ADVANCED STUDIES
IN PURE MATHEMATICS 15

Chief Editor: H. Morikawa (Nagoya University)

Automorphic Forms and Geometry of Arithmetic Varieties

Edited by
K. Hashimoto (Waseda University) and
Y. Namikawa (Nagoya University)

ACADEMIC PRESS, INC.
Harcourt Brace Jovanovich, Publishers
Boston San Diego New York
Berkeley London Sydney
Tokyo Toronto

KINOKUNIYA COMPANY LTD.
TOKYO, JAPAN

Originally Published by Kinokuniya Company Ltd. Tokyo, Japan.
Distributed outside Japan by Academic Press Inc. Orlando, Florida, U.S.A.

ISBN 0-12-330580-2

Library of Congress Catalog Card Number 89-85239

Advanced Studies in Pure Mathematics 15

PRINTED IN JAPAN
by Tokyo Press Co., Ltd.

Foreword

A new era in mathematics is in dawning, as fields continue to cross-fertilize each other and successive challenges consequently mount to cultivate new basic ideas and to formulate and solve new problems.

Our purpose is to bring significant momentum to this new period, by creating conditions that will encourage researchers in the pioneering spirit and action that are needed to explore the mathematical frontiers. Accordingly, symposia will be organized on important topics in pure mathematics, bringing mathematicians of various specialities together in a consciousness of the new directions that are defining themselves.

The organizational activity for the symposia will be provisionally based at the Department of Mathematics of Nagoya University, and the proceedings of the symposia will be published under the title *Advanced Studies in Pure Mathematics*. The board of editors of each volume will consist of the core members listed below plus some of the members of the program committee of each particular symposium.

Efforts are being made at the present time to obtain government approval for establishing a new institute for advanced studies in pure mathematics. At such time as this institute comes into being, the activities described above will be transferred thereto.

H. MORIKAWA and M. NAGATA

This volume is warmly dedicated to

Prof. Ichiro Satake
and
Prof. Friedrich Hirzebruch

on their sixtieth birthdays in 1987
with sincere gratitude
from us all who owe so much to them mathematically
through both their works and their personal guidance

Preface

Since Gauß, automorphic forms have played one of the most important role in number theory together with other but deeply related functicons such as zeta functions.

In this century, the theory of automorphic forms has been extensively enlarged and enriched by relating it to the geometry of symmetric spaces. This generalization allows us to use huge techniques in the Lie group theory, especially the representation theory with harmonic analysis, differential geometry and algebraic geometry. In particular, it has become one of the central problems to study the geometry of the quotients of the Siegel upper half-space or its generalization, the hermitian bounded symmetric domains, by arithmetic subgroups.

The importance of automorphic forms seems to become still greater when we see the very recent development of the theory of strings in quantum field theory where they appear unexpectedly as an essential tool to describe fundamental, physically important quantities. This new tendency would suggest a completely new theory of automorphic forms, namely the one over the moduli space of curves, which would be closely related to non-abelian class field theory.

Here in Japan intensive study on automorphic forms has been made, and it has given, we believe, a not small historical contribution to the theory. It would be our great pleasure if this volume would also contribute something.

This book grew out of a series of symposia held in 1985–86 where the main topic was to study the dimension formulas of various automorphic forms. (For these conferences we obtained financial support from the Research Institute for Mathematical Sciences, Kyoto University, and the Japan Society for the Promotion of Science, to which we together with all participants would express our gratitude.) One of our main concerns was to clarify the relation between two fundamental methods to obtain dimension formulas (for cusp forms), the Selberg trace formula and the index theorem (Riemann-Roch's theorem and the Lefschetz fixed point formula). This relation is well understood when the quotient is compact. But in the case of non-compact quotients, one should analyse the term corresponding to the so-called η-invariants which are related with special values of some zeta functions, and here the relation between the above two methods is not clear. Though this aim is still to be attained (cf. the articles by Satake-Ogata, Sczech,

Tsushima), several important results were obtained which we might consider as fruits of these meetings.

This volume is divided into two parts. Part I consists of survey articles, most of which originated from the lectures given in these meetings but were newly written to be up-to-date. We regret that we could not include here some other interesting survey talks given there, which treat much wider related topics such as those in differential and algebraic geometry.

Part II consists of original papers which have enlarged the subjects treated in the final form of this volume. These subjects cover a large part of those which are now studied in Japan. We also refer the reader to Vol. 13 in this series to understand the whole trend of research in this area in Japan.

This volume is dedicated to Prof. Ichiro Satake and Prof. Friedrich Hirzebruch with our hearty congratulations and thanks on the occasions of their sixtieth birthdays in 1987.

Prof. Satake has long been one of our most important leaders through his work. His theory of compactifications, now known by his name together with Baily and Borel, of quotients of hermitian bounded symmetric domains by arithmetic subgroups, is a fundamental tool to the study of automorphic forms by the geometric method (cf. Tsuyumine's and Tsushima's articles in this volume). Our main aim to clarify the relation between this geometric method and the group-theoretic one (using the trace formula), also originated from his recent work (cf. Satake-Ogata here). Moreover, after his return to Japan, we are enjoying his personal warm encouragements and appropriate advice. The series of meetings mentioned above as the origin of this book was also held under his guidance. Therefore we would like to dedicate this whole volume to Prof. Satake from all the participants of the meetings.

To Prof. Hirzebruch also we owe much, of course, through his Riemann-Roch formula and Lefschetz fixed point formula (which are the main tools to calculate the dimension of cusp forms in geometric method) as well as his deep work on Hilbert modular varieties. But this is not the only reason to dedicate this volume to him. We, the two editors of this volume, ask the reader to allow us to recount our personal experience. Our joint work began with a private seminar at the Max-Planck-Institut für Mathematik in Bonn at the beginning of 1984 by six Japanese members at that time (Prof. Ibukiyama was also one of them) to read Atiyah-Donnelly-Singer's article on the proof of Hirzebruch's conjecture on Hilbert modular varieties. In the course of this seminar we were led to our main motivation for the series of meetings mentioned above. Without this prehistory at MPI, whose director has been Prof. Hirzebruch since its foundation in 1982, we would never have been coeditors of this volume. Already nearly a hundred Japanese mathematicians, including most of the authors in this book, have stayed at Bonn when we consider SFB "Theoretische Mathematik" at Bonn University as MPI's antecedent. Prof. Hirzebruch has always taken the initiative to invite Japanese mathematicians, especially young ones, and given them an opportunity to study in active and international circumstances. His contribution to mathematics in Japan is hence immeasurable, for which we would like to thank him cordially on this occasion.

Lastly, the editors would ask pardon for the long delay of publishing this volume, which is entirely due to their laziness. They are afraid that this delay might have caused the authors some disadvantage.

May 1988

Ki-ichiro Hashimoto
Yukihiko Namikawa

CONTENTS

Advanced Studies in Pure Mathematics 15, 1989
Automorphic Forms and Geometry of Arithmetic Varieties
pp. 1–27

Zeta Functions Associated to Cones and their Special Values

I. Satake and S. Ogata

Introduction

The purpose of this paper is to give a survey on zeta functions associated to (self-dual homogeneous) cones and their special values, including some recent results of ours on this subject.

In § 1 we summarize basic facts on self-dual homogeneous cones and the associated Γ-functions. § 2 is concerned with the zeta functions. Let V be a real vector space, $\mathscr{C}$ a self-dual homogeneous cone in V, and let G be the automorphism group $\mathrm{Aut}(V, \mathscr{C})^{\circ}$. We fix a $\boldsymbol{Q}$-simple $\boldsymbol{Q}$-structure on $(V, \mathscr{C})$. As is well-known, the pair (G, V) is a "prehomogeneous vector space" in the sense of Sato-Shintani [SS]. Following the general idea in [SS], we define a set of zeta functions $\{\xi_I\}$, each one of which is associated to a connected component V_I of $V^{\times}=V-S$, S denoting the singular set; in particular, $\xi_{(0)}=Z_{\mathscr{C}}$ is the zeta function associated to the cone $V_{(0)}=\mathscr{C}$. Then we give an explicit expression for the system of functional equations (Theorems 2.2.2, 2.3.3). Under the assumption that *d is even*, taking suitable linear combinations of these zeta functions, we define a new kind of L-functions L_I, which are shown to satisfy individually (or two in a pair, according to the cases) a functional equation of ordinary type (see (2.3.5)). We give some (new) results (Theorems 2.3.9, 2.4.1) on the residues and special values of these zeta and L-functions, where two extreme L-functions $L_{(0)}$ and $L_{(r_1)}$ play an essential role. These extreme L-functions, which generalize the (partial) Dedekind zeta function and the Shimizu L-function in the Hilbert modular case, seem to be of particular importance from the number-theoretic view point.

In § 3, we consider the corresponding (rational) symmetric tube domain $\mathscr{D}=V+\sqrt{-1}\mathscr{C}$ and, under an additional assumption that the ***Q**-rank of G is one*, study the geometric invariants (χ_∞, τ_∞, etc.) associated to the cusp singularities appearing in the (standard) compactification of the arithmetic quotient space $\tilde{\Gamma}\backslash\mathscr{D}$ ([S3, 4]). A typical example is the Hilbert

Received February 16, 1987.
*) Partially supported by NSF Grand DMS-8502341.

modular singularities, which were studied extensively by Hirzebruch and others ([H2], [HG]). In [H2] Hirzebruch gave a conjecture relating the "signature defects" of the cusps with the zero-values of the corresponding Shimizu L-functions, which was later proved by Atiyah-Donnelly-Singer and Müller ([ADS 1, 2], [M 4, 5]). In view of our results on these invariants and special values, we state in 3.3 some conjectures ((C1), (C2), (C3)) which may be regarded as a natural generalization of the Hirzebruch conjecture.

In § 4, we define a more general zeta function $\boldsymbol{Z}_{\mathscr{C}}$ associated to a "Tsuchihashi singularity" and give a formula for the zero-value $\boldsymbol{Z}_{\mathscr{C}}(0)$ (Theorem 4.2.5) by modifying a method due to Zagier [Z]. Recently, using this formula, Ishida [I3] proved the rationality of $\boldsymbol{Z}_{\mathscr{C}}(0)$ in general. It is hoped that our approach might suggest a new possibility of attacking the generalized Hirzebruch conjecture.

Our study on this subject has been largely inspired by the fundamental works of Professor F. Hirzebruch, to whom this paper is respectfully dedicated. The paper was prepared during a stay at the MSRI, Berkeley in 1986–87, of the first-named author, who would like to thank the staffs of the Institute for superb service and hospitalities.

Notations. The symbols $\boldsymbol{Z}$, $\boldsymbol{Q}$, $\boldsymbol{R}$, $\boldsymbol{C}$ are used in the usual sense, e.g. $\boldsymbol{Q}$ is the field of rational numbers. $\boldsymbol{H}$ is the Hamilton quaternion algebra. We use the symbol like $\boldsymbol{R}_{\geq 0}=\{\lambda \in \boldsymbol{R} \mid \lambda \geq 0\}$, and write $\boldsymbol{R}_+$ for $\boldsymbol{R}_{>0}$. For $\xi \in \boldsymbol{C}$, $\boldsymbol{e}(\xi)$ stands for $\exp(2\pi\sqrt{-1}\xi)$. Let V be a real vector space, $v_1, \cdots, v_r \in V$ and let S be a subset of $\boldsymbol{R}$. Then we write $\{v_1, \cdots, v_r\}_S$ for $\{\sum_{i=1}^r \lambda_i v_i \mid \lambda_i \in S\}$; e.g. $\{v_1, \cdots, v_r\}_{\boldsymbol{R}_{\geq 0}}$ is a closed polyhedral cone generated by $v_1, \cdots, v_r$. For a cone $\mathscr{C}$ and a lattice M in V, $\mathscr{C}^*$ and M^* denote, respectively, the dual cone and the dual lattice in the dual space V^*. For a topological group G, G° denotes the identity connected component of G. For a finite set S, $|S|$ denotes the cardinality of S.

Let F be a subfield of $\boldsymbol{R}$, $\mathscr{G}$ a (connected) algebraic group defined over F, and $G=\mathscr{G}(\boldsymbol{R})^\circ$. By an abuse of notations, we write G_F for $\mathscr{G}(F) \cap G$ and F-rk G for F-rk $\mathscr{G}$ (i.e. the dimension of maximal F-split tori in $\mathscr{G}$). If σ is an imbedding, $F \hookrightarrow \boldsymbol{R}$, then G^σ stands for $\mathscr{G}^\sigma(\boldsymbol{R})^\circ$ and, if F_0 is a subfield of F with $[F:F_0]<\infty$, then $R_{F/F_0}(G)$ stands for $R_{F/F_0}(\mathscr{G})(\boldsymbol{R})^\circ$. When $\mathscr{G}$ is reductive, $G=\mathscr{G}(\boldsymbol{R})^\circ$ is called "reductive", and we write G^s for $\mathscr{G}^s(\boldsymbol{R})^\circ$, $\mathscr{G}^s$ denoting the semisimple part of $\mathscr{G}$.

§ 1. Self-dual homogeneous cones ([BK], [S1], [V])

1.1. Let V be a real vector space of dimension $n>0$. By a *convex cone* in V, we mean a subset $\mathscr{C}$ of V with the following property:

$$x, y \in \mathscr{C},\ \lambda, \mu>0 \Longrightarrow \lambda x+\mu y \in \mathscr{C}.$$

The *dual* of $\mathscr{C}$ is defined by

$$\mathscr{C}^*=\{x^* \in V^* \mid \langle x, x^*\rangle>0 \quad \text{for all } x \in \overline{\mathscr{C}}-\{0\}\}.$$

Then $\mathscr{C}^*$ is an open convex cone in the dual space V^*. It is clear that for a (non-empty) convex cone $\mathscr{C}$ the following three conditions are equivalent:

(i) $\mathscr{C}$ does not contain a line in V;
(ii) $\overline{\mathscr{C}}\cap(-\overline{\mathscr{C}})=\{0\}$;
(iii) $\mathscr{C}^*$ is non-empty.

When these conditions are satisfied, $\mathscr{C}$ is called *non-degenerate*. In what follows, a non-degenerate open convex cone will simply be called a "cone". For a cone $\mathscr{C}$, one has $\mathscr{C}^{**}=\mathscr{C}$.

A cone $\mathscr{C}$ is called *self-dual* if there exists a linear isomorphism $S:(V, \mathscr{C})\rightleftarrows(V^*, \mathscr{C}^*)$, which is symmetric and positive definite. A cone $\mathscr{C}$ is called *homogeneous* if the automorphism group

$$G=\mathrm{Aut}(V, \mathscr{C})^\circ=\{g \in GL(V) \mid g\mathscr{C}=\mathscr{C}\}^\circ.$$

($\circ$ denoting the identity connected component) is transitive on $\mathscr{C}$.

In §§ 1–3, unless otherwise specified, we always assume that $\mathscr{C}$ is self-dual and homogeneous, and fix a positive definite inner product $\langle\ \rangle$ on V defining an isomorphism S mentioned above. Then $(V, \mathscr{C})$ is identified with its dual $(V^*, \mathscr{C}^*)$. In this case, the automorphism group G is the identity connected component of a reductive algebraic group and for any $c_0 \in \mathscr{C}$ the isotropy subgroup

$$K=G_{c_0}=\{g \in G \mid gc_0=c_0\}$$

is a maximal compact subgroup of G. Thus $\mathscr{C}\approx G/K$ has a structure of Riemannian symmetric space (with a flat part).

1.2. In 1957–58, M. Koecher made an observation that the category of self-dual homogeneous cones $(V, \mathscr{C})$ with a base point $c_0 \in \mathscr{C}$ is equivalent to that of "formally real" Jordan algebras by the correspondence given as follows ([BK], [S1]). Let $\mathscr{C}$ be a self-dual homogeneous cone in V with a base point c_0 and let G, K be as above. Let $\mathfrak{g}=\mathrm{Lie}\ G$, $\mathfrak{k}=\mathrm{Lie}\ K$ and let $\mathfrak{g}=\mathfrak{k}+\mathfrak{p}$ be the corresponding Cartan decomposition. Then by the homogeneity assumption there exists a unique linear isomorphism

$$V \ni x \longmapsto T_x \in \mathfrak{p}$$

such that $x=T_x c_0$. The Jordan product in V is then defined by

$$x \circ y=T_x y \quad (x, y \in V).$$

In particular, one has $T_{c0}=\mathrm{id}_V$, i.e. c_0 is the unit element of the Jordan algebra.

By virtue of this equivalence, the classification of self-dual homogeneous cones is reduced to that of formally real Jordan algebras, which was given (by a collaboration of phisicists) as early as in 1934 ([JNW]). A self-dual homogeneous cone $\mathscr{C}$ is decomposed uniquely into the direct product of the "irreducible" ones, for which one has $G=\boldsymbol{R}_+\times G^s$ with G^s $\boldsymbol{R}$-simple (or $=\{1\}$). The irreducible self-dual homogeneous cones are classified into the following five types:

$$\begin{cases} \mathscr{P}_1(\boldsymbol{R})=\boldsymbol{R}_+, \quad \mathscr{P}_r(\boldsymbol{F}) \quad (r\geq 2, \boldsymbol{F}=\boldsymbol{R}, \boldsymbol{C}, \boldsymbol{H}), \\ \mathscr{P}_3(\boldsymbol{O}) \quad (\boldsymbol{O} \text{ denotes the Cayley octonion algebra}), \\ \mathscr{P}(1, n-1)=\{(\xi_i)\in \boldsymbol{R}^n \mid \xi_1>0,\ \xi_1^2-\sum_{i=2}^n \xi_i^2>0\} \quad (n\geq 3), \end{cases}$$

where $\mathscr{P}_r(\boldsymbol{F})$ denotes the cone of positive definite hermitian matrices of size r with entries in $\boldsymbol{F}$. $\mathscr{P}_2(\boldsymbol{R})$, $\mathscr{P}_2(\boldsymbol{C})$, $\mathscr{P}_2(\boldsymbol{H})$ are isomorphic to the "quadratic cones" $\mathscr{P}(1, n-1)$ with $n=3, 4, 6$, respectively. For $\mathscr{C}=\mathscr{P}_3(\boldsymbol{O})$, G^s is an excetional group of type (E_6).

A more general study on "homogeneous cones" was done by Vinberg [V] in the early 60's. In the study of general cones, the characteristic function plays an essential role. For any (non-degenerate, open convex) cone $\mathscr{C}$, the *characteristic function* $\phi(x)=\phi_{\mathscr{C}}(x)$ is defined by

$$\phi_{\mathscr{C}}(x)=\int_{\mathscr{C}^*} e^{-\langle x,x^*\rangle}\mathrm{d}x^*.$$

Clearly one has

$$\phi_{\mathscr{C}}(x)>0, \quad \phi_{\mathscr{C}}(gx)=\det(g)^{-1}\phi_{\mathscr{C}}(x) \qquad \text{for } x\in\mathscr{C},\ g\in G,$$

and $\log\phi_c(x)$ is a convex function, which tends to infinity when $x\in\mathscr{C}$ converges to a boundary point of $\mathscr{C}$. The characteristic function will be used later in § 4.

1.3. Quasi-irreducible cones. Let $\mathscr{C}$ be a self-dual homogeneous cone in V. $\mathscr{C}$ is called *quasi-irreducible* if in its irreducible decomposition all irreducible components are isomorphic.

Lemma 1.3.1. *Suppose $(V, \mathscr{C})$ has a $\boldsymbol{Q}$-simple $\boldsymbol{Q}$-structure; this means that there is a $\boldsymbol{Q}$-vector space V_Q such that $V=V_Q\otimes_Q\boldsymbol{R}$, for which G is (the identity connected component of) an algebraic group defined over $\boldsymbol{Q}$ and that, if $(V, \mathscr{C})=\prod_{\mu=1}^m(V_\mu, \mathscr{C}_\mu)$ is the irreducible decomposition, no partial product of V_μ's is defined over $\boldsymbol{Q}$, or equivalently, that the center of G is of $\boldsymbol{Q}$-rank one. Then $\mathscr{C}$ is quasi-irreducible.*

In fact, under this assumption, there exists a totally real number field F_1 of degree m such that

$$G = G_{F_1/Q}(G_1) = \prod_{\mu=1}^{m} G_\mu, \quad G_\mu = G_1^{\sigma_\mu},$$

where $G_1 = \boldsymbol{R}_+ \times G_1^s$ with G_1^s $\boldsymbol{R}$-simple (which may reduce to $\{1\}$), defined over F_1, and $\{\sigma_\mu\ (1 \leq \mu \leq m)\}$ is the totality of the imbeddings $F_1 \hookrightarrow \boldsymbol{R}$. Then the G_μ^s's are all $\boldsymbol{C}$-isomorphic and hence, by the classification theory, are also $\boldsymbol{R}$-isomorphic except for the case when there exists an even integer r_1 such that every G_μ^s is isogeneous either to $SL(r_1, \boldsymbol{R})$ or to $SL(r_1/2, \boldsymbol{H})$ and when both types $SL(r_1, \boldsymbol{R})$ and $SL(r_1/2, \boldsymbol{H})$ occur in the G_μ^s's. But actually such a "mixed type" can not occur for the following reason. Since the $\boldsymbol{Q}$-rational points are dense in $\mathscr{C}$, one may take c_0 to be $\boldsymbol{Q}$-rational; then the maximal compact subgroup K is also defined over $\boldsymbol{Q}$. One then has the corresponding decomposition

$$K = R_{F_1/Q}(K_1) = \prod_{\mu=1}^{m} K_\mu, \quad K_\mu = K_1^{\sigma_\mu}$$

and hence all K_μ's are also $\boldsymbol{C}$-isomorphic. But the dimension of the maximal compact subgroups of $SL(r_1, \boldsymbol{R})$ and $SL(r_1/2, \boldsymbol{H})$ is equal to $\frac{1}{2}r_1(r_1-1)$, $\frac{1}{2}r_1(r_1+1)$, respectively. Therefore no mixture of these two types can occur, which proves our assertion.

1.4. The norm and trace. The rank of a self-dual homogeneous cone $\mathscr{C}$ is by definition the $\boldsymbol{R}$-rank of the Lie algebra $\mathfrak{g}$, which also coincides with the (absolute) rank of the formally real Jordan algebra (V, c_0). Let $n = \dim V$ and $r = \operatorname{rank} \mathscr{C}$. If $\mathscr{C}$ is irreducible, one has (from the Peirce decomposition of (V, c_0))

$$n = r + \frac{d}{2} r(r-1), \tag{1.4.1}$$

where d is a non-negative integer. For $\mathscr{C} = \boldsymbol{R}_+$, one puts $d = 0$. For $\mathscr{C} = \mathscr{P}_r(\boldsymbol{F})$ $(r \geq 2)$, one has actually rank $\mathscr{C} = r$ and $d = \dim_{\boldsymbol{R}} \boldsymbol{F} = 1, 2, 4, 8$ according as $\boldsymbol{F} = \boldsymbol{R}, \boldsymbol{C}, \boldsymbol{H}, \boldsymbol{O}$. For a quadratic cone $\mathscr{C}$, one has rank $\mathscr{C} = 2$ and $d = n-2$. Thus the pair (r, d) is a complete invariant for an isomorphism class of irreducible self-dual homogeneous cones.

In the Jordan algebra (V, c_0), one can define the (reduced) *norm* $N : V \to \boldsymbol{R}$ as the (unique) homogeneous polynomial function of degree r on V such that, for a "general element" x in V, $N(tc_0 - x)\ (\in \boldsymbol{R}[t])$ is the minimal polynomial for x in the usual sense. When $\mathscr{C}$ is irreducible, the norm is uniquely characterized by the property

(1.4.2) $$N(c_0)=1,\quad N(gx)=\det(g)^{r/n}N(x)\quad \text{for } g\in G,\ x\in V.$$

(Note that $\chi(g)=\det(g)^{r/n}$ is a rational character on G.) Hence one has the relation

(1.4.3) $$N(x)=(\phi_{\mathscr{C}}(c_0)^{-1}\phi_{\mathscr{C}}(x))^{-r/n}\quad (x\in\mathscr{C}).$$

The (reduced) trace $\operatorname{tr}(x)$ is defined by

$$N(tc_0-x)=t^r-\operatorname{tr}(x)t^{r-1}+\cdots+(-1)^rN(x).$$

The trace is K-invariant. It follows that, when $\mathscr{C}$ is irreducible, one has

(1.4.4) $$\operatorname{tr}x=\frac{r}{n}\operatorname{tr}(T_x),$$

where $T_x: y\mapsto xy$ is the multiplication in the Jordan algebra (V, c_0). (Note also, putting $P(x)=2T_x^2-T_{x^2}$, one has the relations $P(gx)=gP(x){}^tg$, $\det(P(x))=N(x)^{2n/r}$.)

In what follows, we assume that $\mathscr{C}$ is quasi-irreducible. Let $(V,\mathscr{C})=\prod_{\mu=1}^{m}(V_\mu,\mathscr{C}_\mu)$ be the irreducible decomposition,

$$G=\prod_{\mu=1}^{m}G_\mu,\quad G_\mu=\operatorname{Aut}(V_\mu,\mathscr{C}_\mu)^\circ,$$

and put $n_1=\dim V_1$, $r_1=\operatorname{rank}\mathscr{C}_1=\boldsymbol{R}\text{-rk}\,G_1$. Then one has $n=mn_1$, $r=mr_1$ and

$$\frac{n}{r}=\frac{n_1}{r_1}=1+\frac{d}{2}(r_1-1).$$

Hence the formulae (1.4.2–4) remain valid. We normalize the inner product on V in such a way that $\langle c_{0,\mu}, c_{0,\mu}\rangle=r_1$ $(1\leq\mu\leq m)$, where $c_0=(c_{0,\mu})$. Then one has

(1.4.5) $$\langle x, y\rangle=\operatorname{tr}(xy).$$

The Euclidean (i.e. self-dual) measure on V for this $\langle\ \rangle$ will be denoted as dx. A G-invariant measure on $\mathscr{C}$ is then given by $N(x)^{-n/r}dx$.

1.5. The Γ-function. Let $\mathscr{C}$ be a quasi-irreducible self-dual homogeneous cone in V. The "Γ-function" of $\mathscr{C}$ (introduced by Koecher) is defined by the integral

(1.5.1) $$\Gamma_{\mathscr{C}}(s)=\int_{\mathscr{C}}e^{-\operatorname{tr}(x)}N(x)^{s-(n/r)}dx\quad (s\in\boldsymbol{C}),$$

which converges absolutely for $\mathrm{Re}\, s > n/r - 1$. By a change of variable, one gets

$$N(x)^{-s}\Gamma_{\mathscr{C}}(s) = \int_{\mathscr{C}} e^{-\langle x,y\rangle} N(y)^{s-n/r} dy \quad (x \in \mathscr{C}). \tag{1.5.2}$$

On V one can define a (unique) differential operator of degree r, denoted as $N(\nabla_x)$, with the property

$$N(\nabla_x) e^{\langle x,y\rangle} = N(y) e^{\langle x,y\rangle}$$

(cf. [R], [SS]). Then $N(\nabla_x)$ is relatively invariant in the sense that one has

$$L_g^{-1} N(\nabla_x) L_g = \det(g)^{-r/n} N(\nabla_x) \quad (g \in G,\ x \in V),$$

where $(L_g f)(x) = f(g^{-1}x)$ for any function f on V. The associated "b-function" is defined by

$$N(\nabla_x) N(x)^s = b(s) N(x)^{s-1} \quad (x \in \mathscr{C},\ s \in \boldsymbol{C})$$

(cf. [SS]).*) Then, applying $N(\nabla_x)$ on the both sides of (1.5.2), one gets

$$b(s) = (-1)^r \frac{\Gamma_{\mathscr{C}}(1-s)}{\Gamma_{\mathscr{C}}(-s)}.$$

By a direct computation from (1.5.1) (see e.g. [S2]), one obtains

$$\Gamma_{\mathscr{C}}(s) = (2\pi)^{(n-r)/2} \left(\prod_{i=1}^{r_1} \Gamma\left(s - \frac{d}{2}(i-1)\right)\right)^m, \tag{1.5.3}$$

$$b(s) = \left(\prod_{i=1}^{r_1} \left(s + \frac{d}{2}(i-1)\right)\right)^m. \tag{1.5.4}$$

(For the Γ-function of a more general cone, see [G].)

§ 2. Zeta functions associated to a self-dual homogeneous cone

2.1. We assume in this section that $(V, \mathscr{C})$ is endowed with a $\boldsymbol{Q}$-simple $\boldsymbol{Q}$-structure in the sense stated in Lemma 1.3.1. Then G^s is $\boldsymbol{Q}$-simple (or reduces to $\{1\}$), the center of G is of $\boldsymbol{Q}$-rank one and $\mathscr{C}$ is quasi-irreducible. The $\boldsymbol{Q}$-rank of G, which we denote by r_0, is a divisor of $r_1 = \boldsymbol{R}\text{-rk}\, G_1$: hence we set $\delta = r_1/r_0$. We fix a base point c_0 in $\mathscr{C} \cap V_{\boldsymbol{Q}}$; then the norm, trace and the (normalized) inner product $\langle\ \rangle$ are all defined over $\boldsymbol{Q}$. We choose a lattice M in $V_{\boldsymbol{Q}}$ and an arithmetic subgroup Γ of G such

*) Note that in some recent literature (e.g. [I1]) it has become more customary to denote our $b(s)$ by $b(s-1)$.

that $\Gamma M=M$. We define a zeta function by

(2.1.1) $$Z_{\mathscr{C},c_0}(\Gamma, M; s)=\sum_{x:\Gamma\backslash\mathscr{C}\cap M}|\Gamma_x|^{-1}N(x)^{-s} \quad (s \in \boldsymbol{C}),$$

where $\Gamma_x=\{\gamma \in \Gamma \mid \gamma x \in x\}$ (which is finite) and the summation is taken over a complete set of representatives of the Γ-orbits in $\mathscr{C}\cap M$. When c_0 is kept fixed, we write $Z_{\mathscr{C}}$ for $Z_{\mathscr{C},c_0}$. It is known that the series on the right hand side of (2.1.1) is absolutely convergent for $\mathrm{Re}\, s>n/r$ and has an analytic continuation to a meromorphic function on the whole plane $\boldsymbol{C}$. It is clear that, if Γ' is a subgroup of Γ of finite index, then one has

$$Z_{\mathscr{C}}(\Gamma', M; s)=[\Gamma : \Gamma']Z_{\mathscr{C}}(\Gamma, M; s).$$

Hence it suffices to consider the zeta function for the full stabilizer $\Gamma_M=\{\gamma \in G \mid \gamma M=M\}$. In that case, we write $Z_{\mathscr{C}}(M; s)$ for $Z_{\mathscr{C}}(\Gamma_M, M; s)$.

In the simplest case where G^s reduces to $\{1\}$, one obtains essentially the (partial) Dedekind zeta function of the totally real number field F_1 (see the Example 2.1.2 below). The case where $\mathscr{C}$ is a quadratic cone was studied by Siegel [S9]. Our zeta function is a special case of the zeta function associated to a (real) "prehomogeneous vector space" in the sense of Sato-Shintani [SS], who treated as examples the cases of $\mathscr{P}_r(\boldsymbol{R})$, $\mathscr{P}_r(\boldsymbol{C})$ and the quadratic cones (see [S7], [SS], pp. 160–168, pp. 155–157). For other cases, see [M3] (cf. also [SF]).

Example 2.1.2. Let F_1 be a totally real number field of degree m and let $V=F_1\otimes_{\boldsymbol{Q}}\boldsymbol{R}\cong\boldsymbol{R}^m$. Then the "angular domain" $\mathscr{C}=\boldsymbol{R}_+^m$ is a self-dual homogeneous cone with respect to the standard inner product in $\boldsymbol{R}^m$. G is identified with the multiplicative group $\boldsymbol{R}_+^m$ and G^s and K reduce to the identity. If one takes c_0 to be 1 (the unit element of F_1), then the norm and the trace are given by

$$N(x)=\prod \xi_\mu, \quad \mathrm{tr}(x)=\sum \xi_\mu \qquad \text{for } x=(\xi_\mu) \in V,$$

and the standard inner product in $\boldsymbol{R}^m$ is normalized. V has a natural $\boldsymbol{Q}$-structure for which $V_{\boldsymbol{Q}}=F_1$, G is defined over $\boldsymbol{Q}$, of $\boldsymbol{Q}$-rank 1, and $G_{\boldsymbol{Q}}=\{\alpha \in F_1^\times \mid \alpha^{\sigma_\mu}>0\,(1\leq\mu\leq m)\}$. Hence the above assumptions are all satisfied and one has $n_1=r_1=r_0=1$, $n=r=m$. Let $\mathcal{O}_{F_1}$ be the ring of integers in F_1 and choose M to be an ideal $\mathfrak{a}_1$ in $\mathcal{O}_{F_1}$. Then Γ_M is the group of totally positive units of $\mathcal{O}_{F_1}$ and one has

$$Z_{\mathscr{C}}(M; s)=\sum_{x:\Gamma_M\backslash\mathscr{C}\cap M} N(x)^{-s}=N(\mathfrak{a}_1)^{-s}\sum_{\mathfrak{a}\sim\mathfrak{a}_1^{-1}} N(\mathfrak{a})^{-s},$$

where the summation in the last expression is taken over all integral ideals $\mathfrak{a}$ "equivalent" to $\mathfrak{a}_1^{-1}$ in the narrow sense. Thus essentially $Z_{\mathscr{C}}(M; s)$ is nothing but a "parital" Dedekind zeta function of F_1 corresponding to the "ray class' of $\mathfrak{a}_1^{-1}$.

2.2. Functional equations. According to the general theory of Sato-Shintani, the functional equations for $Z_{\mathscr{C}}(M; s)$ are obtained as follows. Let

$$V^{\times}=\{x \in V \mid N(x)\neq 0\}=\prod_{\mu=1}^{m} V_{\mu}^{\times}$$

and let

$$V_{\mu}^{\times}=\coprod_{i=0}^{r_1} V_{\mu,i}$$

be the decomposition of $V_{\mu}^{\times}$ into the disjoint union of the connected components, or what amounts to the same thing, into that of the G_{μ}-orbits. (If $c_0=(c_{0,\mu})$ and if $c_{0,\mu}=\sum_{i=1}^{r_1} e_i^{(\mu)}$ is primitive decomposition, then $V_{\mu,i}$ is defined to be the G_{μ}-orbit of $-\sum_{k=1}^{i} e_k^{(\mu)}+\sum_{k=i+1}^{r_1} e_k^{(\mu)}$.) Thus one has

$$V^{\times}=\coprod_{I\in\mathscr{I}_{r_1}^m} V_I,$$

where $\mathscr{I}_{r_1}^m$ denotes the set of all m-tuples $I=(i_1, \cdots, i_m)$ with $0\leq i_{\mu}\leq r_1$ $(1\leq\mu\leq m)$ and for $I=(i_{\mu})$, one sets $V_I=\prod_{\mu=1}^{m} V_{\mu,i_{\mu}}$. Hence $V^{\times}$ consists of $(r_1+1)^m$ connected components. We write (k) for $(k, \cdots, k) \in \mathscr{I}_{r_1}^m$; then $V_{(0)}=\mathscr{C}$ and $V_{(r_1)}=-\mathscr{C}$.

For each $I \in \mathscr{I}_{r_1}^m$, we define a zeta function

$$\xi_I(M; s)=\sum_{x:\Gamma_M\backslash V_I\cap M} \frac{\mu(x)}{|N(x)|^s}, \tag{2.2.1}$$

where the summation is taken over a complete set of representatives of Γ_M-orbits in $V_I\cap M$ and $\mu(x)$ is a "density" defined as follows. For $x \in V_I$, let U_x be a relatively compact neighbourhood of x in V_I and let

$$W_x=\{g \in G \mid gx \in U_x\},$$
$$G_x=\{g \in G \mid gx=x\}, \quad \Gamma_x=G_x\cap\Gamma.$$

Then one has

$$\mu(x)=\int_{\Gamma_x\backslash W_x} \mathrm{d}g \Big/ \int_{U_x} |N(x)|^{-n/r}\mathrm{d}x,$$

where dg is a Haar measure on G normalized in such a way that for any non-negative continuous function f on $\mathscr{C}$ one has

$$\int_G f(gc_0)dg = \int_{\mathscr{C}} f(x)N(x)^{-n/r}\,dx.$$

Then, except for the case $r_1=r_0=2$, $d=1$ (treated in [S7], [S9]), $\mu(x)$ is finite and coincides with the volume of $\Gamma_x \backslash G_x$ with respect to a suitably normalized Haar measure on G_x ([SS], Lemma 2.4) and hence depends only on the Γ-equivalence class of x. In what follows, we omit the above-mentioned exceptional case. Then the series (2.2.1) is absolutely convergent for $\mathrm{Re}\, s > n/r$ and has an analytic continuation to a meromorphic function on $\boldsymbol{C}$ ([SS], [S7]). Clearly one has $\xi_{(0)}=\xi_{(r_1)}=Z_{\mathscr{C}}$. For $I=(i_\mu) \in \mathscr{I}^m_{r_1}$, we set $I^*=(r_1-i_\mu)$. Then it is clear that

$$V_{I^*}=-V_I \quad \text{and} \quad \xi_{I^*}=\xi_I.$$

Thus essentially we get $[(r_1+1)^m/2]$ zeta functions.

Theorem 2.2.2. *The functions $\xi_I(M; s)$ satisfy the functional equations of the following form:*

$$\xi_J\left(M^*; \frac{n}{r}-s\right) = v(M)(2\pi)^{-rs}\Gamma_{\mathscr{C}}(s)\boldsymbol{e}\left(\frac{rs}{4}\right) \sum_{I \in \mathscr{I}^m_{r_1}} \xi_I(M; s)u_{IJ}(s),$$

where M^* is the dual lattice of M, $v(M)=\mathrm{vol}(V/M)$ and, for $I=(i_\mu)$, $J=(j_\mu)$, $u_{IJ}(s)=\prod_{\mu=1}^m u_{i_\mu, j_\mu}(s)$, u_{ij} $(0 \leq i, j \leq r_1)$ being integral polynomials in $\boldsymbol{e}(-s/2)$ of degree $\leq r_1$.

For an explicit expression of u_{ij}, see [SF][*].

2.3. To obtain more precise results, we assume in the rest of this section that *$r \geq 2$ and d is even.* (Note that, if $r=1$, $Z_{\mathscr{C}}$ is essentially the Riemann zeta function. If d is odd, then one has either $r_1=2$ (quadratic cones) or $d=1$ ($\mathscr{P}_{r_1}(\boldsymbol{R})^m (r_1 \geq 2)$).)

Under this assumption, n/r is an integer and there are two cases:

(a) $d \equiv 0 \pmod 4$, or $d \equiv 2 \pmod 4$ and r_1 is odd. In this case, n/r is odd.

(a′) $d \equiv 2 \pmod 4$ and r_1 is even. In this case, n/r is even.

Applying the methods in [SS] and [SF], one obtains

Theorem 2.3.1. *Under the above assumption, the function $\xi_I(M; s)$ has at most r_0 simple poles at $s=n/r-(d/2)\rho$ for $0 \leq \rho \leq r_1-1$, $\delta \mid \rho$ $(\delta=r_1/r_0)$ and one has*

---

[*] Note that $u_{ij}(s)$ in [SS] is in our notation (and in [SF]) given by $c(2\pi)^{(n-r)/2}$ $u_{r-i,r-j}(s)$, if the measure on V in [SS] is equal to $c\,dx$ with dx self-dual. For instance, in the case $V=\mathrm{Her}_r(\boldsymbol{C})$ ([SS], pp. 160–168), one has $c=2^{-(n-r)/2}$.

$$\mathrm{Res}_{s=n/r-(d/2)\rho}\xi_I(M;s)=v(M^*)\Big((2\pi)^{-(d/2)(\rho(\rho+1)/2)}\prod_{k=1}^{\rho}\Gamma\Big(\frac{d}{2}k\Big)\Big)^m$$
$$\times(-1)^{(d/2)\rho|I|}\sqrt{-1}^{(d/2)\rho r}\sum_{J\in\mathscr{I}_\rho^m}(-1)^{(d/2)(r_1-\rho-1)|J|}\binom{\rho}{J}\kappa_J^{(\rho)}(M^*),$$

where M^* is the dual lattice of M and, for $J=(j_\mu)\in\mathscr{I}_\rho^m$, we set

$$|J|=\sum_{\mu=1}^m j_\mu,\quad \binom{\rho}{J}=\prod_{\mu=1}^m\binom{\rho}{j_\mu},$$
$$\kappa_J^{(\rho)}(M^*)=\sum_{x:\,{}^t\Gamma\backslash S_J^{(\rho)}\cap M^*}\int_{G_x/{}^t\Gamma_x}\mathrm{d}\nu_x.$$

$S_J^{(\rho)}$ is the G-orbit in $S=V-V^\times$ containing $\sum_{\mu=1}^m(-\sum_{k=1}^{j_\mu}e_k^{(\mu)}+\sum_{k=j_\mu+1}^{\rho}e_k^{(\mu)})$ and, for $x\in S_J^{(\rho)}$, $\mathrm{d}\nu_x$ is a suitably normalized Haar measure on G_x.

In the special cases, this result is due to [S9], [SS], [S7] and [M3]. It is possible to give a unified proof along the line of [SF] (see [S5a]).

Corollary 2.3.2. *Put $\nu=n/r-(d/2)\rho$ with $0\le\rho\le r_1-1$, $\delta\,|\,\rho$. For a fixed ν, the $(r_1+1)^m$-tuple $(\mathrm{Res}_{s=\nu}\xi_I(M;s))_{I\in\mathscr{I}_{r_1}^m}$ is proportional to $((-1)^{(d/2)\rho|I|})_{I\in\mathscr{I}_{r_1}^m}$. For $\nu\equiv1$ (mod 2), one has*

$$\mathrm{Res}_{s=\nu}\xi_I(M;s)\neq0.$$

Also, for $\nu=n/r$, one has

$$\mathrm{Res}_{s=n/r}\xi_I(M;s)=v(M^*)\int_{\Gamma\backslash\mathscr{C}^1}\mathrm{d}x^1>0,$$

where $\mathscr{C}^1=\{x\in\mathscr{C}\,|\,N(x)=1\}$ and $\mathrm{d}x^1$ is the invariant measure on $\mathscr{C}^1$ induced from $N(x)^{-n/r}\mathrm{d}x$.

Let P_{r_1} denote the symmetric tensor representation of $GL(2,\boldsymbol{C})$ defined by

$$(1,y,\cdots,y^{r_1})\mathrm{P}_{r_1}\Big(\begin{pmatrix}a&b\\c&d\end{pmatrix}\Big)=((a+cy)^{r_1},(a+cy)^{r_1-1}(b+dy),$$
$$\cdots,(b+dy)^{r_1}).$$

For $C=(c_{ij})_{0\le i,j\le r_1}\in GL(r_1+1,\boldsymbol{C})$, we write

$$C^{\otimes m}=(c_{IJ})_{I,J\in\mathscr{I}_{r_1}^m}\in GL((r_1+1)^m,\boldsymbol{C}),\quad c_{IJ}=\prod_{\mu=1}^m c_{i_\mu,j_\mu}.$$

In these notations, we have

Theorem 2.3.3. *The polynomials $u_{IJ}(s)$ in Theorem 2.2.2 are given as follows:*

$$(u_{IJ}(s))_{I,J\in\mathscr{I}^m_{r_1}}=\begin{cases}\mathrm{P}_{r_1}\left(\begin{pmatrix}1 & x\\ x & 1\end{pmatrix}\right)^{\otimes m} & \textit{in Case (a)},\\[2ex] \mathrm{P}_{r_1}\left(\begin{pmatrix}1 & x\\ -x & -1\end{pmatrix}\right)^{\otimes m} & \textit{in Case (a')},\end{cases}$$

where $x=\boldsymbol{e}(-s/2)$.

This follows from [SF], Theorem 2. We put

$$U^{(r_1,m)}(x)=(u_{IJ}(s)),\quad A^{(r_1,m)}=(a_{IJ})=\mathrm{P}_{r_1}\left(\begin{pmatrix}1 & 1\\ 1 & -1\end{pmatrix}\right)^{\otimes m}.$$

Then the matrix $U^{(r_1,m)}(x)$ can be diagonalized as follows:

$$A^{(r_1,m)-1}U^{(r_1,m)}(x)A^{(r_1,m)}=\begin{cases}((1+x)^{r-|I|}(1-x)^{|I|}\delta_{IJ}) & \text{in Case (a)},\\ ((1+x)^{r-|I^*|}(1-x)^{|I^*|}\delta_{I^*,J}) & \text{in Case (a')}.\end{cases}$$

Hence, putting

$$L_J(M;s)=\sum_{I\in\mathscr{I}^m_{r_1}}\xi_I(M;s)a_{IJ}, \tag{2.3.4}$$

one has

$$\begin{aligned}L_I\left(M^*;\frac{n}{r}-s\right)&=v(M)(2\pi)^{-rs}\Gamma_{\mathscr{C}}(s)\\ &\times\left(\boldsymbol{e}\left(\frac{s}{4}\right)+\boldsymbol{e}\left(-\frac{s}{4}\right)\right)^{r-|I|}\left(\boldsymbol{e}\left(\frac{s}{4}\right)-\boldsymbol{e}\left(-\frac{s}{4}\right)\right)^{|I|}\begin{cases}L_I(M;s),\\ L_{I^*}(M;s),\end{cases}\end{aligned} \tag{2.3.5}$$

according to the case (a) and (a').*)

It is easy to see that the matrix $A^{(r_1,m)}=(a_{IJ})$ has the following properties:

$$\begin{cases}a_{(0),J}=1, & a_{I,(0)}=\binom{r_1}{I},\\ a_{I^*,J}=(-1)^{|J|}a_{IJ}, & a_{I,J^*}=(-1)^{|I|}a_{IJ},\\ \sum_{I\in\mathscr{I}^m_{r_1}}a_{IJ}=\delta_{(0),J}2^r, & \sum_{I\in\mathscr{I}^m_{r_1}}(-1)^{|I|}a_{IJ}=\delta_{(r_1),J}2^r.\end{cases} \tag{2.3.6}$$

---

*) This corresponds to the formula (30) in [SF], which was printed erroneously. It should read

$$\begin{aligned}&\Phi_i'(\hat{f},s-n/r)\ (\text{resp. } \Phi'_{r-i}(\hat{f},s-n/r))\\ &\qquad=(2\pi)^{-rs}\Gamma_{\Omega_0}(s)(\boldsymbol{e}(s/4)+\boldsymbol{e}(-s/4))^{i-r}(\boldsymbol{e}(s/4)-\boldsymbol{e}(-s/4))^i\Phi_i'(f,-s)\end{aligned}$$

in Case (a) (resp. (a')).

We use the last relations in the following form.

Lemma 2.3.7. *For* $(x_I) \in \boldsymbol{C}^{(r_1+1)^m}$ *and* $b \in \boldsymbol{C}$, *one has*

$$\sum_{I \in \mathscr{I}^m_{r_1}} x_I a_{IJ} = \delta_{(0),J} b \quad (\textit{resp. } \delta_{(r_1),J} b)$$

if and only if

$$x_I = 2^{-r} b \quad (\textit{resp. } (-1)^{|I|} 2^{-r} b) \qquad \textit{for } I \in \mathscr{I}^m_{r_1}.$$

As another consequence of (2.3.6), one has

Corollary 2.3.8. *If* $|I|$ *is odd, one has* $L_I(M; s) \equiv 0$. *In particular, if* r *is odd, one has* $L_{(r_1)}(M; s) \equiv 0$.

In fact, if $|J|$ is odd, one has by (2.3.4), (2.3.6)

$$\begin{aligned} L_J(s) &= \sum_I \xi_I(s) a_{IJ} = \sum_I \xi_{I^*}(s) a_{I^*,J} \\ &= (-1)^{|J|} \sum_I \xi_I(s) a_{IJ} = -L_J(s), \end{aligned}$$

whence follows that $L_J(s) \equiv 0$.

Theorem 2.3.9. *Let* $\nu = n/r - (d/2)\rho$ *with* $0 \leq \rho \leq r_1 - 1$, $\delta \mid \rho$. *Then*

$$\operatorname{Res}_{s=\nu} L_I(M; s) = \begin{cases} 2^r \operatorname{Res}_{s=\nu} Z_{\mathscr{C}}(M; s) & \textit{if } \nu \equiv \dfrac{n}{r}\ (2) \textit{ and } I = (0), \\ & \textit{or if } \nu \equiv \dfrac{n}{r} - 1(2) \textit{ and } I = (r_1), \\ 0 & \textit{otherwise.} \end{cases}$$

In particular, $L_I(M; s)$ *with* $I \neq (0), (r_1)$ *is entire. If* $(d/2)\delta$ *is even,* $L_{(r_1)}(M; s)$ *is also entire.*

This follows from Corollary 2.3.2 and Lemma 2.3.7.

Corollary 2.3.10. *If* r *is odd, one has*

$$\operatorname{Res}_{s=\nu} Z_{\mathscr{C}}(M; s) = 0 \qquad \textit{for } \nu \equiv \frac{n}{r} - 1(2).$$

This follows from Theorem 2.3.9 and Corollary 2.3.8. In view of the formula in Theorem 2.3.1, this implies that the residue of the zeta functions $\xi_I(M; s)$ is always *real.*

2.4. Special values of zeta and L-functions. We are interested in the special values of the functions $\xi_I(M; s)$ and $L_I(M; s)$ at $s=\nu \in \boldsymbol{Z}$, $\nu \leq 0$. For simplicity, we write $Z_{\mathscr{C}}(s)$, $\xi_I(s)$, $L_I(s)$, $\xi_I^*(s)$ and $L_I^*(s)$ for $Z_{\mathscr{C}}(M; s)$, $\xi_I(M; s)$, $L_I(M; s)$, $\xi_I(M^*, s)$ and $L_I(M^*, s)$, respectively. Putting $s=n/r-\nu$ and replacing M by M^* in (2.3.5), one has for $\nu \equiv n/r-1$ (2) and ≤ 0

$$\operatorname{ord}_{s=\nu} L_I = r-|I|+\begin{cases}\operatorname{ord}_{s=n/r-\nu} L_I^* & \text{(Case (a))},\\ \operatorname{ord}_{s=n/r-\nu} L_{I^*}^* & \text{(Case (a'))},\end{cases}$$

whence follows (noting $r \geq 2$) that

$$L_I(\nu)=0 \qquad \text{for } I \neq (r_1).$$

Similarly, for $\nu \equiv n/r$ (2) and ≤ 0, one has

$$L_I(\nu)=0 \qquad \text{for } I \neq (0).$$

By Lemma 2.3.7, this implies the following

Theorem 2.4.1. *For $\nu \equiv n/r-1$ (2) and ≤ 0, one has*

$$\xi_I(M; \nu)=(-1)^{|I|} Z_{\mathscr{C}}(M; \nu),$$

$$L_I(M; \nu)=\begin{cases}2^r Z_{\mathscr{C}}(M; \nu) & \text{for } I=(r_1),\\ 0 & \text{for } I \neq (r_1).\end{cases}$$

In particular, if r is odd, one has $Z_{\mathscr{C}}(M; \nu)=0$ for all $\nu \equiv 0$ (2) and ≤ 0.

For $\nu \equiv n/r$ (2) and ≤ 0, one has

$$\xi_I(M; \nu)=Z_{\mathscr{C}}(M; \nu),$$

$$L_I(M; \nu)=\begin{cases}2^r Z_{\mathscr{C}}(M; \nu) & \text{for } I=(0),\\ 0 & \text{for } I \neq (0).\end{cases}$$

It has been conjectured in general (including the case d odd) that the special values $Z_{\mathscr{C}}(M; \nu)=\xi_{(0)}(M; \nu)$ ($\nu \in \boldsymbol{Z}$, $\nu \leq 0$) are rational. There are a few evidences supporting this conjecture.

(2.4.2) In the case explained in Example 2.1.2, where one has $r_0=r_1=1$, $r=m$ and $d=0$, the zeta function $Z_{\mathscr{C}}(M; s)$ and the L-function $L_{(0)}=\sum \xi_I$ are essentially the (partial) Dedekind zeta functions of the totally real number field F_1, and the L-function $L_{(1)}=\sum(-1)^{|I|}\xi_I$ coincides with the "Shimizu L-function" (which is a special case of Hecke's zeta function with "Grössencharaktere"). The value $L_{(1)}(M; 1)$, which is related to $L_{(1)}(M^*; 0)$ by the functional equation, appears in the dimension formula for the space of Hilbert modular forms (cf. [S6]). The relation in Theorem 2.4.1 for $\nu=0$, $L_{(1)}(M;0)=2^r Z_{\mathscr{C}}(M;0)$, has been known*). In this case, the rationality of the special values $Z_{\mathscr{C}}(M;\nu)$ ($\nu \in \boldsymbol{Z}$, $\nu \leq 0$) is well-known (see e.g. [S8]).

(2.4.3) In the case where $\mathscr{C}=\mathscr{P}_r(\boldsymbol{R})$ ($r\geq 2$), $M=\mathrm{Sym}_r(\boldsymbol{Z})$ and Γ is a congruence subgroup of $Sp_r(\boldsymbol{Z})$, the special value $Z_{\mathscr{C}}(M^*;\nu)$ appears in the cusp contribution in the dimension formula for the space of Siegel cusp forms for the corresponding congruence subgroup of $Sp_{r-\nu}(\boldsymbol{Z})$ (cf. [M1], [S7]). In this case, one has $F_1=\boldsymbol{Q}$, $r_0=r_1=r$ and $d=1$. (So this is a case excluded here.) For the case $r_1=r=2$, $r_0=1$ and $d=1$ with a $\boldsymbol{Q}$-structure defined by an indefinite quaternion algebra over $\boldsymbol{Q}$, a similar observation was made by Arakawa [A].

(2.4.4) In the case of quadratic cones with $r_0=1$, $r_1=r=2$, Kurihara [K] showed that the values $Z_{\mathscr{C}}(\nu)$ ($\nu \in \boldsymbol{Z}$, $\nu\leq 0$) are rational for $n\leq 3$ and gave an example with $n=4$ for which the rationality holds.

(2.4.5) For the case $r_0=1$, $m=1$, one of the authors ([S2]) has made the following observation, generalizing Shintani's method in [S8]. Suppose Γ is torsion-free and let

$$\mathscr{C}=\coprod_{\substack{\gamma\in\Gamma\\ \alpha\in A}} \gamma\,\mathrm{Int}(\sigma_\alpha)$$

be a Γ-invariant non-singular "r.p.p. decomposition" in the sense to be explained in 4.2. Here $\sigma_\alpha=\{v_1^{(\alpha)},\cdots,v_{l_\alpha}^{(\alpha)}\}_{\boldsymbol{R}\geq 0}$ is a simplicial cone generated by $v_i^{(\alpha)}$ ($1\leq i\leq l_\alpha$) such that $\{v_i^{(\alpha)}\ (1\leq i\leq l_\alpha)\}$ can be extended to a $\boldsymbol{Z}$-basis of M.

Following Ishida [I], we put

$$f(u)=\sum_{\alpha\in A}\prod_{j=1}^{l_\alpha}\frac{1}{e^{\langle v_j{}^{(\alpha)},u\rangle}-1} \qquad (u\in\mathscr{C}),$$

$$f^K(t_1,\cdots,t_r)=\int_K f\Big(k\sum_{i=1}^r t_i e_i\Big)dk \qquad (t_i\in\boldsymbol{R}_+),$$

where $c_0=\sum_{i=1}^r e_i$ is a primitive decomposition of the unit element. Then, it was shown in [S2] that the special values $Z_{\mathscr{C}}(\nu)$ ($\nu\in\boldsymbol{Z}$, $\nu\leq 0$) are a $\boldsymbol{Q}$-linear combination of the Laurent coefficients of $f^K(t_1,\cdots,t_r)$ after a suitable change of variables and hence are a $\boldsymbol{Q}$-linear combination of the integrals of the form

*) We are grateful to R. Sczech for informing us of this relation along with a sketch of proof.

$$\int_K \prod_{i,j} \langle v_j^{(\alpha)}, ke_i\rangle^{\nu_{ij}^{(\alpha)}} \mathrm{d}k,$$

where $\alpha \in A$ and, for a given ν, $(\nu_{ij}^{(\alpha)})$ ranges over a certain finite set of $r\times l_\alpha$ integral matrices.

(2.4.6) In the case of $r_0=1$, the zero-value $Z_{\mathscr{C}}(0)$, which coincides with $2^{-r}L_{(0)}(0)$ or $2^{-r}L_{(r_1)}(0)$ according as $n/r\equiv 0$ or $1 \pmod 2$, seems to have a close connection with some geometric invariants of the corresponding cusp singularity, as we shall explain in the next section. The rationality of $Z_{\mathscr{C}}(0)$ in a more general context of "Tsuchihashi singularities" was proved by Ishida [I3] (see 4.2),

§ 3. Geometric invariants of cusp singularities

3.1. Rational symmetric tube domains. Let V, $\mathscr{C}$, G, $\cdots$ be as before. We consider the tube domain $\mathscr{D}=V+\sqrt{-1}\mathscr{C}$ in $V_{\boldsymbol{C}}\cong \boldsymbol{C}^n$ and let $\tilde{G}=(\mathrm{Hol}\,\mathscr{D})^\circ$ denote the identity connected component of the group of holomorphic automorphisms of $\mathscr{D}$. As is well-known, $\mathscr{D}$ is a symmetric domain (hermitian symmetric space of non-compact type) and $\tilde{G}$ is a semi-simple Lie group of hermitian type with center reduced to the identity, of $\boldsymbol{R}$-rank equal to $r=\boldsymbol{R}\text{-rk}\, G$. The group of affine automorphisms of $\mathscr{D}$, $P=(\mathrm{Aff}\,\mathscr{D})^\circ$, which may naturally be identified with the semi-direct product $G\cdot V$, is a parabolic subgroup of $\tilde{G}$ corresponding to a point boundary component of $\mathscr{D}$, which we denote symbolically by $\sqrt{-1}\infty$. The given $\boldsymbol{Q}$-structure on V determines uniquely a $\boldsymbol{Q}$-structure on $\tilde{G}$ such that P and $G(\subset P)$ are subgroups defined over $\boldsymbol{Q}$; then the $\boldsymbol{Q}$-rank of $\tilde{G}$ is equal to $r_0=\boldsymbol{Q}\text{-rk}\, G$. ($\tilde{\mathfrak{g}}=\mathrm{Lie}\,\tilde{G}$ is the so-called "superstructure algebra" of the Jordan algebra (V, c_0), see [S1].) A symmetric tube domain $\mathscr{D}$ with a $\boldsymbol{Q}$-structure on $\tilde{G}$ determined in this manner is called a "rational symmetric tube domain".

Let $\tilde{\Gamma}$ be a neat arithmetic subgroup of $\tilde{G}$. Here $\tilde{\Gamma}$ being "neat" means that for $\gamma\in\tilde{\Gamma}$ if γ^ν is unipotent for some positive integer ν then γ itself is unipotent; in particular, $\tilde{\Gamma}$ is torsion-free. Then $M=\tilde{\Gamma}\cap V$ is a lattice in V and $\Gamma=(\tilde{\Gamma}\cap P)/(\tilde{\Gamma}\cap V)$ may be regarded as a (torsion-free) arithmetic subgroup of G. We assume that our M and Γ (in § 2) are obtained in this manner. Thus one has an exact sequence

$$1\longrightarrow M\longrightarrow \tilde{\Gamma}\cap P\longrightarrow \Gamma\longrightarrow 1.$$

In general, this group extension may not split, but $\tilde{\Gamma}\cap P$ is a subgroup of finite index in the semi-direct product $\Gamma_M\cdot M$.

In what follows, we restrict ourselves to the *case* $r_0=1$. Then it is well-known that the quotient space $Y=\tilde{\Gamma}\backslash\mathscr{D}$ can be compactified to a normal projective variety Y^* by adding a finitely many points p_ν $(1 \leq \nu \leq h)$, called "cusps":

$$Y^* = (\tilde{\Gamma}\backslash\mathscr{D}) \cup \{p_1, \cdots, p_h\}.$$

This is so-to-speak the minimal compactification. Each cusp corresponds to a $\tilde{\Gamma}$-equivalence class of (point) rational boundary components of $\mathscr{D}$, or equivalently to a $\tilde{\Gamma}$-conjugacy class of (proper) $\boldsymbol{Q}$-parabolic subgroups of $\tilde{G}$. We assume that p_ν corresponds to the class of a $\boldsymbol{Q}$-parabolic subgroup P_ν and, in particular, $P_1 = P$, i.e. p_1 is the class of $\sqrt{-1}\infty$.

By the classification theory, it can be seen that the rational symmetric tube domain with $r_0 = 1$ occurs only in the following cases

	$\mathscr{C}$	$\mathscr{D}$	$\tilde{G}_Q$
(Case 1)	$\boldsymbol{R}_+^m$	$(\mathrm{III}_1)^m$	$SL_2(F_1)/\{\pm 1\}$
(Case 2)	$\mathscr{P}_2(\boldsymbol{R})^m$	$(\mathrm{III}_2)^m$	$SU_2(\boldsymbol{D}_1/F_1)/\{\pm 1\}$
(Case 3)	$\mathscr{P}_\delta(\boldsymbol{C})^m$	$(\mathrm{I}_{\delta,\delta})^m$	$SU_2(\boldsymbol{D}_1'/F_1'/F_1)/\{\pm 1\}$

(Case 1) is usually referred to as the "Hilbert modular case". In this list, F_1 is a totally real number field of degree m, $\boldsymbol{D}_1$ is a totally indefinite quaternion algebra over F_1, F_1' is a totally imaginary quadratic extension of F_1, $\boldsymbol{D}_1'$ is a central division algebra over F_1' with involution of the second kind relative to F_1'/F_1, and SU_2 denotes the special unitary group for an "isotropic" hermitian form of 2 variables (i.e. the hermitian form with matrix $\begin{pmatrix} 0 & 1 \\ 1 & 0 \end{pmatrix}$) in $\boldsymbol{D}_1$ or $\boldsymbol{D}_1'$.

3.2. Geometric invariants of cusps. In general, the cusp p_ν is a singular point on Y^*. A neighborhood of p_ν in Y^* is analytically isomorphic to a neighbourhood of p_ν in the local compactification

$$(\tilde{\Gamma} \cap P_\nu\backslash\mathscr{D}) \cup \{p_\nu\}.$$

(Note that the P_ν are all G_Q-conjugate to $P_1 = G \cdot V$.) Resolving these cusp singularities by the method of toroidal embeddings [AMRT], one obtains a smooth compactification:

$$X \xrightarrow{\pi} Y^*,$$

$$X = (\tilde{\Gamma}\backslash\mathscr{D}) \cup D, \quad D = \textstyle\sum_{\nu=1}^h D^{(\nu)},$$

where D is a divisor with simple normal crossings and $D^{(\nu)} = \pi^{-1}(p_\nu)$ is a connected component of D. Let $D = \sum_{i \in I} D_i$ be the irreducible decomposition of D and put

$$I^{(\nu)}=\{i \in I \mid \pi(D_i)=p_\nu\};$$

then $D^{(\nu)}=\sum_{i\in I^{(\nu)}} D_i$. From the construction, $D^{(\nu)}$ is a "toric divisor" (see [E], [S4, 5]); in particular, each D_i is a toric variety.

We denote by γ_i, $\bar{\gamma}_i$ $(1\leq i\leq n)$ the Chern roots of X and the "logarithmic Chern roots" of X relative to D, respectively (see [H1], [S5]). Let δ_i $(i \in I)$ denotes the 2-cohomology class on X defined by D_i. Then in the cohomology ring $H^*(X, \boldsymbol{Z})$ one has the relation

$$\prod_{i=1}^{n}(1+\gamma_i)=\prod_{i=1}^{n}(1+\bar{\gamma}_i)\prod_{i\in I}(1+\delta_i) \tag{3.2.1}$$

and, since D is toric,

$$\bar{\gamma}_i\cdot\delta_j=0 \quad (1\leq i\leq n,\ j\in I) \tag{3.2.2}$$

([S4], Lemma 2). The arithmetic genus of X and the "logarithmic arithmetic genus" of X relative to D are defined, respectively, by

$$\begin{aligned}\chi(X)&=\left(\prod_{i=1}^{n}\frac{\gamma_i}{(1-e^{-\gamma_i})}\right)_n[X],\\ \bar{\chi}(X, D)&=\left(\prod_{i=1}^{n}\frac{\bar{\gamma}_i}{(1-e^{-\bar{\gamma}_i})}\right)_n[X],\end{aligned} \tag{3.2.3}$$

where $(\cdots)_n$ denotes the homogeneous part of degree $2n$ in $H^*(X, \boldsymbol{Z})$ and $(\cdots)_n[X]$ is its evaluation on the fundamental $2n$-cycle $[X]$. One defines the "cusp contribution" to χ at p_ν by

$$\chi_\infty(p_\nu)=\left(\prod_{i\in I^{(\nu)}}\frac{\delta_i}{1-e^{-\delta_i}}\right)_n[X]. \tag{3.2.4}$$

Then from (3.2.1)–(3.2.4) one obtains the relation

$$\chi(X)=\bar{\chi}(X, D)+\sum_{\nu=1}^{h}\chi_\infty(p_\nu). \tag{3.2.5}$$

Note that, in our case, $\bar{\chi}(X, D)$ can easily be computed by the "proportionality theorem" of Hirzebruch-Mumford ([M2]); in particular, it is independent of the toroidal compactification.

Let $\mathcal{N}(D^{(\nu)})$ denote the nerve (or "dual graph") of $D^{(\nu)}$:

$$\mathcal{N}(D^{(\nu)})=\{J\subset I^{(\nu)} \mid J\neq\emptyset,\ X_J=\textstyle\bigcap_{i\in J}D_i\neq\emptyset\}. \tag{3.2.6}$$

Then $\mathcal{N}(D^{(\nu)})$ is a simplicial complex of dimension $n-1$, which, for a toric divisor $D^{(\nu)}$, is an "Euler complex" ([S4], Lemma 3). It follows that, if one puts

$$q^{(\nu)}(t)=\sum_{J\in\mathcal{N}(D^{(\nu)})} t^{n-|J|},$$

then $q^{(\nu)}(t)$ is a polynomial of degree $n-1$ in $\boldsymbol{Z}[t]$ satisfying the functional equation

(3.2.7) $$t^n q^{(\nu)}\left(\frac{1}{t}-1\right)=q^{(\nu)}(t-1)+q^{(\nu)}(-1)(t-1)^n;$$

in particular, one has

(3.2.8) $$q^{(\nu)}(-1)=(1-(-1)^n)2^{-n}q^{(\nu)}(-2).$$

It is easy to see that $q^{(1)}(-1)$ coincides with the Euler number $e(\Gamma\backslash\mathscr{C}^1)$, which vanishes for a self-dual homogeneous cone $\mathscr{C}$ except for (Cases 1, 2) with $m=1$,

Theorem 3.2.9. *When n is odd, one has*

$$\chi_\infty(p_1)=\frac{1}{2}q^{(1)}(-1)\left(=\frac{1}{2}e(\Gamma\backslash\mathscr{C}^1)\right),$$

where $\mathscr{C}^1=\{x\in\mathscr{C}\mid N(x)=1\}\approx\mathscr{C}/\boldsymbol{R}_+$.

(This is essentially due to Ehlers [E], who proved it in the Hilbert modular case. The general case is given in [S4].)

When n is even, the signature $\tau(X)$, the logarithmic signature $\bar{\tau}(X, D)$ and the cusp contribution to the signature $\tau_\infty(p_\nu)$ are defined similarly to the above; for instance,

(3.2.10) $$\tau_\infty(p_\nu)=2^n\left(\prod_{i\in I^{(\nu)}}\frac{1+e^{-\delta_i}}{1-e^{-\delta_i}}\frac{\delta_i}{2}\right)_n[X].$$

Then from (3.2.1), (3.2.2) and from the fact that D is toric one has

(3.2.11) $$\tau(X)=\bar{\tau}(X, D)+\sum_{\nu=1}^{h}\tau_\infty(p_\nu),$$

(3.2.12) $$\tau_\infty(p_\nu)=2^n\chi_\infty(p_\nu)+q^{(\nu)}(-2)$$

(cf. [S4, 5]).

3.3. Generalized Hirzebruch conjecture. Here we assume that *n is even*. In the Hilbert modular case, the cusp p_1 is "rationally pararelizable" in the sense that p_1 has a compact neighbourhood U in Y^* such that $U-\{p_1\}$ does not contain any other singularities and is retractable to ∂U, and that all Chern classes of the tangent bundle $T(U-\{p_1\})|_{\partial U}$ (restricted to ∂U) in $H^*(\partial U, \boldsymbol{Q})$ vanish. In that case, the "signature defect" $\delta(p_1)$ was defined by Hirzebruch [H2], which in our notation is equal to $\tau_\infty(p_1) - \tau(\tilde{U})$, where $\tilde{U} \to U$ is a desingularization of (U, p_1) and the signature $\tau(\tilde{U})$ is computed in $H^*(\tilde{U}, \partial\tilde{U}, \boldsymbol{Q})$. In the general case, we define the signature defect by $\delta(p_1) = \tau_\infty(p_1) - \tau(\tilde{U})$. (For a direct generalization of the definition, see Looijenga [L]). Then it seems likely that one has

(C1) $$\delta(p_1) = 2^n \chi_\infty(p_1),$$

or equivalently in view of (3.2.12),

(C1′) $$\tau(\tilde{U}) = q^{(1)}(-2).$$

In the Hilbert modular case, Hirzebruch ([H2], p. 230) conjectured that

$(\mathrm{C2})_{r_1=1}$ $$\delta(p_1) = L_{(1)}({}^t\Gamma, M^*; 0),$$

which was proved by Atiyah-Donnelly-Singer [ADS1, 2] and Müller [M4, 5]. (In the case $n=2$, this relation and (C1) were already proved in [H2].) In general (at least for (Case 1, 3)), in view of Theorem 2.4.1 and (C1), it seems natural to conjecture

(C2) $$\delta(p_1) = 2^{n-r} L_{I_1}({}^t\Gamma, M^*; 0),$$

where $I_1 = (0)$ or (r_1) according as $n/r \equiv 0$ or $1 \pmod 2$, and

(C3) $$\chi_\infty(p_1) = Z_{\mathscr{C}}({}^t\Gamma, M^*; 0),$$

or equivalently

(C3′) $$\chi_\infty(p_1) = 2^{-r} L_{I_1}({}^t\Gamma, M^*; 0).$$

The relation similar to (C3) or (C3′) for n odd, where $\chi_\infty(p_1)$ should be replaced by $-\chi_\infty(p_1)$, was proved by Ogata [O], as we shall see in the next section (Theorem 4.2.3, note that, in (Cases 1, 3) with n odd>1, one has $Z_{\mathscr{C}}(0) = \chi_\infty(p_1) = 0$ by Theorems 2.4.1 and 3.2.9). In the Hilbert modular case, the conjecture $(\mathrm{C3}')_{r_1=1}$ was mentioned in [E] and [HG] (p. 95). In this case, comparing the cusp contribution in the dimension formulae for the space of Hilbert cusp forms obtained by Selberg trace formula and by Riemann-Roch-Hirzebruch Theorem ([H1]), one obtains a "weaker form" of $(\mathrm{C3}')_{r_1=1}$:

$$\sum_{\nu=1}^{h} \chi_\infty(p_\nu) = 2^{-n} \sum_{\nu=1}^{h} L_{(1)}({}^t\Gamma_\nu, M_\nu^*; 0)$$

([S6], [F], [S3]; cf. also [A] for (Case 2)).

The relation between these conjectures is shown in the following diagram:

$$\begin{array}{ccc} \chi_\infty(p_1) & \overset{(C3)}{=\!=\!=} & Z_{\mathscr{C}}({}^t\Gamma, M^*; 0) \\ (C1)\Vert & \overset{(C3')}{\diagdown\!\diagdown} & \Vert(\text{Th. } 2.4.1) \\ 2^{-n}\delta(p_1) & \underset{(C2)}{=\!=\!=} & 2^{-r}L_{I_1}({}^t\Gamma, M^*; 0) \end{array}$$

Since the proofs of $(C2)_{r_1=1}$ given in [ADS1, 2] and [M4, 5] are both rather complicated, depending on differential geometry and hard analysis, it seems desirable to give more direct proofs for (C1) and (C3) or (C3′)*⁾.

§ 4. Zeta functions associated to Tsuchihashi singularities

4.1. Tsuchihashi singularities. We consider a normal isolated singularity called "Tsuchihashi cusp" and define the cusp contribution χ_∞ for this kind of singularities.

As before, let V be a real vector space of dimension n and M a lattice (of rank n) in V. Consider a pair $(\mathscr{C}, \Gamma)$ consisting of a (non-degenerate, open convex) cone $\mathscr{C}$ in V, which may not be self-dual nor homogeneous, and a subgroup Γ of $GL(V)$ satisfying the following conditions:

(i) Γ leaves M invariant and is torsion-free;

(ii) Γ leaves $\mathscr{C}$ invariant;

(iii) the quotient space $\Gamma\backslash\mathscr{C}/\boldsymbol{R}_+$ is compact.

Let $\mathscr{D}=V+\sqrt{-1}\mathscr{C}$ be the corresponding (not necessarily symmetric) tube domain. Tsuchihashi [T] constructed a normal isolated singularity associated to the pair $(\mathscr{C}, \Gamma)$, which is the singularity at "infinity" p_1 of $(\Gamma M\backslash\mathscr{D})\cup\{p_1\}$. Let U be a suitable (open) neighbourhood of p_1 and π: $\tilde{U}\to U$ a toroidal desingularization. Let $\pi^{-1}(p_1)=D^{(1)}=\sum_{i\in I^{(1)}} D_i$ be the decomposition of the exceptional set into the union of irreducible components, and let δ_i be the cohomology class determined by D_i in $H^2_c(\tilde{U}, \boldsymbol{Z})$ (the integral 2-cohomology group with compact support) and $[\tilde{U}]$ the fundamental class of $\tilde{U}$. We define $\chi_\infty(p_1)$ by

$$\chi_\infty(p_1)=\left(\prod_{i\in I^{(1)}} \frac{\delta_i}{1-e^{-\delta_i}}\right)_n [\tilde{U}].$$

*⁾ Some statements in [S4] and [O] on Hirzebruch conjecture were incorrect or misleading. In [S4], p. 366, *l.* 18, "χ_∞" should read "$4\chi_\infty$" and the "Hirzebruch conjecture" there should be understood in a weaker sense that the sums of each side of (C2) over all cusps are equal. The "conjecture" on [O], p. 370, *l.* 14 is correct for $n=2$ but should be rectified by (C3) for $n\geq 4$. Note also that our invariants δ and χ_∞ are written in [E] as $2^n\phi$ and ψ.

Then, Theorem 3.2.9 and (3.2.12) remain true, $\tau_\infty(p_1)$ being defined similarly to (3.2.10). We want to relate this cusp contribution to a special value of a zeta function associated to $(\mathscr{C}, \Gamma)$.

4.2. Zeta function associated to $(\mathscr{C}, \Gamma)$. Let $(\mathscr{C}, \Gamma)$ be a pair satisfying (i)–(iii). Let $\mathscr{C}^*$ be the dual cone of $\mathscr{C}$ and M^* the dual lattice of M inthe dual space V^*. $\phi_{\mathscr{C}}(x)$ denotes the characteristic function of the cone $\mathscr{C}$ defined in 1.2. We define the zeta function associated to $(\mathscr{C}, \Gamma)$ by

$$\text{(4.2.1)} \qquad \boldsymbol{Z}_{\mathscr{C}}(\Gamma, M; s) = \sum_{u : \Gamma\backslash\mathscr{C}\cap M} \phi_{\mathscr{C}}(u)^s \quad (\operatorname{Re} s > 1).$$

Note that, when $\mathscr{C}$ is self-dual and homogeneous, we have by (1.4.3)

$$Z_{\mathscr{C},c_0}(\Gamma, M; s) = \phi_{\mathscr{C}}(c_0)^{-(r/n)s} \boldsymbol{Z}_{\mathscr{C}}\left(\Gamma, M; \frac{r}{n} s\right).$$

Theorem 4.2.2. *The function $\boldsymbol{Z}_{\mathscr{C}}(\Gamma, M; s)$ admits a meromorphic continuation to the whole plane and is holomorphic at $s=0$.*

Theorem 4.2.3. *When n is odd, one has*

$$\boldsymbol{Z}_{\mathscr{C}}(\Gamma, M; 0) = -\frac{1}{2} e(\Gamma\backslash\mathscr{C}/\boldsymbol{R}_+) = -\chi_\infty(p_1).$$

A sketch of proofs of these theorems will be given in 4.3.

First, in order to describe the zero-value of $\boldsymbol{Z}_{\mathscr{C}}$, we need the notion of "rational partial polyhedral decomposition" (r.p.p. decomposition, for short) of $\mathscr{C} \cup \{0\}$.

Definition 4.2.4. A (non-empty) collection Σ of closed rational polyhedral cones in V is called an *r.p.p. decomposition* of $\mathscr{C} \cup \{0\}$ if it satisfies the following conditions:

(1) If $\sigma \in \Sigma$ and $\tau \prec \sigma$ (i.e. τ is a face of σ), then $\tau \in \Sigma$. In particular, one has $\{0\} \in \Sigma$. We set $\Sigma^\times = \Sigma - \{0\}$.

(2) If $\sigma, \tau \in \Sigma$, then $\sigma \cap \tau \prec \sigma$.

(3) One has $\mathscr{C} = \coprod_{\sigma \in \Sigma^\times} \operatorname{Int}(\sigma)$.

(4) For any compact subset K contained in $\mathscr{C}$, the set $\{\sigma \in \Sigma \mid \sigma \cap K \neq \emptyset\}$ is finite.

In what follows, we further assume that Σ is "Γ-invariant" and "non-singular", i.e. the following additional conditions are satisfied:

(5) Γ leaves Σ invariant, and acts freely on $\Sigma^\times$.

(6) For every σ in Σ, there exists a $\boldsymbol{Z}$-basis $\{u_1, \cdots, u_n\}$ of M and r with $0 \leq r \leq n$ such that $\sigma = \{u_1, \cdots, u_r\}_{\boldsymbol{R}_{\geq 0}}$.

We now assume that the toroidal desingularization $\pi : \tilde{U} \to U$ is defined by a Γ-invariant and non-singular r.p.p. decomposition Σ of $\mathscr{C}\cup\{0\}$. As before, let $\mathscr{N}(D^{(1)})$ be the nerve of $D^{(1)}$. Then there exists an injective map

$$\mathscr{N}(D^{(1)}) \ni J \longmapsto \sigma_J \in \Sigma^\times$$

such that $\{\sigma_J \mid J \in \mathscr{N}(D^{(1)})\}$ is a complete set of representatives of $\Gamma\backslash\Sigma^\times$, $\dim \sigma_J = |J|$, and that for any $J, J' \in \mathscr{N}(D^{(1)})$ one has $J \subset J'$ if and only if $\gamma(\sigma_J) \subset \sigma_{J'}$ for some $\gamma \in \Gamma$. For $\sigma \in \Sigma^\times$, we put

$$\sigma(1) = \{\tau \in \Sigma \mid \dim \tau = 1 \text{ and } \tau \prec \sigma\}.$$

We denote by dx_σ the Lebesgue measure on the linear subspace $\sigma+(-\sigma)$ of V normalized so that for a $\boldsymbol{Z}$-basis $\{u_1, \cdots, u_n\}$ of M with $\sigma = \{u_1, \cdots, u_r\}_{\boldsymbol{R}\geq 0}$ the volume of the parallelotope spanned by $\{u_1, \cdots, u_r\}$ is one. For $\rho \in \Sigma$ with $\dim \rho = 1$, the symbol ∂_ρ denotes the derivation

$$(\partial_\rho F)(x) = \lim_{t\to 0} \frac{1}{t}\{F(x+tu) - F(x)\}$$

for any differentiable function $F(x)$ on V, where u is the unique primitive element in $\rho \cap M$.

In these notations, we have

Theorem 4.2.5. *For any integer $\nu \geq 2$, one has*

$$\begin{aligned} Z_{\mathscr{C}}(\Gamma, M; 0) &= \sum_{\tau: \Gamma\backslash\Sigma^\times} \int_\tau \left[\prod_{\rho\in\tau(1)} \frac{\partial_\rho}{1-e^{-\partial_\rho}}\right]_{\dim \tau} G_\nu(x)dx_\tau \\ &= (-1)^n 2^{-n} q^{(1)}(-2) + \sum_{J\in\mathscr{N}(D^{(1)})} (-2)^{-(n-|J|)} \tau(X_J) \\ &\quad \times \int_{\sigma_J} \left[\prod_{\rho\in\sigma_J(1)} \frac{\partial_\rho}{2} \frac{1+e^{-\partial_\rho}}{1-e^{-\partial_\rho}}\right]_{|J|} G_\nu(x)dx_{\sigma_J}, \end{aligned}$$

where $G_\nu(x) = \exp(-\phi_{\mathscr{C}}(x)^{-\nu})$ and, for any rational function $\Phi(t) \in \boldsymbol{Q}(t)$, $[\prod_{\rho\in\sigma(1)} \Phi(\partial_\rho)]_k$ is the differential operator of degree k on V obtained as the homogeneous part of total degree k in the formal power series expansion of $\prod_{\rho\in\sigma(1)} \Phi(\partial_\rho)$.

Ishida's proof ([I3]) of the rationality of $Z_{\mathscr{C}}(\Gamma, M; 0)$ is based on the above formula. We note a remarkable similarity of it to (3.2.4) and (3.2.12).

4.3. Outline of proofs ([O]). For $\sigma \in \Sigma^\times$ let

$$Z(\sigma, M; s) = \sum_{u\in M\cap \mathrm{Int}(\sigma)} \phi_{\mathscr{C}}(u)^s, \quad \mathrm{Re}\, s > 1.$$

We prove that $Z(\sigma, M; s)$ can be continued to a meromorphic function on $\boldsymbol{C}$ and calculate the value $Z(\sigma, M; 0)$.

Let $\dim \sigma = r \geq 1$ and $u_1, \cdots, u_r$ a part of a $\boldsymbol{Z}$-basis of M with $\sigma = \{u_1, \cdots, u_r\}_{\boldsymbol{R}_{\geq 0}}$. Then we may write as

$$Z(\sigma, M; s) = \sum_{(l_1, \cdots, l_r) \in (\boldsymbol{Z}_{>0})^r} \phi_{\mathscr{C}}(l_1 u_1 + \cdots + l_r u_r)^s.$$

We employ the method of Zagier [Z] who calculated the values of the zeta functions of real quadratic fields. For simplicity we consider the case $r=2$. The following proposition is well-known.

Proposition 4.3.1. *Let $\psi(s) = \sum_{k=0}^{\infty} a_k \lambda_k^{-s}$ with $\lambda_k > 0$ be a Dirichlet series absolutely convergent for* Re $s > 1$. *Assume that the function $h(t) = \sum_{k=0}^{\infty} a_k \exp(-\lambda_k t)$ has an asymptotic expansion of the form*

$$h(t) \sim b_{-1} \cdot \frac{1}{t} + b_0 + b_1 t + \cdots \quad (t \to 0).$$

Then $\psi(s)$ admits a meromorphic continuation to $\boldsymbol{C}$ and is holomorphic at $s=0$ with the value $\psi(0) = b_0$.

In order to apply this proposition we need an asymptotic expansion of $h(t)$. It is easy to derive the following proposition from the Euler-Maclaurin summation formula.

Proposition 4.3.2. *Let $f(t)$ be a real-valued C^∞-function on* $[0, \infty)$. *Assume that $\int_0^\infty f(t)\,\mathrm{d}t$ is finite. Then $g(t) = \sum_{l=1}^{\infty} f(tl)$ has an asymptotic expansion of the form*

$$g(t) \sim \frac{1}{t} \cdot \int_0^\infty f(t)\,\mathrm{d}t + \sum_{l=0}^{\infty} t^l \beta_l f^{(l)}(0) \quad (t \to 0),$$

where β_l's are the coefficients in the expansion $t/(1-e^{-t}) = \sum_{l=0}^{\infty} \beta_l t^l$.

Applying this to a function to a function $F(x, y)$ of two variables, we have under a certain condition

$$\sum_{m, n \in \boldsymbol{Z}>0} F(mt, nt) \sim \frac{1}{t^2}\left(\int_0^\infty \int_0^\infty F(x, y)\,\mathrm{d}x\mathrm{d}y\right)$$
$$+ \sum_{i=1}^{\infty} \beta_i t^{i-1} \int_0^\infty F^{(i,0)}(0, y)\,\mathrm{d}y + \sum_{j=1}^{\infty} \beta_j t^{j-1} \int_0^\infty F^{(0,j)}(x, 0)\,\mathrm{d}x$$
$$+ \sum_{i,j \in \boldsymbol{Z}_{>0}} \beta_i \beta_j F^{(i,j)}(0, 0) t^{i+j} \quad (t \to 0),$$

where $F^{(i,j)}(x, y) = (\partial^{i+j}/\partial x^i \partial y^j) F(x, y)$. Unfortunately this can not be applied directly to $F(x, y)=\exp(-\phi_{\mathscr{C}}(x\cdot u_1+y\cdot u_2)^{-1})$, because the derivatives of $\exp(-\phi_{\mathscr{C}}(x)^{-1})$ behave badly near $x=0$. In order to surmount this difficulty, we need the following

Lemma 4.3.3. *For a positive integer* ν *and an r-dimensional cone* $\sigma=\{u_1, \cdots, u_r\}_{\boldsymbol{R}_{\geq 0}}$ *contained in* $\mathscr{C}$, *the function*

$$G_{\sigma,\nu}(x_1, \cdots, x_r)=\exp(-\phi_{\mathscr{C}}(x_1u_1+\cdots+x_ru_r)^{-\nu})$$

and its partial derivatives of total order up to $n\nu-1$ *have limits at the origin and the partial derivatives of total order* $n\nu$ *are bounded.*

Modifying Zagier's method and using this lemma, we can prove that the function $\boldsymbol{Z}(\sigma, M; \nu s)$ can be continued to the half plane $\operatorname{Re} s>-1+1/\nu$. Thus $\boldsymbol{Z}(\sigma, M; s)$ can be continued to $\operatorname{Re} s>-\nu+1$, and hence to the whole complex plane. And we get the value of $\boldsymbol{Z}(\sigma, M; s)$ at $s=0$:

Proposition 4.3.4. *For any integer* $\nu\geq 2$ *we have*

$$\boldsymbol{Z}(\sigma, M; 0)=\sum_{|k|=r}\int_{(\boldsymbol{R}_{\geq 0})^r}\beta_k\left(\frac{\partial}{\partial x}\right)^k G_{\sigma,\nu}(x)\,\mathrm{d}x,$$

where $k=(k_1, \cdots, k_r)\in(\boldsymbol{Z}_{\geq 0})^r$, $\beta_k=\beta_{k_1}\cdots\beta_{k_r}$, *and* $(\partial/\partial x)^k=(\partial/\partial x_1)^{k_1}\cdots(\partial/\partial x_r)^{k_r}$.

We rewrite this as

$$\boldsymbol{Z}(\sigma, M; 0)=\int_\sigma\left[\prod_{\rho\in\sigma(1)}\frac{\partial_\rho}{1-e^{-\partial_\rho}}\right]_{\dim\sigma} G_\nu(x)\,\mathrm{d}x_\sigma,$$

where $G_\nu(x)=\exp(-\phi_{\mathscr{C}}(x)^{-\nu})$. By summing this equality side by side over $\sigma\in\Sigma^\times \bmod \Gamma$ we get the first equality in Theorem 4.2.5. By a calculation similar to that leading to (3.2.12) we get the second expression in Theorem 4.2.5, whence follows Theorem 4.2.3.

References

[A] T. Arakawa, The dimension of the space of cusp forms on the Siegel upper half plane of degree two related to a quaternion unitary group, J. Math. Soc. Japan, **33** (1981), 125–145.

[ADS1] M. F. Atiyah, H. Donnelly and I. M. Singer, Eta invariants, signature defects of cusps, and values of L-functions, Ann. of Math., **118** (1983), 131–177.

[ADS2] ——, Signature defects of cusps and values of L-functions, The non-split case, ibid., **119** (1984), 635–637.

[AMRT] A. Ash, D. Mumford, M. Rapoport and Y. Tai, Smooth Compactification of Locally Symmetric Varieties, Math. Sci. Press, Brookline,

1975.

[BK] H. Braun and M. Koecher, Jordan-Algebren, Springer-Verlag, Berlin-Heidelberg-New York, 1966.

[E] F. Ehlers, Eine Klasse komplexer Mannigfaltigkeiten und die Auflösung einiger isolierter Singularitäten, Math. Ann., **218** (1975), 127–156.

[F] E. Freitag, Lokale und globale Invarianten der Hilbertschen Modulgruppe, Invent. Math., **17** (1972), 106–134.

[G] S. G. Gindikin, Analysis in homogeneous domains, Uspehi Mat. Nauk., **19** (1964), 3–92; = Russian Math. Surveys, **19** (1964), 1–89.

[H1] F. Hirzebruch, Topological Methods in Algebraic Geometry (3-rd ed.), Springer-Verlag, Berlin-Heidelberg-New York, 1978.

[H2] ——, Hilbert modular surfaces, Enseign. Math., **19** (1974), 183–281; =Monographie **21**.

[HG] F. Hirzebruch and G. van der Geer, Lectures on Hilbert Modular Surfaces, Presses de l'Université de Montréal, Montréal, 1981.

[I1] J.-I. Igusa, On functional equations of complex powers, Invent. math., **85** (1986), 1–29.

[I2] ——, Zeta distributions associated with some invariants, Amer. J. Math., **109** (1987), 1–34.

[I3] M.-N. Ishida, *T*-complexes and Ogata's zeta zero values, Adv. St. in Pure Math., this volume.

[JNW] P. Jordan, J. von Neumann and E. Wigner, On an algebraic generalization of the quantum mechanical formalism, Ann. of Math., **35** (1934), 29–64.

[K] A. Kurihara, On the values at non-positive integers of Siegal's zeta functions of Q-anisotropic quadratic forms with signature $(1, n-1)$, J. Fac. Sci., Univ. Tokyo, **28** (1982), 567–584.

[KKMS] G. Kempf, F. Knudson, D. Mumford and B. Saint-Donat, Toroidal Embeddings I, Lecture Notes in Math. **339**, Springer-Verlag, Berlin-Heidelberg-New York, 1972.

[L] E. Looijenga, Riemann-Roch and smoothings of singularities, Topology, **25** (1986), 293–302.

[M1] Y. Morita, An explicit formula for the dimension of spaces of Siegel modular forms of degree two, J. Fac. Sci., Univ. Tokyo, **21** (1974), 167–248.

[M2] D. Mumford, Hirzebruch's proportionality theorem in the non-compact case, Invent. Math., **42** (1977), 239–272.

[M3] M. Muro, Micro-local analysis and calculations of functional equations and residues of zeta functions associated with the vector spaces of quadratic forms, Preprint, 1982, partially published in [M3a].

[M3a] ——, Microlocal analysis and calculations on some relatively invariant hyperfunctions related to zeta functions associated with the vector spaces of quadratic forms, Publ. RIMS, Kyoto Univ., **22** (1986), 395–463.

[M4] W. Müller, Signature defects of cusps of Hilbert modular varieties and values of L-series at $s=1$, J. Diff. Geom., **20** (1984), 55–119.

[M4a] ——, L^2-index of elliptic operators on manifolds with cusps of rank one, Akad. Wiss. DDR, Berlin, 1985.

[M5] ——, Manifolds with Cusps of Rank One, Spectral Theory and L^2-Index Theorem, Lect. Notes in Math. **1244**, Springer-Verlag, 1987.

[O] S. Ogata, Special values of zeta functions associated to cusp singularities, Tôhoku Math. J., **37** (1985), 367–384.

[R] H. L. Resnikoff, On a class of linear differential equations for automorphic forms in several complex variables, Amer. J. Math., **95** (1973), 321–331.

[S1] I. Satake, Algebraic Structures of Symmetric Domains, Iwanami-Shoten and Princeton Univ. Press, 1980.

[S2] ——, Special values of zeta functions associated with self-dual homogeneous cones, in Manifolds and Lie Groups (Notre Dame, Ind., 1980), Progress in Math. **14**, Birkhäuser, Basel-Boston-Stuttgart, 1981, 359–384.

[S3] ——, On numerical invariants of arithmetic varieties (the case of Q-rank one) (in Japanese), Sugaku, **35** (1983), 210–220; English transl., Sugaku Expositions, **1** (1988), 1–16.

[S4] ——, On numerical invariants of arithmetic varieties of Q-rank one, in Automorphic Forms of Several Variables, Taniguchi Symposium, Katata, 1983 (I. Satake and Y. Morita, eds.), Progress in Math. **46**, Birkhäuser, Basel-Boston-Stuttgart, 1984, 353–369.

[S5] ——, On the y-genera of arithmetic varieties of Q-rank one, Advances in Math. (Chinese Math. Soc.), **16** (1987), 269–276.

[S5a] ——, On Zeta functions associated with self-dual homogeneous cones, to appear in the Proceedings of S. Ramanujan Birth Centenary Int. Coll., TIFR, Bombay, 1988.

[S6] H. Shimizu, On discontinuous groups operating on the product of upper half planes, Ann. of Math., **77** (1963), 33–71.

[S7] T. Shintani, On zeta-functions associated with the vector space of quadratic forms, J. Fac. Sci., Univ. Tokyo, **22** (1975), 25–65.

[S8] ——, On evaluation of zeta functions of totally real algebraic number fields at non-positive integers, J. Fac. Sci., Univ. Tokyo, **23** (1976), 393–417.

[S9] C. L. Siegel, Über die Zetafunktionen indefiniter quadratischer Formen, Math. Z., **43** (1938), 393–417.

[SF] I. Satake and J. Faraut, The functional equation of zeta distributions associated with formally real Jordan algebras, Tôhoku Math. J., **36** (1984), 469–482.

[SS] M. Sato and T. Shintani, On zeta functions associated with prehomogeneous vector spaces, Ann. of Math., **100** (1974), 131–170.

[T] H. Tsuchihashi, Higher dimensional analogues of periodic continued fractions and cusp singularities, Tôhoku Math. J., **35** (1983), 607–639.

[V] E. B. Vinberg, The theory of convex homogeneous cones, Trudy Moskov. Mat. Obšč., **12** (1963), 303–358; = Trans. Moscow Math. Soc. 1963, 340–403.

[Z] D. Zagier, Valeurs des fonctions zêta des corps quadratiques réels aux entiers négatifs, in Journées Arithmétiques de Caen (Univ. Caen, 1976), Astérisque Nos. **41–43**, Soc. Math. France, Paris, 1977, 135–151.

Department of Mathematics
Tohoku University
Sendai 980
Japan

Advanced Studies in Pure Mathematics 15, 1989
Automorphic Forms and Geometry of Arithmetic Varieties
pp. 29–40

Cusps on Hilbert Modular Varieties and Values of L-Functions

Robert Sczech

§ 1.

Let s be a cusp, and $D=\sum S_\tau$ the corresponding cusp divisor on a Hilbert modular variety X. Every such a cusp belongs to a pair (M, V) where M is a lattice (isomorphic to $\boldsymbol{Z}^n$), and V a group of units (isomorphic to $\boldsymbol{Z}^{n-1}$) in a totally real number field F of degree n over $\boldsymbol{Q}$, subject to the restriction that all elements in V are totally positive, and that V acts on M by multiplication, $VM=M$. However, the cusp divisor D is not unique for a given pair (M, V).

The divisor D is a normal crossing divisor, i.e. the irreducible components S_τ (hypersurfaces on X) intersect only in simple normal crossings. The complicated intersection behavior of the S_τ can be described in terms of a triangulation of the $(n-1)$-torus $\boldsymbol{R}^{n-1}/V$. Every hypersurface S_τ corresponds to a vertex τ of this triangulation, and k different hypersurfaces S_{τ_j} $(1\leq j\leq k)$ intersect either in a $(n-k)$-dimensional submanifold S_σ, or the intersection set is empty. In the first case, σ is the unique simplex of the triangulation having the τ_j as vertices.

This description of the cusp divisor D was given for the first time by Hirzebruch [4] in the case of a real quadratic field F $(n=2)$. He showed in particular that the corresponding triangulation of the torus $S^1=\boldsymbol{R}/V$ is given by the continued fraction expansion of a quadratic irrationality associated with M. In the same paper, Hirzebruch defined a rational number $\varphi(s)=\varphi(M, V)$ called the signature defect of s, in the following way: let Y be a small closed neighbourhood in X of the cusp s. Then Y is a manifold with boundary ∂Y which is a $T^n=\boldsymbol{R}^n/M$ bundle over the torus $T^{n-1}=\boldsymbol{R}^{n-1}/V$ completely determined by the pair (M, V). Let $L(Y)$ be the L-polynomial in the relative Chern classes of Y, and $\operatorname{sign}(Y)$ the signature of Y. From the signature theorem [1], it follows that

$$\varphi(M, V) := L(Y) - \operatorname{sign}(Y)$$

Received March 28, 1987.
Received April 19, 1988.

depends only on the boundary ∂Y of Y, but not on Y itself (which depends on the particular divisor D). This number is called the signature defect of s because in the hypothetical case $\partial Y=\phi$ we would have $\varphi(M, V)=0$. Although easy to define, this number is not easy to compute. But in the case $n=2$, using the explicit description of D, Hirzebruch found the beautiful formula

$$\varphi(M, V)=-\frac{1}{3}\sum_k (b_k-3), \tag{1}$$

where the b_k are the integers arising from the continued fraction expansion of a quadratic irrationality associated with M (the number of terms depends on V).

On the other hand, the pair (M, V) determines a Hecke L-function

$$L(M, V, s)=\sum_{m\in M/V}{}' \frac{\operatorname{sign} N(m)}{|N(m)|^s}, \qquad \operatorname{Re}(s)>1.$$

Here $Y(m)$ is the norm of m, and $m\neq 0$ runs through a set of representatives for the equivalence classes of M modulo V. The special value $L(M, V, 1)$ is well defined because $L(M, V, s)$ has an analytic continuation to the whole complex s-plane. In the quadratic case $n=2$, the calculation of this number was begun by Hecke and completed by C. Meyer and Siegel. Using their results, Hirzebruch proved the relation

$$\varphi(M, V)=\frac{d(M)}{(\pi i)^n}L(M, V, 1) \tag{2}$$

for $n=2$, and then conjectured it for all $n>2$. Here $d(M)$ denotes the volume of a fundamental domain for the action of M on $\boldsymbol{R}^n$. The attraction of this conjecture stems from the fact that the left-hand side reflects only pure topological properties of the cusp whereas the right-hand side is a pure arithmetic quantity. However, at the time of Hirzebruch's paper, it was not even known that the right hand side is always a rational number.

Today, this conjecture is a theorem of Atiyah, Donnelly and Singer who gave a proof in the important recent paper [2]. But in comparison to Hirzebruch's original proof for $n=2$, their proof is not constructive in the sense that it does not answer the question how to calculate both sides of (2) independently. This question makes sense since Ehlers [3] generalized Hirzebruch's construction of the cusp divisor D to all totally real number fields which leads to a closed expression for $\varphi(M, V)$ in terms of the divisor D, and Shintani [8] established a closed formula for $L(M, V, 1)$

which makes it evident that the right hand side of (2) is a rational number. Indeed, a key element in Shintani's formula is also a triangulation of the torus $\boldsymbol{R}^{n-1}/V$. But unfortunately, a proof of (2) along these lines is not easy to accomplish, mostly because of an unexpected relation between $\varphi(M, V)$ and $\varphi(M^*, V)$ which we explain in section 5.

In this paper, we give an explicit formula for $\varphi(M, V)$ in terms of the triangulation of $\boldsymbol{R}^{n-1}/V$ generalizing Hirzebruch's formula (1) in the case $n=2$. In addition, we discuss a new idea for calculating $L(M, V, 1)$ which would lead to the same closed formula for $L(M, V, 1)$ as for $\varphi(M, V)$. The idea is essentially to identify (the conditionally convergent series) $L(M, V, 1)$ as a partial fraction decomposition of $\varphi(M, V)$ using Euler's formula

$$\pi \cot \pi x = \sum_{m \in \boldsymbol{Z}+x}' \frac{1}{m},$$

which is in some sense the case $n=1$ of Hirzebruch's conjecture. If successful, this idea would lead to an elementary proof of Hirzebruch's conjecture. However, at the time of this report, not all of the technical difficulties are solved which arise in the case $n>2$.

§ 2.

We explain our notation. Let $C(k)$ be the set of all $(k-1)$-simplices σ $(1 \leq k \leq n)$ in the triangulation of $\boldsymbol{R}^{n-1}/V$, and denote by C the set of all simplices. In this paper, the letter τ will be used exclusively to denote an element of $C(1)$, and the letter t will always stand for an element in $C(n)$. For example, $\tau \in t$ means τ is a vertex of t. The complex C associated with the divisor D has a second characteristic property. For every $\sigma \in C(k)$, there are k linearly independent lattice points $A_\tau \in M$ (determined only up to a unit in V) such that

$$M \cap \sum_{\tau \in \sigma} \boldsymbol{R} A_\tau = \sum_{\tau \in \sigma} \boldsymbol{Z} A_\tau.$$

(In writing this equation, we assume that M is embedded in $\boldsymbol{R}^n$ through the n different embeddings $F \hookrightarrow \boldsymbol{R}$.) In particular, for every $t \in C(n)$ there is a distinguished $\boldsymbol{Z}$-basis $\{A_\tau, \tau \in t\}$ for M,

$$M = \sum_{\tau \in t} \boldsymbol{Z} A_\tau.$$

Let $\{B_\tau^t\}$ be the dual basis of $M^* = \sum_{\tau \in t} \boldsymbol{Z} B_\tau^t$ given by

$$\operatorname{tr}(A_{\tau'} B_\tau^t) = \begin{cases} 1, & \text{if } \tau = \tau' \in t \\ 0, & \text{if } \tau, \tau' \in t,\ \tau \neq \tau'. \end{cases} \tag{3}$$

The compact submanifolds S_σ define homology classes in $H_{2n-2k}(Y, \boldsymbol{Z})$ if $\sigma \in C(k)$. We want to compute the intersection product ΠS_σ of these homology classes in the homology ring $H_*(Y, \boldsymbol{Z})$. From the construction of the divisor D, c.f. [3], it is clear that

$$\prod_{\tau \in t} S_\tau = 1$$

because the n different hypersurfaces corresponding to the n vertices of $t \in C(n)$ intersect transversally in one point. More generally, we have

$$\prod_{\tau \in \sigma} S_\tau = S_\sigma.$$

To allow multiplicities, we consider a partition p of n,

$$n = \sum_{\tau \in \sigma} p_\tau, \qquad p_\tau \geq 1,$$

where the indices of p run over the vertices of σ. Then the intersection numbers

$$S_\sigma^p := \prod_{\tau \in \sigma} S_\tau^{p_\tau} \in H_0(Y, \boldsymbol{Z}) = \boldsymbol{Z}$$

are well defined integers. Provided that all hypersurfaces S_τ do not have any self-intersections, we can calculate these intersection numbers using the following

Theorem 1. $$S_\sigma^p = \sum_{t \in St(\sigma)} \left(\prod_{\tau \in t} B_\tau^t\right)^{-1} \prod_{\tau \in \sigma} (B_\tau^t)^{p_\tau}.$$

Here t runs over all $(n-1)$-simplices in $\mathrm{St}(\sigma) = \{\sigma' \in C \mid \sigma \subseteq \sigma'\}$, the star of σ, and $B_\tau^t \in F \subseteq \boldsymbol{R}$ denotes the algebraic number defined by (3). The stated formula is a special case of the more general formula

(4) $$S_\sigma^p = \sum_{t \in \mathrm{St}(\sigma)} \left(\prod_{\tau \in t} \mathrm{tr}\,(xB_\tau^t)\right)^{-1} \prod_{\tau \in \sigma} \mathrm{tr}\,(xB_\tau^t)^{p_\tau}$$

valid for all $x \in F$ as long as none of the denominators vanishes. After embedding F in $\boldsymbol{R}^n$, the formula extends (by continuity) to all $x \in \boldsymbol{R}^n$ outside the finite set of hyperplanes defined by $\mathrm{tr}\,(xB_\tau^t) = 0$, $\tau \in t$, $t \in \mathrm{St}\,(\sigma)$. In particular, taking a unit vector for x, we get the special case stated in Theorem 1.

All these formulas can be proved by using induction over k if $\sigma \in C(k)$, starting with $k = n$ which is easy since $\mathrm{St}\,(\sigma) = \{\sigma\}$ and all $p_\tau = 1$ in this case. For the induction step $k \to k-1$, we use the fact that the intersection number $d \circ S$ of

$$S = \prod_{\tau \in \sigma} S_\tau^{q_\tau}, \quad \sigma \in C(k-1), \qquad \sum_{\tau \in \sigma} q_\tau = n-1, \quad q_\tau \geq 1,$$

with the divisor d of a meromorphic function on Y is zero. In particular, for

$$d=\sum_{\tau\in C(1)} c_\tau S_\tau, \qquad c_\tau \in \boldsymbol{Z}$$

we get

$$(\textstyle\sum_1+\sum_2)c_\tau S_\tau \circ S=0$$

where $\sum_1$, runs over all $\tau \in \sigma$, and $\sum_2$ runs over all $\tau \notin \sigma$. By induction hypothesis, we know all the intersection numbers in $\sum_2$. Varying the divisor d, we get in this way a system of linear equations for the intersection numbers $S_\tau \circ S$, $\tau \in \sigma$, which has a unique solution given by (4).

The formula stated in Theorem 1 looks complicated, but it has the following nice property: assume n is even, and let $\sigma, \sigma' \in C(n/2)$. Defining $B_\tau^t=0$ for τ not contained in t, we can write the intersection number $S_\sigma \circ S_{\sigma'}$ as

$$S_\sigma \circ S_{\sigma'}=\sum_t \Big(\prod_{\tau\in\sigma} B_\tau^t\Big)\Big(\prod_{\tau\in t} B_\tau^t\Big)^{-1}\Big(\prod_{\tau\in\sigma} B_\tau^t\Big),$$

where now t runs over all simplices in $C(n)$. To put this formula into the right perspective, write

$$x_{\sigma,t} := \prod_{\tau\in\sigma} B_\tau^t$$

for $\sigma \in C$, $t \in C(n)$, and consider the matrices

$$X=(x_{\sigma,t}), \quad T=(x_{t,t'}), \quad t,\ t' \in C(n).$$

In particular, T is a diagonal matrix whose signature sign (T) equals

$$\operatorname{sign}(T)=\sum_t \operatorname{sign}\Big(\prod_{\tau\in t} B_\tau^t\Big).$$

The formula for the intersection number $S_\sigma \circ S_{\sigma'}$ given above means that the intersection matrix $I=(S_\sigma \circ S_{\sigma'})$, $\sigma, \sigma' \in C(n/2)$, can be factorized as

$$I=(S_\sigma \circ S_{\sigma'})=XT^{-1}X'.$$

If we would know that the homology classes $[S_\sigma]$, $\sigma \in C(n/2)$, generate the middle homology group $H_n(Y, \boldsymbol{Z})$, and in addition, that the signatures of I and T are equal, then by the definition of the signature of Y we could conclude that

Conjecture. (5) $$\operatorname{sign}(Y)=\sum_t \operatorname{sign}\Big(\prod_{\tau\in t} B_\tau^t\Big).$$

Unfortunately, at this point we do not know whether the last two assumptions are true, so the last equation is only a conjecture (which is true for $n=2$). The next result was proved in [6]. Modulo the conjecture (5), it solves the problem (posed by Hirzebruch in [4]) to find an arithmetic expression for $\varphi(M, V)$.

Theorem 2. $$\varphi(M, V) = -\operatorname{sign}(Y) + 2^n \sum_{\sigma \in C} \sum_{p_\tau > 1} S_\sigma^p \prod_{\tau \in \sigma} \frac{B_{p_\tau}}{p_\tau !}.$$

For given σ, p runs here over all partitions of n with length $= \#$ of vertices in σ,

$$n = \sum_{\tau \in \sigma} p_\tau, \qquad p_\tau > 1.$$

The condition $p_\tau > 1$ implies that only the $\sigma \in C(k)$ with $1 \leq k \leq n/2$ contribute to the sum. The numbers B_{p_τ} are the classical Bernoulli numbers

$$B_0 = 1, \quad B_1 = -\frac{1}{2}, \quad B_2 = \frac{1}{6}, \quad B_3 = 0, \cdots$$

§ 3.

In this section we want to discuss two further geometric invariants of the cusp (M, V) which are closely related to the signature defect $\varphi(M, V)$. The first one, $\varphi(M, V, x)$, is the equivariant version of $\varphi(M, V)$ arising from the equivariant signature theorem [1, p. 589] (it is usually called α-invariant, although in [1] it is called σ-invariant). The second one, $\psi(M, V, x)$, could be called equivariant Todd genus because it comes from the holomorphic Lefschetz theorem [1, p. 566]. To define these numbers, let $x \in F$ be a solution of the congruence $V(M+x) = M+x$. (For a given group of units V there are only finitely many different residue classes $M+x$ which will satisfy this equation. Conversely, for every x in F there is some power $V^k = \{v^k / v \in V\}$ such that $V^k(M+x) = M+x$.) Then x induces an action on Y as follows: Recall that the boundary ∂Y is a $\boldsymbol{R}^n/M$-bundle over the base space $\boldsymbol{R}^{n-1}/V$. On this boundary x acts by translation in the fiber $\boldsymbol{R}^n/M$, $w \mapsto w+x$, and this action has a natural extension to Y. For $x \notin M$ (not the identity action) the fixed-point set Y^x is concentrated in the divisor D, so we can define

$$\varphi(M, V, x) := L(Y, x)[Y^x] - \operatorname{sign}(Y, x),$$

where $\operatorname{sign}(Y, x)$ is the equivariant signature, and $L(Y, x)$ is the corresponding cohomology class in $H^*(Y^x, \boldsymbol{C})$. Because x acts trivially on the cohomology group $H^n(Y, \boldsymbol{C})$, we have actually $\operatorname{sign}(Y, x) = \operatorname{sign}(Y)$. Again, $\varphi(M, V, x)$ depends only on the boundary ∂Y and the action of x on it, but not on Y and the particular divisor D. For a proof of the following theorem, see [6].

Theorem 3. $\varphi(M, V, x) = -\operatorname{sign}(Y) + i^n \sum_{\sigma} (-1)^k \sum_{p_\tau > 0} S_\sigma^p \prod_{\tau \in \sigma} c_{p_\tau}(x_\tau)$.

Here the first sum runs only over those $\sigma \in C(k)$, $1 \leq k \leq n$, which admit a representation of x as

$$x = \sum_{\tau \in \sigma} x_\tau A_\tau$$

with rational coordinates x_τ (which are determined only up to an integer because the A_τ are determined only up to a unit in V). The corresponding submanifolds S_σ belong then to the fixed-point set Y^x. The product is taken over the values of the trigonometric functions

$$c_k(u) = \pi^{-k} \sum_{m \in \boldsymbol{Z}+u} m^{-k}, \qquad k = 1, 2, 3, \cdots$$

(if $k=1$, the series has to be ordered according to increasing values of $|m|$). Note that the c_k are essentially the derivatives of the cotangent function

$$c_1(u) = \begin{Bmatrix} 0, & u \in \boldsymbol{Z} \\ \cot \pi u, & u \in \boldsymbol{C} \backslash \boldsymbol{Z} \end{Bmatrix}, \quad \frac{d}{du} c_k(u) = -\pi k c_{k+1}(u), \quad u \notin \boldsymbol{Z}.$$

By Euler, we have for integral values of u,

$$c_k(u) = -(-1)^{k/2} 2^k \frac{B_k}{k!}, \qquad k \geq 2$$

with the Bernoulli numbers B_k. In particular, if $x \in M$, then formally $\varphi(M, V, x) = \varphi(M, V)$. But the definition given for $\varphi(M, V, x)$ makes sense only for $x \notin M$.

The second invariant, the equivariant Todd genus $\psi(M, V, x)$, is the value

$$\psi(M, V, x) := \operatorname{td}(Y, x)[Y^x]$$

of the cohomology class $\operatorname{td}(Y, x) \in H^*(Y^x, \boldsymbol{C})$ contributing to the Lefschetz number of x (acting on the trivial line bundle over Y) in the holomorphic Lefschetz formula [1, p. 566, 6, p. 57]. Again, this makes sense only for $x \notin M$. In the case $x \in M$ we define $\psi(M, V, x) = \psi(M, V)$ as the coefficient of z^n in the formal power series

$$\prod_{\tau} \frac{S_\tau}{1-\exp(-S_\tau z)}.$$

This number was studied, among others by Ehlers [3] and Satake [5]. In [6], we established the following explicit expression for $\psi(M, V, x)$.

Theorem 4. $$\psi(M, V, x)=\left(\frac{i}{2}\right)^n \sum_{\sigma} (-1)^k \sum_{p_\tau>0} S_\sigma^p \prod_{\tau\in\sigma} C_{p_\tau}(x_\tau).$$

Aside from the missing signature term and the additional factor 2^{-n}, the only difference between this expression and the corresponding expression in Theorem 3 is the slightly different definition of the trigonometric functions $C_k(u)$:

$$C_1(u)=c_1(u)+i, \quad C_k(u)=c_k(u) \quad \text{for } k>1.$$

Despite these differences, we conjecture that

Conjecture. (6) $$\varphi(M, V, x)=2^n\psi(M, V, x).$$

This equality, if true, would reflect a very special geometric property of Y. It is equivalent to our conjecture (5) about the signature of Y. This follows from the following result which we found by a very complicated calculation.

Theorem 5. $$\begin{aligned}\varphi(M, V, x)&-2^n\psi(M, V, x)\\ &=\operatorname{sign}(Y)-\sum_{\sigma\in C}(-2)^{n-k}, \quad k=\sharp \textit{ of vertices in } \sigma\\ &=\operatorname{sign}(Y)-\sum_{t\in C(n)}\operatorname{sign}\Big(\prod_{\tau\in\sigma} B_\tau^t\Big).\end{aligned}$$

If n is odd, then $\operatorname{sign}(Y)=0$ by definition. The expressions

$$\sum_{\sigma\in C}(-2)^{n-k}=\sum_{t\in C(n)}\operatorname{sign}\Big(\prod_{\tau\in t} B_\tau^t\Big)$$

also vanish in this case (although this is not obvious). Therefore, for n odd, the equality (6) is indeed true.

§ 4.

To the triple (M, V, x) there is associated the Hecke L-series

$$L(M, V, x, s)=\sum_{m\in M+x/V}{}' \frac{\operatorname{sign} N(m)}{|N(m)|^s}, \qquad \operatorname{Re}(s)>1$$

where m runs through all $m\neq 0$ in $M\neq x$ which are not equivalent under the action of V. As a natural generalization of Hirzebruch's conjecture we propose

Conjecture. (7) $$d(M)L(M, V, x, 1)=(\pi i)^n\varphi(M, V, x) =(2\pi i)^n\psi(M, V, x).$$

Hirzebruch's original conjecture (now a theorem of Atiyah, Donnelly and Singer) is the case $x=0$. Using the functional equation of the L-function at $s=0$,

$$d(M)L(M, V, 0, 1)=(\pi i)^n L(M^*, V, 0, 0),$$

his conjecture can be reformulated as

$$L(M^*, V, 0, 0)=\varphi(M, V, 0). \tag{8}$$

This is exactly the relation Atiyah, Donnelly and Singer have proved in [2]. At $s=1$, the L-series converges conditionally,

$$L(M, V, x, 1)=\lim_{r\to\infty} \sum_{\substack{m\in M+x/V \\ |N(m)|<r}}' \frac{1}{N(m)}.$$

It is this conditionally convergent series which is at the center of our considerations. We want to transform the series into another one where m runs over all elements in $M+x$ instead over the classes in $M+x/V$ only. To this end we apply an old trick of Hecke, and consider an absolutely convergent series

$$\sum_{v\in V} f(vm)=\frac{d(M)}{N(m)} \tag{9}$$

with some (at the moment unspecified) terms $f(vm)$. Then

$$\sum_{m\in M+x/V}' \frac{d(M)}{N(m)}=\sum_{m\in M+x/V}' \sum_{v\in V} f(vm)=\sum_{m\in M+x}' f(m),$$

and therefore,

$$d(M)L(M, V, x, 1)=\lim_{r\to\infty} \sum_{\substack{m\in M+x \\ |N(m)|<r}}' f(m).$$

Clearly, $f(m)$ is not determined by this equation. Although we do not have a complete proof that the following choice of $f(m)$ satisfies (9), it would lead (as we will see in a moment) to an elementary proof of the conjecture (7). For $\sigma \in C$ let

$$M(\sigma)=\{m \in M+x \mid m=\sum_{\tau\in\sigma} m_\tau A_\tau,\ m_\tau\neq 0\}.$$

Then

$$f(m)=\sum_{\sigma\in C} f(\sigma, m)$$

with $f(\sigma, m)=0$ unless $m \in M(\sigma)$ where

$$f(\sigma, m)=(-1)^{n-k}\sum_{p_\tau>0} S_\sigma^p \prod_{\tau\in\sigma} m_\tau^{-p_\tau} \qquad \text{for } \sigma \in C(k).$$

Notice that the coordinates m_τ are just the numbers $m_\tau=\operatorname{tr}(mB_\tau^t)$ for any $t \in \operatorname{St}(\sigma)$. In the special case where $\operatorname{tr}(mB_\tau^t)$ does not vanish for all t and $\tau \in t$, $f(m)$ simplifies to

$$f(m)=\sum_{t\in C}\prod_{\tau\in t} \operatorname{tr}(mB_\tau^t)^{-1}.$$

Every term in this sum has a simple geometric meaning: it is the projective volume of the simplex $t \in C$ (with respect to a certain differential form depending on m), and the identity (9) says then basically that the projective volume of all simplices t in $V\cdot C$ equals the projective volume of the totally positive chamber of the $\boldsymbol{R}^n$. The following theorem follows from a new limit formula, c.f. [7].

Theorem 6. $\displaystyle\lim_{r\to\infty}\sum_{\substack{m\in M+x\\ |N(m)|<r}} f(m)=(2\pi i)^n\psi(M, V, x).$

This identity can be viewed as a partial fraction decomposition of the right hand sice, generalizing Euler's formula

$$\lim_{r\to\infty}\sum_{\substack{m\in \boldsymbol{Z}+x\\ |m|<r}}\frac{1}{m}=\pi\cot\pi x. \tag{10}$$

(In fact, the theorem remains true in the case $n=1$ if we add πi to the right hand side; it is then equivalent to (8)). By definition of $f(m)$, we can write the left hand side as

$$\sum_{\sigma\in C}(-1)^{n-k}\sum_{p_\tau>0} S_\sigma^p R(p, \sigma)$$

with

$$R(p, \sigma)=\lim_{r\to\infty}\sum_{\substack{m\in M(\sigma)\\ |N(m)|<r}}\prod_{\tau\in\sigma} m_\tau^{-p_\tau}.$$

If all p_τ are >1, then this last series converges absolutely, and we have by definition of the $C_k(u)$,

$$R(p, \sigma) = \pi^n \prod_{\tau \in \sigma} C_{p_\tau}(\mathrm{tr}\,(xB_\tau^t)).$$

But if some $p_\tau = 1$, then the series converges only conditionally, and special considerations are necessary to determine its value. For example, if all $p_\tau = 1$, we prove in [7] that

$$R(p, \sigma) = \text{real part of } \pi^n \prod_{\tau \in \sigma} C_1(\mathrm{tr}\,(xB_\tau^t)).$$

It is remarkable that all the contributions coming from the special ordering of $R(p, \sigma)$ according to the increasing values of $|N(m)|$ add up to the expression (5) which we conjecture to be the signature of Y.

§ 5.

If we combine the theorem of Atiyah, Donnelly and Singer with our formula for $\varphi(M, V)$, we get the equality

$$L(M^*, V, 0, 0) = -\mathrm{sign}\,(Y) + 2^n \sum_{\sigma \in C} \sum_{p_\tau > 1} S_\sigma^p \prod_{\tau \in \sigma} \frac{B_{p_\tau}}{p_\tau!}$$

where for given σ, the inner sum runs over all partitions p of n and length $= \#$ of vertices of σ. Assuming our conjecture (5), we can rewrite this as

$$L(M^*, V, 0, 0) = 2^n \sum_{\sigma \in C} \sum_{p_\tau > 0} S_\sigma^p \prod_{\tau \in \sigma} \frac{B_{p_\tau}}{p_\tau!}.$$

On the other hand, by a result of Shintani [8], we have

$$L(M, V, 0, 0) = 2^n \sum_{\sigma \in C} \sum_{q_\tau \geq 0} \mathrm{tr}\left(\frac{1}{n} \prod_{\tau \in \sigma} \frac{B_{q_\tau}}{q_\tau!} A_\tau^{q_\tau - 1}\right)$$

where for given σ, the inner sum runs now over all partition q of k and length $k = \#$ of vertices in σ.

The striking similarity between these two formulas raises the possibility that there is a relation between the rational numbers $L(M, V, 0, 0)$ and $L(M^*, V, 0, 0)$. If $n=2$, then it follows from these formulas that

Conjecture. (11) $L(M, V, 0, 0) = i^n L(M^*, V, 0, 0)$.

Notice that this relation is trivial if the multiplier ring of M in F contains units of negative norm (for example -1 if n is odd) because both sides vanish then. Moreover, if M is fractional ideal in a field of class number 1 whose units have only positive norms, then this relation follows from Hecke's result that

$$\operatorname{sign} N(d) = i^n$$

for every generator d of the different of the field. But beyond this evidence, we do not know whether (11), or equivalently

$$\varphi(M, V) = i^n \varphi(M^*, V),$$

is always true.

References

[1] Atiyah, M. F. and Singer, I. M., The index of elliptic operators III, Ann. of Math., **87** (1968), 546–604.
[2] Atiyah, M. F., Donnelly, H. and Singer, I. M., Eta invariants, signature defects of cusps, and values of L-functions, Ann. of Math., **118** (1983), 131–177.
[3] Ehlers, F., Eine Klasse komplexer Mannigfaltigkeiten und die Auflösung einiger isolierter Singularitäten, Math. Ann., **28** (1975), 127–156.
[4] Hirzebruch, F., Hilbert modular surfaces, Enseign. Math., **19** (1973), 183–281.
[5] Satake, I., On numerical invariants of arithmetic varieties of Q-rank one, in: Automorphic forms of several variables, Kataka 1983, I. Satake and Y. Morita (eds.), Progress in Math. 46, Birkhäuser, Basel, Boston, Stuttgart 1984.
[6] Sczech, R., Zur Summation von L-Reihen, Bonner Mathematische Schriften 141, Bonn 1982.
[7] ——, A new limit formula, in preparation.
[8] Shintani, T., On evaluation of zeta functions of totally real algebraic number fields at non-positive integers, J. Fac. Sci. Univ. Tokyo, **23** (1976), 393–417.

Department of Mathematics
Rutgers University
Newark, NJ 07102
U.S.A.

Advanced Studies in Pure Mathematics 15, 1989
Automorphic Forms and Geometry of Arithmetic Varieties
pp. 41–64

On Dimension Formula for Siegel Modular Forms

Ryuji Tsushima

§ 0. Introduction

Let $\mathfrak{S}_g$ be the Siegel upper half plane of degree g, and let $Sp(g, \boldsymbol{Z})$ be the Siegel modular group of degree g. Let Γ be a subgroup of $Sp(g, \boldsymbol{Z})$ of finite index, and let μ be a holomorphic representation of $GL(g, \boldsymbol{C})$ into $GL(r, \boldsymbol{C})$. By an *automorphic form of type* μ with respect to Γ, we mean a holomorphic mapping f of $\mathfrak{S}_g$ to the r-dimensional complex vector space $\boldsymbol{C}^r$ which satisfies the following equalities:

$$f(M\langle Z\rangle)=\mu(CZ+D)f(Z),$$

for any $M=\begin{pmatrix} A & B \\ C & D \end{pmatrix} \in \Gamma$ and $Z \in \mathfrak{S}_g$, where $M\langle Z\rangle$ is defined to be $(AZ+B)(CZ+D)^{-1}$. (We need to assume the holomorphy of f at "cusps" if $g=1$.) We denote by $A_\mu(\Gamma)$ the complex vector space of automorphic forms of type μ with respect to Γ. It is known that $A_\mu(\Gamma)$ is finite dimensional ([8]). In case $\mu(CZ+D)=\det(CZ+D)^k$, an automorphic form of type μ is also called an automorphic form of *weight* k, and $A_\mu(\Gamma)$ is also denoted by $A_k(\Gamma)$. In case the degree of μ is greater than one, an automorphic form of type μ is called a *vector-valued* automorphic form.

Our main problem is to find a formula for $\dim A_\mu(\Gamma)$ as a function in the signature of μ. The first result to our main problem was obtained by J.-I. Igusa in case $g=2$. (The case $g=1$ is classical.) Let $\Gamma_g(N)$ be the principal congruence subgroup of $Sp(g, \boldsymbol{Z})$ of level N, i.e.,

$$\Gamma_g(N)=\{M \in Sp(g, \boldsymbol{Z}) \mid M\equiv \mathbf{1}_{2g} \bmod N\}.$$

$A(\Gamma):=\oplus_{k\geq 0} A_k(\Gamma)$ has a structure of a graded ring. By using the theory of theta series, he explicitly determined the generators of $A(\Gamma)$ and represented these generators by theta constants for some cases such as $\Gamma=\Gamma_2(1)$ or $\Gamma_2(2)$ ([29], [30]). Especially $\dim A_k(\Gamma_2(1))$ and $\dim A_k(\Gamma_2(2))$ were known in this work. In [31], he constructed a graded ring homomorphism ρ_g from a subring R_g of $A(\Gamma_g(1))$ to the graded ring $S(2, 2g+2)$ of binary

Received November 11, 1987.
Revised June 23, 1988.

$(2g+2)$-forms. R_g contains $A(\Gamma_g(1))^{(2)} := \oplus_{k\geq 0} A_{2k}(\Gamma_g(1))$ and coincides with $A(\Gamma_g(1))$ if g is odd or $g=2, 4$. ρ_1 is bijective and ρ_2 is injective. He reproved the structure theorem of $A(\Gamma_2(1))$ by using this homomorphism ρ_2 and the structure theorem of $S(2, 6)$. But these methods were restricted to special cases of Γ and the case of weight k.

Foı general group Γ and representation μ, two main approaches are known. The first is a geometric one which uses Riemann-Roch-Hirzebruch's formula and the holomorphic Lefschetz fixed points formula (when Γ has fixed points). The second is a group-theoretical one which uses Selberg's trace formula.

If $N\geq 3$, then $\Gamma_g(N)$ acts on $\mathfrak{S}_g$ without fixed points. So $X_g := \Gamma_g(N)\backslash\mathfrak{S}_g$ is a manifold in this case. But X_g is not compact. I. Satake constructed a compactification $\bar{X}_g$ of X_g which is a projective variety ([43]). But if $g\geq 2$, then $\bar{X}_g$ has bad singularities along its cusps: $\bar{X}_g - X_g$. J.-I. Igusa constructed a desingularization $\tilde{X}_2$, $\tilde{X}_3$ of $\bar{X}_2$, $\bar{X}_3$ which is a blowing up of $\bar{X}_2$, $\bar{X}_3$ along its cusps, respectively ([32]). For general g, a smooth compactification $\tilde{X}_g$ of X_g was constructed in [3] and in [41]. When we need to specify the level N, we denote X_g, $\bar{X}_g$ and $\tilde{X}_g$ by $X_g(N)$, $\bar{X}_g(N)$ and $\tilde{X}_g(N)$, respectively.

The first result to our main problem from the first approach was obtained by T. Yamazaki ([57]). He applied Riemann-Roch-Hirzebruch's formula to $\tilde{X}_2$ and calculated $\dim A_k(\Gamma_2(N))$ with $N\geq 3$. The first result to our main problem from the second approach was obtained by Y. Morita and U. Christian independently ([39], [9] and [10]). They calculated $\dim A_k(\Gamma_2(N))$ with $N\geq 3$ by Selberg's trace formula. The next result would be the author's one. In 1979, he calculated $\dim A_k(\Gamma_3(N))$ with $N\geq 3$ by applying Riemann-Roch-Hirzebruch's formula to $\tilde{X}_3$ ([50]). Next in 1980, the author applied the holomorphic Lefschetz fixed points formula to the action on $\tilde{X}_2(N)$ of $\Gamma_2(1)/\Gamma_2(N)$ with $N\geq 3$ and calculated $\dim A_k(\Gamma_2(1))$, and then $\dim A_k(\Gamma_2(2))$ simiiarly ([51]). About the same time K. Hashimoto calculated $\dim A_k(\Gamma_2(1))$ and $\dim A_k(\Gamma_2(2))$ by Selberg's trace formula ([19] I). These results can be regarded as the third proof of the structure theorem of $A(\Gamma_2(1))$. He (and T. Ibukiyama) also explicitly calculated $\dim A_k(\Gamma)$ for more general congruence subgroups such as $\Gamma=\Gamma_0(p)$, where p is a prime number and $\Gamma_0(p) = \left\{M=\begin{pmatrix} A & B \\ C & D\end{pmatrix} \mid C\equiv 0 \bmod p\right\}$ ([19] I, [22] and [28]). Next in 1982, the author explicitly calculated the dimension of the spaces of vector-valued automorphic forms $A_\mu(\Gamma_2(N))$ with $N\geq 1$ by geometric method ([52]).

There are discrete subgroups of $Sp(g, \boldsymbol{R})$ which have quite different properties. For discrete subgroups Γ of $Sp(g, \boldsymbol{R})$ such that $\Gamma\backslash\mathfrak{S}_g$ is compact, $\dim A_\mu(\Gamma)$ was calculated in [26], [33] and [27] by geometric method and in [37] by Selberg's trace formula. They expressed the dimension of $A_\mu(\Gamma)$ as a finite sum of invariants attached to each elliptic fixed point of Γ, or what is the same, elliptic conjugacy classes of Γ. So the dimension can be calculated once a complete list of such data is known. (They solved this problem for general bounded symmetric domains.) $Sp(2, \boldsymbol{R})$ has discrete arithmetic subgroups which are related to a quaternion unitary groups. The quotient space of $\mathfrak{S}_2$ by a group of this type is not compact but it is compactified by adding a finite number of points to it. As to such a group Γ, $\dim A_k(\Gamma)$ was calculated in [56], [46] by geometric method and in [1], [19] II by Selberg's trace formula. We refer the reader to a note by I. Satake [45] and Satake-Ogata's article in this volume.

So our next problem should be to calculate $\dim A_k(\Gamma_3(1))$. The author was studying this problem by geometric method and K. Hashimoto was studying by Selberg's trace formula. But there were some difficulties unsolved on both sides (see § 2). So recently we were studying this problem jointly ([23]). But this work has not been completed now. While we were studying this problem, S. Tsuyumine succeeded to solve it ([54]). His method is similar (but far more complicated) to the Igusa's second proof for the structure theorem of $A(\Gamma_2(1))$. The kernel of Igusa's graded ring homomorphism ρ_3: $A(\Gamma_3(1))\to S(2, 8)$ was known to be an ideal which is generated by a cusp form χ_{18} of weight eighteen ([31]), and the structure of $S(2, 8)$ was known by T. Shioda ([48]). Tsuyumine explicitly determined the structure and the generators of the graded ring $A(\Gamma_3(1))$ by using these results. So especially $\dim A_k(\Gamma_3(1))$ was known.

Although our problem was already solved, it is still not meaningless to solve this problem by our method, because Tsuyumine's method is restricted to the case of the full modular group $\Gamma_3(1)$ and of weight k. If $\dim A_k(\Gamma_3(1))$ is calculated by our method, it is easy to extend this result to the case of general congruence subgroups or the case of vector-valued automorphic forms, since the process to reach there has been already reduced to a routine work (modulo the vanishing theorem in the vector-valued case).

To solve our problem by geometric method, we need to classify the fixed points sets of the action of the finite group $H:=\Gamma_3(1)/\Gamma_3(N)$ with $N\geq 3$ on the smooth compactification $\tilde{X}_3(N)$ and to classify all conjugacy classes of H which have fixed points on $\tilde{X}_3(N)$. We need to calculate the contributions to the dimension formula of these conjugacy classes one by one by the holomorphic Lefschetz fixed points formula. We define that two fixed points sets of H are equivalent to each other if there is an element of H which maps one fixed points set to another. In this classification there are more than one hundred kinds of fixed points sets (in the case of degree two, there were *only* twenty-five) and there are about three hundreds conjugacy classes which have fixed points on $\tilde{X}_3(N)$.

To solve our problem by Selberg's trace formula, we need to classify all conjugacy classes of $\Gamma_3(1)$ which have non-zero contributions to the dimension formula. Although in the case of geometric method, we study the finite group $\Gamma_3(1)/\Gamma_3(N)$ and in the case of Selberg's trace formula, we study the infinite group $\Gamma_3(1)$, the classifications of these conjugacy classes go almost in parallel with each other. So in both methods, we need to calculate the contributions of about three hundred conjugacy classes. To execute such a hard calculation, it is very effective to compare the results from these two approaches with each other.

Let $f(t)$ be the generating function of $A(\Gamma_3(1))$, i.e., $f(t)=\sum_{k\geq 0} t^k \dim A_k(\Gamma_3(1))$. Then this is a rational function and the degree of the denominator of $f(t)$ is one hundred and ten. (There is a misprint in [54] p. 832, $(1-T^{12})^3$ in the denominator should be $(1-T^{12})^2$.) So the expansion of $f(t)$ to partial fractions has essentially fifty five terms, since $f(t)$ is an even function. We are calculating the coefficients of these fifty five terms one by one and we have calculated these coefficients except the following four terms: $1/(1-t)$, $1/(1-t)^2$, $1/(\rho-t)$ and $1/(\rho^2-t)$, where $\rho=\exp(2\pi i/3)$. Among them, the coefficient of $1/(1-t)$ is the most difficult one. To determine this exactly, we need to calculate the contributions of about one hundred conjugacy classes. Perhaps we need to spend much time and effort such as mathematicians of old times spent to calculate the circular constant.

The joint work of K. Hashimoto and the author was expected to be one of main parts of this volume. But unfortunately it is still incomplete, and therefore instead the author presents here an expository note.

In 1986 a paper of M. Eie and C.-Y. Lin was published in American Journal of Mathematics which states that they found the formula for $\dim A_k(\Gamma_3(1))$ by Selberg's trace formula ([15]). If this were true, our effort will be of no use. But the author should confess that this paper seems very deceptive to him. In the end of this note, we comment on this paper.

§ 1. Dimension formula for $\Gamma_2(N)$ and $\Gamma_3(N)$ with $N\geq 3$

For the sake of simplicity we mainly study the case of weight k in this note. In this section we present the calculation of $\dim A_k(\Gamma_2(N))$ with $N\geq 3$. The method we employ now was developed by the author in [50] and is more systematic than the original method of T. Yamazaki.

In both methods (geometric and Selberg), we do not calculate the dimension of $A_\mu(\Gamma)$ directly, but calculate the dimension of the space of "cusp forms" $S_\mu(\Gamma)$. $S_\mu(\Gamma)$ is a subspace of $A_\mu(\Gamma)$, and $\dim A_\mu(\Gamma)$ is calculated by a theory of Eisenstein series from dim $S_\mu(\Gamma)$. So we need to define the space $S_\mu(\Gamma)$.

We recall the construction of the compactification $\overline{X}_g$ of the quotient space X_g. Since $\mathfrak{S}_g$ is not a bounded domain, a part of its boundary is in "infinite place", so we need to consider a bounded domain $\mathscr{D}_g$ which is biholomorphic to $\mathfrak{S}_g$. We put

$$\mathscr{D}_g := \{Z \in M(g, \boldsymbol{C}) \,|\, {}^tZ = Z,\ Z^*Z < \mathbf{1}_g\}.$$

Then $\mathscr{D}_g$ is a bounded domain and the Cayley transformation $c\colon \mathfrak{S}_g \to \mathscr{D}_g$ is defined to be

$$c(Z) = (Z - \sqrt{-1}\mathbf{1}_g)(Z + \sqrt{-1}\mathbf{1}_g)^{-1},$$

and this is a biholomorphic mapping. The action of $Sp(g, \boldsymbol{R})$ on $\mathscr{D}_g$ extends to its closure $\overline{\mathscr{D}}_g$ in $\boldsymbol{C}^{g(g+1)/2}$.

We put

$$F_{g'} := \left\{ \begin{pmatrix} Z & 0 \\ 0 & \mathbf{1}_{g-g'} \end{pmatrix} \,\middle|\, Z \in \mathscr{D}_{g'} \right\},$$

where $0 \leq g' < g$. Then $F_{g'}$ is in $\overline{\mathscr{D}}_g - \mathscr{D}_g$ and biholomorphic to $\mathscr{D}_{g'}$. $F_{g'}$ and its images by the action of $Sp(g, \boldsymbol{R})$ are called *boundary components* of degree g' of $\mathscr{D}_g$. $\overline{\mathscr{D}}_g - \mathscr{D}_g$ is devided into a disjoint union of boundary components. $F_{g'}$ and is images by the action of $Sp(g, \boldsymbol{Z})$ are called *rational* boundary components of degree g' (for the definition of a (rational) boundary component in general, see a text book for example [44]). We define a certain topology on $\mathscr{D}_g^s := \mathscr{D}_g \cup \{\text{rational boundary components}\}$ ([43]), and Satake's compactification $\overline{X}_g(N)$ is the quotient space of $\mathscr{D}_g^s$ by $\Gamma_g(N)$. $\overline{X}_g(N) - X_g(N)$ is a disjoint union of a finite number of copies of $X_{g'}(N)$ $(0 \leq g' < g)$ and a copy of $X_{g'}(N)$ in $\overline{X}_g(N) - X_g(N)$ is called a *cusp* of degree g'. Any smooth compactification $\tilde{X}_g$ has a morphism $\pi_g\colon \tilde{X}_g \to \overline{X}_g$ which is the identity on the quotient space X_g. Here we take a so-called toroidal compactification due to D. Mumford ([3]). Then $E := \tilde{X}_g - X_g$ is a divisor with normal crossings and each irreducible component in E has a structure of a fiber space over a cusp of degree $g-1$ through π_g whose general fiber is an abelian variety of dimension $g-1$.

Let $\mu\colon GL(g, \boldsymbol{C}) \to GL(r, \boldsymbol{C})$ be a holomorphic representation. We define an action of $Sp(g, \boldsymbol{R})$ on the product space $\mathfrak{S}_g \times \boldsymbol{C}^r$ by

$$M(Z, u) := (M\langle Z\rangle, \mu(CZ + D)u),$$

for $M = \begin{pmatrix} A & B \\ C & D \end{pmatrix} \in Sp(g, \boldsymbol{R})$, $Z \in \mathfrak{S}_g$ and $u \in \boldsymbol{C}^r$. If $N \geq 3$, then the quotient space V_μ of $\mathfrak{S}_g \times C^r$ by $\Gamma_g(N)$ has a structure of a vector bundle on X_g. From the construction of the smooth compactification $\tilde{X}_g$ ([3], [41]), it is proved that this vector bundle V_μ has a natural extension to a vector bundle $\tilde{V}_\mu$ on $\tilde{X}_g$. (V_μ is not extended to a vector bundle on $\bar{X}_g$ in general.) In case $\mu(CZ+D)=\det(CZ+D)$, V_μ and $\tilde{V}_\mu$ are line bundles and denoted by L_g and $\tilde{L}_g$, respectively. L_g is extended to a line bundle $\bar{L}_g$ on $\bar{X}_g$ such that the restriction of $\bar{L}_g$ to a cusp of degree g' is isomorphic to $L_{g'}$ and $\tilde{L}_g$ is the pullback of $\bar{L}_g$ by π_g. This line bundle $\bar{L}_g$ is ample and the projectivity of $\bar{X}_g$ follows from this fact ([5]). The space $A_\mu(\Gamma_g(N))$ is naturally identified with $\Gamma(\tilde{X}_g, \mathcal{O}(\tilde{V}_\mu))$.

Definition (1.1). The space of *cusp forms* $S_\mu(\Gamma_g(N))$ is defined to be $\Gamma(\tilde{X}_g, \mathcal{O}(\tilde{V}_\mu - E))$, where $\mathcal{O}(\tilde{V}_\mu - E)$ means the subsheaf of $\mathcal{O}(\tilde{V}_\mu)$ consisting of the germs of sections of $\tilde{V}_\mu$ which vanish along E. If $g \geq 2$, then any subgroup Γ of $Sp(g, Z)$ of finite index contains $\Gamma_g(N)$ for some $N \geq 3$ ([6] and [38]). $S_\mu(\Gamma)$ is defined to be $A_\mu(\Gamma) \cap S_\mu(\Gamma_g(N))$. In case $g=1$, a subgroup Γ of $SL(2, Z)$ of finite index does not contain $\Gamma_1(N)$ in general but always contains a normal subgroup Γ' of finite index which acts fixed points free, so $S_\mu(\Gamma)$ is defined by using Γ' instead of $\Gamma_1(N)$. In case $\mu(CZ+D)=\det(CZ+D)^k$, $S_\mu(\Gamma)$ is also denoted by $S_k(\Gamma)$.

Let $\omega = h(Z)\prod_{i \leq j} dZ_{ij}$ be a holomorphic n-form on $\mathfrak{S}_g$ ($n = \dim \mathfrak{S}_g = g(g+1)/2$). Then ω is invariant under $\Gamma_g(N)$, i.e., $f^*(\omega)=\omega$ for any $f \in \Gamma_g(N)$, if and only if h belongs to $A_{g+1}(\Gamma_g(N))$. So the line bundle $L_g^{\otimes(g+1)}$ on X_g is isomorphic to the canonical line bundle K_{X_g} of X_g. We denote by $\tilde{\omega}$ the n-form on X_g which is induced from ω.

Proposition (1.2). *$\tilde{\omega}$ may have a single pole along E, so the sheaf $\mathcal{O}(\tilde{L}_g^{\otimes(g+1)})$ is isomorphic to $\mathcal{O}(K_{\tilde{X}_g}+E)$.*

Outline of the proof. Let E_1 be an irreducible component of E. Then E_1 consists of the limits of the points in X_g, and E_1 is defined by an equation for example $\operatorname{Im}(Z_{11})=\infty$. In this case the coordinates of the generic point on E_1 consist of $W_{11} := \exp(2\pi i Z_{11}/N)$ and $(Z_{ij})_{1<i \leq j}$, and E_1 is defined by $W_{11}=0$. By these coordinates, $\tilde{\omega}$ is represented as

$$h(Z)N dW_{11}/2\pi i W_{11} \prod_{1<i \leq j} dZ_{ij},$$

since $dZ_{11} = N dW_{11}/2\pi i W_{11}$. So $\tilde{\omega}$ may have a single pole along E.

The reason to calculate $\dim S_k(\Gamma)$ is that we can apply the theorem of Riemann-Roch-Hirzebruch by the following

Theorem (1.3). *For $i>0$ and $k>g+1$, we have*

$$H^i(\tilde{X}_g, \mathcal{O}(\tilde{L}_g^{\otimes k}-E))=0.$$

Hence $\dim S_k(\Gamma_g(N))=\chi(\tilde{X}_g(N), \mathcal{O}(\tilde{L}_g^{\otimes k}-E))$ (*the Euler-Poincaré characteristic*).

Proof. Let $[E]$ be the line bundle associated with the divisor E. Then the sheaf $\mathcal{O}(\tilde{L}_g^{\otimes k}-E)$ is isomorphic to $\mathcal{O}(\tilde{L}_g^{\otimes k}\otimes[E]^{\otimes(-1)})$, and by the above proposition this is isomorphic to $\mathcal{O}(\tilde{L}_g^{\otimes(k-g-1)}\otimes K_{\tilde{X}_g})$. Since $\tilde{L}_g$ is a pull-back of the ample line bundle $\bar{L}_g$, our assertion is essentially a consequence of the Kodaira vanishing theorem ([36]).

Let c_i $(1\leq i\leq n)$ be the i-th Chern class of $\tilde{X}_g$ and let Q_n be the Riemann-Roch polynomial of dimension n. By definition the space of cusp form $S_k(\Gamma_g(N))$ is $\Gamma(\tilde{X}_g, \mathcal{O}(\tilde{L}_g^{\otimes k}-E))$, so if $k\geq g+2$, then from the above theorem the dimension of this space is equal to

$$Q_n(c_1(\tilde{L}_g^{\otimes k}\otimes[E]^{\otimes(-1)});\, c_1, \cdots, c_n),$$

where $c_1(L)$ means the first Chern class of L.

So our problem is reduced to the calculation of this polynomial. To execute this calculation we introduce the notion of logarithmic Chern class ([50]).

Definition (1.4). Let X be a compact complex manifold of dimension n and E a reduced divisor on X with simple normal crossings. Then for $p\in E$, one may take a coordinate system $(z_1, \cdots, z_n)$ around p such that E is defined by $z_1\cdots z_l=0$. Let Θ_X be the sheaf of germs of local holomorphic vector fields on X and let $\Theta_X(\log E)$ denote the (locally free) subsheaf of Θ_X consisting of germs of those local holomorphic vector fields which can be expressed in the form

$$\sum_{i=1}^{l} f_i(z)z_i\frac{\partial}{\partial z_i}+\sum_{i=l+1}^{n} f_i(z)\frac{\partial}{\partial z_i} \qquad (f_i(z) \text{ holomorphic}).$$

$\Theta_X(\log E)$ is the dual sheaf of $\Omega_X^1(\log E)$ which was defined in [11]. We denote by $T_X(\log E)$ the vector bundle which corresponds to $\Theta_X(\log E)$. Then the i-th *logarithmic Chern class* $\bar{c}_i$ of X *relative* to E is defined to be $c_i(T_X(\log E))$.

We denote the total Chern class of X by $c(X)=1+c_1+\cdots+c_n$ and the total logarithmic Chern class of X relative to E by $\bar{c}(X, E)=1+\bar{c}_1+\cdots+\bar{c}_n$. Let $E=\bigcup_{i\in I}E_i$ be the decomposition of E into a union of irreducible components and let $\varepsilon_i:=c_1([E_i])$ be the dual cohomology class of E_i. Then there is the following relation between Chern classes and logarithmic Chern classes.

Theorem (1.5) ([50]). *We have*

$$c(X)=\bar{c}(X, E)\cdot \prod_{i\in I}(1+\varepsilon_i).$$

Now we calculate the dimension of $S_k(\Gamma_2(N))$. In the three dimensional case, the Riemann-Roch polynomial $Q_3(K; c_1, c_2, c_3)$ is equal to

$$K^3/6+c_1K^2/4+(c_1^2+c_2)K/12+c_1c_2/24.$$

So we need to calculate this polynomial replacing K by $c_1(\tilde{L}_2^{\otimes k}\otimes[E]^{\otimes(-1)})$ and c_1, c_2 by the first, second Chern class of $\tilde{X}_2$, respectively. By Proposition (1.2), $c_1(\tilde{L}_2^{\otimes k}\otimes[E]^{\otimes(-1)})$ is equal to

$$-k\bar{c}_1/3-\textstyle\sum_i \varepsilon_i,$$

and by Theorem (1.5), we have

$$c_1=\bar{c}_1+\textstyle\sum_i \varepsilon_i,$$
$$c_2=\bar{c}_2+\bar{c}_1(\textstyle\sum_i \varepsilon_i)+\sum_{i<j}\varepsilon_i\varepsilon_j.$$

Therefore what we need to calculate is (the value at the fundamental class of) the following polynomial:

$$(*)\qquad \begin{aligned}&2^{-3}3^{-4}(-4k^3\bar{c}_1^3+18k^2\bar{c}_1^3-18k\bar{c}_1^3-18k\bar{c}_1\bar{c}_2+27\bar{c}_1\bar{c}_2)\\ &+2^{-3}3^{-2}(-2k^2\bar{c}_1^2+6k\bar{c}_1^2-3\bar{c}_1^2-3\bar{c}_2)(\textstyle\sum_i \varepsilon_i)\\ &+2^{-3}3^{-2}(-2k+3)\bar{c}_1((\textstyle\sum_i \varepsilon_i)^2+\sum_{i<j}\varepsilon_i\varepsilon_j)\\ &-2^{-3}3^{-1}(\textstyle\sum_i \varepsilon_i)(\sum_{i<j}\varepsilon_i\varepsilon_j)+(0)(\sum_i \varepsilon_i)^3.\end{aligned}$$

These intersection numbers in this polynomial are calculated by the following five methods.

a) The intersection numbers in the first line are proportional to the volume of the fundamental domain of $\Gamma_2(N)$ and calculated by the Hirzebruch's proportionality principle ([40]).

b) The intersection numbers in the second line vanish.

c) The first term in the fourth line is calculated by using the theory of torus embedding ([35]).

d) An irreducible component E_1 of E has a structure of an elliptic surface over a cusp of degree one in $\bar{X}_2$ by the restriction of the morphism π_2: $\tilde{X}_2\to\bar{X}_2$ to E_1:

$$\pi\colon E_1\longrightarrow\bar{X}_1.$$

The intersection numbers in the third line are calculated by using the structure of this fiber space.

e) We need not calculate the second term in the fourth line because its coefficient is 0. But in the original work of T. Yamazaki, he did not use the logarithmic Chern classes and in his method the coefficient of this term did not vanish. So he needed to calculate this term. He calculated this term by using Igusa's theory on theta constants. On the other hand in the case of degree three, the coefficient of the term $(\sum_i \varepsilon_i)^6$ does not vanish. So we need to use his method in the case of degree three in fact.

a) Now we return to the general case of degree g. Let $Z \in \mathfrak{S}_g$ and put $Z=(Z_{ij})$. Then for $M=\begin{pmatrix} A & B \\ C & D \end{pmatrix} \in Sp(g, \boldsymbol{R})$ we have

$$(d(M\langle Z\rangle)_{ij})={}^t(CZ+D)^{-1}(dZ_{ij})(CZ+D)^{-1}.$$

So if s_2 is the symmetric tensor representation of $GL(g, \boldsymbol{C})$ of degree two, then the vector bundle V_{s_2} is isomorphic to the cotangent bundle $T^*_{X_g}$ on X_g. We can prove that the vector bundle $\tilde{V}_{s_2}$ is isomorphic to $T_{\tilde{X}_g}(\log E)^*$ similarly as in the proof of Proposition (1.2). For the values of the products of Chern classes of $\tilde{V}_\mu$, the extended Hirzebruch proportionality principle holds ([40]). In particular we have the following

Theorem (1.6). *Let $\check{\mathscr{D}}$ be the compact dual of $\mathfrak{S}_g$ and let $\check{c}_i$ be the i-th Chern class of $\check{\mathscr{D}}$. Let ν_i $(1 \leq i \leq n)$ be non-negative integers with $\sum_i i\nu_i = n$. Then we have*

$$(\textstyle\prod_i \bar{c}_i^{\nu_i})[\tilde{X}_g(N)] = c(\textstyle\prod_i \check{c}_i^{\nu_i})[\check{\mathscr{D}}],$$

where $[\check{\mathscr{D}}]$ (resp. $[\tilde{X}_g(N)]$) is the fundamental class in $H_{2n}(\check{\mathscr{D}}, \boldsymbol{Z})$ (resp. $H_{2n}(\tilde{X}_g(N), \boldsymbol{Z})$), This constant c is rational and expressed as

$$c=(-1)^n \frac{\operatorname{vol}(\Gamma_g(N)\backslash\mathfrak{S}_g)}{\operatorname{vol}(\check{\mathscr{D}})}.$$

We do not present here the definition of the compact dual and the relation between the invariant measures on $\mathfrak{S}_g$ and on $\check{\mathscr{D}}$. It is essential that the intersection number$(\prod_i \bar{c}_i^{\nu_i})$ $[\tilde{X}_g(N)]$ is proportional to the volume of the fundamental domain. Therefore the value of the first line in (∗) is expressed by a polynomial in k and the volume. This polynomial is determined uniquely by the symmetric domain. In the case of Siegel upper half plane of degree g, this is equal to

$$\prod_{0<i\leq j\leq g}(2k+i+j-2g-2),$$

especially this is equal to $(2k-2)(2k-3)(2k-4)$ in the case of degree two ([26]). So the value of the first line is expressed as

$$d(2k-2)(2k-3)(2k-4),$$

where d is a constant. Comparing the coefficient of k^3, we have

$$2^3 d=-2^{-1}3^{-4}\bar{c}_1^{\,3}[\tilde{X}_2(N)].$$

Let $n=\dim \mathfrak{S}_g=g(g+1)/2$ and let X_{ij} (resp. Y_{ij}) be the real (resp. imaginary) part of Z_{ij}. Then the invariant measure vol on $\mathfrak{S}_g$ is defined by $(\det Y)^{-(g+1)}\prod_{i\leq j}dX_{ij}\prod_{i\leq j}dY_{ij}$. By [7] and [25], it is known that

$$\bar{c}_1^{\,n}[\tilde{X}_g(N)]=(-1)^n\pi^{-n}(g+1)^n n!\,2^{-g(g+3)/2}\operatorname{vol}(\Gamma_g(N)\backslash\mathfrak{S}_g).$$

By [8] and by [49], it is known that

$$[\Gamma_g(1)\colon \Gamma_g(N)]=N^{g(2g+1)}\prod_{p\mid N}\prod_{1\leq h\leq g}(1-p^{-2h}),$$
$$\operatorname{vol}(\Gamma_2(1)\backslash\mathfrak{S}_2)=2^{-1}3^{-3}5^{-1}\pi^3.$$

By using the above results, we derive that

$$d=2^{-10}3^{-3}5^{-1}N^{10}\prod_{p\mid N}(1-p^{-2})(1-p^{-4}).$$

So the first line in $(*)$ is calculated.

b) Let E_1 be an irreducible component of $E=\tilde{X}_g-X_g$. Then the morphism $\pi_g\colon \tilde{X}_g\to\bar{X}_g$ maps E_1 to a cusp of degree $g-1$. We assume that the restriction of π_g to E_1:

$$\pi_g\mid E_1\colon E_1\longrightarrow\bar{X}_{g-1}$$

factors through $\pi_{g-1}\colon \tilde{X}_{g-1}\to\bar{X}_{g-1}$. If $g\leq 4$, then Namikawa's smooth compactification satisfies this assumption ([41]). We denote the morphism of E_1 to $\tilde{X}_{g-1}$ by π and the inclusion of E_1 to $\tilde{X}_g$ by i. In case the representation μ is the standard representation of $GL(g, \boldsymbol{C})$, we denote V_μ and $\tilde{V}_\mu$ by V_g and $\tilde{V}_g$, respectively. Then from the construction of $\tilde{V}_{s_2}$ $(\simeq T_{\tilde{X}_g}(\log E)^*)$, it is easily seen that there exist the following short exact sequences of vector bundles:

$$0\longrightarrow\pi^*(T_{\tilde{X}_{g-1}}(\log E)^*)\longrightarrow i^*(T_{\tilde{X}_g}(\log E)^*)\longrightarrow i^*(\tilde{V}_g)\longrightarrow 0,$$
$$0\longrightarrow\pi^*(\tilde{V}_{g-1})\longrightarrow i^*(\tilde{V}_g)\longrightarrow \boldsymbol{C}_{E_1}\longrightarrow 0,$$

where $\boldsymbol{C}_{E_1}$ means the trivial line bundle with fiber $\boldsymbol{C}$ on E_1. We denote by $c(V)$ the total Chern class of a vector bundle V.

Proposition (1.7). *From the above exact sequences, we have*

$$i^*(c(T_{\tilde{X}_g}(\log E)^*)) = \pi^*(c(T_{\tilde{X}_{g-1}}(\log E)^*) \cdot c(\tilde{V}_{g-1})).$$

Thus we proved that the pullback by i of the logarithmic Chern classes of $\tilde{X}_g$ relative to E is represented as a pullback by π of Chern classes of vector bundles on $\tilde{X}_{g-1}$. This property holds for the vector bundle $\tilde{V}_\mu$ for general representation μ.

Let us return to the case of degree two. From the above proposition it is clear that the terms in the second line in $(*)$ vanish, since it holds that

$$\bar{c}_1{}^2 \varepsilon_1[\tilde{X}_2] = (i^*(\bar{c}_1))^2[E_1].$$

$(i^*(\bar{c}_1))^2$ is a pullback by π of a cohomology class on $\bar{X}_1$. But $\bar{X}_1$ is one-dimensional, so this is zero. Similarly $\bar{c}_2\varepsilon_1[\tilde{X}_2]$ vanishes.

The validity of Hirzebruch's proportionality is one of the reasons to use logarithmic Chern classes instead of usual Chern classes. To hold the above property is another reason. In the case of degree three, the author proved a similar proposition as above by using the structure of a group scheme of $E_1 - W$ over $\tilde{X}_2$, where W is the set of points where π is not smooth and used it to prove the vanishing of many intersection numbers. If one uses the usual Chern classes, then the calculation would be much harder because such intersection numbers do not vanish.

c) Now we return to the case of degree two. Let E_1 be an irreducible component of E. The fiber space $\pi: E_1 \to \bar{X}_1(N)$ has a structure of a compactification of the universal family of elliptic curves with level N. If $p \in X_1(N)$, then $\pi^{-1}(p)$ is an elliptic curve and if q is a cusp, then $\pi^{-1}(q)$ consists of N rational curves $l_1, l_2, \cdots, l_N$ such that l_i and l_{i+1} meet at a point. (We put $l_{N+1} = l_1$.) Since $\pi^{-1}(p)$ and $\pi^{-1}(q)$ are algebraically equivalent and do not intersect, there exists the following relation among intersection numbers on E_1:

$$l_1 \cdot \pi^{-1}(q) = l_1 \cdot \pi^{-1}(p) = 0.$$

Moreover we have

$$l_1 \cdot \pi^{-1}(q) = l_1^2 + l_1 l_2 + l_1 l_N,$$

hence

$$l_1^2 = -2.$$

l_1 is an intersection of E_1 and another irreducible component in E which we denote by E_2. So we have

$$i^*(\varepsilon_2^2)[E_1]=\varepsilon_1\varepsilon_2^2[\tilde{X}_2(N)]=-2.$$

Since $\bar{X}_2(N)$ has $(1/2)N^4\prod_{p|N}(1-p^{-4})$ cusps of degree one and $\bar{X}_1(N)$ has $(1/2)N^2\prod_{p|N}(1-p^{-2})$ cusps, there are $(1/12)N^7\prod_{p|N}(1-p^{-2})(1-p^{-4})$ points in $\tilde{X}_2(N)$ where three irreducible components of E intersect and there are $(1/8)N^7\prod_{p|N}(1-p^{-2})(1-p^{-4})$ rational curves where two irreducible components of E intersect. Therefore the first term in the fourth line in $(*)$ is calculated as

$$\begin{aligned}(\textstyle\sum_i\varepsilon_i)(\sum_{i<j}\varepsilon_i\varepsilon_j)&=\textstyle\sum_{i<j}(\varepsilon_i\varepsilon_j^2+\varepsilon_i^2\varepsilon_j)+3\sum_{i<j<k}\varepsilon_i\varepsilon_j\varepsilon_k\\&=(-4/8+3/12)N^7\textstyle\prod_{p|N}(1-p^{-2})(1-p^{-4}).\end{aligned}$$

$\varepsilon_1\varepsilon_2^2[\tilde{X}_2(N)]$ was easily calculated from the structure of E_1 as above. But in the case of degree three, the structure of E_1 is very complicated so we use the theory of torus embedding ([35]) to calculate such intersection numbers (see [50] § 2).

d) From Proposition (1.7) and Proposition (1.2), we have

$$i^*(\bar{c}_1)=-3\pi^*(c_1(\bar{L}_1)).$$

Since $\bar{c}_1\varepsilon_1\varepsilon_2[X_2(N)]=-3\pi^*(c_1(\bar{L}_1))[l_1]$ and $l_1=E_1\cap E_2$ is mapped to a point by π, this number vanishes. So to calculate the third line in $(*)$, it suffices to calculate $\bar{c}_1\varepsilon_1^2[\tilde{X}_2(N)]$.

We have

$$c_1([l_1])i^*(\varepsilon_1)[E_1]=\varepsilon_2\varepsilon_1^2[\tilde{X}_2(N)]=-2.$$

Let q be as before. Then $[q]$ is a line bundle on $\bar{X}_1(N)$ of degree one and we have

$$\pi^*(c_1([q]))i^*(\varepsilon_1)[E_1]=\textstyle\sum_{i=1}^N c_1([l_i])i^*(\varepsilon_1)[E_1]=-2N.$$

So the term $\bar{c}_1\varepsilon_1^2[\tilde{X}_2(N)]=-3\pi^*(c_1(\bar{L}_1))i^*(\varepsilon_1)[E_1]$ is equal to $6N\cdot\deg(\bar{L}_1)$. Since $\mathrm{vol}(\Gamma_1(1)\backslash\mathfrak{S}_1)$ is equal to $\pi/3$, it is similarly proved as before that the degree of $\bar{L}_1$ is equal to $(1/24)N^3\prod_{p|N}(1-p^{-2})$. So we have

$$\begin{aligned}\bar{c}_1(\textstyle\sum_i\varepsilon_i)^2[\tilde{X}_2]&=\bar{c}_1(\textstyle\sum_i\varepsilon_i^2)[\tilde{X}_2]\\&=(1/8)N^8\textstyle\prod_{p|N}(1-p^{-2})(1-p^{-4}).\end{aligned}$$

Thus we calculated the third line and proved the following

Theorem (1.8). *If $N \geq 3$ and $k \geq 4$, then the dimension of $S_k(\Gamma_2(N))$ is equal to*

$$(2^{-10}3^{-3}5^{-1}(2k-2)(2k-3)(2k-4) \\ -2^{-6}3^{-2}(2k-3)N^{-2}+2^{-5}3^{-1}N^{-3})[\Gamma_2(1): \Gamma_2(N)].$$

e) Now we calculate the second term of the fourth line in $(*)$. The following method is due to T. Yamazaki. To calculate this term, it suffices to calculate $\varepsilon_1^3[\tilde{X}_2(N)]$ and to calculate such a self-intersection number we need to replace the cohomology class ε_1 by another cohomology class. We recall the definition of theta constants with half-integral characteristics.

Definition (1.9). Let $m=(m', m'')$ be in $\boldsymbol{Z}^{2g}$ and let m' and m'' be the first and the last g components of m, respectively. For (τ, z) in $\mathfrak{S}_g \times \boldsymbol{C}^g$, we put

$$\theta_m(\tau, z)= \sum_{p \in \boldsymbol{Z}^g} \boldsymbol{e}[(p+m'/2)\tau^t(p+m'/2)/2+(p+m'/2)^t(z+m''/2)],$$

where $\boldsymbol{e}[x]$ is defined to be $e^{2\pi i x}$. This series converges absolutely and uniformly on every compact subset of $\mathfrak{S}_g \times \boldsymbol{C}^g$. We put $\theta_m(\tau)=\theta_m(\tau, 0)$ and call this *theta constant of characteristic m.*

$\theta_m(\tau)$ depends only on m mod 2 and this is identically zero if $m'^t m'' =$ odd. Therefore there are $2^{g-1}(2^g+1)$ theta constants which are not identically zero. $\tau \in \mathfrak{S}_g$ is said to be a *reducible* point if τ is equivalent, with respect to $\Gamma_g(1)$, to a point in the form

$$\begin{pmatrix} \tau_1 & 0 \\ 0 & \tau_2 \end{pmatrix},$$

where $\tau_1 \in \mathfrak{S}_{g'}$ and $\tau_2 \in \mathfrak{S}_{g-g'}$ $(0<g'<g)$. Then we have the following

Theorem (1.10) ([29] and [18]). *In case $g=2$, there are ten non-zero theta constants. Let χ_{10} be the product of the squares of them. Then $\chi_{10} \in S_{10}(\Gamma_2(1))$ and $p \in \mathfrak{S}_2$ is reducible if and only if $\chi_{10}(p)=0$.*

Let R be the zeros of χ_{10} in X_2 and $\bar{R}$ its closure in $\tilde{X}_2$. From the above theorem we have

$$\tilde{L}_2^{\otimes 10} \simeq [\bar{R}]^{\otimes 2} \otimes [E]^{\otimes N}.$$

So it follows that

$$10c_1(\tilde{L}_2)=2c_1([\bar{R}])+Nc_1([E]).$$

We multiply this equation by ε_1^2 and calculate (the values of) both sides. The left hand side is calculated as $10c_1(\tilde{L}_2)\varepsilon_1^2[\tilde{X}_2]=(-10/3)\bar{c}_1\varepsilon_1^2[\tilde{X}_3]=(-5/6)N^4\prod_{p|N}(1-p^2)$. Let R_1 be an irreducible component of $\bar{R}$ which intersects E_1. Then R_1 is biholomorphic to $\bar{X}_1\times\bar{X}_1$ and $E_1\cap R_1$ has a form: $\{p\}\times\bar{X}_1$ (or $\bar{X}_1\times\{p\}$), where p is a cusp of $\bar{X}_1$. $\{p\}\times\bar{X}_1$ and $\{q\}\times\bar{X}_1$ are algebraically equivalent on R_1 and they do not intersect if $p\neq q$. Therefore we have

$$c_1([R_1])\varepsilon_1^2[\tilde{X}_2]=j^*(\varepsilon_1)^2[R_1]=0,$$

where $j\colon R_1\to\tilde{X}_2$ is the inclusion. So it follows that

$$c_1([\bar{R}])\varepsilon_1^2[\tilde{X}_2]=0.$$

Next $c_1([E])\varepsilon_1^2=\varepsilon_1^3+\sum_{1<i}\varepsilon_1^2\varepsilon_i$. So the value of this is equal to $\varepsilon_1^3[\tilde{X}_2]-N^3\prod_{p|N}(1-p^{-2})$. Thus we conclude that

$$\varepsilon_1^3[\tilde{X}_2]=(1/6)N^3\prod_{p|N}(1-p^{-2}).$$

In the above five methods, we have used only the structure of $\tilde{X}_2$ and did not use special results except in the last method. Since the intersection number $\varepsilon_1^3[\tilde{X}_2]$ is of topological nature, it is desirable to calculate this without using Theorem (1.10). So we present here the following

Problem (1.11). *Calculate the intersection number $\varepsilon_1^3([\tilde{X}_2])$ without using J.-I. Igusa's results on theta constants.*

In the case of degree three, dim $S_k(\Gamma_3(N))$ with $N\geq 3$ was calculated by the combinations of the above five methods. The results corresponding to Theorem (1.10) is the following

Theorem (1.12) ([31]). *In case $g=3$, there are thirty six non-zero theta constants. Let χ_{18} be the product of them and Σ_{140} the thirty fifth fundamental symmetric polynomial of the eighth powers of them. The χ_{18} and Σ_{140} belong to $S_{18}(\Gamma_3(1))$ and $S_{140}(\Gamma_3(1))$, respectively and $p\in\mathfrak{S}_3$ is reducible if and only if $\chi_{18}(p)=\Sigma_{140}(p)=0$.*

By using this theorem we can replace the cohomology class $c_1([E])^2$ by another cohomology class and calculate such a self-intersection number as $c_1([E])^6[\tilde{X}_3]$. And the result is as follows.

Theorem (1.13). *If $N\geq 3$ and $k\geq 5$, then the dimension of $S_k(\Gamma_3(N))$ is equal to*

$$(2^{-16}3^{-6}5^{-2}7^{-1}(2k-2)(2k-3)(2k-4)^2(2k-5)(2k-6)$$
$$-2^{-10}3^{-2}5^{-1}(2k-6)N^{-5}+2^{-8}3^{-3}N^{-6})[\Gamma_3(1):\Gamma_3(N)].$$

§ 2. Dimension formula for $\Gamma_2(1)$ and $\Gamma_3(1)$

Since $\Gamma_g(1)$ has fixed points on $\mathfrak{S}_g$, $\Gamma_g(1)\backslash\mathfrak{S}_g$ has singularities. So we need to use the holomorphic Lefschetz fixed points formula to calculate the dimension formula for $\Gamma_g(1)$. First we recall the holomorphic Lefschetz formula. Let X be a compact complex manifold and V a holomorphic vector bundle on X, and let H be a finite group of automorphisms of the pair (X, V). For any $h \in H$, let X^h be the fixed points set of h and let

$$N^h=\sum_\theta N^h(\theta)$$

denote the normal bundle of X^h decomposed according to the eigenvalues $e^{i\theta}$ of h. We put

$$\mathcal{U}^\theta(N^h(\theta))=\prod_\beta\left(\frac{1-e^{-x_\beta-i\theta}}{1-e^{-i\theta}}\right)^{-1},$$

where the (total) Chern class of $N^h(\theta)$ is

$$c(N^h(\theta))=\prod_\beta(1+x_\beta).$$

Let $\mathcal{T}(X^h)$ be the Todd class of X^h and $\mathrm{ch}\,(V\,|\,X^h)(h)$ the Chern character of $V\,|\,X^h$ with h-action ([4]). Put

$$\mu(h)=\left(\frac{\mathrm{ch}\,(V\,|\,X^h)(h)\cdot\prod_\theta\mathcal{U}^\theta(N^h(\theta))\cdot\mathcal{T}(X^h)}{\det\,(1-h\,|\,(N^h)^*)}\right)[X^h].$$

Then we have

Theorem (2.1) ([4]).

$$\sum_{p\geq 0}(-1)^p\,\mathrm{Trace}\,(h\,|\,H^p(X,\mathcal{O}(V)))=\mu(h).$$

Let $H^p(X,\mathcal{O}(V))^H$ be the H-invariant subspace of $H^p(X,\mathcal{O}(V))$. Then by this theorem we have

$$\sum_{p\geq 0}(-1)^p\dim H^p(X,\mathcal{O}(V))^H=\frac{1}{|H|}\sum_{h\in H}\mu(h).$$

Let $N\geq 3$. The group $H:=\Gamma_g(1)/\Gamma_g(N)$ acts on the pair $(\tilde{X}_g(N), \tilde{L}_g^{\otimes k})$, so it acts on $\Gamma(\tilde{X}_g(N), \mathcal{O}(\tilde{L}_g^{\otimes k}(-E)))$. $S_k(\Gamma_g(1))$ is identified with the invariant subspace $\Gamma(\tilde{X}_g(N), \mathcal{O}(\tilde{L}_g^{\otimes k}-(E)))^H$. Therefore dim $S_k(\Gamma_g(1))$ is calculated as

$$\frac{1}{|H|}\sum_{h\in H}\mu(h),$$

under the assumption that the vanishing theorem holds.

Now we consider the case of degree two. To execute this calculation, we need to classify the fixed points sets of H in $\tilde{X}_2$ in the sence we defined in the Introduction. The fixed points sets of $\Gamma_2(1)$ on $\mathfrak{S}_2$ were classified by E. Gottschling in [16] and [17]. By this result the fixed points sets of $H := \Gamma_2(1)/\Gamma_2(N)$ in the quotient space X_2 are classified. Let E_1 be an irreducible component of $E=\tilde{X}_2-X_2$ Then E_1 has a structure of an elliptic surface over X_1. The action of an element h of H which maps E_1 to itself is decomposed into an action of horizontal direction and an action of vertical direction. If h fixes a point in E_1, then the action of horizontal direction of h fixes a point in $\bar{X}_1$. The fixed points in $\bar{X}_1$ come from the fixed points of $\Gamma_1(1)$ on $\mathfrak{S}_1$ which are well known. Since the action of vertical direction of h is decomposed into a translation and the involution with respect to the origin of the elliptic curve, its fixed points are easily classified. Thus we can classify the fixed points over cusps of degree one. Let p be a cusp of degree zero in $\bar{X}_2$ and $\pi_2: \tilde{X}_2 \to \bar{X}_2$ as before. Then $\pi_2^{-1}(p)$ is a reducible rational variety composed of $(1/4)N^3\prod_{p|N}(1-p^{-2})$ projective lines meeting three at each one of the $(1/6)N^3\prod_{p|N}(1-p^{-2})$ points ([32]) and fixed points in $\pi_2^{-1}(p)$ are easily classified.

What we need to do next is to describe exactly the structure of the fixed points sets and to calculate the intersection numbers which appear in the holomorphic Lefschetz fixed points theorem. We refer the reader to the original paper by the author [51].

Now we consider the case of degree three. First we recall Selberg's trace formula. For $M=\begin{pmatrix} A & B \\ C & D \end{pmatrix} \in Sp(g, \boldsymbol{R})$ and $Z \in \mathfrak{S}_g$, we put $J(M, Z)=\det(CZ+D)$ and $Z_{ij}=X_{ij}+\sqrt{-1}\,Y_{ij}$. Then we have

Theorem (2.2) ([8] Exposé 10). *Let Γ be a subgroup of $Sp(g, \boldsymbol{Z})$ of finite index. If $k>2g$, then*

$$\dim S_k(\Gamma)=q\int_{\Gamma\backslash\mathfrak{S}_g}\Sigma_{M\in\Gamma}J(M, Z)^{-k}\det\left(\frac{M\langle Z\rangle-\bar{Z}}{2\sqrt{-1}}\right)^{-k}(\det Y)^k dZ,$$

where q is a polynomial in k of degree $g(g+1)/2$ and dZ is the invariant measure defined by $(\det Y)^{-(g+1)}\prod_{i\le j}dX_{ij}dY_{ij}$.

If C is a conjugacy class or subset of Γ, then the following integral

$$q\int_{\Gamma\backslash\mathfrak{S}_g}\sum_{M\in C}J(M,Z)^{-k}\det\left(\frac{M\langle Z\rangle-\bar{Z}}{2\sqrt{-1}}\right)^{-k}(\det Y)^k dZ,$$

is called the contribution of C to dim $S_k(\Gamma)$.

In the case of degree three, unsolved problems remain on both sides (geometric and Selberg, or Tsushima and Hashimoto) to calculate dim $S_k(\Gamma_3(1))$. On the side of the geometric method, we need to classify the fixed points sets of H on $\tilde{X}_3$. The fixed points sets in the boundary: $\tilde{X}_3-X_3$ are easily classified as before. To classify the fixed points in the quotient space X_3, we need to classify the fixed points sets of $\Gamma_3(1)$ on $\mathfrak{S}_3$. This problem was solved by K. Hashimoto. In [20], he introduced a principle of Milnor and Springer-Steinberg, by which the classification of elliptic conjugacy classes is reduced to a problem of Hermitian forms. By this method, the elliptic conjugacy classes of $\Gamma_3(1)$ were enumerated completely, and their contributions have been determined explicitly ([23]).

On the side of Selberg's trace formula, unsolved problem is to calculate special values of zeta-functions or L-functions of various kinds. These special values appear as the contributions of parabolic of elliptic/parabolic conjugacy classes of $\Gamma_3(1)$ (see [19]). The conjugacy classes of $H=\Gamma_3(1)/\Gamma_3(N)$ corresponding to these conjugacy classes of $\Gamma_3(1)$ fix points in $E=\tilde{X}_3-X_3$. As to the contributions of these conjugacy classes, the geometric method is more effective than Selberg's trace formula and these contributions are calculated by the geometric method. So if we use the geometric method and the Selberg's trace formula jointly, there would be no problem unsolved. This was the starting point of the joint work of K. Hashimoto and the author.

Recently T. Arakawa [2] established the method to calculate these special values of zeta-functions except the value at the origin of Shintani's zeta-function of degree three which appears as the contribution of the conjugacy classes of $\begin{pmatrix}\mathbf{1}_3 & S\\ 0 & \mathbf{1}_3\end{pmatrix}$ with rank $S=3$. So the difficulties on the side of Selberg's trace formula were solved for the most part. But to determine the value of Shintani's zeta-function, we need to use the result of the geometric method.

We recall the definition of Shintani's zeta-function.

Definition (2.3) ([47]). We put

$$\zeta_r^*(s)=\sum\varepsilon(T)^{-1}(\det T)^{-s},$$

where the summation is taken over all the $SL(r, \boldsymbol{Z})$-equivalence classes of positive definite half-integeral symmetric matrices and

$$\varepsilon(T)=\#\{S\in SL(r,Z)\,|\,{}^tSTS=T\}.$$

It is known that the summation in the above definition converges if $\mathrm{Re}(s)>(r+1)/2$. Furthermore $\zeta_r^*(s)$ has an analytic continuation to the whole complex plane with possible simple poles at $s=j/2$ $(j=1,\cdots,r+1)$. Let Π_r be the subset of $\Gamma_g(N)$ consisting of elements which are conjugate, in $\Gamma_g(1)$, to matrices of the form $\begin{pmatrix}\mathbf{1}_g & S\\ 0 & \mathbf{1}_g\end{pmatrix}$, where S is an integral symmetric matrix of size g and of rank r. The importance of Shintani's zeta-function is based on the following theorem and conjecture.

Theorem (2.4) ([47]). *Let $\zeta(s)$ be the Riemannian zeta-function and put*

$$P_{g,g'}(k)=\prod_{i=1}^{g'}(2k-g-i)(2k-g-i+2)\cdots(2k-g+i-2),$$
$$\omega_r=\zeta(2)\zeta(4)\cdots\zeta(2r)\qquad(\omega_0=1),$$
$$u_r=\prod_{i=1}^{r}2\pi^i/i!\qquad(u_0=1).$$

Then the contribution $I_g(\Pi_r,N,k)$ of Π_r to $\dim S_k(\Gamma_g(N))$ $(N\geq 3)$ *is equal to*

$$[\Gamma_g(1):\Gamma_g(N)]N^{-r(2g-r+1)/2}\frac{2^{r(n-r)-1}\omega_{n-r}\zeta_r^*(-g+r)}{u_{n-r}(4\pi)^{(n-r)(n-r+1)/2}}P_{g,g-r}(k).$$

Conjecture (2.5). If $N\geq 3$ and $k>2g$, then

$$\dim S_k(\Gamma_g(N))=\textstyle\sum_{r=0}^{g}I_g(\Pi_r,N,k).$$

This means that the conjugacy classes of $\Gamma_g(N)$ $(N\geq 3)$ other than in Π_r $(0\leq r\leq g)$ have no contributions to $\dim S_k(\Gamma_g(N))$. This conjecture is true if $g=1$ or 2, and in case $g=3$, it was proved by P. Ploch, K. Hashimoto (and M. Eie), independently. $\zeta_1^*(s)$ is the Riemannian zeta-function $\zeta(s)$, so its special values at negative integers are wellknown. The special values of $\zeta_2^*(s)$ at negative integers were calculated by T. Shintani and others.

The special values of $\zeta_3^*(s)$ at the origin and negative integers have not been calculated. So we determine the value of $\zeta_3^*(0)$ by Theorem (1.13), Theorem (2.4), and Conjecture (2.5). Since Conjecture (2.5) was verified in case $g=3$, we have the following

Theorem (2.6). *We have*

$$\zeta_3^*(0)=2^{-7}3^{-3}.$$

In [12] M. Eie claims to have "calculated" all the contributions of the conjugacy classes of $\Gamma_3(2)$ to $\dim S_k(\Gamma_3(2))$ except that of $\Pi_3 \cap \Gamma_3(2)$. By using the fact that the contribution of $\Pi_3 \cap \Gamma_3(2)$ to $\dim S_k(\Gamma_3(2))$ is $2^{-6}[\Gamma_3(1): \Gamma_3(2)]\zeta_3^*(0)$ and $\dim S_k(\Gamma_3(2))$ is an integer, he proved that

$$\zeta_3^*(0)=2^{-7}3^{-3}+2^{-3}3^{-4}5^{-1}7^{-1}l,$$

where l is an integer. So Theorem (2.6) is proved modulo integers by a different method. He also used Theorem (1.13) to determine the value of $\zeta_3^*(0)$ exactly and used it to calculate $\dim S_k(\Gamma_3(2))$.

Thus we solved all the problems to calculate $\dim S_k(\Gamma_3(1))$. No essential problem remains unsolved. But if the calculation is too much, too-muchness may be an essential problem.

In the end we shall make a brief comment on M. Eie and C.-Y. Lin's paper [15] published in American Journal of Mathematics in 1986 and M. Eie's [13] in Memoirs AMS in 1987. In the first paper they claim that they calculated $\dim S_k(\Gamma_3(1))$ (Main Theorem I), $\dim S_k(\Gamma_3(2))$ (Main Theorem II), and $\dim S_k(\Gamma_3(N))$, $N \geq 3$ (Main Theorem III). But we can admit none of them because of its incompleteness or lack of originality.

First concerning Theorem I, in the Acknowledgement in [15], they assert that they obtained their results independently of S. Tsuyumine. It is, however, hard to believe as we show below.

As mentioned in Introduction, our problem is to calculate the coefficients of the partial fractions of the generating function of $A(\Gamma_3(1))$. There are two different standpoints to solve this problem. One is to calculate all the contributions of the conjugacy classes of $\Gamma_3(1)$ and this standpoint is the same as ours. The other is, as written in [15] p. 1064, to calculate only contributions of conjugacy classes which are easy to calculate and determine the coefficients of some part of partial fractions, and to determine the coefficients of the remaining partial fractions by the knowledge of modular forms of lower weights instead of the calculation of the contributions of difficult conjugacy classes.

Their "proof" of Main Theorem I in [15] is not complete from either point of view. From the first standpoint they refer in [15] Proposition 10 to their papers [12], [14] and a paper titled "Conjugacy classes of the modular group $Sp(3, Z)$, Manuscript (1984)" instead of detailed calculations. Since the last paper has not been published, we cannot say anything on it. It might be substituted by [13], on which we comment later. On the other hand their papers [12], [14] are not enough complete to be able to serve as references. First [12] contains many elementary errors. For example in Chapter III Lemma 2, the case when $P(X)=(X^2+X+1)(X^4-X^2+1)$ is missing. But we shall stop listing up all the errors since this was written several years ago.

Now we comment on [14] which was published in Transactions of the American Mathematical Society in 1985. In this paper they "classified" the conjugacy classes of elliptic elements of $Sp(3, \boldsymbol{Z})$ which fix a single point in $\mathfrak{S}_3$. But their result is wrong, since some conjugacy classes are missing. We give an example due to K. Hashimoto. Put

$$M=\begin{pmatrix} 0 & -1 & 1 & -2 & -1 & 0 \\ -1 & 1 & 0 & 0 & -1 & 0 \\ 1 & 0 & -1 & 1 & 1 & -1 \\ 2 & -1 & 0 & 0 & 1 & -1 \\ 0 & 1 & 0 & 1 & 1 & 0 \\ 1 & -1 & 1 & -1 & 0 & -1 \end{pmatrix}$$

and

$${}^tP=(12)^{-1}\begin{pmatrix} -2\sqrt{6} & -3\sqrt{6}-3\sqrt{2} & 3\sqrt{6}-3\sqrt{2} & 2\sqrt{6} & \sqrt{6}+3\sqrt{2} & -\sqrt{6}+3\sqrt{2} \\ 2\sqrt{6} & 3\sqrt{6}-3\sqrt{2} & -3\sqrt{6}-3\sqrt{2} & -2\sqrt{6} & -\sqrt{6}+3\sqrt{2} & \sqrt{6}+3\sqrt{2} \\ 8\sqrt{3} & 0 & 0 & 4\sqrt{3} & -4\sqrt{3} & 4\sqrt{3} \\ 2\sqrt{6} & -\sqrt{6}-3\sqrt{2} & \sqrt{6}-3\sqrt{2} & 2\sqrt{6} & -3\sqrt{6}-3\sqrt{2} & 3\sqrt{6}-3\sqrt{2} \\ -2\sqrt{6} & \sqrt{6}-3\sqrt{2} & -\sqrt{6}-3\sqrt{2} & -2\sqrt{6} & 3\sqrt{6}-3\sqrt{2} & -3\sqrt{6}-3\sqrt{2} \\ -4\sqrt{3} & -4\sqrt{3} & 4\sqrt{3} & 8\sqrt{3} & 0 & 0 \end{pmatrix}.$$

Then $M \in Sp(3, \boldsymbol{Z})$ and $P \in Sp(3, \boldsymbol{R})$. The characteristic polynomial of M is $(X^2+1)(X^4-X^2+1)$. And we have

$$P^{-1}MP=(2)^{-1}\begin{pmatrix} \sqrt{3} & 0 & 0 & 1 & 0 & 0 \\ 0 & -\sqrt{3} & 0 & 0 & 1 & 0 \\ 0 & 0 & 0 & 0 & 0 & 2 \\ -1 & 0 & 0 & \sqrt{3} & 0 & 0 \\ 0 & -1 & 0 & 0 & -\sqrt{3} & 0 \\ 0 & 0 & -2 & 0 & 0 & 0 \end{pmatrix}.$$

So M is conjugate to $e[-1/6, 5/6, 1/2]$ (in the sense of [14]) in $Sp(3, \boldsymbol{R})$ and its conjugacy class is missing in [14], since elliptic elements which are conjugate to $e[1/6, 5/6, 1/2]$ or $e[7/6, 11/6, 3/2]$ do not appear in the Table II in [14].

Therefore their paper cannot be complete from the first standpoint since their paper uses a wrong result.

If they wish their paper to be complete from the second standpoint, they need to make clear the references of *their knowledge of modular forms of lower weights*. As far as we know the dimension of $S_k(\Gamma_3(1))$ with $k \geq 12$ was known by Tsuyumine's work for the first time. (dim $S_k(\Gamma_3(1))$ with

$0 \leq k \leq 10$ was known to be 0 in [31].) In [15] they claim that they have obtained the same result for dim $S_k(\Gamma_3(1))$ as Tsuyumine's, by using the knowledge of dim $S_k(\Gamma_3(1))$ with $10 \leq k \leq 44$, which determines the coefficients of eighteen partial fractions that are difficult to calculate, without giving any reference. We do not know, however, any literature on it except for Tsuyumine's paper.

Now we consider their Main Theorems II and III, calculating dim $S_k(\Gamma_3(2))$ and dim $S_k(\Gamma_3(N))$, respectively. In both cases the essential point of the proof is to determine the exact value of $\zeta_3^*(0)$, for which they use the author's Theorem (1.13) without giving any reference. Theorem III is then even a tautology, while Theorem II could be said their new "result" aside from gaps in the proof.

Next we give some comments on M. Eie's recent paper [13].

In the second chapter he "classified" all the conjugacy classes of $Sp(3, \boldsymbol{Z})$ which have non-zero contributions to dimension formula and in the third chapter he "gave" their contributions. They are very detailed results and may give an impression of a completed and elaborate work. But he still uses the wrong result [14] and there exists another conjugacy class which is missing. In Lemma 1 in the second chapter, he states (without proof) that a conjugacy class which has non-zero contribution fixes a single point on $\mathfrak{S}_3$ or has a representative in Γ_3^2, Γ_3^1 or Γ_3^0, where Γ_3^i is the stabilizer in $Sp(3, \boldsymbol{Z})$ of a rational boundary component of degree i. But this is wrong. In fact, put

$$N=\begin{pmatrix} 0 & 0 & 0 & 1 & 0 & 0 \\ 0 & 0 & 0 & 1 & 1 & 0 \\ 0 & 0 & -1 & 0 & 0 & -1 \\ -1 & 1 & 0 & 0 & 0 & 0 \\ 0 & -1 & 0 & 0 & 0 & 0 \\ 0 & 0 & 1 & 0 & 0 & 0 \end{pmatrix}.$$

Then N is a torsion element of $Sp(3, \boldsymbol{Z})$. The fixed points set of N is a one-dimensional submanifold F of $\mathfrak{S}_3$ and N has non-zero contribution to the dimension formula. Of course, F is biholomorphic to $\mathfrak{S}_1$. The subgroup Δ of $\Gamma_3(1)$ consisting of elements which map F into itself is isomorphic to a discrete subgroup Δ' of $SL(2, \boldsymbol{R})$. We can prove that Δ' is commensurable with the unit group of a maximal order of an indefinite division quaternion algebra over $\boldsymbol{Q}$ which is generated by

$$\begin{pmatrix} 0 & \sqrt{2} \\ \sqrt{2} & 0 \end{pmatrix} \quad \text{and} \quad \begin{pmatrix} 0 & \sqrt{3} \\ -\sqrt{3} & 0 \end{pmatrix}.$$

Therefore $\Delta \backslash F$ is compact. This means that F does not intersect rational boundary components of $\mathfrak{S}_3$, so N does not fix rational boundary components.

Therefore it is impossible that they can reach the same result as Tsuyumine's without adding the contributions for these missing conjugacy classes.

References

[1] T. Arakawa, The dimension of the space of cusp forms on Siegel upper half plane of degree two related to a quaternion unitary group, J. Math. Soc. Japan, **33** (1981), 125–145.

[2] ——, Special values of L-functions associated with the space of quadratic forms and the representation of $Sp(2n, F_p)$ in the space of Siegel cusp forms, in this volume.

[3] A. Ash, D. Mumford, M. Rapoport and Y. Tai, Smooth Compactification of Locally Symmetric Varieties (Lie Groups: History, Frontiers and Applications. Vol. 4), Math. Sci. Press, Brookline, 1975.

[4] M. G. Atiyah and I. M. Singer, The index of elliptic operator, III, Ann. of Math., **87** (1968), 546–608.

[5] W. L. Baily, Jr., On Satake's compactification of V_n, Amer. J. Math., **80** (1958), 348–364.

[6] H. Bass, M. Lazard and J.-P. Serre, Sous-groupes d'indice fini dans $SL(n, \boldsymbol{Z})$, Bull. Amer. Math. Soc., **70** (1964), 385–392.

[7] A. Borel and F. Hirzebruch, Characteristic classes and homogeneous spaces, I, II, III, Amer. J. Math., **80** (1958), 458–538; **81** (1959), 315–382; **82** (1960), 491–504.

[8] H. Cartan, et al., Séminaire Henri Cartan, 1957/1958, Fonctions Automorphes.

[9] U. Christian, Berechnung des Ranges der Schar der Spitzenformen zur Modulgruppe zweiten Grades und Stufe $q > 2$, J. reine angew. Math., **277** (1975), 130–154.

[10] ——, Zur Berechnung des Ranges der Schar der Spitzenformen zur Modulgruppe zweiten Grades und Stufe $q > 2$, J. reine angew. Math., **296** (1977), 108–118.

[11] P. Deligne, Théorie de Hodge, II, III, Publ. Math. Inst. Hautes Etudes Sci., **40** (1971), 5–58; **44** (1974), 5–77.

[12] M. Eie, Siegel cusp forms of degree two and three, Mem. Amer. Math. Soc., **304** (1984), 1–186.

[13] ——, Dimension formulae for the vector spaces of Siegel cusp forms of degree three (II), Mem. Amer. Math. Soc., **373** (1987), 1–124.

[14] M. Eie and C.-Y. Lin, Fixed points and conjugacy classes of regular elliptic elements in $Sp(3, \boldsymbol{Z})$ Trans. Amer. Math. Soc., **289** (1985), 485–496.

[15] ——, Dimension formulae for the vector spaces of Siegel cusp forms of degree three, Amer. J. Math., **108** (1986), 1059–1088.

[16] E. Gottschling, Über die Fixpunkte der Siegelschen Modulgruppe, Math. Ann., **143** (1961), 111–149.

[17] ——, Über die Fixpunktergruppe der Siegelschen Modulgruppe, Math. Ann., **143** (1961), 399–430.

[18] W. F. Hammond, On the graded ring of Siegel modular forms of genus two, Amer. J. Math., **87** (1965), 502–506.

[19] K. Hashimoto, A formula of the spaces of cusp forms on Siegel upper half plane of degree two, I, J. Fac. Sci. Univ. Tokyo Sec. IA, **30** (1983), 403–488; II, Math. Ann., **266** (1984), 539–559.

[20] ——, Elliptic conjugacy classes of the Siegel modular group and unimodular hermitian forms over the ring of cyclotomic fields, J. Fac. Sci. Univ. Tokyo, **33** (1986), 57–82.

[21] ——, Representation of the finite symplectic group $Sp(4, \boldsymbol{F}_p)$ in the spaces of Siegel modular form, The Selberg trace formula and related topics, Contemp. Math., **53**, Amer. Math. Soc., Providence, 1986, 253–276.

[22] K. Hashimoto and T. Ibukiyama, On relations of dimensions of automorphic forms of $Sp(2, \boldsymbol{R})$ and its compact twist $Sp(2)$, (II), Automorphic Forms and Number Theory, Advanced Studies in Pure Math., **7**, Kinokuniya, Tokyo, 1984, 31–102.

[23] K. Hashimoto and R. Tsushima, Dimension formula for Siegel cusp forms of degree three (in preparation).

[24] F. Hirzebruch, Neue Topologische Methoden in der Algebraichen Geometrie, Springer-Verlag, Berlin-Heidelberg-New York 1956.

[25] ——, Characteristic numbers of homogeneous domain, Seminars on Analytic Functions, vol. 2, Institute for Advanced Study, 1957, 92–104.

[26] ——, Automorphe Formen und der Satz von Riemann-Roch, Symposium de Topologia Algebraica, Univ. de Mexico, 1958, 129–143.

[27] ——, Elliptische Differentialoperatoren auf Mannigfaltigkeiten, Arbeitsgemeinschaft für Forschung des Landes Nordrhein-Westfalen, **33** (1965), 583–608.

[28] T. Ibukiyama, On relations of dimensions of automorphic forms of $Sp(2, \boldsymbol{R})$ and its compact twist $Sp(2)$, (I), Automorphic Forms and Number Theory, Advanced Studies in Pure Math., **7**, Kinokuniya, Tokyo, 1984, 7–29.

[29] J.-I. Igusa, On the graded ring of theta-constants, Amer. J. Math., **86** (1964), 219–246; (II), ibid., **88** (1966), 221–236.

[30] ——, On Siegel modular forms of genus two, Amer. J. Math., **84** (1962), 175–200; (II), ibid., **86** (1964), 392–412.

[31] ——, Modular forms and projective invariants, Amer. J. Math., **89** (1967), 817–855.

[32] ——, A desingularization problem in the theory of Siegel modular functions, Math. Ann., **168** (1967), 228–260.

[33] M. Ise, Generalized automorphic forms and certain holomorphic vector bundles, Amer. J. Math., **86** (1964), 70–108.

[34] Y. Kawamata, A generalization of Kodaira-Ramanujan's vanishing theorem, Math. Ann., **261** (1982), 43–46.

[35] G. Kempf, F. Knudsen, D. Mumford and B. Saint-Donat, Toroidal Embeddings, I, Lecture Notes in Math., **339**, Springer, Berlin-Heidelberg-New York, 1978.

[36] K. Kodaira, On a differential geometric method in the theory of analytic stacks, Proc. Nat. Acad. Sci. USA, **39** (1953), 1268–1273.

[37] R. P. Langlands, The dimension of spaces of automorphic forms, Amer. J. Math., **85** (1963), 99–125.

[38] J. Mennicke, Zur Theorie der Siegelschen Modulgruppe, Math. Ann., **159** (1965), 115–129.

[39] Y. Morita, An explicit formula for the dimension of spaces of Siegel modular forms of degree two, J. Fac. Sci. Univ. Tokyo Sec, IA, **21** (1974), 167–248.

[40] D. Mumford, Hirzebruch's proportionality theorem in the noncompact case, Invent. Math., **42** (1977), 239–272.

[41] Y. Namikawa, A new compactification of the Siegel space and degeneration of abelian varieties, I, II, Math. Ann., **221** (1976), 97–141, 201–241.

[42] ——, Toroidal Compactification of Siegel Spaces, Lecture Notes in Math., **812**, Springer, Berlin-Heidelberg-New York, 1980.

[43] I. Satake, On the compactification of Siegel space, J. Indian Math. Soc., **20** (1956), 259–281.

[44] ——, Algebraic Structures of Symmetric Domains, Publ. Math. Soc. Japan

14, Iwanami, Tokyo, 1980.
[45] ——, On numerical invariants of arithmetic varieties—the case of Q-rank 1 (in Japanese), Sugaku, **35** (1983), 210–220.
[46] ——, On numerical invariants of arithmetic varieties of Q-rank one, Automorphic Forms of Several Complex Variables, Progress in Math., **46**, Birkhäuser, Boston-Basel-Stuttgart, 1984, 353–369.
[47] T. Shintani, On zeta-functions associated with the vector space of quadratic forms, J. Fac. Sci. Univ. Tokyo, Sec. IA, **22** (1975), 25–65.
[48] T. Shioda, On the graded ring of invariants of binary octavics, Amer. J. Math., **89** (1967), 1022–1046.
[49] C. L. Siegel, Symplectic geometry, Amer. J. Math., **65** (1943), 1–86.
[50] R. Tsushima, A formula for the dimension of spaces of Siegel cusp forms of degree three, Amer. J. Math., **102** (1980), 937–977.
[51] ——, On the spaces of Siegel cusp forms of degree two, Amer. J. Math., **104** (1982), 843–885.
[52] ——, An explicit dimension formula for the spaces of generalized automorphic forms with respect to $Sp(2, Z)$, Proc. Japan Acad. Ser. A, **59** (1983), 139–142.
[53] ——, The space of Siegel cusp forms of degree two and the representation of $Sp(2, F_p)$, Proc. Japan Acad. Ser. A, **60** (1984), 209–211.
[54] S. Tsuyumine, On Siegel modular forms of degree three, Amer. J. Math., **108** (1986), 755–862.
[55] E. Viehweg, Vanishing theorems, J. reine angew. Math., **335** (1982), 1–8.
[56] H. Yamaguchi, The parabolic contribution to the dimension of the space of cusp forms on Siegel space of degree two, unpublished, 1976.
[57] T. Yamazaki, On Siegel modular forms of degree two, Amer. J. Math., **98** (1976), 39–53.

School of Science and Technology
Meiji University
Kawasaki, Kanagawa 214
Japan

Advanced Studies in Pure Mathematics 15, 1989
Automorphic Forms and Geometry of Arithmetic Varieties
pp. 65–87

On the Graded Rings of Modular Forms in Several Variables

Shigeaki Tsuyumine

The theory of modular forms in one variable has a long history. Gauß already knew modular functions in one variable in connection with elliptic functions (Takagi [49]). The theory of modular forms in several variables is younger than it, but is now nearly a century old. It was noted by Hilbert at the end of the last century, and first treated by his student Blumenthal [2]. In 1935 Siegel introduced a new class of modular forms in several variables in a celebrated paper [47]. Nowadays the theory of automorphic forms on a bounded symmetric domain has been developed in several directions as well as the theory of automorphic forms in a rather wider class. However, in spite of this long history, there is still only a small number of examples of rings of automorphic forms in several variables for which much is known about the structure. Here we give a list of papers concerning the structure of the graded ring, the generators or the relations among them, which is arranged nearly chronologically in each case:

Siegel modular forms; Igusa [28, 29, 30], Hammond [15], Freitag [6], Tsuyumine [56], Satoh [41].

Hilbert modular forms; Gundlach [13, 14], Hammond [16], Fomenko [5], Resnikoff [36], Hirzebruch [21, 22, 23], Van der Geer [11], Van der Geer-Zagier [12], Hermann [17, 18], Nagaoka [35], Müller [32, 33].

Hermitian modular forms; Freitag [7].

Automorphic forms on the complex 2-ball; Resnikoff-Tai [37, 48]. There are some other papers concerning the structure of a ring of modular forms over $\boldsymbol{Z}$. Here we should mention that the dimension formulas for spaces of modular forms, or rather cusp forms, are known for wider class. They have helped, and will keep to help, the work of determining the structure of a graded ring. Shimizu's formula [45] has been continually employed to study a graded ring of Hilbert modular forms (mainly in two dimensional case yet), and the formula by Busam [3] in the symmetric case. The works by Resnikoff-Tai [37, 48] were done by

Received March 16, 1987.

relying on the formula by Cohn [4], and the work by Satoh [41] which is concerned with the module structure of the space of vector valued Siegel modular forms of degree two, was done by relying on Tsushima's formula [51, 52].

The works of this kind have been done after algebraic geometry was fully developed, by which we never mean that all the important methods and ingredients have come from algebraic geometry. The theory of compactification introduced by Satake [39, 40] (cf. [43]) makes us apply algebraic geometry effectively to arithmetic quotients of the Siegel spaces. The works by Igusa [28, 30] have, more or less, flavor of the moduli theory of hyperelliptic curves and of abelian varieties. In the Hilbert modular case, Hirzebruch [19–24] (see also Hirzebruch-Van de Ven [25], Hirzebruch-Zagier [26, 27]) has developed an elegant and powerful geometric method. However it can be said that although we are now equipped with several excellent tools particularly for low dimensional cases, there is still plenty of difficulty because graded rings of modular forms have a character of their own and each of them requires us to treat it in distinct way.

We do not try in the present survey paper such an awful thing as looking round all the tools and the technique to determine the graded rings which have ever been developed. We introduce the method first employed by Gundlach [13] which, or whose modification, is used repeatedly by several authors (Siegel modular case; Hammond [15], Freitag [6], Tsuyumine [56]: Hilbert modular case; Hammond [16], Hermann [17, 18]). After preliminaries in § 1, we introduce it in § 2. In the remaining sections, its applications are shown. In § 3, Gundlach's original argument is introduced from our point of view. In § 4, the structure theorem for the graded ring of Siegel modular forms of degree two is discussed. It is based on the papers by Freitag and by Hammond. In § 5, Siegel modular forms of degree three is discussed by sketching [56], and in the last section we give a small observation about Siegel modular forms of degree four. We expect that the method will be still useful in more cases, and wish that the present paper would be of benefit to people trying further investigation.

The present work was done while the author was staying at Göttingen. He wishes to express his heartfelt gratitude to Sonderforschungsbereich 170 "Géométrie und Analysis" in Göttingen and especially to Prof. U. Christian for his hospitality and support during the author's his visit.

§ 1. A graded ring

We start with a quasi-projective variety X over C and a quasi-coherent sheaf $\mathscr{L}$ on X on which the following assumption is made:

Assumption I. X has a projective compactification $\bar{X}$ such that $\operatorname{codim}(\bar{X}-X)\geq 2$. $\mathscr{L}$ is a torsion-free graded $\mathcal{O}_X$-module $\mathscr{L}=\oplus_k \mathscr{L}(k)$, $k \in \boldsymbol{Z}, \geq 0$, with $\mathscr{L}(0)=\mathcal{O}_X$ where $\mathscr{L}(k)$ is coherent and is locally free of rank one except on proper subvarieties. There is a positive integer d such that if $k>0$ is divisible by d, then $\mathscr{L}(k)$ is an ample invertible sheaf on X which has an extension to an ample invertible sheaf on $\bar{X}$.

For k, $k'\geq 0$, $\mathscr{L}(k)\otimes\mathscr{L}(k')$ is contained in $\mathscr{L}(k+k')$ as an $\mathcal{O}_X$-submodule. The equality holds if k, $k'\equiv 0 \pmod{d}$. It is true also for extensions of $\mathscr{L}(k)$'s, $k\equiv 0 \pmod{d}$, for the extended invertible sheaves are unique from our assumption that $\operatorname{codim}(\bar{X}-X)\geq 2$. The ampleness of $\mathscr{L}(d)$ implies in particular that $\mathcal{O}_X\subset\mathscr{L}(k)$ if k is large enough. Let $d_k\equiv 0 \pmod{d}$ be such that $k\leq d_k$ and $\mathcal{O}_X\subset\mathscr{L}(d_k-k)$. Then for any non-negative integer m, we have $\mathscr{L}(k+md)\subset\mathscr{L}(d_k+md)=\mathscr{L}(d_k)\otimes\mathscr{L}(d)^m$. There is an ascending sequence of $\mathcal{O}_X$-submodules of $\mathscr{L}(d_k)$; $\mathscr{L}(k)\subset \mathscr{L}(k+d)\otimes\mathscr{L}(d)^{-1}\subset\mathscr{L}(k+2d)\otimes\mathscr{L}(d)^{-2}\subset\cdots\subset\mathscr{L}(d_k)$. The coherency shows that $\mathscr{L}(k+md)\otimes\mathscr{L}(d)^{-m}$ are all equal for large m. We have shown that $\mathscr{L}(k)\otimes\mathscr{L}(k')=\mathscr{L}(k+k')$ if $k'\geq 0$ is $\equiv 0 \pmod{d}$ and if k is large enough, or $\equiv 0 \pmod{d}$. Hence $\mathscr{L}$ is finite over the submodule $\oplus_k \mathscr{L}(k)$, $k\equiv 0 \pmod{d}$.

By assumption $H^0(X, \mathscr{L}(k))$ is finite dimensional for any $k\geq 0$. Since by a standard argument (see, for instance, Mumford [34]) $H^0(X, \mathscr{L}(k))\otimes H^0(X, \mathscr{L}(k'))\rightarrow H^0(X, \mathscr{L}(k+k'))$ is surjective for sufficiently large k, k' which are $\equiv 0 \pmod{d}$, a graded ring

$$A^{(d)}:=\bigoplus_k H^0(X, \mathscr{L}(k))$$

is noetherian. Let

$$A:=\bigoplus_k H^0(X, \mathscr{L}(k)),$$

which we call *the graded ring associated with X and $\mathscr{L}$*, or *the graded ring associated with X* when the quasi-coherent sheaf under consideration is obvious. Then by the above argument the graded ring A is finite over $A^{(d)}$ as a module, and hence noetherian. Let

$$X^*:=\operatorname{Proj}(A).$$

X^* is a compactification of X satisfying the condition in the Assumption I, namely it is a projective variety satisfying

$$\operatorname{codim}(X^*-X)\geq 2.$$

Let $i: X \hookrightarrow X^*$ be the natural injection. We write $\mathcal{L}(k)^* = i_*(\mathcal{L}(k))$, which is a coherent sheaf on X^* and which is an ample invertible sheaf if $k \equiv 0 \pmod{d}$. $H^0(X^*, \mathcal{L}(k)^*)$ is equal to $H^0(X, \mathcal{L}(k))$ for any k. There is a finite morphism of X^* onto $\bar{X}$ which is the identity on X. X^*, $\bar{X}$ are not necessarily equal, however they are when $\bar{X}$ is normal.

Let $A_k = H^0(X, \mathcal{L}(k))$, which is the graded part in A of degree k, so that $A = \oplus_k A$. A formal power series $P_A(t)$ in a variable t is called *the generating function of* A if

$$P_A(t) = \sum_k (\dim_C A_k) t^k.$$

Since A is noetherian, $P_A(t)$ is a rational function of t by the Hilbert theorem.

Let V be an irreducible subvariety in X, and let $\bar{V}$ be the closure of V in X^*. We note that if $\bar{V} - V$ is of codimension at least two in $\bar{V}$, then V and $\mathcal{L}|_V$ satisfy the Assumption I.

Example 1. Let $\mathcal{D}$ be a bounded symmetric domain. $\mathcal{D}$ is written as a quotient $G(\boldsymbol{R})^\circ/K$ where $G(\boldsymbol{R})^\circ$ is the (topological) identity component of the group of real points of a semi-simple algebraic group G over $\boldsymbol{Q}$, and K is a maximal compact subgroup of $G(\boldsymbol{R})^\circ$. Let Γ be an arithmetic subgroup of G, which acts properly discontinuously on $\mathcal{D}$. By Baily-Borel [1], there is a natural compactification $(\mathcal{D}/\Gamma)^*$ which is called a Baily-Borel-Satake compactification. It is a normal projective algebraic variety and has $\mathcal{D}/\Gamma$ as an open subvariety. Now let us suppose that G has no normal $\boldsymbol{Q}$-subgroup of dimension three with $\boldsymbol{Q}$-rank one. Then $\operatorname{codim}((\mathcal{D}/\Gamma)^* - \mathcal{D}/\Gamma)$ is at least two. So if we put $X = \mathcal{D}/\Gamma$, then X satisfies the first condition in the Assumption I. We take as $\mathcal{L}$, the quasi-coherent sheaf corresponding to automorphic forms for Γ on $\mathcal{D}$, whose details are given in the following.

Let ρ be a (holomorphic) automorphy factor, i.e., the function on $\Gamma \times \mathcal{D}$ with values in $\boldsymbol{C} - \{0\}$ such that (i) $\rho(\gamma, z)$ is holomorphic in z for any fixed $\gamma \in \Gamma$, (ii) $\rho(\gamma\gamma', z) = \rho(\gamma, \gamma' z) \cdot \rho(\gamma', z)$ for $\gamma, \gamma' \in \Gamma$, and (iii) $\rho(\gamma, z) = \rho(\gamma', z)$ if γ, γ' induce the same automorphism of $\mathcal{D}$. Let $j(\gamma, z)$ be the jacobian at a point $z \in \mathcal{D}$, of an automorphism of $\mathcal{D}$ induced by $\gamma \in \Gamma$. $j(\gamma, z)$ is an example of automorphy factors. Let us consider the automorphy factor ρ which is of the form

$$\rho(\gamma, z) = v(\gamma) j(\gamma, z)^{-r}$$

where r is a positive rational number and v is a multiplier whose value for $\gamma \in \Gamma$ is an m-th root of unity for some fixed $m \in \boldsymbol{Z}$. v is depending on the choice of the branches of $j(\gamma, z)^{-r}$ if $r \in \boldsymbol{Q} - \boldsymbol{Z}$. Let π denote the canonical projection of $\mathscr{D}$ onto $\mathscr{D}/\Gamma$. We define $\mathscr{L}(k)$ to be the coherent sheaf corresponding to ρ^k-automorhpic forms, i.e., the sheaf defined by

$$H^0(U, \mathscr{L}(k)) = \{f \in \mathcal{O}_{\pi^{-1}(U)} \mid f(\gamma z) = \rho(\gamma, z)^k f(z) \quad \text{for } z \in \pi^{-1}(U), \gamma \in \Gamma\},$$

where U is any analytic open subset of $\mathscr{D}/\Gamma$, and $\mathcal{O}_{\pi^{-1}(U)}$ denotes the structure sheaf of $\pi^{-1}(U)$ in analytic sense. $\mathscr{L}(k)$ extends to a coherent sheaf on $(\mathscr{D}/\Gamma)^*$ (Serre [44]), and in particular it is algebraic. By Baily-Borel [1], $\mathscr{L}(k)$ extends to an ample invertible sheaf on $(\mathscr{D}/\Gamma)^*$ if k is sufficiently divisible.

So $X = \mathscr{D}/\Gamma$ and $\mathscr{L} = \bigoplus_k \mathscr{L}(k)$ satisfy the conditions in the Assumption I, if G has no normal $\boldsymbol{Q}$-subgroup of dimension three with $\boldsymbol{Q}$-rank one. $A_k = H^0(X, \mathscr{L}(k))$ is the space of ρ^k-automorphic forms, i.e., holomorphic functions f on $\mathscr{D}$ satisfying $f(\gamma z) = \rho(\gamma, z)^k f(z)$ for $\gamma \in \Gamma$. $A = \bigoplus_k A_k$ is the graded ring of such automorphic forms.

Example 2. Let $\mathscr{D}$, Γ be as above. Let X be an irreducible subvariety of $\mathscr{D}/\Gamma$ such that the closure $\bar{X}$ of X in $(\mathscr{D}/\Gamma)^*$ satisfies that $\operatorname{codim}(\bar{X} - X) \geq 2$. Then X and the sheaf given by restricting to X, the quasi-coherent sheaf on $\mathscr{D}/\Gamma$ corresponding to automorphic forms, satisfy the conditions in the Assumption I. Such is the moduli space $\mathfrak{M}_g$ of curves of genus $g \geq 3$, which we discuss later.

Remark. Let the notation be as in the Example 1. To investigate the structure of A, usually we had better to take $r > 0$ as small at possible. If otherwise, it might make the structure of A unnecessarily complicated. For instance, if we compare a polynomial ring and its subring consisting of polynomials of degree $\equiv 0 \pmod{d}$ for $d > 1$, then the former is easier to handle than the latter.

§ 2. A graded ring and a subring

Let X, $\mathscr{L}$ be as in the preceding section. Let D be an irreducible subvariety in X of codimension one, and let D^* be the closure of D in X^*. Let

$$B := \bigoplus_k B_k, \qquad B_k := H^0(D^*, \mathscr{L}(k)^*|_{D^*}).$$

B is noetherian and is a homogeneous coordinate ring of D^*, hence $D^* = \operatorname{Proj}(B)$. A global section f of $\mathscr{L}(k)$ on X can be regarded as that of $\mathscr{L}(k)^*$ on X^*, and hence $f|_D$ determines the unique element of B_k whose restriction to D equals $f|_D$. We have a homomorphism of graded rings

$$\Psi: A \longrightarrow B$$
$$f \longrightarrow f|_D$$

which is associated with the inclusion map of D into X, or D^* into X^*. We write

$$\bar{A} := \operatorname{Im} \Psi$$

which is a graded subring of B. Now we make the following assumption on D.

Assumption II. There is a homogeneous element χ in A such that χ vanishes only at D. Any such element is equal to a power of χ up to a constant factor.

Under the first condition in the Assumption II, the existence of χ satisfying the second is equivalent to the assertion that if $\varphi: \tilde{X} \to X$ is the normalization, then $\tilde{D} = \varphi^{-1}(D)$ is irreducible. For a positive rational number r we call $D(r)$ the following condition on D; $D(r)$: χ defines $r\tilde{D}$ in $\tilde{X}$ where χ is regarded as an element of $\oplus_k H^0(\tilde{X}, \varphi^*(\mathscr{L}(k)))$. For simplicity, we shall deal only with the case r integral, because the similar argument is applicable to the general case. We note that $D(1)$ is the case that D is defined ideal-theoretically by χ in X. χ defines also a subscheme of X^*, which equals D^* set-theoretically. All irreducible subvarieties in X of codimension one satisfy $D(r)$, $r \leq r_X$ for some fixed $r_X \in \boldsymbol{Z}$, if $\operatorname{Pic}(X) \simeq \boldsymbol{Z} + \{\text{finite torsion}\}$, which is the case if $X = \mathscr{D}/\Gamma$ and $\mathscr{L}$ are as in the Example 1 where $\mathscr{D}$ is an irreducible bounded symmetric domain under a certain condition and Γ has a finite commutator factor group (cf. Tsuyumine [55]). All of them satisfy in $D(1)$ particular if $\operatorname{Pic}(X) \simeq \boldsymbol{Z}$ and if an automorphy factor ρ is taken suitably (loc. cit., see also Freitag [8, 9]).

The Assumption II implies that the divisor nD, for some positive integer n, corresponding to an ample invertible sheaf, and hence that Ψ is surjective for graded parts of sufficiently large degree which is $\equiv 0 \pmod{d'}$, d' being sufficiently divisible.

Let us consider the primitive case $D(1)$. Suppose that the structure of $\bar{A}$ is known, e.g., B is known and Ψ is surjective which is often the case (but not always) in the study of rings of automorphic forms. Then we can deduce some information on A from $\bar{A}$. *In the case of* $D(1)$, *the generating function* $P_A(t)$ *of* A *is given by* $(1 - t^{\deg(\chi)})P_{\bar{A}}(t)$, $\deg(\chi)$ *denoting the degree of* χ *in* A. *If* $\{f_i\}$ *is any system of homogeneous elements in* A *whose images by* Ψ *generate* $\bar{A}$, *then* A *is generated by* χ *and the* f_i*'s.*

In particular if the ring $\bar{A}$ is isomorphic to a polynomial ring, then A is too, and furthermore A and $\bar{A}[\chi]$ are isomorphic as graded rings. The first assertion follows from an exact sequence

$$0 \longrightarrow (\chi) \longrightarrow A \xrightarrow{\Psi} \bar{A} \longrightarrow 0.$$

Let f be any homogeneous element in A, and let $k = \deg(f)$. We are able to prove by induction on k, that f is contained in the ring generated by χ and the f_i's. If $k<0$, then the assertion is trivial. Suppose that it is true for an element of degree $<k$. $\Psi(f)$ is written as a polynomial $Q(\Psi(f_1), \cdots)$ of $\Psi(f_i)$'s. Then the image of $f-Q(f_1, \cdots)$ by Ψ vanishes, and hence it is divisible by χ. If $f-Q(f_1, \cdots)=\chi g$, then g is of degree $<k$, and by the induction hypothesis g is written by χ and the f_i's, and hence f is. The third assertion is immediate from the fact that in the above exact sequence Ψ has a section by assumption.

The general case is to be treated somewhat delicately. Let $\varphi'\colon \tilde{X}^* \to X^*$ be the normalization, which is an extension of $\varphi\colon \tilde{X} \to X$. Let $\tilde{D}^*$ be the closure of $\tilde{D}$ in $\tilde{X}^*$, and let $\bar{A}' := \oplus_k H^0(\tilde{D}^*, \varphi'^*(\mathscr{L}(k)^*)|_{\tilde{D}^*})$, which is a homogeneous coordinate ring of $\tilde{D}^*$. Since $\varphi'|_{\tilde{D}^*}\colon \tilde{D}^* \to D^*$ is a finite morphism, it is shown that $\bar{A}'$ is finite over $\bar{A}$ as a module. Now let us define a valuation on A in terms of $\tilde{D}$. Let $f \in A$ be a homogeneous element. Then for some integer $m>0$, f^m can be regarded as a global section of the invertible sheaf $\mathscr{L}(k)$, $k \equiv 0 \pmod{d}$, and hence that of the invertible sheaf $\varphi^*(\mathscr{L}(k))$ on $\tilde{X}$. Then $\nu(f^m)$ is defined to be the vanishing order of f^m at $\tilde{D}$, and $\nu(f)$ is defined to be $\nu(f^m)/m$ ($\nu(0)=+\infty$). It is easy to check that ν is well-defined. If $\nu(f)=k'r$ for an integer k', then $f/\chi^{k'}$ defines a global section of $\varphi^*(\mathscr{L}(k))$ for $k=\deg(f)-\deg(\chi^{k'})$ (not necessarily that of $\mathscr{L}(k)$), and so $(f/\chi^{k'})|_{\tilde{D}} \in \bar{A}'$. We denote again by Ψ the map $f/\chi^{k'} \to (f/\chi^{k'})|_{\tilde{D}}$, which is an extension of the previous Ψ. For an integer i, let $A(i)$ be the ideal of A generated by homogeneous elements f with $\nu(f) \geq i$. There is a filtration

$$A = A(0) \supset A(1) \supset A(2) \supset \cdots.$$

Let

$$\bar{A}(i) := A(i)/A(i+1).$$

$\bar{A}(0)$ equals $\bar{A}$, which is a noetherian graded ring, and $\bar{A}(i)$'s are noetherian graded modules over $\bar{A}$. Let us fix some i'. If k' is large enough, then there is a homogeneous element g in $A(k'r-i')-A(k'r-i'+1)$, where g is taken to be 1 if $i' \equiv 0 \pmod{r}$. Fixing such g, we define a map $\Psi(i)$ of $A(i)$ of $\bar{A}'$ for $i \equiv i' \pmod{r}$, by

$$\begin{array}{ccccc} A(i) & \longrightarrow & A(i+\nu(g)) & \longrightarrow & \bar{A}' \\ f & \longrightarrow & fg & \longrightarrow & (fg/\chi^{(i+\nu(g))/r})|_{\bar{D}}. \end{array}$$

$\Psi(0)$ equals Ψ by definition. The kernel of $\Psi(i)$ is just $A(i+1)$ and hence $\Psi(i)$ is regarded also as an injective map of $\bar{A}(i)$ into $\bar{A}'$ which is an $\bar{A}$-module homomorphism. By definition, the map $A(i)$ to $\bar{A}'$ given by $f\to\Psi(i+r)(\chi f)$ equals $\Psi(i)$, and in particular there is an inclusion $\Psi(i)(\bar{A}(i))\subset\Psi(i+r)(\bar{A}(i+r))$. If we identify $\bar{A}(i)$ with its $\Psi(i)$-image, then we have r ascending sequences of $\bar{A}$-modules;

$$\begin{array}{l} \bar{A}(0)\subset\bar{A}(r)\subset\bar{A}(2r)\subset\cdots \\ \bar{A}(1)\subset\bar{A}(r+1)\subset\bar{A}(2r+1)\subset\cdots \\ \vdots \\ \bar{A}(r-1)\subset\bar{A}(2r-1)\subset\bar{A}(3r-1)\subset\cdots. \end{array}$$

Since all modules are submodules of $\bar{A}'$ and since $\bar{A}'$ is finite over $\bar{A}$, there is an integer i_0 such that

$$\bar{A}(i)=\bar{A}(i+r) \qquad \text{for any } i\geq i_0,$$

in other words, $\chi\bar{A}(i)=\bar{A}(i+r)$. Then if S is a graded subring of $\bar{A}$ isomorphic to a polynomial ring, then A is isomorphic as graded S-modules, to $\bar{A}(0)\oplus\bar{A}(1)\oplus\cdots\oplus\bar{A}(i_0-1)\oplus\bigoplus_{\mu=0}^{\infty}(\bar{A}(i_0)\oplus\cdots\oplus\bar{A}(i_0+r-1))\chi^{\mu}$, where A is regarded as an S-module by taking some homomorphism of S into A for which $S\to A\overset{\Psi}{\to}\bar{A}$ is the identity on S. Now suppose that the structures of $\bar{A}$ and of a finite number of $\bar{A}$-submodules $\bar{A}(i)$, $1\leq i\leq i_0+r-1$, in $\bar{A}'$ are determined. Then *the generating function of A is given by*

$$\sum_{i=0}^{i_0-1}P_{\bar{A}(i)}(t)+(1-t^{\deg(\chi)})\left(\sum_{i=i_0}^{i_0+r-1}P_{\bar{A}(i)}(t)\right).$$

If $\{f_{j,0}\}$ (resp. $\{f_{j,i}\}$, $i\geq 1$) is any system of homogeneous elements in A (resp. $A(i)$) whose images by Ψ (resp. $\Psi(i)$) generate $\bar{A}$ (resp. $\bar{A}(i)$ over $\bar{A}$), then A is generated by χ and $f_{j,i}$'s with $i\leq i_0+r-1$. To this assertion, a proof similar to the case $D(1)$ can be given, but we omit it.

Suppose that X, $\mathscr{L}$ satisfy the Assumption I. Let

$$X_0\subset X_1\subset\cdots\subset X_m=X$$

be a sequence of inclusions of irreducible subvarieties of X such that the codimensions of X_j in X_{j+1} is one for every $j=0,\cdots,m-1$. Furthermore suppose that the closure $\bar{X}_j$ of X_j in X^* satisfies a condition that the codimension of $\bar{X}_j - X_j$ in $\bar{X}_j$ is at least two for $j \geq 1$. Then X_j and $\mathscr{L}|_{X_j}$, $j \geq 1$, satisfy the Assumption I. If a divisor X_j in X_{j+1} satisfies the Assumption II for each $j=0, \cdots, m-1$ and if the above trick is successful for each j, then we may have some information about the graded ring associated with X, $\mathscr{L}$ from the graded ring $\oplus_k H^0(X_0^*, \mathscr{L}(k)^*|_{X_0^*})$, X_0^* denoting the closure of X_0 in X^*.

§ 3. Hilbert modular forms

Let K be a totally real algebraic number filed of degree $n>1$, and let O_K be the maximal order of K. $SL_2(K)$ acts on the product H^n on n copies of the upper half plane $H=\{z_1 \in \boldsymbol{C} \mid \operatorname{Im} z_1>0\}$ by the modular substitution;

$$z=(z_1, \cdots, z_n) \longrightarrow Mz=\left(\frac{\alpha^{(1)}z_1+\beta^{(1)}}{\gamma^{(1)}z_1+\delta^{(1)}}, \cdots, \frac{\alpha^{(n)}z_n+\beta^{(n)}}{\gamma^{(n)}z_n+\delta^{(n)}}\right)$$

where $M=\begin{pmatrix}\alpha & \beta \\ \gamma & \delta\end{pmatrix} \in SL_2(K)$ and $\alpha^{(1)}, \cdots, \alpha^{(n)}$ denote the conjugates of $\alpha \in K$. The jacobian $j(M, z)$ at z, of the automorphism of H^n induced by M, is $N(\gamma z+\delta)^{-2}=\prod_{i=1}^{n}(\gamma^{(i)}z+\delta^{(i)})^{-2}$. Let Γ_K denote the Hilbert modular group $SL_2(O_K)$. Γ_K acts properly discontinuously on H^n. Let $X_K:=H^n/\Gamma_K$. X_K has a natural compactification X_K^*, which is given by adding h points called cusps to X_K, h denoting the class number of K, and X_K^* is normal and projective. A holomorphic function f on H^n is called a *Hilbert modular form* for Γ_K of weight $k \in \boldsymbol{Z}$, ≥ 0 if it satisfies

$$f(Mz)=N(\gamma z+\delta)^k f(z), \qquad M=\begin{pmatrix}\alpha & \beta \\ \gamma & \delta\end{pmatrix} \in \Gamma_K.$$

We denote by $A(\Gamma_K)=\oplus_k A(\Gamma_K)_k$ the graded ring of Hilbert modular forms. X_K^* is canonically isomorphic to Proj $(A(\Gamma_K))$.

For later use, we fix some notations of elliptic modular functions. Let $\Gamma_1=SL_2(\boldsymbol{Z})$. $A(\Gamma_1)$ denotes the graded ring of elliptic modular forms for Γ_1. $g_4(\tau)$, $g_6(\tau)$, $\tau \in H$, denote the Eisenstein series of weight 4, 6 respectively, and $\Delta(\tau)$ denotes the cusp form of weight 12 which is written as a polynomial of g_4, g_6. $A(\Gamma_1)$ is generated by g_4, g_6, and $A(\Gamma_1)^{(4)}$ is generated by g_4 and g_6^2.

We embed H diagonally into H^n;

$$\begin{aligned} H &\longrightarrow H^n \\ \tau &\longrightarrow (\tau, \cdots, \tau). \end{aligned}$$

Then the stabilizer subgroup of Γ_K at H is equal to $\Gamma_1 \subset \Gamma_K$. Hence we have an embedding $H/\Gamma_1 \to X_K$. If $f(z)$ is a Hilbert modular form of weight k, then $f(\tau, \cdots, \tau)$ is an elliptic modular form of weight nk.

Let $K = \boldsymbol{Q}(\sqrt{5})$. We introduce Gundlach [13] from our point of view. Let D be the modular curve in X_K given as above. We have a homomorphism

$$\Psi : A(\Gamma_K) \longrightarrow A(\Gamma_1).$$
$$f \longrightarrow f|_H$$

Let us denote by $G_2(z)$, $G_6(z)$ the Eisenstein series for Γ_K of weight 2, 6 respectively. Then it can be shown that $\Psi(G_2(z))$ and $\Psi(G_6(z))$ are algebraically independent. Hence $\Psi|_{A(\Gamma_K)^{(2)}}$ gives rise to a surjective map of $A(\Gamma_K)^{(2)}$ onto $A(\Gamma_1)^{(4)}$. Let $f(z)$ be of odd weight. Then the weight of $\Psi(f)$ is $\not\equiv 0 \pmod 4$, and so it is a multiple of g_6. If f is a unit with negative norm, then we have $f(\varepsilon z) = -f(z)$. The usual Fourier expansion argument shows that $\Psi(f)$ vanishes at the cusp of order at least two, and hence $\Psi(f)$ is divisible by Δ^2. So $\Psi(f)$ is a multiple of $g_6\Delta^2$ if f is of odd weight. Gundlach [13, p. 246] found out a Hilbert modular form of weight 15, which we denote by $h(z)$, such that $\Psi(h)$ equals $g_6\Delta^2$ up to a constant factor. Now the image of $A(\Gamma_K)\Psi$ by is determined as

$$\Psi(A(\Gamma_K)) = A(\Gamma_1)^{(4)}[g_6\Delta^2].$$

We define a theta constant with characteristic $\binom{\alpha}{\beta}$ $\alpha, \beta \in O_K$, by setting

$$\theta\begin{bmatrix}\alpha\\ \beta\end{bmatrix}(z) = \sum_{\nu} \boldsymbol{e}\left(\mathrm{tr}\left(\left(\frac{1}{2\sqrt{5}}\left(\nu + \frac{\alpha}{2}\right)^2 z + \frac{1}{2\sqrt{5}}\left(\nu + \frac{\alpha}{2}\right)\beta\right)\right)\right)$$

where $\boldsymbol{e}(\ ) := \exp(2\pi\sqrt{-1}(\ ))$ and where ν runs over O_K, and $\mathrm{tr}(\nu z) := \nu^{(1)}z_1 + \cdots + \nu^{(n)}z_n$. A theta characteristic $\binom{\alpha}{\beta}$ is said to be *even* or *odd* according as $\boldsymbol{e}(\mathrm{tr}(\alpha\beta)/4) = 1$ or -1, and $\theta\begin{bmatrix}\alpha\\ \beta\end{bmatrix}$ is not identically zero if and only if $\binom{\alpha}{\beta}$ is even. There are ten even theta characteristics mod 2, and the corresponding product

$$\Theta_0(z) := \prod_{\text{even}} \theta\begin{bmatrix}\alpha\\ \beta\end{bmatrix}(z)$$

is a moldular form of weight 5. With the aid of Götsky's observation, Gundlach obtained after some calculation, that

$$\mathrm{div}(\Theta_0)=D,$$

which is obtained also by the modular embedding argument as well as the result of Siegel modular forms of degree two (Hammond [16]). So D satisfies the Assumption II and is in the case $D(1)$. By the argument of the preceding section we have the following:

Theorem (Gundlach). *Let* $K=\boldsymbol{Q}(\sqrt{5})$. *Then* $A(\Gamma_K)=\boldsymbol{C}[G_2, G_6, \Theta_0, h]$. *The generating function* $P_{A(\Gamma_K)}(t)$ *equals* $(1+t^{15})/(1-t^2)(1-t^5)(1-t^6)$.

Let K be an arbitrary real quadratic field. Let σ be the automorphism of H^2 given by

$$z=(z_1, z_2)\longrightarrow \sigma z=(z_2, z_1).$$

A Hilbert modular form f is said to be *symmetric* if $f(z_1, z_2)=f(z_2, z_1)$. Let $\hat{\Gamma}_K$ be the composite of Γ_K and $\langle\sigma\rangle$ as groups acting on H^2, and let $\hat{X}_K=H^2/\hat{\Gamma}_K\simeq X_K/\langle\sigma\rangle$. We denote by $A(\hat{\Gamma}_K)$ the graded ring of symmetric Hilbert modular forms. $\hat{X}_K$ has a natural compactification $\hat{X}_K^*:=X_K^*/\langle\sigma\rangle$ which equals $\mathrm{Proj}\,(A(\hat{\Gamma}_K))$.

We return to the case $K=\boldsymbol{Q}(\sqrt{5})$. Let D' be the irreducible divisor of $\hat{X}_K$ defined by $z_1=z_2$, in other words, the image of D by the natural surjective map of X_K onto $\hat{X}_K$. In the above-mentioned system of generators of $A(\Gamma_K)$, G_2, G_6, h are symmetric and Θ_0 is anti-symmetric, i.e., $\Theta_0(z_2, z_1)=-\Theta(z_1, z_2)$. Then Θ_0^2 is symmetric and

$$\mathrm{div}\,(\Theta_0^2)=D'.$$

So D' is in the case $D(1)$, and we have the following:

Theorem (Gundlach). *Let* $K=\boldsymbol{Q}(\sqrt{5})$. *Then* $A(\hat{\Gamma}_K)=\boldsymbol{C}[G_2, G_6, \Theta_0^2, h]$. *The generating function* $P_{A(\hat{\Gamma})}(t)$ *equals* $(1+t^{15})/(1-t^2)(1-t^6)(1-t^{10})$.

The image of $A(\hat{\Gamma}_K)^{(2)}$ by Ψ equals $A(\Gamma_1)^{(4)}=\boldsymbol{C}[g_4, g_6^2]$, and the kernel is the ideal generated by Θ_0^2. By the argument in § 2, $A(\hat{\Gamma}_K)^{(2)}$ is shown to be equal to $\boldsymbol{C}[G_2, G_6, \Theta_0^2]$ which is isomorphic to a polynomial ring. This shows in particular that the symmetric Hilbert modular function field for $\boldsymbol{Q}(\sqrt{5})$ is rational. h^2 is written as a polynomial of G_2, G_6, Θ_0^2, which has been explicitly done in Resnikoff [36], Hirzebruch [21].

Success in this line depends on finding out a "good" divisor of X_K or $\hat{X}_K$. We refer the reader to Hammond [16], Hermann [17, 18] for the case of real quadratic field K other than $\boldsymbol{Q}(\sqrt{5})$. Although Hermann [18] might look like the case when D is not irreducible, it is actually the case $D(8)$. It gives one good example which shows how we should treat the case $D(r)$, $r \geq 2$.

§ 4. Siegel modular forms of degree two

Let H_n be the Siegel space of degree n; $\{Z \in M_n(\boldsymbol{C}) | {}^tZ = Z, \operatorname{Im} Z > 0\}$. Let Γ_n be the modular group $\{M \in M_{2n}(\boldsymbol{Z}) | {}^tMJM = J\}$ where J denotes $\begin{bmatrix} 0 & 1_n \\ -1_n & 0 \end{bmatrix}$, 1_n being the identity matrix of size n. Γ_n acts on H_n by the usual modular transformation;

$$Z \longrightarrow MZ = (AZ+B)(CZ+D)^{-1}, \qquad M = \begin{pmatrix} AB \\ CD \end{pmatrix} \in \Gamma_n.$$

The jacobian $j(M, Z)$ at Z, of the automorphism of H_n induced by M, is $|CZ+D|^{-n-1}$. Let $\mathscr{A}_n := H_n/\Gamma_n$, and let $\mathscr{A}_n^*$ be the Satake compactification which is normal and projective. $\mathscr{A}_n$ is the moduli space of principally polarized abelian varieties over $\boldsymbol{C}$ of dimension n. $\operatorname{codim}(\mathscr{A}_n^* - \mathscr{A}_n)$ equals n, and hence the variety $\mathscr{A}_n$ satisfies the first condition in the Assumption I if $n > 1$. Let Γ be a congruence subgroup of Γ_n. A holomorphic function f on H_n is called a *Siegel modular form* for Γ of weight k if it satisfies

$$f(MZ) = |CZ+D|^k f(Z) \qquad \text{for } M = \begin{pmatrix} AB \\ CD \end{pmatrix} \in \Gamma$$

where for $n=1$ we need the additional condition that f is holomorphic also at the cusps, which is automatic in the case $n > 1$. We denote by $A(\Gamma)$ the graded ring of Siegel modular froms for Γ. $\mathscr{A}_n^*$ equals $\operatorname{Proj}(A(\Gamma_n))$. We denote by $E_k(Z)$ the Eisenstein series of even weight k, which is an example of Siegel modular forms. We define a theta constant by setting

$$\theta\begin{bmatrix} u \\ v \end{bmatrix}(Z) = \sum_{g \in \boldsymbol{Z}^n} \boldsymbol{e}\left(\frac{1}{2}\left(g+\frac{u}{2}\right)Z^t\left(g+\frac{u}{2}\right)+\frac{1}{2}\left(g+\frac{u}{2}\right)^t v\right)$$

where u, v as well as g, are integral row vectors of size n. $\begin{pmatrix} u \\ v \end{pmatrix}$ is called a *theta characteristic of degree n*, and is said to be *even* or *odd* according as $\boldsymbol{e}(u^t v/4) = 1$ or -1. $\theta\begin{bmatrix} u \\ v \end{bmatrix}$ is not identically zero if and only if $\begin{pmatrix} u \\ v \end{pmatrix}$ is even. A point Z of H_n is called *reducible* if it is equivalent to a matrix of the form $\begin{pmatrix} * & 0 \\ 0 & * \end{pmatrix}$ under Γ_n. A *reducible locus* in $\mathscr{A}_n$ is the subset consisting of all points corresponding to reducible points, which is closed in $\mathscr{A}_n$.

Now let us assume $n=2$. We introduce Freitag [6] (cf. [10]) and Hammond [15] from our point of view, in which Igusa's structure theorem for $A(\Gamma_2)^{(2)}$ was reproved. We take as D the reducible locus of $\mathscr{A}_2$, or equivalently, the divisor corresponding to the image of the embedding;

$$H^2 \longrightarrow H_2.$$

$$(z_1, z_2) \longrightarrow \begin{bmatrix} z_1 & 0 \\ 0 & z_2 \end{bmatrix}$$

The stabilizer subgroup of Γ_2 at the image, is generated by the image of $\Gamma_1^{\times 2}$ by

$$\Gamma_1^{\times 2} \longrightarrow \Gamma_2$$

$$\left\{\begin{pmatrix} a & b \\ c & d \end{pmatrix}, \begin{pmatrix} a' & b' \\ c' & d' \end{pmatrix}\right\} \longrightarrow \begin{bmatrix} a & 0 & b & 0 \\ 0 & a' & 0 & b' \\ c & 0 & d & 0 \\ 0 & c' & 0 & d' \end{bmatrix}$$

and by a matrix

$$\begin{bmatrix} 0 & 1 & 0 & 0 \\ 1 & 0 & 0 & 0 \\ 0 & 0 & 0 & 1 \\ 0 & 0 & 1 & 0 \end{bmatrix}.$$

The latter matrix induces the transposition σ of z_1 and z_2. So D is isomorphic to $(H_1/\Gamma_1)^{\times 2}/\langle\sigma\rangle$. Since $(H/\Gamma_1)^* \simeq \mathrm{Proj}\,(\boldsymbol{C}[g_4, g_6])$, D is associated with a graded ring $B := \boldsymbol{C}[g_4 \otimes g_4, g_6 \otimes g_6, g_4^3 \otimes g_6^2, g_6^2 \otimes g_4^3]^{\langle\sigma\rangle}$ where σ changes the first and second components of a tensor. It is easy to see that

$$B = \boldsymbol{C}[g_4 \otimes g_4, g_6 \otimes g_6, g_4^3 \otimes g_6^2 + g_6^2 \otimes g_4^3].$$

Let

$$\Psi : A(\Gamma_2)^{(2)} \longrightarrow B$$

$$f \longrightarrow f|_{H^2}$$

It is shown that $\Psi(E_4)$, $\Psi(E_6)$, $\Psi(E_{12})$ are algebraically independent. Thus Ψ is surjective. There are ten even theta characteristics of degree two. Let Θ be the square of the corresponding product of theta constants;

$$\Theta(Z) := \prod_{\text{even}} \theta \begin{bmatrix} u \\ v \end{bmatrix} (Z)^2.$$

$\Theta(Z)$ is a Siegel modular form for Γ_2 of weight ten, and

$$D = \operatorname{div}(\Theta).$$

The key to prove this is to show that $Z \in H_2$ is reducible if and only if one theta constant $\theta\begin{bmatrix} u \\ v \end{bmatrix}$ vanishes at Z. Hammond [15] has proved this by using moduli theory, that is, if $\theta\begin{bmatrix} u \\ v \end{bmatrix}(Z)=0$, then the principally polarized abelian variety corresponding to $Z \in H_2$ is decomposable. Freitag [6, 10] has done in a way similar to what Gundlach [13] did, namely by the function-theoretic argument of theta constants.

D is in the case $D(1)$, and hence $A(\Gamma_2)^{(2)}$ equals $\mathbf{C}[E_4, E_6, E_{12}, \Theta]$ by the argument of § 2. Θ is equal to $E_4E_6 - E_{10}$ up to a constant factor. So we have the following:

Theorem (Igusa). *$A(\Gamma_2)^{(2)} = \mathbf{C}[E_4, E_6, E_{10}, E_{12}]$, and $E_4, \cdots, E_{12}$ are algebraically independent. The generating function is given by $1/(1-t^4)(1-t^6)(1-t^{10})(1-t^{12})$.*

By this, the Siegel modular function field of degree two is shown to be rational. $A(\Gamma_2)$ is generated by $A(\Gamma_2)^{(2)}$ and a cusp form χ_{35} of weight 35. For this fact and for the definition of χ_{35} we refer the reader to Igusa [30].

Remark. So far, a graded ring of modular forms is written as a finite *free* module over its graded subring isomorphic to a polynomial ring. Such a ring is called Cohen-Macaulay. In the case of real quadratic field K, $A(\Gamma_K)^{(r)}$ is shown to be Cohen-Macualay for any integer $r>1$ ([53]), and moreover it is very likely that $A(\Gamma_K)$ itself is Cohen-Macaulay, which is not generally proved yet. However, unfortunately this nice property does not hold for a general graded ring of modular forms. For instance, neither the graded ring of Hilbert modular forms for K of degree ≥ 3 nor the ring $A(\Gamma_n)$ of Siegel modular forms of degree $n \geq 3$ is Cohen-Macaulay ([53, 54, 58]).

§ 5. Siegel modular forms of degree three

Let $\mathfrak{M}_g$ be the moduli space of smooth curves of genus g. The Torelli map

$$\mathfrak{M}_g \longrightarrow \mathscr{A}_g$$

gives an embedding, and we identify $\mathfrak{M}_g$ with its image. $\mathfrak{M}_g$ is open in

$\mathscr{A}_g$ if $g \leq 3$, and is of codimension one if $g=4$. Let $\mathscr{H}_g$ be the locus in $\mathfrak{M}_g$ consisting of hyperelliptic points, and let $\bar{\mathscr{H}}_g$ be its closure in $\mathscr{A}_g$.

A hyperelliptic curve of genus g is given, as an affine curve, by an equation

$$y^2 = \prod_{i=1}^{2g+2} (x - \xi_i)$$

where $\xi_1, \cdots, \xi_{2g+2}$ are mutually distinct complex numbers. Let W_{2g+2} be an open subvariety of $\boldsymbol{C}^{2g+2}$ consisting of points with distinct coordinates, which parametrizes "all" hyperelliptic curves of genus g. There is a surjective morphism of W_{2g+2} onto $\mathscr{H}_g$ by sending $(\xi_1, \cdots, \xi_{2g+2})$ to the isomorphism class of the corresponding hyperelliptic curve. $\mathscr{H}_g$ is a quotient of W_{2g+2} by the composite of $SL_2(\boldsymbol{C})$ and of the symmetric group $\mathfrak{S}_{2g+2}$ of degree $2g+2$. Let $S(2, 2g+2)$ be the graded ring of invariants of a binary $(2g+2)$-form, which comes out as the invariant subring of $\boldsymbol{C}[\xi_1, \cdots, \xi_{2g+2}]$ under the above-mentioned group where we refer the reader to Schur [42], Tsuyumine [56, Chap. I, II] for detail. $\mathscr{H}_g \to \operatorname{Proj}(S(2, 2g+2))$ is a compactification of $\mathscr{H}_g$. Let $S(2g+2)$ denote the invariant subring of $\boldsymbol{C}[\xi_1, \cdots, \xi_{2g+2}]$ under $SL_2(\boldsymbol{C})$, so that the invariant subring of $S(2g+2)$ under $\mathfrak{S}_{2g+2}$ equals $S(2, 2g+2)$. If $\Gamma_g(2)$ denotes the principal congruence subgroup of level two, then Igusa [30] has obtained the homomorphism

$$\rho_g \colon A(\Gamma_g(2)) \longrightarrow S(2g+2)$$

of graded rings which is described explicitly in terms of theta constants. ρ_g maps the graded part of degree k to that of degree $gk/2$. He showed also that the restriction of ρ_g to $A(\Gamma_g)$, which we denote also by ρ_g, is a homomorphism of $A(\Gamma_g)$ to $S(2, 2g+2)$;

$$\rho_g \colon A(\Gamma_g) \longrightarrow S(2, 2g+2)$$

which is associated with an embedding $\mathscr{H}_g \longrightarrow \mathscr{A}_g$. The kernel of ρ_g is the ideal generated by modular forms vanishing identically on $\mathscr{H}_g \subset \mathscr{A}_g$. ρ_g is injective if $g \leq 2$ because all curves of genus $g \leq 2$ are hyperelliptic.

Now let us consider the case $g=3$. There are 36 even theta characteristics of degree three. The corresponding product of theta constants

$$\chi_{18} := \prod_{\text{even}} \theta \begin{bmatrix} u \\ v \end{bmatrix}$$

is a cusp form for Γ_3 of weight 18. It was shown by Igusa [30] that

$$\bar{\mathscr{H}}_3=\operatorname{div}(\chi_{18}).$$

We have an exact sequence

$$0\longrightarrow(\chi_{18})\longrightarrow A(\Gamma_3)\xrightarrow{\rho_3}S(2, 8)$$

where for the structure of $S(2, 8)$, we refer the reader to Shioda [46].

Let R denote the reducible locus of $\mathscr{A}_3$. Then we have a sequence of inclusions

$$R\subset\bar{\mathscr{H}}_3\subset\mathscr{A}_3$$

where $\bar{\mathscr{H}}_3$ (resp. R) is irreducible of codimension one in $\mathscr{A}_3$ (resp. $\bar{\mathscr{H}}_3$). The sequence satisfies the conditions which are mentioned at the end of § 2. R corresponds to the subset

$$R_0:=\left\{\begin{bmatrix}Z_1 & 0\\ 0 & z_3\end{bmatrix}\in H_3 \mid Z_1\in H_2,\ z_3\in H\right\}$$

of H_3. The stabilizer subgroup of Γ_3 at R_0 is the image of

$$\Gamma_2\times\Gamma_1\longrightarrow\Gamma_3$$

$$\left\{\begin{pmatrix}A & B\\ C & D\end{pmatrix}, \begin{pmatrix}a & b\\ c & d\end{pmatrix}\right\}\longrightarrow\begin{bmatrix}A & 0 & B & 0\\ 0 & a & 0 & b\\ C & 0 & D & 0\\ 0 & c & 0 & d\end{bmatrix}.$$

The subset of R_0 consisting of diagonal matrices, is stable under the matrix $\begin{pmatrix}U & 0\\ 0 & U\end{pmatrix}\in\Gamma_3$ with

$$U=\begin{bmatrix}0 & 0 & 1\\ 0 & 1 & 0\\ 1 & 0 & 0\end{bmatrix}.$$

The graded ring associated with R, which we denote by $\bar{A}(0)$, is given by

$$\left\{\sum\psi\otimes j\in A(\Gamma_2)\otimes A(\Gamma_1) \mid \sum\psi\begin{pmatrix}z_1 & 0\\ 0 & z_2\end{pmatrix}j(z_3)\ \text{is symmetric in}\ z_1, z_2, z_3\right\}$$

where for two graded modules $M=\oplus_k M_k$, $N=\oplus_k N_k$, $M\otimes N$ always denotes a graded module $\oplus_k M_k\otimes N_k$. $\bar{A}(0)$ is easily understood because the structures of $A(\Gamma_1)$, $A(\Gamma_2)$ are known. So if the trick mentioned in § 2 is applicable to $R\subset\bar{\mathscr{H}}_3$ and to $\bar{\mathscr{H}}_3\subset\mathscr{A}_3$, then we get some information about the structure of $A(\Gamma_3)$. It has been done in [56], whose sketch is given in the following (see also [57]).

$\bar{\mathscr{H}}_3 \subset \mathscr{A}_3$ is in the case $D(1)$, indeed $\bar{\mathscr{H}}_3$ is defined by χ_{18} as stated above. So the problem is reduced to determine the graded ring $A(\Gamma_3)/(x_{18})$, whose projective spectrum equals $\bar{\mathscr{H}}_3$. Let us denote by $\nu(f), f \in A(\Gamma_3)$, the vanishing order of $f|_{\bar{\mathscr{H}}_3}$ at R. There is a cusp form χ_{28} of weight 28 whose restriction to $\bar{\mathscr{H}}_3$ vanishes only at R, where for the definition of χ_{28} we refer the reader to [56]. χ_{28} is the modular form of lowest weight satisfying such a property, and $\nu(\chi_{28})=4$. Hence $R \subset \bar{\mathscr{H}}_3$ is in the case $D(4)$.

Remark. In [56], we have defined the order of vanishing to be twice as much as in the present paper in order to unify notations. So the vanishing order $\nu(\chi_{28})$ has been written as 8 there.

$\bar{\mathscr{H}}_3 \subset \mathscr{A}_3$ is corresponding to the union of loci in H_3 of theta constants $\theta\begin{bmatrix} u \\ v \end{bmatrix}$, $\begin{pmatrix} u \\ v \end{pmatrix}$ being even. There are just six theta constants which vanish identically on $R_0 \subset H_3$, e.g., $\theta\begin{bmatrix} 1 & 1 & 1 \\ 1 & 0 & 1 \end{bmatrix}(Z)$. Let V be the irreducible component of the analytic subvariety in H_3 defined by $\theta\begin{bmatrix} 1 & 1 & 1 \\ 1 & 0 & 1 \end{bmatrix}(Z)=0$ which contains R_0. Let f be a modular form for Γ_3. Suppose that $\nu(f) \equiv 0 \pmod 4$. Then $(f/\chi_{28}^{\nu(f)/4})|_{V-\Gamma_3 R_0}$ is holomorphic since χ_{28} vanishes nowhere on $V-\Gamma_3 R_0$, and it extends holomorphically to V by Riemann's removable singularity theorem. Thus

$$\Psi(f/\chi_{28}^{\nu(f)/4})(Z_1, z_3) = \lim_{\substack{Z \to Z_0 \\ Z \in V}} (f/\chi_{28}^{\nu(f)/4})(Z), \qquad Z_0 = \begin{bmatrix} Z_1 & 0 \\ 0 & z_3 \end{bmatrix} \in R_0,$$

is well defined. Let Γ_2' be the subgroup in Γ_2 of index six which leaves a theta characteristic $\begin{pmatrix} 1 & 1 \\ 1 & 0 \end{pmatrix}$ stable mod 2. Then $\Psi(f/\chi_{28}^{\nu(f)/4})$ is shown to be contained in $\bar{A}' := \left\{ \sum \psi \otimes j \in A(\Gamma_2') \otimes A(\Gamma_1) \mid \sum \psi \begin{pmatrix} z_1 & 0 \\ 0 & z_2 \end{pmatrix} j(z_3) \text{ is symmetric in } z_1, z_3 \right\}$. The homomorphism Ψ and the graded ring $\bar{A}'$ can be constructed as the one which we introduced in § 2 where $X = \bar{\mathscr{H}}_3$, and $D=R$.

Let F be a meromorphic modular form for Γ_3 whose restriction to V is, however, holomorphic such as $f/\chi_{28}^{\nu(f)/4}$ above. It is not easy to calculate $\Psi(F)$ directly from definition, particularly if F is not holomorphic globally on H_3. $\rho_3(F)$ is obviously well-defined and is contained in $S(2, 8) \subset \boldsymbol{C}[\xi_1, \cdots, \xi_8]$. Let $A'(\Gamma_3)$ be the graded ring of such modular forms. A hyperelliptic point Z in V moves toward $Z_0 \in R_0$ if the corresponding curve degenerates to the union of a hyperelliptic curve of genus two and of an elliptic curve, which is, for instance, written in terms of $\xi_1, \cdots, \xi_8$, as

$$(\xi_1, \cdots, \xi_5, t+\xi_6, t+\xi_7, t+\xi_8), \qquad t \longrightarrow \infty.$$

Let $S(2, 8)_0$ be the subring of $S(2, 8)$ which is generated by homogeneous elements $I(\xi_1, \cdots, \xi_8)$ of degree s, $s \geq 0$ such that

$$\lim_{t\to\infty} t^{-7s/3} I(\xi_1, \cdots, \xi_5, t+\xi_6, t+\xi_7, t+\xi_8)$$

makes sense. Then the image of $A'(\Gamma_3)$ by ρ_3 is contained in $S(2, 8)_0$. We obtain a homomorphism of graded rings

$$\Psi: S(2, 8)_0 \longrightarrow S(6) \otimes S(2, 4)$$

by using the above limit. There is a commutative diagram

$$\begin{array}{ccc} A'(\Gamma_3) & \xrightarrow{\rho_3} & S(2, 8) \\ \Psi \downarrow & & \Psi \downarrow \\ A(\Gamma_2) \otimes A(\Gamma_1) & \xrightarrow{\rho_2 \otimes \rho_1} & S(6) \otimes S(2, 4) \end{array}$$

The Ψ of right hand side is a map given by polynomial calculation, and is, although sometimes not very easy, surely computable. Thus since $\rho_2 \otimes \rho_1$ is injective, the Ψ of left hand side is computable (at least in principle).

Let $A(i)$ be the ideal of $A(\Gamma_3)/(\chi_{18})$ generated by the modular forms f with $\nu(f) \geq i$, and let $\bar{A}(i) = A(i)/A(i+1)$. Here we note that $A(i)$ is defined differently from [56] (the present $A(i)$ equals $A(i)/(\chi_{18})$ in [56]), but $\bar{A}(i)$'s are the same except for what we mentioned in the above remark. $\bar{A}(i)$'s are $\bar{A}(0)$-modules. We fix three modular forms β, γ, δ with $\nu(\beta)=1$, $\nu(\gamma)=2$, $\nu(\delta)=3$ respectively. If $f \in A(\Gamma_3)$ is of order $i \equiv 0 \pmod 4$ (resp. 1, 2, 3 (mod 4)), then $f/\chi_{28}^{i/4}$ (resp. $f\delta/\chi_{28}^{(i+3)/4}$, $f\gamma/\chi_{28}^{(i+2)/4}$, $f\beta/\chi_{28}^{(i+1)/4}$) is contained in $A'(\Gamma_3)$ and hence the Ψ-image is defined. For $i \equiv 0 \pmod 4$ (resp. 1, 2, 3 (mod 4)) we denote by $\Psi(i)$ the map given by

$$f \longrightarrow \Psi(f/\chi_{28}^{i/4}) \text{ (resp. } \Psi(f\delta/\chi_{28}^{(i+3)/4}),\ \Psi(f\gamma/\chi_{28}^{(i+2)/4}),\ \Psi(f\beta/\chi_{28}^{(i+1)/4})),$$

where in [56], some specific modular forms are taken as β, γ, δ. $\Psi(i)$ induces an injective homomorphism

$$\bar{A}(i) \longrightarrow \bar{A}',$$

which we denote also by $\Psi(i)$. As mentioned above, this map is computable. Let us identify $\bar{A}(i)$ with its $\Psi(i)$-image. Then it is shown that

$$\bar{A}(i)=\bar{A}(i+4) \qquad \text{for } i\geq 3,$$

and so that

$$A(\Gamma_3)/(\chi_{18})\simeq \bar{A}(0)\oplus\bar{A}(1)\oplus\bar{A}(2)\oplus \bigoplus_{\mu=0}^{\infty} (\bar{A}(3)\oplus\bar{A}(4)\oplus\bar{A}(5)\oplus\bar{A}(6))\chi_{28}^{\mu}$$

as modules over a graded subring of $\bar{A}(0)$ isomorphic to a polynomial ring. The structure of $\bar{A}(i)$, $i\leq 6$, could be determined, and hence some of the structure of $A(\Gamma_3)$ too. We refer the reader to [56, pp. 831–2] for the theorem about the structure of $A(\Gamma_3)$. Here we note that *the problems about a minimal system of generators, relations among them, are untouched yet as well as the rationality problem of the variety H_3/Γ_3 which seems one of the outstanding problems.*

We shall close this section with Taniyama's words ([50, Letter to M. Sugiura]), which appears in context, to talk about the theory of complex multiplication. Perhaps the author takes them in a wider sense than Taniyama's original view, which, however, seems still true: When Siegel modular functions can be handled as "easily" as elliptic functions, number theory will have developed in many directions.

§ 6. Siegel modular forms of degree four

Let C be a non-hyperelliptic curve of genus four. We identify C with its canonical curve in $\boldsymbol{P}^3$. Then C is a complete intersection of a quadric and a cubic. There are two types of quadrics in $\boldsymbol{P}^3$, namely, smooth quadrics and quadric cones. A curve C exhibited as a complete intersection of a quadric cone and a cubic is said to *have a vanishing theta constant* because one of the theta constants with even characteristics vanishes at the jacobian point corresponding to C if and only if C is such a curve. Let $\mathfrak{M}_4'$ be the moduli of such curves.

Let R denote the irreducible subvariety of the reducible locus of $\mathscr{A}_4$ corresponding to the direct products of 3-dimensional abelian varieties and elliptic curves, and let $\overline{\mathfrak{M}}_4$, $\overline{\mathfrak{M}}_4'$ be the closures in $\mathscr{A}_4$ of $\mathfrak{M}_4$, $\mathfrak{M}_4'$. There is a sequence of inclusions;

$$R\subset\overline{\mathfrak{M}}_4'\subset\overline{\mathfrak{M}}_4\subset\mathscr{A}_4.$$

$\overline{\mathfrak{M}}_4$ is an irreducible divisor of $\mathscr{A}_4$, and is in the case $D(1)$, indeed it is defined by the single modular form J_8 of weight eight which is called *the Schottky invariant*. It is Schottky's theorem, however whose rigorous proof has been first given by Igusa [31]. Let χ_{68} be the modular form for Γ_4 defined to be the product of all theta constants with even characteristics (mod 2). Then $\overline{\mathfrak{M}}_4'$ is defined in $\overline{\mathfrak{M}}_4$, by $\sqrt{\chi_{68}|_{\overline{\mathfrak{M}}_4}}$ (Tsuyumine [55]). It is not known if there is Siegel modular form of weight 34 whose restriction to $\overline{\mathfrak{M}}_4$ equals $\sqrt{\chi_{68}|_{\overline{\mathfrak{M}}_4}}$. If this is the case, $\overline{\mathfrak{M}}_4' \subset \overline{\mathfrak{M}}_4$ is in the case $D(1)$. However even if we are not lucky, we have only the case $D(2)$. Here we pose the following question:

Question. Is there a Siegel modular form f for Γ_4 such that the ideal (J_8, χ_{68}, f) defines R set-theoretically?

Suppose that the answer is affirmative. Then it can be shown that $R \subset \overline{\mathfrak{M}}_4'$ satisfies the Assumption II, and hence that the above sequence if inclusions satisfies the conditions mentioned at the end of § 2. The result in the preceding section will help to determine the graded ring associated with R which corresponds to the image of the map;

$$H_3 \times H \longrightarrow H_4$$
$$(Z_1, z_4) \longrightarrow \begin{bmatrix} Z_1 & 0 \\ 0 & z_4 \end{bmatrix}.$$

So there will be a chance we get some information about the structure of $A(\Gamma_4)$.

We introduce two facts concerning the above question. In his paper [31], Igusa constructed the modular form of weight 540 which defines $R \cup \mathscr{H}_4$ in $\overline{\mathfrak{M}}_4'$ set-theoretically, $\mathscr{H}_4$ denoting the hyperelliptic locus. Sasaki [38] gave for an arbitrary degree, a system of modular forms which defines set-theoretically the reducible locus. However its number is not small in general.

Though without evidence, the author believes that there should exist such a modular form as in the question. However it is hard to manage the step "$R \subset \overline{\mathfrak{M}}_4'$" if one follows the method in § 5. We shall need to use it jointly with some other method, e.g., an asymptotic dimension formula for spaces of cusp forms for Γ_4 (up to the first or the second degree terms? $\cdots$, I don't know), of course, the perfect one is better. Or is it possible to attack directly the graded ring associated with $\overline{\mathfrak{M}}_4'$ by geometric method? In this section we have shown only that the best card is still in a pile.

References

[1] Baily, W. L. and Borel, A., Compactification of arithmetic quotients of bounded symmetric domains, Ann. Math. (2), **84** (1966), 442–528.

[2] Blumenthal, O., Über Modulfunktionen von mehreren Veränderlichen, (Erste Hälfte), Math. Ann., **56** (1903), 509–548; (Zweite Hälfte), *ibid.*, **58** (1904), 497–527.

[3] Busam, R., Eine Verallgemeinerung gewisser Dimensionsformeln von Shimizu, Invent. Math., **11** (1970), 110–149.
[4] Cohn, L., The dimension of spaces of automorphic forms on a certain two dimensional complex domain, Mem. Amer. Math. Soc., **158** (1975).
[5] Fomenko, O. M., Modular forms and Hilbert functions for the field $Q(\sqrt{2})$ (Russian), Math. Zametki, **4** (1968), 129–136; (English transl.) Math. Notes, **4** (1968), 567–571.
[6] Freitag, E., Zur Theorie der Modulformen zweiten Grades, Nachr. Akad. Wiss. Göttingen Math.-Phys. Kl. II, (1965), 151–157.
[7] ——, Modulformen zweiten Grades zur rationalen und Gaußschen Zahlkörper, Sitz. der Heidelberg Akad. Wiss. No. 1, **49** (1967).
[8] ——, Stabile Modulformen, Math. Ann., **230** (1977), 197–211.
[9] ——, Die Irreduzibilität der Schottkyrelation (Bemerkung zu einem Satz von J. Igusa), Arch. Math., **40** (1983), 255–259.
[10] ——, Siegelsche Modulfunktionen, Grundlehren, **254**, Springer-Verlag, 1983.
[11] Van der Geer, G., Hilbert modular forms for the field $Q(\sqrt{6})$, Math. Ann., **233** (1978), 163–179.
[12] Van der Geer, G. and Zagier, D., The Hilbert modular group for the field $Q(\sqrt{13})$, Invent. Math., **42** (1977), 93–133.
[13] Gundlach, K.-B., Die Bestimmung der Funktionen zur Hilbertschen Modulgruppe des Zahlkörpers $Q(\sqrt{5})$, Math. Ann., **152** (1963), 226–256.
[14] ——, Die Bestimmung der Funktionen zu einigen Hilbertschen Modulgruppen, J. Reine Angew. Math., **220** (1965) 109–153.
[15] Hammond, W. F., On the graded ring of Siegel modular forms of genus two, Amer. J. Math., **87** (1965), 502–506.
[16] ——, The modular groups of Hilbert and Siegel, Amer. J. Math., **85** (1966), 497–516.
[17] Hermann, C. F., Symmetrische Hilbertsche Modulformen und Modulfunktionen zu $Q(\sqrt{17})$, Math. Ann., **256** (1981), 191–197.
[18] ——, Thetafunktionen und symmetrische Hilbertsche Modulformen zu $Q(\sqrt{65})$, J. Reine Angew. Math., **339** (1983), 147–162.
[19] Hirzebruch, F., Hilbert modular surfaces, L'Enseigment Math. (2), **21** (1973).
[20] ——, Kurven auf den Hilbertschen Modulflächen und Klassenzahlrelationen, Lect. Notes in Math., **412**, 75–93, Springer-Verlag, 1974.
[21] ——, Hilbert's modular group for the field $Q(\sqrt{5})$ and the cubic diagonal surface of Clebsch and Klein (Russian), Uspekhi Mat. Nauk, **31:5** (1976), 153–166; (English transl.) Russian Math. Surveys, **31:5** (1976), 96–110.
[22] ——, The ring of Hilbert modular forms for real quadratic fields in small discriminant, In: Modular functions of one variable, IV, Lect. Notes in Math., **627**, 287–323, Springer-Verlag, 1977.
[23] ——, Überlagerungen der projektiven Ebene und Hilbertsche Modulflächen, L'Enseigment Math., **24** (1978), 63–78.
[24] ——, Modulflächen und Modulkurven zur symmetrischen Hilbertschen Modulgruppe, Ann. Sci. École Nor. Sup. (4), **11** (1978), 101–165.
[25] Hirzebruch, F. and Van de Ven, A., Hilbert modular surfaces and the classification of algebraic surfaces, Invent. Math., **23** (1974), 1–29.
[26] Hirzebruch, F. and Zagier, D., Intersection number of curves on Hilbert modular surfaces and modular forms of Neben type, Invent. Math., **36** (1976), 57–113.
[27] ——, Classification of Hilbert modular surfaces, In: Complex analysis and algebraic geometry, 43–77, Iwanami Shoten, Tokyo, 1977.
[28] Igusa, J., On Siegel modular forms of genus two, Amer. J. Math., **84** (1962), 175–200.

[29] ——, On Siegel modular forms of genus two (II), Amer. J. Math., **86** (1964), 817–855.
[30] ——, Modular forms and projective invariants, Amer. J. Math., **89** (1967), 817–855.
[31] ——, On the irreducibility of Schottky's divisor, J. Fac. Sci. Univ. Tokyo, Sect. IA, Math., **28** (1981), 531–545.
[32] Müller, R., Hilbertsche Modulformen und Modulfunktionen zu $Q(\sqrt{8})$, Math. Ann., **266** (1983), 83–103.
[33] ——, Hilbertsche Modulformen und Modulfunktionen zu $Q(\sqrt{5})$, Arch. Math., **45** (1985), 239–251.
[34] Mumford, D., Varieties defined by quadratic equations, In: Questions on algebraic varieties, (C.I.M.E., III Ciclo, Verenna, 1969), 29–100, Edizioni Cremonese, Rome, 1970.
[35] Nagaoka, S., On Hilbert modular forms, III, Proc. Japan. Acad., **59** Ser. A (1983), 346–348.
[36] Resnikoff, H. L., On the graded ring of Hilbert modular forms associated with $Q(\sqrt{5})$, Math. Ann., **208** (1974), 161–170.
[37] Resnikoff, H. L. and Tai, Y.-S., On the structure of a graded ring of automorphic forms on the 2-dimensional complex ball, Math. Ann., **238** (1978), 97–117.
[38] Sasaki, R., Modular forms vanishing at the reducible points of the Siegel upper-half space, J. Reine Angew. Math., **345** (1983), 111–121.
[39] Satake, I., On Siegel's modular functions, In: International symposium on algebraic number theory, 107–129, Tokyo and Nikko, 1955.
[40] ——, On the compactification of the Siegel space, J. Ind. Math. Soc. (NS), **20** (1956), 259–281.
[41] Satoh, T., On certain vector valued Siegel modular forms of degree two, Math. Ann., **274** (1986), 335–352.
[42] Schur, I., Vorlesungen über Invariantentheorie, Grundlehren 143, Springer-Verlag, 1968.
[43] Séminaire Henri Cartan, Fonctions automorphes, École Norm. Sup. 1957/58.
[44] Serre, J.-P., Prolongement de faiceaux analytique cohérents, Ann. Inst. Fourier (Grenoble), **16** (1966), 363–374.
[45] Shimizu, H., On discontinuous groups operating on the product of the upper half planes, Ann. Math., **77** (1963), 33–71.
[46] Shioda, T., On the graded ring of invariants of binary octavics, Amer. J. Math., **89** (1967), 1022–1046.
[47] Siegel, C. L., Einführung in die Theorie der Modulfunktionen *n*-ten Grades, Math. Ann., **116** (1939), 617–657.
[48] Tai, Y.-S. and Resnikoff, H. L., On the structure of a graded ring of automorphic forms on the 2-dimensional complex ball II, Math. Ann., **258** (1982), 367–382.
[49] Takagi, T., Kinsei Sûgaku Shidan (Japanese), Kyôritsu-shuppan, Tokyo, 1970.
[50] Taniyama, Y., The complete works of Yutaka Taniyama.
[51] Tsushima, R., An explicit dimension formula for the space of generalized automorphic forms with respect to $Sp(2, Z)$, Proc. Japan Acad., **59**A, (1983), 139–142.
[52] ——, An explicit dimension formula for the space of generalized automorphic forms with respect to $Sp(2, Z)$, preprint, 1983.
[53] Tsuyumine, S., Rings of modular forms—On Eichler's problem, Nagoya Math. J., **99** (1985), 31–44.
[54] ——, Rings of automorphic forms which are not Cohen-Macaulay, I, J. Math. Soc. Japan, **38** (1986), 147–162.
[55] ——, Factorial property of a ring of automorphic forms, Trans. Amer.

Math. Soc., **296** (1986), 111–123.
[56] ——, On Siegel modular forms of degree three, Amer. J. Math., **108** (1986), 755–862; Addendum, *ibid.*, 1001–1003.
[57] ——, On the modular function field of degree three, Comp. Math., **63** (1987), 83–98.
[58] ——, Rings of automorphic forms which are not Cohen-Macaulay, II, J. Math. Soc. Japan, **40** (1988), 369–381.

Department of Mathematics
Mie University
Tsu 514
Japan

Advanced Studies in Pure Mathematics 15, 1989
Automorphic Forms and Geometry of Arithmetic Varieties
pp. 89–97

Vector Valued Modular Forms of Degree Two and their Application to Triple L-functions

Takakazu Satoh

Abstract

We report some properties of vector valued Siegel modular forms of degree two and triple L-functions of different weight. Their relation is discussed in the section 3. This is motivated by an observation in the section 2. In both sections, certain differential operators are relevant. They are defined in the section 1.

§ 1. Differential operators

In this section we introduce some notations and construct differential operators. Let n be a positive integer and H_n the Siegel upper half space of degree n. Let ρ be a rational representation of $GL(n, \boldsymbol{C})$ and V_ρ the representation space of ρ. We denote by St the standard representation of $GL(n, \boldsymbol{C})$. A C^∞-Siegel modular form f of degree n and of 'weight' ρ is by definition a V_ρ-valued C^∞-function on H_n satisfying the equation

$$f(M\langle Z\rangle) = \rho(J(M, Z))f(Z)$$

for all $Z \in H_n$ and $M=\begin{pmatrix} A & B \\ C & D \end{pmatrix} \in \Gamma_n = Sp(n, \boldsymbol{Z})$ where $M\langle Z\rangle = (AZ+B)(CZ+D)^{-1}$ and $J(M, Z)=CZ+D$. We denote by $M_\rho^\infty(\Gamma_n)$ the $\boldsymbol{C}$-vector space of all such functions. We also denote by $M_\rho(\Gamma_n)$ and $S_\rho(\Gamma_n)$ the subspaces of $M_\rho^\infty(\Gamma_n)$ consisting of all holomorphic forms and of all holomorphic cusp forms, respectively. Instead of $M_\rho^\infty(\Gamma_n)$, we write $M_k^\infty(\Gamma_n)$ if $\rho=\det^k$ and $M_{k,r}^\infty(\Gamma_n)$ if $\rho=\det^k \otimes \mathrm{Sym}^r\,\mathrm{St}$ for simplicity. We define $M_k(\Gamma_n)$, $S_k(\Gamma_n)$, $M_{k,r}(\Gamma_n)$ and $S_{k,r}(\Gamma_n)$ similarly. If either $f \in M_k(\Gamma_1)$ or $g \in M_k(\Gamma_1)$ is a cusp form, we define the Petersson inner product by

$$\langle f, g\rangle_k = \int_{\Gamma_1 \backslash H_1} f(x+iy)\overline{g(x+iy)}\, y^{k-2} dxdy.$$

For a variable $Z=(z_{jl})$ on H_n, we put $\eta=(\eta_{jl})$ with $\eta_{jl}=-4\pi\,\mathrm{Im}(z_{jl})$ and

Received December 22, 1986.

$$\theta^{\pm}=\left(d_{jl}\frac{1}{2\pi i}\frac{\partial}{\partial z_{jl}^{\pm}}\right)_{1\leq j,l\leq n}$$

where $d_{jj}=1$, $d_{jl}=1/2$ for $j\neq l$, $z_{jl}^{+}=z_{jl}$, and $z_{jl}^{-}=\overline{z_{jl}}$. In what follows, we consider the degree two case. Let S_2 be the $\boldsymbol{C}$-vector space of all complex symmetric matrices of size two. The action of $GL(2, \boldsymbol{C})$ on S_2 defined by

$$A\longrightarrow GA{}^tG \quad (A\in S_2 \text{ and } G\in GL(2, \boldsymbol{C})) \tag{1.1}$$

is isomorphic to $\mathrm{Sym}^2\,\mathrm{St}$. We construct differential operators $\nabla^{+}=\nabla_k^{+}$ and $\nabla^{-}=\nabla_{k,r}^{-}$ acting on $M_k^{\infty}(\Gamma_2)$ and $M_{k,r}^{\infty}(\Gamma_2)$ respectively by

$$(\nabla^{+}f)(Z)=k\eta^{-1}f+\theta^{+}f$$

for $f\in M_k^{\infty}(\Gamma_2)$ and

$$(\nabla^{-}f)(Z)=(\mathrm{CFA}\,(\mathrm{Im}\,Z(\theta^{-}f_j)(Z)\,\mathrm{Im}\,Z))_{j=0}^{r}$$

for $f=(f_j)_{j=0}^{r}\in M_{k,r}^{\infty}(\Gamma_2)$, where $\mathrm{CFA}(A)$ is the cofactor matrix of A. Using Shimura [15, (4.2) and (4.4)] and their anti-holomorphic version we see that $\nabla^{\pm}$ define maps

$$\nabla^{+}\colon M_k^{\infty}(\Gamma_2)\longrightarrow M_{k,2}^{\infty}(\Gamma_2)$$

and

$$\nabla^{-}\colon M_{k,r}^{\infty}(\Gamma_2)\longrightarrow M^{\infty}_{\det^{k-2}\otimes\mathrm{Sym}^r\mathrm{St}\otimes\mathrm{Sym}^2\mathrm{St}}(\Gamma_2). \tag{1.2}$$

We note that degree n analogue of (1.2) does not hold since $A\rightarrow\mathrm{CFA}(A)$ is a linear map only for A of size two. These operators have a rather bad property on holomorphy. Holomorphic modular forms are, in general, mapped to non-holomorphic ones by ∇^{+} and annihilated by ∇^{-}. If $r\geq 2$, we have a decomposition

$$\begin{aligned}&\det^k\otimes\mathrm{Sym}^r\mathrm{St}\otimes\mathrm{Sym}^2\mathrm{St}\\&\quad\simeq\det^k\otimes\mathrm{Sym}^{r+2}\mathrm{St}\oplus\det^{k+1}\otimes\mathrm{Sym}^r\mathrm{St}\oplus\det^{k+2}\otimes\mathrm{Sym}^{r-2}\mathrm{St}.\end{aligned} \tag{1.3}$$

In application, we must frequently show the non-vanishing of a desired component. Generally this is not trivial. If $r=0$, there is no such problem. Moreover, in the section 2, we construct holomorphic modular forms using ∇^{+}. The operator ∇^{-} has a nice action on vector valued non-holomorphic Klingen type Eisenstein series. This is used for proving the meromorphy of triple L-functions in the section 3.

§ 2. Construction of certain vector valued modular forms

To cancel the non-holomorphic term, we put

$$[f, g]=\frac{1}{j} f\nabla^{+}g-\frac{1}{k} g\nabla^{+}f$$

for $f \in M_k(\Gamma_2)$ and $g \in M_j(\Gamma_2)$. Then $[f, g] \in M_{k+j,2}(\Gamma_2)$ since it is holomorphic. We show the structure theorem of $M_{k,2}(\Gamma_2)$ and $S_{k,2}(\Gamma_2)$ by using this differential operator.

Recall that the graded $\boldsymbol{C}$-algebra $\bigoplus_k M_k(\Gamma_2)$ where k runs over even integers is generated over $\boldsymbol{C}$ by four algebraically independent elements. (We understand that $M_k(\Gamma_2)=\{0\}$ for a negative k.) They are $\varphi_4 \in M_4(\Gamma_2)$, $\varphi_6 \in M_6(\Gamma_2)$, $\chi_{10} \in S_{10}(\Gamma_2)$ and $\chi_{12} \in S_{12}(\Gamma_2)$. (See Igusa [4] and Maass [9].)

Theorem 1. *For each even integer k, we have (as a $\boldsymbol{C}$-vector space)*

$$\begin{aligned} M_{k,2}(\Gamma_2)=&M_{k-10}(\Gamma_2)[\varphi_4, \varphi_6]\oplus M_{k-14}(\Gamma_2)[\varphi_4, \chi_{10}]\\ &\oplus M_{k-16}(\Gamma_2)[\varphi_4, \chi_{12}]\oplus V_{k-16}(\Gamma_2)[\varphi_6, \chi_{10}]\\ &\oplus V_{k-18}(\Gamma_2)[\varphi_6, \chi_{12}]\oplus W_{k-22}(\Gamma_2)[\chi_{10}, \chi_{12}] \end{aligned} \tag{2.1}$$

and

$$\begin{aligned} S_{k,2}(\Gamma_2)=&S_{k-10}(\Gamma_2)[\varphi_4, \varphi_6]\oplus M_{k-14}(\Gamma_2)[\varphi_4, \chi_{10}]\\ &\oplus M_{k-16}(\Gamma_2)[\varphi_4, \chi_{12}]\oplus V_{k-16}(\Gamma_2)[\varphi_6, \chi_{10}]\\ &\oplus V_{k-18}(\Gamma_2)[\varphi_6, \chi_{12}]\oplus W_{k-22}(\Gamma_2)[\chi_{10}, \chi_{12}] \end{aligned} \tag{2.2}$$

where

$$V_k(\Gamma_2)=M_k(\Gamma_2)\cap \boldsymbol{C}[\varphi_6, \chi_{10}, \chi_{12}] \quad \text{and} \quad W_k(\Gamma_2)=M_k(\Gamma_2)\cap \boldsymbol{C}[\chi_{10}, \chi_{12}].$$

Corollary 2. *Let k be an even integer. Then $M_{k,2}(\Gamma_2)$ and $S_{k,2}(\Gamma_2)$ have a $\boldsymbol{C}$-basis consisting of modular forms whose all Fourier coefficients lie in $\boldsymbol{Z}$.*

The proof proceed as follows. The inclusion $\supset$ is clear in both (2.1) and (2.2). First, we show that subspaces appearing in the right hand side of (2.1) are mutually linearly independent. This is shown by the following lemma.

Lemma ([13, Lemma2.1]). *Let k be an integer. For $j=4$, 6, 10 and 12, let $f_j \in M_{k-j}(\Gamma_2)$. If*

$$f_4\theta^{+}\varphi_4+f_6\theta^{+}\varphi_6+f_{10}\theta^{+}\chi_{10}+f_{12}\theta^{+}\chi_{12}=0,$$

then we have

$$f_4=f_6=f_{10}=f_{12}=0.$$

The dimension formula of $S_{k,r}(\Gamma_2)$ for $r=0$ and $k\geq 4$ or $r\geq 1$ and $k\geq 5$ is obtained by Tsushima [16, 17]. For $k\leq 6$, we have dim $S_{k,2}(\Gamma_2)=0$. This is proved by a method similar to Maass [8, pp. 189–196]. Comparison of dimension shows equality. For details, see [13, §§1–2]. Corollary follows from Theorem 1 and Igusa [5].

Our method is so explicit that we can obtain some interesting numerical values. For $f\in S_{k+r}(\Gamma_1)$, we denote by $[f]_r\in M_{k,r}(\Gamma_2)$ the Klingen type Eisenstein series attached to f. (Precise definition is given in (3.2) below.) As an example, let $f=\Delta_{16}\in S_{16}(\Gamma_1)$ be the normalized eigenform. By Theorem 1, we have

$$M_{14,2}(\Gamma_2)=\boldsymbol{C}[\Delta_{16}]_2\oplus\boldsymbol{C}[\chi_{10},\varphi_4].$$

Using Resnikoff and Saldaña [12, Tables III–V], we have

$$\frac{1}{144}[\varphi_6,\varphi_4^2]=[\Delta_{16}]_2-\frac{403200}{373}[\chi_{10},\varphi_4].$$

On the other hand, let $L_2(s,\Delta_{16})$ be the second L-function of Δ_{16}. By the method of Zagier [18], we obtain

$$\frac{L_2(28,\Delta_{16})(2\pi)^{-41}\Gamma(28)}{\langle\Delta_{16},\Delta_{16}\rangle_{16}}=\frac{2^{10}\cdot 373}{3\cdot 5^2\cdot 7^2\cdot 11}.$$

Here we note $28=2(k+r)-2-r$ with $k=14$ and $r=2$. Let $f\in S_{k+r}(\Gamma_1)$ be a normalized eigenform. The above fact suggests that $L_2(2(k+r)-2-r,f)$ appears in the denominator of Fourier coefficients of $[f]_r$. We note that the case $r=0$ is proved in Mizumoto [10]. (Cf. Kurokawa [6].) In the next section, we show a modified statement: the special value $L_2(2(k+r)-2-r,f)$ appears in the denominator of the pullback of $[f]_r$.

§ 3. Triple L-functions

We study relation between vector valued Klingen type Eisenstein series of degree two and triple L-functions of different weight. Properties of triple L-functions of the same weight are investigated by Garrett [3], where scalar valued degree three Siegel's Eisenstein series are used. For a non-negative integer r, let $V^{(r)}\cong\boldsymbol{C}^{r+1}$ be the representation space of $\rho=\det^k\otimes\mathrm{Sym}^r\mathrm{St}$. We realize a representation ρ on $V^{(r)}$ as

(3.1) $$\rho\left(\begin{pmatrix}a&b\\c&d\end{pmatrix}\right)\begin{pmatrix}x\\y\end{pmatrix}_r=\det\begin{pmatrix}a&b\\c&d\end{pmatrix}^k\left(\begin{pmatrix}a&b\\c&d\end{pmatrix}\begin{pmatrix}x\\y\end{pmatrix}\right)_r$$

where $\begin{pmatrix} x \\ y \end{pmatrix}_r = {}^t(x^r, x^{r-1}y, \cdots, y^r)$. For an integer q with $0 \leq q \leq r$, we put $v_q^{(r)} = {}^t(0, \cdots, 0, 1, 0, \cdots, 0)$ where 1 lies in the $(q+1)$-th column. This is compatible with the definition of v_0 in Arakawa [1, (0.2)]. We omit superscript (r) if there is no ambiguity. By (1.1) and (3.1) we identify S_2 with $V^{(2)}$ as a representation space of $\mathrm{Sym}^2\mathrm{St}$. Especially, $\begin{pmatrix} 1 & 0 \\ 0 & 0 \end{pmatrix} \in S_2$ corresponds to $v_0^{(2)}$. We use $\{v_q \mid 0 \leq q \leq r\}$ as a base of $V^{(r)}$. Set $\begin{pmatrix} a & b \\ c & d \end{pmatrix}^* = a$. For an integer t with $0 \leq t \leq n$, we define $\Gamma_{n,t}$ as the subgroup of Γ_n consisting of the elements whose entries in the first $n+t$ columns and the last $n-t$ rows are zero. For $\rho = \det^k \otimes \mathrm{Sym}^r \mathrm{St}$ and $f \in S_{k+r}(\Gamma_1)$ and $s \in \boldsymbol{C}$, we define the (vector valued non-holomorphic) Klingen type Eisenstein series attached to f by

$$(3.2) \quad [f]_r(Z, s) = \sum_{M \in \Gamma_{2,1} \backslash \Gamma_2} \left(\frac{\det(\operatorname{Im} M\langle Z \rangle)}{\operatorname{Im} M\langle Z \rangle^*} \right)^s f(M\langle Z \rangle^*) \rho(J(M, Z)^{-1}) v_0.$$

By a method similar to Arakawa [1, Proposition 1.2], we see that $[f]_r(Z, s)$ converges absolutely if $\operatorname{Re}(k+2s) > 4$, and it belongs to $M_\rho^\infty(\Gamma_2)$. For Siegel modular forms $f_1, \cdots, f_m$ (of various degree) and a field K, we denote by $K(f_1, \cdots, f_m)$ the field generated by all the Fourier coefficients of $f_1, \cdots, f_m$ over K. For simplicity, we put $[f]_r(Z) = [f]_r(Z, 0)$ and $\boldsymbol{Q}([f]_r) = \boldsymbol{Q}([f]_r(Z, 0))$. We note that under our normalization of v_0,

$$(3.3) \qquad \boldsymbol{Q}(f) \subset \boldsymbol{Q}([f]_r)$$

by Arakawa [1, Proposition 1.2]. For even integers $r \geq 0$ and k and $f \in S_{k+r}(\Gamma_1)$, we have

$$(3.4) \qquad \nabla^- [f]_r(Z, s) = \frac{s}{4\pi} [f]_{r+2}(Z, s+1).$$

for $\operatorname{Re}(k+2s) > 4$ ([14, Proposition 1.1]). Here the right hand side of (3.4) means that the second and the third components in the decomposition (1.3) are zero. By (3.4), the functional equations of scalar valued Klingen type Eisenstein series, which is obtained by Böcherer [2, (16) and (22)], are generalized for vector valued ones as follows.

Proposition 3. *Let k and r be even integers and let $f \in S_{k+r}(\Gamma_1)$. Put*

$$K(s, f) = \pi^{-3s} 2^{-2s} \frac{\Gamma_3\left(s + \frac{k+r}{2}\right) \Gamma(s) \Gamma(2s-1)}{\Gamma_3(s)\left(s + \frac{k+r}{2} - 1\right)} L_2(2s - 2 + k + r, f)$$

and

$$\boldsymbol{E}_r(Z, s, f) = \prod_{j=1}^{r/2} (s-j) K(s+(k/2), f)[f]_r(Z, s)$$

where $\Gamma_3(s) = \prod_{j=0}^{2} \Gamma(s-(j/2))$ *and* $L_2(s, f)$ *is the second L-function of f. Then* $[f]_r(Z, s)$ *extends meromorphically to the whole s-plane and satisfies the functional equation*

$$\boldsymbol{E}_r(Z, s, f) = \boldsymbol{E}_r(Z, 2-k-s, f). \tag{3.5}$$

We write Fourier expansion of $f \in S_k(\Gamma_1)$ as $f(z) = \sum_{n=1}^{\infty} a(n, f) e^{2\pi i n z}$. Let I_n be the identity matrix of size n. If $f \in S_k(\Gamma_1)$ is a normalized Hecke eigenform and p is a prime, we define semi-simple $M_p(f) \in GL(2, \boldsymbol{C})$ (up to the conjugacy class) by $\det(I_2 - tM_p(f)) = 1 - a(p, f)t + p^{k-1}t^2$. For normalized eigen cuspforms f, g and h, define the 'triple L-function' $L(s; f, g, h)$ by

$$L(s; f, g, h) = \prod_{p:\text{prime}} \det(I_8 - p^{-s} M_p(f) \otimes M_p(g) \otimes M_p(h))^{-1}.$$

Theorem 4. *Let f, g and h be normalized eigen cuspforms of degree one and of weight* $\nu(f)$, $\nu(g)$ *and* $\nu(h)$ *respectively satisfying* $\nu(f) \geq \nu(g) \geq \nu(h)$. *Put*

$$\begin{aligned}\tilde{L}(s; f, g, h)&\\ &= \Gamma_{\boldsymbol{C}}(s)\Gamma_{\boldsymbol{C}}(s-\nu(f)+1)\Gamma_{\boldsymbol{C}}(s-\nu(g)+1)\Gamma_{\boldsymbol{C}}(s-\nu(h)+1)L(s; f, g, h)\end{aligned}$$

where $\Gamma_{\boldsymbol{C}}(s) = 2(2\pi)^{-s}\Gamma(s)$. *Then* $\tilde{L}(s; f, g, h)$ *extends meromorphically to the whole s-plane and satisfies the functional equation*

$$\tilde{L}(s; f, g, h) = -\tilde{L}(\nu(f)+\nu(g)+\nu(h)-2-s; f, g, h). \tag{3.6}$$

Moreover, if $\nu(g)+\nu(h)-\nu(f) > 0$ *and* $L(s; f, g, h)$ *is holomorphic at* $s = \dfrac{\nu(f)+\nu(g)+\nu(h)}{2} - 1$, *then*

$$L\left(\frac{\nu(f)+\nu(g)+\nu(h)}{2} - 1; f, g, h\right) = 0, \tag{3.7}$$

and if $\nu(g)+\nu(h)-\nu(f) > 4$, *then*

$$\frac{\pi^{5+\nu(f)-3\nu(g)-3\nu(h)} L(\nu(g)+\nu(h)-2; f, g, h)}{\langle f, f\rangle_{\nu(f)} \langle g, g\rangle_{\nu(g)} \langle h, h\rangle_{\nu(h)}} \in \boldsymbol{Q}([f]_{2\nu(f)-\nu(g)-\nu(h)}, g, h). \tag{3.8}$$

Corollary 5. *Let* $f \in S_k(\Gamma_1)$ *be a normalized Hecke eigenform. Let*

$$L_3(s,f)=\prod_{p:\text{prime}} \det(I_4-p^{-s}\,\mathrm{Sym}^3 M_p(f))^{-1}$$

be the third L-function of f. Put

$$\tilde{L}_3(s,f)=\Gamma_C(s)\Gamma_C(s-k+1)L_3(s,f).$$

Then $\tilde{L}_3(s,f)$ satisfies the functional equation

$$\tilde{L}_3(s,f)=-\tilde{L}_3(3k-2-s,f).$$

In particular we have $L_3((3k/2)-1,f)=0$.

We sketch the proof. For details, see [14, § 2]. Put

$$k=\nu(g)+\nu(h)-\nu(f),$$
$$q=\nu(f)-\nu(g),$$
$$r=2\nu(f)-\nu(g)-\nu(h).$$

Then k, q and r are even integers satisfying $0\leq q\leq r$ and $k+r=\nu(f)\geq 12$. Let $F_q(z, w, s)$ be the component at v_q of $[f]_r\left(\begin{pmatrix} z & 0\\ 0 & w\end{pmatrix}, s\right)$. After some computation, we obtain

$$\begin{aligned}&\langle\langle F_q(z, w, s), g(z)\rangle_{k+r-q}, h(w)\rangle_{k+q}\\ (3.9)\quad &=2(4\pi)^{2-s-2k-r}\frac{\Gamma(2k+r-2+s)\Gamma(k+r-q-1+s)\Gamma(k+q-1+s)}{\Gamma(2k+r-2+2s)}\\ &\times\frac{L(s+2k+r-2;f,g,h)}{L_2(2s+2k+r-2,f)},\end{aligned}$$

where $L_2(s,f)$ is the second L-function of f. Hence by Proposition 3,

$$\begin{aligned}&\left\langle\left\langle K\left(s+\frac{k}{2}\right)F_q(z, w, s), g(z)\right\rangle_{k+r-q}, h(w)\right\rangle_{k+q}\\ &=\pi^{4-4s-r-(7k/2)}2^{9-4s-2k-(3r/2)}\left(s+\frac{k}{2}-1\right)\Gamma(s+k-1)\Gamma(s+k+r-q-1)\\ &\times\Gamma(s+k+q-1)\Gamma(2k+r-2+s)L(s+2k+r-2;f,g,h)\end{aligned}$$

is invariant under $s\leftrightarrow 2-k-s$. This yields (3.6). Using properties of Eisenstein series, $L(s;f,g,h)$ is shown to be meromorphically extended to whole s-plane. Noting a minus sign in (3.6), we have (3.7) under the holomorphy of $L(s;f,g,h)$ at $s=(k/2)+r-1$. By (3.9) and the same argument as in Garrett [3, § 6], we have

$$(3.10)\qquad \frac{\pi^{2-2k-r}L(2k+r-2;f,g,h)}{L_2(2k+r-2,f)\langle g,g\rangle_{k+q-r}\langle h,h\rangle_{k+q}} \in Q([f]_r,g,h).$$

Now (3.8) follows from (3.3), (3.10) and

$$L_2(2k+r-2,f)\pi^{-(3k+r-3)}\langle f,f\rangle_{k+r}^{-1} \in Q(f)$$

due to Zagier [18, Corollary to Theorem 2]. Corollary 5 is a direct consequence of

$$L(s;f,f,f)=L_3(s,f)L_1(s-k+1,f)^2$$

and Moreno and Shahidi [11, Lemma 2].

Remark 6. One can expect that $L(s;f,g,h)$ extends holomorphically to the whole s-plane. At least, this holds for $L(s;f,f,f)$ where $f\in S_k(\Gamma_1)$ with $k\leq 50$ due to Moreno and Shahidi [11, Remark 1].

Remark 7. In fact $Q([f]_r)=Q(f)$ for $r=0$ and $r=2$. For $r=0$, see e.g. Kurokawa [7, Theorem 3] or Mizumoto [10, Theorem 2]. For $r=2$, this is a consequence of Corollary 2.

Note added in proof. I. Piatetski-Shapiro and S. Rallis (Rankin triple L functions: Compositio Math., 64 (1987), 311–115) and T. Ikeda (private communication) obtained information on where poles of triple L functions locate and when they are holomorphic respectively. By virtue of these results, we see that triple L functions are holomorphic at $s=(\nu(f)+\nu(g)+\nu(h))/2-1$ in our full modular case. Hence, the holomorphy condition in Theorem 4 is unnecessary.

References

[1] T. Arakawa, Vector valued Siegel's modular forms of degree two and the associated Andrianov L-functions, Manuscripta Math., **44** (1983), 155–185.

[2] S. Böcherer, Über die Funktionalgleichung automorpher L-Funktionen zur Siegelschen Modulgruppe, J. Reine Angew. Math., **362** (1985), 146–168.

[3] P. B. Garrett, Decomposition of Eisenstein series: Rankin triple products, Ann. of Math., **125** (1987), 209–235.

[4] J. Igusa, On Siegel modular forms of genus two. II, Amer. J. Math., **86** (1964), 392–412.

[5] ——, On the ring of modular forms of degree two over Z, Amer. J. Math., **101** (1979), 149–193.

[6] N. Kurokawa, Congruences between Siegel modular forms of degree two, Proc. Japan Acad., **55A** (1979), 417–422.

[7] ——, Eisenstein series for Siegel modular groups, Proc. Japan Acad., **57A** (1981), 51–55.

[8] H. Maass, Siegel's Modular forms and Dirichlet Series, Lecture Notes in Mathematics, **216**, Springer, Berlin-Heidelberg-New York, 1971.

[9] ——, Lineare Relationen für die Fourierkoeffizienten einiger Modulformen zweiten Grades, Math. Ann., **232** (1978), 163–175.

[10] S. Mizumoto, Fourier coefficients of generalized Eisenstein series of degree two. I, Invent. Math., **65** (1981), 115–135.

[11] C. J. Moreno and F. Shahidi, The L-function $L_3(s, \pi_\Delta)$ is entire, Invent. Math., **79** (1985), 247–251.

[12] H. L. Resnikoff and R. L. Saldaña, Some properties of Fourier coefficients of Eisenstein series of degree two, J. Reine Angew. Math., **265** (1974), 90–109.

[13] T. Satoh, On certain vector valued Siegel modular forms of degree two, Math. Ann., **274** (1986), 335–352.

[14] ——, Some remarks on triple L-function, Math. Ann., **276** (1987), 687–698.

[15] G. Shimura, On the derivatives of theta functions and modular forms, Duke Math. J., **44** (1977), 365–387.

[16] R. Tsushima, An explicit dimension formula for the spaces of generalized automorphic forms with respect to $Sp(2, \boldsymbol{Z})$ Proc. Japan Acad., **59A** (1983), 139–142.

[17] ——, An explicit dimension formula for the spaces of generalized automorphic forms with respect to $Sp(2, \boldsymbol{Z})$ (preprint)

[18] D. Zagier, Modular forms whose Fourier coefficients involve zeta-functions of quadratic fields, Lecture Notes in Math., **627**, Springer, Berlin-Heidelberg-New York, 1977, 105–169.

Department of Mathematics
Tokyo Institute of Technology
Oh-okayama, Meguroku, Tokyo 152
Japan

Advanced Studies in Pure Mathematics 15, 1989
Automorphic Forms and Geometry of Arithmetic Varieties
pp. 99–169

Special Values of L-functions Associated with the Space of Quadratic Forms and the Representation of $Sp(2n, F_p)$ in the Space of Siegel Cusp Forms

Tsuneo Arakawa

Dedicated to Professor I. Satake and Professor F. Hirzebruch for their sixtieth birthdays

Contents

Introduction

0.1. To evaluate special values of various kinds of zeta functions and L-functions and to interpret the meaning of them have been providing fruitful problems to number theory.

Siegel [21], as an initiative work, established an ingenious method of

---

Received March 27, 1987.

evaluating the special values at non-positive integers of partial zeta functions for totally real fields and Klingen-Siegel proved that they are rational numbers (cf. [8]). Shintani [16] presented a more direct method of evaluating them, giving remarkable expressions of partial zeta functions by integrals taken over complex contour paths. Following Shintani's method, Satake [14], introducing zeta functions of self-dual homogeneous cones, studied a general method of obtaining nice expressions of the zeta functions by integrals over contour paths. In some cases, he succeeded in representing the special values at non-positive integers of the zeta functions of cones as a finite sum of certain integrals over some compact Lie group. Kurihara [9], also following Shintani, evaluated the special values at non-positive integers of Siegel zeta functions of $\boldsymbol{Q}$-anisotropic quadratic forms (non-zero forms) with signature $(1, n-1)$ $(n=3, 4)$. However, their methods are not applicable to the zeta functions of cones such that some of edge vectors of cones are contained in the boundary of the self-dual homogeneous cone Ω (see Introduction of [14]).

On the other hand, Hecke [6], [7], more than fifty years before, studied the decomposition into irreducible components of the representation μ of $SL_2(\boldsymbol{F}_p)$ in a certain space of elliptic cusp forms and obtained remarkable relations among multiplicities of irreducible representations in μ. Recently, Yamazaki [24], Tsushima [22], Lee-Weintraub [11], and Hashimoto [5] studied similar problems in the case of the representation μ_k of $Sp(4, \boldsymbol{F}_p)$ in a certain space of Siegel cusp forms of degree two and weight k. The former four authors employed algebro-geometric methods including the Hirzebruch-Riemann-Roch theorem, the holomorphic Lefschetz theorem, and Hashimoto used the Selberg trace formula. To attack the problems, they calculated the traces of μ_k for various elements of $Sp(4, \boldsymbol{F}_p)$. From the viewpoint of the Selberg trace formula, there appear special values of various kinds of zeta functions and L-functions in calculating the dimensions of the spaces of cusp forms and the traces of $\mu_k(\bar{\alpha})$ $(\bar{\alpha} \in Sp(4, \boldsymbol{F}_p))$, as is observed in [16], [1], [13], [4], [5]. In his lecture at Kyoto in 1985, Hashimoto introduced an interesting L-function attached to the ternary zero form $x_1x_2-x_{12}^2$ and expressed the traces of $\mu_k(\bar{\alpha})$ for certain unipotent elements $\bar{\alpha} \in Sp(4, \boldsymbol{F}_p)$ using the special value at $s=3/2$ of that L-function (see the identity (0.5) in 0.2).

The Main purpose of the present paper is to evaluate special values at non-positive integers of two kinds of L-functions, one of which is the one introduced by Hashimoto, associated with the ternary zero form $x_1x_2-x_{12}^2$. We shall follow the method of Satake-Kurihara basically. However, since the quadratic form $x_1x_2-x_{12}^2$ is a zero form (which represents zero non-trivially), we have to deal with partial zeta functions of cones whose edge vectors are not necessarily in the interior of $\mathscr{P}_2$, $\mathscr{P}_2$ being the self-dual homogeneous cone of positive definite symmetric matrices of size two. Because of this reason, Satake-Kurihara's method cannot be applied directly to our situation. We need some original ideas to obtain useful integral representations of partial zeta functions (see Chap. II). The special values of our L-functions are expressed explicitly by (generalized) Bernoulli numbers and special values of Bernoulli polynomials.

Moreover, we shall introduce certain zeta functions with a kind of Gauss sums attached to the space of quadratic forms and express the values of a certain class of integrals appearing in the Selberg trace formula for the trace of $\mu_k(\bar{\alpha})$ ($\bar{\alpha} \in Sp(2n, \boldsymbol{F}_p)$) by using special values at non-positive integers of such zeta functions, where μ_k is the representation of $Sp(2n, \boldsymbol{F}_p)$ in the space of cusp forms of degree n, weight k.

As an application of our results, in the case of $n=2$, we can obtain explicit formulae expressing the traces of $\mu_k(\bar{\alpha})$ by the special values at $s=0$ of our L-functions, which are explicitly evaluated as we stated above.

0.2. We fix the notation and explain our results more precisely. Take an odd prime p and fix it. Let L_n (resp. L_n^*) denote the lattice formed by integral symmetric (resp. half-integral symmetric) matrices of size n, and let $L_{n,+}$, $L_{n,+}^*$ be the subsets consisting of all positive definite matrices of L_n, L_n^*, respectively. Denote by $L_{n,+}^*/SL_n(\boldsymbol{Z})$ (resp. $L_{n,+}/SL_n(\boldsymbol{Z})$) the set of $SL_n(\boldsymbol{Z})$-equivalence classes in $L_{n,+}^*$ (resp. $L_{n,+}$). For an integral symmetric matrix S of size ν ($1 \leqq \nu \leqq n$) with $\det(S) \not\equiv 0 \bmod p$, let $\mathscr{L}_n(S)$ denote the subset of L_n consisting of all $x \in L_n$ such that $x \equiv U\begin{pmatrix} S & 0 \\ 0 & 0 \end{pmatrix}{}^t U \bmod p$ with some $U \in GL_n(\boldsymbol{Z}/p\boldsymbol{Z})$. Define a kind of Gauss sum on L_n^* by $\tau_S^{(n)}(T) = \sum_x \exp(2\pi i \operatorname{tr}(Tx/p))$ ($T \in L_n^*$), x running over all residue classes of elements in $\mathscr{L}_n(S) \bmod pL_n$. Set

$$\xi_n^*(s, \tau_S^{(n)}) = \sum_{T \in L_{n,+}^*/SL_n(\boldsymbol{Z})} \tau_S^{(n)}(T)\varepsilon(T)^{-1} \det(T)^{-s},$$

$$L_n^*(s, \chi_{\det}) = \sum_{T \in L_{n,+}^*/SL_n(\boldsymbol{Z})} \chi(\det(T))\varepsilon(T)^{-1} \det(T)^{-s},$$

where $\varepsilon(T)$ is the order of the unit group $\{U \in SL_n(\boldsymbol{Z}) \mid UT{}^tU = T\}$, and χ is a Dirichlet character mod p. These Dirichlet series, which are typical examples of Hurwitz-type zeta functions and L-functions of prehomogeneous vector spaces, are absolutely convergent for $\operatorname{Re}(s) > (n+1)/2$. As for a general theory of L-functions of prehomogeneous vector spaces, we refer to Sato [15], in a part of which the functional equations of L-functions (in a general situation) are derived from the works of Gyoja-Kawanaka [3] on prehomogeneous vector spaces over finite fields. In the case of $n=2$, Hashimoto introduced the following L-functions $L_2^*(s, \psi_{H,p})$, $L_2(s, \psi_{H,p})$.

Let ψ be the unique non-trivial quadratic character mod p, and let $\psi_{H,p}$ be a mapping from L_2^* to $\boldsymbol{R}$ given as follows; put $\psi_{H,p}(T)=\psi(t)$ if $T\equiv U\begin{pmatrix} t & 0 \\ 0 & 0 \end{pmatrix}{}^tU \bmod p$ with some $U\in GL_2(\boldsymbol{Z}/p\boldsymbol{Z})$, $t\in\boldsymbol{Z}$, and otherwise, put $\psi_{H,q}(T)=0$. Set

$$L_2^*(s,\psi_{H,p})=\sum_{T\in L_{2,+}^*/SL_2(\boldsymbol{Z})}\psi_{H,p}(T)\varepsilon(T)^{-1}\det(T)^{-s},$$

$$L_2(s,\psi_{H,p})=\sum_{T\in L_{2,+}/SL_2(\boldsymbol{Z})}\psi_{H,p}(T)\varepsilon(T)^{-1}\det(T)^{-s}.$$

The L-functions $L_2^*(s,\psi_{H,p})$, $L_2(s,\psi_{H,p})$ are absolutely convergent for $\mathrm{Re}(s)>3/2$. The zeta functions and L-functions given above can be continued analytically to meromorphic functions of s in the whole complex plane which are holomorphic at non-positive integers (see Prop. 1.10, Theorem 1.11 in this paper). The L-functions $L_2^*(s,\chi_{\det})$, $L_2^*(s,\psi_{H,p})$, $L_2(s,\psi_{H,p})$ will be regarded as Siegel L-functions associated with the ternary zero form $x_1x_2-x_{12}^2$ (cf. [20]). Let B_n (resp. $B_n(x)$) be the n-th Bernoulli number (resp. n-th Bernoulli polynomial). Denote by $B_{n,\chi}$ the n-th generalized Bernoulli number attached to a Dirichlet character χ. For a real number x, $\langle x\rangle$ denotes the number satisfying $0<\langle x\rangle\leqq 1$ and $x-\langle x\rangle\in\boldsymbol{Z}$. Set

$$(0.1)\qquad \mathscr{A}=-\sum_{\alpha,\gamma}{}' B_1(\langle(\alpha^2-2\alpha\gamma)/p\rangle)B_1(\langle 2\alpha\gamma/p\rangle)B_1(\langle(\gamma^2-\alpha^2)/p\rangle)$$
$$+\frac{1}{12}(3+\delta_{p,3})B_{1,\psi},$$

where α,γ run over all residue classes mod p satisfying $\alpha^2\not\equiv 2\alpha\gamma \bmod p$, $\alpha\gamma\not\equiv 0 \bmod p$, $\alpha^2\not\equiv\gamma^2 \bmod p$, and $\delta_{p,3}$ is the Kronecker symbol ($\delta_{p,3}=1$ if $p=3$, and otherwise, $\delta_{p,3}=0$). Moreover, set

$$(0.2)\qquad \mathscr{B}=-\frac{1}{3}\sum_{\alpha,\gamma}{}'' B_2(\langle(\alpha^2-2\alpha\gamma)/p\rangle)B_1(\langle(\gamma^2-\alpha^2)/p\rangle),$$

where α,γ run over all residue classes mod p satisfying $\alpha^2\not\equiv\gamma^2 \bmod p$.

Theorem 1. *Let p be an odd prime.*

(i) *The special values $L_2^*(1-m,\psi_{H,p})$ $(m=1,2,\cdots)$ are rational numbers.*

(ii) *If $p\equiv 1$ mod 4, then $L_2^*(0,\psi_{H,p})=0$.*

(iii) *If $p\equiv 3$ mod 4, then,*

$$L_2^*(0,\psi_{H,p})=\mathscr{A}+\mathscr{B}+\frac{11}{36p}B_{3,\psi}-\frac{1}{24p}B_{1,\psi}.$$

We present a conjecture on the value $L_2^*(0, \psi_{H,p})$.

Conjecture I. *If* $p\equiv 3 \bmod 4$, $L_2^*(0, \psi_{H,p})=\frac{1}{24}B_{1,\psi}$.

(Note that $h(-p)=-B_{1,\psi}$ for $p>3$ and $p\equiv 3 \bmod 4$, where $h(-p)$ is the class number of the quadratic field $\boldsymbol{Q}(\sqrt{-p})$.)

Conjecture I is true for $p<500$ by the numerical calculation using a computer.

Theorem 2. *Let* χ *be a primitive character* mod p $(p>2)$. *Then, for* $m=1, 2, \cdots$,

$$L_2^*(1-m, \chi_{\det})=\begin{cases} \dfrac{\chi(-4)^{-1}(-1)^m}{2^{2m+1}m}B_{2m,\chi^2}\cdots & \text{if } \chi\neq\psi, \\ \dfrac{\chi(-1)(-1)^{m-1}}{2^{2m+1}m}(p^{2m-1}-1)B_{2m}\cdots & \text{if } \chi=\psi. \end{cases}$$

Theorems 1, 2 will be regarded as a kind of generalization of the well-known formula $L(1-m, \chi)=-B_{m,\chi}/m$ for Dirichlet L-function $L(s, \chi)$.

Let $\Gamma_{2n}(p)$ be the principal congruence subgroup of $\Gamma_{2n}=Sp(2n, \boldsymbol{Z})$ with level p. The quotient group $\Gamma_{2n}/\Gamma_{2n}(p)$ is isomorphic to the finite symplectic group $Sp(2n, \boldsymbol{F}_p)$ of degree $2n$, and the surjection $\Gamma_{2n}\longrightarrow Sp(2n, \boldsymbol{F}_p)$ is denoted by $\alpha\longrightarrow\bar{\alpha}$ $(\alpha\in\Gamma_{2n})$. Let $\mathfrak{H}_n$ be the Siegel upper half plane of degree n, on which the real symplectic group $\mathfrak{G}_{2n}=Sp(2n, \boldsymbol{R})$ acts in a usual manner; for $Z=X+iY\in\mathfrak{H}_n$ and for $\gamma=\begin{pmatrix} A & B \\ C & D\end{pmatrix}\in\mathfrak{G}_{2n}$, $\gamma\langle Z\rangle=(AZ+B)(CZ+D)^{-1}$. Set $J(\gamma, Z)=\det(CZ+D)$, and

$$H(\gamma; Z)=J(\gamma, Z)^{-k}\det\left(\frac{\gamma\langle Z\rangle-\bar{Z}}{2i}\right)^{-k}\det(Y)^k,$$

$$dZ=\det(Y)^{-n-1}\prod_{1\leq i\leq j\leq n} dX_{ij}dY_{ij}.$$

For any subset Π of Γ_{2n} which is invariant by the conjugation of any elements in $\Gamma_{2n}(p)$, set

$$I_n(\Pi; k)=a(k)\int_{\Gamma_{2n}(p)\backslash\mathfrak{H}_n}\sum_{\gamma\in\Pi} H(\gamma; Z)dZ,$$

where $a(k)$ is a constant given by (3.1.3) in Chap. III. Denote by $\mathfrak{S}_k(\Gamma_{2n}(p))$ the space of Siegel cusp forms of degree n, weight k with respect to $\Gamma_{2n}(p)$. The representation μ_k of the group $Sp(2n, \boldsymbol{F}_p)$ in the space $\mathfrak{S}_k(\Gamma_{2n}(p))$ is given by $\mu_k(\bar{\alpha}^{-1})f(Z)=J(\alpha, Z)^{-k}f(\alpha\langle Z\rangle)(\alpha\in\Gamma_{2n}, f\in$

$\mathfrak{S}_k(\Gamma_{2n}(p))$, and see (3.1.2)). It is known by [2] that $\mathrm{tr}(\mu_k(\bar{\alpha}^{-1}))= I_n(\Gamma_{2n}(p)\alpha; k)$ for $k>2n$. For a symmetric matrix x of size ν, denote by $t_{n,\nu}(x)=\begin{pmatrix}1_n & \tilde{x}\\ 0 & 1_n\end{pmatrix}$, where $\tilde{x}=\begin{pmatrix}x & 0\\ 0 & 0\end{pmatrix}$ (see (3.1.4)). For $\alpha\in\Gamma_{2n}$, let $\Pi_r(\alpha)$ $(1\leqq r\leqq n)$ be the set consisting of all elements in $\Gamma_{2n}(p)\alpha$ that are conjugate to some elements $t_{n,r}(x)$ with $x\in L_r$, $\det(x)\neq 0$ ($\Pi_r(\alpha)$ depends only on $\bar{\alpha}$). Following Shintani [16, Chap. 2], we will show that, if $k\geqq 2n+3$, then, for each $\alpha=t_{n,\nu}(S)$ with $S\in L_\nu$, $\det(S)\not\equiv 0 \bmod p$,

$$I_n(\Pi_r(\alpha); k)=[\Gamma_{\alpha,p}:\Gamma_{2n}(p)]p^{-r(n-(r-1)/2)}b(n,k,r)\Omega_{n,r}\xi_r^*(r-n,\tau_S^{(r)}) \qquad (\nu\leqq r\leqq n),$$

where $b(n,k,r)$, $\Omega_{n,r}$ are rational numbers given by (3.2.7), (3.2.8), respectively, and $\Gamma_{\alpha,p}=\{\gamma\in\Gamma_{2n}\,|\,\gamma^{-1}\alpha\gamma\equiv\alpha \bmod p\}$ (a subgroup of Γ_{2n}), of which $\Gamma_{2n}(p)$ is a normal subgroup with index finite. If $\alpha=1_{2n}$, the results above coincide with those of [16]. In the case of $n=2$, $\nu=1$, put, for any integer μ prime to p, $\alpha_\mu=t_{2,1}(\mu)$ $(\in\Gamma_4)$. It is essentially known by [13], [1], [4], [5], and can be proved in a similar manner that, if $k\geqq 7$, $\mathrm{tr}(\mu_k(\bar{\alpha}_\mu))=\sum_{r=1}^2 I_2(\Pi_r(\alpha_{-\mu}); k)$. It will be shown that, if $k\geqq 7$,

$$\begin{aligned}(0.3)\quad \mathrm{tr}(\mu_k(\bar{\alpha}_\mu))= & -2^{-5}3^{-1}p^2(p^2-1)\{\psi(-\mu)\tau_\psi B_{2,\psi}+(p^2-1)/6\}(2k-3)\\ & +2^{-1}p(p^2-1)\{\psi(-1)pL_2^*(0,\psi_{\det})+\psi(-\mu)\tau_\psi pL_2^*(0,\psi_{H,p})\\ & +2^{-4}3^{-1}(p^2-1)\},\end{aligned}$$

where τ_ψ is the Gauss sum associated with ψ.

Theorem 3. *Let p be an odd prime and let κ be a non-quadratic residue $\bmod p$. Let $k\geqq 7$. The difference of the two traces* $\mathrm{tr}(\mu_k(\bar{\alpha}_\mu))$ $(\mu=1,\kappa)$ *is given by*

$$(0.4)\quad \mathrm{tr}(\mu_k(\bar{\alpha}_1))-\mathrm{tr}(\mu_k(\bar{\alpha}_\kappa))=\begin{cases}-2^{-4}3^{-1}p^2(p^2-1)\sqrt{p}\,B_{2,\psi}(2k-3) & \cdots p\equiv 1 \bmod 4,\\ -p^2(p^2-1)\sqrt{-p}L_2^*(0,\psi_{H,p}) & \cdots p\equiv 3 \bmod 4.\end{cases}$$

We present another conjecture, which is based on Conjecture I.

Conjecture II. *If $p>3$, $p\equiv 3 \bmod 4$, and $k\geqq 7$, then,*

$$\mathrm{tr}(\mu_k(\bar{\alpha}_1))-\mathrm{tr}(\mu_k(\bar{\alpha}_\kappa))=\frac{1}{24}p^2(p^2-1)\sqrt{-p}h(-p).$$

Tsushima calculated the traces of $\mu_k(\bar{\alpha}_\mu)$ in his private notes, which are based on the results of [22]. Lee-Weintraub [11] announced the explicit values of the imaginary parts of $\mathrm{tr}(\mu_k(\bar{\alpha}_\mu))$ and presented a conjecture analogous to our Conjecture II. Their methods are purely geometric and the results do not involve special values of L-functions. We are much concerned how the traces can be expressed by special values of L-functions and how they can be explicitly evaluated.

In his lecture at Kyoto in 1985, Hashimoto announced that the difference of the traces of $\mu_k(\bar{\alpha}_\mu)$ ($\mu=1, \kappa$) is given as follows:

$$\text{(0.5)} \quad \mathrm{tr}\,(\mu_k(\bar{\alpha}_1))-\mathrm{tr}\,(\mu_k(\bar{\alpha}_\kappa))=\frac{-p^4(p^2-1)i}{4\pi^2}L_2(3/2, \psi_{H,p})\cdots p\equiv 3 \bmod 4.$$

It follows from (0.4), (0.5) that

$$\text{(0.6)} \qquad L_2(3/2, \psi_{H,p})=4\pi^2 p^{-3/2}L_2^*(0, \psi_{H,p})\cdots \text{if } p\equiv 3 \bmod 4,$$

which, together with (iii) of Theorem 1, will be regarded as a Kronecker limit formula for $L_2(s, \psi_{H,p})$. It is guessed that $L_2^*(s, \psi_{H,p})$, $L_2(s, \psi_{H,p})$ are related with each other by a functional equation under $s\to 3/2-s$. The relation (0.6) will be derived from the functional equation.

The author wishes to express his hearty gratitude to Professor K. Hashimoto for informing him of his precious unpublished results and to Professor F. Sato for several valuable comments. He is also very grateful to Professor A. Kurihara and Professor I. Makino for verifying Conjecture I numerically by computer. It is a pleasure for the author to mention that this paper is an answer to the problems presented by Professor Hashimoto at Kyoto, 1985.

Notation

Let $\boldsymbol{N}$, $\boldsymbol{Z}$, $\boldsymbol{Q}$, $\boldsymbol{R}$, and $\boldsymbol{C}$ denote the set of natural numbers, the ring of rational integers, the rational number field, the real number field, and the complex number field, respectively. For any commutative ring S, $M(m, n; S)$, $M_n(S)$, $GL_n(S)$, and $SL_n(S)$ denote the module of $m\times n$ matrices with entries in S, the ring of matrices of size n with entries in S, the group of invertible elements in $M_n(S)$, and the group of elements in $M_n(S)$ whose determinants are one, respectively. For any element A of $M_n(S)$, let ${}^tA$, $\mathrm{tr}\,(A)$, and $\det(A)$ denote the transposed matrix of A, the trace of A, and the determinant of A, respectively. We denote by 1_n the unit matrix of $M_n(S)$. Moreover, we put $S^\times=GL_1(S)$.

For any element Z of $M_n(\boldsymbol{C})$, we denote by $\mathrm{Re}\,(Z)$, $\mathrm{Im}\,(Z)$, and $\bar{Z}$, the real part of Z, the imaginary part of Z, and the complex conjugate of Z, respectively. For real symmetric matrices A, B of the same size, $A>B$ means that $A-B$ is positive definite. For any $x\in\boldsymbol{R}$, $\langle x\rangle$ denotes the real number with $x-\langle x\rangle\in\boldsymbol{Z}$, $0<\langle x\rangle\leqq 1$. Let $\Gamma(s)$ and $\zeta(s)$ be the gamma function and the Riemann zeta function, respectively. Finally, the symbol $e[w]$ ($w \in \boldsymbol{C}$) is used as an abbreviation for $\exp(2\pi i w)$.

Chapter I. *L*-functions of quadratic forms

1.1. Definition of zeta functions and *L*-functions

Following Shintani [16], we shall define certain zeta functions with Gauss sums and certain *L*-functions which are associated with the vector space of symmetric matrices. We shall not discuss a general theory of those functions but only some properties that will be needed in later chapters.

Let $G_{\boldsymbol{R}}^{(n)}$ be $GL_n(\boldsymbol{R})$, and let $V_{\boldsymbol{R}}^{(n)}$ be the $\boldsymbol{R}$-vector sapce of real symmetric matrices of size n. Then, the group $G_{\boldsymbol{R}}^{(n)}$ acts on $V_{\boldsymbol{R}}^{(n)}$ in a usual manner; for $g \in G_{\boldsymbol{R}}^{(n)}$, $x \in V_{\boldsymbol{R}}^{(n)}$, we put $\rho(g)x = gx{}^tg$. Let L_n be the lattice of $V_{\boldsymbol{R}}^{(n)}$ consisting of all integral symmetric matrices of size n, and let L_n^* be its dual with respect to the bilinear form $(x, y) = \operatorname{tr}(xy)$ $(x, y \in V_{\boldsymbol{R}}^{(n)})$. Namely, L_n^* is the lattice consisting of half-integral symmetric matrices of size n.

Let p be an odd prime and fix it once and for all. Denote by $\boldsymbol{Z}_{(2)}$ the $\boldsymbol{Z}$-module consisting of all rational numbers of the form $2^{-m}x$ with $m, x \in \boldsymbol{Z}$. For $a, b \in \boldsymbol{Z}_{(2)}$, $a \equiv b \bmod p$ means that $(a-b)/p$ is a p-adic integer. Let $1 \leqq \nu \leqq n$ and let $S \in L_\nu$ with $\det(S) \not\equiv 0 \bmod p$. We set

$$\mathscr{L}_n(S) = \left\{ x \in L_n \mid x \equiv U \begin{pmatrix} S & 0 \\ 0 & 0 \end{pmatrix} {}^tU \bmod p \text{ with some } U \in GL_n(\boldsymbol{Z}/p\boldsymbol{Z}) \right\},$$

$\boldsymbol{Z}/p\boldsymbol{Z}$ being the ring of residue classes mod p. Then, $\mathscr{L}_n(S)$ is invariant under the action of $SL_n(\boldsymbol{Z})$; namely, $U\mathscr{L}_n(S){}^tU = \mathscr{L}_n(S)$ for any $U \in SL_n(\boldsymbol{Z})$. Moreover, $\mathscr{L}_n(VS{}^tV) = \mathscr{L}_n(S)$ for any $V \in GL_\nu(\boldsymbol{Z}/p\boldsymbol{Z})$. If $\nu < n$, this definition of $\mathscr{L}_n(S)$ amounts to saying that

$$\mathscr{L}_n(S) = \left\{ x \in L_n \mid x \equiv U \begin{pmatrix} S & 0 \\ 0 & 0 \end{pmatrix} {}^tU \bmod p \text{ with some } U \in SL_n(\boldsymbol{Z}) \right\}.$$

We denote by $\mathscr{L}_n(S)/pL_n$ all residue classes of elements in $\mathscr{L}_n(S)$ mod pL_n. Define a subgroup $\tilde{\Gamma}_{S,\infty}$ of $GL_n(\boldsymbol{Z}/p\boldsymbol{Z})$ by

$$\tilde{\Gamma}_{S,\infty} = \left\{ \begin{pmatrix} A & B \\ C & D \end{pmatrix} \in GL_n(\boldsymbol{Z}/p\boldsymbol{Z}) \mid AS{}^tA \equiv S \bmod p,\ C \equiv 0 \bmod p \right\}.$$

A complete set of representatives of $\mathscr{L}_n(S)/pL_n$ is given by the set

$$\left\{ U \begin{pmatrix} S & 0 \\ 0 & 0 \end{pmatrix} {}^tU \mid U \in GL_n(\boldsymbol{Z}/p\boldsymbol{Z})/\tilde{\Gamma}_{S,\infty} \right\}. \tag{1.1.1}$$

A kind of Gauss sum $\tau_S^{(n)}(T)$ $(T \in L_n^*)$ is defined as follows:

(1.1.2) $$\tau_S^{(n)}(T)=\sum_{x\in \mathscr{L}_n(S)/pL_n} e[\mathrm{tr}\,(Tx)/p],$$

which depends only on $T \bmod p$. Then we have $\tau_S^{(n)}(UT{}^tU)=\tau_S^{(n)}(T)$ for any $U \in GL_n(\boldsymbol{Z}/p\boldsymbol{Z})$. Denote by $L_{n,+}^*$ the subset of L_n^* consisting of all positive definite symmetric matrices. For each $T \in L_{n,+}^*$, let $\varepsilon(T)$ be the order of the unit group $\{U \in SL_n(\boldsymbol{Z}) \mid UT{}^tU=T\}$. Two matrices T_1, T_2 of L_n^* are called $SL_n(\boldsymbol{Z})$-equivalent, if there exists some $U \in SL_n(\boldsymbol{Z})$ with $T_2=UT_1^tU$. Denote by $L_{n,+}^*/SL_n(\boldsymbol{Z})$ the set of $SL_n(\boldsymbol{Z})$-equivalence classes in $L_{n,+}^*$. We define a zeta function $\xi_n^*(s, \tau_S^{(n)})$ with the Gauss sum $\tau_S^{(n)}$ as follows:

$$\xi_n^*(s, \tau_S^{(n)})=\sum_{T\in L_{n,+}^*/SL_n(\boldsymbol{Z})} \tau_S^{(n)}(T)\varepsilon(T)^{-1}\det(T)^{-s}.$$

For a primitive character $\chi \bmod p$, we also define an L-function $L_n^*(s, \chi_{\det})$ by putting

$$L_n^*(s, \chi_{\det})=\sum_{T\in L_{n,+}^*/SL_n(\boldsymbol{Z})} \chi(\det(T))\varepsilon(T)^{-1}\det(T)^{-s},$$

where χ is naturally extended over $\boldsymbol{Z}_{(2)}$. Moreover, set

$$\xi_n^*(s)=\sum_{T\in L_{n,+}^*/SL_n(\boldsymbol{Z})} \varepsilon(T)^{-1}\det(T)^{-s}.$$

It is well-known that $\xi_n^*(s)$ is absolutely convergent for $\mathrm{Re}(s)>(n+1)/2$, Therefore, $\xi_n^*(s, \tau_S^{(n)})$, $L_n^*(s, \chi_{\det})$ also converge absolutely if $\mathrm{Re}(s)>(n+1)/2$. The zeta function $\xi_n^*(s)$ is one of the zeta functions intensively studied by Shintani in [16].

Denote by ψ the unique non-trivial quadratic character mod p, which is characterized by $\psi(a)=(a/p)$ for any integer a prime to p, (a/p) being the Legendre symbol. Let $\psi_{H,p}$ be a mapping from L_2^* to $\boldsymbol{R}$ given by

$$\psi_{H,p}(T)=\begin{cases}\psi(t)\cdots \text{if } T\equiv U\begin{pmatrix} t & 0\\ 0 & 0\end{pmatrix}{}^tU \bmod p \text{ with some } U\in GL_2(\boldsymbol{Z}/p\boldsymbol{Z}) \text{ and}\\ \qquad\qquad t\in \boldsymbol{Z} \text{ prime to } p,\\ 0\cdots\cdots \text{otherwise.}\end{cases}$$

In the case of $n=2$, Hashimoto introduced the following L-function:

$$L_2^*(s, \psi_{H,p})=\sum_{T\in L_{2,+}^*/SL_2(\boldsymbol{Z})} \psi_{H,p}(T)\varepsilon(T)^{-1}\det(T)^{-s},$$

which is absolutely convergent for $\mathrm{Re}(s)>3/2$. We have a relation among $\xi_2^*(s, \tau_S^{(2)})$, $L_2^*(s, \psi_{H,p})$, $L_2^*(s, \psi_{\det})$, and $\xi_2^*(s)$, which is given in Proposition 1.2.

Lemma 1.1. *Let μ be any integer prime to p and take μ as S in* (1.1.2) *in the case of $n=2$, $\nu=1$. Then, for $T\in L_2^*$, we have*

(i) $\tau_\mu^{(2)}(T)=\{\psi(-1)\psi(\det(T))p-1\}/2\cdots$ *if* $\det(T)\not\equiv 0 \bmod p$,

(ii) $\tau_\mu^{(2)}(T)=\{\psi(\mu)\psi_{H,p}(T)p\tau_\psi-1\}/2\cdots$ *if* $\det(T)\equiv 0 \bmod p$, $T\not\equiv 0 \bmod p$,

(iii) $\tau_\mu^{(2)}(T)=(p^2-1)/2\cdots$ *if* $T\equiv 0 \bmod p$,

where τ_ψ is the ordinary Gauss sum associated with ψ: $\tau_\psi=\sum_{a\not\equiv 0 \bmod p}\psi(a)\cdot e[a/p]$.

Proof. Writing $U=\begin{pmatrix}\alpha & \beta\\ \gamma & \delta\end{pmatrix}\in GL_2(Z/pZ)$, we have $U\begin{pmatrix}\mu & 0\\ 0 & 0\end{pmatrix}{}^tU=\mu\begin{pmatrix}\alpha^2 & \alpha\gamma\\ \alpha\gamma & \gamma^2\end{pmatrix}$. We set

$$\mathcal{M}(p)=\{(\alpha,\gamma)\in Z/pZ\times Z/pZ \mid (\alpha,\gamma)\not\equiv(0,0) \bmod p\}. \tag{1.1.3}$$

It is easy to see from (1.1.1) that, for each $T\in L_2^*$,

$$\tau_\mu^{(2)}(T)=\frac{1}{2}\sum_{(\alpha,\gamma)\in\mathcal{M}(p)} e\left[\mu\,\mathrm{tr}\left(\begin{pmatrix}\alpha^2 & \alpha\gamma\\ \alpha\gamma & \gamma^2\end{pmatrix}T\right)/p\right].$$

If $\det(T)\not\equiv 0 \bmod p$, we may assume that $T\equiv\begin{pmatrix}t_1 & 0\\ 0 & t_2\end{pmatrix} \bmod p$ with $(t_j, p)=1$ $(j=1, 2)$. Then we get

$$\tau_\mu^{(2)}(T)=\frac{1}{2}\left\{\sum_{(\alpha,\gamma)\in\mathcal{M}(p)} e[\mu(t_1\alpha^2+t_2\gamma^2)/p]\right\}.$$

As is well-known, if $(a, p)=1$, then, $\sum_{u \bmod p} e[au^2/p]=\psi(a)\tau_\psi$. Thus we have, with the help of the property $\tau_\psi^2=\psi(-1)p$,

$$\tau_\mu^{(2)}(T)=\frac{1}{2}\{\psi(-1)\psi(t_1t_2)p-1\}.$$

Next let $\det(T)\equiv 0 \bmod p$ and $T\not\equiv 0 \bmod p$. We may assume that $T\equiv\begin{pmatrix}t & 0\\ 0 & 0\end{pmatrix} \bmod p$ with $(t, p)=1$. Then we get

$$\tau_\mu^{(2)}(T)=\frac{1}{2}\{\psi(\mu t)p\tau_\psi-1\},$$

from which the assertion (ii) follows. The assertion (iii) is clear.

Proposition 1.2. *Let μ be any integer prime to p. Then,*

$$\xi_2^*(s, \tau_\mu^{(2)}) = \frac{1}{2}\{\psi(-1)pL_2^*(s, \psi_{\det}) + \psi(\mu)p\tau_\psi L_2^*(s, \psi_{H,p}) - (1-p^{2-2s})\xi_2^*(s)\}.$$

We omit the proof of Proposition 1.2, which is immediate from Lemma 1.1.

Let κ be a non-quadratic residue mod p. Taking 1, κ as μ in Proposition 1.2, we get the following corollary.

Corollary to Proposition 1.2. *We have*

$$p\tau_\psi L_2^*(s, \psi_{H,p}) = \xi_2^*(s, \tau_1^{(2)}) - \xi_2^*(s, \tau_\kappa^{(2)}),$$
$$\psi(-1)pL_2^*(s, \psi_{\det}) = \xi_2^*(s, \tau_1^{(2)}) + \xi_2^*(s, \tau_\kappa^{(2)}) + (1-p^{2-2s})\xi_2^*(s).$$

1.2. Some properties of $\xi_n^*(s, \tau_S^{(n)})$, $L_2^*(s, \psi_{\det})$, **and** $L_2^*(s, \psi_{H,p})$ (analytic continuations, poles, residues)

We follow Section 2 of Chapter 2 in [16]. Let $\mathscr{P}_n$ be the symmetric space formed by positive definite real symmetric matrices of size n. For each $\lambda \in \boldsymbol{N}$, the functions $f_n(x, \lambda)$, $f_n^*(x, \lambda)$ ($x \in V_{\boldsymbol{R}}^{(n)}$) on $V_{\boldsymbol{R}}^{(n)}$ are defined as follows:

$$f_n(x, \lambda) = \begin{cases} \det(x)^{\lambda-(n+1)/2} \exp(-2\pi \operatorname{tr}(x)) \cdots \text{if } x \in \mathscr{P}_n, \\ 0 \cdots \text{if } x \in V_{\boldsymbol{R}}^{(n)},\ x \notin \mathscr{P}_n, \end{cases}$$
$$f_n^*(x, \lambda) = \det(1_n - ix)^{-\lambda}.$$

An Euclidean measure dx on $V_{\boldsymbol{R}}^{(n)}$ is normalized by $dx = \prod_{1 \le i \le j \le n} dx_{ij}$. Put $\chi(g) = \det(g)^2$. Moreover, we set

$$\gamma_n(s) = \prod_{i=0}^{n-1} \Gamma(s+1+i/2), \qquad C_n = \prod_{k=1}^{n} \frac{2\pi^{k/2}}{\Gamma(k/2)},$$

and

$$v_n = \begin{cases} \dfrac{1}{2}\zeta(2)\zeta(3)\cdots\zeta(n) & (n \geqq 2) \\ \dfrac{1}{2} & (n=1). \end{cases}$$

Let p be an odd prime and fix it.

Lemma 1.3. *Let* $1 \leqq \nu \leqq n$, *and let* $S \in L_\nu$ *with* $\det(S) \not\equiv 0 \bmod p$. *If* $\lambda > n+1$, *then,*

$$\sum_{T \in L^*} \tau_S^{(n)}(T) f_n(\rho({}^t g^{-1})(T/p), \lambda) = \mu_{n,\lambda} \chi(g)^{(n+1)/2} \sum_{x \in \mathscr{L}_n(S)} f_n^*(\rho(g)x, \lambda),$$

where $g\in G_{\boldsymbol{R}}^{(n)}$, *and* $\mu_{n,\lambda}=(4\pi)^{n(n-1)/4}p^{n(n+1)/2}(2\pi)^{-\lambda n}\gamma_n(\lambda-(n+1)/2)$ *(both sides are absolutely convergent).*

Proof. The Fourier transform of $f_n(\rho({}^tg^{-1})x,\lambda)$ on $V_{\boldsymbol{R}}^{(n)}$ is given by

$$\int_{V_{\boldsymbol{R}}^{(n)}} f_n(\rho({}^tg^{-1})x,\lambda)\,e[\mathrm{tr}\,(xy)]dx$$
$$=\pi^{n(n-1)/4}(2\pi)^{-\lambda n}\gamma_n(\lambda-(n+1)/2)\chi(g)^{(n+1)/2}f_n^*(\rho(g)y,\lambda)$$
$(g\in G_{\boldsymbol{R}}^{(n)},\ y\in V_{\boldsymbol{R}}^{(n)}$, and the integral converges absolutely).

This identity is nothing but Hilfssatz 37 of [18]. Replacing y with $u+py$ ($u\in V_{\boldsymbol{R}}^{(n)}$) and changing the variables by $x\to x/p$, we get

$$\int_{V_{\boldsymbol{R}}^{(n)}} f_n(\rho({}^tg^{-1})(x/p),\lambda)\,e[\mathrm{tr}\,(ux/p)]\,e[\mathrm{tr}\,(xy)]dx$$
$$=2^{-n(n-1)/2}\mu_{n,\lambda}\chi(g)^{(n+1)/2}f_n^*(\rho(g)(u+py),\lambda).$$

By virtue of the Poisson summation formula (see (iii) of Lemma 19 in [16]), we have

$$\sum_{x\in L_n^*} f_n(\rho({}^tg^{-1})(x/p),\lambda)\,e[\mathrm{tr}\,(ux/p)]=\mu_{n,\lambda}(g)^{(n+1)/2}\sum_{y\in L_n} f_n^*(\rho(g)(u+py),\lambda)$$
(the both sides are absolutely convergent for $\lambda>n+1$).

If we let u run over $\mathscr{L}_n(S)/pL_n$, we obtain the formula in Lemma 1.3.

Let d_ng be a Haar measure on $G_{\boldsymbol{R}}^{(n)}$ normalized by

$$d_ng=\det(g)^{-n}\prod_{1\leq i,\,j\leq n} dg_{ij}.$$

Set $G_{\boldsymbol{R},+}^{(n)}=\{g\in G_{\boldsymbol{R}}^{(n)}\,|\,\det(g)>0\}$. We put

$$(1.2.1)\quad Z^*(s,\tau_S^{(n)})=\int_{G_{\boldsymbol{R},+}^{(n)}/SL_n(\boldsymbol{Z})}\chi(g)^s\sum_{T\in L_n^*}\tau_S^{(n)}(T)f_n(\rho(g)(T/p),\lambda)d_ng.$$

We denote by $Z_+^*(s,\tau_S^{(n)})$ the integral obtained by replacing the region of integration with the set $\{g\in G_{\boldsymbol{R},+}^{(n)}/SL_n(\boldsymbol{Z})\,|\,\chi(g)\geqq 1\}$ in the right side of (1.2.1). As is shown in Lemma 21, (i) in [16], the integral $Z^*(s,\tau_S^{(n)})$ is absolutely convergent if $\mathrm{Re}\,(s)>(n+1)/2$, $\mathrm{Re}\,(\lambda+s)>n$, and then,

$$(1.2.2)\quad Z^*(s,\tau_S^{(n)})$$
$$=p^{ns}\pi^{n(n-1)/4}(2\pi)^{-(\lambda+s-(n+1)/2)n}2^{-n-1}C_n\gamma_n(\lambda+s-n-1)\xi_n^*(s,\tau_S^{(n)}).$$

We keep the conditions: $S\in L_\nu$, $\det(S)\not\equiv 0 \bmod p$. Set, for $r\in N$ $(r\leqq n)$,

$$(1.2.3)\qquad \mathscr{L}_n^{(r)}(S)=\{x\in\mathscr{L}_n(S)\,|\,\mathrm{rank}\,(x)=r\}.$$

We see immediately that, if $\mathscr{L}_n^{(r)}(S)\neq\emptyset$, then, of necessity, $\nu\leqq r\leqq n$.

Lemma 1.4. *Let $\nu\leqq r\leqq n$. Each $x\in\mathscr{L}_n^{(r)}(S)$ can be written as $x=U\begin{pmatrix}x_1&0\\0&0\end{pmatrix}{}^tU$ with some $x_1\in\mathscr{L}_r^{(r)}(S)$, $U\in SL_n(\boldsymbol{Z})$.*

Proof. We assume that $r<n$, otherwise we have nothing to do. Take $x\in\mathscr{L}_n^{(r)}(S)$. Then there exist $U\in SL_n(\boldsymbol{Z})$ and $x_1\in L_r$ with $\det(x_1)\neq 0$ such that $x=U\begin{pmatrix}x_1&0\\0&0\end{pmatrix}{}^tU$. Since $x\in\mathscr{L}_n(S)$, the rank of x_1 mod p as a matrix of $M_r(\boldsymbol{Z}/p\boldsymbol{Z})$ is ν. Thus there exist some $W_1\in GL_r(\boldsymbol{Z}/p\boldsymbol{Z})$ and $x_2\in L_\nu$, $\det(x_2)\not\equiv 0$ mod p with the condition

$$x_1\equiv W_1\begin{pmatrix}x_2&0\\0&0\end{pmatrix}{}^tW_1 \bmod p. \tag{1.2.4}$$

Since $x\in\mathscr{L}_n(S)$, so is $\begin{pmatrix}x_1&0\\0&0\end{pmatrix}$. Hence we see easily from (1.2.4) that there exists some $V\in GL_n(\boldsymbol{Z}/p\boldsymbol{Z})$ such that $\begin{pmatrix}x_2&0\\0&0\end{pmatrix}\equiv V\begin{pmatrix}S&0\\0&0\end{pmatrix}{}^tV$ mod p. Writing $V=\begin{pmatrix}V_1&V_2\\V_3&V_4\end{pmatrix}$ with $V_1\in M_\nu(\boldsymbol{Z}/p\boldsymbol{Z})$, we have $x_2\equiv V_1S\,{}^tV_1$ mod p, which implies that $V_1\in GL_\nu(\boldsymbol{Z}/p\boldsymbol{Z})$. We get, again by (1.2.4), $x_1\equiv W\begin{pmatrix}S&0\\0&0\end{pmatrix}{}^tW$ mod p with some $W\in GL_r(\boldsymbol{Z}/p\boldsymbol{Z})$. Thus, $x_1\in\mathscr{L}_r^{(r)}(S)$. q.e.d.

Let P_n^r denote the subgroup of $G_{\boldsymbol{R}}^{(n)}$ formed by all matrices whose left lower $(n-r)\times r$ blocks are zero. Denote by $P_{n,+}^r$ the connected component of 1_n in P_n^r.

Lemma 1.5. *The following decompositions hold.*
(i) $\mathscr{L}_n(S)=\bigcup_{r=\nu}^{n}\mathscr{L}_n^{(r)}(S)$ *(disjoint union).*
(ii) *For each r ($\nu\leqq r\leqq n$),*

$$\mathscr{L}_n^{(r)}(S)=\bigcup_{U\in SL_n(\boldsymbol{Z})/SL_n(\boldsymbol{Z})\cap P_n^r}\left\{U\begin{pmatrix}x&0\\0&0\end{pmatrix}{}^tU \,\middle|\, x\in\mathscr{L}_r^{(r)}(S)\right\}$$

(disjoint union).

The proof is due to Lemma 1.4 and is immediate.
We set

$$Z(f_n^*(x,\lambda),\mathscr{L}_n(S),s)=\int_{G_{\boldsymbol{R},+}^{(n)}/SL_n(\boldsymbol{Z})}\chi(g)^s\sum_{x\in\mathscr{L}_n^{(n)}(S)}f_n^*(\rho(g)x,\lambda)d_ng. \tag{1.2.5}$$

By virtue in Lemma 21, (ii) in [16] and Lemma 1.5 above, one has the following proposition (or one can prove it in the same manner).

Proposition 1.6. *Assume that* λ, s *satisfy the following inequalities*:

$$(1.2.6)\qquad \begin{cases} \lambda>1,\ \mathrm{Re}(s)<\lambda & \textit{for } n=1, \\ \lambda>\mathrm{Max}\,(13/2,\ 2\,\mathrm{Re}(s)+7/2) & \textit{for } n=2, \\ \lambda>n+7/2,\ \mathrm{Re}(s)<\lambda-(n-1)/2 & \textit{for } n\geqq 3. \end{cases}$$

In addition to (1.2.6), *if* s *satisfies* $\mathrm{Re}(s)>(n-1)/2$, *then the integral* $Z(f_n^*(x,\lambda), \mathscr{L}_n(S), s)$ *is absolutely convergent.*

We impose the following assumption on λ:

(1.2.7) $\lambda>1$ for $n=1$, $\lambda>13/2$ for $n=2$, and $\lambda>n+7/2$ for $n\geqq 3$.

For $\lambda\in \boldsymbol{N}$ satisfying (1.2.7), we put

$$(1.2.8)\qquad a_\lambda=\begin{cases} \lambda-1 & \text{for } n=1 \\ (\lambda-13/2)/2 & \text{for } n=2 \\ \lambda-n & \textit{for } n\geqq 3. \end{cases}$$

We define the integral $Z_+(f_n^*(x,\lambda), \mathscr{L}_n(S), s)$ by restricting the region of integration to the set $\{gG\in \mathscr{P}^{(n)}_{\boldsymbol{R},+}/SL_n(\boldsymbol{Z})\,|\,\chi(g)\geqq 1\}$ in the definition (1.2.5) of $Z(f_n^*(x,\lambda), \mathscr{L}_n(S), s)$. We see easily from Proposition 1.6 that $Z(f_n^*(x,\lambda), \mathscr{L}_n(S), (n+1)/2-s)$ is absolutely convergent if $-a_\lambda<\mathrm{Re}(s)<1$, and hence that $Z_+(f_n^*(x,\lambda), \mathscr{L}(S), (n+1)/2-s)$ is absolutely convergent for $-a_\lambda<\mathrm{Re}(s)$.

Proposition 1.7. *Assume that* λ *satisfies the condition* (1.2.7). *If* $\mathrm{Re}(s)>(n+1)/2$, *then the following identity holds*:

$$(1.2.9)\qquad Z^*(s,\tau_S^{(n)})=Z_+^*(s,\tau_S^{(n)})+\mu_{n,\lambda}\Big\{Z_+(f_n^*(x,\lambda), \mathscr{L}_n(S), (n+1)/2-s)$$
$$+\sum_{r=\nu}^{n-1}\frac{C_n v_{n-r}}{C_r C_{n-r}(s-(n+1-r)/2)}Z(f_r^*(x,\lambda), \mathscr{L}_r(S), n/2)\Big\}.$$

Proposition 1.7 can be proved quite in a similar manner as in the proof of Lemma 21, (iii) of [16]. For the convenience of the reader, we give a proof, which is based on Proposition 1.6.

Proof of Proposition 1.7. First we notice that the integral $Z(f_r^*(x,\lambda), \mathscr{L}_r(S), n/2)$ is absolutely convergent by Proposition 1.6. Using Lemma 1.3, we have, if $\mathrm{Re}(s)>(n+1)/2$,

$$(1.2.10)\quad Z^*(s,\tau_S^{(n)})=Z_+^*(s,\tau_S^{(n)})$$
$$+\mu_{n,\lambda}\int_{G_{\boldsymbol{R},+}^{(n)}/SL_n(\boldsymbol{Z}),\chi(g)\geqq 1}\chi(g)^{(n+1)/2-s}\sum_{x\in\mathscr{L}_n(S)}f_n^*(\rho(g)x,\lambda)d_ng.$$

We need the next lemma. For the proof, see [19] and also Lemma 17 in [16].

Lemma 1.8. *If* $\mathrm{Re}(s)>0$ *and* $t>0$, *then*

$$\int_{G_{\boldsymbol{R},+}^{(n)}/SL_n(\boldsymbol{Z}),\chi(g)\geqq 1/t}\chi(g)^{-s}d_ng=\frac{t^s}{s}v_n.$$

Let dp be the right invariant measure on $P_{n,+}^r$ normalized by

$$dp=\det(p_1)^{n-r}\det(p_2)^{-r}d_rp_1d_{n-r}p_2dq,$$

where $p=\begin{pmatrix}p_1&0\\0&p_2\end{pmatrix}\begin{pmatrix}1_r&q\\0&1_{n-r}\end{pmatrix}$ with $p_1\in G_{\boldsymbol{R},+}^{(r)}$, $p_2\in G_{\boldsymbol{R},+}^{(n-r)}$, $q\in M(r,n-r;\boldsymbol{R})$, and $dq=\prod_{1\leqq i\leqq r,1\leqq j\leqq n-r}dq_{ij}$. Let $\nu\leqq r\leqq n-1$. With the help of Lemma 1.5, Lemma 18 in [16], and Lemma 1.8, we get, if $\mathrm{Re}(s)>(n+1)/2$,

$$\int_{G_{\boldsymbol{R},+}^{(n)}/SL_n(\boldsymbol{Z}),\chi(g)\geqq 1}\chi(g)^{(n+1)/2-s}\sum_{x\in\mathscr{L}_n^{(r)}(s)}f_n^*(\rho(g)x,\lambda)d_ng$$
$$=2^{-1}\int_{G_{\boldsymbol{R},+}^{(n)}/SL_n(\boldsymbol{Z})\cap P_{n,+}^r,\chi(g)\geqq 1}\chi(g)^{(n+1)/2-s}\sum_{x\in\mathscr{L}_r^{(r)}(S)}f_n^*\left(\rho(g)\begin{pmatrix}x&0\\0&0\end{pmatrix},\lambda\right)d_ng$$
$$=\frac{C_n}{C_rC_{n-r}}\int_{\substack{P_{n,+}^r/SL_n(\boldsymbol{Z})\cap P_{n,+}^r\\ \chi(p_1)\chi(p_2)\geqq 1}}\chi(g)^{(n+1)/2-s}\sum_{x\in\mathscr{L}_r^{(r)}(S)}f_r^*(\rho(p_1)x,\lambda)dp$$
$$=\frac{C_nv_{n-r}}{C_rC_{n-r}(s-(n+1-r)/2)}Z(f_r^*(x,\lambda),\mathscr{L}_r(S),n/2).$$

Thus we see from (1.2.10) that the expression (1.2.9) holds for $\mathrm{Re}(s)>(n+1)/2$. We have completed the proof of Proposition 1.7.

We notice that $Z_+^*(s,\tau_S^{(n)})$ is a holomorphic function of s in the whole complex plane and that $Z_+(f_n^*(x,\lambda),\mathscr{L}_n(S),(n+1)/2-s)$ is holomorphic for $\mathrm{Re}(s)>-a_\lambda$. Taking the identities (1.2.2), (1.2.9) into account of, we see that $\xi_n^*(s,\tau_S^{(n)})$ can be continued to a meromorphic function in the region $\mathrm{Re}(s)>-a_\lambda$. Since λ can be taken sufficiently large, $\xi_n^*(s,\tau_S^{(n)})$ is extended to a meromorphic function in the whole complex plane. Moreover, we see that the poles of $\xi_n^*(s,\tau_S^{(n)})$ are located only at $s=(n+1-r)/2$ $(\nu\leqq r\leqq n)$ and that they are simple poles.

The next Proposition 1.9 corresponds to Corollary to Lemma 21 in [16].

Proposition 1.9. *Let λ satisfy the condition* (1.2.7). *The following functional equation holds:*

$$Z(f_n^*(x,\lambda),\mathscr{L}_n(S),(n+1)/2-s)$$
$$=\frac{\pi^{n(n+1)/2}p^{n(s-(n+1)/2)}C_n\gamma_n(\lambda+s-n-1)}{2(2\pi)^{ns}\gamma_n(\lambda-(n+1)/2)}\xi_n^*(s,\tau_S^{(n)})$$

(*the left side is defined at least if* $-a_\lambda<\mathrm{Re}(s)<1$).

Proof. We start from the identity (1.2.9). If $-a_\lambda<\mathrm{Re}((s)<1$, then we can get, similarly as in the proof of Proposition 1.7,

$$Z_+^*(s,\tau_S^{(n)})=\mu_{n,\lambda}\int_{G_{\mathbf{R},+}^{(n)}/SL_n(\mathbf{Z}),\chi(g)\leqq 1}\chi(g)^{(n+1)/2-s}\sum_{x\in\mathscr{L}_n^{(n)}(S)}f_n^*(\rho(g)x,\lambda)d_ng$$
$$-\mu_{n,\lambda}\sum_{r=\nu}^{n-1}\frac{C_nv_{n-r}}{C_rC_{n-r}(s-(n+1-r)/2)}Z(f_r^*(x,\lambda),\mathscr{L}_r(S),n/2),$$

from which we obtain

$$Z(s,\tau_S^{(n)})=\mu_{n,\lambda}Z(f_n^*(x,\lambda),\mathscr{L}_n(S),(n+1)/2-s).$$

Thus, by (1.2.2), we get the functional equation in Proposition 1.9,

Proposition 1.10. *The zeta function $\xi_n^*(s,\tau_S^{(n)})$ can be continued analytically to a meromorphic function in the whole complex plane which has simple poles only at $s=(n+1-r)/2$ ($\nu\leqq r\leqq n-1$). The residue of the pole at $s=(n+1-r)/2$ is given by*

$$\frac{2^{(n-r)(n+r+1)/2}v_{n-r}}{C_{n-r}}\xi_r^*((r+1-n)/2,\tau_S^{(r)})\qquad(\nu\leqq r\leqq n-1).$$

Proof. The former part of Proposition 1.10 has been verified. We have only to calculate the residues of the poles. We see easily from Proposition 1.7, Proposition 1.9 that the residue of the pole at $s=(n+1-r)/2$ ($\nu\leqq r\leqq n$) is given by

$$\frac{(2\pi)^{rn/2}C_nv_{n-r}\gamma_r(\lambda-(r+1+n)/2)}{2^{1+r(r+1)/2}p^{nr/2}\gamma_r(\lambda-(r+1)/2)}\mu_{n,\lambda}\xi_r^*((r+1-n)/2,\tau_S^{(r)}).$$

Thus, using the identity (1.2.2), we get the explicit residue at $s=(n+1-r)/2$ of $\xi_n^*(s,\tau_S^{(n)})$ as in Proposition 1.10.

For a primitive character $\chi \bmod p$, $B_{k,\chi}$ denotes the k-th generalized Bernoulli number given by

$$B_{k,\chi}=p^{k-1}\sum_{a=1}^{p-1}\chi(a)B_k(a/p). \tag{1.2.11}$$

In the case of $n=2$, one can derive some information of $L_2^*(s, \psi_{H,p})$, $L_2^*(s, \psi_{\det})$ from Proposition 1.10.

Theorem 1.11. *Let p be an odd prime and let ψ be the unique non-trivial quadracic character mod p. The L-functions $L_2^*(s, \psi_{H,p})$, $L_2^*(s, \psi_{\det})$ are continued analytically to meromorphic functions in the whole complex plane which are holomorphic except at $s=3/2$, 1. Then, $L_2^*(s, \psi_{H,p})$ has the unique simple poles at $s=1$ with residue $-B_{1,\psi}/p$, and $L_2^*(s, \psi_{\det})$ has simple poles at $s=3/2$, 1. The residue of $L_2^*(s, \psi_{\det})$ at $s=3/2$ (resp. $s=1$) is given by $\psi(-1)3^{-1}p^{-2}(p-1)\pi$ (resp. $-\psi(-1)2^{-1}p^{-1}(p-1)$).*

Proof. The former part is clear from Corollary to Proposition 1.2 and Proposition 1.10. Let $\mu \in \boldsymbol{Z}$ with $(\mu, p)=1$. An elementary computation shows that, in the case of $n=1$,

$$\xi_1^*(s, \tau_\mu^{(1)})=\frac{1}{2}\{\psi(\mu)\tau_\psi L(s, \psi)-(1-p^{1-s})\zeta(s)\}, \tag{1.2.12}$$

where $L(s, \psi)$ is the Dirichlet L-function associated with ψ. Since $L(0, \psi) = -B_{1,\psi}$, $\zeta(0)=-1/2$. We have

$$\xi_1^*(0, \tau_\mu^{(1)})=-\frac{1}{2}\psi(\mu)\tau_\psi B_{1,\psi}+\frac{1}{4}(1-p).$$

We see immediately from Proposition 1.10 that the residue of $\xi_2^*(s, \tau_\mu^{(2)})$ at the pole $s=1$ is given by $\xi_1^*(0, \tau_\mu^{(1)})$. It is known by [16, Theorem 2 or Corollary to Lemma 21] that $\xi_2^*(s)$ has simple poles only at $s=3/2$, 1 with residues $\pi/3$, -1 respectively. Since $\xi_2^*(s, \tau_\mu^{(2)})$ is holomorphic at $s=3/2$, so is $L_2^*(s, \psi_{H,p})$. Thus we get, by Corollary to Proposition 1.2, the assertion of Theorem 1.11.

Chapter II. Evaluation of special values of L-functions (the cases of degree two)

2.1. L-functions, and partial zeta functions

Let $\partial\mathscr{P}_2$ denote the boundary of the domain $\mathscr{P}_2$ in $V_{\boldsymbol{R}}^{(2)}$, that is, $\partial\mathscr{P}_2$ is the set of positive semi-definite symmetric matrices of size two. Let $\{W_1, W_2, \cdots, W_r\}$ be an r-tuple of elements in $\mathscr{P}_2 \cup \partial\mathscr{P}_2$ such that $W_1, W_2, \cdots, W_r$ are linearly independent over $\boldsymbol{R}$. Then, necessarily, $r \leqq 3$. For any r-tuple $\xi=(\xi_1, \cdots, \xi_r)$ of positive numbers, we define a partial zeta function $\zeta(s;\{W_1,\cdots,W_r\},\xi)$ as follows (see (2.2) of [14] and (1.4) of [9]):

(2.1.1) $$\zeta(s;\{W_1,\cdots,W_r\},\xi)=\sum_{m_1,\cdots,m_r=0}^{\infty}\det\Big(\sum_{j=1}^{r}(\xi_j+m_j)W_j\Big)^{-s}.$$

Let $C=C(W_1,\cdots,W_r)$ be a simplicial cone spanned by $W_1,\cdots,W_r$:

$$C=C(W_1,\cdots,W_r)=\Big\{\sum_{j=1}^{r}\lambda_j W_j\,|\,\lambda_j>0\ (1\leqq j\leqq r)\Big\}.$$

We assume that the cone $C(W_1,\cdots,W_r)$ is contained in $\mathscr{P}_2$. Then it is easily shown that the zeta functions $\zeta(s;\{W_1,\cdots,W_r\},\xi)$ is absolutely convergent for $\mathrm{Re}(s)>r/2$. For any subset M of $V_{\boldsymbol{R}}^{(2)}$, the zeta function $\zeta(s;C,M)$, if it converges absolutely, is defined by

(2.1.2) $$\zeta(s;C,M)=\sum_{T\in C\cap M}\det(T)^{-s}.$$

It is well-known that, as a fundamental domain of $\mathscr{P}_n$ under the usual action of the group $GL_n(\boldsymbol{Z})$, one can take the so-called Minkowski domain $\mathscr{R}_n$ of reduced matrices (see, for instance, § 9 of [12]). In the case of $n=2$, the domain $\mathscr{R}_2$ has a simple form:

$$\mathscr{R}_2=\Big\{\begin{pmatrix}y_1 & y_{12}\\ y_{12} & y_2\end{pmatrix}\,|\,0\leqq 2y_{12}\leqq y_1\leqq y_2,\ 0<y_1\Big\}.$$

We fix three special elements V_1, V_2, V_3 in $\mathscr{P}_2\cup\partial\mathscr{P}_2$ throughout Chapter II; put

$$V_1=\begin{pmatrix}1 & 0\\ 0 & 1\end{pmatrix},\quad V_2=\begin{pmatrix}1 & 1/2\\ 1/2 & 1\end{pmatrix},\quad \text{and}\quad V_3=\begin{pmatrix}0 & 0\\ 0 & 1\end{pmatrix}.$$

We set, for simplicity,

(2.1.3) $$\begin{aligned}&C_{123}=C(V_1,V_2,V_3), &&C_{ij}=C(V_i,V_j)\quad(1\leqq i<j\leqq 3),\\ &C_j=C(V_j)\quad(j=1,2),\end{aligned}$$

which are simplicial cones contained in $\mathscr{P}_2$. Then the domain $\mathscr{R}_2$ has the decomposition

(2.1.4) $$\mathscr{R}_2=C_{123}\cup C_{12}\cup C_{13}\cup C_{23}\cup C_1\cup C_2\qquad\text{(disjoint union)}.$$

For each cone C in (2.1.3) and any $Y\in C$, the order $\varepsilon^*(Y)$ of the group $\{U\in GL_2(\boldsymbol{Z})\,|\,UY{}^tU=Y\}$ takes the same value independent of Y belonging to C, and one can put

$$\varepsilon^*(C)=\varepsilon^*(Y)\qquad(Y\in C).$$

It is easily verified that

(2.1.5) $\varepsilon^*(C_{123})=2, \quad \varepsilon^*(C_{ij})=4 \ (1\leqq i<j\leqq 3), \quad \varepsilon^*(C_1)=8, \quad \varepsilon^*(C_2)=12.$

For a real number x, we denote by $\langle x\rangle$ the unique real number which satisfies $0<\langle x\rangle\leqq 1$ and $x-\langle x\rangle \in \boldsymbol{Z}$. Let p be an odd prime.

The aim of this section is to represent the L-functions $L_2^*(s, \psi_{H,p})$, $L_2^*(s, \chi_{\mathrm{det}})$ and the zeta function $\xi_2^*(s)$ as a finite linear combination of partial zeta functions (2.1.1).

First we shall discuss the L-function $L_2^*(s, \psi_{H,p})$. Let $\mathscr{M}(p)$ be the set given by (1.1.3). For each integer μ prime to p, let $L^*(\mu)$ be the set consisting of all elements $T \in L_2^*$ satisfying $\psi_{H,p}(T)=\psi(\mu)$. Then it immediately follows that

$$L^*(\mu)=\left\{T \in L_2^* \mid T\equiv\mu\begin{pmatrix}\alpha^2 & \alpha\gamma\\ \alpha\gamma & \gamma^2\end{pmatrix} \bmod p \text{ for some } (\alpha, \gamma) \in \mathscr{M}(p)\right\}$$

and that $L^*(\mu l^2)=L^*(\mu)$ for any integer l prime to p. For each $(\alpha, \lambda) \in \mathscr{M}(p)$ and for each integer μ prime to p, we put

(2.1.6) $$\xi_{\alpha,\gamma,\mu}=(\langle\mu(\alpha^2-2\alpha\gamma)/p\rangle, \langle 2\mu\alpha\gamma/p\rangle, \langle\mu(\gamma^2-\alpha^2)/p\rangle).$$

Let $\Xi_{H,\mu}$ be the set of all triples $\xi_{\alpha\gamma\mu}$: $\Xi_{H,\mu}=\{\xi_{\alpha,\gamma,\mu} \mid (\alpha, \gamma) \in \mathscr{M}(p)\}$. Then, $\mathscr{M}(p)/\{\pm 1\}$ corresponds to $\Xi_{H,\mu}$ bijectively by $\pm(\alpha, \gamma)\rightarrow\xi_{\alpha,\gamma,\mu}(=\xi_{-\alpha,-\gamma,\mu})$. For any integers i, j with $1\leqq i<j\leqq 3$, we set

$$\Xi_{H,\mu}^{(i,j)}=\{\xi=(\xi_1, \xi_2, \xi_3) \in \Xi_{H,\mu} \mid \xi_k=1\},$$

where k is the unique integer of 1, 2, 3 satisfying $\{i, j, k\}=\{1, 2, 3\}$. We notice that

(2.1.7) $$\begin{cases}\Xi_{H,\mu}^{(1,2)}=\{\xi_{\alpha,\gamma,\mu} \mid (\alpha, \gamma) \in \mathscr{M}(p),\ \alpha^2\equiv\gamma^2 \bmod p\},\\ \Xi_{H,\mu}^{(1,3)}=\{\xi_{\alpha,\gamma,\mu} \mid (\alpha, \gamma) \in \mathscr{M}(p),\ \alpha\gamma\equiv 0 \bmod p\},\\ \Xi_{H,\mu}^{(2,3)}=\{\xi_{\alpha,\gamma,\mu} \mid (\alpha, \gamma) \in \mathscr{M}(p),\ \alpha^2\equiv 2\alpha\gamma \bmod p\}.\end{cases}$$

For each cone C of the form (2.1.3), the zeta function $\zeta(s; C, L^*(\mu))$ given by (2.1.2) is absolutely convergent at least for $\mathrm{Re}(s)>3/2$.

Proposition 2.1. *The following expressions for the zeta functions* $\zeta(s; C, L^*(\mu))$ *hold:*

$$\zeta(s; C_{123}, L^*(\mu))=p^{-2s}\sum_{\xi\in\Xi_{H,\mu}}\zeta(s; \{V_1, V_2, V_3\}, \xi),$$

$$\zeta(s; C_{ij}, L^*(\mu))=p^{-2s}\sum_{(\xi_1,\xi_2,\xi_3)\in\Xi_{H,\mu}^{(i,j)}}\zeta(s; \{V_i, V_j\}, (\xi_i, \xi_j)) \qquad (1\leqq i<j\leqq 3),$$

$\zeta(s;\ C_1,\ L^*(\mu))=0,$

$$\zeta(s;\ C_2,\ L^*(\mu))=\begin{cases}0 & \cdots \textit{if } p>3,\\ p^{-2s}\zeta(s;\ \{V_2\},\ \langle\mu/p\rangle)\cdots & \textit{if } p=3.\end{cases}$$

Proof. Take $T\in C_{123}\cap L^*(\mu)$ and write $T=\sum_{j=1}^{3} m_j V_j$ with all $m_j\in \boldsymbol{N}$. If we take a pair $(\alpha,\gamma)\in\mathcal{M}(p)$ such that

$$T\equiv\mu\begin{pmatrix}\alpha^2 & \alpha\gamma\\ \alpha\gamma & \gamma^2\end{pmatrix} \bmod p, \tag{2.1.8}$$

then, m_j's satisfy the congruences:

$$\begin{cases}m_1\equiv\mu(\alpha^2-2\alpha\gamma) \bmod p, \qquad m_2\equiv 2\mu\alpha\gamma \bmod p,\\ m_3\equiv\mu(\gamma^2-\alpha^2) \bmod p.\end{cases}$$

Therefore, there exists a triple $l=(l_1,\ l_2,\ l_3)$, l_j being nonnegative integers, such that $(m_1,\ m_2,\ m_3)=p(\xi_{\alpha,\gamma,\mu}+l)$. Each $T\in C_{123}\cap L^*(\mu)$ determines a triple l uniquely and also $(\alpha,\gamma)\in\mathcal{M}(p)$ uniquely up to (± 1)-multiplication. Thus the first identity of Proposition 2.1 follows. Next, for instance, let $T\in C_{12}\cap L^*(\mu)$ and write $T=\sum_{j=1}^{2} m_j V_j$ $(m_j\in \boldsymbol{N})$. A pair (α,γ) can be so taken as in (2.1.8). Then the congruences $m_1\equiv\mu(\alpha^2-2\alpha\gamma) \bmod p$, $m_2\equiv 2\mu\alpha\gamma \bmod p$ follow, and necessarily, the relation $\alpha^2\equiv\gamma^2 \bmod p$ has to hold. Therefore, the identity for $\zeta(s;\ C_{12},\ L^*(\mu))$ immediately follows. Other identities left are quite similarly verified. So the proof is omitted.

q.e.d.

Let κ be a non-quadratic residue mod p as in Chapter I.

Proposition 2.2. *Let ψ be the unique non-trivial quadratic character* mod p. *Then we have*

$$L_2^*(s,\ \psi_{H,p})=\sum_{\mu}\psi(\mu)\Big\{\zeta(s;\ C_{123},\ L^*(\mu))+\frac{1}{2}\sum_{i<j}\zeta(s;\ C_{ij},\ L^*(\mu))$$
$$+\frac{1}{6}\delta_{p,3}\zeta(s;\ C_2,\ L^*(\mu))\Big\},$$

where μ is taken over 1 *and κ, and the summation $\sum_{i<j}$ indicates that i, j run over all integers with $1\leqq i<j\leqq 3$. Moreover, $\delta_{p,3}=0$ if $p\neq 3$, and $\delta_{p,3}=1$ if $p=3$.*

Proof. Only in this proof, we introduce the L-function $M_2^*(s,\ \psi_{H,p})$ which is quite similar to $L_2^*(s,\ \psi_{H,p})$. We set

$$M_2^*(s,\ \psi_{H,p})=\sum_{T\in L_{2,+}^*/GL_2(\boldsymbol{Z})}\psi_{H,p}(T)\varepsilon^*(T)^{-1}\det(T)^{-s}$$

where T is taken over $GL_2(Z)$-equivalence classes of positive definite half-integral symmetric matrices of size two, and $\varepsilon^*(T)$ is the order of the unit group $\{U \in GL_2(Z) \mid UT{}^tU=T\}$ of T. Then an elementary observation shows that $L_2^*(s, \psi_{H,p})=2M_2^*(s, \psi_{H,p})$. In view of the decomposition (2.1.4) of $\mathscr{R}_2$, we may take a disjoint union $\cup_C (C \cap L_2^*)$, C varying all simplicial cones in (2.1.3), as a complete set of $GL_2(Z)$-equivalence classes of all elements in $L_{2,+}^*$. Thus we get, with the help of the decomposition $L_2^*=L^*(1) \cup L^*(\kappa)$ (disjoint union),

$$L_2^*(s, \psi_{H,p})=2\sum_C \varepsilon^*(C)^{-1} \sum_\mu \psi(\mu)\zeta(s; C, L^*(\mu)),$$

which, together with (2.1.5) and Proposition 2.1, completes the proof of Proposition 2.2.

Let χ be a primitive character mod p. Secondly, we treat the L-function $L_2^*(s, \chi_{\det})$. For each integer δ prime to p, we set

$$M^*(\delta)=\{T \in L_2^* \mid \det(T) \equiv \delta \bmod p\}.$$

For each $T=\begin{pmatrix} t_1 & t_{12} \\ t_{12} & t_2 \end{pmatrix} \in L_2^*$, a triple ξ_T is defined by

$$\xi_T=(\langle (t_1-2t_{12})/p \rangle, \langle 2t_{12}/p \rangle, \langle (t_2-t_1)/p \rangle),$$

which depends only on T mod p. Let Ξ_δ be the set of all triples ξ_T, T varying all elements of L_2^*/pL_2^* with $\det(T) \equiv \delta \bmod p$: $\Xi_\delta=\{\xi_T \mid T \in L_2^* \bmod pL_2^*, \det(T) \equiv \delta \bmod p\}$. For integers i, j $(1 \leqq i < j \leqq 3)$, we set

$$\Xi_\delta^{(i,j)}=\{\xi=(\xi_1, \xi_2, \xi_3) \in \Xi_\delta \mid \xi_k=1\},$$

k being the the unique integer of 1, 2, 3 with $\{i, j, k\}=\{1, 2, 3\}$.

For any cone C of the form (2.1.3), the zeta function $\zeta(s; C, M^*(\delta))$, which is absolutely convergent for $\mathrm{Re}(s) > 3/2$, has the following expression.

Proposition 2.3. *Let δ be any integer prime to p. Then,*

$$\zeta(s; C_{123}, M^*(\delta))=p^{-2s} \sum_{\xi \in \Xi_\delta} \zeta(s; \{V_1, V_2, V_3\}, \xi),$$

$$\zeta(s; C_{ij}, M^*(\delta))=p^{-2s} \sum_{(\xi_1, \xi_2, \xi_3) \in \Xi_\delta^{(i,j)}} \zeta(s; \{V_i, V_j\}, (\xi_i, \xi_j)) \qquad (1 \leqq i < j \leqq 3).$$

For the cone C_j $(j=1, 2)$, we have

$$\sum_{\delta \not\equiv 0 \bmod p} \chi(\delta)\zeta(s; C_1, M^*(\delta))=\sum_{t_1 \not\equiv 0 \bmod p} \chi(t_1^2)\zeta(s; \{V_1\}, \langle t_1/p \rangle),$$

$$\sum_{\delta \not\equiv 0 \bmod p} \chi(\delta)\zeta(s;\, C_2,\, M^*(\delta)) = \sum_{t_{12} \not\equiv 0 \bmod p} \chi(3t_{12}^2)\zeta(s;\, \{V_2\},\, \langle 2t_{12}/p \rangle)$$

(note that, if $p=3$, then, the right side of the last equality coincides identically with zero).

Proposition 2.4. *Let χ be a primitive character* mod p. *Then,*

$$L_2^*(s, \chi_{\det}) = \sum_{\delta \not\equiv 0 \bmod p} \chi(\delta)\Big\{\zeta(s:\, C_{123},\, M^*(\delta)) + \frac{1}{2}\sum_{i<j}\zeta(s;\, C_{ij},\, M^*(\delta))$$
$$+\frac{1}{4}\zeta(s;\, C_1,\, M^*(\delta)) + \frac{1}{6}\,\zeta(s;\, C_2,\, M^*(\delta))\Big\}.$$

We omit the proofs of Proposition 2.3 and Proposition 2.4, which are quite similar to those of Proposition 2.1 and Proposition 2.2.

Finally, we obtain Proposition 2.5, which asserts that the zeta function $\xi_2^*(s)$ can be represented as a finite linear combination of partial zeta functions.

Proposition 2.5. *We have*

$$\xi_2^*(s) = \zeta(s;\, \{V_1, V_2, V_3\},\, (1, 1, 1)) + \frac{1}{2}\sum_{i<j}\zeta(s;\, \{V_i, V_j\},\, (1, 1))$$
$$+\frac{1}{4}\zeta(s;\, \{V_1\},\, 1) + \frac{1}{6}\zeta(s;\, \{V_2\},\, 1).$$

The proof of Proposition 2.5 is omitted as well.

2.2. Integral representations of partial zeta functions I

The aim of subsequent two sections is to obtain convenient expressions of partial zeta functions as integrals over contour paths, and then to evaluate special values of them at non-positive integers.

Let $\zeta(s;\, \{W_1, \cdots, W_r\},\, \xi)$ be a partial zeta function as is defined in (2.1.1). We assume that the cone $C(W_1, \cdots, W_r)$ is contained in $\mathscr{P}_2$. The following formula is well-known (see for instance [12] and also Lemma 1 of [14]):

$$\text{(2.2.1)}\quad \det(T)^{-s} = \frac{1}{\Gamma_2(s)}\int_{\mathscr{P}_2} \det(Y)^s e^{-\operatorname{tr}(TY)} dv(Y) \quad (T \in \mathscr{P}_2,\ \operatorname{Re}(s) > 1/2),$$

where we put

$$\Gamma_2(s) = \pi^{1/2}\Gamma(s)\Gamma(s-1/2) \quad \text{and} \quad dv(Y) = \det(Y)^{-3/2}\prod_{1 \le i \le j \le 2} dY_{ij}.$$

We set, for $t \in \boldsymbol{C}$, $x \in \boldsymbol{R}$,

$$\phi(t; x)=\frac{e^{tx}}{e^t-1},$$

which is the generating function of Bernoulli polynomials $B_k(x)$. Namely, the Laurent expansion at $t=0$ of $\phi(t; x)$ is given by

$$\phi(t; x)=\sum_{k=0}^{\infty}\frac{B_k(x)}{k!}t^{k-1} \qquad (|t|<2\pi). \tag{2.2.2}$$

By a usual argument which uses the formula (2.2.1), we get an expression of $\zeta(s; \{W_1, \cdots, W_r\}, \xi)$ for $\mathrm{Re}(s)>r/2$ by the integral taken over $\mathscr{P}_2$:

$$\zeta(s; \{W_1, \cdots, W_r\}, \xi)=\frac{1}{\Gamma_2(s)}\int_{\mathscr{P}_2}\det(Y)^s\prod_{j=1}^{r}\phi(\mathrm{tr}(W_jY); 1-\xi_j)dv(Y).$$

We set, for $\theta \in \boldsymbol{R}$, $k_\theta=\begin{pmatrix}\cos\theta & \sin\theta\\ -\sin\theta & \cos\theta\end{pmatrix}$.

Following Satake [14, 2.2], we make a change of variables $Y\to(t, u, \theta)$ with $Y=tk_\theta\begin{pmatrix}u & 0\\ 0 & 1\end{pmatrix}{}^tk_\theta$ $(0<t, 0<u\leqq 1, 0\leqq\theta\leqq\pi)$. We thus obtain, using the relation $dv(Y)=t^{-1}u^{-3/2}(1-u)dtdud\theta$,

$$\begin{aligned}\zeta(s; \{W_1, \cdots, W_r\}, \xi)&=\frac{1}{\Gamma_2(s)}\\ &\times\int_0^\infty dt\int_0^1 du\int_0^\pi d\theta\cdot t^{2s-1}u^{s-3/2}(1-u)\Phi((t, u, \theta); \{W_1, \cdots, W_r\}, \xi),\end{aligned} \tag{2.2.3}$$

where we put

$$\Phi((t, u, \theta), \{W_1, \cdots, W_r\}, \xi)=\prod_{j=1}^{r}\phi(t\lambda((u, \theta), W_j); 1-\xi_j)$$

and

$$\lambda((u, \theta), W)=\mathrm{tr}\left(Wk_\theta\begin{pmatrix}u & 0\\ 0 & 1\end{pmatrix}{}^tk_\theta\right) \qquad \text{for any } W\in\mathscr{P}_2\cup\partial\mathscr{P}_2.$$

The following condition (2.2.4) on vectors W_j;

$$\text{all } W_j\ (1\leqq j\leqq r) \text{ are contained in } \mathscr{P}_2, \tag{2.2.4}$$

being imposed, then, the integral (2.2.3) has been studied in a full generality by Satake [14] and by Kurihra [9] in a special but significant case.

However, for our aim to evaluate special values of L-functions discussed in 2.1, it is indispensable to get rid of the condition (2.2.4). In view of Proposition 2.1, Proposition 2.3, and Proposition 2.5, we have only to consider the cases in which, with respect to an r-tuple $\{W_1, \cdots, W_r\}$, the vectors $W_1, \cdots, W_{r-1}$ are all in $\mathscr{P}_2$, and W_r coincides with the special vector V_3 in $\partial\mathscr{P}_2$.

Now we set

$$\psi(t; x) = \phi(t; x) - \frac{1}{t},$$

which is a holomorphic function of t in the region $|t| < 2\pi$. Let $\{W_1, \cdots, W_{r-1}, V_3\}$ ($r=2$ or 3) be an r-tuple of vectors in $\mathscr{P}_2 \cup \partial\mathscr{P}_2$ such that $W_1, \cdots, W_{r-1}$ are all in $\mathscr{P}_2$. We set, for an r-tuple $\xi = (\xi_1, \cdots, \xi_r)$ of positive numbers,

$$\begin{aligned}\Phi_P((t, u, \theta); \{W_1, \cdots, W_{r-1}, V_3\}, \xi)\\ = \prod_{j=1}^{r-1} \phi(t\lambda((u, \theta), W_j); 1-\xi_j)\psi(t\lambda((u, \theta), V_3); 1-\xi_3),\end{aligned}$$

and, for an $(r-1)$-tuple $\xi' = (\xi_1, \cdots, \xi_{r-1})$,

$$\begin{aligned}\Phi_S((t, u, \theta); \{W_1, \cdots, W_{r-1}, V_3\}, \xi')\\ = \frac{1}{\lambda((u, \theta), V_3)} \prod_{j=1}^{r-1} \phi(t\lambda((u, \theta), W_j); 1-\xi_j).\end{aligned}$$

Moreover, we set

$$\begin{aligned}(2.2.5)\qquad \zeta_P(s; \{W_1, \cdots, W_{r-1}, V_3\}, \xi)\\ = \frac{1}{\Gamma_2(s)} \int_0^\infty dt \int_0^1 du \int_0^\pi d\theta \cdot t^{2s-1} u^{s-3/2}(1-u)\\ \times \Phi_P((t, u, \theta); \{W_1, \cdots, W_{r-1}, V_3\}, \xi),\end{aligned}$$

$$\begin{aligned}(2.2.6)\qquad \zeta_S(s; \{W_1, \cdots, W_{r-1}, V_3\}, \xi')\\ = \frac{1}{\Gamma_2(s)} \int_0^\infty dt \int_0^1 du \int_0^\pi d\theta \cdot t^{2s-1} u^{s-3/2}(1-u)\\ \times \Phi_S((t, u, \theta); \{W_1, \cdots, W_{r-1}, V_3\}, \xi')\end{aligned}$$

(the letter P (resp. S) is used to intend that the function given in (2.2.5) (resp. in (2.2.6)) is a principal (resp. singular) part of $\zeta(s; \{W_1, \cdots, W_{r-1}, V_3\}, \xi)$).

The integrals in (2.2.5), (2.2.6) are absolutely convergent at least for $\mathrm{Re}(s) > 3/2$. Obviously,

$$\zeta(s;\{W_1,\cdots,V_3\},\xi)=\zeta_P(s;\{W_1,\cdots,V_3\},\xi)+\zeta_S(s;\{W_1,\cdots,V_3\},\xi').$$

For our later use, we prepare some symbols. For a positive number ε, let $I_\varepsilon(\infty)$ (resp. $I_\varepsilon(1)$) be the contour path consisting of the oriented half line $(+\infty,\varepsilon)$ (resp. $(1,\varepsilon)$), a counterclockwise circle of radius ε around the origin, and the oriented half line $(\varepsilon,+\infty)$ (resp. $(\varepsilon,1)$). We would like to modify the integral in (2.2.3) directly into the integral taken over contour paths $I_\varepsilon(\infty)$ and $I_\varepsilon(1)$ (for a small ε) with respect to t and u, respectively. However, the function $\phi(t\lambda((u,\theta),V_3);1-\xi_3)$ has serious singularities as a function of t and u on the paths $I_\varepsilon(\infty)$, $I_\varepsilon(1)$, because of the form of $\lambda((u,\theta),V_3)=u\sin^2\theta+\cos^2\theta$, and therefore such a modification cannot be done easily. To avoid the difficulties derived from such singularities, we divide zeta function $\zeta(s;W_1,\cdots,W_{r-1},V_3\},\xi)$ into two parts as above. In the rest of this section, we shall mainly discuss the function $\zeta_P(s:\{W_1,\cdots,W_{r-1},V_3\},\xi)$ and its expression by an integral over contour paths. The singular part $\zeta_S(s;\{W_1,\cdots,W_{r-1},V_3\},\xi)$ will be dealed with in the next section.

For a positive number δ, we denote by $D_\delta(\infty)$ and $D_\delta(1)$ the regions given as follows:

$$D_\delta(\infty)=\{z\in \boldsymbol{C}\,|\,|z|<\delta\}\cup\{z\in \boldsymbol{C}\,|\,\mathrm{Re}\,(z)>0 \text{ and } |\mathrm{Im}\,(z)|<\delta\},$$
$$D_\delta(1)=D_\delta(\infty)\cap\{z\in \boldsymbol{C}\,|\,|z|\leqq 1\}.$$

If W is in $\mathscr{P}_2$, we can take a positive constant a, b satisfying

$$a1_2<W<b1_2. \tag{2.2.7}$$

We may write $\lambda((u,\theta),W)=\alpha_1u+\alpha_2$ with $a<\alpha_1,\alpha_2<b$. It then follows that

$$\begin{aligned}&|\lambda((u,\theta),W)|<b(1+|u|),\\ &\mathrm{Re}\,(\lambda((u,\theta),W))>a-b\delta \quad \text{if } \mathrm{Re}\,(u)>-\delta \quad (\delta>0).\end{aligned} \tag{2.2.8}$$

We need some analytic properties of the functions $\phi(t\lambda((u,\theta),W);1-\xi)$ and $\psi(t\lambda((u,\theta),V_3);1-\xi)$ $(\xi>0)$.

Lemma 2.6. *Suppose that $W\in\mathscr{P}_2$ satisfies the condition* (2.2.7). *Let $\xi>0$ and $0<\delta<a/b$.*

(i) *If $|t|<\pi/2b$, and $u\in D_\delta(1)$, then, $t\phi(t\lambda((u,\theta),W);1-\xi)$, which is a holomorphic function of (t,u) for each θ in that region of (t,u), has the power series expansion with respect to t:*

$$t\phi(t\lambda((u,\theta),W);1-\xi)=\frac{1}{\lambda((u,\theta),W)}+\sum_{k=1}^{\infty}\frac{B_k(1-\xi)}{k!}\{\lambda((u,\theta),W)\}^{k-1}t^k,$$

(ii) *If* $t>0$ *and* $0\leqq u\leqq 1$, *then,*

$$t\phi(t\lambda((u,\theta),W);1-\xi)<\frac{te^{-t\xi a}}{1-e^{-ta}}.$$

Proof. If $|t|<\pi/2b$, $u\in D_\delta(1)$, then, we get, by (2.2.8),

$$|t\lambda((u,\theta),W)|<\pi(1+|u|)/2<2\pi.$$

Therefore, the Laurent expansion (2.2.2) implies the assertion (i). The assertion (ii) is immediately derived from the inequality $\lambda((u,\theta),W)>a$ $(0\leqq u\leqq 1)$.

Lemma 2.7. *Let* $\xi>0$ *and* $0<\delta<1$. *Then* $\psi(t\lambda((u,\theta),V_3);1-\xi)$ *is a holomorphic function of* t, u *for each* θ *in the region* $\{(t,u)\,|\,|t|<2\pi,\ u\in D_\delta(1)\}$, *and has a Taylor expansion with respect to* t:

$$\psi(t\lambda((u,\theta),V_3);1-\xi)=\sum_{k=1}^{\infty}\frac{B_k(1-\xi)}{k!}(u\sin^2\theta+\cos^2\theta)^{k-1}t^{k-1}.$$

Proof. Recalling that $\lambda((u,\theta),V_3)=u\sin^2\theta+\cos^2\theta$, and moreover that $|t\lambda((u,\theta),V_3)|<2\pi$ if $|t|<2\pi$, $u\in D_\delta(1)$, we immediately get the assertion of Lemma 2.7.

The function $\psi(t;1-\xi)$ has a preferable property which will be used in the proof of Proposition 2.9.

Lemma 2.8. *Let* $\xi>0$. *There exist positive constants* M_k $(k=1,2,\cdots)$ *independent of* t *such that, if* $0\leqq t<+\infty$,

$$|\psi^{(k)}(t;1-\xi)|<M_k,$$

where $\psi^{(k)}(t;1-\xi)$ *denotes the k-th derivative of* $\psi(t;1-\xi)$ *as a function of* t.

We omit the proof of Lemma 2.8, which is an easy exercise of differential calculus.

It is easy to see from (2.2.8) and Lemma 2.6 that, if δ is taken sufficiently small, then, $t\phi(t\lambda((u,\theta),W);1-\xi)$ $(W\in\mathscr{P}_2,\ \xi>0)$ is holomorphic as a function of t, u for each $\theta\in\boldsymbol{R}$ in the region $D_\delta(\infty)\times D_\delta(1)$. Moreover, taking (2.2.8), Lemma 2.6, and Lemma 2.7 into account, we see without difficulty that the integral

$$t^2\int_0^{\pi}\Phi_P((t,u,\theta);\{W_1,\cdots,W_{r-1},V_3\},\xi)d\theta$$

indicates a holomorphic function of t, u in the region $D_\delta(1)\times D_\delta(1)$ for a sufficiently small δ. We notice here that the range of t is the region $D_\delta(1)$ (not $D_\delta(\infty)$).

To define the function $t^s=e^{s\log t}$, we take the branch of $\log t$ with $0<\arg t<2\pi$.

Proposition 2.9. *The function $\zeta_P(s;\{W_1,\cdots,W_{r-1},V_3\},\xi)$ is analytically continued to a meromorphic function in the whole complex plane which is holomorphic at $s=1-m$ $(m=1,2,\cdots)$. Moreover, the special value at $s=1-m$ is given by*

$$\begin{aligned}&\zeta_P(1-m;\{W_1,\cdots,W_{r-1},V_3\},\xi)\\&\quad=C(m)\int_{\Gamma_\varepsilon}dt\int_{I_\varepsilon(\infty)}du\int_0^\pi d\theta t^{1-2m}u^{-m-1/2}(1-u)\\&\qquad\times\phi_P((t,u,\theta);\{W_1,\cdots,W_{r-1},V_3\},\xi),\end{aligned}$$

where $C(m)=(2m-1)!/2^{2m+2}\pi^2 i$ and Γ_ε denotes a circle of radius ε around the origin oriented counterclockwise, ε being taken sufficiently small.

Proof. We set, only in the proof of this proposition,

$$f(t,u,\theta)=\Phi_P((t,u,\theta);\{W_1,\cdots,W_{r-1},V_3\},\xi).$$

We divide the integral in (2.2.5) into two parts by the range of the variable t. We set

$$\zeta_P(s;\{W_1,\cdots,W_{r-1},V_3\},\xi)=\frac{1}{\Gamma_2(s)}(I_1(s)+I_2(s)),$$

where

$$I_1(s)=\int_0^1 dt\int_0^1 du\int_0^\pi d\theta\cdot t^{2s-1}u^{s-3/2}(1-u)f(t,u,\theta),$$

$$I_2(s)=\int_1^\infty dt\int_0^1 du\int_0^\pi d\theta\cdot t^{2s-1}u^{s-3/2}(1-u)f(t,u,\theta).$$

It is easy to see from the remark just before the statement of Proposition 2.9 that $I_1(s)$ has the following expression by an integral over contour paths:

$$\begin{aligned}I_1(s)=&\frac{1}{(e[2s]-1)(e[s-3/2]-1)}\int_{I_\varepsilon(1)}dt\int_{I_\varepsilon(1)}du\int_0^\pi d\theta\\&\times t^{2s-1}u^{s-3/2}(1-u)f(t,u,\theta),\end{aligned}\tag{2.2.9}$$

where ε is taken sufficiently small. Since the integral in (2.2.9) indicates an entire function of s, the function $I_1(s)$ can be continued analytically to a meromorphic function in the whole complex plane. Thus we easily obtain

$$(2.2.10)\quad \lim_{s\to 1-m}\frac{I_1(s)}{\Gamma_2(s)}=C(m)\int_{\Gamma_\varepsilon}dt\int_{I_\varepsilon(1)}du\int_0^\pi d\theta\cdot t^{1-2m}u^{-m-1/2}(1-u)f(t,u,\theta)$$
$$(m=1,2,\cdots).$$

On the one hand, we see easily from Lemma 2.8 and so on that the function $f(t, u, \theta)$ is a C^∞-function of (t, u, θ) in the region $(0, +\infty)\times[0, 1]\times[0, \pi]$, and especially that the partial derivatives $(\partial^k f/\partial u^k)(t, u, \theta)$ $(k=0, 1, 2, \cdots)$ are bounded in the region $[1, +\infty]\times[0, 1]\times[0, \pi]$ of (t, u, θ). We set, for $\operatorname{Re}(s)>0$,

$$F(s;(t,\theta))=\int_0^1 u^{s-1}f(t,u,\theta)du.$$

Then we have

$$(2.2.11)\quad I_2(s)=\int_1^\infty dt\int_0^\pi d\theta\cdot t^{2s-1}\{F(s-1/2;(t,\theta))-F(s+1/2;(t,\theta))\}.$$

Using the integration by parts recurrently, we obtain

$$(2.2.12)\quad F(s;(t,\theta))=\sum_{j=0}^{m-1}\frac{(-1)^j}{s(s+1)\cdots(s+j)}\frac{\partial^j f}{\partial u^j}(t,1,\theta)$$
$$+\frac{(-1)^m}{s(s+1)\cdots(s+m-1)}\int_0^1 u^{s+m-1}\frac{\partial^m f}{\partial u^m}(t,u,\theta)du$$
$$(\operatorname{Re}(s)>-m).$$

Since any $m\in\boldsymbol{N}$ can be taken, it follows from (2.2.11), (2.2.12) that $I_2(s)$ can be continued to an entire function of s. Thus we get

$$(2.2.13)\quad \left[\frac{I_2(s)}{\Gamma_2(s)}\right]_{s=1-m}=0.$$

The analytic continuation of $\zeta_P(s;\{W_1,\cdots,W_{r-1},V_3\},\xi)$ immediately follows from those of $I_1(s)$, $I_2(s)$. The last assertion of Proposition 2.9 is derived from (2.2.10), (2.2.13). q.e.d.

The following proposition which deals with partial zeta functions whose edge vectors are all in $\mathscr{P}_2$ is only a small part of the results obtained by Satake in [14].

Proposition 2.10. *Let all vectors W_j $(1 \leqq j \leqq r)$, which are linearly independent over* $\boldsymbol{R}$, *be in* $\mathscr{P}_2$. *Then the zeta function* $\zeta(s; \{W_1, \cdots, W_r\}, \xi)$ *has the analytic continuation to a meromorphic function of s in the whole complex plane, which is holomorphic at* $s=1-m$ $(m=1, 2, \cdots)$. *The special value at* $s=1-m$ *is then given by*

$$\zeta(1-m; \{W_1, \cdots, W_r\}, \xi) = C(m) \int_{\Gamma_\varepsilon} dt \int_{I_\varepsilon(1)} du \int_0^\pi d\theta \cdot t^{1-2m} u^{-m-1/2}(1-u) \times \Phi((t, u, \theta); \{W_1, \cdots, W_r\}, \xi).$$

Proof. For the completeness, we give a proof. Lemma 2.6 implies that the integral

$$t^2 \int_0^\pi \Phi((t, u, \theta); \{W_1, \cdots, W_r\}, \xi) d\theta$$

indicates a holomorphic function of (t, u) in some region $D_\delta(\infty) \times D_\delta(1)$. Thus one obtains, for a sufficiently small ε,

$$\zeta(s; \{W_1, \cdots, W_r\}, \xi) = \frac{1}{\Gamma_2(s)(e[2s]-1)(e[s-3/2]-1)} \int_{I_\varepsilon(\infty)} dt \int_{I_\varepsilon(1)} du \int_0^\pi d\theta \times t^{2s-1} u^{s-3/2}(1-u) \Phi((t, u, \theta); \{W_1, \cdots, W_r\}, \xi).$$

Since the integral in the right side of the equality is absolutely convergent, this identity gives the analytic continuation of $\zeta(s; \{W_1, \cdots, W_r\}, \xi)$ to a meromorphic function of s in the whole complex plane. Substituting $s = 1-m$, we get the identity in Proposition 2.10.

In view of Proposition 2.1, Proposition 2.3, and Proposition 2.5, we need only partial zeta functions of the form

$$\zeta(s; \{V_1, V_2, V_3\}, (\xi_1, \xi_2, \xi_3)), \quad \zeta(s; \{V_i, V_j\}, (\xi_i, \xi_j)) \quad (1 \leqq i < j \leqq 3)$$
$$\zeta(s; \{V_j\}, \xi) \quad (j=1, 2, \text{ and } \xi > 0).$$

Now we discuss the evaluation of $\zeta_P(1-m; \{V_1, V_2, V_3\}, (\xi_1, \xi_2, \xi_3))$ $(m \in N)$ and so on as a continuation of Proposition 2.9, Proposition 2.10. For each triple (k_1, k_2, k_3) of integers such that $k_1, k_2 \geqq -1$, $k_3 \geqq 0$, and $k_1 + k_2 + k_3 = 2(m-1)$, we define a number $\Lambda_{(k_1, k_2, k_3)}$ by putting

$$(2.2.14) \quad \Lambda_{(k_1, k_2, k_3)} = \int_{I_\varepsilon(1)} du \int_0^\pi d\theta \cdot u^{-m-1/2}(1-u) \prod_{j=1}^3 \lambda((u, \theta), V_j)^{k_j},$$

where the integral in the right side is independent of the choice of a small positive number ε.

Remark. The numbers $\Lambda_{(k_1,k_2,k_3)}$ essentially coincide with $N(1-m; k_1+1, k_2+1, k_3+1; v_1, v_2, v_3)$ in Kurihara [9], though his definition is different from ours.

Proposition 2.11. *Let* $\xi=(\xi_1, \xi_2, \xi_3)$ *be a triple of positive numbers, and* $m \in \boldsymbol{N}$. *Then we have*

$$\zeta_P(1-m; \{V_1, V_2, V_3\}, \xi)=-2\pi i C(m) \sum_{k_1,k_2,k_3}{}' \left\{\prod_{j=1}^{3} \frac{B_{k_j+1}(\xi_j)}{(k_j+1)!}\right\} \Lambda_{(k_1,k_2,k_3)},$$

where k_1, k_2, k_3 *run over all integers satisfying the conditions* $k_1, k_2 \geqq -1$, $k_3 \geqq 0$, *and* $k_1+k_2+k_3=2(m-1)$.

Proof. We take δ sufficiently small so that Lemma 2.6 for V_j $(j=1, 2)$ and Lemma 2.7 hold. Then we get the following power series expansion, if $|t|<\delta$, $u \in D_\delta(1)$,

$$\Phi_P((t, u, \theta); \{V_1, V_2, V_3\}, \xi)=\sum_{k_j} \prod_{j=1}^{3} \left\{\frac{B_{k_j+1}(1-\xi_j)}{(k_j+1)!} \lambda((u, \theta), V_j)^{k_j}\right\} \cdot t^{k_1+k_2+k_3},$$

where k_1, k_2, k_3 run over all integers satisfying $k_1, k_2 \geqq -1$ and $k_3 \geqq 0$. Applying Proposition 2.9, we obtain the expression for $\zeta_P(1-m; \{V_1, V_2, V_3\}, \xi)$ in Proposition 2.11.

Quite in the same manner as in the proof of Proposition 2.11, one can evaluate special values at $s=1-m$ of the functions $\zeta_P(s; \{V_j, V_3\}, (\xi_j, \xi_3))$ $(j=1, 2)$, $\zeta(s; \{V_1, V_2\}, (\xi_1, \xi_2))$ and $\zeta(s; \{V_j\}, \xi)$ $(j=1, 2, \xi>0)$. So we omit the proof of the following proposition.

Proposition 2.12. *Let* ξ, ξ_j $(j=1, 2, 3)$ *be positive numbers and* $m \in \boldsymbol{N}$. *Then the following expressions hold.*

(a) $$\zeta_P(1-m; \{V_1, V_3\}, (\xi_1, \xi_3))=2\pi i C(m) \sum_{k_1,k_3}{}' \frac{B_{k_1+1}(\xi_1) B_{k_3+1}(\xi_3)}{(k_1+1)!(k_3+1)!} \Lambda_{(k_1,0,k_3)},$$

where k_1, k_3 *run over all integers with* $k_1 \geqq -1$, $k_3 \geqq 0$, $k_1+k_3=2(m-1)$.

(b) $$\zeta_P(1-m; \{V_2, V_3\}, (\xi_2, \xi_3))=2\pi i C(m) \sum_{k_2,k_3}{}' \frac{B_{k_2+1}(\xi_2) B_{k_3+1}(\xi_3)}{(k_2+1)!(k_3+1)!} \Lambda_{(0,k_2,k_3)},$$

where k_2, k_3 *run over all integers with* $k_2 \geqq -1$, $k_3 \geqq 0$, $k_2+k_3=2(m-1)$.

(c) $$\zeta(1-m; \{V_1, V_2\}, (\xi_1, \xi_2))=2\pi i C(m) \sum_{k_1,k_2}{}' \frac{B_{k_1+1}(\xi_1) B_{k_2+1}(\xi_2)}{(k_1+1)!(k_2+1)!} \Lambda_{(k_1,k_2,0)},$$

where k_1, k_2 *run over all integers with* $k_1, k_2 \geqq -1$, $k_1+k_2=2(m-1)$.

(d) $$\zeta(1-m;\{V_1\},\xi)=-2\pi i C(m)\frac{B_{2m-1}(\xi)}{(2m-1)!}\Lambda_{(2m-2,0,0)}.$$

(e) $$\zeta(1-m;\{V_2\},\xi)=-2\pi i C(m)\frac{B_{2m-1}(\xi)}{(2m-1)!}\Lambda_{(0,2m-2,0)}.$$

To complete the evaluation of special values of zeta functions above, we have to study some properties of the numbers $\Lambda_{(k_1,k_2,k_3)}$.

Proposition 2.13. *Let k_1, k_2, k_3 be integers with $k_1, k_2 \geqq -1$, $k_3 \geqq 0$, and $k_1+k_2+k_3=2(m-1)$ ($m \in N$). If k_1, k_2, k_3 satisfy one of the following three conditions, then, $(1/\pi)\Lambda_{(k_1,k_2,k_3)}$ is a rational number.*

(i) *k_1, k_2, k_3 are non-negative integers,*

(ii) *$k_1=-1$ and k_2 is a positive odd integer,*

(iii) *$k_2=-1$ and k_1, k_3 are non-negative integers.*

Proof. A straightforward computation shows that

(2.2.15) $$\begin{cases}\lambda((u,\theta),V_1)=1+u, \quad \lambda((u,\theta),V_2)=1+u+(1-u)\sin\theta\cos\theta,\\ \lambda((u,\theta),V_3)=u\sin^2\theta+\cos^2\theta.\end{cases}$$

Changing the variable by $\cot\theta=x$ in (2.2.14), we obtain

(2.2.16) $$\frac{1}{2\pi}\Lambda_{(k_1,k_2,k_3)}=\int_{I_\varepsilon(1)}u^{-m-1/2}(1-u)P_{k_1,k_2,k_3}(u)du,$$

where ε is a sufficiently small number and

(2.2.17) $$P_{k_1,k_2,k_3}(u) =\frac{1}{2\pi}(1+u)^{k_1}\int_R\left\{1+u+(1-u)\frac{x}{1+x^2}\right\}^{k_2}\left(\frac{x^2+u}{1+x^2}\right)^{k_3}\frac{dx}{1+x^2}.$$

The computation of the integral in (2.2.17) easily shows that, if k_1, k_2, k_3 satisfy either of the conditions (i), (ii), then, $P_{k_1,k_2,k_3}(u)$ is a polynomial of u with rational coefficients. On the one hand,

(2.2.18) $$\int_{I_\varepsilon(1)}u^{k-1/2}du=-\frac{4}{2k+1} \qquad \text{for any } k \in Z.$$

Therefore, according to (2.2.16), the value $(1/2\pi)\Lambda_{(k_1,k_2,k_3)}$ is a rational number, if each triple (k_1, k_2, k_3) satisfies either of the conditions (i), (ii).

Suppose that $k_2=-1$ and $k_1, k_3 \geqq 0$. Set

$$Q(u)=3u^2+10u+3,$$

and

$$\omega(u)=\frac{-(1-u)+i\sqrt{Q(u)}}{2(1+u)} \qquad (0\leqq u\leqq 1).$$

Then, $\omega(u)$, $\overline{\omega(u)}$ (the complex conjugate of $\omega(u)$) are the distinct roots of the quadratic equation: $(1+u)x^2+(1-u)x+1+u=0$. We write simply ω for $\omega(u)$. Applying the residue theorem in computing the integral in (2.2.17), we obtain

$$\text{(2.2.19)} \quad P_{k_1,-1,k_3}(u)=i\frac{(1+u)^{k_1}}{(1+u)(\omega-\bar{\omega})}\left(1-\frac{1-u}{1+\omega^2}\right)^{k_3}+R_{k_1,k_3}(u),$$

where we put

$$R_{k_1,k_3}(u)=i(1+u)^{k_1}\operatorname*{Res}_{x=i}\{(1+u)x^2+(1-u)x+(1+u)\}^{-1}\left(1-\frac{1-u}{1+x^2}\right)^{k_3}.$$

Since the residue at $x=i$ of the function

$$\{(1+u)x^2+(1-u)x+1+u\}^{-1}\left(\frac{1-u}{1+x^2}\right)^l \qquad (l\in \boldsymbol{Z},\ l>0),$$

is a polynomial of u with coefficients in the Gaussian field $\boldsymbol{Q}(i)$, so is $R_{k_1,k_3}(u)$. Therefore, the real part of $R_{k_1,k_3}(u)$ is a polynomial of u with rational coefficients. An elementary calculation shows that

$$(1+u)(\omega-\bar{\omega})=i\sqrt{Q(u)}, \quad 1+\omega^2=-\frac{1-u}{1+u}\omega, \quad \omega\bar{\omega}=1, \quad \text{and}$$

$$1-\frac{1-u}{1+\omega^2}=\frac{1+u-i\sqrt{Q(u)}}{2}.$$

Then, we get, by (2.2.19),

$$\text{(2.2.20)} \quad P_{k_1,-1,k_3}(u)=2^{-k_3}(1+u)^{k_1}\{1+u-i\sqrt{Q(u)}\}^{k_3}Q(u)^{-1/2}+R_{k_1,k_3}(u).$$

The function $u^{-m+1}(1+u)^{2m-2-2j}Q(u)^j$ $(0\leqq j\leqq m-1)$ is invariant under the transformation $u\to 1/u$, and consequently, is a polynomial of $(u+1/u)$ with degree $m-1$. Thus one can write

$$\text{(2.2.21)} \quad u^{-m+1}(1+u)^{2m-2-2j}Q(u)^j=\sum_{k=0}^{m-1}b_{j,m,k}(u^k+u^{-k}) \qquad (0\leqq j\leqq m-1)$$

with some $b_{j,m,k}\in\boldsymbol{Q}$. We set, for a sufficiently small $\varepsilon>0$,

$$g(s)=\int_{I_\varepsilon(1)} u^{s-1/2}Q(u)^{-1/2}du,$$

where the branch of $Q(u)^{1/2}$ is so taken that $Q(u)^{1/2}>0$, if $u\in \boldsymbol{R}$. The integral in the right side is independent of the choice of ε and converges for arbitrary $s\in \boldsymbol{C}$. Accordingly, $g(s)$ stands for an entire function of s. We define a sequence $\{\alpha_n\}$ by putting

$$\alpha_n=g(n)-g(-n) \qquad (n=0,1,2,\cdots).$$

Lemma 2.14. *The sequence $\{\alpha_n\}$ satisfies the recursive formula*

$$3(n-1/2)\alpha_n+10(n-1)\alpha_{n-1}+3(n-3/2)\alpha_{n-2}=-16 \qquad (n\geqq 2)$$

with $\alpha_0=0$, $\alpha_1=-16/3$. Consequently, all α_n are rational numbers.

Proof. We begin with, if $\mathrm{Re}\,(s)>-1/2$.

$$\begin{aligned}3g(s+1)+5g(s)&=\int_{I_\varepsilon(1)} u^{s-1/2}(3u+5)Q(u)^{-1/2}du\\&=(e[s-1/2]-1)\int_0^1 u^{s-1/2}\frac{d}{du}Q(u)^{1/2}du.\end{aligned}$$

The integration by parts then implies that, if $\mathrm{Re}\,(s)>1/2$,

$$3g(s+1)+5g(s)=-4(1+e[s])-(s-1/2)\int_{I_\varepsilon(1)} u^{s-3/2}Q(u)^{1/2}du.$$

Writing $Q(u)^{1/2}=Q(u)\cdot Q(u)^{-1/2}$, we get the following functional equation:

$$\text{(2.2.22)}\quad 3(s+1/2)g(s+1)+10sg(s)+3(s-1/2)g(s-1)=-4(1+e[s]),$$

which is valid for arbitrary $s\in \boldsymbol{C}$ by the analytic continuation. Putting $s=0$, one gets $\alpha_1=g(1)-g(-1)=-16/3$. Moreover, if we substitute $s=m-1$ and $s=1-m$, respectively in (2.2.22), and add the both equalities so obtained, then we have the recursive formula in Lemma 2.14.

We continue the proof of Proposition 2.13. We see from (2.2.17) that $P_{k_1,-1,k_3}(u)$ is real valued, if $0\leqq u\leqq 1$, and hence from (2.2.20) that $P_{k_1,-1,k_3}(u)$ is a polynomial of u with rational coefficients plus a $\boldsymbol{Q}$-linear sum of the functions $(1+u)^{2m-1-2j}Q(u)^{j-1/2}$ $(0\leqq j\leqq m-1)$. Moreover, we find from (2.2.21) that

$$\begin{aligned}&\int_{I_\varepsilon(1)} u^{-m-1/2}(1-u)(1+u)^{2m-1-2j}Q(u)^{j-1/2}du\\&\qquad=-2b_{j,m,0}\alpha_1+\sum_{k=1}^{m-1} b_{j,m,k}(\alpha_{k-1}-\alpha_{k+1}) \qquad (0\leqq j\leqq m-1),\end{aligned}$$

which is a rational number owing to Lemma 2.14. Thus, taking (2.2.16), (2.2.18) into account of, we can conclude that $(1/2\pi)\Lambda_{(k_1,-1,k_3)}(k_1, k_3 \geqq 0)$ is a rational number. q.e.d.

Remark For triples (k_1, k_2, k_3) not satisfying the conditions of Proposition 2.13, it will be hard to compute $\Lambda_{(k_1,k_2,k_3)}$ in an elementary manner. However we do not need the explicit values of them (see 2.4, 2.5 of this paper).

To evaluate the special values at $s=0$ of L-functions, we need the following explicit values of $\Lambda_{(k_1,k_2,k_3)}$.

Proposition 2.15. *We have*

$$\Lambda_{(0,0,0)}=\Lambda_{(-1,1,0)}=8\pi, \quad \Lambda_{(1,-1,0)}=32\pi/3, \quad \textit{and} \quad \Lambda_{(0,-1,1)}=16\pi/3.$$

Proof. The first two identities are straightforward. Then, the identity (2.2.20) implies that $P_{1,-1,0}(u)=2P_{0,-1,1}(u)=(1+u)Q(u)^{-1/2}$. Therefore, we get

$$\Lambda_{(1,-1,0)}=2\Lambda_{(0,-1,1)}=2\pi(g(-1)-g(1))=-2\pi\alpha_1=32\pi/3.$$

2.3. Integral representations of partial zeta functions II

We keep the notation used in 2.2. We shall study the analytic continuations of the functions $\zeta_S(s, \{V_1, V_2, V_3\}, (\xi_1, \xi_2))$, $\zeta_S(s: \{V_j, V_3\}, \xi)$ $(j=1, 2)$, and determine the first and, if possible, the second term of the Laurent expansions at $s=1-m$ $(m \in N)$ of them.

For simplicity we write λ_j for $\lambda((u, \theta), V_j)$ $(j=1, 2, 3)$, if there is no fear of confusion. We see easily from (2.2.6) that, for positive numbers ξ_1, ξ_2, ξ, and for $\mathrm{Re}(s)>3/2$,

$$\text{(2.3.1)} \qquad \begin{aligned} &\zeta_S(s; \{V_1, V_2, V_3\}, (\xi_1, \xi_2))=\frac{1}{\Gamma_2(s)(e[2s]-1)}I(s; (\xi_1, \xi_2)) \\ &\zeta_S(s; \{V_j, V_3\}, \xi)=\frac{1}{\Gamma_2(s)(e[2s]-1)}I_j(s; \xi) \qquad (j=1, 2), \end{aligned}$$

where we put

$$I(s; (\xi_1, \xi_2))=\int_{I_\varepsilon(\infty)} dt\int_0^1 du\int_0^\pi d\theta\cdot t^{2s-2}u^{s-3/2}(1-u)\,\frac{1}{\lambda_3}\prod_{j=1}^{2}\phi(t\lambda_j; 1-\xi_j),$$

$$I_j(s; \xi)=\int_{I_\varepsilon(\infty)} dt\int_0^1 du\int_0^\pi d\theta\cdot t^{2s-2}u^{s-3/2}(1-u)\,\frac{1}{\lambda_3}\phi(t\lambda_j; 1-\xi) \quad (j=1, 2),$$

ε being taken sufficiently small. The absolutely convergence for $\mathrm{Re}(s)>3/2$ of the integrals above is easily verified by Lemma 2.6. We shall first integrate with respect to θ. Changing the variable by $\cot\theta=x$ $(-\infty<x<+\infty)$, we get, by (2.2.15),

$$\text{(2.3.2)} \qquad \lambda_1=1+u,\quad \lambda_2=1+u+(1-u)\frac{x}{1+x^2},\quad \lambda_3=\frac{u+x^2}{1+x^2}.$$

As is easily seen, for each positive number $\beta<1$, there exists a small positive number $\delta=\delta(\beta)$ such that $\phi(t(1+z);\,1-\xi)$ $(\xi>0)$, as a function of t, z, is holomorphic in the region $\{(t,z)\in \boldsymbol{C}^2\,|\,\{t\in D_\delta(\infty),\ t\neq 0,\ |z|\leqq\beta\}$. Then, $\phi(t(1+z);\,1-\xi)$, as a function of z, has the power series expansion

$$\text{(2.3.3)} \qquad \phi(t(1+z);\,1-\xi)=\sum_{k=0}^{\infty}\frac{\phi^{(k)}(t;\,1-\xi)t^k}{k!}\cdot z^k$$

$$(t\in D_\delta(\infty),\ t\neq 0,\ |z|\leqq\beta).$$

It follows from (2.3.3) that, if $t\in D_\delta(\infty)$, $t\neq 0$, and $0\leqq u\leqq 1$,

$$\text{(2.3.4)} \qquad \phi(t\lambda_2;\,1-\xi)=\sum_{k=0}^{\infty}\frac{\phi^{(k)}(t(1+u);\,1-\xi)t^k(1-u)^k}{k!}\cdot\left(\frac{x}{1+x^2}\right)^k,$$

where δ is taken sufficiently small. We then define a function $\mathscr{H}_k(u)$ for each non-negative integer k:

$$\text{(2.3.5)} \qquad \mathscr{H}_k(u)=(1-u)^k\int_R\frac{1}{x^2+u}\left(\frac{x}{1+x^2}\right)^k dx \qquad (u>0).$$

Obviously, we have $\mathscr{H}_k(u)=0$ for any odd k. Applying the residue theorem in calculating the integral in (2.3.5), one can divide $\mathscr{H}_{2k}(u)$ $(k\in\boldsymbol{Z},\ k\geqq 0)$ into two parts as follows:

$$\text{(2.3.6)} \qquad \mathscr{H}_{2k}(u)=\pi u^{-1/2}(-u)^k+\pi\mathscr{A}_{2k}(u),$$

where

$$\mathscr{A}_{2k}(u)=2i(1-u)^{2k}\operatorname*{Res}_{x=i}\left(\frac{1}{x^2+u}\left(\frac{x}{1+x^2}\right)^{2k}\right).$$

An elementary computation shows that $\mathscr{A}_0(u)=0$, $\mathscr{A}_2(u)=(1+u)/2$. Moreover, we observe that each $\mathscr{A}_{2k}(u)$ is a polynomial of u with rational coefficients. We set, for $\xi>0$,

$$F_1(t,u;\,1-\xi)=\sum_{k=0}^{\infty}\frac{\phi^{(2k)}(t(1+u);\,1-\xi)}{(2k)!}\cdot t^{2k}(-u)^k,$$

$$F_2(t, u; 1-\xi)=\sum_{k=0}^{\infty}\frac{\phi^{(2k)}(t(1+u); 1-\xi)}{(2k)!}\cdot t^{2k}\mathscr{A}_{2k}(u).$$

We shall discuss the convergence and the regularity of $F_j(t, u; 1-\xi)$ $(j=1, 2)$ as functions of t, u. For that purpose, some preparations will be needed. We put, for $a \in \boldsymbol{R}$, and $j \in \boldsymbol{N}$,

$$\phi_j(t; a)=\frac{e^{ta}}{(e^t-1)^j}.$$

Lemma 2.16. *Let $n \in \boldsymbol{N}$ and $a<1$. If we write, in a unique way,*

$$\phi^{(n)}(t; a)=\sum_{j=0}^{n}\lambda_{j,n}(a)\phi_{j+1}(t; a) \qquad \text{with some } \lambda_{j,n}(a)\in \boldsymbol{R}, \tag{2.3.7}$$

then, we have $(-1)^n\lambda_{j,n}(a)>0$ for each j $(0\leqq j\leqq n)$.

Proof. Differentiating the both sides of (2.3.7) with respect to t, and using the identity $\phi'_{j+1}(t; a)=(a-j-1)\phi_{j+1}(t; a)-(j+1)\phi_{j+2}(t; a)$, we get recursive relations:

$$\begin{cases}\lambda_{0,n+1}(a)=(a-1)\lambda_{0,n}(a),\\ \lambda_{j,n+1}(a)=(a-j-1)\lambda_{j,n}(a)-j\lambda_{j-1,n}(a) \qquad (1\leqq j\leqq n),\\ \lambda_{n+1,n+1}(a)=-(n+1)\lambda_{n,n}(a).\end{cases} \tag{2.3.8}$$

In the case of $n=1$, we have, trivially, $(-1)\lambda_{j,1}(a)>0$ $(j=0, 1)$. Thus the assertion follows by induction on n from (2.3.8).

Taking the k-th derivative of (2.2.2), one gets, if $|t|<2\pi$,

$$\frac{\phi^{(k)}(t; a)}{k!}=(-1)^k t^{-k-1}+\sum_{n=k+1}^{\infty}\frac{B_n(a)}{n!}\binom{n-1}{k}t^{n-1-k}. \tag{2.3.9}$$

For $x \in \boldsymbol{R}$, $[x]$ denotes the largest integer less than or equal to x.

Lemma 2.17. *Let $0<\beta<1$ and $a \in \boldsymbol{R}$. If $|t|\leqq\pi/2$, $|w|\leqq\beta$, then,*

$$t\sum_{k=0}^{\infty}\frac{\phi^{(2k)}(t; a)t^{2k}}{(2k)!}\cdot w^k=\frac{1}{1-w}+\sum_{n=1}^{\infty}\frac{B_n(a)}{n!}\left(\sum_{k=0}^{\kappa_n}\binom{n-1}{2k}w^k\right)t^n, \tag{2.3.10}$$

where we put $\kappa_n=[(n-1)/2]$, and the infinite series of both sides are absolutely convergent. Moreover, the function defined by the infinite series (2.3.10) indicates a holomorphic function of t, w in the region $\{(t, w)|\ |t|\leqq\pi/2, |w|\leqq\beta\}$.

Proof. By virtue of the fact that $t\phi(t;a)=\sum_{k=0}^{\infty}(B_k(a)/k!)t^k$ is absolutely convergent for $|t|<2\pi$, there exists a positive constant C_1 independent of k which satisfies

$$|B_k(a)/k!|<C_1(3\pi/2)^{-k} \qquad (k=1,2,\cdots).$$

Using (2.3.9), we get

$$\begin{aligned}\sum_{k=0}^{\infty}\left|\frac{\phi^{(2k)}(t;a)t^{2k+1}}{(2k)!}\cdot w^k\right| &\leqq \sum_{k=0}^{\infty}|w|^k+C_1\sum_{k=0}^{\infty}\sum_{n=2k+1}^{\infty}\binom{n-1}{2k}|w|^k\left(\frac{2|t|}{3\pi}\right)^n \\ &\leqq \frac{1}{1-|w|}+C_1\sum_{n=1}^{\infty}\left(\sum_{k=0}^{\kappa_n}\binom{n-1}{2k}\beta^k\right)\left(\frac{2|t|}{3\pi}\right)^n \\ &\leqq \frac{1}{1-|w|}+C_1\sum_{n=1}^{\infty}\left(\frac{4|t|}{3\pi}\right)^n,\end{aligned}$$

where the last infinite series is convergent for $|t|\leqq\pi/2$. Thus the infinite series in the both sides of (2.3.10) are absolutely and uniformly convergent for $|t|\leqq\pi/2$, $|w|\leqq\beta$. In a similar manner, the identity (2.3.10) is easily shown to hold.

Proposition 2.18. *If we take δ sufficiently small, then, the infinite series $tF_j(t,u;1-\xi)$ $(\xi>0, j=1,2)$ are absolutely and uniformly convergent in the region $D_\delta(\infty)\times D_\delta(1)$. Consequently, $tF_j(t,u;1-\xi)$ $(j=1,2)$ are holomorphic in the same region. Moreover, $tF_j(t,u;1-\xi)$, as functions of t, have the following power series expansions; if $|t|<\delta$, $u\in D_\delta(1)$, then, we have*

$$tF_1(t,u;1-\xi)=\frac{1+u}{1+3u+u^2}+\sum_{n=1}^{\infty}\frac{B_n(1-\xi)}{n!}\cdot\mu_n(u)t^n, \tag{2.3.11}$$

$$tF_2(t,u;1-\xi)=\sum_{n=0}^{\infty}\frac{\mathscr{A}_{2n}(u)}{(1+u)^{2n+1}}+\sum_{n=1}^{\infty}\frac{B_n(1-\xi)}{n!}\cdot\nu_n(u)t^n, \tag{2.3.12}$$

where we put

$$\begin{cases}\mu_n(u)=\sum_{k=0}^{\kappa_n}\binom{n-1}{2k}(1+u)^{n-1-2k}(-u)^k, \\ \nu_n(u)=\sum_{k=0}^{\kappa_n}\binom{n-1}{2k}(1+u)^{n-1-2k}\mathscr{A}_{2k}(u) \quad \left(\kappa_n=\left[\frac{n-1}{2}\right],\ n\geqq1\right).\end{cases} \tag{2.3.13}$$

Proof. First we consider the infinite series $tF_1(t,u;1-\xi)$. We put $t'=t(1+u)$, $w=-u/(1+u)^2$. Obviously,

$$tF_1(t, u; 1-\xi)=\frac{t'}{1+u}\sum_{k=0}^{\infty}\frac{\phi^{(2k)}(t'; 1-\xi)t'^{2k}}{(2k)!}\cdot w^k.$$

By the inequality $|-u/(1+u)^2|\leqq 1/4$ for $0\leqq u\leqq 1$, there exists a small positive constant δ_1 such that $|-u/(1+u)^2|\leqq 1/2$ for $u\in D_{\delta_1}(1)$. If $|t|<\pi/4$, $u\in D_{\delta_1}(1)$, then we get $|t'|<\pi/2, |w|\leqq 1/2$. It follows easily from Lemma 2.17 that $tF_1(t, u; 1-\xi)$ converges absolutely and is holomorphic in the region $\{(t, u)|\ |t|<\pi/4, u\in D_{\delta_1}(1)\}$, and moreover that the power series expansion (2.3.11) holds in the same region. We take δ sufficiently small, comparing with δ_1. Let $|t|\geqq\pi/4$, $t\in D_\delta(\infty)$, and moreover, $u\in D_\delta(1)$. Set $\tau'=\mathrm{Re}(t')$. Then we may have $\tau'>|t'|/\sqrt{2}$, δ being taken sufficiently small. An elementary observation shows that

$$|\phi_j(t'; a)|\leqq\phi_j(\tau'; a)\qquad(a\in \boldsymbol{R}, j=0, 1, 2, \cdots),$$

from which, in addition to Lemma 2.16, we get

$$|\phi^{(2k)}(t', 1-\xi)|\leqq\phi^{(2k)}(\tau', 1-\xi)\qquad(k=0, 1, 2, \cdots).$$

Hence we see from Lemma 2.16 and the expansion (2.3.4) that

$$\begin{aligned}\sum_{k=0}^{\infty}\left|\frac{\phi^{(2k)}(t'; 1-\xi)t'^{2k}}{(2k)!}\cdot w^k\right| &\leqq\sum_{k=0}^{\infty}\frac{\phi^{(2k)}(\tau'; 1-\xi)|t'|^{2k}}{(2k)!}\cdot(1/2)^k\\ &\leqq\sum_{k=0}^{\infty}\frac{\phi^{(k)}(\tau'; 1-\xi)\tau'^k}{k!}\cdot\left(\frac{-|t'|}{\sqrt{2}\,\tau'}\right)^k\\ &=\phi\left(\tau'-\frac{|t'|}{\sqrt{2}}; 1-\xi\right).\end{aligned}$$

Thus, $F_1(t, u; 1-\xi)$ is absolutely convergent in the region $\{(t, u)|\ |t|\geqq\pi/4, t\in D_\delta(\infty), u\in D_\delta(1)\}$. Moreover, we see from the observation above that $F_1(t, u; 1-\xi)$ is uniformly convergent in some small neighbourhood of (t, u) contained in the region above. Consequently, $F_1(t, u; 1-\xi)$ is holomorphic in that region.

Next we consider the series $tF_2(t, u; 1-\xi)$. We have to estimate $\mathscr{A}_{2k}(u)$ from the above. The definition of $\mathscr{A}_{2k}(u)$ implies that

$$\pi\mathscr{A}_{2k}(u)=(1-u)^{2k}\int_{\boldsymbol{R}}\frac{1}{x^2+u}\left\{\left(\frac{x}{1+x^2}\right)^{2k}-\left(\frac{-u}{(1-u)^2}\right)^k\right\}dx.$$

Putting $s(x)=(x/1+x^2)^2$ for simplicity, we get the expression

$$\pi\mathscr{A}_{2k}(u)=\int_{\boldsymbol{R}}\frac{ux^2+1}{(1+x^2)^2}\sum_{j=0}^{k-1}\{s(x)(1-u)^2\}^{k-1-j}(-u)^j dx, \tag{2.3.14}$$

which holds for any $u \in \boldsymbol{C}$. We take a positive number δ_2 in such a manner that, if $|u|<\delta_2$, then, $|-u/(1-u)^2|<1/16$. Accordingly, $\delta_2 \leqq 9-4\sqrt{5}=0.0557\cdots$. Using the inequalities $s(x) \leqq 1/4$, we see easily from (2.3.14) that, if $|u|<\delta_2$,

$$\left|\frac{\mathscr{A}_{2k}(u)}{(1+u)^{2k}}\right|<(1/4)^{k-1}\frac{|1-u|^{2(k-1)}}{|1+u|^{2k}}\sum_{j=0}^{k-1}\left|\frac{-4u}{(1-u)^2}\right|^j$$
$$<\frac{1}{(1-\delta_2)^2}(1/4)^{k-1}\left(\frac{1+\delta_2}{1-\delta_2}\right)^{2(k-1)}\cdot\frac{4}{3}.$$

Since we have the inequality $(1+\delta_2)/2(1-\delta_2)<3/5$, there exists a positive constant C_2 independent of k such that

$$\left|\frac{\mathscr{A}_{2k}(u)}{(1+u)^{2k}}\right|<C_2(3/5)^{2k} \quad \text{if } |u|<\delta_2 \quad (k=0, 1, 2, \cdots). \tag{2.3.15}$$

We put $\delta_3=\delta_2/2$. On the other hand, if $|u| \geqq \delta_2$, and $u \in D_{\delta_3}(1)$, then,

$$|1+u| \geqq 1+\mathrm{Re}\,(u)>1+\delta_3, \quad |1-u| \leqq 1, \quad \text{and} \quad |u| \leqq 1.$$

Thus the inequalities just above and (2.3.14) imply that, for any non-negative integer k,

$$\left|\frac{\mathscr{A}_{2k}(u)}{(1+u)^{2k}}\right|<\frac{4}{3}\left(\frac{1}{1+\delta_3}\right)^{2k} \qquad \text{if } |u| \geqq \delta_2 \text{ and } u \in D_{\delta_3}(1). \tag{2.3.16}$$

Putting $t(1+u)=t'$, we get, formally,

$$tF_2(t, u; 1-\xi)=\frac{t'}{1+u}\sum_{k=0}^{\infty}\frac{\Phi^{(2k)}(t'; 1-\xi)t'^{2k}}{(2k)!}\cdot\frac{\mathscr{A}_{2k}(u)}{(1+u)^{2k}}. \tag{2.3.17}$$

With the help of (2.3.15), (2.3.16), we can prove that the right side of (2.3.17) is absolutely convergent if $|t|<\pi/4$, $u \in D_{\delta_3}(1)$, and we obtain, similarly as in the proof of (2.3.11), the identity (2.3.12). Since $\mathscr{A}_{2k}(u)$ is a polynomial of u, we see easily from the expression (2.3.12) that $tF_2(t, u; 1-\xi)$ is holomorphic in the region $\{t \in \boldsymbol{C} \mid |t|<\pi/4\} \times D_{\delta_3}(1)$. The rest of assertions for $tF_2(t, u; 1-\xi)$ can be verified in the same manner as in the case of $tF_1(t, u; 1-\xi)$ by using the inequlities (2.3.15), (2.3.16). q.e.d.

We take δ sufficiently small so that the identity (2.3.4) and Proposition 2.18 simultaneously hold. Then, taking the identities (2.3.4), (2.3.5), (2.3.6) and the inequality $|x/(1+x^2)| \leqq 1/2$ into account of, we obtain

$$(2.3.18)\qquad \int_0^{\pi}\frac{1}{\lambda_3}\phi(t\lambda_2;\,1-\xi)d\theta=\pi u^{-1/2}F_1(t,u;\,1-\xi)+\pi F_2(t,u;\,1-\xi)$$
$$(t\in D_\delta(\infty),\ 0<u\leqq 1).$$

We set, for positive numbers ξ_1, ξ_2, ξ,

$$(2.3.19)\qquad \begin{cases}\Phi^{(0)}(t,u;\,\xi_1,\xi_2)=\phi(t(1+u);\,1-\xi_1)F_1(t,u;\,1-\xi_2),\\ \Phi^{(1)}(t,u;\,\xi)=\phi(t(1+u);\,1-\xi),\\ \Phi^{(2)}(t,u;\,\xi)=F_1(t,u;\,1-\xi),\end{cases}$$

$$(2.3.20)\qquad \begin{cases}\Psi^{(0)}(t,u;\,\xi_1,\xi_2)=\phi(t(1+u);\,1-\xi_1)F_2(t,u;\,1-\xi_2),\\ \Psi^{(1)}(t,u;\,\xi)=0,\\ \Psi^{(2)}(t,u;\,\xi)=F_2(t,u;\,1-\xi).\end{cases}$$

Let $\Phi(t,u)$ (resp. $\Psi(t,u)$) be one of the three functions given in (2.3.19) (resp. in (2.3.20)). We set

$$(2.3.21)\qquad \begin{aligned}J(s;\Phi)&=\int_{I_\varepsilon(\infty)}dt\int_{I_\varepsilon(1)}du\cdot t^{2s-2}u^{s-2}(1-u)\Phi(t,u),\\ K(s;\Psi)&=\int_{I_\varepsilon(\infty)}dt\int_{I_\varepsilon(1)}du\cdot t^{2s-2}u^{s-3/2}(1-u)\Psi(t,u),\end{aligned}$$

where ε is taken sufficiently small with $\varepsilon<\delta$, δ being the same as in (2.3.18). Then, by virtue of Proposition 2.18 and its proof, the integrals $J(s;\Phi)$, $K(s;\Psi)$ are independent of the choice of ε and absolutely convergent for arbitrary $s\in \boldsymbol{C}$. Moreover, they indicate entire functions of s. We write, for convenience,

$$(2.3.22)\qquad \begin{cases}J(s;\,(\xi_1,\xi_2))=J(s;\,\Phi^{(0)}(t,u;\,\xi_1,\xi_2)),\\ J_j(s;\,\xi)=J(s;\,\Phi^{(j)}(t,u;\,\xi)) \qquad (j=1,2),\\ K(s;\,(\xi_1,\xi_2))=K(s;\,\Psi^{(0)}(t,u;\,\xi_1,\xi_2)),\\ K_j(s;\,\xi)=K(s;\,\Psi^{(j)}(t,u;\,\xi)) \qquad (j=1,2).\end{cases}$$

Trivially, $K_1(s;\xi)=0$. Thus, using (2.3.18), we obtain convenient expressions for $I(s;(\xi_1,\xi_2))$, $I_j(s;\xi)$ $(j=1,2)$ by the integrals (2.3.22) over contour paths $I_\varepsilon(\infty)$, $I_\varepsilon(1)$.

Proposition 2.19. *Let ξ_1, ξ_2, ξ be positive numbers. We have*

$$I(s;\,(\xi_1,\xi_2))=\frac{\pi}{(e[s]-1)}J(s;\,(\xi_1,\xi_2))+\frac{\pi}{(e[s-3/2]-1)}K(s;\,(\xi_1,\xi_2)),$$

$$I_j(s;\xi)=\frac{\pi}{(e[s]-1)}J_j(s;\xi)+\frac{\pi}{(e[s-3/2]-1)}K_j(s;\xi) \qquad (j=1,2),$$

which give the analytic continuation to meromorphic functions of s in the whole complex plane.

Corollary to Proposition 2.19. *The functions* $\zeta_S(s;\{V_1,V_2,V_3\},(\xi_1,\xi_2))$, $\zeta_S(s;\{V_j,V_3\},\xi)\ (j=1,2)$ *can be continued analytically to meromorphic functions of s in the whole complex plane.*

The corollary is an immediate consequence of Proposition 2.19 and (2.3.1).

Proposition 2.19 shows us that the Laurent expansions at $s=1-m$ $(m\in\boldsymbol{N})$ of $I(s;(\xi_1,\xi_2))$, $I_j(s;\xi)\ (j=1,2)$ are given as follows:

$$\begin{aligned}(2.3.23)\quad I(s;(\xi_1,\xi_2))=&\frac{J(1-m;(\xi_1,\xi_2))}{2i(s+m-1)}+\frac{1}{2i}\{J'(1-m;(\xi_1,\xi_2))\\&-\pi iJ(1-m;(\xi_1,\xi_2))-\pi iK(1-m;(\xi_1,\xi_2))\}\\&+\text{higher terms of }(s+m-1),\end{aligned}$$

$$\begin{aligned}(2.3.24)\quad I_j(s;\xi)=&\frac{J_j(1-m;\xi)}{2i(s+m-1)}+\frac{1}{2i}\{J_j'(1-m;\xi)-\pi iJ_j(1-m;\xi)\\&-\pi iK_j(1-m;\xi)\}+\text{higher terms of }(s+m-1)\\&(j=1,2).\end{aligned}$$

Let $C(m)\ (m\in\boldsymbol{N})$ be the constant given in Proposition 2.9. Then, as a Taylor expansion at $s=1-m\ (m\in\boldsymbol{N})$, we have

$$\frac{1}{\Gamma_2(s)(e[2s]-1)}=-2C(m)+\beta_m(s+m-1)+\text{higher terms of }(s+m-1)$$

with some constant $\beta_m\in\boldsymbol{C}$. Thus, by (2.3.1), (2,3.23), we get the Laurent expansion at $s=1-m$ of $\zeta_S(s;\{V_1,V_2,V_3\},(\xi_1,\xi_2))$:

$$\begin{aligned}(2.3.25)\quad &\zeta_S(s;\{V_1,V_2,V_3\},(\xi_1,\xi_2))\\&=\frac{iC(m)J(1-m;(\xi_1,\xi_2))}{(s+m-1)}+\frac{1}{2i}\{(\beta_m+2\pi iC(m))J(1-m;(\xi_1,\xi_2))\\&-2C(m)J'(1-m;(\xi_1,\xi_2))+2\pi iC(m)K(1-m;(\xi_1,\xi_2))\}\\&+\text{higher terms of }(s+m-1).\end{aligned}$$

At this stage we have to evaluate $J(1-m;(\xi_1,\xi_2))$, $J'(1-m;(\xi_1,\xi_2))$, $K(1-m;(\xi_1,\xi_2))$ and so on.

We consider the integral $J(s, \Phi)$ in (2.3.21). Putting $s=1-m$, we get

$$(2.3.26)\qquad J(1-m;\Phi)=\int_{\Gamma_\varepsilon} dt\int_{\Gamma_\varepsilon} du\cdot t^{-2m}u^{-m-1}(1-u)\Phi(t,u)$$

(for the path Γ_ε, see Proposition 2.9).

Furthermore, derivating the integrand of $J(s;\Phi)$ with respect to s, we obtain

$$(2.3.27)\qquad J'(1-m;\Phi)$$
$$=\int_{I_\varepsilon(\infty)} dt\int_{I_\varepsilon(1)} du\cdot t^{-2m}u^{-m-1}(1-u)(2\log t+\log u)\Phi(t,u),$$

where the integral is absolutely convergent again by Proposition 2.18. For non-negative integers n, the functions $\Phi_n(t)$, according to the choice of $\Phi(t, u)$, are defined as follows; We set

$$\begin{cases}\Phi_n(t)=\dfrac{\phi(t;1-\xi_1)\phi^{(2n)}(t;1-\xi_2)(-1)^n}{(2n)!} & \text{if } \Phi(t,u)=\Phi^{(0)}(t,u;\xi_1,\xi_2),\\ \Phi_0(t)=\phi(t;1-\xi),\quad \Phi_n(t)=0\quad (n\geqq 1) & \text{if } \Phi(t,u)=\Phi^{(1)}(t,u;\xi),\\ \Phi_n(t)=\dfrac{\phi^{(2n)}(t;1-\xi)(-1)^n}{(2n)!} & \text{if } \Phi(t,u)=\Phi^{(2)}(t,u;\xi).\end{cases}$$

Moreover, we see from Proposition 2.18 that $\Phi(t, u)$ has a Laurent expansion with respect to t:

$$(2.3.28)\qquad \Phi(t,u)=\sum_{n=-2}^{\infty} b_n(u;\Phi)t^n \qquad \text{if } |t|<\delta,\ u\in D_\delta(1).$$

Proposition 2.20. *Let $m\in N$. We have, for a sufficiently small ε,*

(i) $J(1-m;\Phi)=2\pi i\int_{\Gamma_\varepsilon}\Phi_m(t)dt,$

(ii) $$J'(1-m;\Phi)=4\pi i\int_{I_\varepsilon(\infty)}\log t\cdot\Phi_m(t)dt-4\pi i\sum_{j=0}^{m-1}\frac{(2m-2j-1)!}{\{(m-j)!\}^2}$$
$$\times\int_{\Gamma_\varepsilon} t^{2(j-m)}\Phi_j(t)dt+2\pi i\int_{I_\varepsilon(1)} u^{-m-1}(1-u)\log u\cdot b_{2m-1}(u;\Phi)du.$$

Proof. The function $\Phi(t, u)$, which is holomorphic if $t\in D_\delta(\infty)$, $t\neq 0$, and $|u|<\delta$, can be expanded in a power series of u as follows:

$$(2.3.29)\qquad \Phi(t,u)=\sum_{n=0}^{\infty}\frac{1}{n!}\frac{\partial^n\Phi}{\partial u^n}(t,0)u^n.$$

On the other hand, it is easy to see from the definition of $\Phi_n(t)$ that

$$(2.3.30)\quad t^2\Phi(t, u)=t^2\sum_{n=0}^{\infty}\Phi_n(t(1+u))t^{2n}u^n \qquad (t\in D_\delta(\infty),\ u\in D_\delta(1)).$$

Since the infinite series in the right side of (2.3.30) is uniformly convergent in $D_\delta(\infty)\times D_\delta(1)$ by Proposition 2.18, and each term $t^2\Phi_n(t(1+u))t^{2n}u^n$ is a holomorphic function of t, u, we can differentiate it termwise. Thus, taking the k-th derivative of (2.3.30) with respect to u, we get

$$\frac{\partial^k\Phi}{\partial u^k}(t, u)=\sum_{j=0}^{k}\sum_{n=j}^{\infty}\binom{k}{j}n(n-1)\cdots(n-j+1)\Phi_n^{(k-j)}(t(1+u))t^{k-j+2n}u^{n-j}.$$

Therefore,

$$(2.3.31)\qquad \frac{\partial^k\Phi}{\partial u^k}(t, 0)=\sum_{j=0}^{k}\binom{k}{j}j!\,\Phi_j^{(k-j)}(t)t^{k+j}.$$

It follows from (2.3.29), (2.3.31) that

$$\begin{aligned}(2.3.32)\quad &\int_{I_\varepsilon(1)}u^{-m-1}(1-u)\Phi(t, u)du\\ &=2\pi i\left\{\frac{1}{m!}\cdot\frac{\partial^m\Phi}{\partial u^m}(t, 0)-\frac{1}{(m-1)!}\cdot\frac{\partial^{m-1}\Phi}{\partial u^{m-1}}(t, 0)\right\}\\ &=2\pi i\left(\Phi_m(t)t^{2m}+\sum_{j=0}^{m-1}\left\{\frac{t^{m+j}}{(m-j)!}\Phi_j^{(m-j)}(t)-\frac{t^{m-1+j}}{(m-1-j)!}\Phi_j^{(m-1-j)}(t)\right\}\right).\end{aligned}$$

With the use of the integration by parts, the identity

$$(2.3.33)\qquad -s\int_{I_\varepsilon(\infty)}t^{s-1}\Phi_j^{(m-1-j)}(t)\,dt=\int_{I_\varepsilon(\infty)}t^s\Phi_j^{(m-j)}(t)dt$$

holds for each j $(0\leqq j\leqq m-1)$. Putting $s=j-m$, we get

$$(2.3.34)\quad (m-j)\int_{I_\varepsilon(\infty)}t^{j-m-1}\Phi_j^{(m-1-j)}(t)dt=\int_{I_\varepsilon(\infty)}t^{j-m}\Phi_j^{(m-j)}(t)dt.$$

Differentiating the both sides of (2.3.33) with respect to s and then, putting $s=j-m$, we have

$$\begin{aligned}(2.3.35)\quad (m-j)&\int_{I_\varepsilon(\infty)}\log t\cdot t^{j-m-1}\Phi_j^{(m-1-j)}(t)\,dt-\int_{I_\varepsilon(\infty)}t^{j-m-1}\Phi_j^{(m-1-j)}(t)dt\\ &=\int_{I_\varepsilon(\infty)}\log t\cdot t^{j-m}\Phi_j^{(m-j)}(t)dt \qquad (0\leqq j\leqq m-1).\end{aligned}$$

Therefore, the identities (2.3.26), (2.3.32), and (2.3.34) imply the assertion (i). The recurrent use of the integration by parts yields

$$(2.3.36)\quad \int_{I_\varepsilon(\infty)} t^{j-m-1}\Phi_j^{(m-1-j)}(t)dt = \frac{(2m-2j-1)!}{(m-j)!}\int_{\Gamma_\varepsilon} t^{2(j-m)}\Phi_j(t)dt.$$

It is easy to see from (2.3.32), (2.3.35), and (2.3.36) that

$$(2.3.37)\quad 2\int_{I_\varepsilon(\infty)} dt \int_{I_\varepsilon(1)} du\cdot t^{-2m}u^{-m-1}(1-u)\log t\cdot\Phi(t,u)$$

$$=4\pi i\int_{I_\varepsilon(\infty)} \log t\cdot\Phi_m(t)dt - 4\pi i\sum_{j=0}^{m-1}\frac{(2m-2j-1)!}{\{(m-j)!\}^2}\int_{\Gamma_\varepsilon} t^{2(j-m)}\Phi_j(t)dt.$$

Moreover, we get, using the expansion (2.3.28),

$$\int_{I_\varepsilon(\infty)} dt \int_{I_\varepsilon(1)} du\cdot t^{-2m}u^{-m-1}(1-u)\log u\cdot\Phi(t,u)$$

$$=2\pi i\int_{I_\varepsilon(1)} u^{-m-1}(1-u)\log u\cdot b_{2m-1}(u;\Phi)du,$$

which, in addition to (2.3.27), (2.3.37), completes the proof.

Let $m\in\boldsymbol{N}$. For integers k, n with $k, n\geqq 0$, $k+n=2m+1$, we define the numbers $\mathcal{M}_{(k-1,n-1)}$ by putting

$$(2.3.38)\quad \mathcal{M}_{(k-1,n-1)} = \int_{I_\varepsilon(1)} \log u\cdot u^{-m-1}(1-u)(1+u)^{k-1}\mu_n(u)du,$$

where $\mu_n(u)$ ($n\geqq 1$) is a polynomial of u given by (2.3.13), and

$$\mu_0(u)=\frac{1+u}{1+3u+u^2}.$$

The numbers $\mathcal{M}_{(k-1,n-1)}$ are independent of the choice of small ε.

Lemma 2.21. *If $k, n\geqq 1$ with $k+n=2m+1$, then $(1/2\pi i)\mathcal{M}_{(k-1,n-1)}$ are rational numbers.*

Proof. It follows from (2.3.13) that

$$(1-u)(1+u)^{k-1}\mu_n(u)=\sum_{j=0}^{\kappa_n}\binom{n-1}{2j}(1-u)^{2(m-j)-1}(-u)^j.$$

By the conditions $k\geqq 1$ and $k+n=2m+1$, we have $m>j$ for each j ($0\leqq j\leqq \kappa_n=[(n-1)/2]$). As is easily shown, the coefficient of the term u^m

of the polynomial $(1-u)(1+u)^{2(m-j)-1}(-u)^j$ vanishes. Therefore, the assertion of Lemma 2.21 is reduced to the formula

$$\int_{I_\varepsilon(1)} \log u \cdot u^p du = \frac{2\pi i}{p+1} \qquad \text{for } p \in Z,\ p \neq -1, \tag{2.3.39}$$

In the case of $\Phi(t, u)=\Phi^{(0)}(t, u; \xi_1, \xi_2)$, Proposition 2.20 yields

Proposition 2.22. *Let ξ_1, ξ_2 be positive numbers and $m \in N$. Then,*

(i) $$J(1-m; (\xi_1, \xi_2)) = \frac{4\pi^2(-1)^m}{(2m+1)!}\{B_{2m+1}(\xi_1)+B_{2m+1}(\xi_2)\},$$

(ii) $$\begin{aligned} J'(1-m; (\xi_1, \xi_2)) &= \frac{4\pi i(-1)^m}{(2m)!}\int_{I_\varepsilon(\infty)} \log t \cdot \phi(t; 1-\xi_1)\phi^{(2m)}(t; 1-\xi_2)dt \\ &-8\pi^2 \sum_{j=0}^{m-1} \frac{(2m-2j-1)!(-1)^j}{\{(m-j)!\}^2} \\ &\times \left\{\frac{B_{2m+1}(\xi_1)}{(2m+1)!} + \sum_{n=2j+1}^{2m+1} \frac{B_{2m+1-n}(\xi_1)B_n(\xi_2)}{(2m+1-n)!n!} \cdot \binom{n-1}{2j}\right\} \\ &-2\pi i \sum_{n=0}^{2m+1} \frac{B_{2m+1-n}(\xi_1)B_n(\xi_2)}{(2m+1-n)!n!} \cdot \mathcal{M}_{(2m-n, n-1)}. \end{aligned}$$

Proof. In the proof we have $\Phi(t, u)=\Phi^{(0)}(t, u; \xi_1, \xi_2)$, and

$$\Phi_n(t) = \frac{\phi(t; 1-\xi_1)\phi^{(2n)}(t; 1-\xi_2)(-1)^n}{(2n)!} \qquad (n=0, 1, 2, \cdots).$$

The expansions (2.2.2), (2.3.9) show that the coefficient of the term t^{-1} in the Laurent expansion at $t=0$ of $\Phi_m(t)$ is given by

$$\frac{(-1)^m}{(2m+1)!}\{B_{2m+1}(1-\xi_1)+B_{2m+1}(1-\xi_2)\}.$$

Thus, by (i) of Proposition 2.18, the assertion (i) follows. In view of the expansions (2.2.2), (2.3.11) of Proposition 2.18, the coefficient $b_{2m-1}(u; \Phi)$ in (2.3.28) is given as follows:

$$\begin{aligned} b_{2m-1}(u; \Phi) &= \frac{B_{2m+1}(1-\xi_1)(1+u)^{2m+1}}{(2m+1)!(1+3u+u^2)} + \sum_{n=1}^{2m+1} \frac{B_{2m+1-n}(1-\xi_1)B_n(1-\xi_2)}{(2m+1-n)!n!} \\ &\times (1+u)^{2m-n}\mu_n(u). \end{aligned}$$

Therefore, we see from (2.3.38) that

$$\int_{I_\varepsilon(1)} u^{-m-1}(1-u)\log u\cdot b_{2m-1}(u;\Phi)du$$
$$=-\sum_{n=0}^{2m+1}{}' \frac{B_{2m+1-n}(\xi_1)B_n(\xi_2)}{(2m+1-n)!n!}\cdot\mathscr{M}_{(2m-n,n-1)}.$$

Computing the coefficient of the term $t^{2(m-j)-1}$ $(0\leqq j\leqq m-1)$ in the Laurent expansion at $t=0$ of $\Phi_j(t)$, we have

$$\int_{\Gamma_\varepsilon} t^{2(j-m)}\Phi_j(t)dt$$
$$=2\pi i(-1)^j\left\{\frac{B_{2m+1}(1-\xi_1)}{(2m+1)!}+\sum_{n=2j+1}^{2m+1}\frac{B_{2m+1-n}(1-\xi_1)B_n(1-\xi_2)}{(2m+1-n)!n!}\cdot\binom{n-1}{2j}\right\}.$$

Summing up the results above, we obtain the assertion (ii).

In another two cases of $\Phi(t, u)$, we obtain the following

Proposition 2.23. *Let $m\in\boldsymbol{N}$ and $\xi>0$. Then,*

(i) $J_j(1-m;\xi)=0\qquad(j=1,2)$,

(ii) $$J_1'(1-m;\xi)=\frac{4\pi^2B_{2m}(\xi)}{m(m!)^2}+2\pi i\frac{B_{2m}(\xi)}{(2m)!}\mathscr{M}_{(2m-1,0)},$$
$$J_2'(1-m;\xi)=\frac{4\pi^2(-1)^mB_{2m}(\xi)}{m(2m!)}+\frac{4\pi^2B_{2m}(\xi)}{m}\sum_{j=0}^{m-1}\frac{(-1)^j}{\{(m-j)!\}^2(2j)!}$$
$$+2\pi i\frac{B_{2m}(\xi)}{(2m)!}\mathscr{M}_{(0,2m-1)}.$$

(iii) *In particular, $J_j'(1-m;\xi)\in\pi^2\boldsymbol{Q}$ $(j=1,2)$.*

Proof. If $\Phi(t,u)=\phi(t(1+u);1-\xi)$, then, we have $\Phi_0(t)=\phi(t;1-\xi)$, $\Phi_n(t)=0$ $(n\geqq1)$, and $b_{2m-1}(u;\Phi)=B_{2m}(1-\xi)(1+u)^{2m-1}/(2m)!$. Hence, we see immediately from Proposition 2.20 that $J_1(1-m;\xi)=0$, and that

$$J_1'(1-m;\xi)=-\frac{4\pi i(2m-1)!}{(m!)^2}\int_{\Gamma_\varepsilon} t^{-2m}\phi(t;1-\xi)dt$$
$$+\frac{2\pi iB_{2m}(1-\xi)}{(2m)!}\int_{I_\varepsilon(\infty)}\log u\cdot u^{-m-1}(1-u)(1+u)^{2m-1}du,$$

from which we get the expression for $J_1'(1-m;\xi)$ in the assertion (ii) (note that $\mu_1(u)=1$). In the case of $\Phi(t,u)=F_1(t,u;1-\xi)$, we have

$$\Phi_n(t)=\frac{\phi^{(2n)}(t;1-\xi)(-1)^n}{(2n)!}\quad\text{and}\quad b_{2m-1}(u;\Phi)=\frac{B_{2m}(1-\xi)}{(2m)!}\cdot\mu_{2m}(u).$$

Since the integration by parts implies that

$$-s\int_{I_\varepsilon(\infty)} t^{s-1}\phi^{(2m-1)}(t;\,1-\xi)dt=\int_{I_\varepsilon(\infty)} t^{s}\phi^{(2m)}(t;\,1-\xi)dt,$$

we get

$$\begin{aligned}(2.3.40)\qquad \int_{I_\varepsilon(\infty)} \log t\cdot\phi^{(2m)}(t;\,1-\xi)dt &= -\int_{I_\varepsilon(\infty)} t^{-1}\phi^{(2m-1)}(t;\,1-\xi)dt\\ &= -\frac{\pi i B_{2m}(1-\xi)}{m}.\end{aligned}$$

Thus,

$$4\pi i\int_{I_\varepsilon(\infty)} \log t\cdot\Phi_m(t)dt=\frac{4\pi^2(-1)^m B_{2m}(\xi)}{m(2m)!}.$$

Moreover, we have, by a usual argument,

$$\int_{\Gamma_\varepsilon} t^{2(j-m)}\Phi_j(t)dt=\frac{2\pi i(-1)^j B_{2m}(1-\xi)}{(2m)!}\cdot\binom{2m-1}{2j}.$$

Hence, the assertion (ii) of Proposition 2.20 implies the expression for $J_2'(1-m;\,\xi)$ in (ii). The equality $J_2(1-m;\,\xi)=0$ is clear.

Finally, we evaluate the special values at $s=1-m$ of $K(s;\,(\xi_1,\,\xi_2))$ $K_2(s;\,\xi)$ (for the definition, see (2.3.22)).

Let $\nu_n(u)$ $(n\geqq 1)$ be the polynomials of u defined by (2.3.13). For instance, $\nu_1(u)=\nu_2(u)=0$, $\nu_3(u)=(1+u)/2$. We put, for convenience,

$$\nu_0(u)=\sum_{j=0}^{\infty}\frac{\mathscr{A}_{2j}(u)}{(1+u)^{2j+1}}\qquad \text{for } |u| \text{ sufficiently small.}$$

For any pairs $(k,\,n)$ of non-negative integers with $k+n=2m+1$ $(m\in \boldsymbol{N})$, we define the numbers $\mathscr{N}_{(k-1,\,n-1)}$ by putting

$$(2.3.41)\qquad \mathscr{N}_{(k-1,\,n-1)}=\int_{I_\varepsilon(1)} u^{-1/2-m}(1-u)(1+u)^{k-1}\nu_n(u)du,$$

where ε is taken sufficiently small. The integral in the right side of (2.3.41) is independent of the choice of small ε. Then the identity (2.2.18) implies that

$$(2.3.42)\qquad \mathscr{N}_{(k-1,\,n-1)}\in \boldsymbol{Q}\qquad \text{for } k,\,n\geqq 1.$$

Proposition 2.2.4. *Let* $\xi_1,\,\xi_2,\,\xi>0$ *and* $m\in \boldsymbol{N}$. *Then,*

$$\text{(i)}\qquad K(1-m;\,(\xi_1,\,\xi_2))=-2\pi i\sum_{n=0}^{2m+1}\frac{B_{2m+1-n}(\xi_1)B_n(\xi_2)}{(2m+1-n)!\,n!}\cdot\mathscr{N}_{(2m-n,\,n-1)},$$

(ii)
$$K_2(1-m;\,\xi)=\frac{2\pi i B_{2m}(\xi)}{(2m)!}\cdot\mathcal{N}_{(0,2m-1)}.$$

Proof. Let $\Psi(t, u)$ be one of the functions given in (2.3.20). Recalling the definition (2.3.21) of $K(s;\,\Psi)$, we have by the Fubini theorem,

$$K(1-m;\,\Psi)=\int_{I\varepsilon(1)} u^{-m-1/2}(1-u)du\int_{\Gamma_\varepsilon} dt\cdot t^{-2m}\Psi(t, u).$$

If $\psi(t, u)=\psi^{(0)}(t, u;\,\xi_1, \xi_2)$, then, observing the expansions (2.2.2), (2.3.13), we get

$$\int_{\Gamma_\varepsilon} t^{-2m}\psi(t, u)dt=2\pi i\sum_{n=0}^{\infty}\frac{B_{2m+1-n}(1-\xi_1)B_n(1-\xi_2)}{(2m+1-n)!n!}\cdot(1+u)^{2m-n}\nu_n(u).$$

Thus we obtain the assertion (i). The assertion (ii) is similarly verified.

As is observed in (2.3.25), the function $\zeta_S(s;\,\{V_1, V_2, V_3\}, (\xi_1, \xi_2))$ has a possible simple pole at $s=1-m$ ($m\in\mathbf{N}$). On the other hand, it is easy to see from (2.3.1), Proposition 2.19, and (i) of Proposition 2.23 that the functions $\zeta_S(s;\,\{V_j, V_3\}, \xi)$ ($j=1, 2$) are holomorphic at $s=1-m$. The special values at $s=1-m$ of them can be given explicitly by applying Proposition 2.23, Proposition 2.24.

Proposition 2.25. *Let $m\in\mathbf{N}$ and $\xi>0$. Then,*
(i) $\zeta_S(1-m;\,\{V_j, V_3\}, \xi)\in\mathbf{Q}$ ($j=1, 2$).
(ii) *In particular,*

$$\zeta_S(0;\,\{V_1, V_3\}, \xi)=\frac{1}{2}B_2(\xi),$$

$$\zeta_S(0;\,\{V_2, V_3\}, \xi)=\frac{3}{8}B_2(\xi).$$

Proof. Proposition 2.19 and (i) of Proposition 2.23 show that

$$I_j(1-m;\,\xi)=\frac{1}{2i}\,J_j'(1-m;\,\xi)-\frac{\pi}{2}\,K_j(1-m;\,\xi)\qquad(j=1, 2),$$

We observe from (2.3.1) that

$$\zeta_S(1-m;\,\{V_j, V_3\}, \xi)=-2C(m)I_j(1-m;\,\xi)\qquad(j=1, 2).$$

Thus the assertion (i) follows from Proposition 2.23, Lemma 2.21, Proposition 2.24, and (2.3.42) (for the explicit value of $C(m)$, see Proposition 2.9). Let $m=1$. We have, by (2.3.39),

(2.3.43) $$\mathcal{M}_{(1,0)}=\mathcal{M}_{(0,1)}=\int_{I_\varepsilon(1)}\log u\cdot u^{-2}(1-u^2)du=-4\pi i.$$

Therefore, we find from (ii) of Proposition 2.23 that $J_1'(0;\xi)=8\pi^2 B_2(\xi)$ and $J_2'(0;\xi)=6\pi^2 B_2(\xi)$. Since $\mathcal{N}_{(0,1)}=0$ and accordingly, $K_2(0;\xi)=0$, we get the assertion (ii) (notice that $K_1(s;\xi)$ is identically zero).

2.4. Evaluation of special values of $L_2^*(s,\psi_{H,p})$

The aim of the subsequent two sections is to prove the rationality of special values of L-functions at non-positive integers and in particular to obtain the explicit special values at $s=0$ of them. We keep the notation used in the previous sections.

Suppose that p is an odd prime. For any integers μ, δ prime to p, let $L^*(\mu)$, $M^*(\delta)$ be the same as in 2.1. Corresponding to $M=L^*(\mu)$, $M^*(\delta)$, we shall define the principal part $\zeta_P(s;C,M)$ and the singular part $\zeta_S(s;C,M)$ of the zeta function $\zeta(s;C,M)$, C being the simplicial cones C_{123}, C_{j3} $(j=1,2)$. In view of Proposition 2.1, Proposition 2.3, we set, for $M=L^*(\mu)$ or $M^*(\delta)$,

(2.4.1)
$$\zeta_P(s;C_{123},M)=p^{-2s}\sum_{\xi\in\Xi_M}\zeta_P(s;\{V_1,V_2,V_3\},\xi),$$
$$\zeta_S(s;C_{123},M)=p^{-2s}\sum_{\xi\in\Xi_M}\zeta_S(s;\{V_1,V_2,V_3\},(\xi_1,\xi_2)),$$
$$\zeta_P(s;C_{j3},M)=p^{-2s}\sum_{\xi\in\Xi_M^{(j,3)}}\zeta_P(s;\{V_j\,V_3\},(\xi_j,\xi_3)),$$
$$\zeta_S(s;C_{j3},M)=p^{-2s}\sum_{\xi\in\Xi_M^{(j,3)}}\zeta_S(s;\{V_j,V_3\},\xi_j)\qquad(j=1,2),$$

(ξ being denoted by (ξ_1,ξ_2,ξ_3)),

where we set $\Xi_M=\Xi_{H,\mu}$, $\Xi_M^{(j,3)}=\Xi_{H,\mu}^{(j,3)}$ (resp. $\Xi_M=\Xi_\delta$, $\Xi_M^{(j,3)}=\Xi_\delta^{(j,3)}$), if $M=L^*(\mu)$ (resp. if $M=M^*(\delta)$). Proposition 2.2 then makes it possible to define the principal and singular parts of the L-function $L_2^*(s,\psi_{H,p})$. We set

(2.4.2)
$$L_{2,P}^*(s,\psi_{H,p})=\sum_\mu\psi(\mu)\Big\{\{\zeta_P(s;C_{123},L^*(\mu))+\frac{1}{2}\sum_{j=1}^{2}\zeta_P(s;C_{j3},L^*(\mu))$$
$$+\frac{1}{2}\zeta(s;C_{12},L^*(\mu))+\frac{\delta_{p,3}}{6}\zeta(s;C_2,L^*(\mu))\Big\},$$
$$L_{2,S}^*(s,\psi_{H,p})=\sum_\mu\psi(\mu)\Big\{\zeta_S(s;C_{123},L^*(\mu))+\frac{1}{2}\sum_{j=1}^{2}\zeta_S(s;C_{i3},L^*(\mu))\Big\}.$$

where μ runs over 1 and κ, (κ being a non-quadratic residue mod p). Let

χ be a primitive character mod p. Viewing Proposition 2.4, Proposition 2.5, we define P- nad S-parts of $L_2^*(s, \chi_{\det})$, $\xi_2^*(s)$ as follows; we set

$$\begin{aligned} (2.4.3)\quad & L_{2,P}^*(s, \chi_{\det}) \\ &= \sum_{\delta \not\equiv 0 \bmod p} \chi(\delta)\Big\{\zeta_P(s; C_{123}, M^*(\delta)) + \frac{1}{2}\sum_{j=1}^{2}\zeta_P(s; C_{j3}, M^*(\delta)) \\ &\quad + \frac{1}{2}\zeta(s; C_{12}, M^*(\delta)) + \frac{1}{4}\zeta(s; C_1, M^*(\delta)) + \frac{1}{6}\zeta(s; C_2, M^*(\delta))\Big\}, \end{aligned}$$

$$\begin{aligned} & L_{2,S}^*(s, \chi_{\det}) \\ &= \sum_{\delta \not\equiv 0 \bmod p} \chi(\delta)\Big\{\zeta_S(s; C_{123}, M^*(\delta)) + \frac{1}{2}\sum_{j=1}^{2}\zeta_S(s; C_{j3}, M^*(\delta))\Big\}, \end{aligned}$$

$$\begin{aligned} \xi_{2,P}^*(s) &= \zeta_P(s; \{V_1, V_2, V_3\}, (1, 1, 1)) + \frac{1}{2}\sum_{j=1}^{2}\zeta_P(s; \{V_j, V_3\}, (1, 1)) \\ &\quad + \frac{1}{2}\zeta(s; \{V_1, V_2\}, (1, 1)) + \frac{1}{4}\zeta(s; \{V_1\}, 1) + \frac{1}{6}\zeta(s; \{V_2\}, 1), \end{aligned}$$

$$\xi_{2,S}^*(s) = \zeta_S(s; \{V_1, V_2, V_3\}, (1, 1)) + \frac{1}{2}\sum_{j=1}^{2}\zeta_S(s; \{V_j, V_3\}, 1\}.$$

The following obvious identities hold:

$$(2.4.4)\qquad \begin{cases} L_2^*(s, \psi_{H,p}) = L_{2,P}^*(s, \psi_{H,p}) + L_{2,S}^*(s; \psi_{H,p}) \\ L_2^*(s, \chi_{\det}) = L_{2,P}^*(s, \chi_{\det}) + L_{2,S}^*(s, \chi_{\det}) \\ \xi_2^*(s) = \xi_{2,P}^*(s) + \xi_{2,S}^*(s). \end{cases}$$

It is easy to see from Proposition 2.9, Proposition 2.10 and Corollary to Proposition 2.19 that $L_{2,P}^*(s, \psi_{H,p})$, $L_{2,S}^*(s, \psi_{H,p})$, $L_{2,P}^*(s, \chi_{\det})$, $L_{2,S}^*(s, \chi_{\det})$, $\xi_{2,P}^*(s)$, and $\xi_{2,S}^*(s)$ can be continued analytically to meromorphic functions of s in the whole complex plane.

In the rest of this section we shall discuss the evaluation of special values at $s=1-m$ ($m \in N$) of $L_2^*(s, \psi_{H,p})$. First of all, we need two lemmas related to the Bernoulli polynomials.

Lemma 2.26. *Let $m \in N$. Then,*

$$\sum_{\xi \in \Xi_{H,\mu}} B_{2m+1}(\xi_j) = 0 \qquad (j=1, 2, 3),$$

where ξ_j is the j-component of $\xi \in \Xi_{H,\mu}$.

Proof. The proof is based on the distribution property of $B_k(x)$:

$$(2.4.5)\qquad B_k(x) = p^{k-1}\sum_{r=0}^{p-1} B_k\left(\frac{x+r}{p}\right) \qquad \text{(see for instance [10, p. 35]).}$$

Moreover, if we note that $B_{2m+1}(0)=B_{2m+1}(1)=0$, then, the assertion immediately follows.

Lemma 2.27. *Let $m \in \boldsymbol{N}$ and let k_2, k_3 be non-negative integers with $k_2+k_3=2m-1$.*

(i) *If k_2 is an even positive integer, then,*

$$\sum_{\xi \in \Xi_{H,\mu}} B_{k_2+1}(\xi_2)B_{k_3+1}(\xi_3)=0.$$

(ii) *If $k_2=0$, then,*

$$\sum_{\xi \in \Xi_{H,\mu},\, \xi_2 \neq 1} B_1(\xi_2)B_{2m}(\xi_3)=0$$

where $\xi=(\xi_1, \xi_2, \xi_3)$ runs over all triples of $\Xi_{H,\mu}$ with $\xi_2 \neq 1$.

Proof. Let k_2 be an even positive integer. We note that

$$\langle x\rangle+\langle -x\rangle=1 \qquad \text{if } x \in \boldsymbol{R}-\boldsymbol{Z}. \tag{2.4.6}$$

The identity $B_{k_2+1}(\langle x\rangle)=-B_{k_2+1}(\langle -x\rangle)$ $(x \in \boldsymbol{R}-\boldsymbol{Z})$ shows that

$$\sum_{(\alpha,\gamma)\in \mathcal{M}(p)} B_{k_2+1}(\langle 2\mu\alpha\gamma/p\rangle)B_{k_3+1}(\langle \mu(\gamma^2-\alpha^2)/p\rangle)$$
$$=-\sum_{(\alpha,\gamma)\in \mathcal{M}(p)} B_{k_2+1}(\langle -2\mu\alpha\gamma/p\rangle)B_{k_3+1}(\langle \mu(\gamma^2-\alpha^2)/p\rangle).$$

Then, (α, γ) being replaced with $(-\alpha, \gamma)$, the assertion (i) follows. The proof of the assertion (ii) is quite similar to that of (i).

Let $m \in \boldsymbol{N}$. In the below, let k_1, k_2, k_3 be integers satisfying $k_1, k_2 \geqq -1$, $k_3 \geqq 0$ and $k_1+k_2+k_3=2(m-1)$. For any triple $\xi=(\xi_1, \xi_2, \xi_3)$ of positive numbers, we write, for convenience,

$$B(k_1, k_2, k_3; \xi)=\prod_{j=1}^{3} \frac{B_{k_j+1}(\xi_j)}{(k_j+1)!}. \tag{2.4.7}$$

Let μ be any integer prime to p. Viewing Proposition 2.11, Proposition 2.12, we define the numbers $\mathcal{A}_{(k_1,k_2,k_3)}(\mu)$ as follows;

(i) If $k_1k_2k_3 \neq 0$, we set

$$\mathcal{A}_{(k_1,k_2,k_3)}(\mu)=-2\pi i C(m)\Lambda_{(k_1,k_2,k_3)} \sum_{\xi \in \Xi_{H,\mu}} B(k_1, k_2, k_3; \xi).$$

(ii) Let r be an integer with $1 \leqq r \leqq 3$. If $k_r=0$ and other k_j's $(j \neq r)$ are non-zero, we set

$$\mathcal{A}_{(k_1,k_2,k_3)}(\mu)=-2\pi i C(m)\Lambda_{(k_1,k_2,k_3)} \sum_{\xi \in \Xi_{H,\mu},\, \xi_r \neq 1} B(k_1, k_2, k_3; \xi).$$

(iii) Let r, n be integers with $1 \leqq r < n \leqq 3$. If $k_r = k_n = 0$, and the rest of k_j's is non-zero (then, necessarily, $m > 1$), we set

$$\mathscr{A}_{(k_1,k_2,k_3)}(\mu) = -2\pi i C(m) \Lambda_{(k_1,k_2,k_3)} \{ \sum_{\xi \in \Xi_{H,\mu}, \xi_r, \xi_n \neq 1} B(k_1, k_2, k_3; \xi) - s(k_1, k_2, k_3; \mu) \},$$

where

$$s(k_1, k_2, k_3; \mu) = \begin{cases} \frac{1}{8} g_{2m-1}(\mu) & (k_1, k_2, k_3) = (0, 0, 2m-2), \\ \frac{1}{24} \delta_{p,3} g_{2m-1}(\mu) & (k_1, k_2, k_3) = (0, 2m-2, 0), \\ 0 & (k_1, k_2, k_3) = (2m-2, 0, 0), \end{cases}$$

$g_d(\mu)$ ($d \in \boldsymbol{Z}$, $d \geqq 0$) being $\sum_{\alpha \not\equiv 0 \bmod p} B_d(\langle \mu\alpha^2/p \rangle)/d!$.

(iv) In the case of $(k_1, k_2, k_3) = (0, 0, 0)$, we set

$$\mathscr{A}_{(0,0,0)}(\mu) = -2\pi i C(1) \Lambda_{(0,0,0)} \left\{ \sum_{\xi \in \Xi_{H,\mu}, \xi_j \neq 1 \ (j=1,2,3)} B(0, 0, 0; \xi) - \frac{1}{8} g_1(\mu) - \frac{1}{24} \delta_{p,3} g_1(\mu) \right\}.$$

Note that, if d is prime to p, then, $\mathscr{A}_{(k_1,k_2,k_3)}(\mu d^2) = \mathscr{A}_{(k_1,k_2,k_3)}(\mu)$. The special values at $s = 1-m$ ($m \in \boldsymbol{N}$) of $L^*_{2,P}(s, \psi_{H,p})$ can be evaluated with the use of the numbers defined above.

Proposition 2.28. *Let $m \in \boldsymbol{N}$. Then,*

(i) $L^*_{2,P}(1-m, \psi_{H,p}) = p^{2(m-1)} \sum_{\mu} \sum'_{k_1,k_2,k_3} \psi(\mu) \mathscr{A}_{(k_1,k_2,k_3)}(\mu)$,

where μ is over 1 *and κ, and (k_1, k_2, k_3) runs over all triples of integers with $k_1, k_2 \geqq -1$, $k_3 \geqq 0$, and $k_1 + k_2 + k_3 = 2(m-1)$.*

(ii) *Accordingly, $L^*_{2,P}(1-m, \psi_{H,p}) \in \boldsymbol{Q}$.*

(iii) *If $p \equiv 1$ mod 4, then, $L^*_{2,P}(1-m, \psi_{H,p}) = 0$,*
If $p \equiv 3$ mod 4, then, in particular,

$$L^*_{2,P}(0, \psi_{H,p}) = \mathscr{A} + \mathscr{B},$$

where $\mathscr{A}$, $\mathscr{B}$ are the constants given by (0.1), (0.2) *in the introduction.*

Proof. Let $\xi_{\alpha,\gamma,\mu}$ be the triple of $\Xi_{H,\mu}$ given by (2.1.6). We notice that $B_1(1) = 1/2$, and moreover that

$$(2.4.8)\quad \begin{cases} \Xi_{H,\mu}^{(2,3)} \cap \Xi_{H,\mu}^{(1,3)} = \{\xi_{\alpha,\gamma,\mu} \mid \alpha \equiv 0 \bmod p,\ \gamma \not\equiv 0 \bmod p\}, \\ \Xi_{H,\mu}^{(2,3)} \cap \Xi_{H,\mu}^{(1,2)} = \begin{cases} \phi & (p>3), \\ \{\xi_{\alpha,\gamma,\mu} \mid \gamma \equiv -\alpha \bmod p,\ \alpha \not\equiv 0 \bmod p\} & (p=3), \end{cases} \\ \Xi_{H,\mu}^{(1,3)} \cap \Xi_{H,\mu}^{(1,2)} = \phi. \end{cases}$$

Taking very carefully (2.4.1), (2.4.2), Proposition 2.11, Proposition 2.12, and (2.4.8) into account of, we obtain the expression (i) for $L_{2,P}^*(1-m, \psi_{H,p})$.

If a triple (k_1, k_2, k_3) satisfies any of the conditions (i), (ii), (iii) in Proposition 2.13, then, $(1/\pi)\Lambda_{(k_1,k_2,k_3)}$ is a rational number. Therefore, $\mathscr{A}_{(k_1,k_2,k_3)}(\mu)$ is also a rational number. If $k_1=k_2=-1$ (resp. If $k_1=-1$ and k_2 is even), then, Lemma 2.26 (resp. Lemma 2.27) shows that $\mathscr{A}_{(-1,-1,2m)}(\mu)=0$ (resp. $\mathscr{A}_{(-1,k_2,k_3)}(\mu)=0$). Therefore, the assertion (ii) follows from (i).

For convenience, we write $\xi_{\alpha,\gamma,\mu}=(\xi_{\alpha,\gamma,\mu}^{(1)}, \xi_{\alpha,\gamma,\mu}^{(2)}, \xi_{\alpha,\gamma,\mu}^{(3)})$. Exchanging μ for $-\mu$, we observe from (2.1.6), (2.4.6) that

$$(2.4.9)\quad \begin{cases} \xi_{\alpha,\gamma,-\mu}^{(j)} = 1-\xi_{\alpha,\gamma,\mu}^{(j)} & \text{if } \xi_{\alpha,\gamma,\mu}^{(j)} \neq 1, \\ \xi_{\alpha,\gamma,-\mu}^{(j)} = 1 & \text{if } \xi_{\alpha,\gamma,\mu}^{(j)} = 1 \quad (j=1,2,3) \end{cases}$$

Further, we note that $B_k(1)=(-1)^k B_k(1)$ for $k\neq 1$. Then we see easily from (2.4.9) that, for any triple (k_1, k_2, k_3),

$$(2.4.10)\quad \mathscr{A}_{(k_1,k_2,k_3)}(-\mu) = -\mathscr{A}_{(k_1,k_2,k_3)}(\mu).$$

If $p\equiv 1 \bmod 4$, then, -1 is a quadratic residue mod p. Hence, $\mathscr{A}_{(k_1,k_2,k_3)}(-\mu)=\mathscr{A}_{(k_1,k_2,k_3)}(\mu)$, which implies that $\mathscr{A}_{(k_1,k_2,k_3)}(\mu)=0$. Thus the first assertion of (iii) follows.

Suppose that $p\equiv 3 \bmod 4$. Then we may take -1 as κ. The assertion (i), together with (2.4.10) and the property $\psi(-1)=-1$, show, that

$$L_{2,P}^*(0, \psi_{H,p}) = 2\{\mathscr{A}_{(-1,1,0)}(1)+\mathscr{A}_{(1,-1,0)}(1)+\mathscr{A}_{(0,-1,1)}(1)+\mathscr{A}_{(0,0,0)}(1)\}.$$

We have, by Proposition 2.15,

$$(2.4.11)\quad \begin{aligned} \mathscr{A}_{(-1,1,0)}(1) &= \frac{-1}{4} \sum_{\substack{(\alpha,\gamma)\in\mathscr{M}(p), \\ \alpha^2\not\equiv\gamma^2 \bmod p}} B_2(\langle 2\alpha\gamma/p\rangle) B_1(\langle(\gamma^2-\alpha^2)/p\rangle), \\ \mathscr{A}_{(1,-1,0)}(1) &= \mathscr{B}, \\ \mathscr{A}_{(0,-1,1)}(1) &= \frac{-1}{6} \sum_{\substack{(\alpha,\gamma)\in\mathscr{M}(p), \\ \alpha^2\not\equiv 2\alpha\gamma \bmod p}} B_1(\langle(\alpha^2-2\alpha\gamma)/p\rangle) B_2(\langle\gamma^2-\alpha^2)/p\rangle), \\ \mathscr{A}_{(0,0,0)}(1) &= \mathscr{A}/2, \end{aligned}$$

where, to show the last equality, we used the fact that $B_{1,\psi}=g_1(1)$. In the first equality of (2.4.11), if we replace (α, γ) with (γ, α), then, we get, with the help of (2.4.6), $\mathscr{A}_{(-1,1,0)}(1)=0$. In the third equality of (2.4.11), replacing (α, γ) with $(\alpha-\gamma, -\gamma)$, we have

$$\mathscr{A}_{(0,-1,1)}(1)=\frac{-1}{6}\sum_{\substack{(\alpha,\gamma)\in\mathscr{M}(p)\\ \alpha^2\not\equiv\gamma^2 \bmod p}} B_2(\langle(2\alpha\gamma-\alpha^2)/p\rangle)B_1(\langle(\alpha^2-\gamma^2)/p\rangle)=-\mathscr{B}/2.$$

Thus the second assertion of (iii) follows. q.e.d.

Now we study the singular part $L^*_{2,S}(s, \psi_{H,p})$.

Proposition 2.29. *Let $m \in \boldsymbol{N}$ and let μ be an integer prime to p. The function $\zeta_S(s; C_{123}, L^*(\mu))$ is holomorphic at $s=1-m$, and the special value at $s=1-m$ is given by*

$$\begin{aligned}&\zeta_S(1-m; C_{123}, L^*(\mu))\\ &\quad=iC(m)p^{2(m-1)}\sum_{\xi\in\Xi_{H,\mu}}\{J'(1-m; (\xi_1, \xi_2))-\pi iK(1-m; (\xi_1, \xi_2))\}.\end{aligned}$$

Proof. Proposition 2.22 and Lemma 2.26 show that

$$\sum_{\xi\in\Xi_{H,\mu}} J(1-m; (\xi_1, \xi_2))=0.$$

Thus we see immediately from (2.4.1) and the expansion (2.3.25) that $\zeta_S(s; C_{123}, L^*(\mu))$ is holomorphic at $s=1-m$, and that the special value at $s=1-m$ is expressed as in the proposition.

If follows from Proposition 2.25, Proposition 2.29, and (2.4.2) that $L^*_{2,S}(s, \psi_{H,p})$ is holomorphic at $s=1-m$ $(m \in \boldsymbol{N})$. The following proposition plays a key role to evaluate its special value at $s=1-m$.

Proposition 2.30. *Let $m \in \boldsymbol{N}$. Then,*

$$\begin{aligned}&\frac{4\pi i(-1)^m}{(2m)!}\sum_{\mu}\psi(\mu)\sum_{\xi\in\Xi_{H,\mu}}\int_{I_\varepsilon(\infty)}\log t\cdot\phi(t; 1-\xi_1)\phi^{(2m)}(t; 1-\xi_2)dt\\ &\quad=\frac{8\pi^2(-1)^m}{p^{2m-1}}\sum_{j=1}^{2m+1}\frac{B_{2m+1-j}B_{j,\psi}}{j!(2m+1-j)!j},\end{aligned}$$

where μ runs over 1 and κ, and ε is taken sufficiently small.

Proof. As in the proof of Proposition 2.28, let $\xi^{(j)}_{\alpha,\gamma,\mu}$ $(j=1, 2, 3)$ be the j-component of $\xi_{\alpha,\gamma,\mu}\in\Xi_{H,\mu}$ (see (2.1.6)). We set $\mu\alpha^2=x$, $2\mu\alpha\gamma=u$. If (α, γ) runs over all elements of $\mathscr{M}(p)$ with $\alpha\not\equiv 0 \bmod p$, and μ is over 1 and κ, then, $(x, u)=(\mu\alpha^2, 2\mu\alpha\gamma)$ just doubly covers all elements of $\mathscr{M}(p)$ with $x\not\equiv 0 \bmod p$. If $\alpha\equiv 0 \bmod p$, then, $\xi^{(1)}_{\alpha,\gamma,\mu}=\xi^{(2)}_{\alpha,\gamma,\mu}=1$. Thus,

$$(2.4.12)\quad \frac{1}{2}\sum_{\mu}\psi(\mu)\sum_{(\alpha,\gamma)\in\mathscr{M}(p)}\int_{I_\varepsilon(\infty)}\log t\cdot\phi(t;1-\xi^{(1)}_{\alpha,\gamma,\mu})\phi^{(2m)}(t;1-\xi^{(2)}_{\alpha,\gamma,\mu})dt$$

$$=\sum_{x\not\equiv 0 \bmod p}\sum_{u \bmod p}\psi(x)\int_{I_\varepsilon(\infty)}\log t\cdot\phi(t;1-\langle(x-u)/p\rangle)\phi^{(2m)}(t;1-\langle u/p\rangle)dt.$$

We need the following lemma.

Lemma 2.31. *Let x be any integer prime to p. Then,*
(i)

$$\sum_{u \bmod p}\phi(t;1-\langle(x-u)/p\rangle)\phi^{(2m)}(t;1-\langle u/p\rangle)=\sum_{j=1}^{2m+1}\gamma_{j,2m}\phi^{(j)}(t;1-\langle x/p\rangle)$$

with

$$\gamma_{j,2m}=\frac{(-1)^j(2m)!}{j!(2m+1-j)!p^{2m-j}}\cdot B_{2m+1-j}\qquad(1\leqq j\leqq 2m+1),$$

(ii)
$$\int_{I_\varepsilon(\infty)}\log t\cdot\sum_{u \bmod p}\phi(t;1-\langle(x-u)/p\rangle)\phi^{(2m)}(t;1-\langle u/p\rangle)dt$$

$$=-2\pi i\sum_{j=1}^{2m+1}\gamma_{j,2m}\frac{(-1)^jB_j(\langle x/p\rangle)}{j}.$$

Proof. We may take x so that $1\leqq x\leqq p-1$. We get, with the help of Lemma 2.16,

$$\sum_{u \bmod p}\phi(t;1-\langle(x-u)/p\rangle)\phi^{(2m)}(t;1-\langle u/p\rangle)$$

$$=\sum_{j=0}^{2m}\sum_{u \bmod p}\lambda_{j,2m}(1-\langle u/p\rangle)\frac{\exp t(2-\langle u/p\rangle-\langle(x-u)/p\rangle)}{(e^t-1)^{j+2}}$$

$$=\sum_{j=0}^{2m}\frac{1}{(e^t-1)^{j+2}}\left\{\sum_{u=1}^{x-1}\lambda_{j,2m}(1-u/p)e^{t(2-x/p)}+\sum_{u=x}^{p}\lambda_{j,2m}(1-u/p)\,e^{t(1-x/p)}\right\}$$

$$=\sum_{j=0}^{2m}\left\{\sum_{u=1}^{x-1}\lambda_{j,2m}(1-u/p)\phi_{j+1}(t;1-x/p)+\sum_{u=1}^{p}\lambda_{j,2m}(1-u/p)\phi_{j+2}(t;1-x/p)\right\}.$$

Then, using the formula $\phi_{j+1}(t;a)=-(1/j)\phi_j'(t;a)+((a-j)/j)\phi_j(t;a)$ $(j\geqq 1)$ recurrently, we obtain the expression

$$(2.4.13)\qquad \sum_{u \bmod p}\phi(t;1-\langle(x-u)/p\rangle)\phi^{(2m)}(t;1-\langle u/p\rangle)$$

$$=\sum_{j=0}^{2m+1}\gamma_{j,2m}(x)\phi^{(j)}(t;1-x/p),$$

where $\gamma_{j,2m}(x)$ $(0\leqq j\leqq 2m+1)$ are certain rational numbers (they may depend on x). Recalling the Laurent expansion (2.3.9) of $\phi^{(k)}(t;a)$, and comparing the coefficients of the term t^{-j-1} in the Laurent expansions at $t=0$ of the both sides of (2.4.13), we get the explicit values of $\gamma_{j,2m}(x)$:

$$(-1)^j j!\gamma_{j,2m}(x)=\sum_{u \bmod p}\frac{(2m)!\,B_{2m+1-j}(1-\langle(x-u)/p\rangle)}{(2m+1-j)!}\qquad(1\leqq j\leqq 2m+1),$$

$$\gamma_{0,2m}(x)=\frac{1}{2m+1}\sum_{u \bmod p}\{B_{2m+1}(1-\langle(x-u)/p\rangle)+B_{2m+1}(1-\langle u/p\rangle)\}.$$

Then, the property (2.4.5) of $B_k(x)$ shows that

$$\gamma_{0,2m}(x)=0,\qquad \gamma_{j,2m}(x)=\frac{(-1)^j(2m)!\,B_{2m+1-j}}{j!(2m+1-j)!\,p^{2m-j}}\qquad(1\leqq j\leqq 2m+1).$$

Thus the assertion (i) follows. Similarly as in (2.3.40), we have

$$\int_{I\varepsilon(\infty)}\log t\cdot\phi^{(j)}(t;\,1-x/p)dt=-2\pi i\frac{B_j(1-x/p)}{j},$$

which, together with (i), completes the proof of the assertion (ii).

Thus, (2.4.6) and Lemma 2.31 with the definition (1.2.11) of the generalized Bernoulli numbers imply Proposition 2.30.

Proposition 2.32. *Let* $m\in \boldsymbol{N}$. *Then,*
(i) $\sum_\mu \psi(\mu)\zeta_S(1-m;\,C_{123},\,L^*(\mu))\in \boldsymbol{Q}$.
(ii) *In particular, if* $m=1$,

$$\sum_\mu \psi(\mu)\zeta_S(0;\,C_{123},\,L^*(\mu))=\frac{1}{p}\left\{\frac{11}{36}B_{3,\psi}-\frac{7}{16}B_{2,\psi}-\frac{1}{24}B_{1,\psi}\right\}.$$

Proof. Taking (ii) of Proposition 2.22, Lemma 2.21, Lemma 2.26, and Proposition 2.30 into account of, we observe that

$$\sum_\mu \psi(\mu)\sum_{\xi\in\Xi_{H,\mu}} J'(1-m;\,(\xi_1,\,\xi_2))\in \pi^2\boldsymbol{Q}.$$

We get, immediately by (i) of Proposition 2.24, (2.3.42), and Lemma 2.26,

$$\sum_{\xi\in\Xi_{H,\mu}} K(1-m;\,(\xi_1,\,\xi_2))\in \pi i\boldsymbol{Q}.$$

Thus the assertion (i) follows from Proposition 2.29. Next suppose that $m=1$. Since obviously, $\mathcal{N}_{(0,1)}=\mathcal{N}_{(1,0)}=0$, we see easily again from Proposition 2.24, Lemma 2.26 that

$$\sum_{\xi\in\Xi_{H,\mu}} K(0;\,(\xi_1,\,\xi_2))=0.$$

Therefore we find from Proposition 2.29, Proposition 2.22, (2.3.43), Lemma 2.26, and Proposition 2.30 that

$$\begin{aligned}\sum_{\mu}\psi(\mu)\zeta_S(0;\,C_{123},\,L^*(\mu))&=\frac{1}{16\pi^2}\sum_{\mu}\psi(\mu)\sum_{\xi\in\Xi_{H,\mu}}J'(1-m;\,(\xi_1,\,\xi_2))\\&=\frac{1}{p}\left\{-\frac{1}{24}B_{1,\psi}+\frac{1}{16}B_{2,\psi}-\frac{1}{36}B_{3,\psi}\right\}\\&\quad-\frac{1}{2}\sum_{\mu}\psi(\mu)\sum_{\xi\in\Xi_{H,\mu}}\{B_1(\xi_1)B_2(\xi_2)+B_2(\xi_1)B_1(\xi_2)\}.\end{aligned}$$

Hence the following lemma completes the proof of Proposition 2.32.

Lemma 2.33. *We have*

$$\sum_{\mu}\psi(\mu)\sum_{\xi\in\Xi_{H,\mu}}\{B_1(\xi_1)B_2(\xi_2)+B_2(\xi_1)B_1(\xi_2)\}=\frac{1}{p}B_{2,\psi}-\frac{2}{3p}B_{3,\psi}.$$

Proof. We notice that

$$\int_{\Gamma_\varepsilon}t^{-2}\phi(t;\,1-\xi_1)\phi(t;\,1-\xi_2)dt=-2\pi i\left\{\frac{1}{6}\sum_{j=1}^{2}B_3(\xi_j)+\frac{1}{2}\sum_{j=1}^{2}B_1(\xi_j)B_2(\xi_{3-j})\right\}.$$

Then it follows from Lemma 2.26 that the left side of the equality in the lemma is equal to

$$\begin{aligned}&\frac{-1}{\pi i}\int_{\Gamma_\varepsilon}t^{-2}\sum_{\mu}\psi(\mu)\sum_{\xi\in\Xi_{H,\mu}}\phi(t;\,1-\xi_1)\phi(t;\,1-\xi_2)dt\\&=\frac{-1}{\pi i}\int_{\Gamma_\varepsilon}t^{-2}\sum_{x\not\equiv 0\bmod p}\psi(x)\sum_{u\bmod p}\phi(t;\,1-\langle(x-u)/p\rangle)\phi(t;\,1-\langle u/p\rangle)dt\\&=\frac{1}{\pi i}\int_{\Gamma_\varepsilon}t^{-2}\sum_{x\not\equiv 0\bmod p}\psi(x)\{\phi(t;\,1-\langle x/p\rangle)+p\phi'(t;\,1-\langle x/p\rangle)\}dt\\&=2\sum_{x\not\equiv 0\bmod p}\psi(x)\{B_2(1-\langle x/p\rangle)/2+pB_3(1-\langle x/p\rangle)/3\},\end{aligned}$$

which, in addition to (1.2.11), completes the proof of Lemma 2.33.

Finally, we evaluate the special values at $s=0$ of the zeta functions attached to the cones C_{j3} ($j=1,\,2$).

Proposition 2.34. *The following identities hold;*

$$\sum\psi(\mu)\zeta_S(0;\,C_{23},\,L^*(\mu))=\frac{3}{8p}B_{2,\psi},$$

$$\sum \psi(\mu)\zeta_S(0;\, C_{13},\, L^*(\mu)) = \frac{1}{2p} B_{2,\psi}.$$

Proof. Only the first identity will be proved. Proposition 2.25 and (2.4.1) show that

$$\begin{aligned}\sum \psi(\mu)\zeta_S(0;\, C_{23},\, L^*(\mu)) &= \frac{3}{16} \sum \psi(\mu) \sum_{\substack{(\alpha,\gamma)\in\mathcal{M}(p)\\ \alpha^2\equiv 2\alpha\gamma \bmod p}} B_2(\langle 2\mu\alpha\gamma\rangle/p)\\ &= \frac{3}{8p} B_{2,\psi}. \qquad \text{q.e.d.}\end{aligned}$$

Gathering (2.4.2), Proposition 2.25, Proposition 2.32, and Proposition 2.34 together, we obtain the following proposition.

Proposition 2.35. *We have*

(i) $L^*_{2,S}(1-m,\, \psi_{H,p}) \in \boldsymbol{Q}$ $(m \in \boldsymbol{N})$.

(ii) $L^*_{2,S}(0,\, \psi_{H,p}) = (11/36p)B_{3,\psi} - (1/24p)B_{1,\psi}$.

Now Theorem 1 (the main theorem) in the introduction follows immediately from Proposition 2.28, Proposition 2.35, (2.4.4), and the fact that if $p \equiv 1 \bmod 4$, $B_{k,\psi} = 0$ for any odd k.

2.5. Evaluation of special values of $L^*_{2,P}(s, \chi_{\det})$, $\xi^*_2(s)$.

Let χ be a primitive character mod p, p being an odd prime. We have defined the functions $L^*_{2,P}(s, \chi_{\det})$, $L^*_{2,S}(s, \chi_{\det})$, $\xi^*_{2,P}(s)$, $\xi^*_{2,S}(s)$ in (2.4.3). As we see from Proposition 2.9, Proposition 2.10, the principal parts $L^*_{2,P}(s, \chi_{\det})$, $\xi^*_{2,P}(s)$ are holomorphic at $s = 1-m$ $(m \in \boldsymbol{N})$.

We need a lemma concerning the Bernoulli polynomials.

Lemma 2.36. *Let* $m \in \boldsymbol{N}$. *Then,*

(i) $\sum_{a \not\equiv 0 \bmod p} \chi(a^2) B_{2m-1}(\langle a/p\rangle) = 0$.

(ii) *Let* k *be any integer with* $0 \leq k \leq 2m+1$. *Then,*

$$\sum_{a \not\equiv 0 \bmod p} \chi(a^2) B_{2m+1-k}(\langle -a/p\rangle) B_k(\langle a/p\rangle) = 0.$$

Proof. Replacing a with $-a$, and then, using the property $B_k(1-x) = (-1)^k B_k(x)$ and (2.4.6), we have the assertions (i), (ii).

Then, Proposition 2.3, (d), (e) of Proposition 2.12, and Lemma 2.36 show that

$$\text{(2.5.1)} \qquad \sum_{\delta \not\equiv 0 \bmod p} \chi(\delta)\zeta(1-m;\, C_j,\, M^*(\delta)) = 0 \qquad (j = 1, 2).$$

For each triple (k_1, k_2, k_3) of integers satisfying $k_1, k_2 \geqq -1$, $k_3 \geqq 0$, and $k_1+k_2+k_3=2(m-1)$ $(m \in N)$, we define the numbers $\mathscr{B}_{(k_1,k_2,k_3)}$ as follows;

(i) If $k_1k_2k_3 \neq 0$, we set

$$\mathscr{B}_{(k_1,k_2,k_3)} = -2\pi i C(m) \Lambda_{(k_1,k_2,k_3)} \sum_{\delta \not\equiv 0 \bmod p} \chi(\delta) \sum_{\xi \in \Xi_\delta} B(k_1, k_2, k_3; \xi).$$

(ii) Let r be an integer with $1 \leqq r \leqq 3$. If $k_r=0$ and other k_j's $(j \neq r)$ are non-zero, we set

$$\mathscr{B}_{(k_1,k_2,k_3)} = -2\pi i C(m) \Lambda_{(k_1,k_2,k_3)} \sum_{\delta \not\equiv 0 \bmod p} \chi(\delta) \sum_{\xi \in \Xi_\delta, \xi_r \neq 1} B(k_1, k_2, k_3; \xi).$$

(iii) Let r, n be integers with $1 \leqq r < n \leqq 3$. If $k_r=k_n=0$ and the rest of k_j's is non-zero $(m>1)$, we set

$$\mathscr{B}_{((k_1,k_2,k_3)} = -2\pi i C(m) \Lambda_{(k_1,k_2,k_3)} \sum_{\delta \not\equiv 0 \bmod p} \chi(\delta) \sum_{\substack{\xi \in \Xi_\delta \\ \xi_r, \xi_n \neq 1}} B(k_1, k_2, k_3, \xi).$$

(iv) If $(k_1, k_2, k_3)=(0, 0, 0)$, we set

$$\mathscr{B}_{(0,0,0)} = -2\pi i C(1) \Lambda_{(0,0,0)} \sum_{\delta \not\equiv 0 \bmod p} \chi(\delta) \sum_{\substack{\xi \in \Xi_\delta \\ \xi_j \neq 1 \ (j=1,2,3)}} B(0, 0, 0; \xi).$$

As the following Lemma 2.37 shows, all the numbers $\mathscr{B}_{(k_1,k_2,k_3)}$ are proved to be zero. However, these numbers are useful to evaluate the special values $L^*_{2,P}(1-m, \chi_{\det})$.

Lemma 2.37. *For any triple (k_1, k_2, k_3) as above, $\mathscr{B}_{(k_1,k_2,k_3)}=0$.*

Proof. For each $T=\begin{pmatrix} t_1 & t_{12} \\ t_{12} & t_2 \end{pmatrix} \in L_2^*$, we write

$$\xi_T = (\langle (t_1-2t_{12})/p \rangle, \langle 2t_{12}/p \rangle, \langle (t_2-t_1)/p \rangle) = (\xi_T^{(1)}, \xi_T^{(2)}, \xi_T^{(3)}).$$

Replacing T with $-T$, we observe that, if $\xi_T^{(j)} \neq 1$, then, $\xi_{-T}^{(j)}=1-\xi_T^{(j)}$, and that, if $\xi_T^{(j)}=1$, then, $\xi_{-T}^{(j)}=1 (1 \leqq j \leqq 3)$. Thus it follows from the definition of $\mathscr{B}_{(k_1,k_2,k_3)}$ and some properties of $B_k(x)$ that $\mathscr{B}_{(k_1,k_2,k_3)} = -\mathscr{B}_{(k_1,k_2,k_3)}$.

q.e.d.

The special values at $s=1-m$ $(m \in N)$ of $L^*_{2,P}(s, \chi_{\det})$ are obtained in a similar manner as in the proof of Proposition 2.28. The result is very simple.

Proposition 2.38. *We have $L^*_{2,P}(1-m, \chi_{\det})=0$ $(m=1, 2, \cdots)$.*

Proof. We notice that, for each integer δ prime to p,

$$\Xi_\delta^{(1,2)} \cap \Xi_\delta^{(1,3)} = \left\{ \xi_T \mid T = \begin{pmatrix} t_1 & 0 \\ 0 & t_1 \end{pmatrix}, \ t_1 \in Z/pZ, \ t_1^2 \equiv \delta \bmod p \right\},$$
$$\Xi_\delta^{(1,3)} \cap \Xi_\delta^{(2,3)} = \phi,$$
$$\Xi_\delta^{(2,3)} \cap \Xi_\delta^{(1,2)} = \left\{ \xi_T \mid T = \begin{pmatrix} 2t_{12} & t_{12} \\ t_{12} & 2t_{12} \end{pmatrix}, \ t_{12} \in Z/pZ, \ 3t_{12}^2 \equiv \delta \bmod p \right\}.$$

Then we have, by (i) of Lemma 2.36,

$$(2.5.2) \qquad \begin{cases} \sum_{\delta \not\equiv 0 \bmod p} \chi(\delta) \sum_{\xi \in \Xi_\delta^{(1,2)} \cap \Xi_\delta^{(1,3)}} B_{2m-1}(\xi_1) = 0, \\ \sum_{\delta \not\equiv 0 \bmod p} \chi(\delta) \sum_{\xi \in \Xi_\delta^{(2,3)} \cap \Xi_\delta^{(1,2)}} B_{2m-1}(\xi_2) = 0. \end{cases}$$

Thus, making use of the definition (2.4.3) of $L_{2,P}^*(s, \chi_{\det})$, Proposition 2.11, Proposition 2.12, and the identities (2.5.1), (2.5.2), we obtain

$$L_{2,P}^*(1-m, \chi_{\det}) = p^{2(m-1)} \sum_{(k_1,k_2,k_3)}' \mathscr{B}_{(k_1,k_2,k_3)},$$

where (k_1, k_2, k_3) is taken over all triples of integers satisfying $k_1, k_2 \geqq -1$, $k_3 \geqq 0$, and $k_1 + k_2 + k_3 = 2(m-1)$. Therefore, Lemma 2.37 completes the proof of Proposition 2.38.

Now we consider the function $L_{2,S}^*(s, \chi_{\det})$. We have

$$(2.5.3) \qquad \sum_{\delta \not\equiv 0 \bmod p} \chi(\delta) \zeta_S(s; C_{123}, M^*(\delta))$$
$$= p^{-2s} \sum_{\delta \not\equiv 0 \bmod p} \chi(\delta) \sum_{\substack{T \in L_2^*/pL_2^* \\ \det(T) \equiv \delta \bmod p}} \zeta_S(s; \{V_1, V_2, V_3\}, (\xi_T^{(1)}, \xi_T^{(2)})).$$

Dividing the summation into two parts according to $t_1 \not\equiv 0 \bmod p$ or not in $T = \begin{pmatrix} t_1 & t_{12} \\ t_{12} & t_2 \end{pmatrix}$, we see that the left side of (2.5.3) is equal to

$$p^{-2s} \sum_{\substack{\delta, t_1 \not\equiv 0 \bmod p \\ t_{12} \bmod p}} \chi(\delta) \zeta_S(s; \{V_1, V_2, V_3\}, (\langle (t_1 - 2t_{12})/p \rangle, \langle 2t_{12}/p \rangle))$$
$$+ p^{-2s} \sum_{\substack{t_{12} \not\equiv 0 \bmod p \\ t_2 \bmod p}} \chi(-t_{12}^2) \zeta_S(s; \{V_1, V_2, V_3\}, (\langle -2t_{12}/p \rangle, \langle 2t_{12}/p \rangle)).$$

Using the well-known fact that $\sum_{\delta \not\equiv 0 \bmod p} \chi(\delta) = 0$, we get

$$(2.5.4) \qquad \sum_{\delta \not\equiv 0 \bmod p} \chi(\delta) \zeta_S(s; C_{123}, M^*(\delta))$$
$$= p^{1-2s} \chi(-4)^{-1} \sum_{a \not\equiv 0 \bmod p} \chi(a^2) \zeta_S(s; \{V_1, V_2, V_3\}, (\langle -a/p \rangle, \langle a/p \rangle)).$$

Since Proposition 2.22, Lemma 2.36 show that

$$\sum_{a\not\equiv 0 \bmod p} \chi(a^2)J(1-m; (\langle -a/p\rangle, \langle a/p\rangle))=0 \qquad (m\in \boldsymbol{N}),$$

it is not difficult to see from (2.5.4), (2.3.25), Proposition 2.22, Proposition 2.24, and Lemma 2.36 that the function in the left side of the equality (2.5.4) is holomorphic at $s=1-m$, and moreover that its special value at $s=1-m$ is given by

$$(2.5.5) \qquad \sum_{\delta\not\equiv 0 \bmod p} \chi(\delta)\zeta_S(s; C_{123}, M^*(\delta))\Big|_{s=1-m} = \frac{\chi(-4)^{-1}4\pi(-1)^{m+1}p^{2m-1}C(m)}{(2m)!}$$

$$\times\int_{I_\varepsilon(\infty)} \log t\cdot \sum_{a\not\equiv 0 \bmod p} \chi(a^2)\phi(t; \langle a/p\rangle)\phi^{(2m)}(t; 1-\langle a/p\rangle)dt,$$

where ε is taken sufficiently small. In a similar manner as in Lemma 2.31, we can write

$$(2.5.6) \qquad \sum_{a\not\equiv 0 \bmod p} \chi(a^2)\phi(t; \langle a/p\rangle)\phi^{(2m)}(t; 1-\langle a/p\rangle)=\sum_{j=1}^{2m+1}\lambda_j\phi^{(j)}(t),$$

where $\phi(t)=\phi(t; 0)=1/(e^t-1)$, and λ_j $(1\leqq j\leqq 2m+1)$ are given by

$$(2.5.7) \qquad (-1)^j j!\,\lambda_j=\frac{(2m)!}{(2m+1-j)!}\sum_{a\not\equiv 0 \bmod p} \chi^2(a)B_{2m+1-j}(\langle a/p\rangle).$$

We notice that $\lambda_j=0$ if j is even. By a similar argument as in the proof of Lemma 2.31, we have

$$(2.5.8) \qquad \int_{I_\varepsilon(\infty)} \log t\cdot\sum_{j=1}^{2m+1}\lambda_j\phi^{(j)}(t)dt=-2\pi i\sum_{j=1}^{2m+1}\frac{\lambda_jB_j}{j}=\pi i\lambda_1$$

(note that $\lambda_jB_j=0$ if $j\neq 1$). Computing the number λ_1 from (2.5.7), we get, by (1.2.11), (2.4.5),

$$(2.5.9) \qquad \lambda_1=\begin{cases} -p^{1-2m}B_{2m,\chi^2} & \text{if } \chi^2 \text{ is non-trivial,}\\ \left(1-\dfrac{1}{p^{2m-1}}\right)B_{2m} & \text{if } \chi \text{ is quadratic.}\end{cases}$$

Moreover, we find that

$$\begin{aligned}&\sum_{\delta\not\equiv 0 \bmod p} \chi(\delta)\zeta_S(s; C_{13}, M^*(\delta))\\ &\qquad =p^{-2s}\sum_{t_1,t_2\not\equiv 0 \bmod p} \chi(t_1t_2)\zeta_S(s; \{V_1, V_2\}, \langle t_1/p\rangle)\\ &\qquad =0.\end{aligned}$$

The case of the cone C_{23} is similar. Thus we have, identically,

(2.5.10) $$\sum_{\delta\not\equiv \bmod p} \chi(\delta)\zeta_S(s; C_{j3}, M^*(\delta))=0 \qquad (j=1, 2).$$

In view of the identities from (2.5.4) to (2.5.10), we obtain the following proposition.

Proposition 2.39. *The sepcial values at* $s=1-m$ $(m=1, 2, \cdots)$ *of* $L^*_{2,S}(s, \chi_{\det})$ *are given as follows:*

$$L^*_{2,S}(1-m, \chi_{\det})=\begin{cases} \dfrac{\chi(-4)^{-1}(-1)^m}{2^{2m+1}m} B_{2m,\chi^2} & \text{if } \chi\neq\psi, \\ \dfrac{\chi(-1)(-1)^{m-1}}{2^{2m+1}m}(p^{2m-1}-1)B_{2m} & \text{if } \chi=\psi. \end{cases}$$

ψ *being the unique non-trivial quadratic character mod* p.

By virtue of Proposition 2.38 and Proposition 2.39, we can evaluate the special values $L^*_2(1-m, \chi_{\det})$ explicitly. The result is given in Theorem 2 in the introduction.

Finally, we shall evaluate the value $\xi^*_2(0)$. Using (2.4.3), Proposition 2.11, Proposition 2.12, and Proposition 2.15, we get

$$\xi^*_{2,P}(0)=1/24.$$

Further, we see from Proposition 2.22 that $J(1-m; (1, 1))=0$ $(m \in \boldsymbol{N})$ and that

$$J'(0; (1, 1))=-2\pi i\int_{I_\varepsilon(\infty)} \log t\cdot\phi(t)\phi^{(2)}(t)dt+16\pi^2 B_1B_2.$$

Since we can write $\phi(t)\phi^{(2)}(t)=\mu_1\phi'(t)+\mu_2\phi^{(2)}(t)+\mu_3\phi^{(3)}(t)$ with $\mu_1=-1/6$, $\mu_2=-1/2$, $\mu_3=-1/3$, we have

$$\int_{I_\varepsilon(\infty)} \log t\cdot\phi(t)\phi^{(2)}(t)dt=-2\pi i\sum_{j=1}^{3}\frac{\mu_jB_j}{j}=-\pi i/12,$$

which shows that $J'(0; (1, 1))=-3\pi^2/2$. We see from Proposition 2.24 that $K(0; (1, 1))=0$ (note that $\mathcal{N}_{(0,1)}=\mathcal{N}_{(1,0)}=0$). Therefore, the function $\zeta_S(s; \{V_1, V_2, V_3\}, (1, 1))$ is holomorphic at $s=1-m$ $(m \in \boldsymbol{N})$, and its special value at $s=0$ is given by

(2.5.11) $$\zeta_S(0; \{V_1, V_2, V_3\}, (1, 1))=iC(1)J'(0; (1, 1))=-3/32.$$

Moreover, we have by Proposition 2.25,

$$\zeta_S(0; \{V_1, V_3\}, 1)=1/12, \qquad \zeta_S(0; \{V_2, V_3\}, 1)=1/16,$$

which, in addition to (2.5.11), imply that $\xi_{2,S}^*(0) = -1/48$. Accordingly,

$$\xi_2^*(0) = \xi_{2,P}^*(0) + \xi_{2,S}^*(0) = 1/48.$$

Thus we are successful in giving another proof of the following theorem due to Siegel [20, Satz 3], Shintani [16, Theorem 2].

Theorem 2.40 (Siegel-Shintani). *The special value at $s=0$ of the zeta function $\xi_2^*(s)$ is given by*

$$\xi_2^*(0) = 1/48.$$

Chapter III. Some applications to the representation of $Sp(2n, \boldsymbol{F}_p)$ in the space of Siegel cusp forms

3.1. The representatinn μ_k of $Sp(2n, \boldsymbol{F}_p)$ in the space of cusp forms

Let $\mathfrak{H}_n$ be the Siegel upper half plane of degree n: $\mathfrak{H}_n = \{Z \in M_n(\boldsymbol{C}) \mid Z = {}^tZ \ \mathrm{Im}\,(Z) > 0\}$. The real symplectic group $\mathfrak{G}_{2n} = Sp(2n, \boldsymbol{R})$ of degree 2n acts on $\mathfrak{H}_n$ in a usual manner:

$$Z \longrightarrow g\langle Z\rangle = (AZ+B)(CZ+D)^{-1} \qquad \left(Z \in \mathfrak{H}_n,\ g = \begin{pmatrix} A & B \\ C & D \end{pmatrix} \in \mathfrak{G}_{2n}\right).$$

By this action, $\mathfrak{G}_{2n}/\{\pm 1\}$ gives the group of biholomorphic automorphisms of $\mathfrak{H}_n$. We put

$$J(g, Z) = \det\,(CZ+D) \qquad \text{for } g = \begin{pmatrix} A & B \\ C & D \end{pmatrix} \in \mathfrak{G}_{2n}.$$

Denote by $\Gamma_{2n}(l)$ the principal congruence subgroup with level $l\,(l \in \boldsymbol{N})$ of the Siegel modular group $Sp(2n, \boldsymbol{Z})$:

$$\Gamma_{2n}(l) = \{\gamma \in Sp(2n, \boldsymbol{Z}) \mid \gamma \equiv 1_{2n} \bmod l\}.$$

For a positive integer k, let $\mathfrak{S}_k(\Gamma_{2n}(l))$ be the space of Siegel cusp forms of degree n, weight k with respect to $\Gamma_{2n}(l)$; namely, it consists of all holomorphic functions $f(Z)$ on $\mathfrak{H}_n$ which satisfy the following two conditions:

$$\begin{cases} \text{(i)} \ \ f(\gamma\langle Z\rangle) = J(\gamma, Z)^k f(Z) & \text{for any } \gamma \in \Gamma_{2n}(l), \\ \text{(ii)} \ \ \det\,(\mathrm{Im}\,(Z))^{k/2}\,|f(Z)| & \text{is bounded on } \mathfrak{H}_n. \end{cases}$$

We write, simply, $\Gamma_{2n} = \Gamma_{2n}(1)\ (= Sp(2n, \boldsymbol{Z}))$. The group $\Gamma_{2n}(l)$ is a normal subgroup of Γ_{2n}. For $\alpha \in \Gamma_{2n}$ and $f \in \mathfrak{S}_k(\Gamma_{2n}(l))$, we put

(3.1.1) $$(f|[\alpha]_k)(Z)=J(\alpha, Z)^{-k}f(\alpha\langle Z\rangle).$$

Obviously, $f|[\alpha]_k \in \mathfrak{S}_k(\Gamma_{2n}(l))$. In the following we suppose that p is an odd prime. It is well-known that the quotient group $\Gamma_{2n}/\Gamma_{2n}(p)$ is isomorphic to the finite symplectic group $Sp(2n, \boldsymbol{F}_p)$ of degree $2n$ over the finite field $\boldsymbol{F}_p$ of p-elements. For $\alpha \in \Gamma_{2n}$, $\bar{\alpha}=\alpha \bmod p$ is regarded as an element of $Sp(2n, \boldsymbol{F}_p)$ via the isomorphism $\Gamma_{2n}/\Gamma_{2n}(p)\cong Sp(2n, \boldsymbol{F}_p)$. By means of (3.1.1), one can construct a representation μ_k of $Sp(2n, \boldsymbol{F}_p)$ in the space $\mathfrak{S}_k(\Gamma_{2n}(p))$; for $\bar{\alpha} \in Sp(2n, \boldsymbol{F}_p)$ $(\alpha \in \Gamma_{2n})$ and $f \in \mathfrak{S}_k(\Gamma_{2n}(p))$, we set

(3.1.2) $$\mu_k(\bar{\alpha})f=f|[\alpha^{-1}]_k.$$

It is easily checked that (3.1.2) is well-defined. We put, for $\gamma \in \mathfrak{G}_{2n}$, $Z \in \mathfrak{H}_n$,

$$H(\gamma; Z)=\det(\operatorname{Im}(Z))^k \det\left(\frac{\gamma\langle Z\rangle-\bar{Z}}{2i}\right)^{-k} J(\gamma, Z)^{-k}.$$

We have $H(\gamma; g\langle Z\rangle)=H(g^{-1}\gamma g; Z)$ for $g, \gamma \in \mathfrak{G}_{2n}$. Denote by dZ the invariant measure on $\mathfrak{H}_n$ given by

$$dZ=\det(Y)^{-n-1}\prod_{1\leq i\leq j\leq n} dX_{ij}dY_{ij} \qquad \text{for } Z=X+iY.$$

The following theorem is due to Godement [2] (see also Lemma 1 of [5]).

Theorem 3.1 (Godement). *Let $k>2n$. The trace of $\mu_k(\bar{\alpha}^{-1})$ $(\alpha \in \Gamma_{2n})$ is given by the following formula:*

$$\operatorname{tr}(\mu_k(\bar{\alpha}^{-1}))=a(k)\int_{\Gamma_{2n}(p)\backslash\mathfrak{H}_n}\sum_{\gamma\in\Gamma_{2n}(p)\alpha} H(\gamma; Z)dZ,$$

where we put

(3.1.3) $$a(k)=\frac{\gamma_n(k-(n+1)/2)}{2^n(2\pi)^{n(n+1)/2}\gamma_n(k-n-1)} \qquad (\textit{for } \gamma_n(s), \textit{ see Chap. I}).$$

For a symmetric matirx x of size ν $(1\leq\nu\leq n)$, we write

(3.1.4) $$t_{n,\nu}(x)=\begin{pmatrix} 1_n & \begin{matrix} x & 0 \\ 0 & 0 \end{matrix} \\ 0 & 1_n \end{pmatrix}.$$

Suppose that $\bar{\alpha}$ $(\alpha \in \Gamma_{2n})$ is $Sp(2n, \boldsymbol{F}_p)$-conjugate to some element $t_{n,\nu}(S)$ with $S \in L_\nu$, $\det(S)\not\equiv 0 \bmod p$. For each integer r $(1\leq r\leq n)$, let $\Pi_r(\alpha)$ be the set consisting of all elements $\gamma \in \Gamma_{2n}(p)\alpha$ that are Γ_{2n}-conjugate to some elements $t_{n,r}(x)$ with $x \in L_r$, $\det(x) \neq 0$. Set, following [16, §3 of Chap. 2],

$$(3.1.5) \qquad I_n(\Pi_r(\alpha); k) = a(k) \int_{\Gamma_{2n}(p)\backslash \mathfrak{H}_n} \sum_{\gamma \in \Pi_r(\alpha)} H(\gamma; Z) dZ.$$

Put
$$U_n = \prod_{k=1}^{n} (2\pi^k / \Gamma(k)).$$

We denote by $\delta_n g$ the invariant measure on $\mathfrak{G}_{2n}$ which satisfies

$$\int_{\mathfrak{G}_{2n}} f(g\langle i1_n \rangle)\delta_n g = 2^{-n} U_n \int_{\mathfrak{H}_n} f(Z) dZ$$

for any integrable function f on $\mathfrak{H}_n$. Then we have

$$(3.1.6) \qquad I_n(\Pi_r(\alpha); k) = a'(k) \int_{\Gamma_{2n}(p)\backslash \mathfrak{G}_{2n}} \sum_{\gamma \in \Pi_r(\alpha)} H(g^{-1}\gamma g; i1_n)\delta_n g$$

where $a'(k) = 2^n a(k)/U_n$.

3.2. On the integrals $I_n(\Pi_r(\alpha); k)$

In the case of $\alpha = 1_{2n}$, Shintani [16] proved the absolutely convergence of the integrals $I_n(\Pi_r(\alpha); k)$ under a certain condition for k and expressed the values of them as elementary constant multiples of the special values at non-positive integers of the zeta function $\xi_r^*(s)$. In the following, as quite an analogy of Shintani's results, we evaluate the values of the integrals $I_n(\Pi_r(\alpha); k)$ for a general α by using the special values of $\xi_r^*(s; \tau_S^{(r)})$. We keep the notation used in Chap. I.

Let $1 \leqq \nu \leqq n$ and let $S \in L_\nu$ with $\det(S) \not\equiv 0 \bmod p$. We may assume that $\alpha = t_{n,\nu}(S)$. Then we observe that $\Pi_r(\alpha) \neq \phi$ only if $\nu \leqq r \leqq n$. Set

$$\Gamma_{\alpha,p} = \{\sigma \in \Gamma_{2n} \mid \sigma^{-1}\alpha\sigma \equiv \alpha \bmod p\}.$$

Then, $\Gamma_{\alpha,p}$ is a subgroup of Γ_{2n}, and $\Gamma_{2n}(p)$ is a normal subgroup of $\Gamma_{\alpha,p}$. In a formal manner, we get, by (3.1.6),

$$(3.2.1) \qquad I_n(\Pi_r(\alpha); k) = a'(k)[\Gamma_{\alpha,p} : \Gamma_{2n}(p)]$$
$$\times \int_{\Gamma_{2n}\backslash \mathfrak{G}_{2n}} \sum_{\sigma \in \Gamma_{\alpha,p}\backslash \Gamma_{2n}} \sum_{\gamma \in \sigma^{-1}\Pi_r(\alpha)\sigma} H(g^{-1}\gamma g; i1_n)\delta_n g,$$

where $[\Gamma_{\alpha,p} : \Gamma_{2n}(p)]$ denotes the group index of $\Gamma_{\alpha,p}$ to $\Gamma_{2n}(p)$. Let $\mathscr{L}_n^{(r)}(S)$ be the same as in (1.2.3). Denote by $\mathfrak{G}_{2n,r}$ the subgroup of $\mathfrak{G}_{2n}$ consisting of all matrices whose left lower $(2n-r)\times r$ blocks are zero. Every element q of $\mathfrak{G}_{2n,r}$ has the following block decomposition:

$$(3.2.2)\qquad q=\begin{pmatrix}1_n & \begin{matrix} y_1 & y_{12} \\ {}^t y_{12} & 0\end{matrix} \\ & 1_n\end{pmatrix}\begin{pmatrix}\begin{matrix}1_r & x_{12}\\ 0 & 1_{n-r}\end{matrix} & \\ & \begin{matrix}1_r & 0 \\ -{}^t x_{12} & 1_{n-r}\end{matrix}\end{pmatrix}\begin{pmatrix}a & & & \\ & \alpha & & \beta \\ & & {}^t a^{-1} & \\ & \gamma & & \delta\end{pmatrix},$$

where $a \in GL_r(\boldsymbol{R})$, $h=\begin{bmatrix}\alpha & \beta \\ \gamma & \delta\end{bmatrix} \in \mathfrak{G}_{2(n-r)}$, $x_{12}, y_{12} \in M(r, n-r; \boldsymbol{R})$, $y_1 \in M_r(\boldsymbol{R})$ (${}^t y_1 = y_1$). Let $\delta_{n,r}q$ be the left invariant measure of $\mathfrak{G}_{2n,r}$ given by

$$\delta_{n,r}q = |\det(a)|^{-(2n-r+1)} d_r a \delta_{n-r} h dx_{12} dy_1 dy_{12}.$$

Lemma 3.2. *Let* $\alpha = t_{n,\nu}(S)$ *with* $S \in L_\nu$, $\det(S) \not\equiv 0 \bmod p$. *For each integer* r *with* $\nu \leq r \leq n$, *we have*

$$\bigcup_{\sigma \in \Gamma_{\alpha,p} \backslash \Gamma_{2n}} \sigma^{-1} \Pi_r(\alpha) \sigma = \bigcup_{\gamma \in \Gamma_{2n} \cap \mathfrak{G}_{2n,r} \backslash \Gamma_{2n}} \gamma^{-1} \{t_{n,r}(x) \mid x \in \mathscr{L}_r^{(r)}(S)\} \gamma$$

(*the both sides are disjoint unions*).

Proof. In the proof we use, implicitly, the fact that p is an odd prime. The disjointness of the unions is clear. First, let $x \in \mathscr{L}_r^{(r)}(S)$, $\gamma \in \Gamma_{2n}$. There exists $U \in GL_r(\boldsymbol{Z}/p\boldsymbol{Z})$ with $x \equiv U\begin{pmatrix}S & 0 \\ 0 & 0\end{pmatrix}{}^t U \bmod p$. Since the map $\beta \to \bar{\beta}$ of Γ_{2n} to $Sp(2n, \boldsymbol{F}_p)$ is surjective, there exists $\sigma_1 \in \Gamma_{2n}$ such that $t_{n,r}(x) \equiv \sigma_1^{-1} \alpha \sigma_1 \bmod p$. Thus, $\gamma^{-1} t_{n,r}(x) \gamma = (\sigma_1 \gamma)^{-1} \tau_1 \alpha (\sigma_1 \gamma)$ with some $\tau_1 \in \Gamma_{2n}(p)$. Write $\sigma_1 \gamma = \rho \sigma$ with $\rho \in \Gamma_{\alpha,p}$ and $\sigma \in \Gamma_{\alpha,p} \backslash \Gamma_{2n}$. Then, $\gamma^{-1} t_{n,r}(x) \gamma = \sigma^{-1} \rho^{-1} \tau_1 \alpha \rho \sigma$. Since $\rho^{-1} \alpha \rho \equiv \alpha \bmod p$, we have $\rho^{-1} \alpha \rho = \tau_2 \alpha$ with some $\tau_2 \in \Gamma_{2n}(p)$. Hence, $\rho^{-1} \tau_1 \alpha \rho \in \Pi_r(\alpha)$.

Conversely, let $\sigma \in \Gamma_{\alpha,p} \backslash \Gamma_{2n}$, and let $\tau \in \Gamma_{2n}(p)$ with $\tau\alpha \in \Pi_r(\alpha)$. Then we have $\sigma^{-1} \tau \alpha \sigma = \gamma^{-1} t_{n,r}(x) \gamma$ with some $\gamma \in \Gamma_{2n} \cap \mathfrak{G}_{2n,r} \backslash \Gamma_{2n}$ and $x \in L_r$, $\operatorname{rank}(x) = r$. The task we have to do is to prove that $x \in \mathscr{L}_r^{(r)}(S)$. If we put $\gamma \sigma^{-1} = \gamma_1$, then, $\gamma_1^{-1} \alpha \gamma_1 \equiv t_{n,r}(x) \bmod p$. Writing $\gamma_1 = \begin{pmatrix}A & B \\ C & D\end{pmatrix} \in \Gamma_{2n}$ we have

$$(3.2.3)\qquad \begin{aligned}&\begin{pmatrix}S & 0 \\ 0 & 0\end{pmatrix} C \equiv 0 \bmod p, \qquad \begin{pmatrix}S & 0 \\ 0 & 0\end{pmatrix} D \equiv A \begin{pmatrix}x & 0 \\ 0 & 0\end{pmatrix} \bmod p,\\ &C\begin{pmatrix}x & 0 \\ 0 & 0\end{pmatrix} \equiv 0 \bmod p.\end{aligned}$$

Since ${}^t DA - {}^t BC = 1_n$, we get, by (3.2.3),

$$(3.2.4)\qquad {}^t D \begin{pmatrix}S & 0 \\ 0 & 0\end{pmatrix} D \equiv {}^t DA \begin{pmatrix}x & 0 \\ 0 & 0\end{pmatrix} \equiv (1_n + {}^t BC)\begin{pmatrix}x & 0 \\ 0 & 0\end{pmatrix} \equiv \begin{pmatrix}x & 0 \\ 0 & 0\end{pmatrix} \bmod p.$$

We decompose A, C, D as follows:

$$A=\begin{pmatrix} a_1 & a_2 \\ a_3 & a_4 \end{pmatrix},\quad C=\begin{pmatrix} c_1 & c_2 \\ c_3 & c_4 \end{pmatrix},\quad D=\begin{pmatrix} d_1 & d_2 \\ d_3 & d_4 \end{pmatrix} \qquad \text{with } a_1,\ c_1,\ d_1 \in M_\nu(\boldsymbol{Z}).$$

Since $\det(S)\not\equiv 0 \bmod p$, we have, again by (3.2.3), $c_1\equiv 0 \bmod p$, $c_2\equiv 0 \bmod p$. Thus we see easily from the relation $A^tD-B^tC=1_n$ that $a_1^td_1+a_2^td_2\equiv 1_\nu \bmod p$. Accordingly, there exists some $V\in GL_n(\boldsymbol{Z}/p\boldsymbol{Z})$ of the form $V=\begin{pmatrix} d_1 & d_2 \\ * & * \end{pmatrix}$. We have, by (3.2.4), ${}^tV\begin{pmatrix} S & 0 \\ 0 & 0 \end{pmatrix}V\equiv\begin{pmatrix} x & 0 \\ 0 & 0 \end{pmatrix} \bmod p$. Thus, $\begin{pmatrix} x & 0 \\ 0 & 0 \end{pmatrix}\in\mathscr{L}_n^{(r)}(S)$. In the same manner as in the proof of Lemma 1.4, we can conclude that $x\in\mathscr{L}_r^{(r)}(S)$. q.e.d.

We set

$$\omega_n=\zeta(2)\zeta(4)\cdots\zeta(2n).$$

Let $\nu\leq r\leq n$. Using the decomposition in Lemma 3.2, we get, by the equality (3.2.1),

$$\text{(3.2.5)}\quad I_n(\Pi_r(\alpha);k)$$
$$=a'(k)[\Gamma_{\alpha,p}:\Gamma_{2n}(p)]\int_{\Gamma_{2n}\cap\mathfrak{G}_{2n,r}\backslash\mathfrak{G}_{2n}}\sum_{x\in\mathscr{L}_r^{(r)}(S)}H(g^{-1}t_{n,r}(x)g;\,i1_n)\delta_n g.$$

Making use of Lemma 22 of [16] and the decomposition (3.2.2) of $q\in\mathfrak{G}_{2n,r}$, we have

$$\text{(3.2.6)}\quad I_n(\Pi_r(\alpha);k)=a'(k)[\Gamma_{\alpha,p}:\Gamma_{2n}(p)]\frac{U_n}{U_{n-r}C_r}$$
$$\times\int_{\Gamma_{2n}\cap\mathfrak{G}_{2n,r}\backslash\mathfrak{G}_{2n,r}}\sum_{x\in\mathscr{L}_r^{(r)}(S)}H(t_{n,r}(a^{-1}x^ta^{-1});\,i1_n)\delta_{n,r}q$$
$$=a'(k)[\Gamma_{\alpha,p}:\Gamma_{2n}(p)]2^{r(2n-r+1)/2}\frac{U_n\omega_{n-r}}{U_{n-r}C_r}$$
$$\times\int_{G_{\boldsymbol{R},+}^{(r)}/SL_r(\boldsymbol{Z})}\chi(g)^{n-(r-1)/2}\sum_{x\in\mathscr{L}_r^{(r)}(S)}\det(1_r-igx^tg)^{-k}d_rg$$
$$=a'(k)[\Gamma_{\alpha,p}:\Gamma_{2n}(p)]2^{r(2n-r+1)/2}\frac{U_n\omega_{n-r}}{U_{n-r}C_r}$$
$$\times Z(f_r^*(x,k),\,\mathscr{L}_r(S),\,n-(r-1)/2).$$

By virtue of Proposition 1.6, the integral $Z(f_r^*(x,k),\mathscr{L}_r(S),n-(r-1)/2)$ is absolutely convergent for $k\geq 2n+3$, and hence the equalities (3.1.6), (3.2.1), (3.2.5), (3.2.6) can be justified definitely. Thus the absolutely convergence of the integral $I_n(\Pi_r(\alpha);k)$ follows for $k\geq 2n+3$. We obtain

Proposition 3.3.

Proposition 3.3. *Let* $1\leqq\nu\leqq n$ *and take* $S\in L_\nu$ *with* $\det(S)\not\equiv 0 \bmod p$. *Put* $\alpha=t_{n,\nu}(S)$. *For each* r $(\nu\leqq r\leqq n)$, *the integral* $I_n(\Pi_r(\alpha); k)$ *given by* (3.1.5) *is absolutely convergent for* $k\geqq 2n+3$ *and is equal to*

$$[\Gamma_{\alpha,p}: \Gamma_{2n}(p)]p^{-r(n-(r-1)/2)}b(n, k, r)\Omega_{n,r}\xi_r^*(r-n, \tau_S^{(r)}),$$

where we put

$$(3.2.7)\qquad b(n, k, r)=\prod_{j=1}^{n-r}(2k-n-j)(2k-n-j+2)\cdots(2k-n+j-2),$$

$$(3.2.8)\qquad \Omega_{n,r}=\frac{\omega_{n-r}2^{r(n-r)-1}}{U_{n-r}(4\pi)^{(n-r)(n-r+1)/2}}$$

(*we understand* $b(n, k, r)=1$, $\Omega_{n,r}=1$ *for* $r=n$).

Proof. The absolutely convergence of the integral $I_n(\Pi_r(\alpha); k)$ has already been verified. The functional equation in Proposition 1.9 shows that

$$(3.2.9)\qquad Z(f_r^*(x, k), \mathscr{L}_r(S), n-(r-1)/2)$$
$$=\frac{\pi^{r(r+1)/2}p^{-r(n-(r-1)/2)}C_r\gamma_r(k-n-1)}{2(2\pi)^{r(r-n)}\gamma_r(k-(r+1)/2)}\cdot\xi_r^*(r-n, \tau_S^{(r)}).$$

The latter half of the assertion is a direct consequence of (3.2.6), (3.2.9).

3.3. Traces of $\mu_k(\bar{\alpha})$ in the case of degree 4 $(n=2)$

We consider the case of $n=2$, $\nu=1$. Take a non-quadratic residue κ mod p and fix it. For any integer μ prime to p, we put

$$\alpha_\mu=t_{2,1}(\mu)=\begin{pmatrix} 1_2 & \begin{matrix}\mu & 0\\ 0 & 0\end{matrix} \\ 0 & 1_2\end{pmatrix}\in\Gamma_4.$$

It is easy to see that

$$[\Gamma_{\alpha_\mu,p}: \Gamma_4(p)]=2p^4(p^2-1).$$

Since $L(-1, \psi)=-B_{2,\psi}/2$, $\zeta(-1)=-1/12$, the identity (1.2.12) shows that

$$\xi_1^*(-1, \tau_\mu^{(1)})=-\frac{1}{4}\psi(\mu)\tau_\psi B_{2,\psi}+\frac{1}{24}(1-p^2).$$

It follows from Proposition 3.3 that, if $r=1$.

$$I_2(\Pi_1(\alpha_\mu);\ k) = -2^{-5}3^{-1}p^2(p^2-1)\{\psi(\mu)\tau_\psi B_{2,\psi}+(p^2-1)/6\}(2k-3) \quad (k\geqq 7).$$

It is immediate to see from Proposition 1.2, Proposition 3.3 that, if $r=2$,

$$I_2(\Pi_2(\alpha_\mu);\ k) = \frac{p(p^2-1)}{2}\{p\psi(-1)L_2^*(0,\ \psi_{\det})+p\psi(\mu)\tau_\psi L_2^*(0,\ \psi_{H,p})$$
$$+(p^2-1)\xi_2^*(0)\} \qquad (k\geqq 7).$$

Thus, by virtue of Theorem 2 in the introduction and Theorem 2.40, we obtain

$$I_2(\Pi_2(\alpha_\mu);\ k) = \frac{p(p^2-1)}{2}\{p\psi(\mu)\tau_\psi L_2^*(0,\ \psi_{H,p})+2^{-4}3^{-1}(2p^2-p-1)\}.$$

It is essentially known by [1], [13], [14], [5] and verified in a similar manner that, if $n=2$, in the trace formula for $\operatorname{tr}(\mu_k(\bar{\alpha}_\mu^{-1}))$ in Theorem 3.1, the contributions from any other conjugacy classes except from $\Pi_r(\alpha_\mu)$ $(r=1, 2)$ vanish identically. Thus one obtains, if $k\geqq 7$,

$$\operatorname{tr}(\mu_k(\bar{\alpha}_\mu)) = \sum_{j=1}^{2} I_2(\Pi_j(\alpha_{-\mu});\ k).$$

Summing up the results above, we obtain the following theorem (and also (0.3) in the introduction).

Theorem 3.4. *Let μ be any integer prime to p. If $k\geqq 7$, then,*

$$\operatorname{tr}(\mu_k(\bar{\alpha}_\mu)) = -2^{-5}3^{-1}p^2(p^2-1)\{\psi(-\mu)\tau_\psi B_{2,\psi}+(p^2-1)/6\}(2k-3)$$
$$+2^{-1}p(p^2-1)\{p\psi(-\mu)\tau_\psi L_2^*(0,\ \psi_{H,p})+2^{-4}3^{-1}(2p^2-p-1)\}.$$

It is well-known that

$$\tau_\psi = \begin{cases} \sqrt{p} & p\equiv 1 \bmod 4, \\ i\sqrt{p} & p\equiv 3 \bmod 4. \end{cases}$$

Substituting 1, κ for μ and subtracting $\operatorname{tr}(\mu_k(\bar{\alpha}_\kappa))$ from $\operatorname{tr}(\mu_k(\bar{\alpha}_1))$ in Theorem 3.4, we obtain Theorem 3 in the introduction. As a direct corollary of Theorem 3.4, the imaginary part of $\operatorname{tr}(\mu_k(\bar{\alpha}_\mu))$ is given as follows:

$$\operatorname{Im}(\operatorname{tr}(\mu_k(\bar{\alpha}_\mu))) = \begin{cases} 0 & p\equiv 1 \bmod 4, \\ 2^{-1}p^{5/2}(p^2-1)\psi(-\mu)L_2^*(0,\ \psi_{H,p}) & p\equiv 3 \bmod 4. \end{cases}$$

References

[1] U. Christian, Berechnung des Ranges der Schar der Spitzenformen zur Modulgruppe zweiten Grades und Stufe $q>2$, J. Reine Angew. Math., **277** (1975), 130–154; Zur Berechnung des Ranges der Schar der Spitzenformen zur Modulgruppe zweiten Grades und Stufe $q>2$, ibid., **296** (1977), 108–118.

[2] R. Godement, Seminaire H. Cartan, 1957–58, Fonctions automorphes, Exposé 5–10.

[3] A. Gyoja and N. Kawanaka, Gauss sums of prehomogeneous vector spaces, Proc. Japan Acad., **61** Ser. A (1985), 19–22.

[4] K. Hashimoto, The dimension of the space of cusp forms on Siegel upper half plane of degree two (I), J. Fac. Sci. Univ. Tokyo, Sect. IA, **30** (1983), 403–488.

[5] ——, Representation of the finite symplectic group $Sp(4, \boldsymbol{F}_p)$ in the space of Siegel modular forms, Contemporary Math., **53** (1986), 253–276.

[6] E. Hecke Über ein Fundamentalproblem aus der Theorie der elliptischen Modulfunktionen, Abh. Math. Seminar der Univ. Hamburg, **6** (1928), 235–257.

[7] ——, Über das Verhalten der Integrale 1. Gattung bei Abbildungen, insbesondere in der Theorie der elliptischen Modulfunktionen, ibid., **8** (1930), 271–281.

[8] H. Klingen, Über die Werte der Dedekindschen Zetafunktion, Math. Ann., **145** (1962), 265–272.

[9] A. Kurihara, On the values at non-positive integers of Siegel's zeta functions of Q-anisotropic quadratic forms with signature $(1, n-1)$, J. Fac. Sci. Univ. Tokyo, Sect. IA, **28** (1982), 567–584.

[10] S. Lang, Cyclotomic Fields, Springer Verlag, New York-Heidelberg-Berlin, 1978.

[11] R. Lee and S. H. Weintraub, On a generalization of a theorem of Erich Hecke, Proc. Nat. Acad. Sci. U.S.A., **79** (1982), 7955–7957.

[12] H. Maass, Siegel's Modular Forms and Dirichlet Series, Lecture Notes in Math., **216**, Springer Verlag, 1971.

[13] Y. Morita, An explicit formula for the dimension of spaces of Siegel modular forms of degree two, J. Fac. Sci. Univ. Tokyo, Sect. IA, **21** (1974), 167–248.

[14] I. Satake, Special values of zeta functions associated with self-dual homogeneous cones, Progress in Math., **14**, Birkhäuser, (1981), 359–384.

[15] F. Sato, Remarks on functional equations of zeta distributions, Advanced Studies in pure Math., (this volume).

[16] T. Shintani, On zeta-functions associated with the vector space of quadratic forms, J. Fac. Sci. Univ. Tokyo, Sect. IA, **22** (1975), 25–65.

[17] ——, On evaluation of zeta functions of totally real algebraic number fields at non-positive integers, J. Fac. Sci. Univ. Tokyo, Sect. IA, **23** (1976), 393–417.

[18] C. L. Siegel, Über die analytische Theorie der quadratischen Formen, Ann. of Math., **36** (1935), 527–606.

[19] ——, The volume of the fundamental domain for some infinite groups, Trans. Amer. Math. Soc., **39** (1936), 209–218.

[20] ——, Über die Zetafunktionen indefiniter quadratischer Formen, Math. Z., **43** (1938), 682–708.

[21] ——, Berechnung von Zetafunktionen an ganzzähligen Stellen, Nachr. Akad. Wiss. Göttingen, Math.-Phys. Kl, (1969), 87–102.

[22] R. Tsushima, On the spaces of Siegel cusp forms of degree two, Amer. J. Math., **104** (1982), 843–885.

[23] ——, The spaces of Siegel cusp forms of degree two and the representation of $Sp(2, \boldsymbol{F}_p)$, Proc. Japan Acad., **60** (1984), 209–211.
[24] T. Yamazaki, On the representation of $Sp(2, \boldsymbol{F}_p)$ in the space of Siegel cusp forms, (preprint, 1974).

Department of Mathematics
Rikkyo University
Nishi-Ikebukuro, Tokyo 171
Japan

Advanced Studies in Pure Mathematics 15, 1989
Automorphic Forms and Geometry of Arithmetic Varieties
pp. 171–210

Selberg-Ihara's Zeta function for p-adic Discrete Groups

Ki-ichiro Hashimoto*) and Akira Hori

Dedicated to Prof. Friedrich Hirzebruch and Prof. Ichiro Satake on their sixtieth birthdays

Contents

Introduction

0-1. Let G be $\mathrm{PSL}(2, \boldsymbol{R})$ and let Γ ($\subset G$) be a Fuchsian group of the first kind. In [Sel], a zetafunction $Z_\Gamma(s)$ was introduced and proved to have many important properties which resemble those of usual L-functions, such as Euler product, functional equation, and analogue of Riemann Hypothesis. This function, now called with the name of Selberg, is generalized to any discrete subgroup Γ of a semi-simple Lie group of $\boldsymbol{R}$-rank one, when G/Γ is compact by Gangolli [Gan], and later by Gangolli-Warner [G-W] to the case when G/Γ has a finite volume. Meanwhile, an analogue of $Z_\Gamma(s)$ was introduced by Ihara [I-1], for a cocompact torsion-free discrete subgroup Γ of $\mathrm{PSL}(2, K)$ or $\mathrm{PL}(2, K)$, where K is a p-adic field. Especially it was shown that Ihara's zeta function $Z_\Gamma(u)$ is a

Received November 30, 1987.
Revised April 1, 1988.

*) The first author has been partially supported by Sonderforschungsbereich 170, 'Geometrie und Analysis', at Univ. Göttingen.

rational function of the indeterminate u. This result is based on a remarkable structure theorem of such Γ, which in particular asserts that it is a free group with explicitly constructible basis (i.e. minimal set of generators).

0-2. In this paper, we shall extend Ihara's results to the case when G is a semi-simple algebraic group over a $\mathfrak{p}$-adic field K and Γ is a discrete subgroup of G such that G/Γ is compact, under the assumption that G has K-rank one, and that it has an affine Tits system (G, B, N, S). In this case, S consists of two elements, say, s_1, s_2, and Bruhat-Tits' building X attached to (G, B, N, S) is a tree with two distinct kinds of vertices ([T-1, 2]). It is known that Γ acts on X freely and that Γ is again a free group ([Ser-1]). As in Ihara [I-1], our zetafunction $Z_\Gamma(u; \rho)$ is defined, for an n-dimensional unitary representation ρ of Γ, by the infinite (formal) product

$$Z_\Gamma(u; \rho) := \prod_{\{\gamma\}_\Gamma} \det\{I_n - \rho(\gamma) u^{\deg\{\gamma\}_\Gamma}\}^{-1} \tag{0.1}$$

where the product is taken over the set of "primitive hyperbolic" Γ-conjugacy classes $\{\gamma\}_\Gamma$, and its degree is defined as follows. Recall that $U_1 = B \cup Bs_1B$, $U_2 = B \cup Bs_2B$ are the subgroups of G which contain B properly and which represent the G-conjugacy classes of the maximal (open) compact subgroups. We define the length function $l: G \to \boldsymbol{N} \cup \{0\}$ by

$$l(g) = m \Longleftrightarrow g \in G_m := U_1(s_1s_2)^m U_1, \tag{0.2}$$

and put

$$\deg\{\gamma\}_\Gamma := \operatorname*{Min}_{x \in G} l(x^{-1}\gamma x). \tag{0.3}$$

0-3. To state our main result, put

$$q_i := \#(B\backslash Bs_iB) \quad \text{and} \quad h_i := \#(U_i\backslash G/\Gamma) \quad (i=1, 2).$$

It follows from our assumption on the compactness of G/Γ that h_i (the class number of Γ with respect to U_i) is finite. Let $\mathscr{H}(G, U_1)$ be the Hecke algebra of the pair (G, U_1), and let $A_{1,\rho}$ be the following matrix given by the Brandt representation of $\mathscr{H}(G, U_1)$ attached to ρ:

$$A_{1,\rho} := (a_{ij}) \in M(nh_1, \boldsymbol{C}), \text{ with } a_{ij} = \sum_{\gamma} \rho(\gamma), \tag{0.4}$$

where the last sum is taken over the set $\Gamma \cap x_i^{-1}G_1x_j$, $\{x_i \ (1 \leq i \leq h_1)\}$ being a complete set of representatives of $U_1\backslash G/\Gamma$. See the text (§ 7) for another interpretation of $A_{1,\rho}$. The purpose of this paper is to prove the following:

Theorem (0.5). *Let the notation and assumptions be as above. Then* Γ *is a free group of rank* $r = q_1h_1 - h_2 + 1$ $(= q_2h_2 - h_1 + 1)$, *and we have*

(0.6) $$Z_\Gamma(u;\rho)^{-1}=(1-u)^{n(r-1)}(1+q_2u)^{n(h_2-h_1)}\times\det\{I_{nh_1}-(A_{1,\rho}-q_2+1)u+q_1q_2u^2\}.$$

Let $\rho=\mathbf{1}$ be the trivial representation, and put $Z_\Gamma(u)=Z_\Gamma(u;\mathbf{1})$. Combining the above result and that of Garland [Gar] (see also [Cas]), we get an information on the spectral decomposition of $L^2(G/\Gamma)$, which was first found by Ihara [I-1] for $G=\mathrm{PSL}_2(K)$:

Corollary (0.7). *We have the following equality*

$$-\operatorname*{Res}_{u=1}\frac{d}{du}\log Z_\Gamma(u)=r=\textit{the multiplicity of the Steinberg representation in } L^2(G/\Gamma).$$

There are at least three different ways to prove the above theorem. The first, which is based on the combinatorial argument on the structure of Γ, is a reformulation of the ideas of Ihara [I-1] in terms of the Tits system. We shall describe this proof in some detail. The second proof, which uses the graph-theoretic interpretation of the Γ-conjugacy classes, will be sketched briefly. The third method, which uses the Selberg's original idea of trace formula, is not complete yet. It would be an interesting problem to give an interpretation of our results and some questions arising therefrom, in terms of the spectral decomposition of $L^2(G/\Gamma)$. This will be treated in the subsequent paper [Ha].

0-4. As a matter of fact, our zetafunction $Z_\Gamma(u;\rho)$ can be defined for subgropus Γ of a much wider class of groups G; and the above theorem holds in such general setting. In the Appendix, we shall study the class of groups for which the whole procedure of the evaluation of $Z_\Gamma(u;\rho)$ developed in this paper can be applied without any change of notation.

Throughout the following, for any set S, $\sharp(S)$ will denote its cardinal number. For any ring A and a positive integer n, $M(n, A)$ will denote the ring of all n by n matrices whose entries are elements of A. The symbol $[x]$ ($x\in\boldsymbol{R}$) will be the largest integer n such that $n\leq x$.

§ 1. Groups with axiom (G, l, I), (G, l, II)

Let G be an abstract group. Assume that we are given a map

$$l: G\longrightarrow \boldsymbol{N}\cup\{0\}$$

called *the length*, satisfying the following conditions (G, l, I), (G, l, II), where G_l denotes the set of all elements of G with length l ($l=0, 1, 2, \cdots$) and $U:=G_0$.

(G, l, I) *For any l, G_l is non-empty, U is a subgroup of G, and*

$$G_l^{-1}=G_l, \quad UG_lU=G_l, \quad \sharp(U\backslash G_l)<\infty.$$

According to this condition we can define the Hecke algebra $\mathscr{H}(G, U)$ with respect to the pair (G, U) as in Shimura [Sh]. Since each G_l is a union of finite number of U-double cosets, it can be considered as an element of $\mathscr{H}(G, U)$ (by taking formal sum instead of disjoint union).

(G, l, II) *There exist two natural numbers q_1, q_2 such that*

$$G_1^2=G_2+(q_2-1)G_1+q_2(q_1+1)U \tag{1.1}$$

$$G_1G_l=G_{l+1}+(q_2-1)G_l+q_1q_2G_{l-1} \quad (2\leq l), \tag{1.2}$$

where the products G_1^2, G_1G_l are taken in $\mathscr{H}(G, U)$.

From this it follows immediately that

$$\sharp(U\backslash G_l)=(q_1q_2)^{l-1}q_2(q_1+1) \qquad \text{for } 1\leq l. \tag{1.3}$$

The class of groups satisfying our axiom (G, l, I), (G, l, II) has been studied in Appendix. Here we collect some of the results. The proofs are given in Appendix. Let $\Omega:=\{\omega_j;\ 1\leq j\leq t:=q_2(1+q_1)\}$ be a complete set of representatives of $U\backslash G_1$. Then one has the following

Lemma (1.4).
(i) *For each $\omega\in\Omega$, there is a unique $\rho\in\Omega$ such that $\rho\omega\in U$.*
(ii) *For each $\omega\in\Omega$, $\sharp\{\rho\in\Omega;\ \rho\omega\in G_1\}=q_2-1$.*
(iii) *For all other $\rho\in\Omega$, one has $\rho\omega\in G_2$.*

Lemma (1.5). *Any element $g\in G$ is expressed as a product*

$$g=u\omega_{j_1}\omega_{j_2}\cdots\omega_{j_l} \quad (u\in U,\ \omega_{j_1},\ \cdots,\ \omega_{j_l}\in\Omega),$$

for which one has $l(g)\leq l$, with equality if and only if $g\in G_l$. Moreover, the last condition is equivalent to

$$\omega_{j_n}\omega_{j_{n+1}}\in G_2 \qquad \text{for } n=1, 2, \cdots, l-1. \tag{*}$$

Lemma (1.6). *Suppose that $x, y, z\in G_1$.*
(i) *If $xy\in G_2$ and $yz\in G_i$, then $xyz\in G_{i+1}$ $(i=0, 1, 2)$.*
(ii) *If $xy\in G_i$ and $yz\in G_2$, then $xyz\in G_{i+1}$ $(i=0, 1, 2)$.*

Corollary (1.7). *Suppose that $x=x_1x_2\cdots x_m\in G_m$, and $y=y_1y_2\cdots y_n\in G_n$ are (free) products of $x_1, \cdots, x_m\in G_1$, $y_1, \cdots, y_n\in G_1$ respectively. If $x_my_1\in G_1$, then one has $l(xy)=m+n-1$, and vice versa.*

These results have simple interpretations in terms of trees (cf. Appendix). We now describe it briefly. Recall first the following:

Definition (1.8). Let q_1, q_2 be two positive integers. By a semi-regular bipartite tree of valency (q_1+1, q_2+1), we mean a connected tree $X(q_1, q_2)=(V_1, V_2; E)$ whose set of vertices is a disjoint union of two subsets V_1, V_2, and each vertex $P \in V_1$ (resp. $Q \in V_2$) is adjacent exactly to q_1+1 (resp. q_2+1) vertices of V_2 (resp. V_1) (see Fig. 1 in §2).

Let V_1 be the coset space $U\backslash G$, on which G acts by the right multiplication. For each $P=Ux \in V_1$, the set $V(P; 2)$ consisting of the points $Q=Uy$ such that $l(xy^{-1})=1$ has the cardinality $t=q_2(1+q_1)$(=independent of P), as one sees immediately from (G, l, I) and (1.3).

Lemma (1.9). *The set $V(P; 2)$ is divided into a disjoint union of q_1+1 subset $E_j(P)$ $(0 \leq j \leq q_1)$; and for any two distinct points $Q=Uy$, $R=Uz$ of $E_j(P)$, one has $l(yz^{-1})=1$.*

Proof. See Appendix Lemma (10.9).

Now we consider the (disjoint) union E^* of all pairs $(P, E_j(P))$, $P \in V_1$:

$$E^* := \bigcup_{P \in V_1} \bigcup_{j=0}^{q_1} (P, E_j(P)) \tag{1.10}$$

One can regard E^* as a fibre space over V_1, whose fibre over the point P is a set of q_1+1 subsets $E_j(P)$. This fibre space structure is compatible with the action of G. On the other hand one can introduce an equivalence relation on E^*:

$$(P, E_j(P)) \simeq (Q, E_k(Q)) \Longleftrightarrow E_j(P) \cup \{P\} = E_k(Q) \cup \{Q\}. \tag{1.11}$$

One can denote the equivalence class of $(P, E_j(P))$ by the q_2+1 points set $E_j(P) \cup \{P\}$. Note that each set $E_j(P) \cup \{P\}$ also has the same property as in the Lemma (1.9). Call V_2 the set of equivalence classes: $V_2 := E^*/\simeq$. Now we define a bipartite graph X, whose set of vertices consists of the disjoint union of V_1 and V_2; and we require that two vertices $P \in V_1$ and $E_j \in V_2$ are ajacent (i.e., joined by an edge $e=[P, E_j]$) iff $P \in E_j$. It is not difficult to see that X is a semi-regular bipartite tree of valency (q_1+1, q_2+1). Moreover, the action of G is naturally extended to X, in such a way that is transitive on V_1. This gives the first half of the following: (cf. Theorem (11.10))

Theorem (1.12). *Let q_1, q_2 be the given positive integers. There exists a bijection between the following objects:*

(i) *The group G with length function which satisfy the axiom* (G, l, I), (G, l, II).

(ii) *The groups G which acts on the semiregular bipartite tree $X(q_1, q_2)$, of valency (q_1+1, q_2+1), whose action is transitive on the first set V_1 of its vertices.*

Moreover, one can identify the coset space $U\backslash G$ and the set V_1; the length function l on $U\backslash G$ is then equal to the half of the distance d_X defined by the tree $X(q_1, q_2)$: $l(Ux, Uy)=d_X(Ux, Uy)/2$.

Note that we are assuming that G acts on $X(q_1, q_2)$ from the right, so that on $V_1 \simeq U\backslash G$, the action of G is that of the right multiplication. Let $P_0 \in V_1$ be the distinguished point which is corresponding to U. Then one has

$$g \in G_l \Longleftrightarrow d_X(P_0, P_0 g)=2l. \tag{1.13}$$

§ 2. Tits system and building

The most important class of groups G and the length function l, which satisfy the conditions (G, l, I), (G, l, II) in the preceding paragraph, is supplied by the socalled Tits system, or B-N pair. We shall review the definition and some of the basic facts on Tits system that we need for our study. Recall that a Tits system is a quadruplet (G, B, N, S) consisting of a group G, subgroups B, N of G, and a subset S of $W:=N/(B\cap N)$ which satisfy the following conditions:

(T. 1) $B\cup N$ generates G and $B\cap N$ is a normal subgroup of N.

(T. 2) S consists of a finite number of elements of order 2, and W is generated by them.

(T. 3) $sBw\subset BwB\cup BswB$ $(s \in S,\ w \in W)$.

(T. 4) For any $s \in S$, we have $sBs\not\subset B$.

In (T. 3) and (T. 4), expressions like sBw, BwB make sense since s, w are defined modulo $T:=N\cap B\subset B$. The group W is called the Weyl group of the Tits system.

Fact (2.1). (W, S) is a Coxeter system.

Fact (2.2). $G=BWB$, and the mapping $w\to BwB$ is a bijection of W onto $B\backslash G/B$ (Bruhat decomposition).

For $X\subset S$, let W_X be the subgroup of W generated by X. Put $G_X:=BW_XB$.

Fact (2.3). G_X is a subgroup of G and is generated by $\bigcup_{s\in X} BsB$.

Moreover, the assignment $X\to G_X$ is a bijection between the power set of S and the set of subgroups of G containing B.

Fact (2.4). Let $(X_i)_{i\in I}$ is a family of subset of X; if $X=\bigcap_i X_i$ then $G_X=\bigcap_i G_{X_i}$. And we have $G_{X_i}\subset G_{X_j}$ if and only if $X_i\subset X_j$.

We call G_X the standard parabolic subgroup of type X. It is known that any subgroup containing B is of this type.

Now we assume, throughout this paper, that

(2.5) (i) S consists of two elements, say, s_1, s_2.
(ii) $s_1 s_2$ is of infinite order.

Then $W=\langle s_1, s_2; s_1^2=s_2^2=1\rangle$ is isomorphic to the infinite dihedral group. Let U_1, U_2 be the standard parabolic subgroups corresponding to the subsets $\{s_1\}$, $\{s_2\}$ of S respectively:

(2.6) $$U_1:=B\cup Bs_1B, \qquad U_2:=B\cup Bs_2B.$$

Moreover, we assume that the cardinalities

(2.7) $$q_i:=\#(B\backslash Bs_iB) \quad (i=1, 2)$$

are finite. Then it follows immediately that

(2.8) $$\#(B\backslash U_i)=1+q_i \quad (i=1, 2).$$

Let $\mathscr{H}(G, B)$ be the Hecke algebra of (G, B). By a result of Iwahori-Matsumoto [I-M], we have

(2.9) $$\mathscr{H}(G, B)\simeq \boldsymbol{Z}[T_1, T_2]_{nc}$$
(non-commutative ring generated by T_1, T_2)

with $Bs_1B\to T_1$, $Bs_2B\to T_2$, and with the fundamental relations

(2.10) $$T_i^2=(q_i-1)T_i+q_i \quad (i=1, 2).$$

We have also the Hecke algebras $\mathscr{H}(G, U_i)$ for which we have

(2.11) $$\mathscr{H}(G, U_i)\simeq \boldsymbol{Z}[T] \ (i=1, 2), \quad \text{with } U_i s_1 s_2 U_i \longrightarrow T.$$

Now we recover the length function $l: G\to \boldsymbol{N}\cup\{0\}$ as follows. First we define $l(w)\,(w\in W)$ to be the number of s_2 in the reduced expression of w. Then we put

(2.12) $$l(g)=l(w), \qquad \text{if } g\in BwB.$$

It is immediate to see that the condition (G, l, I) is satisfied. Notice that $U=l^{-1}(0)=U_1$. We shall show that it also satisfy (G, l, II).

Lemma (2.13). *Let* $w=w_1w_2\cdots w_r$ $(w_i\in S)$ *be the reduced expression of* $w\in W$. *Then one has*

$$\sharp(B\backslash BwB)=\prod_{i=1}^{r}\sharp(B\backslash Bw_iB).$$

Proof. See [I-M], and [Mac], Proposition (3.1.7). Q.E.D.

Lemma (2.14). *As an element of* $\mathscr{H}(G, B)$, *we have*

$$G_l=(1+T_1)T_2(T_1T_2)^{l-1}(1+T_1)\quad (l\geq 1).$$

The relations (1.1), (1.2) now follow easily from this and (2.10). Thus we see that a Tits system of affine type of rank two, together with a choice of one of its two generators in S, gives a group G satisfying (G, l, I), (G, l, II).

Next we review the Bruhat-Tits building X attached to the Tits system (G, B, N, S) of affine type of rank two. By this assumption on the rank, it is a tree described as follows. The set V of its vertices is a disjoint union of V_1 and V_2, each of which is in one-to-one correspondence with $U_1\backslash G$ and $U_2\backslash G$ respectively. The set E of its edges is in one-to-one correspondence with $B\backslash G$. Let $\varphi_i: B\backslash G\to U_iG$ be the natural projection $(i=1, 2)$. Then two vertices P, Q are joined by a (non-oriented) edge e if and only if $P=\varphi_1(e)$, $Q=\varphi_2(e)$, where we regard P, Q, e as elements of $U_1\backslash G$, $U_2\backslash G$, and $B\backslash G$ respectively.

One can describe it also as follows. First notice that the normalizer of any parabolic subgroup of G coincides with itself. So one can identify the cosets U_ig in $U_i\backslash G$ (resp. Bg in $B\backslash G$) with the set of conjugates $g^{-1}U_ig$ (resp. $g^{-1}Bg$) of the parabolic subgroups of each type. Then the two vertices are adjacent (i.e., joined by an edge) if and only if the intersection of the corresponding parabolic subgroups is a conjugate of B.

(2.15) **Example.** $P_0\longleftrightarrow U_1$, $Q_0\longleftrightarrow U_2$, $e_0\longleftrightarrow B$

$q_1=4$, $q_2=2$

In the figure below, black vertices $\{\bullet\}$ and white ones $\{\circ\}$ correspond to the conjugates $\{gU_1g^{-1}\}$ and $\{gU_2g^{-1}\}$ respectively. Each black vertex has $1+q_1$ adjacent (white) vertices, and each white one has $1+q_2$ adjacent (black) vertices.

Let P_0 and Q_0 be, as in Fig-1, the adjacent pair of vertices corresponding to U_1, U_2 respectively. For the sake of simplicity, we identify V_i with $U_i\backslash G$, E with $B\backslash G$, and denote them simply by V_i, E respectively $(i=1, 2)$.

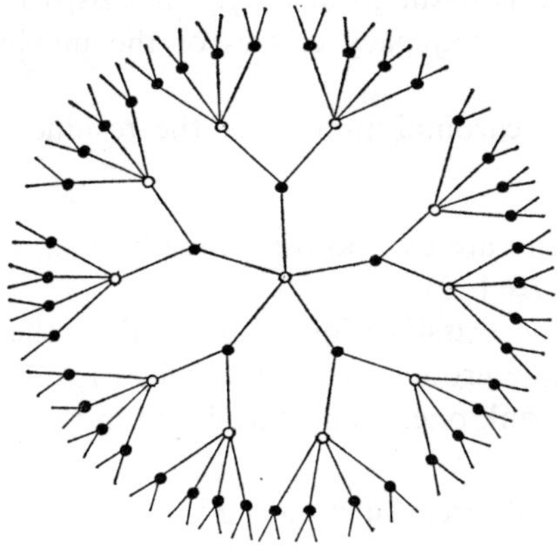

(Figure-1)

§ 3. *P*-adic algebraic groups

Let G be a semi-simple algebraic group defined over a local field K, and assume that G has K-rank one and has an affine Tits system (G, B, N, S). Note that the last condition is satisfied if G is simply connected. In this case B is called the Iwahori subgroup. Let W be the (affine) Weyl group of our Tits system. The assumption on the K-rank implies that our Tits system has rank two, so that, writing $S=\{s_1, s_2\}$, all the conditions of the preceding paragraph are satisfied.

♯	Type	simply connected groups	(d_1, d_2)
(1)	A_1	$SL(2, K)$	(1, 1)
(2)	$C\text{-}BC_1$	$SU(3, L/K)$; $[L:K]=2$, $L/K=$ramified	(1, 1)
(3)	${}^dA_{2d-1}$ $(d\geq 2)$	$SL(2, D)$; $D=K$-simple div. algebra $[D:K]=d^2$	(d, d)
(4)	${}^2A_3''$	$SU(4, L/K)$; $[L:K]=2$, $L/K=$unramified	(3, 3)
(5)	${}^2A_2'$	$SU(3, L/K)$; $[L:K]=2$, $L/K=$unramified	(3, 1)
(6)	${}^2C_3$	$Sp(3)=U(3, D)$; $D=$div. quaternion$/K$	(3, 2)
(7)	${}^2C_2$	$Sp(2)=U(2, D)$; $D=$div. quaternion$/K$	(1, 2)
(8)	${}^2C\text{-}B_3$	$SU_{\text{skew}}(4, D/K)$; $[L:K]=2$, $L/K=$ramified	(3, 2)
(9)	${}^2C\text{-}B_2$	$SU_{\text{skew}}(3, D/K)$; $[L:K]=2$, $L/K=$ramified	(1, 2)
(10)	${}^4D_4$	$SU_{\text{skew}}(4, D/K)$; $[L:K]=2$, $L/K=$unramified	(1, 4)
(11)	${}^4D_5$	$SU_{\text{skew}}(5, D/K)$; $L=K\oplus K$	(3, 4)

In (8)-(11), $SU_{\text{skew}}(r, D/K)$ denotes the special unitary group of a quaternion skew-hermitian form $q(x)$ of rank r, with Witt index$=1$, and L denotes the center of the even Clifford algebra of $q(x)$.

Fact (3.1). The two subgroups $U_i := B \cup Bs_iB$ $(i=1, 2)$ form a representatives of the G-conjugacy classes of the maximal open compact subgroups of G.

Let q denote the cardinal number of the residue field of K, i.e. $O/\mathfrak{p} \simeq \boldsymbol{F}_p$.

Fact (3.2). There are two positive integers d_1, d_2 such that $\#(B\backslash Bs_iB) = q^{d_i}$ $(=q_i$, say, for $i=1, 2)$.

According to the classification of Tits [T-2], there are 11 types of simple and semi-simple groups over a local field K, up to the central isogeny, which have K-rank one. They are listed in the above table.

§ 4 Structure of the discrete subgroups Γ

Let G be a group with a length function $l: G \to \boldsymbol{N} \cup \{0\}$ satisfying the axiom (G, l, I), (G, l, II) in § 1. And let Γ be a subgroup of G which satisfies:

(Γ, I) Γ is torsion free, and $\Gamma \cap x^{-1}Ux = \{1\}$ for any $x \in G$.

(Γ, II) $\#(U\backslash G/\Gamma) < \infty$.

We put $h := \#(U\backslash G/\Gamma)$, and denote by $\{x_1, \cdots, x_h\}$ a complete set of representatives in $U\backslash G/\Gamma$, which is fixed once and for all. Thus we have

$$G = \bigcup_{j=1}^{h} Ux_j\Gamma \quad \text{(disjoint)}. \tag{4.1}$$

Also put, for each $l=1, 2, \cdots$:

$$S_{ij}^{(l)} := \Gamma \cap x_i^{-1} G_l x_j \quad (1 \leq i, j \leq h). \tag{4.2}$$

Note that (Γ, I), (Γ, II) imply that $S_{ij}^{(l)}$ are all finite set. One can easily show, as in [I-1], that the mapping $G_l \to (S_{ij}^{(l)})$ defines a ring homomorphism φ: $\boldsymbol{Z}[G_l; l=0, 1, 2, \cdots] \to M(h, \boldsymbol{Z}[\Gamma])$, where we identify the set $S_{ij}^{(l)}$ and the formal sum of its elements in the group ring $\boldsymbol{Z}[\Gamma]$. And one can show

$$\begin{aligned}\sum_{i=1}^{h} \#(S_{ij}^{(l)}) &= \sum_{j=1}^{h} \#(S_{ij}^{(l)}) \\ &= \#(U\backslash G_l) = (q_1q_2)^{l-1}q_2(1+q_1) \quad (l \geq 1).\end{aligned} \tag{4.3}$$

We shall denote $S_{ij}^{(1)}$ simply by S_{ij}. From the assumption (G, l, II) and the above remark, it follows that Γ is generated by $\bigcup_{i,j} S_{ij}$. More precisely, one has

Lemma (4.4). *Let i, j $(1 \leq i, j \leq h)$ be given. Then*

(i) $\Gamma=\bigcup_{l=0}^{\infty} S_{ij}^{(l)}$ *(disjoint).*

(ii) *Each element* $\gamma \in S_{ij}^{(l)}$ *has a following expression*

$$(*) \qquad \gamma=\sigma_{ii_1}\sigma_{i_1i_2}\cdots\sigma_{i_{l-1}j} \ (\sigma_{i_ki_{k+1}} \in S_{i_ki_{k+1}};\ i_0=i,\ i_l=j).$$

with

$$(**) \qquad \sigma_{i_{k-1}i_k}\sigma_{i_ki_{k+1}} \in S^{(2)}_{i_{k-1}i_{k+1}} \quad (k=1, 2, \cdots, l-1).$$

(iii) *In the above expression, the indices* $(i_1, i_2, \cdots, i_{l-1})$ *and the elements* $\sigma_{i_ki_{k+1}}$ $(0\le k\le l-1)$ *are uniquely determined by* γ *and* (i, j).

Proof. (i) This follows trivially from $G=x_i^{-1}Gx_j=\bigcup_{l=0}^{\infty} x_i^{-1}G_lx_j$ (disjoint). (ii) Let γ be any element of $S_{ij}^{(l)}$. Then one sees that $x_i\gamma x_j^{-1} \in G_l$, so that from Lemma (1.5), there exists a unique expression of $x_i\gamma x_j^{-1}$ of the following form:

$$x_i\gamma x_j^{-1}=u\omega_{j_1}\omega_{j_2}\cdots\omega_{j_l} \ (u \in U,\ \omega_{j_1}, \cdots, \omega_{j_l} \in \Omega),$$

with $\qquad \omega_{j_n}\omega_{j_{n+1}} \in G_2 \qquad$ *for* $n=1, 2, \cdots, l-1.$

Put

$$g_0:=u\omega_{j_1}\omega_{j_2}\cdots\omega_{j_l}x_j,$$
$$g_1:=\omega_{j_2}\omega_{j_3}\cdots\omega_{j_l}x_j,$$
$$\cdots,$$
$$g_l:=x_j.$$

There exists a unique index i_{l-1} such that $g_{l-1} \in Ux_{i_{l-1}}\Gamma$; writing $g_{l-1}=\omega_{j_l}x_j=v_{l-1}x_{i_{l-1}}\sigma_{i_{l-1}j}$ $(v_{l-1} \in U,\ \sigma_{i_{l-1}j} \in \Gamma)$, one sees that $\sigma_{i_{l-1}j} \in S_{i_{l-1}j}$. Next let i_{l-2} be the unique index such that $g_{l-2}\sigma_{i_{l-1}j} \in Ux_{i_{l-2}}\Gamma$; and write

$$g_{l-2}=\omega_{j_{l-1}}\omega_{j_l}x_j=v_{l-2}x_{i_{l-2}}\sigma_{i_{l-2}i_{l-1}}\sigma_{i_{l-1}j} \ (v_{l-2} \in U,\ \sigma_{i_{l-2}i_{l-1}} \in \Gamma),$$

One sees immediately that $\sigma_{i_{l-2}i_{l-1}} \in S_{i_{l-2}i_{l-1}}$. Repeating this procedure, one gets an expression $(*)$ of γ as asserted. The condition $(**)$ follows from the corresponding condition $\omega_{j_n}\omega_{j_{n+1}} \in G_2$, since one has

$$\sigma_{i_{k-1}i_k}\sigma_{i_ki_{k+1}}=x_{i_{k-1}}^{-1}(v_{k-1}^{-1}\omega_{j_k}\omega_{j_{k+1}}v_{k+1})x_{i_{k+1}} \in \Gamma\cap x_{i_{k-1}}^{-1}G_2x_{i_{k+1}}.$$

(iii) The uniqueness follows easily from the above arguments.

Q.E.D.

Note that $S_{ij}=S_{ji}^{-1}$. Let S be a subset of $\cup S_{ij}-\{1\}$ such that $\cup S_{ij}=S\cup S^{-1}\cup\{1\}$.

Corollary (4.5). *The subgroup* Γ *of* G *satisfying* (Γ, I), (Γ, II) *is a free group with free generators* S.

Proof. The fact that Γ is a free group is proved in Appendix, Corollary (10.12). Therefore our assertion follows immediately from Lemm (4.4). Q.E.D.

However, in general, it is difficult to determine the set S explicitly, as has been done in [I-1]. This is partly because the same element σ can belong to many of the subsets S_{ij}.

Remark (4.6). Using the action of G on the tree $X(q_1, q_2)$, the above Lemma (4.4) can be interpreted as follows. We first note that

$$(4.7) \qquad S_{ij}^{(l)}=\{\gamma \in \Gamma;\, d_X((P_0x_j, P_0x_i\gamma)=2l\} \quad (P_0=U \in U\backslash G),$$

and that, if we put $P_j:=P_0x_j$ $(1\leq j\leq h)$, then we have

$$(4.8) \qquad V_1=\bigcup_{\gamma\in\Gamma}\{P_j;\, 1\leq j\leq h\}\gamma \quad \text{(disjoint)}.$$

In particular, each point $P \in V_1$ determines uniquely the index i $(1\leq i\leq h)$ and $\gamma \in \Gamma$ such that $P=P_0x_i\gamma$.

Now for each $\gamma \in S_{ij}^{(l)}=\Gamma\cap x_i^{-1}G_lx_j$, one has $d_X(P_0x_j, P_0x_i\gamma)=2l$. Put $R_0:=P_0x_i\gamma$, $R_l:=P_0x_j$, and let

$$C(R_0, R_l)=[R_0, Q_1, R_1, Q_2, R_2, \cdots, Q_l, R_l]$$

be the geodesic path from R_0 to R_l in $X(q_1, q_2)$. Then each $R_k \in V_1$ determines an index i_k $(1\leq i_k\leq h)$ and an element $\gamma_k \in \Gamma$. Now it is easy to see that

$$\begin{aligned}
&\gamma_l=1,\\
&\gamma_{l-1}=\sigma_{i_{l-1}j},\\
&\gamma_{l-2}=\sigma_{i_{l-2}i_{l-1}}\sigma_{i_{l-1}j},\\
&\cdots\cdot,\\
&\gamma_0=\sigma_{ii_1}\sigma_{i_1i_2}\cdots\sigma_{i_{l-1}j}.
\end{aligned}$$

Lemma (4.9). *For any element* σ *of* S_{ij} $(1\leq i, j\leq h)$, *one has*

$$\sum_{k=1}^{h} \#\{\tau \in S_{ki};\, \tau\sigma \in S_{kj}^{(l)}\}=\begin{cases}1 & \cdots\ l=0\\ q_2-1 & \cdots\ l=1\\ q_1q_2 & \cdots\ l=2.\end{cases}$$

Proof. Note that $\sigma \in S_{ij}$ (resp. $\tau \in S_{ki}$) is equivalent to $d_X(P_0x_i\sigma, P_0x_j)=2$ (resp. $d_X(P_0x_i, P_0x_k\tau)=2$). Now the set $B:=\{R \in V_1;\, d_X(P_0x_i, R)=2\}$ is divided, according to its distance from $P_0x_j\sigma^{-1}$, into three disjoint subsets: $B=B_0\cup B_1\cup B_2$, with

$$B_l=\{R \in V_1;\ d_X(R, P_0x_j\sigma^{-1})=2l\} \quad (l=0, 1, 2).$$

From the above remark, it follows that if $R \in B_l$, then R is expressed as $R=P_0x_k\tau$ with $1\leq k\leq h$, $\tau \in S_{ki}^{(l)}$. Now the assertion follows from Lemma (1.4). Q.E.D.

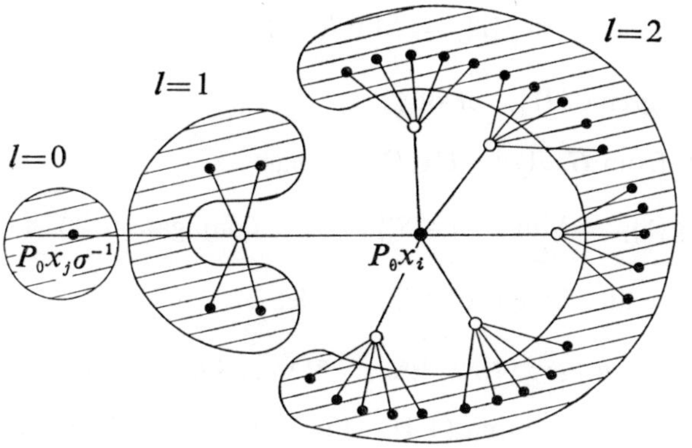

Figure-2

Finally we shall make the following observation.

Lemma (4.10). *Let Γ be a subgroup of G satisfying the conditions* (Γ, I), (Γ, II). *Then one has:*

(i) $h_2:=(1+q_1)h/(1+q_2)$ *is an integer.*

(ii) *The free rank r of Γ is given by* $r=q_1h-h_2+1$.

Proof. Consider the quotient graph $Y=X(q_1, q_2)/\Gamma$, for which one has $\#(V_1Y)=\#(V_1/\Gamma)=h$. Putting $h_2=\#(V_2Y)=\#(V_2/\Gamma)$, one sees that the number of edges in Y is $(q_1+1)h=(q_2+1)h_2$, hence (i). To prove (ii), one notes that the Euler characteristic of Y is

$$\begin{aligned}(4.11)\qquad \chi(Y)&=(h+h_2)-(q_1+1)h\\ &=\dim H^0(\Gamma, \boldsymbol{R})-\dim H^1(\Gamma, \boldsymbol{R})\\ &=1-r.\end{aligned}$$

Q.E.D.

Remark (4.12). It would be interesting to interprete h_2 and prove the equality (i) combinatorially, without using the graph Y. In fact, using the notation of (1.10), h_2 is seen to be the number of the equivalence classes $E_j(P)\cup\{P\}$ modulo Γ; and (i) can be proved by counting the cardinality of E^*/Γ in two different ways.

§ 5. Γ-conjugacy classes of given degree

In this and the next sections, the notation and the assumptions will be the same as in the previous sections. We begin with the same definition as in [I-1]:

Definition (5.1) For any Γ-conjugacy class $\{\gamma\}_\Gamma$ ($\gamma \in \Gamma$), we put

$$\deg\{\gamma\}_\Gamma := \operatorname*{Min}_{x \in G} l(x^{-1}\gamma x),$$

and call it the *degree* of $\{\gamma\}_\Gamma$, or of γ.

Note that, from $G=\bigcup_{i=1}^{h} Ux_i\Gamma$, one has

$$\deg\{\gamma\}_\Gamma := \operatorname*{Min}_{1\leq i\leq h} \{l;\ \delta \in S_{ii}^{(l)},\ \delta \text{ is } \Gamma\text{-conjugate to } \gamma\}. \tag{5.2}$$

Now let γ be an element of $S_{ii}^{(l)}$, and let

$$\gamma=\sigma_{ii_1}\sigma_{i_1i_2}\cdots\sigma_{i_{l-1}i}\ (\sigma_{i_ki_{k+1}} \in S_{i_ki_{k+1}};\ i_0=i_l=i).$$

be the expression as in Lemma (4.4). For the sake of simplicity, put $\sigma_m=\sigma_{i_{m-1}i_m}$ ($m=1, 2, \cdots, l$). Let k ($1\leq k\leq[l/2]$) be such that

$$\begin{aligned}
&i_1=i_{l-1} \quad \text{and} \quad \sigma_l\sigma_1=1,\\
&i_2=i_{l-2} \quad \text{and} \quad \sigma_{l-1}\sigma_2=1,\\
&\qquad \cdots,\\
&i_k=i_{l-k} \quad \text{and} \quad \sigma_{l-k+1}\sigma_k=1,\\
&i_{k+1}\neq i_{l-k-1} \quad \text{or} \quad \sigma_{l-k}\sigma_{k+1}\neq 1.
\end{aligned} \tag{5.3}$$

Lemma (5.4). *Let $\gamma \in S_{ii}^{(l)}$ satisfy the above condition. Then one has*
(i) *If $\sigma_{l-k}\sigma_{k+1} \in S_{i_{l-k-1}i_{k+1}}^{(2)}$ then* $\deg\{\gamma\}_\Gamma=l-2k$.
(ii) *If $\sigma_{l-k}\sigma_{k+1} \in S_{i_{l-k-1}i_{k+1}}^{(1)}$ then* $\deg\{\gamma\}_\Gamma=l-2k-1$.

Proof. By the condition (5.3) one sees that

$$\gamma=\sigma_1\sigma_2\cdots\sigma_k(\sigma_{k+1}\cdots\sigma_{l-k})\sigma_k^{-1}\cdots\sigma_1^{-1},$$

so that γ is Γ-conjugate to $\gamma':=\sigma_{k+1}\cdots\sigma_{l-k}$. Again by (5.3) one sees that $\gamma' \in S_{i_ki_k}^{(l-2k)}$. Also from $i_{l-k}=i_k$, one has $\sigma_{l-k}\sigma_{k+1} \in S_{i_{l-k-1}i_{k+1}}^{(m)}$ with $m=1$ or 2. Consider the element $\gamma'':=\sigma_{k+1}^{-1}\gamma'\sigma_{k+1}=\sigma_{k+2}\cdots\sigma_{l-k-1}\sigma_{l-k}\sigma_{k+1}$, which is Γ-conjugate to γ. Since the condition (**) of Lemma (4.4) is satisfied with only possible exception for $m=1$ in the last factor, we see from (1.7) that $\gamma'' \in S_{i_{k+1}i_{k+1}}^{(l-2k+2-m)}$. Thus we have $\deg\{\gamma\}_\Gamma\geq l-2k+2-m$. Now let $\delta \in \Gamma$ be such that $\delta^{-1}\gamma''\delta \in S_{ff}^{(n)}$, and suppose that $n<l-2k+2-m$. Using Lemma (4.4), one finds $s\geq 1$ such that $\delta \in S_{i_{k+1}f}^{(s)}$, so that δ is ex-

pressed as

$$\delta=\tau_{i_{k+1}f_1}\tau_{f_1f_2}\cdots\tau_{f_{s-1}f}\ (\tau_{f_tf_{t+1}}\in S_{f_tf_{t+1}}),$$

with the similar condition as (**) of Lemma (4.4). We may assume that δ has been so chosen that $s(\geq 1)$ is minimal among such elements. Suppose first that $m=2$. If we have

$$\tau^{-1}_{i_{k+1}f_1}\sigma_{i_{k+1}i_{k+2}}\in S^{(2)}_{f_1i_{k+1}}\cup S^{(1)}_{f_1i_{k+2}},\quad \text{or}$$
$$\sigma_{i_ki_{k+1}}\tau_{i_{k+1}f_1}\in S^{(2)}_{i_kf_1}\cup S^{(1)}_{i_kf_1},$$

then using Lemma (1.4) it follows easily that $n=l(\delta^{-1}\gamma''\delta)\geq l-2k+2s-2\geq l-2k$, which contradicts to our assumption. Thus we have either $\tau_{i_{k+1}f_1}=\sigma_{i_{k+1}i_{k+2}}$ and $f_1=i_{k+2}$, or $\tau_{i_{k+1}f_1}=\sigma^{-1}_{i_ki_{k+1}}$ and $f_1=i_{k+1}$. Replacing γ'' by γ' or by $\sigma_{l-k}\sigma_{k+1}\cdots\sigma_{l-k-1}$, this reduces the problem to the case of δ with $s-1$, which contradicts to the minimality of s. The case $m=1$ can be treated similarly, if we regard the product $\sigma_{l-k}\sigma_{k+1}$ as a single element of $S^{(1)}_{i_{l-k-1}i_{k+1}}$. Q.E.D.

The geometric interpretation of this result is quite simple and helpful to understand the situation. We first note that for $\gamma\in S^{(l)}_{ii}$, the expression

$$(*)\qquad \gamma=\sigma_{ii_1}\sigma_{i_1i_2}\cdots\sigma_{i_{l-1}i}\ (\sigma_{i_ki_{k+1}}\in S_{i_ki_{k+1}};\ i_0=i_l=i),$$

with

$$(**)\qquad \sigma_{i_{k-1}i_k}\sigma_{i_ki_{k+1}}\in S^{(2)}_{i_{k-1}i_{k+1}}\ (k=1,2,\cdots,l-1)$$

corresponds to the geodesic path C (i.e., the path without backtracking) in the tree $X(q_1,q_2)$, joining P_0x_i and $P_0x_i\gamma$. Now, if we project C to the quotient graph $X(q_1,q_2)/\Gamma$, we get a closed path $\overline{C}$ in $X(q_1,q_2)/\Gamma$. In

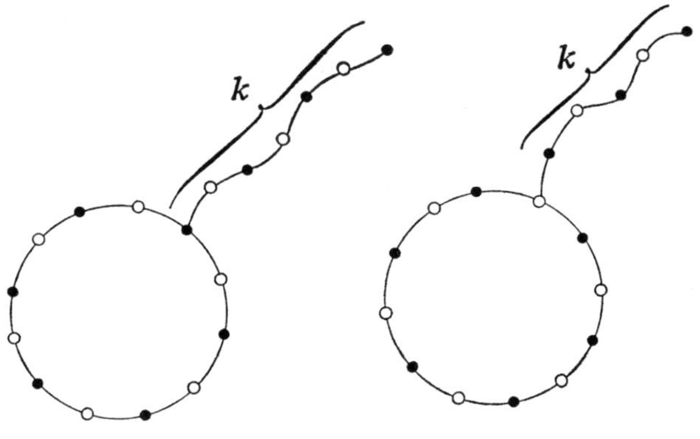

(Figure-3)

general, however, $\bar{C}$ has a backtracking at the origin=terminal, as is illustrated in the above figures.

From the proof of Lemma (5.4) and the above interpretation, we immediately have the following

Lemma (5.5). *Let* $\gamma \in S_{ii}^{(l)}$ *and let*

$$\gamma=\sigma_1\sigma_2\cdots\sigma_l \ (\sigma_k=\sigma_{i_k i_{k+1}} \in S_{i_k i_{k+1}};\ i_0=i_l=i)$$

be the expression as in Lemma (4.4). *Moreover, suppose that* $\sigma_l\sigma_1 \in S_{i_{l-1}i_1}^{(2)}$. *Then*

(i) *The elements* $\gamma_j:=\sigma_{j+1}\sigma_{j+2}\cdots\sigma_l\sigma_1\cdots\sigma_j$ $(0\leq j\leq l-1)$ *are exactly those which are* Γ*-conjugate to* γ *and which are contained in some* $S_{mm}^{(l)}$ $(1\leq m\leq h$; *l is fixed*).

(ii) *The above l elements* γ_j $(0\leq j\leq l-1)$ *are not necessarily distinct. If exactly d of them are mutually distinct, then each one belongs to exactly d different* $S_{mm}^{(l)}$*'s; i.e.,* $\#\{m;\gamma_j \in S_{mm}^{(l)}\}=d$.

§ 6. Zeta function $Z_\Gamma(u;\rho)$

Let γ $(\neq 1)$ be an element of Γ. Since Γ is a free group of finite rank, the centralizer $C_\Gamma(\gamma)$ of γ in Γ is an infinite cyclic group. We call γ or, the conjugacy class $\{\gamma\}_\Gamma$, to be primitive, if $C_\Gamma(\gamma)$ is generated by γ.

Definition. Let $\rho:\Gamma\to U(n)$ be an n-dimensional unitary representation of Γ. Then the zeta function $Z_\Gamma(u;\rho)$ of Γ attached to ρ is defined by an infinite product

(6.1) $$Z_\Gamma(u;\rho):=\prod_{\{\gamma\}_\Gamma}\det(I_n-\rho(\gamma)u^{\deg\{\gamma\}_\Gamma})^{-1}.$$

Let $G=\bigcup_{i=1}^h Ux_i\Gamma$ be a decomposition of G into disjoint union whose set of representatives is fixed throughout the following. Consider the Brandt representation of $\mathscr{H}(G,U)$ attached to ρ:

(6.2) $$\varphi_\rho:\mathscr{H}(G,U)\longrightarrow M(nh,\boldsymbol{C})=M(h,\boldsymbol{C})\otimes M(n,\boldsymbol{C})$$
$$UyU\longrightarrow\Big(\sum_{\gamma\in\Gamma\cap x_i^{-1}UyUx_j}\rho(\gamma)\Big)_{1\leq i,j\leq h}.$$

We put

(6.3) $$A_{l,\rho}:=\varphi_\rho(G_l)\qquad\text{for } l=0,1,2,\cdots$$

Also put $h_1:=h$, $h_2:=(q_1+1)h_1/(q_2+1)$; recall that h_2 is an integer, and that the rank of Γ is equal to $h_1q_1-h_2+1=h_2q_2-h_1+1$ (cf. Lemma (4.10)).

Now our main result of this paper is the following

Theorem (6.4). *The zeta function $Z_\Gamma(u;\rho)$ is a rational function of u which has the following expression:*

$$Z_\Gamma(u;\rho)^{-1}=(1-u)^{n(r-1)}(1+q_2u)^{n(h_2-h_1)}\\ \times\det\{I_{nh_1}-(A_{1,\rho}-q_2+1)u+q_1q_2u^2\}.$$

We shall prove this theorem. Taking the logarithmic derivative of (6.1), one gets

(6.5) $$u\frac{d}{du}\log Z_\Gamma(u;\rho)=\sum_{l=0}^{\infty}N_{l,\rho}u^l,$$

with

(6.6) $$N_{l,\rho}:=\sum_{d\mid l}d\sum_{P=\{\delta\}_\Gamma}\operatorname{tr}\rho(\delta^{l/d})$$

where the second sum in (6.6) is extended over the set of primitive Γ-conjugacy classes P such that $\deg P=d$. Suppose that $\delta\in S_{ii}^{(d)}$ is a representative of P. Then by Lemma (5.4) we see that $\gamma=\delta^{l/d}\in S_{ii}^{(l)}$, and that, if we express γ as in Lemma (5.5) then by the primitiveness of $P=\{\delta\}_\Gamma$, there are exactly d mutually different ones among the l expressions γ_j $(0\leq j\leq l-1)$. It follows that each of them is contained in exactly d distinct $S_{mm}^{(l)}$'s $(1\leq m\leq h)$. Thus we have

(6.7) $$N_{l,\rho}=\sum_{j=1}^{h}\sum_{\substack{\gamma\in S_{jj}^{(l)}\\ \deg\{\gamma\}_\Gamma=l}}\operatorname{tr}\rho(\gamma).$$

For each $l\geq 1$, k $(0\leq k\leq[(l-1)/2])$, and i $(1\leq i\leq h)$, put

$$S_{ii}^{(l,k,+)}:=\{\gamma\in S_{ii}^{(l)};\ (5.3)\text{ with }\sigma_{l-k}\sigma_{k+1}\in S_{i_{l-k-1}i_{k+1}}^{(2)}\}$$
$$S_{ii}^{(l,k,-)}:=\{\gamma\in S_{ii}^{(l)},\ (5.3)\text{ with }\sigma_{l-k}\sigma_{k+1}\in S_{i_{l-k-1}i_{k+1}}^{(1)}\},$$

where we use the abbreviated notation as in Lemma (5.4), and k is determined by (5.3). Then one has for each i $(1\leq i\leq h)$ and $l\geq 1$,

(6.8) $$S_{ii}^{(l)}=\bigcup_{k=0}^{[(l-1)/2]}S_{ii}^{(l,k,+)}\cup\bigcup_{k=0}^{[(l-2)/2]}S_{ii}^{(l,k,-)}\ \text{(disjoint)}.$$

Lemma (6.9). (i) *For $k\geq 1$, the mapping*

(6.10) $$\phi^+:\sum_{i=1}^{h}S_{ii}^{(l,k,+)}\longrightarrow\sum_{j=1}^{h}S_{jj}^{(l-2k,0,+)}$$

$$\gamma=\sigma_1\sigma_2\cdots\sigma_k(\sigma_{k+1}\cdots\sigma_{l-k})\sigma_k^{-1}\cdots\sigma_1^{-1}\longrightarrow\sigma_{k+1}\sigma_{k+2}\cdots\sigma_{l-k}$$

is $(q_1-1)q_2(q_1q_2)^{k-1}$*-to-one.*

(ii) *For* $k\geq 0$, *the mapping*

$$(6.11)\qquad \phi^-:\sum_{i=1}^{h}S_{ii}^{(l,k,-)}\longrightarrow\sum_{j=1}^{h}S_{jj}^{(l-2k-1,0,+)}$$

$$\gamma=\sigma_1\sigma_2\cdots\sigma_k(\sigma_{k+1}\cdots\sigma_{l-k})\sigma_k^{-1}\cdots\sigma_1^{-1}\longrightarrow\sigma_{k+2}\cdots\sigma_{l-k}\sigma_{k+1}$$

is $(q_2-1)(q_1q_2)^k$*-to-one.*

Proof. We first prove (i). Suppose we are given $\gamma'=\sigma_{k+1}\cdots\sigma_{l-k}\in S_{i_ki_k}^{(l-2k)}$, and consider the number of choices for $\sigma_k,\cdots,\sigma_1$. We first show that there are $(q_1-1)q_2$ choices for σ_k. Indeed, σ_k should be chosen under the following conditions:

$$\sigma_k\sigma_{k+1}\in S_{i_ki_{k+1}}^{(2)}\text{ i.e., }d_X(P_0x_{i_{k+1}}\sigma_{k+1}^{-1},\ P_0x_{i_{k-1}}\sigma_k)=2$$
$$\sigma_{l-k}\sigma_k^{-1}\in S_{i_{l-k-1}i_{k-1}}^{(2)}\text{ i.e., }d_X(P_0x_{i_{k-1}}\sigma_k,\ P_0x_{i_{l-k-1}}\sigma_{l-k})=2$$
$$\sigma_{l-k}\sigma_{k+1}\in S_{i_{l-k-1}i_{k+1}}^{(2)}\text{ i.e., }d_X(P_0x_{1_{k+1}}\sigma_{k+1}^{-1},\ P_0x_{i_{l-k-1}}\sigma_{l-k})=2.$$

Noting that the index i_k has been given, we see that to choose σ_k is equivalent to choose the vertex $P_0x_{i_{k-1}}\sigma_k\in V_1$, which is at the distance 4 from the (given) vertices $P_0x_{i_{k+1}}\sigma_{k+1}^{-1}$, $P_0x_{i_{l-k-1}}\sigma_{l-k}$. Now it is easy to see from Lemma (4.9) that there exactly $(q_1-1)q_2$ such vertices (see also Figure 2). Next for each choice of σ_k, one can show by Lemma (4.9) that there are q_1q_2 choices for σ_{k-1}. Similarly for each choice of $\sigma_k,\sigma_{k-1},\cdots,\sigma_{j+1}$, there are q_1q_2 choices for σ_j. This proves (i). To prove (ii), it suffices to note that, for given $\sigma_{k+2},\cdots,\sigma_{l-k}$, there are exactly q_2-1 choices of σ_{k+1} such that $\gamma'=\sigma_{k+2}\cdots\sigma_{l-k}\sigma_{k+1}\in S_{i_{k+1}i_{k+1}}^{(l-2k-1)}$, which follows easily from Lemma (4.9). Then one can proceed as in the case (i). Q.E.D.

Let f be a class function on Γ. For a finite subset S of Γ, we denote by $f(S)$ the sum of $f(\gamma)$, $\gamma\in S$.

Corollary (6.12). *For any class function* f *on* Γ, *one has*

$$\sum_{i=1}^{h}f(S_{ii}^{(l,k,+)})=(q_1-1)q_2(q_1q_2)^{k-1}\sum_{i=1}^{h}f(S_{ii}^{(l-2k,0,+)}),$$
$$\sum_{i=1}^{h}f(S_{ii}^{(l,k,-)})=(q_2-1)(q_1q_2)^{k}\sum_{i=1}^{h}f(S_{ii}^{(l-2k-1,0,+)}).$$

Proof. Let χ be the character of ρ, i.e., $\chi(\gamma)=\mathrm{tr}\,\rho(\gamma)$ ($\gamma\in\Gamma$). Then we have

$$\begin{aligned}
\mathrm{tr}(A_{l,\rho}) &= \sum_{i=1}^{h} \chi(S_{ii}^{(l)}) \\
&= \sum_{k=0}^{[(l-1)/2]} \sum_{i=1}^{h} \chi(S_{ii}^{(l,k,+)}) + \sum_{k=0}^{[(l-2)/2]} \sum_{i=1}^{h} \chi(S_{ii}^{(l,k,-)}) \\
&= \sum_{i=1}^{h} \chi(S_{ii}^{(l,0,+)}) + (q_1-1)q_2 \sum_{k=1}^{[(l-1)/2]} (q_1q_2)^{k-1} \sum_{i=1}^{h} \chi(S_{ii}^{(l-2k,0,+)}) \\
&\quad + (q_2-1) \sum_{k=0}^{[(l-2)/2]} (q_1q_2)^{k} \sum_{i=1}^{h} \chi(S_{ii}^{(l-2k-1,0,+)})
\end{aligned}$$

Thus from (6.7), we obtain

$$\begin{aligned}
(6.13)\qquad \mathrm{tr}(A_{l,\rho}) &= N_{l,\rho} + (q_1-1)q_2 \sum_{k=1}^{[(l-1)/2]} (q_1q_2)^{k-1} N_{l-2k,\rho} \\
&\quad + (q_2-1) \sum_{k=0}^{[(l-2)/2]} (q_1q_2)^{k} N_{l-2k-1,\rho}. \qquad \text{Q.E.D.}
\end{aligned}$$

Lemma (6.14).

$$\begin{aligned}
N_{l,\rho} &= \mathrm{tr}(A_{l,\rho}) - (q_2-1)\mathrm{tr}(A_{l-1,\rho}) \\
&\quad - \sum_{k=1}^{l-2} \left\{ (q_2-1) + (q_1-q_2)q_2 \sum_{m=0}^{l-2-k} (-q_2)^m \right\} \mathrm{tr}(A_{k,\rho}).
\end{aligned}$$

Proof. We prove this by induction on l. Note first that the assertion for $l=1$ follows from (6.13). Suppose that it is true for $l=1, 2, \cdots, p-1$. Using (6.13) for $l=p$, and the induction hypothesis, we have an equality:

$$\begin{aligned}
(6.15)\qquad N_{p,\rho} &= \mathrm{tr}(A_{p,\rho}) - (q_2-1)\mathrm{tr}(A_{p-1,\rho}) \\
&\quad - (q_1-1)q_2 \sum_{k=1}^{[(p-1)/2]} (q_1q_2)^{k-1} \mathrm{tr}(A_{p-2k,\rho}) \\
&\quad + (q_1-1)(q_2-1)q_2 \sum_{k=1}^{[(p-1)/2]} (q_1q_2)^{k-1} \sum_{j=1}^{p-2k-1} \mathrm{tr}(A_{j,\rho}) \\
&\quad + (q_1-1)(q_1-q_2)q_2^2 \sum_{k=1}^{[(p-1)/2]} (q_1q_2)^{k-1} \sum_{j=1}^{p-2k-2} \sum_{m=0}^{p-2k-2-j} (-q_2)^m \mathrm{tr}(A_{j,\rho}) \\
&\quad - (q_2-1) \sum_{k=1}^{[(p-2)/2]} (q_1q_2)^{k} \mathrm{tr}(A_{p-2k-1,\rho}) \\
&\quad + (q_2-1)^2 \sum_{k=0}^{[(p-2)/2]} (q_1q_2)^{k} \sum_{j=1}^{p-2k-2} \mathrm{tr}(A_{j,\rho}) \\
&\quad + (q_2-1)(q_1-q_2)q_2 \sum_{k=0}^{[(p-2)/2]} (q_1q_2)^{k} \sum_{j=1}^{p-2k-3} \sum_{m=0}^{p-2k-3-j} (-q_2)^m \mathrm{tr}(A_{j,\rho}).
\end{aligned}$$

Now a direct calculation shows that the total sum of the right hand side of (6.15) is equal to that of (6.14) for $l=p$. Q.E.D.

Using (6.5) and (6.14), we have

$$(6.16)\quad u\frac{d}{du}\log Z_\Gamma(u;\rho)=\sum_{l=1}^{\infty}\mathrm{tr}(A_{l,\rho})u^l$$

$$-(q_2-1)\sum_{l=2}^{\infty}\mathrm{tr}(A_{l-1,\rho})u^l-(q_2-1)\sum_{l=3}^{\infty}\sum_{k=1}^{l-2}\mathrm{tr}(A_{k,\rho})u^l$$

$$-(q_1-q_2)q_2\sum_{l=3}^{\infty}\sum_{k=1}^{l-2}\sum_{m=0}^{l-2-k}(-q_2)^m\mathrm{tr}(A_{k,\rho})u^l$$

$$=\left\{1-(q_2-1)\sum_{k=1}^{\infty}u^k-(q_1-q_2)q_2\sum_{k=2}^{\infty}\sum_{m=0}^{k-2}(-q_2)^m u^k\right\}\sum_{l=1}^{\infty}\mathrm{tr}(A_{l,\rho})u^l$$

$$=\left\{1-\frac{(q_2-1)u}{1-u}-\frac{(q_1-q_2)q_2u^2}{(1-u)(1+q_2u)}\right\}\sum_{l=1}^{\infty}\mathrm{tr}(A_{l,\rho})u^l$$

$$=\frac{1-q_1q_2u^2}{(1-u)(1+q_2u)}\cdot\sum_{l=1}^{\infty}\mathrm{tr}(A_{l,\rho})u^l.$$

Applying the Brandt representation φ_ρ to the both sides of (1.1), (1.2), we have

$$(6.17)\qquad A_{1,\rho}^2=A_{2,\rho}+(q_2-1)A_{1,\rho}+q_2(q_1+1)A_{0,\rho},$$

$$(6.18)\qquad A_{1,\rho}A_{l,\rho}=A_{l+1,\rho}+(q_2-1)A_{l,\rho}+q_1q_2A_{l-1,\rho}\quad (l\geq 2).$$

It is easy to see that these equalities are equivalent to the following one in the ring $\boldsymbol{Z}[A_{l,\rho};\, l\geq 0]\,[[u]]$ of formal power series.

$$(6.19)\qquad \sum_{l=0}^{\infty}A_{l,\rho}u^l=\frac{(1-u)(1+q_2u)}{1-(A_{1,\rho}-q_2+1)u+q_1q_2u^2}.$$

Taking the trace of both sides of (6.19), we have

$$(6.20)\quad u\frac{d}{du}\log Z_\Gamma(u;\rho)=(1-q_1q_2u^2)\mathrm{tr}\{1-(A_{1,\rho}-q_2+1)u+q_1q_2u^2\}^{-1}$$
$$-nh\cdot\frac{1-q_1q_2u^2}{(1-u)(1+q_2u)}.$$

Now the following equality is easily proved:

$$(6.21)\quad \mathrm{tr}\{1-(A_{1,\rho}-q_2+1)u+q_1q_2u^2\}^{-1}=\frac{u}{1-q_1q_2u^2}$$
$$\times\frac{d}{du}\log\det\{1-(A_{1,\rho}-q_2+1)u+q_1q_2u^2\}^{-1}+\frac{nh}{1-q_1q_2u^2}.$$

Combining (6.20) and (6.21) together with $Z_\Gamma(0;\rho)=1$, we get the following expression of $Z_\Gamma(u;\rho)$:

$$Z_\Gamma(u;\rho)^{-1}=(1-u)^{(q_1q_2-1)nh/(q_2+1)}(1+q_2u)^{(q_1-q_2)nh/(q_2+1)} \times\det\{I_{nh_1}-(A_{1,\rho}-q_2+1)u+q_1q_2u^2\}.$$

Note that, by Lemma (4.10), we have

$$\frac{(q_1q_2-1)h}{q_2+1}=\frac{q_2(q_1+1)h_1}{q_2+1}-\frac{(q_2+1)h_1}{q_2+1}=q_2h_2-h_1=r-1,$$

$$\frac{(q_1-q_2)h}{q_2+1}=\frac{(q_1+1)h_1}{q_2+1}-\frac{(q_2+1)h_1}{q_2+1}=h_2-h_1.$$

This completes the proof of the theorem (6.4).

§ 7. Remarks

Finally we shall give some remarks on our results.

7-1. First, we shall give a graph-theoretic interpretation of our results. The basic ideas have been suggested by [Ser-1] (see also [Su]). Let Y denote the finite quotient graph $X(q_1, q_2)/\Gamma$. Then the group Γ can be identified with the fundamental group of Y, or its realization as a CW-complex, with respect to a (fixed) point. And there is a bijective correspondence between the set of Γ-conjugacy classes and the set of geodesic cycles. Note that, by definition, a geodesic cycle is a closed path in Y without backtracking, modulo the equivalence induced by shifting the origin. A cycle is called primitive, if it is not a power of another cycle. This corresponds to the primitive Γ-conjugacy classes. Thus, denoting by $|P|$ the length of the geodesic cycle P in Y, we get

$$Z_\Gamma(u;\rho)=\prod_{P:\text{primitive}}\det(I_n-\rho(\langle P\rangle)u^{|P|/2})^{-1}, \tag{7.1}$$

where the product is extended over all primitive cycles P of Y, and $\langle P\rangle$ denotes the corresponding conjugacy class of Γ. Taking the logarithmic derivative of (7.1), we get

$$N_{l,\rho}=\sum_C \chi(\langle C\rangle), \tag{7.2}$$

where C runs through the set of geodesic path in Y of length $2l$, with origin a vertex belonging to V_1Y (=image of V_1). In this context, the matrix $A_{l,\rho}$ has the following interpretation. Let $S_\rho(\Gamma)$ be the $\boldsymbol{C}$-vector space

(7.3) $$S_\rho(\Gamma):=\{f\colon V_1\to C^n;\, f(P\gamma)=f(P)\rho(\gamma) \text{ for any } \gamma\in\Gamma\}.$$

Since Γ acts on $X(q_1, q_2)$ without fixed point, it follows that $S_\rho(\Gamma)$ is of finite dimension with $\dim S_\rho(\Gamma)=n\#(V_1Y)=nh$. Note that $S_\rho(\Gamma)$ is regarded as a space of C^n-valued functions on $U\backslash G$, on which one can define the representation of the Hecke algebra $\mathcal{H}(G, U)$ by the convolution. It is immediate to see that the resulting representation is equivalent to the Brandt representation φ_ρ defined by (6.2); and $A_{l,\rho}$ corresponds to the following linear operator on $S_\rho(\Gamma)$:

(7.4) $$(A_{l,\rho}f)(P):=\sum_{\substack{Q\in V_1\\ d_X(P,Q)=2l}} f(Q).$$

Let $V_1^*=\{P_k;\, 1\le k\le h\}$ be a complete set of representatives of $V_1/\Gamma=V_1Y$, and let $e_j:=(0,\cdots,0,1,0,\cdots,0)$ $(1\le j\le n)$ be the standard unit vectors of C^n. Then the functions $f_{k,j}$ $(1\le k\le h,\ 1\le j\le n)$ of $S_\rho(\Gamma)$ which are determined by the condition $f_{k,j}(P_k)=e_j$, $f_{k,j}(P_m)=0$ $(m\ne k)$ from a basis of $S_\rho(\Gamma)$. Using this basis, it is easy to show

(7.5) $$\operatorname{tr}(A_{l,\rho})=\sum_{\substack{\gamma\in\Gamma,\, P\in V_1^*\\ d_X(P,P\gamma)=2l}} \chi(\gamma).$$

Using these interpretations, one can simplify the proofs of the results in § 6.

7-2. Suppose that (G, l) is obtained from a Tits system (G, B, N, S). Then one can proceed the same calculation of $Z_\Gamma(u;\rho)$, with U_2 instead of U_1. Thus one gets a second formula for $Z_\Gamma(u;\rho)$:

(7.6) $$\begin{aligned} Z_\Gamma(u;\rho)^{-1}&=(1-u)^{n(r-1)}(1+q_1u)^{n(h_1-h_2)}\\ &\times\det\{I_{nh_2}-(A'_{1,\rho}-q_1+1)u+q_1q_2u^2\}. \end{aligned}$$

where $A'_{1,\rho}$ is defined similarly as $A_{1,\rho}$. Note that, while the zeta function $Z_\Gamma(u;\rho)$ defined by (7.1) is independent of the choice between U_1 and U_2, the final results (6.4) and (7.6) are not symmetric in q_1, q_2 and h_1, h_2. It is, therefore, an interesting problem to explain this difference in the two expressions. It is also an important problem to ask the possible relation between our results on $Z_\Gamma(u;\rho)$ and the spectral decomposition of $L^2(G/\Gamma)$. We shall study these problems in the subsequent paper [Ha]. Here we content ourselves with the following observation.

Suppose that G is an algebraic group over a local field K satisfying the conditions of § 3. Suppose moreover, that $\rho=\mathbf{1}$, the trivial representation, and put $A_1=A_{1,1}$. Then, by the same argument as in [I-1], we see that A_1 has the eigenvalue $q_2(1+q_1)$ with multiplicity one. It follows that

$Z_\Gamma(u; \mathbf{1})^{-1}$ has the factor $(1-u)$ with multiplicity $r = \dim H^1(\Gamma, \boldsymbol{R})$. This, combined with a result of [Car], [Gas], implies the following result which generalizes that of Ihara [I-1]:

Proposition (7.7). *We have the following equality*

$$-\operatorname*{Res}_{u=1} \frac{d}{du} \log Z_\Gamma(u) = r = \textit{the multiplicity of the Steinberg representation in } L^2(G/\Gamma).$$

We shall give a different proof of this in [Ha], which is independent of [Gar], [Cas].

Appendix. Bipartite trees, Hecke algebras, and flowers of groups (by Ki-ichiro Hashimoto)

Contents

§ 8. Introduction

In [I-1], Ihara studied the discrete subgroups of $SL(2, K)$ over a $\boldsymbol{p}$-adic field K. There he established, among others, a remarkable structure theorem which states that any discrete torsion free subgroup Γ of $SL(2, K)$ is a free group, whose free basis can be constructed in an explicit way. In fact what he did is more; such structure theorem was proved for subgroups of more general groups G satisfying certain axioms, and a zeta function for such Γ has been introduced and studied with a number of applications. The proofs are based on somewhat mysterious combinatorial arguments, as was written in the introduction of [Ser-1]. Later Serre [Ser-1] gave a graph-theoretic interpretation of the first result, generalizing it to much more general class of groups G. In one of the main theorems (Th. 13 in § I.5.4), he established a structure theorem of groups acting on a tree X. G is then recovered as the fundamental group $\pi_1(\boldsymbol{G}, Y, T)$ of the graph of groups $(\boldsymbol{G}, Y)$ at a maximal tree T $(\subset Y)$, where Y is the quotient graph X/G. Ihara's structure theorem is generalized to subgroups Γ of such G.

Here we note that, according to a philosophy of Tits, the tree attached to $G=SL(2, K)$ is regarded as the analogue of the upper half-plane for $SL(2, \boldsymbol{R})$. Based on this idea, a theory of harmonic analysis on trees has been constructed (cf. [Car]). Ihara's main result in [I-1] on the zeta function of Γ ($\subset SL(2, K)$) is particulary interesting if viewed from this point.

However, in such full generality, it seems difficult to extend the results of [I-1] on zeta functions, or to get a deep result on the arithmetic of a subgroup Γ of G.

We note here that in [I-1], an essential role has been played by the relations (G, l, II) (cf. (11.12)) of Hecke operators (a $\boldsymbol{p}$-adic analogue of the Laplacian), which have been given in [Ser-1] a simple interpretation in terms of graph theory, or trees. It is, therefore, natural to study the class of groups acting on trees, *which are not necessarily homogeneous*, for which one can expect a nice relation for elements of its Hecke algebra, that will lead us to the evaluation of zeta functions for its (discrete) subgroups Γ, as well as its application to the spectral decomposition of $L^2(G/\Gamma)$. We require that the Hecke operators satisfy, instead of the relations (G, l, II) of [I-1], one of those which appear in the theory of Iwahori-Matsumoto [I-M] (see (1.1), (1.2)).

The purpose of this note is to prove the equivalence between the following classes of objects consisting of groups and some extra data:

(8.1) *Groups G satisfying similar axioms* (G, l, I), (G, l, II) *as in* [I-1], *the latter describing a structure of the Hecke algebra* $\mathscr{H}(G, U)$ *of G with respect to a subgroup U* (cf. (1.1), (1.2)).

(8.2) *Groups G with an action on a semi-regular bipartite tree* $X(q_1, q_2)=(V_1, V_2; E)$ *of valency* (q_1+1, q_2+1), *which is transitive on* V_1, *one of its two kinds of vertices.*

(8.3) *Fundamental groups* $\pi_1(\boldsymbol{G}, F, T)$ *of the "flowers F", of groups with certain regularity condition* (cf. (11.4) (11.5)); here *a flower* is a finite graph described as in the following:

Fig. (8.4)

The equivalence of (8.2) and (8.3) follows as a special case of Serre's description of the groups acting on a tree, mentioned above. We included it in our main result, not only because of its beautiful characterization of our class, but also because it shows very well the range of it. As a simplest case, our class contains the groups which have the Tits system (G, B, N, S), where S consists of two elements generating the infinite dihedral group, and the flower reduces to ●—○. This is the case for any simply connected groups over the local fields K with K-rank one. In fact our motivation has been to extend the results of [I-1] to such groups. However, (8.3) shows that our class covers much more wide class of groups which in general fail to have a Tits system.

Notation. For a finite set S, $\#(S)$ denotes its cardinality. By a graph, we mean, unless otherwise stated, a *non-oriented* one. If X is a graph, we denote VX (resp. EX) the set of its vertices (resp. edges). If X is a tree, $d_X: VX \times VX \to N \cup \{0\}$ denotes the distance on X. For a vertex $P \in VX$ and $l \in N \cup \{0\}$, we put $VX(P; l) := \{Q \in VX;\ d_X(P, Q) = l\}$.

§ 9. Groups with axioms (G, l, I), (G, l, II)

We shall prove the results described in § 1. Let (G, l) be as in § 1. We begin with the following remark. Our axioms $(G, l, \mathrm{I}, \mathrm{II})$ can be viewed as a generalization of those of Ihara [I-1], in two ways. First, if we put $q_1 = q_2 = q$, then we see that our axioms can be derived from those of [I-1], by considering only G_{2l}'s (see Remark (11.14)). This is what we obtain by replacing $PL_2(K)$ by $PSL_2(K)$ in [I-1]. On the other hand, axioms of [I-1] is recovered also by putting $q_2 = 1$ in our axioms. In terms of trees, this corresponds to considering the barycentric subdivision of the homogeneous tree attached to $PL_2(K)$ (see § 10).

Proof of Lemma (1.4). Since (i) is trivial, and (iii) follows from (i), (ii) and (1.1), we only need to show (ii). From the definition of the product $G_1 \cdot G_1$ in $\mathscr{H}(G, U)$, we have

$$G_1 = \textstyle\sum_i U\omega_i \Longrightarrow G_1^2 = \sum_{i,j} U\omega_i\omega_j,$$

where the formal sum is taken with multiplicities (cf. [Sh], Chap. 3). Comparing this with (1.1), we see that, for any $\xi \in G_1$,

$$q_2 - 1 = \#\{(i, j);\ U\omega_i\omega_j = U\xi\} = \#\{j;\ \xi\omega_j^{-1} \in \textstyle\sum_i U\omega_i = G_1\}.$$

The assertion (ii) follows from this by taking $\xi = \omega^{-1} \in G_1^{-1} = G_1$. Q.E.D.

It follows from (1.1) and Lemma (1.4), (i) that we have $t = \#(U \backslash G_1)$

$=q_2(q_1+1)$; and the induction using (1.2) shows

$$\sharp(U\backslash G_l)=(q_1q_2)^{l-1}q_2(q_1+1) \qquad \text{for } l\geq 1. \tag{1.3}$$

Proof of Lemma (1.5). Repeated application of the argument as above shows that

$$G_1^l=\sum_{j_1,j_2,\cdots,j_l} U\omega_{j_1}\omega_{j_2}\cdots\omega_{j_l}, \tag{9.6}$$

with multiplicity taking into account. On the other hand, we see from (G, l, I), (G, l, II) and the induction on l that

$$G_1^l=G_l+c_1G_{l-1}+\cdots+c_lU \quad (c_i \in \boldsymbol{N}\cup\{0\}).$$

Comparing these two expressions for G_1^l, we get the assertions except for the equivalence; $l(g)=l\Leftrightarrow(*)$. To prove this, we note that this assertion is equivalent to the identity

$$G_l=\sum_{j_1,j_2,\cdots,j_l}^{(*)} U\omega_{j_1}\omega_{j_2}\cdots\omega_{j_l}, \tag{9.7}$$

where $\Sigma^{(*)}$ is the partial sum of (9.6) of those which satisfy the condition $(*)$. From what we have seen, it follows that G_l is contained in the sum $\Sigma^{(*)}$. On the other hand, from Lemma (1.4) we see that the number of U-cosets in this sum is $(q_1q_2)^{l-1}q_2(q_1+1)$, which is equal to $\sharp(U\backslash G_l)$ (cf. (1.3)). This proves our asssertion. Q.E.D.

We remark that, under the condition (G, l, I), (G, l, II) is equivalent to the statements in Lemmas (1.4) and (1.5).

Following [I-1], we call a product xy $(x, y \in G)$ *free*, if $l(xy)=l(x)+l(y)$. The free product of n elements $x_1\cdots x_n$ is defined similarly. It is easy to see from the above lemma that, if xy, yz are free products and $y\notin U$, then xyz is free. In particular for $x_1, x_2, \cdots, x_l \in G_1$, we have

$$l(x_1x_2\cdots x_l)=l \Longleftrightarrow l(x_ix_{i+1})=2 \quad (i=1, 2, \cdots, l-1). \tag{9.8}$$

The following is also an easy consequence of Lemma (1.5).

Lemma (9.9). *If the product xy is free and $xy=u\omega_{j_1}\omega_{j_2}\cdots\omega_{j_l}$ is its unique expression as in Lemma* (1.5), *then one can find $v \in U$ such that $x=u\omega_{j_1}\omega_{j_2}\cdots\omega_{j_m}v$ and $y=v^{-1}\omega_{j_{m+1}}\omega_{j_{m+2}}\cdots\omega_{j_l}$ $(m=l(x))$.*

Now we give a

Proof of Lemma (1.6). The assertions for $i=0$ are trivial, since $UG_lU=G_l$. Also the assertions for $i=2$ follow from the remark preceding Lemma (9.9). So we prove them for $i=1$.

Let $G_1=\sum_j U\omega_j$ and $G_2=\sum^{(*)}_{i_1,i_2} U\omega_{i_1}\omega_{i_2}$ be the decompositions into U-cosets as above. Then the product G_1G_2 in $\mathscr{H}(G, U)$ is expressed as

$$G_1G_2=\sum_j \sum^{(*)}_{i_1,i_2} U\omega_j\omega_{i_1}\omega_{i_2}$$
$$=\sum^{(*)}_{j,i_1,i_2} U\omega_j\omega_{i_1}\omega_{i_2}+\sum^{(1)} U(\omega_j\omega_{i_1})\omega_{i_2}+\sum^{(0)} U(\omega_j\omega_{i_1})\omega_{i_2},$$

where $\sum^{(i)}$ is the sum of $U\omega_j\omega_{i_1}\omega_{i_2}$ such that $\omega_j\omega_{i_1}\in G_i$ $(i=0, 1)$. From (9.7), we have $\sum^{(*)}_{j,i_1,i_2}=G_3$. Also from Lemma (1.4), (i) and (iii), it follows that $\sum^{(0)}=q_1q_2G_1$. Now comparing theabove expression for G_1G_2 with that of (1.2), $l=2$, we get $\sum^{(1)}=(q_2-1)G_2$, from which it follows that $\omega_j\omega_{i_1}\omega_{i_2}\in G_2$.

Thus we have proved the assertion (i) for $i=1$, in the case $x=\omega_j$, $y=\omega_{i_1}$, $z=\omega_{i_2}\in\Omega$. The general case is easily reduced to this and Lemma (9.9), since Ω is an arbitrary set of representatives of $U\backslash G_1$. (ii) follows from (i), by taking inverse $(G_l^{-1}=G_l)$. Q.E.D.

§ 10. Construction of a tree $X(q_1, q_2)$

A tree of complete graphs.

Let (G, l) be a pair of a group and a function l on G taking values in non-negative integers, which satisfy (G, l, I). Following an idea of Serre [Ser-1], we define a graph $Z=Z(q_1, q_2)$ on which G acts. Let $VZ:=U\backslash G$ be the homogeneous space of G consisting of the (left) U-cosets. We define a function d_Z: $VZ\times VZ\to \boldsymbol{N}\cup\{0\}$ by

(10.1) $$d_Z(Ux, Uy):=l(xy^{-1})\quad (x, y\in G).$$

Then it follows immediately from (G, l, I) that d_Z satisfies the following properties.

(10.2) d_Z *is symmetric and G-invariant.*

(10.3) $d_Z(P, R)=0\Longleftrightarrow P=R$ $(P, R\in VZ)$.

(10.4) *For any* $P\in VZ$, *and* $l\in \boldsymbol{N}\cup\{0\}$, *the cardinality* t_l *of the set* $\{R\in VZ; d_Z(P, R)=l\}$ *is finite, and it is independent of* P.

Then we define two points $P, R\in VZ$ to be *adjacent* (or joined by an *edge*), if $d_Z(P, R)=1$. This gives us a combinatorial graph Z on which G acts, having VZ as its set of vertices. Recall that a combinatorial graph is a graph which does not contain any circuit of length $n\leq 2$.

Now we assume that G satisfies the second axiom (G, l, II). Then from (1.3), we have $t_l=(q_1q_2)^{l-1}t$, $t=t_1=q_2(q_1+1)$. Call a circuit of length 3 a *triangle*. Lemma (1.4), (ii) implies that, if $q_2>1$ then the graph Z

contains a triangle; in fact $C:=[U, U\omega, U\rho\omega, U]$ ($\omega, \rho, \rho\omega \in G_1$) is such a triangle. If, on the other hand, $q_2=1$, then our axioms reduce to that of [I-1], in which case (G, l, II) is equivalent to the assertion that Z is a tree (see [Ser-1], p. 117, ex. 2).

We call a circuit $C_n=[P_0, P_1, \cdots, P_n=P_0]$ in a graph Z to be *minimal*, if the only pairs of vertices in $\{P_0, P_1, \cdots, P_{n-1}\}$ adjacent in Z are $\{P_i, P_{i+1}\}$ $(i=0, 1, \cdots, n-1)$; or equivalently, if no proper subset of this is the set of vertices of a circuit in Z.

Lemma (10.5). *The graph $Z(q_1, q_2)$ contains no minimal circuit of length $n\geq 4$.*

Proof. Suppose on the contrary that Z contains a minimal circuit $C_n=[P_0, P_1, \cdots, P_n=P_0]$ of length $n\geq 4$. Since the action of G on the set of vertices of Z is transitive, we may assume that $P_0=U$, and write $P_1=Ux_1$, $P_2=Ux_2x_1$, $\cdots$, $P_{n-1}=Ux_{n-1}\cdots x_2x_1$, where $x_1, \cdots, x_{n-1} \in G_1$. Now the assumptions that $n\geq 4$ and that C_n is minimal, imply that $x_ix_{i+1} \in G_2$ for $i=1, \cdots, n-2$. By (9.8), this implies that $l(x_{n-1}\cdots x_2x_1)=d_Z(P_0, P_{n-1})$ $=n-1>1$, a contradiction. Q.E.D.

Lemma (10.6). *If two distinct triangles $\{P_1, P_2, P_3\}$, $\{P_1, P_2, R_3\}$ in Z share an edge $[P_1, P_2]$, then the remaining two vertices are adjacent: $d_Z(P_3, R_3)=1$.*

Proof. Again we may write $P_1=U$, $P_2=U\omega$, $P_3=U\rho\omega$, and $R_3=U\tau\omega$, where $\omega, \rho, \rho\omega, \tau, \tau\omega \in G_1$. Suppose, on the contrary, that $d_Z(P_3, R_3)=l(\rho\tau^{-1})=2$. Then applying Lemma (1.6), (i) for $x=\rho$, $y=\tau^{-1}$, $z=\tau\omega$, one gets $l(xyz)=l(\rho\omega)=2$, a contradiction. Q.E.D.

Now it is easy to see how our graph $Z(q_1, q_2)$ looks like. The following Lemma shows that it is a '*tree of complete graphs $K(q_2+1)$*'; recall that a complete graph $K(n)$ is a combinatorial graph consisting of n vertices, any two of which are adjacent. Namely, each vertex P in Z is a common vertex of q_1+1 distinct complete graphs isomorphic to $K(q_2+1)$. Call a complete graph isomorphic to $K(4)$ a *tetrahedron* (cf. Fig. (10.7)).

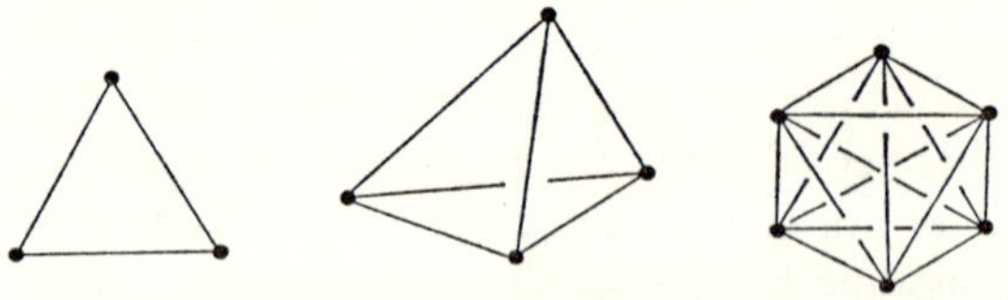

$K(3)$: triangle $K(4)$: tetrahedron $K(n)$: $n=6$

Fig. (10.7)

$Z(q_1, q_2)$: $q_1=1$, $q_2=3$

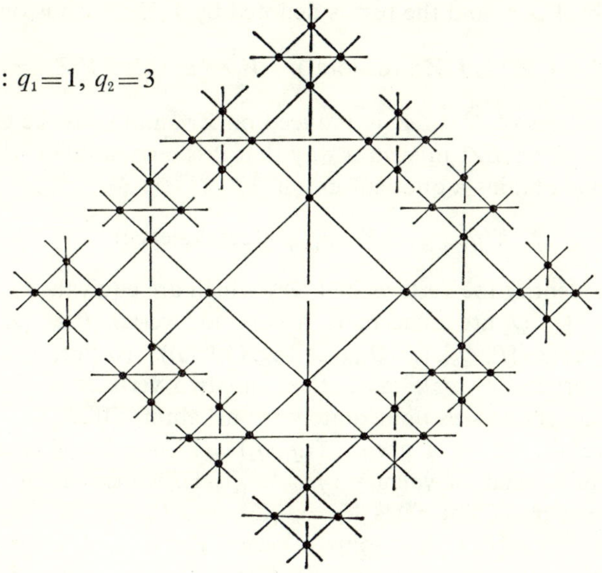

Fig. (10. 8)

We say that two (not necessarily distinct) edges to be *related*, if they belong to a tetrahedron in Z (or a triangle if $q_2=2$, or identical if $q_2=1$).

Lemma (10.9). (i) *The above relation is an equivalence relation in the set EZ of edges of Z, which is compatible with the G-action.*

(ii) *For any vertex $P \in VZ$, it induces an equivalence relation in the set $Z(P)$ of vertices of Z adjacent to P, each equivalence class consisting of q_2 vertices.*

Proof. (i) is easily proved by an application of Lemma (10.6). Two vertices $R_1, R_2 \in Z(P)$ are equivalent if and only if the edges $[P, R_1]$, $[P, R_2]$ are related. To prove (ii), we may assume that $P=P_0=U$. Writing $R_1 = U\omega_1$, $R_2=U\omega_2$ ($\omega_1, \omega_2 \in \Omega$), we see that this is the case if and only if $\omega_1\omega_2^{-1} \in U \cup G_1$. For each R_2, the number of such $R_1 \in Z(P)$ is seen from Lemma (1.4) to be $1+(q_2-1)=q_2$. Q.E.D.

Tree $X(q_1, q_2)$.

We now construct a tree $X(q_1, q_2)$ from $Z(q_1, q_2)$, which is of bipartite type of valency (q_1+1, q_2+1), and which is acted upon by our group G. Let $Z^*(q_1, q_2)$ be the barycentric subdivision of the graph $Z=Z(q_1, q_2)$. This means that we add new vertex Q at the middle point of each edge $[P, R]$ of Z, and call it to be adjacent to P, R. The set of added vertices is denoted by V_2Z^*, and the rest is denoted by V_1Z^*. Thus one has

$$VZ^* = V_1Z^* \cup V_2Z^* \text{ (disjoint)}, \quad V_1Z^* \simeq VZ, \quad V_2Z^* \simeq EZ.$$

Clealy the action of G on Z is induced on Z^*, and so is the equivalence relation on $V_2Z^* \simeq EZ$, in such a way that it is compatible with G-action. Therefore one obtains a quotient graph

$$X(q_1, q_2) := Z^*(q_1, q_2)/(\text{equivalence}).$$

Namely we identify the vertices in V_2Z^* which are equivalent; and if two such vertices Q_1, Q_2 are adjacent to a common vertex $P \in V_1Z^*$, we also identify two edges $[P, Q_1]$, $[P\ Q_2]$ (see Fig. (10.10)). Notice, among others, that by this process of making quotient, no two vertices of V_1Z^* ($\simeq VZ$) are identified, as one sees immediately from Lemma (10.9).

As a result, we get a graph $X(q_1, q_2) = (V_1, V_2; E)$ of bipartite type, which is semi-regular of valency (q_1+1, q_2+1), whose set of vertices are divided into two disjoint parts $V_1 := V_1Z^*$ and $V_2 := V_2Z^*/(\text{equivalence})$, and each edge $y = [P, Q] \in E$ joins a vertex $P \in V_1$ and a vertex $Q \in V_2$. Now the following properties (i), (ii), (iii) are easily proved by what we have seen (cf. Fig. (10.10)).

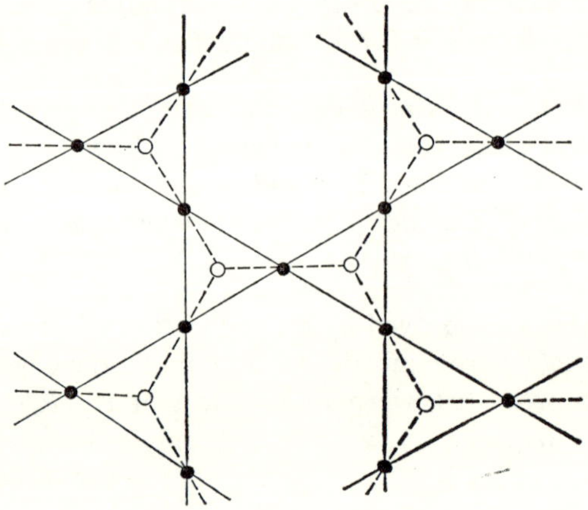

Fig. (10.10)

Theorem (10.11). *Let the notation be as above.*

(i) *The graph $X(q_1, q_2) = (V_1, V_2; E)$ is a connected tree of semi-regular bipartite type with valency (q_1+1, q_2+1): i.e., it is a tree where each vertex $P \in V_1$ is adjacent to q_1+1 vertices in V_2, and each $Q \in V_2$ is adjacent to q_2+1 vertices in V_1.*

(ii) *The distance on $V_1 \simeq VZ$ which is induced from that of the tree*

$X(q_1, q_2)$, coincides with $2d_Z(P, P')$.

(iii) *The group G acts on $X(q_1, q_2)$ in such a way that it has no inversion and that it is transitive on V_1.*

Conversely, suppose that a group G and its action on the connected semi-regular bipartite tree $X=(V_1, V_2, E)$ of valency (q_1+1, q_2+1), satisfying the condition (iii) *above, are given. Then the mapping $l: G \to \boldsymbol{N} \cup \{0\}$ given by $l(g)=(1/2)d_X(P, Pg)$ $(P \in V_1)$ satisfies (G, l, I), (G, l, II).*

Proof. We prove the last assertion. By the assumption we can identify V_1 with $U \backslash G$, where $U:=\mathrm{Stab}(P_0)$, $P_0=U$. The properties (G, l, I) for the mapping l is immediately seen. To show (G, l, II), we note that $\mathscr{H}(G, U)$ is regarded as a subring of $\mathrm{End}(\boldsymbol{Z}[V_1])$, where $\boldsymbol{Z}[V_1]$ is the free $\boldsymbol{Z}$-module over V_1. Namely, a U-double coset $UxU=\bigcup_j Ux_j$ maps $P=Uz$ $(\in V_1)$ to the formal sum $\sum_j Ux_jz$. Then one sees that $G_l : P \to \sum Q$, the sum taken over the set of vertices such that $d_X(P, Q)=2l$. The assertion follows easily from this. Q.E.D.

Corollary (10.12). *Let G be as above.*

(i) *If Γ is a torsion free subgroup of G such that $\Gamma \cap x^{-1}Ux=\{1\}$ for any $x \in G$. Then Γ is a free group.*

(ii) *If, moreover, Γ satisfies $\#(U \backslash G/\Gamma)<\infty$, then it has a finite basis of rank $r_\Gamma=h_1q_1-h_2+1$ $(=h_2q_2-h_1+1)$, where $h_1=\#(U \backslash G/\Gamma)=\#(V_1/\Gamma)$, $h_2=\#(V_2/\Gamma)$ are the number of Γ-orbits in V_1, V_2, respectively.*

Proof. This follows from [Ser-1], I.3.3, Theorem 4.4′, together with the relation $(q_1+1)h_1=(q_2+1)h_2$. In fact the assumptions in (i) imply that the (restricted) action of Γ on the tree $X(q_1\, q_2)$ is a free action. To see this, suppose, on the contrary, that $\gamma \in \Gamma$ $(\gamma \neq 1)$ has a fixed point $Q \in V_2$. Then it induces a permutation on the finite set of vertices adjacent to Q. Hence some power γ^n of γ has a fixed point $P=Ux \in V_1$. This implies that $\gamma^n \in \Gamma \cap x^{-1}Ux$, a contradiction. Q.E.D.

Notice that the assumption on the torsion-freeness of Γ can be weakened to the one that Γ has no torsion element $\gamma \neq 1$ such that $\gamma^n=1$ for $n=(q_2+1)!$.

§ 11. Graph of groups over a flower

Now we consider the quotient graph $F:=X(q_1, q_2)/G$. Thus we assume that G is a group acting on $X(q_1, q_2)$ as described in (10.11). Then it is clear that F is a finite graph; in fact the image of the vectices of V_1 (resp. V_2) consists of a single (resp. at most $1+q_1$) points. Denote them by $V_1F=\{P\}$ and $V_2F=\{Q_1, \cdots, Q_r\}$. Also denote the image of E by EF.

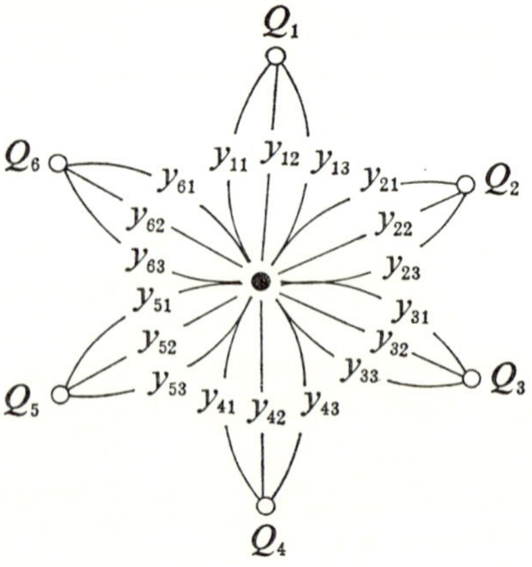

Fig. (11. 1)

From its shape, we shall call F *a flower*, and each of its subgraphs consisting of $F^{(i)}:=\{P, Q_i$, all edges $y_{i,j}$ $(1 \leq j \leq t_i)$ connecting them$\}$, *a petal* of the flower F $(1 \leq i \leq r)$.

Let $P_i = P_0 \in V_1$, and let $Q_{i,j} \in V_2$ be the vertex such that the edge $[P_i, Q_{i,j}]$ project to $y_{i,j}$ $(1 \leq j \leq t_i)$. We denote by $U(Q_{i,j})$ the stabilizer in G of $Q_{i,j}$. Similarly, let $Q_i \in V_2$, and let $P_{i,j} \in V_1$ be the vertex such that the edge $[Q_i, P_{i,j}]$ project to $\bar{y}_{i,j}$ $(1 \leq j \leq t_i)$. We denote by $U(P_{i,j})$ the stabilizer in G of $P_{i,j}$. Note that $U \cap U(Q_{i,j})$ (resp. $U(Q_i) \cap U(P_{i,j})$) is the stabilizer of the edge $[P_i, Q_{i,j}]$ (resp. $[Q_i, P_{i,j}]$). Since the two edges $[P_i. Q_{i,j}]$ and $[P_{i,j}, Q_i]$ have the same image $y_{i,j}$ in F, one finds $g_{i,j} \in G$ such that $[P_i, Q_{i,j}]g_{i,j} = [P_{i,j}, Q_i]$, i.e., $P_i g_{i,j} = P_{i,j}$, $Q_i g_{i,j}^{-1} = Q_{i,j}$. It follows that

$$(11.2) \qquad U(Q_i) \cap U(P_{i,j}) = g_{i,j}^{-1}(U \cap U(Q_{i,j}))g_{i,j}.$$

We refer to the following equalities (11.4), (11.5) as *the regularity condition.*

Proposition (11.3). *Let the notation be as above.*

(i) *The number of petals in* $F = r = \sharp(V_2/G)$.

(ii) *The number of edges in a petal* $F_i = t_i = \sharp(V(Q_i; 1)/U(Q_i))$, *and one has*

$$(11.4) \qquad \sum_{j=1}^{t_i} [U(Q_i): U(Q_i) \cap U(P_{i,j})] = q_2 + 1 \quad (1 \leq i \leq r).$$

(iii) *One has similarly*

$$(11.5) \qquad \sum_{i=1}^{r} \sum_{j=1}^{t_i} [U(P_i): U(P_i) \cap U(Q_{i,j})] = q_1 + 1.$$

Proof. (i) is clear. To prove (ii), one notes first that for two vertices Q, $Q' \in V(P_i; 1)$, the edges $[P_i, Q]$, $[P_i, Q']$ have the same image in F if and only if there exists $g \in G$ such that $P_i g = P_i$ and $Qg = Q'$, i.e., if and only if they belong to the same $U(P_i)$-orbit. The equality $t_i = \sharp(V(P_i; 1)/U(P_i))$ follows from this, and counting the vertices in $V(P_i; 1)$ in two ways, one gets (11.4). (iii) is proved similarly. Q.E.D.

Recall that a graph of groups $(\boldsymbol{G}, Y)$ is a connected graph Y together with an assignment $\boldsymbol{G}$: $Y \to \{\text{groups}\}$ which assigns to each vertex $P \in VY$ (resp. oriented edge $y = [P, Q] \in EY$) a group $\boldsymbol{G}(P)$ (resp. $\boldsymbol{G}(y)$). Moreover, it is required that $\boldsymbol{G}([P, Q]) = \boldsymbol{G}([Q, P])$, and there are monomorphisms

$$(11.6) \qquad \begin{array}{ll} \boldsymbol{G}([P, Q]) \longrightarrow \boldsymbol{G}(Q), & a \to a^y \\ \| & \\ \boldsymbol{G}([Q, P]) \longrightarrow \boldsymbol{G}(P), & a \to a^{\bar{y}} \end{array} \qquad \text{(cf. [Ser-1]).}$$

Suppose that a group G acts on a graph X, and $Y := X/G$ be its quotient graph. Then one gets in the natural way a graph of groups $(\boldsymbol{G}, Y)$, called the *associated graph of groups*. This is exactly what we described above; namely $\boldsymbol{G}(P) = U(P) =$ the stabilizer of P in G, and $\boldsymbol{G}([P, Q]) = \boldsymbol{G}(P) \cap \boldsymbol{G}(Q)$ (note the compatibility condition (11.2)).

A maximal tree T in a graph Y is a subgraph which is a tree, and which is maximal (i.e., containing all vertices of Y). In [Ser-1] the fundamental group $\pi_1(\boldsymbol{G}, Y, T)$ of $(\boldsymbol{G}, Y)$ at T is defined. It is a group with generators $\boldsymbol{G}(P)$ $(P \in VY)$ and $g_y (y \in EY)$ with the relations

$$(11.7) \qquad \begin{array}{l} g_y^{-1} a^y g_y = a^{\bar{y}}, \quad g_{\bar{y}} = g_y^{-1} \ (y \in EY, a \in \boldsymbol{G}(y)), \\ \text{and} \quad g_y = 1 \quad \text{if} \quad y \in ET \ (= \text{edges of } T). \end{array}$$

Then the following fundamental results have been proved in [Ser-1]:

Theorem (11.8) (Serre). *Let $(\boldsymbol{G}, Y)$ be a graph of groups and T a maximal subtree of Y.*

(1) *There exists a tree $\tilde{X}(\boldsymbol{G}, Y, T)$ on which $\pi_1(\boldsymbol{G}, Y, T)$ acts in such a way that $\tilde{X}(\boldsymbol{G}, Y, T)/\pi_1(\boldsymbol{G}, Y, T) \simeq Y$.*

(2) *$\tilde{X}(\boldsymbol{G}, Y, T)$ and $\pi_1(\boldsymbol{G}, Y, T)$ are universal; i.e., for any group G and a connected graph X on which G acts in such a way that $X/G \simeq Y$, and that the associated graph of groups is isomorphic to $(\boldsymbol{G}, Y)$, there are natural surjective morphisms*: $\varphi : \pi_1(\boldsymbol{G}, Y, T) \to \boldsymbol{G}$, $\psi : \tilde{X}(G, Y, T) \to X$.

Theorem (11.9) (Serre). *Suppose that a group G acts on a connected non-empty graph X, and let $(\boldsymbol{G}, Y)$ be its associated graph of groups, where $Y = X/G$. Then the following properties are equivalent:*

(a) *X is a tree.*

(b) *ψ: $\tilde{X}(\boldsymbol{G}, Y, T) \longrightarrow X$ is an isomorphism.*
(c) *φ: $\pi_1(\boldsymbol{G}, Y, T) \longrightarrow G$ is an isomorphism.*

Now it is not difficult to show, by using the explicit construction of $\tilde{X}(\boldsymbol{G}, Y, T)$ described in [Ser-1], § I.5.3, that, if we start from a graph of groups $(\boldsymbol{G}, F)$ over our flower F which satisfies the regularity condition (11.4), (11,5), then $\tilde{X}(\boldsymbol{G}, F, T)$ is isomorphic to $X(q_1, q_2)$. In fact using (11.4) (resp. (11.5)), one can show that any vertex Q (resp. P) lying over a white one $Q_i \in V_2F$ (resp. black one $P_0 \in V_1F$) has q_2+1 edges (resp. q_1+1 edges).

Summing up, we have shown the equivalence of the following classes of groups (with extra structures).

Theorem (11.10) (Main result). *The following three classes of groups with extra data are mutually equivalent:*

(1) *The groups G with a length function $l: G \to \boldsymbol{N} \cup \{0\}$, satisfying axioms (G, l, I), (G, l, II).*

(2) *The groups G which act on a semi-regular bipartite tree $X(q_1, q_2) = (V_1, V_2; E)$ in such a way that the action is transitive on V_1.*

(3) *The fundamental groups $\pi_1(\boldsymbol{G}, F, T)$ of flowers of groups $(\boldsymbol{G}, F)$ with regularity condition* (11.4), (11.5).

Example (11.11). Let G^* be a free group with n generators $\alpha_1, \cdots, \alpha_n$. Denote by $l^*(x)$ the length of the reduced word of $x \in G^*$. Define G to be the subgroup of G^* consisting of x such that $l^*(x) \equiv 0 \pmod 2$, and put $l(x) := (1/2)l^*(x)$. It is well known that G is again a free group, and by Schreier's index theorem, the rank of G is $r_G = (n-1)[G^*: G]+1 = 2n-1$. It is immediately seen that (G, l) satisfies (G, l, I) with $U=\{1\}$. To see the second axiom (G, l, II), we note first that (G^*, l^*) satisfies the corresponding axioms of [I-1], and that $G_l = G^*_{2l}$ for any $l \in \boldsymbol{N} \cup \{0\}$. Namely one has in $\mathscr{H}(G^*, U^*) = \boldsymbol{Z}[G^*]$ (=the group ring of G^*),

$$(11.12) \quad (G_1^*)^2 = G_2^* + (q+1)U^*, \quad G_1^* \cdot G_l^* = G_{l+1} + qG_{l-1}^* \quad (l \geq 2),$$

with $q=2n-1$. Our axiom (G, l, II) with $q_1 = q_2 = 2n-1$ easily follows from these equalities. Let $Z = Z(q_1, q_2)$ be the graph associated with (G, l) (cf. § 10), and let $P_0 = U = 1$ be the origin of Z. It is also easy to see that two vertices $R, R' \in Z$ are related as in Lemma (10.9), (ii) if and only if their reduced words coincide after the first words. From this it follows that the action of G on the set V_2 of new vertices of $X(q_1, q_2)$ is transitive. On the other hand, since $U=\{1\}$, no two edges adjacent to P_0 are G-equivalent. We have shown that the quotient graph $X(q_1, q_2)/G$ looks like a petal as in the following:

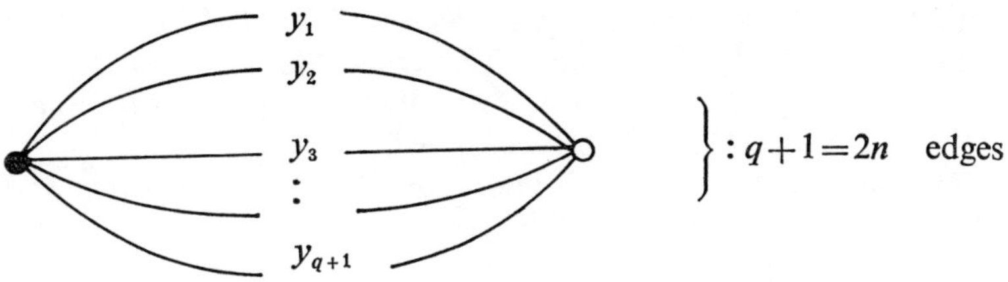

Fig. (11.13)

This example can also be constructed from the graph of groups, associated to the above petal where to both vertices attaches the trivial group (cf. [Ser-1]).

Remark (11.14). As described above, the results of Ihara [I-1] are reproduced if we put $q_1=q_2=q$, and $G_l=G^*_{2l}$, where (G^*, l^*) satisfies the axioms of [I-1]. This explains the relation for $PGL(2, K)$ and $PSL(2, K)$.

§ 12. Tits system and the Hecke algebra

In the rest of this paper, we shall study the steps in which our group G comes to have properties which appear in the theory of Tits system. Thus we assume that G is a group with the conditions described in Theorem (11.10).

Notation being as in § 1, we put $P_0 := U \in V_1$, and choose, once and for all, a vertex $Q_0 \in V_2$ which is adajcent to P_0. Also put, $V := V_1 \cup V_2$, and for any point $P \in V$, call $U(P)$ the stabilizer of P in G. The action of G on $X(q_1, q_2)$ induces the action of $U(P)$ on each $V(P; l)$. Notice that $U(P_0)=U$, and we have the natural bijection $V(P_0; 2l)/U \simeq U \backslash G_l/U$.

Lemma (12.1). *We have the following inequalities:*

$$\#(V_2/G) \leq \#(V(P_0: 1)/U) \leq \#(V(P_0; 2)/U) \leq \cdots \leq \#(V(P_0; l)/U) \leq \cdots$$

Proof. For any vertex $Q \in V_2$, take a vertex $P \in V_1$ which is adjacent to Q. If we write $P=Ux$ $(x \in G)$, then we have $Qx^{-1} \in V(P_0; 1)$. This shows that any G-orbit in V_2 is represented by a vertex in $V(P_0; 1)$, hence the first inequality. The rest follows from the fact that, by mapping the adjacent vertex near to the origin P_0, one has the inverse system

$$V(P_0; 0) \leftarrow V(P_0; 1) \leftarrow V(P_0; 2) \leftarrow \cdots \leftarrow V(P_0; l) \leftarrow \cdots \tag{12.2}$$

where each map is surjective, and U-equivariant. Q.E.D.

Proposition (12 3). *The following conditions are equivalent.*

(a) *G acts transitively on E* ($=$*the set of edges of* $X(q_1, q_2)$)

(b) $\#(V(P_0; 1)/U)=1$, *i.e.*, $V(P_0; 1)=Q_0\cdot U$.

(c) $G=U_1 *_B U_2$ (*product with amalgamated subgroup B*), *where* $U_1:=U$, $U_2:=U(Q_0)$, *and* $B:=U_1\cap U_2$.

Moreover, it follows under these conditions that the action of U_2 *on* $V(Q_0; 1)$ *is transitive*: $\#(V(Q_0; 1)/U_2)=1$.

Proof. The equivalence (a)$\Leftrightarrow$(c) is nothing but Theorems 6, 7 of [Ser-1], § I.4.1. Other part is easily proved as in the above Lemma. (See also the proof of Proposition (12.4)). Q.E.D.

Proposition (12.4). *Suppose that the conditions of the above proposition are satisfied. Then one has the following assertions.*

(i) $\#(U\backslash G_1/U)=1 \Longleftrightarrow \#(B\backslash U_2/B)=2 \Longleftrightarrow \#(V(Q_0; 1)/B)=2$.

(ii) *Under these conditions, one has* $s_2^2\in B$ *for any* $s_2\in U_2-B$. *Similar assertions hold for* U_1, $s_1\in U_1-B$.

Proof. (i) Note first that, in general one has the inequalities

$$[U_1: B]\leq 1+q_1, \quad [U_2: B]\leq 1+q_2, \tag{12.5}$$

where both equalities hold if and only if G acts transitively on E. In fact the index $[U_1: B]$ is equal to the number of vertices of the U_1-orbit $Q_0\cdot U_1$ in $V(P_0; 1)$, the latter having $1+q_1$ vertices. And similarly for $[U_2: B]$. Now suppose that G is transitive on E, so that U_1 is transitive on $V(P_0; 1)$. Let $\omega\in G_1$ be such that the vertex P_ω lies in the same component of $X-\{P_0\}$ as Q_0, i.e., $d_X(Q_0, P_\omega)=1$. Then one has

$$\begin{aligned}\#(U\backslash G_1)\geq\#(U\backslash U\omega U)&=\#(U\backslash U\cdot\omega U\omega^{-1})\\&=\#((U\cap\omega U\omega^{-1})\backslash\omega U\omega^{-1})\\&=[U: B]\cdot[B: U\cap\omega^{-1}U\omega].\end{aligned}$$

From this and the equality in (12.5), one sees that

$$G_1=U\omega U \Longleftrightarrow [B: U\cap\omega^{-1}U\omega]=q_2. \tag{$*$}$$

To prove the last equivalence, we notice again that our assumption implies that U_2 is transitive on $V(Q_0; 1)$. Since the stabilizer in U_2 of the vertex P_0 is B, we see that $B\backslash U_2/B$ is in one-to-one correspondence with the orbit space $V(Q_0; 1)/B$, Moreover, since $U\cap\omega^{-1}U\omega$ is the stabilizer of $P_\omega\in V(Q_0: 1)$, we see that the right hand side of ($*$) is equivalent to the condition that $V(Q_0; 1)-\{P_0\}$ is a single B-orbit, i.e., $\#(B\backslash U_2/B)=2$.

(ii) is an immediate consequence of what we have seen, since s_2 induces the transposition of the two B-orbits in $V(Q_0; 1)$. Q.E.D.

We note that the element s_2 above belongs to G_1. The following Lemma is easily proved by the above arguments and (9.8).

Lemma (12.6). *Suppose that the conditions stated in* (12.4) *are satisfied. Then one has* $(s_1s_2)^l, (s_2s_1)^l \in G_l$ *for any* $l \in \boldsymbol{N}$.

Now we suppose that the conditions in (12.4) are satisfied, and consider the Hecke algebra $\mathscr{H}(G, B)$. Call T_i the element Bs_iB $(i=1, 2)$, and I_B the unit element of $\mathscr{H}(G, B)$.

Proposition (12.7). $T_i^2=(q_i-1)T_i+q_iI_B \quad (i=1, 2)$.

Proof. By our assumption, the set E of the edges of $X(q_1, q_2)$ is identified with $B\backslash G$, and $\mathscr{H}(G, B)$ can be regarded as a subring of $\mathrm{End}(\boldsymbol{Z}[E])$, where $\boldsymbol{Z}[E]$ denotes the free $\boldsymbol{Z}$-module over the set E. Regarding Q_0 as the origin, one can identify E with $V-\{Q_0\}$, by assigning each edge to its end vertex lying on the other side of Q_0. Let $E(Q_0; 1)$ be the set of edges corresponding to $V(Q_0; 1)$. Then from what we have seen in the proof of Proposition (12.4), it follows that $E(Q_0; 1)$ consists of two B-orbits, corresponding to the decomposition $B\backslash U_2/B=B\cup Bs_2B$. This implies that as an element of $\mathrm{End}(\boldsymbol{Z}[E])$, T_2 maps each edge e to the formal sum of the q_2 edges which share the same vertex $Q \in V_2$ with e. The identity for T_2 follows immediately from this fact. The one for T_1 is proved similarly.

Q.E.D.

Notice that $T_1T_2\neq T_2T_1$. It follows from the above proof that the monomials $I_B, T_1T_2, T_2T_1, \cdots, T_{i_1}T_{i_2}\cdots T_{i_l}$ $(i_m\neq i_{m+1}, m=1, \cdots, l-1)$ are pairwise disjoint, hence they are linearly independent. Also it is easy to prove:

Corollary (12.8). *As elements of* $\mathscr{H}(G, B)$, *one has*

$$G_l=(1+T_1)T_2(T_1T_2)^{l-1}(1+T_1) \quad (l\geq 1).$$

Theorem (12.9). *Notation being as above, suppose that* $q_1, q_2\geq 2$, *and that the conditions in* (12.4) *are satisfied. Then the following assertions are all equivalent.*

(i) *If* $i_1, \cdots, i_l \in \{1, 2\}$ *and* $i_m\neq i_{m+1}$ $(m=1, 2, \cdots, l-1)$, *then* $T_{i_1}T_{i_2}\cdots T_{i_l}$ *consists of a single B-double coset.*

(ii) $\mathscr{H}(G, B)=\boldsymbol{Z}[T_1, T_2]$.

(iii) $\sharp(U\backslash G_l/U)=1$ *for any* $l \in \boldsymbol{N}$.

(iv) *U acts transitively on* $V(P_0; l)$ *for any* $l \in \boldsymbol{N}$.

(iv)′ U_2 *acts transitively on* $V(Q_0; l)$ *for any* $l \in \boldsymbol{N}$.

(v) *There exist a subgroup N of G and a set S such that* $(G, \boldsymbol{B}, N, S)$ *is a Tits system.*

Proof. (i)⇒(ii): This follows immediately from (12.8).

(ii)⇒(i): This follows from the obvious relation

$$Bs_{i_1}s_{i_2}\cdots s_{i_l}B \subseteq (Bs_{i_1}B)\cdots(Bs_{i_l}B) = T_{i_1}\cdots T_{i_l},$$

since the left hand side is expressed as a polymomial of T_1, T_2.

(i)⇒(iii): From (12.8), we see that (i) is equivalent to

$$G_l = Bs_2(s_1s_2)^{l-1}B \cup B(s_1s_2)^lB \cup B(s_2s_1)^lB \cup B(s_1s_2)^ls_1B. \tag{12.10}$$

From this and the fact that $U = B \cup Bs_1B$, it follows that $G_l = U(s_1s_2)^lU$.

(iii)⇔(iv): This is immediate from the remark preceding (12.1).

(iv)⇒(i): We shall show that (iv) implies (12.10).

First note that, for any $l \in N$, the set $V(P_0; 2l)$ is decomposed according to the distance from Q_0:

$$V(P_0; 2l) = (V(P_0; 2l) \cap V(Q_0; 2l-1)) \cup (V(P_0; 2l) \cap V(Q_0; 2l+1)), \tag{12.11}$$

where both of these components are stable under B. It is easy to see that $s_1(\in U-B)$ induces the transposition of them. Since U is assumed to be transitive on $V(P_0; 2l)$, this implies that B acts on each components transitively, i.e., $\sharp(V(P_0\ 2l)/B) = 2$, or equivalently $\sharp(U\backslash G_l/B) = 2$. From Lemma (12.6) it follows that

$$G_l = U(s_2s_1)^lB \cup U(s_2s_1)^{l-1}s_2B. \tag{12.12}$$

Next let $EV(P_0; 2l)$ be the set of all edges of $X(q_1, q_2)$ which have as their one of end points the vertices of $V(P_0; 2l)$. From our assumption, it is easy to see that $EV(P_0; 2l)$ consists of two U-orbits:

$EV^{\pm}(P_0; 2l) := \{e \in EV(P_0; 2l);$ the other end point $\in V(Q_0; 2l \pm 1)\}$. Taking P_0 as the origin of $X(q_1, q_2)$, we can naturally identify E with $V-\{P_0\}$, in such a way that the identification is compatible with U-action. It then follows from (12.3) that one has a commutative diagram

$$\begin{array}{ccc} E & \simeq & B\backslash G \\ \cup & & \cup \\ EV(P_0; 2l) & \simeq & B\backslash G_l \end{array} \quad (l \in N), \tag{12.13}$$

which is compatible with U-action. We then see from (12.12) that

$$EV^+(P_0; 2l) \simeq B(s_2s_1)^{-l}U, \quad EV^-(P_0; 2l) \simeq Bs_2^{-1}(s_2s_1)^{-l}U.$$

Now it is easy to describe the B-orbits decomposition of $EV^{\pm}(P_0; 2l)$. Namely, as in (12.11), each of them is divided into two V-orbits according to the distance between their end points in V_2 and Q. Thus we have

$$\sharp(B\backslash B(s_2s_1)^{-l}U/B)=\sharp(B\backslash U(s_2s_1)^lB/B)=2,$$
$$\sharp(B\backslash Bs_2^{-1}(s_2s_1)^{-l}U/B)=\sharp(B\backslash Us_2(s_2s_1)^lB/B)=2,$$

hence $\sharp(B\backslash G_l/B)=4$. Now we see from (12.8) that

(12.14) $$(T_2T_1)^l=B(s_2s_1)^lB,\quad T_1(T_2T_1)^l=Bs_1(s_2s_1)^lB,$$
$$(T_2T_1)^{l-1}T_2=B(s_2s_1)^{l-1}s_2B,\quad (T_1T_2)^2=B(s_1s_2)^lB,$$

for any $l\in N$, hence (12.10).

We have established the equivalence (i)$\Leftrightarrow$(ii)$\Leftrightarrow$(iii)$\Leftrightarrow$(iv). Since the conditions (i), (ii) are symmetric in q_1, q_2, one sees that (iv)$\Leftrightarrow$(iv)$'$.

(iv), (iv)$'\Leftrightarrow$(v): This has been pointed out in [Ser-1], II.1 ex. However, we sketch briefly how one finds N and S, under the assumptions of (iv), (iv)$'$. Let $L_0^+:=(P_0, Q_0, P_1, Q_1, \cdots)\in\varprojlim V(P_0; l)$ an infinite half line (path) starting from P_0, and let $L_0^-:=(Q_0, P_0, Q_{-1}, P_{-1}, \cdots)\in\varprojlim V(Q_0, l)$ be a similar one starting from Q_0. Put $L_0:=L_0^+\cup L_0^-$. From the assumptions (iv), (iv)$'$, one finds element $s_1\in U_1$ (resp. $s_2\in U_2$) which induces on L_0 the reflection at P_0 (resp. Q_0). Call N the subgroup of G consisting of elements which keep L stable, and put $T:=B\cap N$. Then it is easy to see that $T|L=\{\text{id.}\}$, and T is a normal subgroup of N, and that $N/T=\text{Aut}(L)=\langle s_1, s_2\,|\,s_1^2=s_2^2=\text{id.}\rangle$. Putting $S=\{s_1, s_2 \pmod T\}$, it is not difficult to check the axioms of Tits system (or, BN-pair; see [Bou], [Ser-1], [T-2]. In fact, the only non-trivial part is to show (T-4), which follows from the property (12.14) above. Conversely, if (G, B, N, S) is a Tits system, then it is easy to see that the associated Tits building is isomorphic to our tree $X(q_1, q_2)$ with the same action of G. Therefore the assertion reduces to the well known properties of the Tits system. This completes the proof of Theorem (12.9). Q.E.D.

References

[Bou] N. Bourbaki, Groupes et Algèbres de Lie, Chap. IV. Hermann, Paris, 1968.

[B-T] F. Bruhat and J. Tits, Groupes algébriques simples sur un corps local, Proc. Conf. Local Fields, 23–36, Springer-Verlag, 1967.

[Car] P. Cartier, Fonctions harmoniques sur un arbre, Symposia Mathematica, **9** (1972), 203–270.

[Cas] W. Casselman, On a p-adic vanishing theorem of Garland, Bull. Amer. Math. Soc., **80–5** (1974), 1001–1004.

[Gan] R. Gangolli, Zeta function of Selberg's type for compact space forms of symmetric spaces of rank one, Illinois J. Math., **21** (1977), 1–42.

[G-W] R. Gangolli and G. Warner, Zeta functions of Selberg's type for some noncompact quotient of symmetric spaces of rank one, Nagoya Math. J., **78** (1980), 1–44.

[Ha] K. Hashimoto, Zeta functions of finite graphs and representations of p-adic groups, this volume.

[I-1] Y. Ihara, On discrete subgroups of the two by two projective linear group over p-adic fields, J. Math. Soc. Japan, **18** (1966), 219–235.

[I-2] ——, Discrete subgroups of $PL(2, k_p)$ Proc. Symposia Pure Math., **IX**, AMS 1966, 272–278.

[I-M] N. Iwahori and H. Matsumoto, On some Bruhat decomposition and the structure of the Hecke ring of p-adic Chevalley groups, Publ. Math. IHES, **25** (1965), 5–48.

[Mac] I. G. Macdonald, Sperical functions on a group of p-adic type, Publ. of the Ramanujan Inst., **2**, 1971.

[Sel] A. Selberg, Harmonic analysis and discontinuous groups in weakly symmetric Riemannian spaces with applications to Dirichlet series, J. Indian Math. Soc., **20** (1956), 47–87.

[Ser-1] J-P. Serre, Arbres, Amalgames, SL_2, Astérisque, no **46**, Soc. Math. France, 1977.

[Ser-2] ——, Cohomologie des groupes discretes, Ann. Math. Studies, **70**, 77–169, Princeton Univ. Press, 1971.

[Sh] G. Shimura, Introduction to the arithmetic theory of automorphic functions, Publ. Math. Soc. Japan, **11**, 1971.

[Su] T. Sunada, L-functions in geometry and some applications, Springer Lecture Notes, **1201**, 266–284.

[T-1] J. Tits, Sur le groupe des automorphismes d'un arbre. Essays on Topology and related topics, Memoires dediés à Georges de Rham, Springer-Verlag, 1970, 188–211.

[T-2] ——, Reductive groups over local fields. Proc. Symposia Pure Math., **33-1** (1979), 29–69.

K. Hashimoto
Department of Mathematics
Waseda University
Ohkubo 3–4–1, Shinjuku-ku
Tokyo 160, Japan

A. Hori
Department of Mathematics
University of Tokyo
Hongo 7–3–1, Bunkyo-ku
Tokyo 113, Japan

Advanced Studies in Pure Mathematics 15, 1989
Automorphic Forms and Geometry of Arithmetic Varieties
pp. 211–280

Zeta Functions of Finite Graphs and Representations of p-Adic Groups

Ki-ichiro Hashimoto*)

Dedicated to Prof. Friedrich Hirzebruch and Prof. Ichiro Satake on their sixtieth birthdays

Contents

§ 0. Introduction

0-1. In this paper we shall be concerned with the two different subjects, which have been developed separately. One is a combinatorial problem in algebraic graph theory, and the other is an arithmetic of discrete subgroups of $\boldsymbol{p}$-adic groups and their representations.

Suppose that X is a finite (multi) graph, which is not a tree. We always assume that X is *non-oriented.* A closed path C in X is called reduced, if C and $C^2 = C.C$ have no backtracking. Then obviously the set $\mathscr{C}_l^{red}(X)$ of reduced closed paths of length l is finite, and $\sharp(\mathscr{C}_l^{red}(X)) \to \infty$ $(l \to \infty)$ if X is not homotopic to a circuit, i.e., S^1. (See § 1 for

Received November 30, 1987.
Revised April 1, 1988.
*) The author has been supported by Sonderforschungsbereich 170, 'Geometrie und Analysis' at Univ. Göttingen.

precise definitions). The main problems to be studied in this paper can be stated as follows:

(0.1) Count the number N_l of elements of $\mathscr{C}_l^{red}(X)$; can one give a closed formula for it?

(0.2) To which extent does the data $(N_l)_{l \in \boldsymbol{N}}$ reflect basic properties of X?

We are especially interested in studying the properties of the generating function

$$Z_X(u) := \exp\left\{\sum_{l=1}^{\infty} (N_l/l)u^l\right\}, \qquad N_l := \sharp(\mathscr{C}_l^{red}(X)). \tag{0.3}$$

More generally, let f: $EX \to \boldsymbol{C}$ be a function on the set EX of edges of X. For a reduced closed path $C=(e_1, \cdots, e_l)$ $(e_j \in EX)$, we put $f(C) := \prod_{k=1}^{l} f(e_k)$, and define

$$Z_X(f;u) = \exp\left\{\sum_{l=1}^{\infty} \sum_{C \in \mathscr{C}_l^{red}(X)} (f(C)/l)u^l\right\}. \tag{0.4}$$

These functions are expected to keep many properties of X, as in the case of the congruence zeta functions of algebraic varieties over finite fields. One of our main results (cf. (2.22), (2.27)) implies that $Z_X(f; u)$ is always a rational function of u. In fact, the first of which implies that $Z_X(u)^{-1}$ is a polynomial with integral coefficients:

Theorem (0.5). *Suppose that X is a connected multigraph with m edges and n vertices $\{P_j;\ 1 \le j \le n\}$, and denote by q_j+1 the valency of P_j. Then there exist $2m$ complex numbers (in fact algebraic integers) α_i $(1 \le i \le 2m)$, such that*

$$N_l = \sum_{i=1}^{2m} \alpha_i^l \qquad (l = 1, 2, \cdots), \tag{0.6}$$

$$\prod_{j=1}^{2m} \alpha_j = (-1)^n q_1 q_2 \cdots q_n. \tag{0.7}$$

Moreover,

(i) *the number r of indices such that $\alpha_j = 1$ satisfies*

$$r = \dim_{\boldsymbol{C}} H_1(X, \boldsymbol{C}), \tag{0.8}$$

except for the trivial case that X is homotopic to a circuit Cir_n (i.e., $H_1(X, \boldsymbol{C}) \simeq \boldsymbol{C}$); *in such case one has $r=2$.*

(ii) *the number r' of indices such that $\alpha_j = -1$ satisfies*

$$(0.9)\qquad \begin{aligned} r' &= \dim_{\boldsymbol{C}} H_1(X, \boldsymbol{C}) - 1, && \textit{if } X \textit{ is non-bipartite,} \\ &= \dim_{\boldsymbol{C}} H_1(X, \boldsymbol{C}), && \textit{if } X \textit{ is bipartite and } \dim_{\boldsymbol{C}} H_1(X, \boldsymbol{C}) > 1, \\ &= \dim_{\boldsymbol{C}} H_1(X, \boldsymbol{C}) + 1 = 2, && \textit{if } X \textit{ is homotopic to } \mathrm{Cir}_{2n}. \end{aligned}$$

We note, among others, that the last result (0.9) gives a new characterization of finite bipartite multigraphs. If X is assumed to have some regularity, then one has much more strong results:

Theorem (0.10). (i) *Suppose that X is regular, i.e., $q_j = q$ (>1) for $1 \leq j \leq n$. Then exactly one of the α_i's is equal to q. Moreover, those α_i's s.t. $\alpha_i \neq \pm 1$, $\pm q$ are divided into the union of pairs $\{\alpha_k, \alpha'_k\}$ which satisfy*

$$(0.11)\qquad \alpha_k \alpha'_k = q, \quad \alpha_k + \alpha'_k \in \boldsymbol{R} \quad (\alpha_k, \alpha'_k \neq 1, q).$$

(ii) *Suppose that X is a semi-regular bipartite multigraph of valency $(q+1, q'+1)$, with $q \geq q'$, $qq' > 1$; and put $n_1 = \#\{j; q_j = q\}$, $n_2 = \#\{j; q_j = q'\}$. Then one has $\alpha_{i+m} = -\alpha_i$ $(1 \leq i \leq m)$, and*

$$(0.12)\qquad \{\alpha_1^2, \cdots, \alpha_m^2\} = \{\overbrace{1, \cdots, 1}^{r};\ qq';\ \overbrace{-q', \cdots, -q'}^{n_2 - n_1},\ \beta_k, \beta'_k\ (1 \leq k \leq n_1 - 1)\},$$

where each pair $\{\beta_k, \beta'_k\}$ satisfies

$$(0.13)\qquad \beta_k \beta'_k = qq', \quad \beta_k + \beta'_k \in \boldsymbol{R} \quad (\beta_k, \beta'_k \neq 1, qq').$$

These results imply that our "zeta function $Z_X(u)$" of X has in fact many properties which are strikingly analogous to the congruence zeta functions of algebraic curves over a finite field $\boldsymbol{F}_q$.

0-2. On the other hand, an analogue of the Selberg zeta function $Z_\Gamma(u)$ has been introduced by Ihara [I-1], for the discrete subgroups of SL(2, K) over a $\boldsymbol{p}$-adic field K. It is defined by an infinite formal product

$$(0.14)\qquad Z_\Gamma(u) := \prod_{\{\gamma\}_\Gamma} (1 - u^{\deg\{\gamma\}_\Gamma})^{-1},$$

extended over the set of "primitive hyperbolic" Γ-conjugacy classes. As is remarked in [I-2], this $Z_\Gamma(u)$ coincides with a congruence zeta function of a modular curve over $\boldsymbol{F}_p$ for some arithmetically defined Γ. Moreover, as it was pointed out by Serre [Ser], $Z_\Gamma(u)$ has an interpretation as a zeta function of certain regular graph of valency p+1; the above Theorem (0.10), (i) is then a consequence of a result of [I-1]. However, these relations have been seldom taken up seriously until recently. It is an interesting problem to ask for a generalization of such relations to a wider class of groups and graphs, and possibly, varieties (curves).

0–3. In our previous paper [H-H], Ihara's zeta function has been generalized to a class of groups, containing those defined over a $\boldsymbol{p}$-adic field K, with K-rank one. We have given it an expression as a rational function. Actually, $Z_\Gamma(u)^{-1}$ was shown to be a polynomial of u, which is essentially a Hecke polynomial for an element $T(p)$ of $\mathscr{H}(G, U)$, the Hecke algebra of G with respect to a maximal open compact subgroup U. This raises some interesting questions. It is known that a simply connected semisimple group over a local field K with K-rank one has two non-conjugate maximal open compact subgroups, say U_1 and U_2. The formula for $Z_\Gamma(u)$ is [H-H] is not symmetric in U_1, U_2, reflecting that the calculation has been done depending on a choice of one of them. How can one explain the difference between the two expressions? $\cdots$ This has been a motivation for the present paper.

Another problem, which is more important and closely related to it, is the relation between $Z_\Gamma(u)$ and the spectral decomposition of $L^2(G/\Gamma)$. Recall that in the case of real Lie groups of $\boldsymbol{R}$-rank one, to determine the Selberg zeta function is equivalent to know the eigenvalues of the Laplacian on $L^2(G/\Gamma)$.

0–4. In studying these problems we were led to introduce the third expression for $Z_\Gamma(u)$, which was based on $B=U_1\cap U_2$, the Iwahori subgroup of G. The Hecke algebra $\mathscr{H}(G, B)$ is generated by two elements T_1, T_2. Now the main result of the present paper states that our zeta function $Z_\Gamma(u;\rho)$ with additional parameter ρ (= unitary representation of Γ of finite degree), can be expressed:

$$Z_\Gamma(u;\rho)^{-1}=\det\{I-\rho^*(T_1T_2)u\}, \tag{0.15}$$

where ρ^* is a representation of $\mathscr{H}(G, B)$ associated with ρ. A simple argument on the representations of $\mathscr{H}(G, B)$ now makes the situation clear, and one can reproduce quite simply from (0.15) the main result in [H-H]. Moreover, if one combines this result with the theory of Borel [Bo-2] on the admissible representations of $\boldsymbol{p}$-adic groups having fixed vectors under B, one easily gets a complete answer to the above questions. We shall describe it in Section 6. Among others, we shall prove the following result, which is also a generalization of [I-1].

Theorem (0.16). *The zeta function $Z_\Gamma(u)$ describes the spectral decomposition in $L^2(G/\Gamma)$, of those components which have B-fixed vectors. Namely, let (π, V_π) be the irreducible unitary representation of G such that*

$V^B \neq \{0\}$, and let φ be the induced representation of $\mathscr{H}(G, B)$ on V^B. Then the multiplicity of (π, V_π) in $L^2(G/\Gamma)$ is equal to that of the characteristic polynomial of $\varphi(T_1T_2)$ in $Z_\Gamma(u)^{-1}$.

0–5. Quite surprisingly, it turned out that the expression (0.15) holds for an arbitrary finite multigraph X, if one only assume that it is of bipartite type. The non-bipartite case can easily be reduced to this case. Here we need no regularity condition on X. T_1, T_2 are interpreted as the correspondences on the edges EX. Based on this result, one can study various combinatorial properties of finite graphs in terms of our zetafunctions. Especially, we shall describe the relation between the spectrum of a finite multigraph X and $Z_X(u)$. This has a number of interesting applications. Among others, we shall give a formula which relates the complexity $\kappa(X)$ of X and the value of $(1-u)^r Z_X(u)$ at $u=1$; this is an analogue of the class number formula for a number field, or a function field.

Finally, it is a great pleasure for the author to remark that the present paper, as well as the previous one [H-H], grew out of the effort to understand the important (mysterious) paper of Ihara [I-1], and to ask for the possible generalization of his results. He also would like to make the following:

Acknowlegements. The author wishes to express his gratitude to Prof. H. Hijikata for a useful discussion during the preparation of this paper. He is also grateful to Prof. P. Sarnak, who informed him (without proof) the connection between $\kappa(X)$ and the class number of the function field of a modular curve over $\boldsymbol{F}_p$.

Notation. As usual, we denote by $\boldsymbol{Z}$ the ring of integers, and by $\boldsymbol{Q}$, $\boldsymbol{R}$, $\boldsymbol{C}$ the field of rational, real, and complex numbers, respectively. If A is a ring with unit 1, then $M(n, A)$ is the ring of $n \times n$ matrices, and $GL(n, A)$ the group of invertible matrices in $M(n, A)$. The unit matrix is denoted by I_n, or simply by I. For a finite set S, $\sharp(S)$ denotes its cardinality. We shall often write the collection of indexed elements set theoretically as $\{a_1, \cdots, a_n\}$, even if a_j's are not necessarily distinct.

Since there are several different terminology even for basic objects in the graph theory, we collect in Section 1 those which we use, to avoid confusion.

§ 1. Graphs and multigraphs

The purpose of this section is to fix the notation and basic definitions in our graph theory, and to state our main points of interests. We remark first that the (multi)graphs X we are interested are all assumed to be *non-oriented*, unless otherwise stated.

A *multigraph* X is a triple (VX, EX, ε), consisting of two sets VX, EX, whose elements are called *vertices*, and *edges* respectively, and a map $\varepsilon=\varepsilon_X$, called the *incident map* of X:

(1.1) $$\varepsilon\colon EX \longrightarrow VX\times VX, \qquad \varepsilon(y)=(o(y), t(y)),$$

where the vertex $o(y)$ (resp. $t(y)$) is called the *origin* (resp. *terminus*) of y, respectively. Moreover, we require that there is an involution ι_X; $EX\to EX$, $y\to y^{-1}$ $(y\neq y^{-1})$ such that $o(y^{-1})=t(y)$, $t(y^{-1})=o(y)$. We call the pair $e=\{y, y^{-1}\}$ a *non-oriented edge*, or simply an *edge* if there is no fear of confusion, and write $\varepsilon(e)=\{o(y), t(y)\}$. The set of (non-oriented) edges are also denoted by EX. An edge $e\in EX$ is said to *join* its *ends* $P, Q\in VX$, or often called to be *incident* to P, Q if $\varepsilon(e)=\{P, Q\}$; and two vertices are then called to be *joined* by the edge e, or *adjacent*. Also, we call two distinct edges $e_1, e_2\in EX$ to be *adjacent*, if $\varepsilon(e_1)\cap\varepsilon(e_2)\neq\phi$. We do not assume that ends of an edge to be distinct; if $\varepsilon(e)=\{P, P\}$, e is called a *loop*. Also, more than one non-oriented edges may have the same pair $\{P, Q\}$ as their ends, in which case X is said to have a *multiple edge*. If X has no loop and no multiple edge, it is a *combinatorial graph*, or simply a *graph*. The number of the non-oriented edges e which is incident to a vertex $P\in VX$ is called the *valency* of P. The vertex of valency one is called an *end point*. A multigraph X is called to be *s-partite*, if VX is divided into the disjoint union of s subsets V_i $(1\leq i\leq s)$ such that no two vertices of the same V_i are adjacent. 2-partite multigraph is simply called *bipartite*.

Fig. 1

A *morphism* of a multigraph $\varphi\colon X=(VX, EX, \varepsilon_X)\to Y=(VY, EY, \varepsilon_Y)$ is a pair of mappings $\varphi_0\colon VX\to VY$, $\varphi_1\colon EX\to EY\cup VY$ such that the following diagram, together with the involutions ι_X, ι_Y, is commutative, where $\varphi_0^*: VX\times VX\to VY\times VY$ is the map induced naturally from φ.

$$\text{(1.2)} \qquad \begin{array}{ccc} EX & \xrightarrow{\ \varphi_1\ } & EY\cup VY \\ {\scriptstyle \varepsilon_X}\big\downarrow & & \big\downarrow{\scriptstyle \varepsilon_Y\cup\Delta_Y} \\ VX\times VX & \xrightarrow{\ \varphi_0^*\ } & VY\times VY \end{array}$$

Here $\Delta_Y: VY\to VY\times VY$ is the diagonal map. X and Y are called *isomorphic*, if there is a morphism $\varphi=(\varphi_0, \varphi_1)$ such that φ_0 and $\varphi_1: EX\to EY$ are bijective.

A *path C of length n* (≥ 0) on a multigraph X is a sequence

$$\text{(1.3)} \qquad C=(P_0, y_1, P_1, y_2, \cdots, P_{n-1}, y_n, P_n)$$

of $n+1$ vertices and n *oriented* edges such that $\varepsilon(y_i)=(P_{i-1}, P_i)$ for $i=1, 2, \cdots, n$. Here we do not require that the vertices or edges are distinct. The length of C is denoted by $|C|$. We write $P_0=o(C)$, $P_n=t(C)$, and call them *the origin, the terminus* of C, respectively. The inverse of a path C is defined by

$$C^{-1}=(P_n, y_n^{-1}, P_{n-1}, y_{n-1}^{-1}, \cdots, P_1, y_1^{-1}, P_0).$$

Also the composition $C.C'$ of two paths C, C' satisfying $t(C)=o(C')$ is defined by an obvious way. If C satisfies $y_i\neq y_{i+1}^{-1}$ $(1\leq i\leq n-1)$, it is called a *proper path*, or to have *no backtracking*. We always regard a path of length 1 to be proper. A path C can be written without any ambiguity as $C=(y_1, y_2, \cdots, y_n)$. If X is a graph, then C is determined also by giving a sequence of $n+1$ vertices $(P_0, P_1, \cdots, P_n)$ where P_i and P_{i+1} are adjacent for $0\leq i\leq n-1$. One sees in this case that C is a proper path if and only if $P_{i-1}\neq P_{i+1}$ for $1\leq i\leq n-1$. We shall often use these simplified notation. A multigraph is called *connected*, if for any two distinct vertices P, Q, there exists a path C such that $o(C)=P$, $t(C)=Q$. We assume, unless otherwise stated, that all multigraphs we treat are connected. Then we can define a distance d_X on VX; $d_X(P, Q)$ is the length of the shortest (proper) path C with origin P, and terminus Q, where we note that $d_X(P, Q)=0$ if $P=Q$, and conversely. The maximum of $d_X(P, Q)$ $(P, Q\in VX)$ is called the *diameter* of X, and denoted by $d(X)$. A path C is called *closed*, if $o(C)=t(C)$.

A typical example of the closed proper path of length n is an *n-circuit*; it is a closed path where the vertices $P_0, P_1, \cdots, P_{n-1}$ are all distinct. In other words, n-circuit is an isomorphic image of the following graph Cir_n:

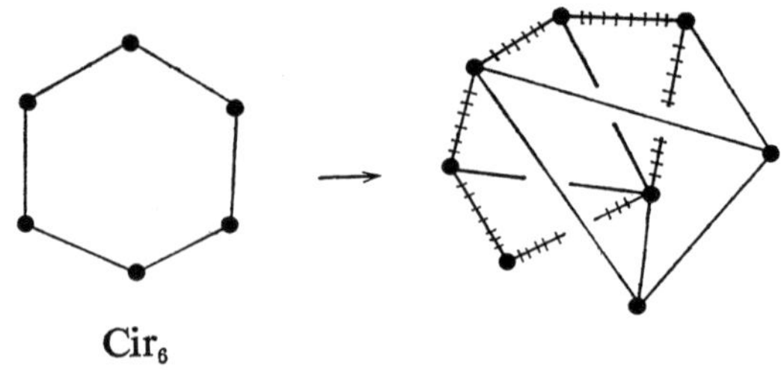

Fig. 2

The minimum length of the circuits of X is called, if it exists, the *girth* of X, and written by $g(X)$. X is called a *tree*, if it has no circuit, and we put $g(X)=\infty$. Note that $g(X)\geq 2$ unless X has a loop, and $g(X)\geq 3$ if X is a graph.

Definition. A closed proper path $C=(y_1, \cdots, y_n)$ on X is called *reduced*, if either $n=1$, or $y_1\neq y_n^{-1}$. In other words, C is reduced if and only if C and $C.C=C^2$ are both proper. We put

(1.4) $$\mathscr{C}_n=\mathscr{C}_n(X):=\{C; C=\text{a proper closed path}, |C|=n\},$$

(1.5) $$\mathscr{C}_n^{red}=\mathscr{C}_n^{red}(X):=\{C\in\mathscr{C}_n; C=\text{reduced}\}.$$

As illustrated in the following figure, to each closed proper path $C\in\mathscr{C}_n$ is assigned a unique reduced path $C^*\in\mathscr{C}_{n-2k}^{red}$ by removing the edges $\{y_1, y_n\}, \{y_2, y_{n-1}\}, \cdots, \{y_k, y_{n-k+1}\}$, when $y_i^{-1}=y_{n-i+1}$ $(1\leq i\leq k)$ and $y_{k+1}\neq y_{n-k}^{-1}$.

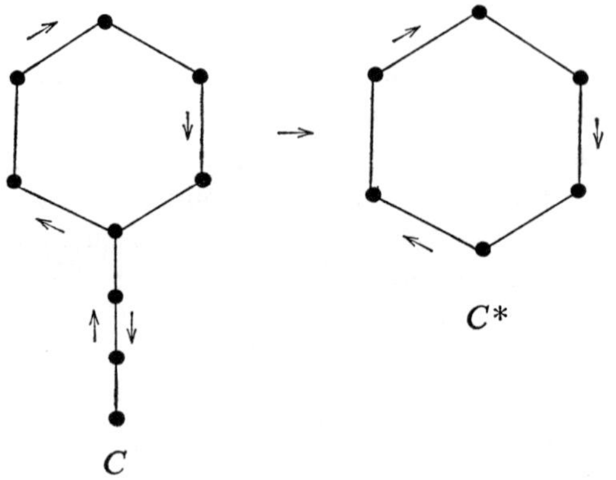

Fig. 3

If $C\in\mathscr{C}_n^{red}(X)$ is reduced, then for each k $(1\leq k\leq n-1)$, the closed path

$$C_k:=(y_{k+1}, y_{k+2}, \cdots, y_n, y_1, \cdots, y_k),$$

which is derived from C by shifting the origin by k steps, is also reduced. We call C_k to be conjugate to C, and write $C_k \sim C$. This defines an equivalence relation in the set $\mathscr{C}_n^{red}$ of the reduced closed paths of length n. We call an equivalence class of C *a cycle* of X, and denote it by [C]. A cycle, or a reduced closed path C of length n which represents it, is called *non-primitive*, if there exists a positive integer k $(1 \le k < n)$ such that $C = C_k$; and otherwise it is called *primitive*. In other words, C is primitive if and only if it is not of the form $C = D^m$ with $m > 1$. We denote by $\mathscr{C}_n^{red,pr}$ the set of all primitive reduced closed paths of length n. Then one sees that there is an obvious decomposition:

$$(1.6) \qquad \mathscr{C}_n^{red} = \bigcup_{d|n} (\mathscr{C}_d^{red,pr})^{n/d} \qquad \text{(disjoint).}$$

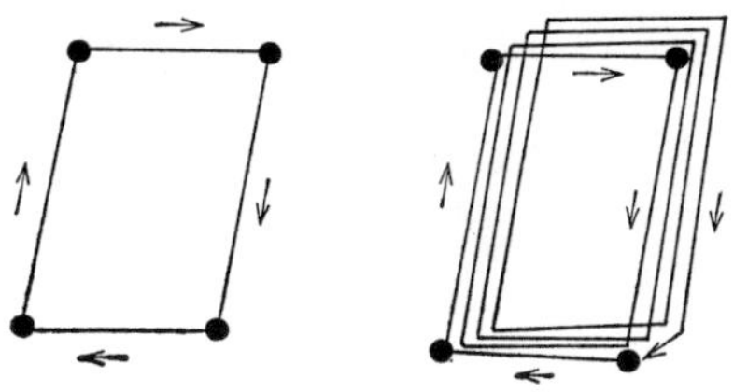

Fig. 4

Finally we recall the basic properties of the fundamental group. Let $P_0 \in VX$ be a vertex which we fix once and for all. Then, as usual, the fundamental group $\pi_1(X, P_0)$ of X with base point P_0 is defined as the group of homotopy classes $\langle C \rangle$ of the closed paths C in X such that $o(C) = t(C) = P_0$, where the product in $\pi_1(X, P_0)$ is induced from the composition of the paths. As illustrated in Fig. 3, each C determines the cycle $[C^*]$ of reduced path C^* which is homotopic to C. Moreover, it is easy to see that the mapping $\langle C \rangle \to [C^*]$ induces the bijection

$$(1.7) \qquad \{\text{conjugacy classes in } \pi_1(X, P_0)\} \simeq \{\text{cylces in } X\}.$$

Note that $\pi_1(X, P_0)$ is a free group of rank $r = \dim_C H_1(X, C)$. Hence the centralizer $Z(\gamma)$ of any element $\gamma \neq 1$ of $\pi_1(X, P_0)$ is an infinite cyclic group. It is called *primitive*, if $Z(\gamma)$ is generated by itself. It is also easy to see that, in the bijection (1.7), the primitive conjugacy classes correspond to the primitive cycles.

It is also known that there exists a tree $\tilde{X}$ and an action of $\pi_1(X, P_0)$ on $\tilde{X}$ such that the quotient $\tilde{X}/\Gamma$ is isomorphic to X. In other words, $\tilde{X}$ together with the natural morphism $p\colon \tilde{X} \to X$ is a universal covering of X.

Let $\tilde{P}_0$ be a vertex of $\tilde{X}$ such that $p(\tilde{P}_0)=P_0$, and let, for each $\gamma \in \pi_1(X, P_0)$, $\tilde{C}_\gamma$ be the unique proper path in $\tilde{X}$ with origin $\tilde{P}_0$ and terminus $\tilde{P}_0 \cdot \gamma$. Then the image $p(\tilde{C}_\gamma)$ of $\tilde{C}_\gamma$ is a closed proper path in X with origin P_0, and we recover γ as the class of $p(\tilde{C}_\gamma)$.

For the more detailed description of $\pi_1(X, P_0)$ and alternative definitions, we refer to [Ser].

§ 2. Zeta functions of finite multigraphs

We begin with a generalization of the definition of zeta function. Let X be a finite connected multigraph, and let $\Gamma := \pi_1(X, P_0)$ be the fundamental group of X. As remarked in Section 1, Γ is a free group, and its rank r is given by

$$r = \#(EX) - \#(VX) + 1 = \dim_C H_1(X, C). \tag{2.1}$$

Let $EX = \{e_1, \cdots, e_m\}$ be the set of non-oriented edges of X. By a *labelling* on EX, we mean an assignment $e_j \to u_j$ $(1 \le j \le m)$, where $u_1, \cdots, u_m$ are (independent) variables. We denote them simply by $\boldsymbol{u} = (u_1, u_2, \cdots, u_m)$.

Definition. For each Γ-conjugacy class $\boldsymbol{P} = \{\gamma\}_\Gamma$, we put

$$\boldsymbol{u}^{\boldsymbol{P}} = \boldsymbol{u}^{C_\gamma} := \prod_{k=1}^{d} u_{i_k} \tag{2.2}$$

where $C_\gamma = (y_{i_1}, y_{i_2}, \cdots, y_{i_d})$ is a reduced closed path corresponding to $\boldsymbol{P}$, such that $y_{i_k} \in e_{i_k}$ $(1 \le k \le d)$. Also we define the degree of $\boldsymbol{P}$ by

$$\begin{aligned} \deg \boldsymbol{P} &:= d = \deg \boldsymbol{u}^{\boldsymbol{P}} \\ &= |C_\gamma| \quad (= \text{the length of the reduced path } C_\gamma). \end{aligned} \tag{2.3}$$

We denote by $\langle C \rangle$ the Γ-conjugacy class, or an element in the class, corresponding to the cycle $[C]$, with $C \in \mathscr{C}_d^{red}(X)$. Let $\rho \colon \Gamma \to U(n)$ be an n-dimensional unitary representation of Γ. Then the zeta function of X attached to $(\rho, \boldsymbol{u})$ is defined by either one of the following equivalent (formal) infinite products:

$$\begin{aligned} Z_X(\boldsymbol{u}; \rho) &:= \prod_{\boldsymbol{P} = \{\gamma\}_\Gamma : \text{primitive}} \det\{I_n - \rho(\gamma)\boldsymbol{u}^{\boldsymbol{P}}\}^{-1} \\ &= \prod_{[C] : \text{primitive}} \det\{I_n - \rho(\langle C \rangle)\boldsymbol{u}^{C}\}^{-1} \end{aligned} \tag{2.4}$$

where $\boldsymbol{P} = \{\gamma\}_\Gamma$ (resp. $[C]$) runs over the set of primitive Γ-conjugacy classes (resp. cycles). Taking the logarithm of (2.4), we have

$$\begin{aligned}\log Z_X(\boldsymbol{u};\rho)&=\sum_{[C]}-\log\det\{I_n-\rho(\langle C\rangle)\boldsymbol{u}^C\}\\&=\sum_{[C]}\sum_{i=1}^{n}-\log(1-\alpha_i\boldsymbol{u}^C)\\&=\sum_{[C]}\sum_{i=1}^{n}\sum_{k=1}^{\infty}(\alpha_i^k/k)\boldsymbol{u}^{Ck},\\&=\sum_{k=1}^{\infty}\sum_{d=1}^{\infty}\sum_{\substack{[C]\\|C|=d}}\left\{\sum_{i=1}^{n}(\alpha_i^k/k)\boldsymbol{u}^{Ck}\right\},\end{aligned}$$

where $\{\alpha_i;\,(1\leq i\leq n)\}$ are the eigenvalues of $\rho(\langle C\rangle)$. Notice that the last sum in the bracket is homogeneous of degree kd. Applying the Euler operator, we obtain

$$\sum_{j=1}^{m}u_j(\partial/\partial u_j)\log Z_X(\boldsymbol{u};\rho)=\sum_{k=1}^{\infty}\sum_{d=1}^{\infty}d\sum_{\substack{[C]\\|C|=d}}\left\{\sum_{i=1}^{n}\alpha_i^k\right\}\boldsymbol{u}^{Ck}. \tag{2.5}$$

We note, from the definition of the cycles, that the natural map $C\to[C]$, from $\mathscr{C}_d^{red,pr}$ to the set of primitive cycles, is d-to-one. Putting $kd=l$, we see that the right hand side of (2.5) is reformed to

$$\sum_{l=1}^{\infty}\sum_{d|l}\sum_{C\in\mathscr{C}_d^{red,pr}}\operatorname{tr}\rho(\langle C\rangle^{l/d})\boldsymbol{u}^{C(l/d)}=\sum_{l=1}^{\infty}\sum_{C\in\mathscr{C}_l^{red}}\operatorname{tr}\rho(\langle C\rangle)\boldsymbol{u}^C.$$

The last equality follows from the obvious decomposition (1.6). Put

$$N_{l,\rho}(\boldsymbol{u}):=\sum_{C\in\mathscr{C}_l^{red}}\operatorname{tr}\rho(\langle C\rangle)\boldsymbol{u}^C, \tag{2.6}$$

$$N_{l,\rho}:=\sum_{C\in\mathscr{C}_l^{red}}\operatorname{tr}\rho(\langle C\rangle)=N_{l,\rho}(1,\cdots,1). \tag{2.7}$$

Then we can express our zetafunction as follows:

$$\sum_{j=1}^{m}u_j(\partial/\partial u_j)\log Z_X(\boldsymbol{u};\rho)=\sum_{l=1}^{\infty}N_{l,\rho}(\boldsymbol{u}). \tag{2.8}$$

Note that $Z_X(u,\cdots,u;\rho)$ agrees with $Z_X(u;\rho)$ defined by (0.3). Also, for $\rho=\mathbf{1}$ (= trivial representation) and for a function f on EX, we get $Z_X(f(e_1),\cdots,f(e_m);\mathbf{1})=Z_X(f;u)$ (cf. (0.4)). Finally we have

$$N_l=N_{l,\mathbf{1}}=\#(\mathscr{C}_l^{red}), \tag{2.9}$$

hence $Z_X(u,\cdots,u;\mathbf{1})=Z_X(u)$ (cf. (0.3)). Here we note that in (2.9), we count, if any, a loop C twice, distinguishing C and C^{-1}.

To proceed further, we first assume that X is a *bipartite* multigraph. Then it is easy to see that any closed path has an even length. Let $\tilde{X}$ be the universal covering tree of X as above. Since X is assumed to be of bipartite, so is $\tilde{X}$, and one has a partition $V\tilde{X}=\tilde{V}_1\cup\tilde{V}_2$ (disjoint), which is preserved by the action of $\Gamma=\pi_1(X, P_0)$. For each edge $e=\{P_1, P_2\}\in E\tilde{X}$, such that $P_i\in\tilde{V}_i$, let $\tilde{E}_i(e)$ be the set of edges which are incident to P_i $(i=1, 2)$.

Definition (2.10). Let $\boldsymbol{Z}[E\tilde{X}]$ be the free $\boldsymbol{Z}$-module over the set $E\tilde{X}$. We define the correspondences T_1, T_2 on $E\tilde{X}$ to be the elements in $\mathrm{End}(\boldsymbol{Z}[E\tilde{X}])$ given by

(2.11) $$T_i(e):=\sum_{\substack{e'\in\tilde{E}_i(e)\\ e'\neq e}} e' \qquad (i=1, 2).$$

The following lemma is immediately seen (see Fig. 5).

Lemma (2.12). *One has for each $e\in E\tilde{X}$, and $l\geq 0$*

$$(T_2T_1)^l(e)=\sum_{(*)} e^{2l+1}.$$

Here the sum is extended over the edges $e^{2l+1}\in E\tilde{X}$ such that there exists $C=(y_{(1)}, y_{(2)}, \cdots, y_{(2l+1)})\in\tilde{\mathscr{C}}^{(+)}_{2l+1}(e)$ with $y_{(1)}\in e$, $o(y_{(1)})\in V_2$, $t(y_{(1)})\in V_1$ and $y_{(2l+1)}\in e^{2l+1}$; $\tilde{\mathscr{C}}^{(+)}_{2l+1}(e)$ is the set of all such proper paths $C=(y_{(1)}, y_{(2)}, \cdots, y_{(2l+1)})$ of length $2l+1$. Similar formula holds for $(T_1T_2)^l(e)$, with $C\in\tilde{\mathscr{C}}^{(-)}_{2l+1}(e)$ running to the other direction, starting with $y^{-1}_{(1)}$.

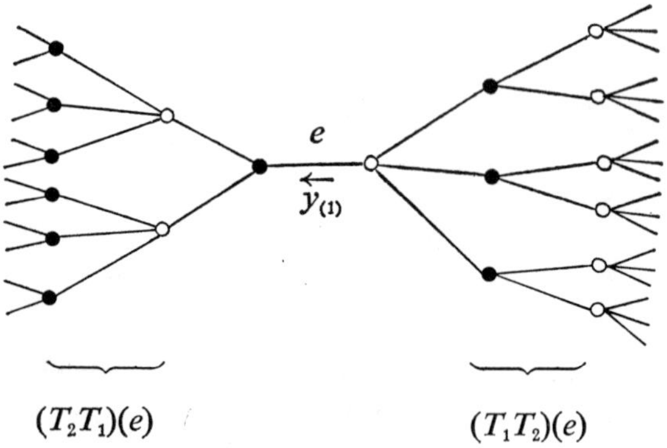

Fig. 5

Now we consider the $\boldsymbol{C}$-vector space M_ρ consisting of the $\boldsymbol{C}^n$-valued functions on $E\tilde{X}$ satisfying

(2.13) $$f(e\cdot\gamma)=f(e)\cdot\rho(\gamma) \qquad \text{for any } \gamma\in\Gamma=\pi_1(X, P_0).$$

It is easy to see that $\dim_{C} M_{\rho}=n.\,\sharp(EX)$. In fact, let $\mathscr{E}=\{e^{(i)}; (1\leq i\leq m=\sharp(EX))\}$ be a complete set of representatives of $E\tilde{X}/\Gamma$, and let $\{\boldsymbol{v}_j\ (1\leq j\leq n)\}$ an orthonormal basis of $\boldsymbol{C}^n$, with respect to the standard inner product. Then from the fact that Γ acts on $\tilde{X}$ without fixed point, it follows that the functions $f_{i,j}$ determined by

$$(2.14)\qquad \begin{aligned} f_{i,j}(e)&=\boldsymbol{v}_j\cdot\rho(\gamma) && (\text{if } e=e^{(i)}\cdot\gamma,\ \exists\gamma\in\Gamma),\\ &=0 && (\text{otherwise}), \end{aligned}$$

form a basis of M_{ρ}. Moreover, one can introduce an inner product on M_{ρ} in such a way that these basis form an orthonormal system. In other words, we put

$$(2.15)\qquad (f, f'):=\sum_{e^{(i)}\in\mathscr{E}}(f(e^{(i)}), f'(e^{(i)})),$$

where those in the right hand side denote the standard inner product on $\boldsymbol{C}^n$.

Since the action of Γ on $\tilde{X}$ preserves the incidence relation and the partition of $V\tilde{X}$, the correspondences T_i induce naturally the endomorphisms on M_{ρ},

$$(2.16)\qquad (\rho^{*}(T_i)f)(e):=\sum_{e'\in\tilde{E}_i(e)} f(e')-f(e)\qquad (i=1, 2).$$

In fact it is easy to see that ρ^{*} is an anti-representation of $\mathrm{End}(\boldsymbol{Z}[E\tilde{X}])$. One also sees immediately that $\rho^{*}(T_i)$ are hermitian operators with respect to the inner product (2.15).

Let $A:=\boldsymbol{C}[\boldsymbol{u}]=\boldsymbol{C}[u_1, \cdots, u_m]$ be the polynomial ring of $\boldsymbol{u}$ over $\boldsymbol{C}$, and define the elements $\rho_{\boldsymbol{u}}^{*}(T_i)$ of $\mathrm{End}(M_{\rho}\otimes_{\boldsymbol{C}}A)$:

$$(2.17)\qquad (\rho_{\boldsymbol{u}}^{*}(T_i)F)(e):=\sum_{\substack{e'\in\tilde{E}_i(e)\\ e'\neq e}} F(e')\boldsymbol{u}^{e'}\qquad (F\in M_{\rho}\otimes_{\boldsymbol{C}}A),$$

where $\boldsymbol{u}^{e'}=u_i$ if $p(e')=e^{(i)}(1\leq i\leq m)$, and we regard F naturally as a $\boldsymbol{C}^n\otimes_{\boldsymbol{C}}A=A^n$-valued function on $E\tilde{X}$. Note that these are well-defined endomorphisms over A, since one has $\boldsymbol{u}^{e\cdot\gamma}=\boldsymbol{u}^{e}$ for any $e\in E\tilde{X}$, $\gamma\in\Gamma$. Moreover, one can extend the inner product (2.15) to a nondegenerate hermitian pairing $(M_{\rho}\otimes_{\boldsymbol{C}}A)\times(M_{\rho}\otimes_{\boldsymbol{C}}A)\to A$, and see that $\rho_{\boldsymbol{u}}^{*}(T_i)$ are self-adjoint operators on a finite A-module $M_{\rho}\otimes_{\boldsymbol{C}}A$. From Lemma (2.12), we obtain

$$(2.18)\qquad (\rho_{\boldsymbol{u}}^{*}(T_1)\rho_{\boldsymbol{u}}^{*}(T_2))^{l}F(e)=\sum_{C\in\tilde{\mathscr{C}}_{2l+1}^{(+)}(e)} F(e')\boldsymbol{u}^{C}.$$

Now the following lemma is a key to our study:

Lemma (2.19). *As endomorphisms over* $\boldsymbol{A}$, *one has*

$$\operatorname{tr}(\rho_{\boldsymbol{u}}^*(T_2)\rho_{\boldsymbol{u}}^*(T_1))^l = \operatorname{tr}(\rho_{\boldsymbol{u}}^*(T_1)\rho_{\boldsymbol{u}}^*(T_2))^l = (1/2)\cdot N_{2l,\rho}(\boldsymbol{u}).$$

Proof. First note that the elements $F_{i,j} := f_{i,j}\otimes 1$ $(1\leq i\leq m,\ 1\leq j\leq n)$ form a basis of $M_\rho\otimes_C \boldsymbol{A}$ over $\boldsymbol{A}$. Therefore we have

$$\begin{aligned}(2.20)\qquad \operatorname{tr}(\rho_{\boldsymbol{u}}^*(T_2)\rho_{\boldsymbol{u}}^*(T_1))^l &= \sum_{i=1}^m\sum_{j=1}^n ((\rho_{\boldsymbol{u}}^*(T_2)\rho_{\boldsymbol{u}}^*(T_1))^l(F_{i,j}),\ F_{i,j})\\ &= \sum_{i=1}^m\sum_{j=1}^n\sum_{k=1}^m ((\rho_{\boldsymbol{u}}^*(T_2)\rho_{\boldsymbol{u}}^*(T_1))^l(F_{i,j})(e^{(k)}\otimes 1),\ F_{i,j}(e^{(k)}\otimes 1))\\ &= \sum_{i=1}^m\sum_{j=1}^n ((\rho_{\boldsymbol{u}}^*(T_2)\rho_{\boldsymbol{u}}^*(T_1))^l(F_{i,j})(e^{(i)}\otimes 1),\ \boldsymbol{v}_j\otimes 1).\end{aligned}$$

Using (2.18), we see that

$$(\rho_{\boldsymbol{u}}^*(T_2)\rho_{\boldsymbol{u}}^*(T_1))^l(F_{i,j})(e^{(i)}\otimes 1) = \sum_{C\in\tilde{\mathscr{C}}_{2l+1}^{(+)}(e^{(i)})} f_{i,j}(e')\boldsymbol{u}^C,$$

where the sum $\sum(*)$ is extended over the proper path $C=(y^{(i)},\cdots,y')\in\tilde{\mathscr{C}}_{2l+1}^{(+)}(e^{(i)})$ such that $y'=y^{(i)}\cdot\gamma$ $(y^{(i)}\in e^{(i)})$ for some $\gamma\in\Gamma$. Denote the set of such paths by $\tilde{\mathscr{C}}_{2l+1}^{(+,0)}(e^{(i)})$, and the similar subset of $\tilde{\mathscr{C}}_{2l+1}^{(-)}(e^{(i)})$ by $\tilde{\mathscr{C}}_{2l+1}^{(-,0)}(e^{(i)})$. Since γ is determined uniquely by C, it follows from (2.14) that the last expression of (2.20) can be reformed as

$$\sum_{i=1}^m\sum_{j=1}^n\sum_{C\in\tilde{\mathscr{C}}_{2l+1}^{(+,0)}(e^{(i)})}(\boldsymbol{v}_j\cdot\rho(\gamma)\otimes 1,\ \boldsymbol{v}_j\otimes 1)\boldsymbol{u}^C = \sum_{i=1}^m\sum_{C\in\tilde{\mathscr{C}}_{2l+1}^{(+,0)}(e^{(i)})}\operatorname{tr}\rho(\gamma)\boldsymbol{u}^C$$

Now the morphism $p\colon \tilde{X}\to X$ induces the following bijection:

$$(2.21)\qquad \bigcup_{i=1}^m[\tilde{\mathscr{C}}_{2l+1}^{(+,0)}(e^{(i)})\cup\tilde{\mathscr{C}}_{2l+1}^{(-,0)}(e^{(i)})]\ \underset{p}{\simeq}\ \mathscr{C}_{2l}^{red},$$

where the $\cup$'s in the left are all disjoint. Thus we have

$$\operatorname{tr}(\rho_{\boldsymbol{u}}^*(T_2)\rho_{\boldsymbol{u}}^*(T_1))^l + \operatorname{tr}(\rho_{\boldsymbol{u}}^*(T_1)\rho_{\boldsymbol{u}}^*(T_2))^l = \sum_{C\in\mathscr{C}_{2l}^{red}}\operatorname{tr}\rho(\langle C\rangle)\boldsymbol{u}^C.$$

This completes the proof. Q.E.D.

Now we have the following

Main theorem (I). *Let* X *be a finite connected multigraph of bipartite type, and let* ρ, $\boldsymbol{u}$ *be a unitdry representation of* $\Gamma=\pi_1(X, P_0)$ *of degree* n, *and a labelling of* EX, *respectively. Then the zetafunction* $Z_X(\boldsymbol{u};\rho)$ *attached to* $(\boldsymbol{u},\rho)$ *is a rational function of* $\boldsymbol{u}$, *and it is given by*

(2.22) $$Z_X(\boldsymbol{u};\rho)=\det(I_{nm}-\rho_{\boldsymbol{u}}^*(T_2)\rho_{\boldsymbol{u}}^*(T_1))^{-1} \qquad (m=\sharp(EX)).$$

In particular, one has (putting $\boldsymbol{u}=(u,\cdots,u)$*)*

(2.23) $$Z_X(u;\rho)=\det(I_{nm}-\rho^*(T_2T_1)u^2)^{-1}.$$

Proof. Taking log of the right hand side of (2.22), one get

$$\log\det(I_{nm}-\rho_{\boldsymbol{u}}^*(T_2)\rho_{\boldsymbol{u}}^*(T_1))^{-1}=\sum_{l=1}^{\infty}(1/l)\operatorname{tr}(\rho_{\boldsymbol{u}}^*(T_2)\rho_{\boldsymbol{u}}^*(T_1))^l.$$

Notice that $\operatorname{tr}(\rho_{\boldsymbol{u}}^*(T_2)\rho_{\boldsymbol{u}}^*(T_1))^l$ is a homogeneous polynomial in $\boldsymbol{u}$ of degree $2l$, and apply the Euler's operator. One obtains

$$\sum_{j=1}^{m}u_j(\partial/\partial u_j)\log\det(I_{nm}-\rho_{\boldsymbol{u}}^*(T_2)\rho_{\boldsymbol{u}}^*(T_1))^{-1}=2\sum_{l=1}^{\infty}\operatorname{tr}(\rho_{\boldsymbol{u}}^*(T_2)\rho_{\boldsymbol{u}}^*(T_1))^l.$$

The assertion follows by comparing this and (2.8), (2.19). Q.E.D.

Remark (2.24). (i) We note that, in both (2.19) and (2.22), the operator $\rho^*(T_1T_2)$ plays the same role as the Frobenius endomorphism in the case of congruence zeta functions of algebraic curves over finite fields.

(ii) For a bipartite multigraph X, the zeta function $Z_X(\boldsymbol{u};\rho)^{-1}$ is an even polynomial of $\boldsymbol{u}$ i.e., $Z_X(-\boldsymbol{u};\rho)=Z_X(\boldsymbol{u};\rho)$. For the zeta function of single variable, it is often convenient to modify it and put

(2.25) $$Z_{X,b}(u;\rho):=Z_X(u^{1/2};\rho)=\det(I_{nm}-\rho^*(T_2T_1)u)^{-1}.$$

Next we consider a general multigraph X. Let $X^{(2)}$ be the barycentric subdivision of X. This means that we add a new vertex to the middle point of each edge of X, hence the edge is divided into two edges. It can be seen immediately that $X^{(2)}$ is a graph of bipartite type. Namely $VX^{(2)}=V_1\cup V_2$, where $V_1=VX$ and $V_2\simeq EX$ is the set of new vertices.

Definition (2.26). Let $EX^{(2)}=\{e_1^{(2)},\cdots,e_{2m}^{(2)}\}$ be the edges of $X^{(2)}$ such that $e_{2j-1}^{(2)}$, $e_{2j}^{(2)}$ correspond to the edge e_j of X. For a labelling $\boldsymbol{u}$ of EX, we put $\boldsymbol{u}^{(2)}:=(u_1,u_1,u_2,u_2,\cdots,u_m,u_m)$ and regard it a (reduced) labelling of $EX^{(2)}$. We also put $\boldsymbol{u}^2:=(u_1^2,\cdots,u_m^2)$.

Main theorem (II). *Suppose that X is an arbitrary finite connected multigraph, and let $\boldsymbol{u}$, ρ be as above. Then we have*

(2.27) $$Z_X(\boldsymbol{u}^2;\rho)=Z_{X^{(2)}}(\boldsymbol{u}^{(2)};\rho),$$

where the right hand side is the zeta function of a bipartite multigraph, evaluated in (2.22).

Proof. First it should be noted that the geometric realizations of X, $X^{(2)}$ as CW-complexes are the same, and the fundamental groups of them are identified in the obvious way. Note also that there is a natural bijection $\mathscr{C}_l^{red}(X) \simeq \mathscr{C}_{2l}^{red}(X^{(2)})$, so that, if C and $C^{(2)}$ are the corresponding reduced closed paths, then one has

$$\text{(2.28)} \qquad \boldsymbol{u}^{2C} = (\boldsymbol{u}^{(2)})^{C^{(2)}}.$$

The assertino follows from this and (2.4). Q.E.D.

We conclude this section with the following observation, which follows either from (2.8) or from (2.22). Let X^* be the connected multigraph obtained from X by removing its endpoints. Note that the fundamental groups of X and X^* can be identified. Then we have

Proposition (2.29).

$$Z_{X^*}(\boldsymbol{u};\rho) = Z_X(\boldsymbol{u};\rho).$$

In particular, $Z_X(\boldsymbol{u};\rho)$ *does not involve the variable* u_i, *if the corresponding edge* e_i *is incident to an endpoint of* X.

From (2.22), one sees that $Z_X(\boldsymbol{u};\rho)^{-1}$ is a polynomial of $\boldsymbol{u}$ of degree at most $2n.\sharp(EX)$, where $n = \deg(\rho)$. For a multigraph X $(=X^*)$ having no endpoint, we have the following fact, which follows from (5.23):

Proposition (2.30). *Suppose that* $X = X^*$. *Then* $Z_X(\boldsymbol{u};\rho)^{-1}$ *is a polynomial of* $\boldsymbol{u}$ *of degree* $2n.\sharp(EX)$. *If* X *is bipartite,* $Z_{X,b}(u)^{-1}$ *is a polynomial of degree* $n.\sharp(EX)$.

§ 3. Spectrum of a finite multigraph

In this section we study the relation between our zeta functions and the spectra of the finite multigraphs X, under the assumption that X has certain regularity. We shall be mostly concerned with the simplest zeta function $Z_X(u) = Z_X(u;\mathbf{1})$.

Let $VX = \{P_1, \cdots, P_n\}$, and $EX = \{e_1, \cdots, e_m\}$ be the sets of vertices and edges of a multigraph X.

Definition (3.1). The $n \times n$ matrix $A = A(X) = (a_{ij})$, defined by

$$a_{ij} := \sharp\{y \in EX;\ o(y) = P_i,\ t(y) = P_j\},$$

is called *the adjacency matrix*. Applying the involution $\iota_X\colon EX \to EX$, we see that $A(X)$ is a symmetric matrix. The polynomial

$$\phi_X(z) := \det(z.I_n - A) \tag{3.2}$$

is called the *characteristic polynomial* of X, and the set of its roots, counted with multiplicities, is called *the spectrum* of X, and denoted by $\mathrm{Spec}(X)$.

Note that A being a real symmetric matrix implies that the spectrum consists of real numbers; $a_{ii}=0$ $(1 \leq i \leq n)$ if X has no loop, which implies that $\mathrm{tr}(A)=0$. Moreover, if X is a graph then all nonzero entries of A are equal to 1. Note also that the (i, i)-entry of A^m is equal to the number of closed paths having P_i as their origin, hence we have

$$\mathrm{tr}(A^m) \geq N_m. \tag{3.3}$$

Let λ be the eigenvalue of A with maximal absolute value. Then we see immediately that the power series in (0.3) converges absolutely for $|u| < |\lambda|^{-1}$. This already shows that our zeta function is connected with the spectrum of X. Thus it is interesting to ask how two polynomials $\phi_X(u)$ and $Z_X(u)^{-1}$ are related. We shall give an answer in the case where X satisfies some regularity conditions.

Definition (3.4). Let $P_i \in VX$ be a vertex of a multigraph X. Then one has

$$\sum_{j=1}^{n} a_{ij} = \text{the number of edges incident to } P_i,$$

and this number is called the *valency* of P_i; if the valency, say k, is the same for all vertices P_i, then X is called *regular* of valency k, or *k-valent*. The following fact has been known for a long time:

Proposition (3.5). *Suppose that X is a connected regular k-valent multigraph. Then*

(i) *k belongs to* $\mathrm{Spec}(X)$ *with multiplicity one.*

(ii) *Moreover, k is the eigenvalue of $A(X)$ with maximal absolute value.*

(iii) $-k \in \mathrm{Spec}(X)$ *if and only if X is of bipartite type.*

Proof. See [Bi]. We shall give a proof, based on the same idea as in [Bi], in Section 4 (see (4.26)). Q.E.D.

Also the zeta function $Z_X(u)$ of regular graphs has been known. However, their simple relation, described below, has been seldom noticed until recently.

Theorem (3.6) (Ihara [I-1]). *Suppose that X is a connected regular multigraph with valency $q+1$, and the adjacency matrix A. Then one has*

$$(3.7) \qquad Z_X(u)^{-1}=(1-u^2)^{r-1}\det[I_n-Au+qu^2],$$

where $r=(q-1)n/2$ is the rank of $\Gamma=\pi_1(X, P_0)$, and $n=\sharp(VX)$. In other words,

$$(3.8) \qquad Z_X(u)^{-1}=(1-u^2)^{r-1}u^n\phi_X\left(\frac{1+qu^2}{u}\right).$$

Proof. This can be viewed as a special case of the following Main theorem (III), in two ways. First, if X is of bipartite type, it is directly included. To see it in other way, we take the barycentric subdivison $X^{(2)}$ of X, and apply (2.27). Q.E.D.

Remark (3.9). In [I-1], (3.7) was proved for a slightly different context. It was Serre [Ser] who pointed out that Ihara's result could be interpreted in terms of graphs. The detail has been given by Sunada [Su-1, 2].

The above theorem has a number of interesting consequences, since the spectra of regular graphs have been studied by many authors, and their results can be translated to the properties of zeta function $Z_X(u)$. We shall describe some of them in Sections 8, 9.

Next we proceed to the case of bipartite multigraphs.

Definition (3.10). Let X be an *s-partite* multigraph, and let $VX=\bigcup_{i=1}^{s} V_i$ be the corresponding decomposition of VX. It is also called *bipartite*, if $s=2$. An s-partite graph is called *semi-regular* of valency $(k_1, \cdots, k_s)$, if each vertex $P \in V_i$ is incident to exactly k_i edges for each $i=1, 2, \cdots, s$.

Note that a multi-partite multigraph has no loop. As we have seen in the previous section, the bipartite mutigraphs play an essential role in our study. For such graphs, one has the following

Lemma (3.11). *Suppose that X is a bipartite mutigraph, and let $VX=V_1\cup V_2$ be the corresponding partition, with $\sharp(V_i)=n_i$ $(i=1, 2)$, $n_2\geq n_1$. Then the spectrum of X has the form*

$$\mathrm{Spec}(X)=\{\pm\lambda_1, \pm\lambda_2, \cdots, \pm\lambda_{n_1}, 0, \cdots, 0\ (n_2-n_1\ \textit{times})\},$$

with $\lambda_1\geq\lambda_2\geq\cdots\geq\lambda_{n_1}\geq 0$.

Proof. From the definition, the adjacency matrix A of X has the form

$$A=\left(\begin{array}{c|c} \begin{matrix}0 & \cdots & 0\\ & \cdots & \\ 0 & \cdots & 0\end{matrix} & B \\ \hline {}^tB & \begin{matrix}0 & \cdots & 0\\ & \cdots & \\ 0 & \cdots & 0\end{matrix}\end{array}\right) \quad \text{with } B\in M(n_1, n_2; \boldsymbol{Z}). \tag{3.12}$$

One can find an orthogonal matrix $W\in O(n_2)$ in such a way that

$$BW=\overset{\longleftarrow n_1 \longrightarrow \;\; \longleftarrow n_2-n_1 \longrightarrow}{\left(\begin{array}{cccc|ccc} \mu_1 & 0 & \cdots & 0 & 0 & \cdots & 0\\ & \ddots & & \vdots & \vdots & & \vdots\\ * & \mu_2 & \ddots & \vdots & \vdots & & \vdots\\ \vdots & & \ddots & 0 & \vdots & & \vdots\\ * & & * & \mu_{n_1} & 0 & \cdots & 0\end{array}\right)}=(C|0), \text{ say.}$$

Then one sees that A is conjugate to the matrix

$$\overset{\longleftarrow n_1 \longrightarrow \longleftarrow n_1 \longrightarrow \longleftarrow n_2-n_1 \longrightarrow}{\left(\begin{array}{c|c|c} 0 & C & 0\\ \hline {}^tC & 0 & 0\\ \hline 0 & 0 & 0\end{array}\right)}.$$

Now the last matrix has the characteristic polynomial

$$\begin{aligned} x^{n_2-n_1}\det\begin{pmatrix} x.I & -C\\ -{}^tC & x.I\end{pmatrix} &= x^{n_2-n_1}\det\begin{pmatrix} x.I & -C\\ -{}^tC & x.I\end{pmatrix}\begin{pmatrix} I & x^{-1}C\\ O & I\end{pmatrix}\\ &= x^{n_2-n_1}\det\begin{pmatrix} x.I & O\\ -{}^tC & x.I-x^{-1}\,{}^tC.\,C\end{pmatrix}\\ &= x^{n_2-n_2}\det(x^2 I-{}^tC.C). \end{aligned}$$

Since ${}^tC.C$ is a positive semi-definite symmetric matrix, its eigenvalues are of the form $\{\lambda_i^2;\ \lambda_i\geq 0\ (1\leq i\leq n_1)\}$. Q.E.D.

Now we state a theorem which generalizes Ihara's result (3.7) to the connected regular bipartite multigraphs. First we need some notation. Let X be such a graph, and let $VX=V_1\cup V_2$ as above. Then one can construct a new multigraphs $X^{[i]}$ $(i=1, 2)$ as follows:

$$VX^{[i]}:=V_i,\ EX^{[i]}:=\{C=\text{proper path};\ |C|=2,\ o(C),t(C)\in V_i\}, \tag{3.13}$$
$$\varepsilon\colon C\to\{o(C),\ t(C)\}.$$

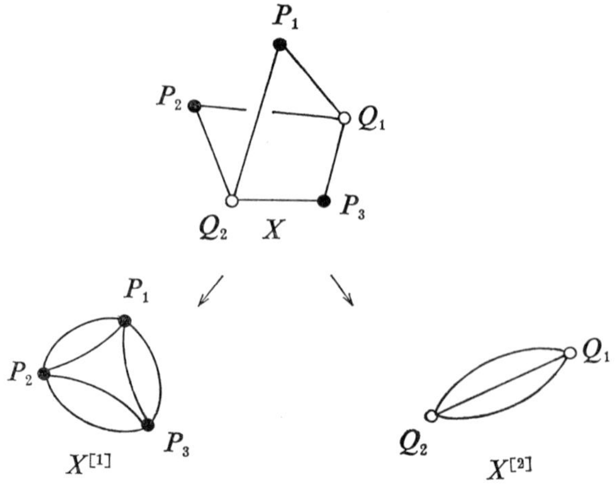

Fig. 6

Main theorem (III). *Suppose that X is a connected semiregular bipartite multigraph of valency (q_1+1, q_2+1), $\sharp(V_i)=n_i$ $(i=1, 2)$, $q_1 \geq q_2$, and let $A^{[i]}$ be the adjacency matrix of the associated multi graph $X^{[i]}$ $(i=1, 2)$. Then one has*

$$(3.14)\quad \begin{aligned} Z_{X,b}(u)^{-1} &= (1-u)^{(r-1)}(1+q_2u)^{(n_2-n_1)} \times \det[I_{n_1}-(A^{[1]}-q_2+1)u+q_1q_2u^2], \\ &= (1-u)^{(r-1)}(1+q_1u)^{(n_1-n_2)} \times \det[I_{n_2}-(A^{[2]}-q_1+1)u+q_1q_2u^2], \end{aligned}$$

where $Z_{X,b}(u)$ is defined by (2.25), *and $r=n_1q_1-n_2+1=n_2q_2-n_1+1$ is the rank of $\Gamma=\pi_1(X, P_0)$.*

In particular, if $\mathrm{Spec}\,(X)=\{\pm\lambda_1, \pm\lambda_2, \cdots, \pm\lambda_{n_1}, 0, \cdots, 0\}$ *are as in Lemma* (3.11), *one has*

$$(3.15)\quad \det[I_{n_1}-(A^{[1]}-q_2+1)u+q_1q_2u^2] = \prod_{j=1}^{n_1}\{1-(\lambda_j^2-q_1-q_2)u+q_1q_2u^2\}.$$

Proof. A proof of (3.14) has been given in [H-H]; see Remark (3.19) below. We shall give in Section 5 a different proof, which describes that it can be derived from our Main Theorem (II). To show (3.15), one notes first that the last equality in (3.14) is equivalent to the following relation between $A^{[1]}$ and $A^{[2]}$:

$$(3.16)\quad A^{[2]} \simeq \left(\begin{array}{c|c} A^{[1]}+(q_1-q_2)I_{n_1} & 0 \\ \hline 0 & -(q_2+1)I_{n_2-n_1} \end{array}\right).$$

On the other hand, one can see easily that

$$A^2=\left(\begin{array}{c|c} A^{[1]}+(q_1+1)I_{n_1} & 0 \\ \hline 0 & A^{[2]}+(q_2+1)I_{n_2} \end{array}\right).$$

From these follows

$$A^{[1]}\simeq\begin{pmatrix} \lambda_1^2-(1+q_1) & & 0 \\ & \ddots & \\ 0 & & \lambda_{n_1}^2-(1+q_1) \end{pmatrix}. \tag{3.17}$$

This proves (3.15). Q.E.D.

From the above proof, one has the following

Corollary (3.18). *Suppose X is a regular connected multigraph of valency $q+1$, with $n=\sharp(VX)$, $m=\sharp(EX)$, and let* $\mathrm{Spec}(X^{(2)})=\{\pm\lambda_1, \pm\lambda_2, \cdots, \pm\lambda_n, 0, \cdots, 0$ *($m-n$ times)$\}$ be the spectrum of the barycentric subdivision $X^{(2)}$ of X (cf.* (3.11)). *Then the spectrum of X is given by*

$$\mathrm{Spec}(X)=\{\lambda_1^2-q-1, \cdots, \lambda_n^2-q-1\}.$$

Remark (3.19). The expression (3.14) has been given in [H-H] for a zeta function $Z_\Gamma(u;\rho)$ attached to a subgroup Γ of G, which satisfies axioms (G, l, I), (G, l, II) (cf. (6.1)). That these two zeta functions are the same thing can be seen as follows. Consider the universal covering tree $\tilde{X}$ and the action of the fundamental group $\Gamma=\pi_1(X, P_0)$, which is faithful so that it can be regarded as a subgroup of $G:=\mathrm{Aut}(\tilde{X})$. Now it is known and easy to see that G acts transitively on $E\tilde{X}$, hence also on $V_1\tilde{X}$, $V_2\tilde{X}$. Thus we can apply the results of our previous paper [H-H] to get (3.14). In fact it suffices to note that the matrix $A_{1,\rho}$ in [H-H] is nothing but our $A^{[1]}$, for $\rho=\mathbf{1}$; and similarly for $A^{[2]}$.

Now one can prove the following generalization of Proposition (3.5):

Proposition (3.20). *Suppose that X is a connected semiregular bipartite multigraph of valency (k_1, k_2). Then $\sqrt{k_1k_2}$ belong to* $\mathrm{Spec}(X)$, *with multiplicity one; moreover, $\sqrt{k_1k_2}$ is the maximal absolute value of the eigenvalues of A.*

Proof. It is easy to see that $X^{[1]}$ (resp. $X^{[2]}$) is a regular multigraph of valency $q_2(q_1+1)$ (resp. $q_1(q_2+1)$), where we put $k_i=q_i+1$ $(i=1, 2)$. Therefore one sees from Proposition (3.5) that $A^{[1]}$ has $q_2(q_1+1)$ as its eigenvalue of maximal absolute value, with multiplicity one. This implies by (3.14), that $\det[I-(A^{[1]}-q_2+1)u+q_1q_2u^2]$ has a simple factor

$$1-\{q_2(q_1+1)-q_2+1\}u+q_1q_2u^2=(1-u)(1-q_1q_2u)$$

which is, by (3.15), equal to $1-\{\lambda_1^2-(q_1+q_2)\}u+q_1q_2u^2$. Hence we get

$$\lambda_1^2=1+q_1+q_2+q_1q_2=(q_1+1)(q_2+1)=k_1k_2,$$

as asserted. Similarly, we have $\lambda_j^2-(q_1+1)<q_2(q_1+1)$ $(j\geq 2)$, hence $\lambda_j^2<k_1k_2$. Q.E.D.

See Proposition (4.26) for an interpretation of the eigenvalue $\sqrt{k_1k_2}$. Finally we make the following

Definition (3.21). Let X be a finite connected semiregular bipartite multigraph of valency (q_1+1, q_2+1), such that $q_1\geq q_2>1$. We call X a *weak Ramanujan graph*, if

(3.22) $$\operatorname{ord}_{(1+q_2u)}Z_{X,b}(u)^{-1}=n_2-n_1,$$

where $n_i=\#(VX_i)$. Note that we have $n_2\geq n_1$. X is called a *Ramanujan graph*, if

(3.23) $$|\lambda^2-q_1-q_2|\leq 2\sqrt{q_1q_2},$$

for any $\lambda\in\operatorname{Spec}(X)$ such that $\lambda^2\neq(1+q_1)(1+q_2)$.

Observe that (3.23) is equivalent to the condition that the nonlinear factors $1-(\lambda^2-q_1-q_2)u+q_1q_2u^2$ of $Z_{X,b}(u)^{-1}$ have the imaginary roots α, α' such that $|\alpha|=|\alpha'|=\sqrt{q_1q_2}$. Also note that this definition of the Ramanujan graph agrees with the one given in [L-P-S], in the case when X is regular bipartite, as one sees easily. Also it is easy to prove the following assertions:

Lemma (3.24). *Suppose that X is as in* (3.21), *and that it is a Ramanujan graph. Then one has*

(i) *X is a weak Ramanujan graph.*

(ii) *If $q_1=q_2$, then $X^{(2)}$ (= the barycentric subdivision of X) is also a Ramanujan graph.*

Conversely, if $q_1=q_2$ and $X^{(2)}$ is a Ramanujan graph, then X is a Ramanujan graph.

Proof. Omitted.

We shall give in Section 9 a number of examples of multigraphs which are *not* Ramanujan graphs.

§ 4. Harmonic functions and the Hodge decomposition

In the previous two sections we gave two essentially different expressions of $Z_X(u)$. While the first one (2.23) (and (2.27)) is completely general, it does not give a direct connection with the spectrum of X. On the other hand the second expression (3.14), which applies only to the semi-regular bipartite graphs, shows that $Z_{X,b}(u)$ is determined by the spectrum of X, together with its topological invariants r, n_2-n_1. Moreover, the latter expression has two distinguished factors, $(1-u)$ and $(1+q_2u)$. In this section, we study the interpretation of these factors.

Throughout this section, we always assume that our multigraph X is of bipartite type, and $VX=V_1\cup V_2$ is the corresponding partition. (This assumption does not cause any loss of generality; take the barycentric subdivision of X if necessary).

We introduce, only in this section, an orientation on X. Thus we call an oriented edge y to be *positive*, if $o(y)\in V_1$ and $t(y)\in V_2$. If $e=\{y, y^{-1}\}$ and y is positive, we write $y=e^+$ and $y^{-1}=e^-$. We define the signature $\mathrm{sgn}(y)$ of an oriented edge y by

$$\mathrm{sgn}(y)=\begin{cases} 1 & \text{if } y \text{ is positive} \\ -1 & \text{if } y \text{ is negative.}\end{cases} \tag{4.1}$$

Recall that if $\rho=\mathbf{1}$ is the trivial representation of $\pi_1(X, P_0)$, the $\boldsymbol{C}$-space $M^1(X):=M_\rho(X)$ is regarded as consisting of $\boldsymbol{C}$-valued functions on EX, and it is equipped with an inner product (2.15).

We first describe an important decomposition of $M^1(X)$, called the *Hodge decomposition*, which distinguishes the subspace corresponding to the factor $(1-u)$. This is the one dimensional version of Garland and Borel [Gar], [Bo-1]. Let $M^0(X)$ denote the space of functions on VX. Then we can write $M^0(X)=M^0(V_1)\oplus M^0(V_2)$, where $M^0(V_i)$ is the space of functions on V_i $(i=1, 2)$. We define an inner product on $M^0(X)$ by putting

$$(f, g):=\sum_{P\in V} k(P)f(P)\overline{g(P)} \qquad (k(P):=\#(E(P))=\text{valency of } P). \tag{4.2}$$

Define the linear maps d, δ by

$$\begin{aligned} &d\colon M^0(X)\to M^1(X),\ (df)(e):=f(t(e^+))-f(o(e^+)).\\ &\delta\colon M^1(X)\to M^0(X),\ (\delta f)(P):=(-1)^i\sum_{e\in E(P)} f(e)/k(P) \quad (P\in V_i).\end{aligned} \tag{4.3}$$

Then it is easily checked that d and δ are the adjoint of each other:

$$(\delta f, g)=(f, dg) \qquad (f\in M^1(X),\ g\in M^0(X)). \tag{4.4}$$

Finally define the Laplace operator by $\Delta = d\delta: M^1(X) \to M^1(X)$, and put $\boldsymbol{H}^1(X) := \mathrm{Ker}(\Delta)$, and call it the space of *harmonic* functions.

Lemma (4.5).
(i) $\mathrm{Ker}(d)\ (:= C^0(X)) = \{$*constant functions on* $VX\}$.
(ii) $\boldsymbol{H}^1(X) = \mathrm{Ker}(\delta)$.

Proof. (i) is trivial. We have for $f \in M^1(X)$,

$$f \in \boldsymbol{H}^1(X) \Longleftrightarrow (\Delta f, g) = (\delta f, \delta g) = 0 \qquad (\forall g \in M^1(X))$$
$$\Longleftrightarrow \delta f = 0 \qquad (\text{Take } g := f).$$
Q.E.D.

From this and (4.4) one one can easily prove the following result, which is called the *Hodge decomposition*:

Corollary (4.6). *We have the orthogonal decompositions*:

$$M^1(X) = \boldsymbol{H}^1(X) \oplus dM^0(X), \qquad M^0(X) = \mathrm{Ker}(d) \oplus \delta M^1(X).$$

Let $C^0(V_i)$ denote the space of constant functions on V_i, and let $M_0^0(V_i) := (M^0(V_i) \cap (C^0(V_i))^\perp$ be the orthogonal complement in $M^0(V_i)$ of $C^0(V_i)$. Then we see immediately that

$$\mathrm{Ker}(d) = C^0(X) \subset C^0(V_1) \oplus C^0(V_2), \qquad d(C^0(V_1) \oplus C^0(V_2)) = C^1(X),$$

and that d is injective on $M_0^0(V_1) \oplus M_0^0(V_2)$.

Definition (4.7). The subspace $d(M_0^0(V_1) \oplus M_0^0(V_2))$ of $M^1(X)$ is called the space of *cusp forms* on X, and denoted by $M^1_{\mathrm{cusp}}(X)$.

Lemma (4.8). *One has the orthogonal decomposition*

$$d(M^0(X)) = M^1_{\mathrm{cusp}}(X) \oplus C^1(X).$$

Proof. This follows from the following formula: for $f, g \in M^0(V)$, one has

$$(4.9) \quad (df, dg) = (f, g) - \sum_{e \in EX} \{f(o(e^+))\overline{g(t(e^+))} + f(o(e^-))\overline{g(t(e^-))}\}.$$

Q.E.D.

Now recall that $M^1(X)$ is acted upon by the hermitian operators $\rho^*(T_i)$ $(i=1, 2)$. For the real numbers $\alpha, \beta \in \boldsymbol{R}$, we denote by $M(\alpha, \beta)$ the subspace of $M^1(X)$ consisting of the functions satisfying

$$(4.10) \qquad \rho^*(T_1)f = \alpha f, \quad \rho^*(T_2)f = \beta f. \qquad (f \in M^1(X)).$$

For each vertex $P \in VX$, we denote by $E(P)$ the set of edges of X which are incident to P, i.e., $E(P):=\{e \in EX;\ P \in \varepsilon(e)\}$.

Lemma (4.11). *The following conditions are equivalent*:
(i) $f \in M(-1, -1)$.
(ii) $f \in \boldsymbol{H}^1(X)$; *i.e., f is a harmonic function.*
(iii) $\sum_{e \in E(P)} f(e)=0$ *for any* $P \in VX$.

Proof. If one puts $\varepsilon(e)=\{P_1, P_2\}$ $(P_i \in V_i)$, one has $E(P_i)=E_i(e)$. From the definitions (2.16), (4.3), it follows

$$(4.12) \qquad (\rho^*(T_i)f)(e)=\sum_{e' \in E(P_i)} f(e')-f(e) \\ =(-1)^i k(P_i)(\delta f)(P_i)-f(e).$$

The assertion follows immediately from this. Q.E.D.

Proposition (4.13). *There is a canonical isomorphism*

$$\eta\colon H_1(X, \boldsymbol{C}) \simeq \boldsymbol{H}^1(X)=M(-1, -1).$$

In particular, one has

$$(4.14) \qquad \dim_{\boldsymbol{C}} \boldsymbol{H}^1(X)=\dim_{\boldsymbol{C}} H_1(X, \boldsymbol{C})=r=\textit{rank of } \pi_1(X, P_0).$$

Proof. Let $C=(y_1, \cdots, y_{2l})$ be a closed path in X, so that $t(y_{j-1})=o(y_j)$ $(1 \leq j \leq 2l)$, with the convention $y_0=y_{2l}$. Define a function $f=f_C \in M^1(X)$ by

$$(4.15) \qquad f_C(e):=\sum_{y_j \in e} \operatorname{sgn}(y_j).$$

It is easily seen that f_C satisfies (4.11), (iii), hence $f_C \in \boldsymbol{H}^1(X)=M(-1, -1)$. It is also clear that f_C depends only on the reduced path to which C corresponds, or the homology class of C. Thus the map $\eta\colon C \to f_C$, is a well defined linear map on $H_1(X, \boldsymbol{Z})$. That it is injective on $H_1(X, \boldsymbol{Z})$ is obvious, since any element of this group is represented by a single closed path. It follows easily that η can be extended to an injective linear map on $H_1(X, \boldsymbol{C})=H_1(X, \boldsymbol{Z}) \otimes_{\boldsymbol{Z}} \boldsymbol{C}$. We shall show that $\dim_{\boldsymbol{C}} H_1(X, \boldsymbol{C})=\dim_{\boldsymbol{C}} M(-1, -1)$. First note that by Lemma (4.11), $M(-1, -1)$ is the orthogonal complement of the subspace generated by the characteristic functions f_P of $E(P)$, $P \in VX$. Call this subspace N. We claim that $\dim_{\boldsymbol{C}} N=\sharp(VX)-1$. To prove this, suppose that $\Sigma c_P f_P=0$ is a linear relation among f_P's. Let $e \in EX$ be an edge such that $o(e^+)=P$, $t(e^+)=Q$. Evaluating at e, we get $c_P+c_Q=0$. Since X is connected, this implies that there exists a constant c such that $c_P=c$,

$c_Q = -c$ for any $P \in V_1$, $Q \in V_2$. Thus, the functions f_P's have exactly one linear relation, which proves our claim. Now we have $\dim_C M(-1, -1) = \dim_C M^1(X) - (\#(VX) - 1) = \dim_C H_1(X, C)$ (cf. 2.1). This completes the proof.

Q.E.D.

Thus the space of harmonic functions is acted upon by $\rho^*(T_i)$ as the scalar -1; and this part of $M^1(X)$ contributes in $Z_X(u)^{-1}$ to the factor $(1-u)$. Namely one has:

Corollary (4.16). *Suppose that X is an arbitrary connected multigraph, which is not necessarily of bipartite type. Then the multiplicity of the factor $(1-u)$ in $Z_X(u)^{-1}$ is at least $r = \text{rank } \pi_1(X, P_0)$; moreover, it is exactly r if X is regular, or semiregular bipartite, with valency (q_1+1, q_2+1), $q_1 q_2 > 1$.*

Proof. The first assertion is an immediate consequence of (2.23). The second statement is already shown in the proof of Proposition (3.20); indeed, by (3.15), we see that the multiplicity of $(1-u)$ in $\det[I - (A^{[1]} - q_2 + 1)u + q_1 q_2 u^2]$ is equal to that of $\sqrt{(1+q_1)(1+q_2)}$ in $\text{Spec}(X)$ (see also Proposition (4.26)).

Q.E.D.

Remark (4.17). In Section 5, we shall prove that the multiplicity of $(1-u)$ in $Z_X(u)^{-1}$ is equal to r also for any non-regular multigraph X which is not homotopic to a circle (cf. (0.8)).

Next we consider the factor $(1+q_2 u)$.

Proposition (4.18). *Suppose that X is a semiregular bipartite multigraph of valency (q_1+1, q_2+1). Then*

(i) $M(q_1, -1) = dM^0(V_1) \cap (dM^0(V_2))^{\perp} \subset M^1_{\text{cusp}}(X)$,

$M(-1, q_2) = dM^0(V_2) \cap (dM^0(V_1))^{\perp} \subset M^1_{\text{cusp}}(X)$.

(ii) *$M(-1, q_2)$ (resp, $M(q_1, -1)$) consists of the functions satisfying the local conditions:*

$$\sum_{e \in E(P)} f(e) = 0 \qquad \textit{for any } P \in V_1 \ (\textit{resp. } P \in V_2), \tag{4.19}$$

$$f \textit{ is constant on each } E(Q),\ Q \in V_2 \ (\textit{resp. } Q \in V_1). \tag{4.20}$$

(iii) *One has the following equalities:*

$$\dim_C M(-1, q_2) = \textit{the multiplicity of } (1+q_2 u) \textit{ in } Z_X(u), \tag{4.21}$$

$$\dim_C M(q_1, -1) = \textit{the multiplicity of } (1+q_1 u) \textit{ in } Z_X(u).$$

Proof. We first prove (i), (ii). We first show the equivalence

$$\rho^*(T_1)f = -f \Longleftrightarrow (4.19) \Longleftrightarrow f \in dM^0(V_1)^{\perp}.$$

The first one is easy from (4.12). On the other hand, one has for any $g \in M^0(V_1)$

$$(f, dg) = \sum_{e \in EX} f(e)\overline{(dg)(e)} = -\sum_{P \in V_1} \overline{g(P)}\Big(\sum_{e \in E(P)} f(e)\Big),$$

from which follows the second equivalence. Similarly one has

$$\rho^*(T_2)f = q_2 f \Longleftrightarrow (4.20) \Longleftrightarrow f \in dM^0(V_2).$$

In fact, one has

$$\rho^*(T_2)f = q_2 f \Longleftrightarrow \sum_{e' \in E(Q)} f(e') = (q_2+1)f(e) \qquad (Q \in V_2,\ t(e^+) = Q).$$

Since the sum in the left side depends only on Q, we see this is equivalent to (4.20). The second one is easily shown. The least assertion (iii) will be proved in Section 5. Q.E.D.

Remark (4.22). Counting the number of equations in (4.19), (4.20), we know, without using (3.14), the following estimate

$$\dim_C M(-1, q_2) \geq \#(EX) - [\#(V_1) + q_2\#(V_2)] = n_2 - n_1.$$

However, with the knowledge of Main theorem (III), one sees that this is a consequence of the following:

Proposition (4.23). *Suppose that* $q_1 \geq q_2$. *Then one has*

$$\dim_C M(-1, q_2) - \dim_C M(q_1, -1) = n_2 - n_1.$$

Proof. From (3.14), (3.15), we see that the extra factors $(1+q_2u)$ of $Z_{X,b}(u)^{-1}$, other than those $n_2 - n_1$ one's, come from

$$\{1 - (\lambda_j^2 - q_1 - q_2)u + q_1q_2u^2\} \qquad \text{for } \lambda_j \in \operatorname{Spec}(X), \quad (1 \leq j \leq n_1).$$

This implies that the quadratic polynomial has a root $-q_2^{-1}$, hence the other root is $-q_1^{-1}$. Q.E.D.

Corollary (4.24). *Let the assumptions be as above. Then the following assertions are equivalent.*

(i) *X is a weak Ramanujan graph.*

(ii) *$Z_{X,b}(u)^{-1}$ is not divisible by* $(1+q_1u)$.

If $q_2 = 1$, one can, in analogy of Proposition (4.8), construct a map

$$\eta^{\#}: H_1^{(0)}(X, Z) \longrightarrow M(-1, 1) \text{ modulo } \{\pm 1\},$$

where $H_1^{(0)}(X, Z)$ is a Z-submodule of $H_1(X, Z)$ generated by the closed paths of length divisible by 4. In fact, if $C=(y_1, \cdots, y_{4l})$ is such a path, one defines $\eta^{\#}(C)=f \in M^1(X)$ by

(4.25) $$f(e):=\sum_{y_j \in e}(-1)^{[(j-1)/2]+j-1}\operatorname{sgn}(y_j).$$

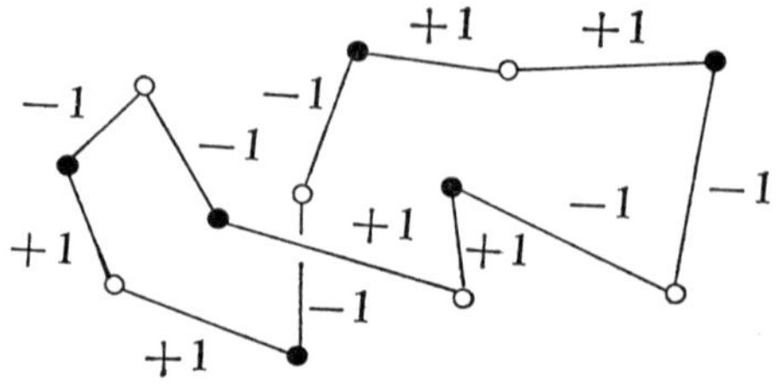

Fig. 7

It is immediately seen that $f=\eta^{\#}(C)$ belongs to $M(-1, 1)$. One can prove, as in the proof of Theorem (5.32), that the image of $\eta^{\#}$ spans $M(-1, 1)$.

We close this section with the following observation:

Proposition (4.26). *Suppose X is a semiregular bipartite multigraph of valency (q_1+1, q_2+1). Then*

$$M(q_1, q_2)=C^1(X) \ (:=\textit{constant functions}),$$

and if $q_1q_2>1$, one has

$$\dim_C M(q_1, q_2)=1=\textit{multiplicity of } (1-q_1q_2u) \textit{ in } Z_{X,b}(u)^{-1}.$$

Proof. The first assertion is proved similarly as in Proposition (4.18). In view of (2.25), in order to prove the last assertion, it suffices to show that q_1q_2-eigenspace of $\rho^*(T_1T_2)$ is $M(q_1, q_2)$.

In fact, let $f \in M^1(X)$ be such that $\rho^*(T_1T_2)f=q_1q_2f, f\neq 0$, Let $e \in EX$ be an edge such that $f(e)$ has a maximal absolute value. By Lemma (2.12), we have

$$(\rho^*(T_1T_2)f)(e)=\sum_{(*)} f(e'),$$

where the sum in the right consists of q_1q_2 terms. It follows that

$$q_1q_2|f(e)|\leq \sum_{(*)}|f(e')|\leq q_1q_2|f(e)|,$$

hence we see that f is a constant function on EX. This proves our assertion. Q.E.D.

§ 5. Representation of $C[T_1, T_2]$; a proof of (3.14)

In this section X is always assumed to be a bipartite multigraph such that $VX = V_1 \cup V_2$. We first treat the case that X is semiregular of valency (q_1+1, q_2+1).

Let $\tilde{X}$ be, as in Section 2, the universal covering tree, and let $T_1, T_2 \in \mathrm{End}(Z[E\tilde{X}])$ be the correspondences on $E\tilde{X}$ as in (2.10). We denote by $C[T_1, T_2]$ the subalgebra over C generated by T_i $(i=1, 2)$.

Lemma (5.1). *$C[T_1, T_2]$ is the non-commutative ring of polynomials of T_1, T_2 with the fundamental relations*

(5.2) $$T_i^2 = (q_i - 1)T_i + q_i \qquad (i=1, 2).$$

In particular, we have

(5.3) $$C[T_1, T_2] = \sum_{(i_1, i_2, \cdots, i_l)} C T_{i_1} T_{i_2} \cdots T_{i_l} \qquad (i_k = 1, 2),$$

where the sum is extended over $(i_1, \cdots, i_l)$, $l \geq 0$, such that $i_k \neq i_{k+1}$.

Proof. The relations (5.2) is an immediate consequence of the definition (2.10). On the other hand, from Lemma (2.12) and (Fig. 5), we see that the monomials $T_{i_1} T_{i_2} \cdots T_{i_l}$; $i_k \neq i_{k+1}$, $l=0, 1, 2, \cdots$, are linearly independent. Q.E.D.

Recall that (2.16) defines an anti-representation ρ^* of $C[T_1, T_2]$ on the space $M^1(X)$ of C-valued functions on EX. Since $\rho^*(T_1)$, $\rho^*(T_2)$ are hermitian operators, we see that ρ^* is semi-simple. Throughout the following, we shall omit the prefix "anti-", and refer ρ^* as a representation of $C[T_1, T_2]$.

Lemma (5.4). *Suppose that φ: $C[T_1, T_2] \rightarrow \mathrm{End}_C(W)$ be an irreducible representation of finite dimension $n = \dim_C(W)$. Then we have $n \leq 2$.*

Proof. From (5.2), we see that either $\varphi(T_1)$ is a scalar, or it has $(X+1)(X-q_1)$ as the minimal polynomial. In the first case, we get $\dim_C(W) = n = 1$. Suppose that $n \geq 2$, in which case we have the following two decompositions of W into eigenspaces:

$$W = W_1 \oplus W_1', \quad \varphi(T_1) \mid W_1 = q_1, \quad \varphi(T_1) \mid W_1' = -1,$$
$$W = W_2 \oplus W_2', \quad \varphi(T_2) \mid W_2 = q_2, \quad \varphi(T_2) \mid W_2' = -1,$$

with $\dim_C(W_i)$, $\dim_C(W_i') \geq 1$. From our assumption that φ is irreducible of dimension $n \geq 2$, it follows that the intersection of any two of the subspaces W_1, W_2, W_1', W_2' is the zero space. Then we see that they all have the same dimension, and that

$$W=W_1\oplus W_2=W_1\oplus W_2'. \tag{5.5}$$

Corresponding to the first decomposition, we see that φ is expressed as

$$\varphi(T_1)=\left(\begin{array}{c|c} q_1 I & 0 \\ \hline C_1 & -I \end{array}\right), \qquad \varphi(T_2)=\left(\begin{array}{c|c} -I & C_2 \\ \hline 0 & q_2 I \end{array}\right). \tag{5.6}$$

Moreover, taking the conjugation with a matrix $\begin{pmatrix} X & 0 \\ 0 & Y \end{pmatrix}$, $X, Y \in GL(n/2, \boldsymbol{C})$, we can transform C_1, C_2 to YC_1X^{-1}, XC_2Y^{-1}, respectively. Now it is easy to see that φ=irreducible implies that $n/2=1$ and $C_1, C_2\neq 0$.

Q.E.D.

Proposition (5.7). *Suppose that $q_1q_2>1$. The irreducible representations φ of $\boldsymbol{C}[T_1, T_2]$ are classified as follows:*
(we put $p_\varphi(u):=\det[I-\varphi(T_1T_2)u]$).

(i) *Degree one*;

$$\varphi(T_1)=q_1,\quad \varphi(T_2)=q_2;\quad \varphi(T_1T_2)=q_1q_2,\quad p_\varphi(u)=1-q_1q_2u,$$
$$\varphi(T_1)=q_1,\quad \varphi(T_2)=-1;\quad \varphi(T_2T_2)=-q_1,\quad p_\varphi(u)=1+q_1u,$$
$$\varphi(T_1)=-1,\quad \varphi(T_2)=q_2;\quad \varphi(T_1T_2)=-q_2,\quad p_\varphi(u)=1+q_2u,$$
$$\varphi(T_1)=-1,\quad \varphi(T_2)=-1;\quad \varphi(T_1T_2)=+1,\quad p_\varphi(u)=1-u.$$

(ii) *Degree two*; φ *is parametrized by* $c\in \boldsymbol{C}$, $c\neq 0$, $(q_1+1)(q_2+1)$, *with*

$$\varphi(T_1)=\left(\begin{array}{c|c} q_1 & 0 \\ \hline c & -1 \end{array}\right), \qquad \varphi(T_2)=\left(\begin{array}{c|c} -1 & 1 \\ \hline 0 & q_2 \end{array}\right); \tag{5.8}$$

$$\varphi(T_1T_2)=\left(\begin{array}{c|c} -q_1 & q_1 \\ \hline -c & c-q_2 \end{array}\right), \qquad p_\varphi(u)=1-(c-q_1-q_2)u+q_1q_2u^2.$$

Proof. (i) follows immediately from (5.2). (ii): From the proof of Lemma (5.4), one sees that φ is equivalent to the representation given by (5.8). A direct computation shows that (5.8) is reducible if and only if $c=0$, or $(1+q_1)(1+q_2)$, or equivalently

$$p_\varphi(u)=(1+q_1u)(1+q_2u), \qquad \text{or } (1-u)(1-q_1q_2u). \tag{5.9}$$

Q.E.D.

Corollary (5.10). *Suppose that* (φ, W) *is a 2-dimensional irreducible* *subspace of* $M^1(X)$ *s.t.* $p_\varphi(u)=1-(c-q_1-q_2)u+q_1q_2u^2$. *Then* W *has a basis* f_1, f_2 *which satisfies the following relations*:

$$(5.11)\qquad f_1(e)=\sum_{e'\in E(o(e+))} f(e'), \qquad cf_2(e)=\sum_{e'\in E(t(e+))} f(e').$$

Proof. This follows easily from (5.8) and (4.12). Q.E.D.

Corollary (5.12). *The irreducible representation* φ *of* $C[T_1, T_2]$ *is determined by the characteristic polynomial* $p_\varphi(u)$ *of* $\varphi(T_1T_2)$.

We note that the above Proposition (5.7), together with (5.9), gives the proof of (4.21).

A Proof of (3.14). Now we shall give a new proof of (3.14), which is a simple consequence of the general formula (2.22), and which is independent of our previous proof in [H-H]. We know already that $Z_{X,b}(u)^{-1}$ is a polynomial of u, which is divisible by

$$(1-u)^r(1-q_1q_2u)(1+q_2u)^{n_2-n_1}.$$

Moreover, we know from the above results, that each of these factors corresponds to one-dimensional subspace of $M^1(X)$, invariant by $C[T_1, T_2]$. Denote by M_* the orthogonal complement in $M^1(X)$ of the direct sum of them.

We have $\dim_C(M_*)=\sharp(EX)-[r+1+n_2-n_1]=2(n_1-1)$.

Lemma (5.13). $\det \rho^*(T_i)=(-1)^{n_iq_i}(q_i)^{n_i}$ $(i=1, 2)$.

Proof. We prove this for T_1. The assumption on X implies that EX is the disjoint union of $E(P)$'s ($P\in V_1$), hence we have the decomposition $M^1(X)=\oplus_P M^1(E(P))$, where $M^1(E(P))$ consists of the functions on $E(P)$. Since T_1 preserves $Z[E(P)]$, it follows that $\rho^*(T_1)$ preserves the above decomposition of $M^1(X)$. Now from (4.18), (5.2), we see easily that on each $M^1(E(P))$, $\rho^*(T)$ has a simple eigenvalue q_1, and the eigenvalue -1 with multiplicity q_1. The assertion follows from this. Q.E.D.

It follows that $\det(\rho^*(T_i)\,|\,M_*)=(-q_i)^{n_1-1}$. Now the comparison of (5.7) and (5.13), together with Proposition (4.13), implies that the restriction of $\rho^*\,|\,M_*$ is decomposed into a direct sum of two-dimensional invariant subspaces on which neither one of $\rho^*(T_i)$ acts as a scalar. Let W be such a subspace. Then as in the proof of Lemma (5.4), we see that W has a one-dimensional subspace W_1 such that $\rho^*(T_1)\,|\,W_1=q_1$. From (5.2), we see that $\rho^*(1+T_1)$ gives the projection of W onto W_1, and that

$$\rho^*(1+T_1)^2=(q_1+1)\cdot\rho^*(1+T_1),$$

Now define an element A of $\boldsymbol{C}[T_1, T_2]$ by

$$A:=(1+q_1)^{-1}(1+T_1)T_2(1+T_1). \tag{5.14}$$

Recall that $X^{[1]}$ is the multigraph derived from X such that $VX^{[1]}=V_1$ and $EX^{[1]}$ consists of the proper paths of the form (P, e_1, Q, e_2, P') of length 2 ($P, P' \in V_1$). We identify $M^0(X^{[1]})$ with $M^0(V_1)$, and regard the adjacency matrix $A^{[1]}$ of $X^{[1]}$ as an endomorphism of $M^0(V_1)$ as follows. Let $VX^{[1]}=V_1=\{P_1, \cdots, P_{n_1}\}$, and let $f \in M^0(V_1)$. Then we have

$$(A^{[1]}f)(P_i):=\sum_j a_{ij}f(P_j), \qquad (A^{[1]}=(a_{ij})).$$

From the above remark, we see that $M_*(\subset dM^0(X))$ is decomposed as $M_*=M_*^{[1]}\oplus M_*^{[2]}$, where $M_*^{[i]}$ is the subspace such that $\rho^*(T_i)\,|\,M_*^{[i]}=q_i$. Now from the proof of Proposition (4.18), we can regard $M_*^{[i]}$ as a space of functions on V_i, i.e., $M_*^{[i]}\subset M^0(V_i)=M^0(X^{[i]})$. Moreover, one can see easily that $M_*^{[1]}$ is stable under $A^{[1]}$, and that $A\,|\,M_*^{[1]}=A^{[1]}\,|\,M_*^{[1]}$. Similar assertion holds for $A^{[2]}$, $M_*^{[2]}$.

Lemma (5.15). *Notation and assumptions being as above, we have*
(i) $W_1\subset dM^0(V_1)$.
(ii) W_1, *regarded as a subspace of* $M^0(X^{[1]})$, *is stable by* $A^{[1]}$, *and*

$$A^{[1]}\,|\,W_1=A\,|\,W_1=\mathrm{tr}\,(\rho^*(T_1T_2)\,|\,W)+q_2-1. \tag{5.16}$$

Proof. Since W_1 is the eigenspace of $\rho^*(T_1)$ with eigenvalue q_1, (i) follows from the proof of Proposition (4.18). From the invariance of W by $\boldsymbol{C}[T_1, T_2]$, and the above remark, we see that W_1 is invariant by $A^{[1]}$, and that $A^{[1]}\,|\,W_1=A\,|\,W_1$. Now we have

$$\begin{aligned} A^{[1]}\,|\,W_1=\rho^*(A)\,|\,W_1&=(q_1+1)^{-1}\rho^*[(1+T_1)T_2(1+T_1)]\,|\,W_1\\ &=(q_1+1)^{-1}\rho^*[(1+T_1)^2T_2]\,|\,W_1\\ &=\mathrm{tr}\,(\rho^*(T_1T_2)\,|\,W)+q_2-1. \end{aligned}$$

Q.E.D.

It follows that

$$\begin{aligned} \det(I-(\rho^*(T_1T_2)\,|\,M_*)u)&=\prod_W \det(I_2-(\rho^*(T_1T_2)\,|\,W)u)\\ &=\prod_W (1-\mathrm{tr}\,(\rho^*(T_1T_2)\,|\,W)u+q_1q_2u^2)\\ &=\prod_W \{1-((A^{[1]}\,|\,W^{[1]})-q_2+1)u+q_1q_2u^2\}. \end{aligned}$$

Now the equality (3.14) follows from this and (2.25).

Next we consider the general case, where X is a bipartite multigraph, which is not necessarily semiregular, such that

$$VX=V_1\cup V_2,\quad V_1=\{P_j;\ 1\le j\le n_1\},\quad V_2=\{Q_j;\ 1\le j\le n_2\},$$

and the valency of P_j (resp. Q_j) is $q_1^{(j)}+1$ (resp. $q_2^{(j)}+1$). In such general case, the algebra $\boldsymbol{C}[T_1, T_2]$ has a complicated structure, and it seems difficult to classify the equivalence classes of irreducible representations of it. Nevertheless, it would be interesting to ask the following questions:

(5.17) Are the irreducible representations of $\boldsymbol{C}[T_1, T_2]$ of degree ≤ 2? (see example (5.20)).

(5.18) Can one classify them, or, if the answer to (5.17) is negative, those of degree one and two, as in Proposition (5.7)?

(5.19) Does the characteristic polynomial $p_\varphi(u)$ of $\varphi(T_1T_2)$ determine the (irreducible) representation φ?

Example (5.20). The following example shows that for $X=Y^{(2)}$, the barycentric subdivision of $Y=X_1(4, 5)$, $(\rho^*, M^1(X))$ has an irreducible component of degree 3.

$Y{:}=X_1(4.5)$

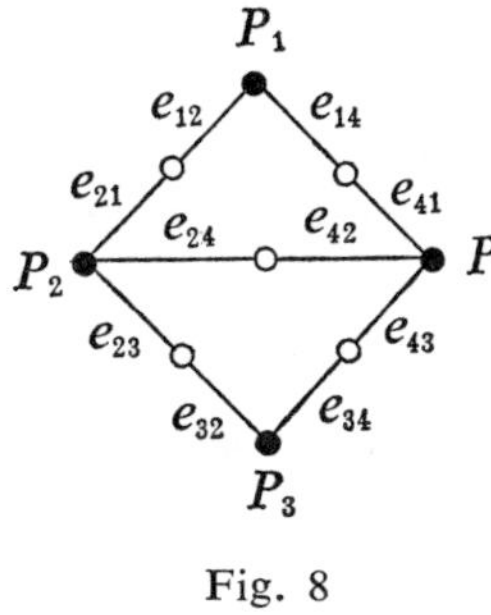

Fig. 8

Arrange the edges of EX is the following order:

$$EX=\{e_{12}, e_{14};\ e_{21}, e_{23}, e_{24};\ e_{32}, e_{34};\ e_{41}, e_{42}, e_{43}\}.$$

We identify $M^1(X)$ with $\boldsymbol{C}$-space spanned by EX. Then taking EX as the basis, we have the following decomposition of $M^1(X)$:

$$M^1(X)=W_1\oplus W_2\oplus W_3\oplus W_4\oplus W_5;\quad \rho^*=\varphi_1\oplus\varphi_2\oplus\varphi_3\oplus\varphi_4\oplus\varphi_5.$$

$$W_1=\{(-a-b, a+b, a+b, -b, -a, b, -b, -a-b, a, b);\ a, b\in \boldsymbol{C}\}$$

$$\varphi_1(T_1)=-1,\quad \varphi_1(T_2)=-1,\quad p_\varphi(u)=(1-u)^2.$$

$$W_2=\{(a, -a, a, -a, 0, -a, a, -a, 0, a);\ a\in \boldsymbol{C}\}$$

$\varphi_2(T_1)=-1,\quad \varphi_2(T_2)=1,\quad p_\varphi(u)=(1+u).$

$W_3=\{(-a,\ -a,\ b,\ -b,\ 0,\ a,\ a,\ b,\ 0,\ -b);\ a,\ b\in \boldsymbol{C}\}$

$\varphi_3(T_1)\simeq\begin{pmatrix}1 & 0\\ 0 & -1\end{pmatrix},\quad \varphi_3(T_2)\simeq\begin{pmatrix}0 & -1\\ -1 & 0\end{pmatrix},\quad p_\varphi(u)=1+u^2.$

$W_4=\{(a-b,\ -a+b,\ -b,\ -b,\ -a,\ a-b,\ -a+b,\ b,\ a,\ b);\ a,\ b\in \boldsymbol{C}\}$

$\varphi_4(T_1)\simeq\begin{pmatrix}0 & 1\\ 2 & 1\end{pmatrix},\quad \varphi_4(T_2)\simeq\begin{pmatrix}-1 & -1\\ 0 & 1\end{pmatrix},\quad p_\varphi(u)=1+u+2u^2.$

$W_5=\{(a,\ a,\ a+b-c,\ a+b-c,\ b+c,\ a,\ a,\ a+b-c,\ b+c,\ a+b-c);$
$a,\ b,\ c\in \boldsymbol{C}\}$

$$\varphi_5(T_1)\simeq\begin{pmatrix}1 & 1 & 1\\ 0 & 2 & 0\\ 0 & -1 & -1\end{pmatrix},\quad \varphi_5(T_2)\simeq\begin{pmatrix}1 & 0 & 0\\ 1 & 0 & 1\\ -1 & 1 & 0\end{pmatrix},\quad p_\varphi(u)=1-u^2-2u^3.$$

$$\begin{aligned}Z_Y(u)^{-1}=Z_{X,b}(u)^{-1}&=1-4u^3-2u^4+4u^6+4u^7+u^8-4u^{10}\\ &=(1-u)^2(1+u)(1+u^2)(1+u+2u^2)(1-u^2-2u^3).\end{aligned}$$

We begin with the following observation. In $\mathrm{End}_{\boldsymbol{Z}}(\boldsymbol{Z}[E\tilde{X}])$, the element T_1 (resp. T_2) preserves the decomposition

$$E\tilde{X}=\bigcup_{p(\tilde{P}_j)=P_j} E(\tilde{P}_j)\qquad(\text{resp. } E\tilde{X}=\bigcup_{p(\tilde{Q}_k)=Q_k} E(\tilde{Q}_k)). \tag{5.21}$$

where $E(\tilde{P}_j)$ denotes the set of edges incident to $\tilde{P}_j$. And as the element of $\mathrm{End}_{\boldsymbol{Z}}(\boldsymbol{Z}[E(\tilde{P}_j)])$ (resp. $\mathrm{End}_{\boldsymbol{Z}}(\boldsymbol{Z}[E(\tilde{Q}_k)])$), it satisfies

$$T_1^2=(q_1^{(j)}-1)T_1+q_1^{(j)}\qquad(\text{resp. } T_2^2=(q_2^{(k)}-1)T_2+q_2^{(k)}). \tag{5.22}$$

It follows from this that the only possible linear representations φ of $\boldsymbol{C}[T_1, T_2]$ are:

$$\begin{aligned}
&\varphi(T_1)=q_1^{(j)},\quad \varphi(T_2)=q_2^{(k)};\quad \varphi(T_1T_2)=q_1^{(j)}q_2^{(k)},\quad p_\varphi(u)=1-q_1^{(j)}q_2^{(k)}u,\\
&\varphi(T_1)=q_1^{(j)},\quad \varphi(T_2)=-1;\quad \varphi(T_1T_2)=-q_1^{(j)},\quad p_\varphi(u)=1+q_1^{(j)}u,\\
&\varphi(T_1)=-1,\quad \varphi(T_2)=q_2^{(k)};\quad \varphi(T_1T_2)=-q_2^{(k)},\quad p_\varphi(u)=1+q_2^{(k)}u,\\
&\varphi(T_1)=-1,\quad \varphi(T_2)=-1;\quad \varphi(T_1T_2)=+1,\quad p_\varphi(u)=1-u.
\end{aligned}$$

Now we consider the representation ρ^* of $\boldsymbol{C}[T_1, T_2]$ on $M^1(X)$. From (5.21), (5.22), we easily get the following equalities:

$$\begin{aligned}
\det\rho^*(T_1)&=\prod_{j=1}^{n_1}(-1)^{q_1^{(j)}}q_1^{(j)}=(-1)^{m-n_1}q_1^{(1)}\cdots q_1^{(n_1)},\\
\det\rho^*(T_2)&=\prod_{k=1}^{n_2}(-1)^{q_2^{(k)}}q_2^{(k)}=(-1)^{m-n_2}q_2^{(1)}\cdots q_2^{(n_2)}.
\end{aligned} \tag{5.23}$$

Proposition (5.24). *Suppose that X is not homotopic to a single circle, and that (φ, W) is an irreducible component of $\rho^*: \mathbf{C}[T_1, T_2] \to \mathrm{End}_{\mathbf{C}}(M^1(X))$, such that $p_\varphi(1)=0$. Then φ is linear (i.e.* $\dim W=1$) *and* $W \subset M(-1, -1)$: $\varphi(T_1)=\varphi(T_2)=-1$.

Proof. We first note that, for any $f \in M^1(X)$, one has

$$(5.25)\quad (\rho^*(T_1T_2)f)(e) = \sum_{Q' \in V(P)} \{ \sum_{e' \in E(Q')} f(e')\} - \sum_{e' \in E(Q)} f(e') - \sum_{e' \in E(P)} f(e') + f(e),$$

where we put $o(e^+)=P$, $t(e^+)=Q$, and $V(P)$ denotes the set of vertices Q' adjacent to P. Now suppose that $f \in W \subset M^1(X)$ is an eigenfunction of $\varphi(T_1T_2)$ such that $\varphi(T_1T_2)f=f$. By (5.25), this implies that the following equality holds for any $P \in V_1$ and $Q \in V(P)$:

$$\sum_{Q' \in V(P)} \{ \sum_{e' \in E(Q')} f(e')\} - \sum_{e' \in E(P)} f(e') = \sum_{e' \in E(Q)} f(e').$$

Since the left hand side depends only on P, it follows from this that, for any $P=P^{(j)} \in V_1$,

$$f^{**}(Q) := \sum_{e' \in E(Q)} f(e') \qquad \text{is a constant function on } V(P^{(j)}),$$

and that

$$f^*(P^{(j)}) := \sum_{e' \in E(P^{(j)})} f(e') = q_1^{(j)} f^{**}(Q) \qquad (Q \in V(P^{(j)})).$$

Since X is connected, one sees that $f^{**}(Q)$ is constant, say c, on V_2. Now from the disjoint union $EX = \cup E(P_j) = \cup E(Q_k)$, one obtain

$$cn_2 = \sum_{e \in EX} f(e) = \sum_{P_j \in V_1} q_1^{(j)} f^{**}(Q) = c \sum_{j=1}^{n_1} q_1^{(j)},$$

hence from (2.1),

$$c\left(\sum_{j=1}^{n_1} (q_1^{(j)}+1) - n_1 - n_2\right) = c(\#(EX) - \#(VX)) = c(r-1) = 0.$$

It follows that, if $r>1$, then $f^*(P)=f^{**}(Q)=0$ for any $P \in V_1$, $Q \in V_2$. From Lemma (4.11), this implies that $f \in M(-1, -1)$. Q.E.D.

As an immediate consequence, we obtain the following result which is a generalization of (4.16).

Theorem (5.26). *Suppose that Y is a connected finite multigraph, and* *let* $r:=\dim_{C}H_1(Y, C)$ *be the number of independent cycles of* Y. *Then one has*

$$\begin{aligned}\operatorname{ord}_{(1-u)} Z_Y(u)^{-1} &= r, \qquad \textit{if } r>1,\\ &= r+1=2, \qquad \textit{if } r=1.\end{aligned}$$

Proof. Let $X:=Y^{(2)}$ be the barycentric subdivision of Y. Then by (2.25), (2.27), we have $Z_Y(u)^{-1}=Z_{X,b}(u)^{-1}=\det(I-\rho^*(T_1T_2)u)$. Therefore, the assertion follows from (5.24), if $r>1$. Suppose that $r=1$, and let $C=(y_1, y_2, \cdots, y_{2l})$ be the circuit contained in X, which is unique up to the orientation and shifting the origin. Then it is easy to see that the two linearly independent functions

$$f_1: f_1(e_i)=(-1)^i, \quad \text{and} \quad f_2: f_2(e_i)=1 \qquad (1\leq i\leq 2l),$$

which vanish outside C, form a basis of 1-eigenspace of $\rho^*(T_1T_2)$. Q.E.D.

In what follows, we keep our assumption that φ is an irreducible component of $\rho^*: C[T_1, T_2]\rightarrow \operatorname{End}_C(M^1(X))$.

Proposition (5.27). *Suppose* X *is as above, and* (φ, W) *is a linear representation such that* $\varphi(T_1)=q_1^{(j)}$, $\varphi(T_2)=q_2^{(k)}$ *for some indices* $1\leq j\leq n_1$, $1\leq k\leq n_2$. *Then* φ *occurs in* $(\rho^*, M^1(X))$ *if and only if* X *is semiregular; moreover, it occurs exactly once.*

Proof. Let $f\in M(q_1^{(j)}, q_2^{(k)})$ be a function which spans W. Then as in the proof of Proposition (4.18), one sees that f is constant on each subsets $E(P)$, $E(Q)$ $(P\in V_1, Q\in V_2)$. Since X is connected, this implies that f is a constant function, say c, on EX. Now for any $P_i\in V_1$, and $e\in E(P_i)$, one has $(\rho^*(T_1)f)(e)=q_1^{(i)}f$, hence $q_1^{(i)}=q_1^{(j)}$; and similarly for $Q_i\in V_2$. Q.E.D.

Next we describe the condition under which the linear representation $\varphi(T_1)=-1$, $\varphi(T_2)=q_2^{(k)}$ occurs in $(\rho^*, M^1(X))$. Let $\{Q_{k_1}, \cdots, Q_{k_t}\}$ be the vertices of V_2 such that $q_2^{(k_i)}=q_2^{(k)}$, and let $X(q_2^{(k)})$ be the subgraph of X whose edges are the union of $E(Q_{k_i})$. Now we remove from $X(q_2^{(k)})$ the edges of $E(Q_{k_i})$, if one of it has an end point. Repeating this procedure, we finally get a sub-multigraph $X_0(q_2^{(k)})$ of X, which is not necessarily connected. The following assertion can be proved by a similar argument as above, and we omit the detail.

Proposition (5.28). *Notation being as above, the linear representation* φ *occurs in* $M^1(X)$ *if and only if it occurs in* $M^1(X_0(q_2^{(k)}))$. *Moreover, the* *multiplicity of φ in $M^1(X)$ is the sum of multiplicities of it in the connected components of $X_0(q_2^{(k)})$.*

Example (5.29). $X=X_{10}(6, 8)\to X(3)=X_0(3)\simeq K(3, 2)$.

$$Z_{X,b}(u)^{-1}=(1-u)^3(1+2u)^2(1-u-4u^2-4u^3)$$
$$Z_{X_0(3),b}(u)=(1-u)^2(1+u)^2(1-2u)(1+2u).$$

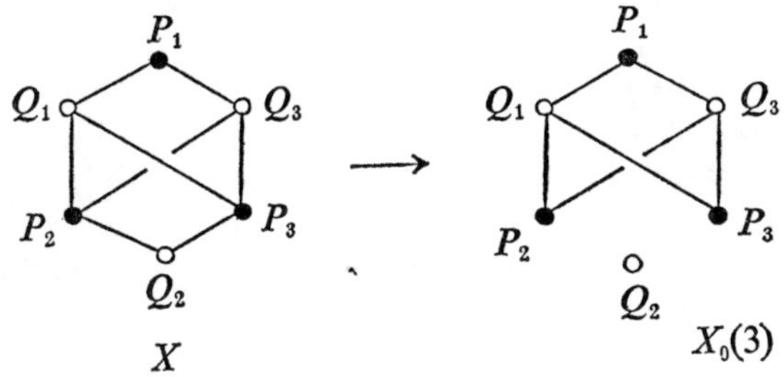

Fig. 9

Next we shall prove the following result, which characterize the factor $(1+u)$ in $Z_X(u)$.

Proposition (5.30). *Suppose that Y is a non-bipartite multigraph, $X:=Y^{(2)}$ is the barycentric subdivision of Y, and (φ, W) is an irreducible component of $(\rho^*, M^1(X))$ such that $p_\varphi(-1)=0$. Then φ is linear, and $W\subset M(-1, 1)$.*

Proof. By the assumption, there exists a nonzero function $f\in W\subset M^1(X)$ such that $\rho^*(T_1T_2)f=-f$. By a similar argument using (5.25), we see that the function $f^*(P):=\sum_{e\in E(P)} f(e)$ of $P\in V_1$ satisfies

$$f^*(P)=f(e')-f(e) \qquad \text{if } e\in E(P),\ \{e, e'\}=E(Q),\ Q\in V_2. \tag{5.31}$$

Now the assumption that Y is nonbipartite implies the existence of a closed path $C=(P_0, \cdots, P_l)$ in Y of *odd* length such that $P_0=P_l=P$. From (5.31), one obtains $f^*(P)=0$. By (4.12), this implies that $\rho^*(T_1)f=-f$, hence also $\rho^*(T_2)f=f$. Since (φ, W) is assumed to be irreducible, we have $W=Cf$, which completes the proof. Q.E.D.

Using the above result, and (2.27), we can now prove the following theorem, which gives a new characterization of the bipartite multigraphs.

Theorem (5.32). *Suppose that Y is a connected finite multigraph, and let $r:=\dim_C H_1(Y, C)$ be the number of independent cycles of Y. Suppose moreover that $r>1$. Then one has*

$$\operatorname{ord}_{(1+u)} Z_Y(u)^{-1} = r-1, \qquad \textit{if } Y \textit{ is non-bipartite,}$$
$$= r, \qquad \textit{if } Y \textit{ is bipartite.}$$

Proof. If Y is bipartite, the assertion is a consequence of (5.26) and the fact $Z_Y(u)^{-1}$ is a polynomial of u^2. So suppose that Y is non-bipartite, and put $X=Y^{(2)}$. The above proposition (5.30) shows that $\operatorname{ord}_{(1+u)} Z_Y(u)^{-1}$ is equal to $\dim_C M(-1, 1)$. By (4.18), (ii), we can identify $M(-1, 1)$ with the space M_Y^0 of functions on EY, which satisfy $\sum_{e \in E(P)} f(e)=0$ for all $P \in VY$. Now with respect to the inner product (4.2), M_Y^0 is the orthogonal complement of the subspace spanned by $\{f_P; P \in VY\}$, where f_P is the characteristic function of $E(P)$. We claim that the f_P's are linearly independent. In fact, let $\sum_P c_P f_P=0$ be a linear relation among them. Let $C=(C_0, \cdots, P_l)$ be as in the proof of Proposition (5.30). Evaluating at the edge $[P_i, P_{i+1}]$, we obtain $c_{P_i}+c_{P_{i+1}}=0$ for $0 \leq i \leq l-1$, from which follows $c_P=0$ for all $P \in VY$. Therefore we obtain

$$\operatorname{ord}_{(1+u)} Z_Y(u)^{-1} = \dim_C M(-1, 1) = \#(EY) - \#(VY) = r-1. \qquad \text{Q.E.D.}$$

Remark (5.33). From the above results, we see that, if $X=Y^{(2)}$ for a non-bipartite multigraph Y, then we always have $M(1, -1)=\{0\}$.

§ 6. Representations of p-adic groups

Let G be an abstract group to which is given a mapping $l: G \to \boldsymbol{N} \cup \{0\}$, called the length function, satisfying the following conditions (G, l, I), (G, l, II).

(G, l, I) For each $l \geq 0$, $G_l := \{x \in G; l(x)=l\}$ is non-empty, and $U := G_0$ is a subgroup; moreover, $UG_lU=G_l=G_l^{-1}$, and $\#(U \backslash G_l) < \infty$.

Under these conditions, one can define the Hecke algebra $\mathscr{H}(G, U)$ of the pair (G, U), and one can regard G_l as an element of $\mathscr{H}(G, U)$.

(G, l, II) There exist two positive integers q_1, q_2 such that one has the following relations in $\mathscr{H}(G, U)$.

$$(6.1) \qquad (G_1)^2 = G_2 + (q_2-1)G_1 + q_2(q_1+1)U,$$
$$G_1G_l = G_{l+1} + (q_2-1)G_l + q_1q_2G_{l-1} \qquad (l \geq 2).$$

We describe two typical classes of groups satisfying these conditions.

(6.2) **Example 1.** Let X be a semiregular bipartite multigraph of valency (q_1+1, q_2+1), and let $\tilde{X}$ be the universal covering tree of X.

Then we know that $\tilde{G}=\mathrm{Aut}(\tilde{X})$ satisfies the above conditions, with $l(x)=(1/2)d_{\tilde{X}}(P_0, P_0x)$ where $d_{\tilde{X}}$ is th distance on $\tilde{X}$ and $P_0 \in \tilde{V}_1$ is a fixed vertex.

Moreover, any subgroup G of $\tilde{G}$ which acts transitively on $\tilde{V}_1$ satisfies (G, l, I), (G, l, II) (c.f. [H-H]).

(6.3) **Example 2.** The algebraic groups G over a local field K (i.e., locally compact field with respect to a discrete valuation of K), which is simply connected and has K-rank one.

In both examples, the group G is a topological group which is locally compact, totally disconnected and unimodular. Moreover, it has a Tits system (G, B, N, S) of affine type, where the Weyl group is the infinite dihedral group, and the associated building is isomorphic to $\tilde{X}$. Here, if one choose two adjacent vertices $P_0 \in \tilde{V}_1$, $Q_0 \in \tilde{V}_2$, their stabilizers $U=U_1:=\mathrm{Stab}_G(P)$ and $U_2:=\mathrm{Stab}_G(Q_0)$ are representatives of the two maximal parahoric subgroups of G, up to G-conjugation. Recall that a subgroup is called parahoric, if it contains a conjugate of B. We may assume that $B=U_1\cap U_2$. Then the set of edges $E\tilde{X}$ can be naturally identified with $B\backslash G$, on which G acts by right translation.

The following facts are well known:

Lemma (6.4). *$\mathscr{H}(G, B)\simeq C[T_1, T_2]$, $U_i\to 1+T_i$ $(i=1, 2)$, where $C[T_1, T_2]$ is as in* (5.2).

Lemma (6.5). *$\mathscr{H}(G, U)=C[G_1]$, and there is a monomorphism $\mathscr{H}(G, U)\to\mathscr{H}(G, B)$ which maps $I_U\to(1+q_1)I_B$.*

Now we recall the basic facts on the representation theory. Let G be a locally compact, totally disconnected unimodular group. A representation (π, V) of G is a homomorphism $\pi\colon G\to GL(V)$. It is called *smooth*, if

(6.6) for each $v \in V$, the stabilizer $\{x \in G; \pi(x)v=v\}$ is an open subgroup of G.

We note that this is equivalent to the assertion $V=\bigcup_B V^B$, where B runs over the open compact subgroups of G, and $V^B:=\{v \in V;\ \pi(g)v=v\,(g \in B)\}$ is the space of B-fixed vectors. (π, V) is called *admissible* if, in addition,

(6.7) $\quad \dim_C V^B<\infty \quad$ for any open compact subgroup B.

If (π, V) is smooth, then one gets a representation of the algebra $\mathscr{H}(G)$ of locally constant and compactly supported functions f on G, the product being defined by the convolution

(6.8) $$(f_1*f_2)(g):=\int_G f_1(x)f_2(x^{-1}g)dx, \qquad (f_1, f_2 \in \mathscr{H}(G)).$$

with respect to a Haar measure of G. Namely it is defined by

$$\pi(f)\cdot v := \int_G f(x)\pi(x)\cdot v\,dx \qquad (f \in \mathscr{H}(G),\ v \in V). \tag{6.9}$$

One of the basic fact here is that the correspondence $(\pi, V) \to (\pi, \mathscr{H}(G))$ induces an equivalence of the category of smooth representations of G and the non-degenerate $\mathscr{H}(G)$-modules. In particular a subspace V_1 of V is G-invariant if and only if it is $\mathscr{H}(G)$-invariant. If, moreover, (π, V) is admissible, then $\pi(f)$ is of finite rank for any $f \in \mathscr{H}(G)$, and conversely.

Let B be an open compact subgroup of G. Then we can define the Hecke algebra $\mathscr{H}(G, B)$ to be the subalgebra of $\mathscr{H}(G)$, which consists of the bi-B-invariant compactly supported functions on G. If (π, V) is a smooth representation, then V^B is stable under the action of $\mathscr{H}(G, B)$. In this way, one obtains a natural map

$$\rho\colon \mathrm{Hom}_G(V, W) \longrightarrow \mathrm{Hom}_{\mathscr{H}(G,B)}(V^B, W^B), \tag{6.10}$$

where (π, V), (π', W) are smooth representations of G. Let (φ, E) be a representation of $\mathscr{H}(G, B)$. Put

$$I(E) = I(\varphi) = C_c(G/B) \otimes_{\mathscr{H}(G,B)} E, \tag{6.11}$$

where $C_c(G/B)$ denotes the space of compactly supported right B-invariant functions on G, which is acted upon by $\mathscr{H}(G, B)$ from the right, and regard it as a G-module by left translation. It is easily seen that $I(E)$ is a smooth G-module.

Proposition (6.12) ([Bo-2], (2.5)). *Suppose that* $\dim_{\mathbf{C}} E < \infty$. *Then the natural map* $\rho_I\colon \mathrm{Hom}_G(I(E), V) \to \mathrm{Hom}_{\mathscr{H}(G,B)}(E, V^B)$ *is bijective.*

Now suppose, in addition, that G has a Tits system (G, B, N, S) of affine type, where B is an open compact subgroup of G, called the Iwahori subgroup. The following facts have been proved in [Bo-2].

Theorem (6.13) ([Bo-2], (4.4) (4.10)). *Assumption being as above, let* (φ, E) *be a finite dimensional* $\mathscr{H}(G, B)$*-module. Then* $I(E)$ *is an admissible* G*-module, and it is irreducible if and only if* E *is an irreducible* $\mathscr{H}(G, B)$*-module. The assignment* $E \to I(E)$ *is an exact functor from finite dimensional* $\mathscr{H}(G, B)$*-modules to admissible* G*-modules.*

From this theorem and (6.12), it follows that the irreducible admissible G-modules V such that $V^B \neq \{0\}$ are parametrized by the irreducible $\mathscr{H}(G, B)$-modules. To apply this to the spectral decomposition of

$L^2(G/\Gamma)$, we need the following fact which is also well known. Let (π, V) be a unitary rerpesentation of G, where V is a Hilbert space. A vector $v \in V$ is called smooth, if the isotropy group of v is an open subgroup of G. The set V_∞ of smooth vectors forms a vector subspace of V, which is stable under G. In this way, one gets a G-module (π_∞, V_∞).

Theorem (6.14) (cf. [Car-2]). *Suppose that (π, V) is an irreducible unitary representation of G. Then V_∞ is dense in V, and the representation (π_∞, V_∞) is admissible.*

Now let G be a locally compact, totally disconnected unimodular group, and assume that G has a Tits system (G, B, N, S) of affine type such that the Weyl group $W := \langle S \rangle$ $(S = \{s_1, s_2\})$ is the infinite dihedral group. Let Γ be a discrete subgroup such that

(6.15) Γ is torsion free, and G/Γ is compact.

Let, as usual, $L^2(G/\Gamma)$ be the Hilbert space of square integrable functions on G, which are right Γ-invariant. Then one can define the left regular representation of G on $L^2(G/\Gamma)$ by

(6.16) $$(\pi(g)f)(x) := f(g^{-1}x) \qquad (g \in G,\ x \in G/\Gamma,\ f \in L^2(G/\Gamma)).$$

This gives a unitary representation of G. It is well known that the assumption of the compactness of G/Γ implies that $L^2(G/\Gamma)$ decomposes into an orthogonal direct sum

(6.17) $$L^2(G/\Gamma) = \bigoplus_{[\pi]} V(\pi),$$

where $[\pi]$ is extended over the set of equivalent classes of irreducible unitary representations of G, and $V(\pi)$ is the closed G-invariant subspace of $L^2(G/\Gamma)$ which is π-isotypic. Moreover, each $V(\pi)$ is isomorphic to a direct sum of a finite copies of an irreducible G-module V_π which belongs to $[\pi]$. Denote by $m_\Gamma(\pi)$ the multiplicity of π in $V(\pi)$:

(6.18) $$V(\pi) \simeq m_\Gamma(\pi) \cdot V_\pi.$$

Now we can state the second half of our main results in this paper. It is concerned with the multiplicities $m_\Gamma(\pi)$ of π in $L^2(G/\Gamma)$. Let G and Γ be as above, and let $\tilde{X}$ be the building associated with the Tits system (G, B, N, S). By our assumption, $\tilde{X}$ is a tree of semiregular bipartite type with valency (q_1+1, q_2+1), and the quotient, say X, is a finite graph with the same property. Moreover, Γ can be identified with the fundamental group of X as a CW-complex. Let $Z_\Gamma(u) = Z_X(u)$ be the zetafunction of Γ or X which has been evaluated in Sections 2, 3.

Main Theorem (IV). *Notation and assumption being as above, let* (π, V) *be an irreducible unitary representation of* G, *and let* (φ, V_∞^B) *be the corresponding irreducible representation of* $\mathscr{H}(G, B)$ *as in Theorem* (6.13). *Assume that* $V_\infty^B \neq \{0\}$. *Then one has*

$$m_\Gamma(\pi) = \text{the multiplicity of } p_\varphi(u) \text{ in } Z_\Gamma(u)^{-1}, \tag{6.19}$$

where $p_\varphi(u) = \det(I - \varphi(T_1 T_2)u)$ *is a linear or quadratic polynomial as in Proposition* (5.7). *In other words, the zeta function describes the spectral decomposition in* $L^2(G/\Gamma)$, *those components of which have* (*nonzero*) *B-fixed vectors.*

Proof. This is a consequence of the above quoted facts and the obvious equality

$$L^2(G/\Gamma)^B = L^2(B\backslash G/\Gamma) \simeq M^1(X) \qquad (X = \tilde{X}/\Gamma). \tag{6.20}$$

Q.E.D.

The above result has a number of applications. First of all, recall that $\mathscr{H}(G, B) = \boldsymbol{C}[T_1, T_2]$ has exactly 4 linear representations (cf. Proposition (5.9)). Each of such representation corresponds to an admissible representation of G. In [Bo-2], Borel determined, in a more general context, the linear representations which correspond to those representations of G which are square integrable. Among such representations, there is a distinguished one, called the Steinberg representation, which corresponds to the linear representation $T_i \to -1$ $(i = 1, 2)$, hence to $p_\varphi(u) = 1 - u$. That this is square integrable was first noted by Matsumoto [Ma]. The following result was first proved by Ihara [I-1] for $G = SL_2(K)$. We note also that the first equality has been proved by Garland [Gar] for a general $\boldsymbol{p}$-adic group.

Corollary (6.21) (cf. [Gar], [Bo-2]). *For the Steinberg representation* π_{St}, *one has*

$$\begin{aligned} m_\Gamma(\pi_{St}) &= r = \dim_{\boldsymbol{C}} H^1(\Gamma, \boldsymbol{C}) \\ &= -\operatorname{ord}_{(1-u)} Z_\Gamma(u). \end{aligned} \tag{6.22}$$

Proof. This follows immediately from (6.19) and (4.16). Q.E.D.

Now suppose that $q_1 > q_2$. Then the linear representation such that $p_\varphi(u) = 1 + q_2 u$ is also known to be square integrable, whereas the one which corresponds to $p_\varphi(u) = 1 + q_1 u$, is not (cf. [Bo-2]. We note that there is a misprint in the statement of [Bo-2], pp 254, (ii)). This fact would be compared with the following result, which is an immediate consequence of Proposition (4.23).

Corollary (6.23). *Suppose that $q_1 > q_2$, and (π, V) corresponds to a linear representation of $\mathcal{H}(G, B)$ such that $p_\varphi(u) = 1 + q_2 u$. Then one has*

$$m_\Gamma(\pi) = -\operatorname{ord}_{(1+q_2u)} Z_\Gamma(u) \geq n_2 - n_1 > 0, \tag{6.24}$$

where for $i = 1, 2$

$$n_i = \#(U_i \backslash G/\Gamma) = \text{the number of vertices of } X \text{ of valency } q_i + 1. \tag{6.25}$$

Moreover, the equality $m_\Gamma(\pi) = n_2 - n_1$ holds if and only if $X = \tilde{X}/\Gamma$ is a weak Ramanujan graph.

§ 7. Special values of zeta functions

As in the cases of zeta functions which appear in number theory, we can expect that the special values of our $Z_X(u)$ are related with properties of the graph X. In fact the results (4.16), (4.21), and (4.26) are regarded as giving such relations. Moreover, recall that in the special case where X can be derived from an arithmetic subgroup of $SL(2, \boldsymbol{Q}_p)$, the essential part of $Z_X(u)$ is nothing but the congruence zeta function of the reduction modulo p of a modular curve $X_0(N)$ (see Ihara [I-2]). In this case, the class number of the function field of $X_0(N) \otimes \boldsymbol{F}_p$ is expressed as the residue of the congruence zeta function.

In this section, we shall extend this fact to $Z_{X,b}(u)$, and give an interpretation of the residue of it at $u = 1$.

Definition (7.1). Let X be an arbitrary connected multigraph. A *spanning tree* T is a subgraph of X, which is a tree and such that $VT = VX$. If X is finite, the number of spanning trees is called the *complexity* of X, and denoted by $\kappa(X)$.

We shall show that the complexity is an analogue of the class numbers of the global fields. Let X be a finite connected multigraph such that $VX = \{P_j;\ 1 \leq j \leq n\}$, and let $A \in M(n, \boldsymbol{Z})$ be its adjacency matrix (cf. (3.1)). Let D be the diagonal matrix such that

$$D := \operatorname{diag}(k_1, \cdots, k_n), \tag{7.2}$$

with $\quad k_j :=$ valency of $P_j = \#\{e \in EX;\ P_j \in \varepsilon(e)\} \qquad (1 \leq j \leq n)$.

Also put

$$J = J_n := \text{the } n \times n \text{ matrix whose entries are all } 1, \tag{7.3}$$

$$Q := D - A. \tag{7.4}$$

Now the following result is fundamental:

Theorem (cf. [Bi]). *With the above notation, we have*

(i) *The matrix of cofactors (adjugate) of Q is a multiple of J, by $\kappa(X)$:*

$$\operatorname{adj}(Q)=\kappa(X)\cdot J. \tag{7.5}$$

(ii) *The complexity of X is given by the formula*

$$\kappa(X)=n^{-2}\det(J+Q). \tag{7.6}$$

Theorem (7.7). *Suppose that X is a regular connected multigraph of valency $q+1$, such that $q>1$, $\sharp(VX)=n$, and let $r=\dim_{C} H_1(X, \boldsymbol{C})$. Then one has the following formula:*

$$\kappa(X)=\frac{-1}{n(q-1)2^{r-1}}\cdot\frac{1}{(1-u)^r Z_X(u)}\bigg|_{u=1}.$$

Proof. The assumption on X implies that $D=(q+1)I_n$, so that $Q=(q+1)I-A$. Moreover from the regularity, we see immediately that $JA=AJ$. Now the constant function $f\in M^0(X)$ is a common eigenfunction of A and J: $Af=(q+1)f$, $Jf=nf$, hence $Qf=0$. Note also that J is 0 on the orthogonal complement of $\boldsymbol{C}f$. Thus one see that

$$\det(J+Q)=\det[J+(q+1)I-A]=n\prod_{j=1}^{n-1}(q+1-\eta_j),$$

where $\eta_1, \cdots, \eta_{n-1}$, and $\eta_n=q+1$ are the eigenvalues of A. Applying (3.7) to the above result, we thus obtain

$$\begin{aligned}\kappa(X)&=n^{-1}\prod_{j=1}^{n-1}(q+1-\eta_j)\\ &=n^{-1}\{\det[I_n-Au+qu^2]/(1-u)(1-qu)\}|_{u=1}\\ &=\frac{-1}{n(q-1)2^{r-1}}\cdot\frac{1}{(1-u)^r Z_X(u)}\bigg|_{u=1}.\end{aligned} \tag{7.8}$$

Q.E.D.

The above result has an interesting application to the class number of the function field of the modular curve $X_0(l)$ over $\boldsymbol{F}_p$. Let l be a prime such that $(p, l)=1$, and let $\boldsymbol{B}$ be the definite quaternion algebra over $\boldsymbol{Q}$, which ramifies exactly at the places ∞, l. For a subset S of $\boldsymbol{B}$, we put $S^{(1)}:=\{s\in S;\ \mathrm{Nr}(s)=1\}$, where $\mathrm{Nr}(s)$ is the reduced norm of s. Let O be a maixmal order of $\boldsymbol{B}$, and put

$$\Gamma:=\boldsymbol{B}^{(1)}\cap[\boldsymbol{B}_p^{(1)}\times\prod_{q\neq p}O_q^{(1)}]\qquad(O_\infty=\boldsymbol{B}_\infty), \tag{7.9}$$

and regard Γ as a subgroup of $G := \boldsymbol{B}_p^{(1)} \simeq SL(2, \boldsymbol{Q}_p)$ through the projection to the first component. Since $O_q^{(1)}$ $(q \neq p)$ are compact groups, Γ is a discrete subgroup of G. Moreover, by the strong approximation theorem, one has

$$\begin{aligned} O_A^{(1)} \backslash \boldsymbol{B}_A^{(1)} / \boldsymbol{B}^{(1)} &= O_A^{(1)} \backslash [\boldsymbol{B}_p^{(1)} \times \prod_{q \neq p} O_q^{(1)}] \cdot \boldsymbol{B}^{(1)} / \boldsymbol{B}^{(1)} \\ &= O_A^{(1)} \backslash [\boldsymbol{B}_p^{(1)} \times \prod_{q \neq p} O_q^{(1)}] / \Gamma \\ &= O_p^{(1)} \backslash \boldsymbol{B}_p^{(1)} / \Gamma \simeq SL(2, \boldsymbol{Z}_p) \backslash G / \Gamma. \end{aligned}$$

Since $\sharp[O_A^{(1)} \backslash \boldsymbol{B}_A^{(1)} / \boldsymbol{B}^{(1)}] = \sharp[O_A^{\times} \backslash \boldsymbol{B}_A^{\times} / \boldsymbol{B}^{\times}] = h =$ the class number of $\boldsymbol{B}$ is known to be finite (see (7.10) below), we see that G/Γ is compact. Now a famous result of Eichler [Ei] states that there is an isomorphism

$$(7.10) \qquad L^2(O_A^{(1)} \backslash \boldsymbol{B}_A^{(1)} / \boldsymbol{B}^{(1)}) = M^0(V_1) \simeq S_2(\Gamma_0(l)) \oplus \boldsymbol{C}$$

as modules over the Hecke algebra $\bigotimes_{q \neq l} \mathscr{H}(\boldsymbol{B}_q^{(1)}, O_q^{(1)})$, where $S_2(\Gamma_0(l))$ is the space of holomorphic cusp forms of weight 2 for the group $\Gamma_0(l)$. In particular we have $n = h = \dim_{\boldsymbol{C}} S_2(\Gamma_0(l)) + 1$. It follows that the main part of our zeta function $Z_X(u)$ attached to the graph $X = \tilde{X}/\Gamma$ is

$$(7.11) \quad \det[I_h - Au + pu^2] = (1-u)(1-pu) \cdot \det[I - uT(p) \,|\, S_2(\Gamma_0(l)) + pu^2].$$

On the other hand, from Eichler-Shimura's congruence relation ([Sh]), it is well known that the congruence zetafunction $Z(X_0(l)/\boldsymbol{F}_p; u)$ of the modular curve $X_0(l)$ over $\boldsymbol{F}_p$ is equal to

$$(7.12) \quad Z(X_0(l)/\boldsymbol{F}_p; u) = \det[I - uT(p) \,|\, S_2(\Gamma_0(l)) + pu^2]/(1-u)(1-pu).$$

It is also well known, and easy to show that

Lemma (7.13). *The class number $H_{l,p}$ of the function field of the modular curve $X_0(l)$ over $\boldsymbol{F}_p$ $(p \neq l)$, is equal to the number of $\boldsymbol{F}_p$-rational points of the jacobian variety of $X_0(l)$, and it is given by*

$$H_{l,p} = (1-p)[(1-u)Z(X_0(l)/\boldsymbol{F}_p; u)]|_{u=1}.$$

Proof. See [Sch].

Comparing this with (7.7), we get the following observation for the class number $H_{l,p}$:

(7.14) *Suppose that $l \equiv 1 \pmod{12}$. Then one has $H_{l,p} = (g+1)\kappa(X)$,*

where $g = \dim_{\boldsymbol{C}}(S_2(\Gamma_0(l))$ is the genus of $X_0(l)$, and is independent of p. In particular $H_{l,p}$ is always a multiple of $g+1 = (l-1)/12$.

Proof. We prove that, under the assumption $l \equiv 1 \pmod{12}$, the group $\bar{\Gamma}=\Gamma/\{\pm 1\}$ is torsion free, hence it is a free group. Let γ be an element of Γ of order n. Then, as an element of $\boldsymbol{B}_\infty^{(1)}=SU(2)$, γ is conjugate to an element of its maximal torus $SO(2)$. It follows that γ has the characteristic polynomial $X^2-2\cos(2\pi/n)X+1$. By the definition of Γ, we see that $2\cos(2\pi/n) \in \boldsymbol{Z}_q$ $(\subset \boldsymbol{Q}_q)$ for any $q \neq l$. In particular, it follows that each q has only the prime divisors of degree 1 in $K=\boldsymbol{Q}(\zeta_n+\zeta_n^{-1})$, where ζ_n denotes a primitive n-th root of unity. This is possible only for $K=\boldsymbol{Q}$, hence for $n=1, 2, 3, 4, 6$. On the other hand, the assumption on l implies that $\boldsymbol{B}_l$ has no element of order 3, 4, 6. This proves that $\{\pm 1\}$ is the only torsion subgroup of Γ. Since this $\{\pm 1\}$ acts trivially on the tree $\tilde{X}$ (= the Tits building of $G=SL_2(\boldsymbol{Q}_p)$), we have $\tilde{X}/\Gamma=\tilde{X}/\bar{\Gamma}=X$. Now applying (7.7) and (7.13), we obtain the result.

Q.E.D.

The above result is a special case of (slightly stronger form of) a theorem of Doi and Brumer ([D-M]):

Theorem (7.15). *Let l be an odd prime. Then for any prime $p \neq l$, one has*

$$H_{l,p}=\det[I-uT(p)\,|\,S_2(\Gamma_0(l))+pu^2]|_{u=1}\equiv 0 \qquad (\mathrm{mod}\ e_0\cdot(l-1)/12),$$

where $e_0=1, 3, 2, 6$ (resp. 6) *according as* $l \equiv 1, 5, 7, 11 \bmod 12$ (resp. $l=3$).

Proof. This is proved as a consequence of the basic properties (7.16) of the Brandt matrix $B(p)$, and was proved in [Ha]. For the convenience of the reader, we quote it as Lemma (7.17) below. Note that we have $\sum_i (1/e_i)=(l-1)12$, by Eichler's mass formula. Q.E.D.

Suppose that $A=(a_{ij}) \in M(n, \boldsymbol{Z})$ is an integral matrix and $e_1, \cdots, e_n$ are positive integers satisfying the conditions:

(7.16) (i) $\sum_{j=1}^{n} a_{ij}=k$ for all $i=1, \cdots, n$.

(ii) $e_i a_{ij}=e_j a_{ji}$ for any i, j.

Then regarding A as an endomorphism of $W=\boldsymbol{R}^n$, one gets $A\boldsymbol{u}=k\boldsymbol{u}$ for $\boldsymbol{u}={}^t(1, \cdots, 1)$. Let W_0 be the orthogonal complement of $\boldsymbol{u}$ in W, and denote by A_0 the restriction of A to W_0. Put

$$M:=\sum_{i=1}^{n} 1/e_i, \qquad e_0:=\mathrm{l.c.m}(e_1, \cdots, e_n).$$

Lemma (7.17). *Assumption being as above, one has*

$$\det(kI_{n-1}-A_0)\equiv 0 \qquad (\mathrm{mod}\ e_0M).$$

Proof. See [Ha-1], Theorem 2. Q.E.D.

It seems to be possible to obtain the above result (7.15) using graphs also in the case when $\Gamma/\{\pm 1\}$ has torsion elements. Now we shall generalize Theorem (7.7) to the case where X is an arbitrary connected semiregular multigraph with valency (q_1+1, q_2+1).

Theorem (7.18). *Suppose X is a connected semiregular bipartite multigraph of valency (q_1+1, q_2+1), such that $\#(V_i)=n_i$, and $q_1q_2>1$. Then we have*

$$\kappa(X)=\frac{1+q_2}{n_1}\cdot\frac{1}{1-q_1q_2}\cdot\frac{1}{(1-u)^r Z_{X,b}(u)}\bigg|_{u=1}.$$

Proof. We have to evaluate $\det(J+Q)$, where $J=J_n$, $n=n_1+n_2$, and $Q=D-A$ is as in (7.4). From (7.5), it follows again that $JQ=QJ$, and the constant function $f\in M^0(X)$ satisfies $Jf=nf$, $Qf=0$. Thus, it suffices to evaluate the product of the eigenvalues of Q, excluding the last one, which is 0. For this we arrange the vertices of X as $\{P_1, \cdots, P_{n_1}, P_{n_1+1}=Q_1, \cdots, P_n=Q_{n_2};\ P_i\in V_1, Q_j\in V_2\}$, so that Q is expressed as

$$Q=\left(\begin{array}{c|c} (q_1+1)I_{n_1} & -B \\ \hline -{}^tB & (q_2+1)I_{n_2}\end{array}\right).$$

One can find an orthogonal matrix $W\in O(n_2)$ in such a way that

$$BW=\left[\begin{array}{cccc|ccc} \mu_1 & 0 & \cdots & 0 & 0 & \cdots & 0 \\ * & \mu_2 & \ddots & \vdots & \vdots & & \vdots \\ \vdots & \ddots & \ddots & 0 & \vdots & & \vdots \\ * & & 0 & \mu_{n_1} & 0 & \cdots & 0 \end{array}\right]=(C\,|\,0),\ \text{say}.$$

(with the first block of n_1 columns and the second block of n_2-n_1 columns)

Then one sees that Q is conjugate to the matrix

$$\left[\begin{array}{c|c|c} (q_1+1)I & -C & 0 \\ \hline -{}^tC & (q_2+1)I & 0 \\ \hline 0 & 0 & (q_2+1)I \end{array}\right].$$

(block columns of widths n_1, n_1, n_2-n_1)

Now the last matrix has the characteristic polynomial

$$\{x-(q_2+1)\}^{n_2-n_1}\det\left(\begin{array}{c|c} (x-(q_1+1))I_{n_1} & C \\ \hline {}^tC & (x-(q_2+1))I_{n_1} \end{array}\right)$$
$$=\{x-(q_2+1)^{n_2-n_1}\det[\{x-(q_1+1)\}\{x-(q_2+1)\}I_{n_1}-C\,{}^tC]$$
$$=\{x-(q_2+1)\}^{n_2-n_1}\prod_{j=1}^{n_1}[\{x-(q_1+1)\}\{x-(q_2+1)\}-\lambda_j^2].$$

From this follows that the eigenvalues of Q are

$$q_2+1 \quad (n_2-n_1 \text{ times}), \qquad \alpha_j,\ \alpha'_j \quad (1\leq j\leq n_1)$$

with $\alpha_j\alpha'_j=(q_1+1)(q_2+1)-\lambda_j^2$. Here $\{\pm\lambda_j\ (1\leq j\leq n_1), 0, \cdots, 0\}$ are the eigenvalues of A, and we know from Proposition (3.19) that

$$\lambda_1^2=(q_1+1)(q_2+1)>\lambda_2^2\geq\cdots\geq\lambda_{n_1}^2.$$

It follows from (7.6), that

$$\kappa(X)=n_1^{-1}(1+q_2)^{n_2-n_1+1}$$
$$\times\{\det[I_{n_1}-(A^{[1]}-q_2+1)u+q_1q_2u^2]/(1-u)(1-q_1q_2u)\}|_{u=1}$$

The assertion now follows from (3.15). Q.E.D.

Remark (7.19). The above formula gives a generalization of (7.7). In fact, if X is a regular multigraph of valency $q+1$, we take the barycentric subdivision $X^{(2)}$, which can be regarded as a semiregular bipartite multigraph of valency $(q+1, 2)$. Then (7.7), (7.18), and (2.27) gives

$$\kappa(X^{(2)})=2^r\cdot\kappa(X). \tag{7.20}$$

This relation can be explained as follows: A spanning tree of X is obtained by removing an edge from each of the r independent set of cycles in X. On the other hand, to each edge $e\in EX$ which are to be removed, we have two choices of edges in $EX^{(2)}$, to get a spanning tree of $X^{(2)}$. Thus, (7.20) holds for any multigraphs X and $X^{(2)}$.

Problem (7.21). Can one show the relation for $\kappa(X)$ and the values of $(1-u)^rZ_X(u)$ at $u=1$, analogous to (7.18), in the general case where no regularity of X is assumed?

Finally we prove the following result which generalize (7.20) partially.

Proposition (7.22). *Suppose that X is a semiregular bipartite multigraph with valency (q_1+1, q_2+1), such that $q_1\geq q_2$, $q_1q_2>1$, and $n_i=\#(V_i)$*

($i=1, 2$). *Them* $\kappa(X)$ *is a multiple of* $(1+q_2)^{n_2-n_1+1}$.

Proof. This follows from the formula for $\kappa(X)$ given in the proof of Theorem (7.18), applying Lemma (7.17) to $A^{[1]}$. Q.E.D.

Note that, if in Lemma (7.17) one assumes that $a_{ij}\geq 0$ for any i, j, and that $e_i=1$ for all i, then A is regarded as the adjacency matrix of a regular multigraph X of valency k. The assertion then is also a consequence of (7.8), since $\kappa(X)$ is an integer.

§ 8. Miscellaneous results

Here we describe some of the general results on the relation between the spectra of two or more finite (multi)-graphs, which are immediately interpreted in terms of our zeta functions. They are the complementary graphs, line graphs, and those derived from the general compositions. The last ones are useful to compute $Z_X(u)$ for various families of graphs, or to construct them with prescribed properties.

(8.1) Complementary graphs.

Let X be a finite connected graph with n vertices. Then there exists an injective morphism $\iota: X\to K(n)$, where $K(n)$ is the complete graph with n vertices (i.e., any two vertices in $K(n)$ are joined by an edge). Since ι is unique up to the permutation of VX, one can identify X with its image $\iota(X)$, and regard it a subgraph of $K(n)$. If one removes from $K(n)$ all edges of X, one gets a new graph with n vertices VX, which is called the *complementary graph* of X, and denoted by X^c. It is in general not connected, and it may have isolated points (i.e., no edge incident to it), or endpoints. If we put $r=\dim_{\mathbf{C}} H_1(X, \mathbf{C})$, $r^c=\dim_{\mathbf{C}} H_1(X^c, \mathbf{C})$, then we have the following relation which follows from (2.1):

(8.2) $$r+r^c=(n-1)(n-4)/2, \qquad \text{if } X^c \text{ is connected.}$$

Proposition (8.3). *Suppose that* X *is a connected regular graph with valency* k, *and let* $\operatorname{Spec}(X)=\{\lambda_1=k, \lambda_2, \cdots, \lambda_n\}$ *be the spectrum of* X. *Then* X^c *is again regular of valency* $n-1-k$, *and one has*

(8.4) $$\operatorname{Spec}(X^c)=\{n-1-k, -\lambda_2-1, -\lambda_3-1, \cdots, -\lambda_n-1\}.$$

If, moreover, X^c *is assumed to be connected, one has*

(8.5) $$Z_{X^c}(u)^{-1}=(1-u^2)^{n(n-3-k)/2}(1-u)[1-(n-1-k)u] \times\prod_{j=2}^{n}[1+(\lambda_j+1)u+(n-2-k)u^2].$$

Proof. This can be proved similarly as Proposition (8.7), and we omit the detail. Q.E.D.

Suppose next that X is a connected bipartite graph such that $\#(V_1) = n_1$, $\#(V_2) = n_2$. Then an analogue of X^c, denoted by X^{bc}, is defined as follows. X^{bc} has the same set $VX = V_1 \cup V_2$ of vertices as X, and an edge joins the vertices $P \in V_1$ and $Q \in V_2$ if and only if they are not adjacent in X. Put $r' := \dim_{\boldsymbol{C}} H_1(X^{bc}, \boldsymbol{C})$, and observe

$$(8.6) \qquad r + r' = n_1 n_2 - 2n_1 - 2n_2 + 2, \qquad \text{if } X^{bc} \text{ is connected.}$$

Proposition (8.7). *Notation being as above, suppose that X is semiregular of valency (k_1, k_2), $k_1 \geq k_2 > 1$, and that X^{bc} is connected. Let*

$$\mathrm{Spec}(X) = \{\pm\sqrt{k_1 k_2}, \pm\lambda_2, \cdots, \pm\lambda_{n_1}, 0, \cdots, 0 \ (n_2 - n_1 \text{ times})\}$$

be the spectrum of X as in (3.11). *Then X^{bc} is again semiregular bipartite of valency $(k_1', k_2') = (n_2 - k_1, n_1 - k_2)$, and one has*

$$(8.8) \quad \mathrm{Spec}(X^{bc}) = \{\pm\sqrt{k_1' k_2'} = \pm\lambda_1, \pm\lambda_2, \cdots, \pm\lambda_{n_1}, 0, \cdots, 0 \ (n_2 - n_1 \ \textit{times})\},$$

$$(8.9) \quad Z_{X^{bc}, b}(u)^{-1} = (1-u)^{r'-1}(1+q_2' u)^{n_2 - n_1} \prod_{j=1}^{n_1} [1 - (\lambda_j^2 + q_1' - q_2')u - q_1' q_2' u^2],$$

where we put $q_i' := k_i' - 1$ $(i = 1, 2)$.

Proof. We shall prove (8.8), since (8.9) follows from it and (3.15). Let $A^{[1]}$, $A_{bc}^{[1]}$ be the adjacency matrices of $X^{[1]}$, $(X^{bc})^{[1]}$ respectively (cf. (3.13)). In view of the equation (3.17), it suffices to evaluate the eigenvalues of $A_{bc}^{[1]}$. Let A, A_{bc} be the adjacency matrices of X, X^{bc}, and express A as in (3.12). Then it follows from the definition that

$$A_{bc} = \left(\begin{array}{c|c} 0 & J \\ \hline {}^t J & 0 \end{array}\right) - A = \left(\begin{array}{c|c} 0 & J - B \\ \hline {}^t(J-B) & 0 \end{array}\right)$$

where $J = J_{n_1, n_2}$ is the $n_1 \times n_2$ matrix whose entries are all 1. We have

$$A_{bc}^2 = \left(\begin{array}{c|c} (J-B)\,{}^t(J-B) & 0 \\ \hline 0 & {}^t(J-B)(J-B) \end{array}\right),$$

and, by the assumption that X is semiregular of valency (k_1, k_2), we have $B\,{}^t J = J\,{}^t B = k_1 J_{n_1}$, hence

$$(J-B)\,{}^t(J-B) = J\,{}^t J - 2k_1 J_{n_1} + B\,{}^t B = (n_2 - 2k_1) J_{n_1} + B\,{}^t B.$$

Similarly we have

$$(B\,{}^tB)J_{n_1}=k_1k_2J_{n_1}=J_{n_1}(B\,{}^tB).$$

Now observing that $(B\,{}^tB)$ and J_{n_1} has the common eigen vector ${}^t(1, \cdots, 1)$, one sees that, from the assumption

$$B\,{}^tB\simeq \mathrm{diag}(\lambda_1^2, \cdots, \lambda_{n_1}^2)$$

it follows that

$$(J-B)\,{}^t(J-B)\simeq \mathrm{diag}(\lambda_1^2+(n_2-2k_1)n_1,\ \lambda_2^2, \cdots, \lambda_{n_1}^2).$$

The assertion (8.8) now follows, if one observes the relations (3.17) and $\lambda_1^2+(n_2-2k_1)n_1=k_1'k_2'$, which follows from

$$k_1+k_1'=n_2, \quad k_2+k_2'=n_1, \quad \text{and} \quad k_1k_2'=k_1'k_2. \qquad \text{Q.E.D.}$$

(8.10) Line graphs.

The line graph $L(X)$ of a (connected) graph X is defined as follows. As the set of vertices, put $VL(X)=EX$, and define two vertices in $L(X)$ to be adjacent, if the corresponding edges in X have a common vertex. The following result is well known:

Proposition (8.11) (Sachs 1967). *Let X be a regular graph of valency k, such that $n=\#(VX)$, $m=nk/2=\#(EX)$. Then $L(X)$ is again regular of valency $2k-2$, and*

(8.12) $\mathrm{Spec}(L(X))$

$$=\{\lambda_j+k-2\ (1\leq j\leq n),\ -2,\ -2, \cdots, -2\ (m-n\ \textit{times})\}.$$

where $\mathrm{Spec}(X)=\{\lambda_j\ (1\leq j\leq n)\}$. *Therefore if X is connected, one has*

(8.13) $Z_{L(X)}(u)^{-1}$

$$=(1-u^2)^{nk(k-2)/2}(1+2u+q'u^2)^{m-n}\prod_{j=1}^{n}[1-(k-2+\lambda_j)u+q'u^2],$$

where $q'=2k-3$.

Proof. For (8.12), see [Bi], Theorem 3.8. (8.13) is a direct consequence of this and (3.7). Q.E.D.

(8.14) Cartesian product $X_1\times X_2$.

The cartesian product of two graphs $X_1\times X_2$ is defined to be a graph X which has $VX_1\times VX_2$ as the set of vertices and two vertices (P_1, P_2),

(Q_1, Q_2) are defined to be adjacent if either (i) $P_1 = Q_1$ and P_2, Q_2 are adjacent, or (ii) $P_2 = Q_2$ and P_1, Q_1 are adjacent. From this definition we see that the valency of $(P, Q) \in V(X_1 \times X_2)$ is the sum of the valencies of P and Q. It is also easy to see that, when X_1, X_2 are connected, then so is $X_1 \times X_2$, and

$$\begin{aligned}(8.15)\qquad R &:= \dim_{C} H_1(X_1 \times X_2, \mathbf{C}) \\ &= \#(E(X_1 \times X_2)) - \#(V(X_1 \times X_2)) + 1 \\ &= \#(VX_1)\#(EX_2) + \#(VX_2)\#(EX_1) - \#(VX_1)\#(VX_2) + 1.\end{aligned}$$

Proposition (8.16). *Suppose that X_i is a connected graph and let $\mathrm{Spec}(X_1) = \{\lambda_j;\ 1 \le j \le n_1\}$, $\mathrm{Spec}(X_2) = \{\mu_j;\ 1 \le j \le n_2\}$ be the spectrum of X_1, X_2 respectively. Then one has*

$$(8.17)\qquad \mathrm{Spec}(X_1 \times X_2) = \{\lambda_i + \mu_j;\ 1 \le i \le n_1,\ 1 \le j \le n_2\}.$$

If, moreover, X_i is assumed to be regular of valency k_i $(i = 1, 2)$, then one has

$$(8.18)\qquad Z_{X_1 \times X_2}(u)^{-1} = (1 - u^2)^{(q-1)n_1 n_2/2} \prod_{i=1}^{n_1} \prod_{j=1}^{n_2} [1 - (\lambda_i + \mu_j)u + qu^2],$$

where $q = k_1 + k_2 - 1$.

Proof. Although (8.17) is well known (see [Scw]), we give here a simple proof. Arrange the vertices of $X_1 \times X_2$ in n_2 blocks:

$$(P_1, Q_1), \cdots, (P_{n_1}, Q_1); (P_1, Q_2), \cdots, (P_{n_1}, Q_2); \cdots; (P_1, Q_{n_2}), \cdots, (P_{n_1}, Q_{n_2}).$$

Then the adjacency matrix C of $X_1 \times X_2$ is written as a matrix with $n_2 \times n_2$ blocks

$$C = \begin{array}{r} \\ i) \\ \\ \\ \\ j) \\ \\ \end{array} \left(\begin{array}{ccccc} A & & & & \\ & & & I & \\ & A & & & \\ & & A & & \\ & & & A & \\ & I & & & \\ & & & & A \end{array}\right) \in M(n_1 n_2, \mathbf{Z}),$$

where A is the adjacency matrix of X_1, and the unit matrix $I = I_{n_1}$ appears at the (i, j) block if and only if Q_i and Q_j are adjacent in X_2. Let $W = (w_{ij}) \in GL_{n_2}(\mathbf{C})$ be a nonsingular matrix which transforms the adjacency matrix B of X_2 into a diagonal matrix: $X^{-1}BX = \mathrm{diag}(\mu_1, \cdots, \mu_{n_2})$, and let $\tilde{W}$ be the matrix having the same type as C, whose (i, j) block is $w_{ij}I_{n_1}$. Then one can easily see that

$$\tilde{W}^{-1}C\tilde{W}=\begin{pmatrix} \mu_1 I+A & & & 0 \\ & \mu_2 I+A & & \\ & & \ddots & \\ 0 & & & \mu_{n_2} I+A \end{pmatrix},$$

and the assertion (8.17) follows. Q.E.D.

(8.19) Conjunction $X_1 \wedge X_2$.

The conjunction $X_1 \wedge X_2$ of two graphs X_1, X_2 is defined as follows. Again we have $V(X_1 \wedge X_2)=V(X_1)\times V(X_2)$, and (P_1, Q_1), (P_2, Q_2) are defined to be adjacent if and only if both pairs $\{P_1, P_2\}$ and $\{Q_1, Q_2\}$ are adjacent. We see easily that if X_1, X_2 are connected, then so is $X_1 \wedge X_2$, and

(8.20)
$$\begin{aligned} R:&=\dim_C H_1(X_1\wedge X_2, C) \\ &=\#(E(X_1\wedge X_2))-\#(V(X_1\wedge X_2))+1 \\ &=2\#(EX_1)\#(EX_2)-\#(VX_1)\#(VX_2)+1. \end{aligned}$$

Proposition (8.21). *Suppose that X_i and* Spec(X_i) *are as above* $(i=1, 2)$. *Then one has*

(8.22)
$$\operatorname{Spec}(X_1\wedge X_2)=\{\lambda_i\mu_j;\ 1\le i\le n_1,\ 1\le j\le n_2\}.$$

If, moreover, X_i is assumed to be regular of valency k_i $(i=1, 2)$, then $X_1\wedge X_2$ is also regular with valency k_1k_2, and one has

(8.23)
$$Z_{X_1\wedge X_2}(u)^{-1}=(1-u^2)^{(q-1)n_1n_2/2}\prod_{i=1}^{n_1}\prod_{j=1}^{n_2}[1-(\lambda_i\mu_j)u+qu^2],$$

where $q=k_1k_2-1$.

Proof. With the same notation as in the proof of Proposition (8.16), we see that the adjacency matrix C of $X_1\wedge X_2$ now is

(8.24)
$$C=\begin{array}{c} \\ i) \\ \\ \\ \\ j) \\ \\ \end{array}\!\!\begin{pmatrix} 0 & & & & \\ & \overset{i}{\smile} & & \overset{j}{\smile}\ A & \\ & & 0 & & \\ & & & 0 & \\ & & & & 0 \\ & A & & & \\ & & & & 0 \end{pmatrix}\in M(n_1, n_2, Z),$$

where we put A in (i, j) block if Q_i and Q_j are adjacent. In other words, $C=A\otimes B$. Now the assertion follows from this by a similar argument as above. Q.E.D.

(8.25) Strong product $X_1 * X_2$.

The strong product $X_1 * X_2$ of two graphs X_1, X_2 is defined as follows. Again we have $V(X_1 * X_2)=V(X_1)\times V(X_2)$, and (P_1, Q_1), (P_2, Q_2) are defined to be adjacent if and only if $d_{X_1}(P_1, P_2)\leq 1$, $d_{X_2}(Q_1, Q_2)\leq 1$. We easily see that if X_1, X_2 are connected, then so is $X_1 * X_2$, and

$$
\begin{aligned}
(8.26)\qquad R &:= \dim_{\boldsymbol{C}} H_1(X_1 * X_2, \boldsymbol{C}) \\
&= \#(E(X_1 * X_2)) - \#(V(X_1 * X_2)) + 1 \\
&= 2\#(EX_1)\#(EX_2) + \#(VX_1)\#(EX_2) + \#(VX_2)\#(EX_1) \\
&\qquad - \#(VX_1)\#(VX_2) + 1.
\end{aligned}
$$

Proposition (8.27). *Suppose that X_i and* Spec(X_i) *are as above* $(i=1, 2)$. *Then one has*

$$(8.28)\qquad \operatorname{Spec}(X_1 * X_2)=\{\lambda_i\mu_j+\lambda_i+\mu_j;\ 1\leq i\leq n_1,\ 1\leq j\leq n_2\}.$$

If, moreover, X_i is assumed to be regular of valency k_i $(i=1, 2)$, then $X_1 * X_2$ is also regular with valency $k_1k_2+k_1+k_2$, and one has

$$(8.29)\qquad Z_{X_1 * X_2}(u)^{-1}=(1-u^2)^{(q-1)n_1n_2/2}\prod_{i=1}^{n_1}\prod_{j=1}^{n_2}[1-(\lambda_i\mu_j+\lambda_i+\mu_j)u+qu^2],$$

where $q=k_1k_2+k_1+k_2-1$.

Proof. This is a combination of (8.15) and (8.21). Q.E.D.

(8.30) Join X_1+X_2.

The join of two graphs X_1 and X_2 is a graph X_1+X_2 such that $V(X_1+X_2)=V(X_1)\cup V(X_2)$ (disjoint), and to $EX_1\cup EX_2$ we add the edges joining any $P\in V(X_1)$ and any $Q\in V(X_2)$. To get a result for this graph similar to the above ones, we have to assume that both X_1, X_2 are regular, say, of valency k_1, k_2 respectively.

Proposition (8.31) ([Scw]). *Assumption being as above, one has*

$$(8.32)\qquad \operatorname{Spec}(X_1+X_2)=\{\xi, \xi', \lambda_i\ (2\leq i\leq n_1),\ \mu_j\ (2\leq j\leq n_2)\},$$

where we assume that $\lambda_1=k_1$ $\mu_1=k_2$, and ξ, ξ' are the roots of $X^2-(k_1+k_2)X+(k_1k_2-n_1n_2)$. If, moreover, X_1+X_2 is assumed to be regular, then one has

$$(8.33)\qquad Z_{X_1+X_2}(u)^{-1}=(1-u^2)^{(q-1)(n_1+n_2)/2}(1-u)(1-qu)(1-cu+qu^2)\times\prod_{i=2}^{n_1}[1-\lambda_i u+qu^2]\prod_{j=2}^{n_2}[1-\mu_j u+qu^2],$$

where we put $q=k_1+n_2-1$ $(=k_2+n_1-1)$, $c=k_1-n_1=k_2-n_2$.

Proof. (8.32) is a special case of (8.35) below. Note that the regularity assumption on X_1+X_2 is equivalent to the equality $k_1+n_2=k_2+n_1$. Q.E.D.

(8.34) Composition $G[X_1, \cdots, X_p]$.

Let G be a graph such that $VG=\{P_1, \cdots, P_p\}$, and let X_i be a graph for $i=1, 2, \cdots, p$. Then one can form a new graph $X:=G[X_1, \cdots, X_p]$, whose set of vertices is the disjoint union of VX_i $(1\le i\le p)$, and as the set of edges, we add to $\cup EX_i$, those joining $Q_i\in VX_i$ and $Q_j\in VX_j$ whenever P_i and P_j are adjacent in G. This supplies a quite general way of constructing new graphs from known ones. Now we assume that each X_i is regular, say, of valency k_i. Also put $n_i:=\#(VX_i)$. The following result, due to Schwenk [Scw] is very useful.

Proposition (8.34) ([Scw]). *Assumption being as above, let*

$$\mathrm{Spec}(X_i)=\{\lambda_1^{(i)}=k_i, \lambda_2^{(i)}, \cdots, \lambda_{n_i}^{(i)}\}$$

be the spectrum of X_i. *Then one has*

$$(8.35)\qquad \mathrm{Spec}(G[X_1, \cdots, X_p])=\mathrm{Spec}(C^*)\cup\bigcup_{i=1}^{p}(\mathrm{Spec}(X_i)\setminus\{\lambda_1^{(i)}\})$$

where in the union we count with multiplicities, and $\mathrm{Spec}(C^*)$ *is the spectrum of the following matrix*

$$(8.36)\qquad C^*:=\begin{matrix} \\ \\ i) \\ j) \\ \\ \end{matrix}\begin{pmatrix} k_1 & & \overset{i}{\smile} & \overset{j}{\smile} & \\ & k_2 & & & \\ & & \ddots & n_j & \\ & & n_i & \ddots & \\ & & & & k_p \end{pmatrix}\in M(p, \boldsymbol{Z}).$$

where we put n_j *in* (i, j) *entry such that* $i\ne j$, *whenever* P_i *and* P_j *are adjacent in* G.

Proof. We shall give a proof along the same idea as [Scw]. We first note that the adjacency matrix C of $X=G[X_1, \cdots, X_p]$ is divided into $p\times p$ blocks according to the partition of $N:=\sharp(VX)=n_1+n_2+\cdots+n_p$, and written as follows:

$$
(8.37)\qquad C=\begin{matrix} \\ \\ i) \\ \\ j) \\ \\ \\ \end{matrix}\begin{bmatrix} A_1 & & & & & & \\ & A_2 & & & & & \\ & & \ddots & & J & & \\ & & & \ddots & & & \\ & & {}^tJ & & \ddots & & \\ & & & & & A_{p-1} & \\ & & & & & & A_p \end{bmatrix} \in M(N, Z),
$$

(with column markers i and j above the matrix)

From the regularity of X_i, we get

$$
(8.38)\qquad A_iJ=k_iJ,\quad {}^tJA_i=k_i{}^tJ\quad (J=J_{n_i,n_j}\in M(n_i, n_j; Z)),
$$

where J is the matrix of prescribed size whose entries are all 1. Let $W_i=C^{n_i}$ be the C-space to which A_i and ${}^tJ$ act. Observe that $u_i:={}^t(1,\cdots,1)$ is a common eigenvector of them: $A_iu_i=k_iu_i$, ${}^tJu_i=n_iu_j$, and let W_i^0 be the orthogonal complement of u_i in W_i, which consists of the vectors whose sum of coordinates is 0. Then from (8.37), (8.38), C induces an endomorphism C^0 on $\bigoplus_{i=1}^p W_i^0$, and C^* on $\bigoplus_{i=1}^p(W_i/W_i^0)\simeq C^p$. From (8.38), it is easy to see that $C^0\simeq\bigoplus_{i=1}^p(A_i\,|\,W_i^0)$, and that C^* is represented by the matrix (8.36). Since $\mathrm{Spec}(C)=\mathrm{Spec}(C^0)\cup\mathrm{Spec}(C^*)$ (with multiplicity), this completes the proof. Q.E.D.

Corollary (3.39). *Suppose that* G (resp. X_i $(1\leq i\leq p)$) *is a connected regular graph of valency* k_0 (resp. k) *and* $\sharp(VG)=p$ (resp. $\sharp(VX_i)=n$). *Then* $G[X_1,\cdots,X_p]$ *is again a regular connected graph of valency* $K=k+nk_0$, *and if* $\mathrm{Spec}(G)=\{\mu_j\ (1\leq j\leq p)\}$ (resp. $\mathrm{Spec}(X_i)=\{\lambda_j^{(i)};\ 1\leq j\leq n\}$) *is the spectrum of* G *with* $\mu_1=k_0$ (resp. $\lambda_1^{(i)}=k$), *one has*

$$
\begin{aligned}
(8.40)\qquad & Z_{G[X_1,\cdots,X_p]}(u)^{-1} \\
&=(1-u^2)^{(q-1)pn/2}(1-u)(1-qu)\prod_{j=2}^{p}[1-(n\mu_j+k)u+qu^2] \\
&\quad\times\prod_{i=1}^{p}\prod_{j=2}^{n}[1-\lambda_j^{(i)}u+qu^2],
\end{aligned}
$$

where $q=K-1=k+nk_0-1$.

Proof. This follows immediately from (8.35) and (3.7). Q.E.D.

§ 9. Zeta functions of well known families of graphs

In this section, we shall compute the zeta functions $Z_X(u)$ for some well known families of graphs. These computations give us many examples of graphs which are not Ramanujan graphs, and also the examples which answer the questions raised in Section 5. Most of them are derived from (3.7), (3.15), and the knowledge of Spec(X), which has been studied by various authors.

(9.1) Cayley graphs of finite groups.

Let G be a group and let S be a symmetric set of generators, i.e., $S=S^{-1}$ and $G=\langle S\rangle$. We assume moreover that $1 \notin S$. Now a connected graph $X=X(G, S)$, called the *Cayley graph* of (G, S), is defined as follows. Put $VX=G$, and define two vertices g, h to be adjacent, if $g=hs$ for an element $s \in S$. This gives a connected regular graph of valency $\sharp(S)$. $X(G, S)$ is a bipartite graph if and only if no product of odd number of element of S is equal to 1. As a matter of fact, many of the graphs which have interesting properties are Cayley graphs. A simple example is given by taking $S=G\setminus\{1\}$, in which case we obtain the complete graph $K(n)$ $(n=\sharp(G))$, which will be treated below.

However, it is a difficult problem to determine the zeta function, or the spectrum of it, without specifying G and S. This is reduced to the following problem. Let f be a function on G, and let $A(G,f) \in M(n, \boldsymbol{C})$ $(n=\sharp(G))$ be the matrix, indexed by G, such that

(9.2) $A(G,f)=(f(g^{-1}h))_{g,h\in G}$.

Can one determine the characteristic polynomial of $A(G,f)$? In fact the adjacency matrix of the Cayley graph $X(G, S)$ is nothing but $A(G,f)$ with f being the characteristic function of the set S.

Here, we restrict ourselves to the case where G is an abelian group. In this case it is easy to determine the eigenvalues of $A(G,f)$, using a well known result on group determinant:

Lemma (9.3). *The eigenvalues of* $A(G,f)$ *are*

$$\{\sum_{g\in G} \chi(g)f(g)\ (\chi \in \hat{G}:=\text{the group of characters of } G)\}.$$

Thus using (3.7), we obtain

Proposition (9.4). *For the Cayley graph* $X=X(G, S)$ *of a finite abelian group* G, *we have*

$$\mathrm{Spec}(X)=\{\sum_{s\in S} \chi(s);\ \chi \in \hat{G}\},$$

$$Z_X(u)^{-1}=(1-u^2)^{(q-1)n/2} \prod_{x\in \hat{G}} [1-(\sum_{s\in S} \chi(s))u+qu^2],$$

where we put $\sharp(G)=n$, $\sharp(S)=q+1$.

Note that $\sum \chi(s) \in \boldsymbol{R}$, because $S=S^{-1}$. This result supplies many examples of graphs with the second eigenvalue λ_2 as close to $\lambda_1=q+1$ as one wants.

For example, take $G_{l,t}:=\boldsymbol{Z}/l\boldsymbol{Z}\oplus\cdots\oplus\boldsymbol{Z}/l\boldsymbol{Z}$, the elementary abelian l-group of rank t $(n=l^t)$, and let

$$S=\{(0,\cdots,0,\overset{i}{\breve{\pm 1}},0,\cdots,0);\ 1\leq i\leq t\},\qquad q=2t-1.$$

Then we have for $X=X(G_{l,t},S)$

$$\operatorname{Spec}(X)=\left\{2\sum_{i=1}^{t}\cos\left(\frac{2\pi a_i}{l}\right);\ \boldsymbol{a}\in G_{l,t}\right\},\qquad (\boldsymbol{a}:=(a_1,\cdots,a_t)),$$

$$\text{(9.5)}\qquad Z_X(u)^{-1}=(1-u^2)^{l^t(t-1)}\prod_{\boldsymbol{a}\in G_{l,t}}\left[1-2u\sum_{i=1}^{t}\cos\left(\frac{2\pi a_i}{l}\right)+qu^2\right].$$

Note that X is bipartite if and only if l is even. Choosing l suitably, one can give examples of regular bipartite graphs X for which $Z_{X,b}(u)^{-1}$ has an irreducible factor over $\boldsymbol{Q}$ with arbitrary prescribed degree. We remark finally that, putting $t=1$, we get a circuit C_l of length l:

$$\text{(9.6)}\qquad Z_{C_l}(u)^{-1}=\prod_{a=0}^{l-1}\left[1-2u\cos\left(\frac{2\pi a}{l}\right)+u^2\right]=(1-u^l)^2.$$

(9.7) Flower of l petals.

X is a semiregular bipartite multigraph, which consists of $l+1$ vertices: $V_1=\{P\}$, $V_2=\{Q_1,\cdots,Q_l\}$, and each Q_j is joined to P by $q+1$ edges.

Fig. 10

Thus we see that the adjacent matrix is

$$A=\begin{bmatrix} 0 & q+1 & q+1 & \cdots & q+1 \\ q+1 & 0 & 0 & \cdots & 0 \\ q+1 & 0 & 0 & \cdots & 0 \\ \vdots & \vdots & \vdots & & \vdots \\ q+1 & 0 & 0 & & 0 \end{bmatrix} \in M(n, \boldsymbol{Z}), \qquad n=l+1.$$

It is easy to calculate the eigenvalues of A, and we obtain

$$\text{(9.8)} \qquad \operatorname{Spec}(X)=\{\pm(q+1)\sqrt{l},\ 0,\ \cdots,\ 0\ (l-1 \text{ times})\}.$$

Arranging the order of the edges as

$$EX=\{e_0^{(1)}, e_1^{(1)}, \cdots, e_q^{(1)}, e_0^{(2)}, \cdots, e_q^{(2)}, \cdots, e_0^{(l)}, \cdots, e_q^{(l)}\},$$

we find the following matrix representation of $\rho^*(T_i)$:

$$\rho^*(T_1)=\begin{bmatrix} J-I & I & I & \cdots & I \\ I & J-I & I & \cdots & I \\ I & I & J-I & \cdots & I \\ \vdots & \vdots & \vdots & & \vdots \\ I & I & I & \cdots & J-I \end{bmatrix} \in M(m, \boldsymbol{Z}), \qquad m=l(q+1), \quad I=I_{q+1}, \quad J=J_{q+1},$$

$$\rho^*(T_2)=\begin{bmatrix} J-I & 0 & 0 & \cdots & 0 \\ 0 & J-I & 0 & \cdots & 0 \\ 0 & 0 & J-I & \cdots & 0 \\ \vdots & \vdots & \vdots & & \vdots \\ 0 & 0 & 0 & \cdots & J-I \end{bmatrix} \in M(m, \boldsymbol{Z}), \qquad m=l(q+1),$$

$$\rho^*(T_1T_2)=\begin{bmatrix} (q-1)J+I & qJ & qJ & \cdots & qJ \\ qJ & (q-1)J+I & qJ & \cdots & qJ \\ qJ & qJ & (q-1)J+I & \cdots & qJ \\ \vdots & \vdots & \vdots & & \vdots \\ qJ & qJ & qJ & \cdots & (q-1)J+I \end{bmatrix}.$$

Using the fact that $\det(xI-J)=x^q(x-(q+1))$, or from (3.15), we get

$$\text{(9.9)} \qquad Z_{X,b}^{-1}(u)=(1-u)^{ql}(1+qu)^{l-1}(1-q(lq+l-1)u).$$

Note that X is semiregular bipartite of valency $(q_1+1, q_2+1)=(lq+l, q)$, hence $r=\dim_{\boldsymbol{C}} H_1(X, \boldsymbol{C})=ql$. Thus we have

$$M^1(X)=M(-1,\ -1)\oplus M(-1,q)\oplus M(lq+l-1,q),$$
$$\text{dim:}\quad (q+1)l= \quad ql \quad + \quad l-1 \quad + \quad 1$$

and see that the space of cuspidal functions vanishes.

(9.11) Complete bipartite graph $K(q_1+1,q_2+1)$.

This is a typical example of semiregular bipartite graph of valency (q_1+1,q_2+1). The vertices are divided into two classes V_1, V_2 such that $\sharp(V_1)=q_2+1$, $\sharp(V_2)=q_1+1$, and each $P\in V_1$ (resp. $Q\in V_2$) is joined to all vertices of V_2 (resp. V_1). To determine the spectrum of this graph, we consider the adjacency matrix $A^{[1]}$ of the multigraph $X^{[1]}$ derived from $X=K(q_1+1,q_2+1)$ as in (3.13).

Since there are exactly (q_1+1) proper paths C of length two such that $o(C)=P_i$, $t(C)=P_j$ for any distinct vertices P_i, $P_j\in V_1$, we get $A^{[1]}=(q_1+1)(J-I)$, where $J,I\in M(q_2+1,\boldsymbol{Z})$ are as above. Hence

$$I-(A^{[1]}-q_2+1)u+q_1q_2u^2=(1+(q_1+q_2)u+q_1q_2u^2)I-(q_1+1)Ju,$$

and we easily get

$$\text{(9.12)}\qquad Z_{X,b}^{-1}(u)=(1-u)^{q_1q_2}(1-q_1q_2u)(1+q_1u)^{q_2}(1+q_2u)^{q_1}.$$

(9.13) Complete graph $K(q+2)$ $(q\geq 1)$.

This is the most elementary graph, having $n=q+2$ vertices, any two of which are joined by an edge. One sees immediately that $K(q+2)$ is a regular graph with valency $q+1$. The zetafunction $Z_X(u)$ can be simply derived as above. Namely, the adjacent matrix of $K(n)$ is $A=J_n-I_n$, hence we have $I-Au+qu^2=(1+u+qu^2)I-uJ$. From (3.7), one obtain

$$\text{(9.14)}\qquad Z_{K(q+2)}(u)^{-1}=(1-u)^r(1+u)^{r-1}(1-qu)(1+u+qu^2)^{q+1},$$
$$r=q(q+1)/2.$$

This result is also a consequence of the well known fact that

(9.15) $\operatorname{Spec}(K(q+2))=\{q+1,-1,\cdots,-1\ (q+1 \text{ times})\}$ (cf. [Bi]).

Note that (9.14) implies that the representation of $\boldsymbol{C}[T_1,T_2]$ on the space $M^1_{\text{cusp}}(X^{(2)})$ of cusp forms on the edges of $X^{(2)}$, the barycentric subdivision of $X=K(q+2)$, is a direct sum of $M(-1,1)$, which is the $(q+1)(q-2)/2$ copies of a linear representation, and $q+1$ copies of a 2-dimensional irreducible representation, say φ (cf. (4.7)). This fact has an interesting interpretation in terms of the representations of the symmetric group $\mathfrak{S}_{q+2}$ of degree $q+2$, as follows.

The group $\mathfrak{S}_{q+2}$ acts on $K(q+2)$ through the permutation of its vertices, hence it acts also on $X^{(2)}$, and one gets a representation of $\mathfrak{S}_{q+2}$ on our space $M^1(X^{(2)})$. To consider how it decomposes into the irreducible ones, let us identify $M^1(X^{(2)})$ with the space M of bilinear forms $F: \boldsymbol{C}^{q+2}\times\boldsymbol{C}^{q+2}\to\boldsymbol{C}$ which satisfy $F(\boldsymbol{v}_i, \boldsymbol{v}_i)=0$ $(1\leq i\leq q+2)$, where $\{\boldsymbol{v}_i\ (1\leq i\leq q+2)\}$ is the standard orthonormal basis of $\boldsymbol{C}^{q+2}$. This can be done by identifying the function $f\in M^1(X^{(2)})$ with the bilinear form $F(x, y):=\sum_{i,j} f([P_i Q_j])x_i y_j$, where $[P_i, Q_j]$ denotes the edge of $X^{(2)}$ joining $P_i\in VX$ and Q_j, the middle point of $[P_i, P_j]\in EX$. Now if we add M the space of diagonal forms, we get $W^*\otimes W^*$, where $W=\boldsymbol{C}^{q+2}$, and W^* is its dual space:

$$\text{(9.16)}\qquad M^1(X^{(2)})\oplus\boldsymbol{C}^{q+2}\simeq M\oplus\boldsymbol{C}^{q+2}\simeq W^*\otimes W^*,\qquad \text{as } \mathfrak{S}_{q+2}\text{-modules.}$$

On the other hand, as is well known, the canonical representation of $\mathfrak{S}_{q+2}$ in W^* decomposes as

$$\text{(9.17)}\qquad (\rho, W^*)\simeq\rho_{q+2}\oplus\rho_{q+1,1},$$

where, in general, $\rho_{n_1,n_2,\cdots,n_t}$ denotes the irreducible representation of $\mathfrak{S}_{q+2}$ corresponding to the partition $q+2=n_1+n_2+\cdots+n_t$, hence, in particular, ρ_{q+2} denotes the trivial one. Now from (9.16), (9.17), to decompose $M^1(X^{(2)})$ into irreducible $\mathfrak{S}_{q+2}$-submodules is equivalent to decompose $\rho_{q+1,1}\otimes\rho_{q+1,1}$. To this, the following answer has been known:

Lemma (9.18) (Murnaghan [Mu]).

$$\rho_{q+1,1}\otimes\rho_{q+1,1}\simeq\rho_{q+2}\oplus\rho_{q+1,1}\oplus\rho_{q,2}\oplus\rho_{q,1,1}.$$

Using this result, one can reproduce (9.14) without using (3.7), as follows. We first note that the above irreducible components have degrees

$$\deg(\rho_{q+1,1})=q+1,\quad \deg(\rho_{q,2})=(q-1)(q+2)/2,\quad \deg(\rho_{q,1,1})=q(q+1)/2.$$

Next we notice that the actions of $\boldsymbol{C}[T_1, T_2]$ and $\mathfrak{S}_{q+2}$ are commutative. This implies that T_1, T_2 acts as scalars on each irreducible components of $M^1(X^{(2)})$. Also note that $M(q, 1)\simeq\rho_{q+2}$ as $\mathfrak{S}_{q+2}$-modules.

Lemma (9.19). *One has*

$$M(-1, -1)\simeq\rho_{q,1,1},\quad M(-1, 1)\simeq\rho_{q,2}\qquad \textit{as } \mathfrak{S}_{q+2}\textit{-modules.}$$

Moreover, the orthogonal complement M_0 *of* $M(-1, 1)$ *in* $M^1_{\text{cusp}}(X)$ *is isomorphic to* $\rho_{q+1,1}\oplus\rho_{q+1,1}$.

Proof. Under the above identification, the $+1$ (resp. -1)-eigen-space of T_2 corresponds to the space of symmetric (resp. skew-symmetric) bilinear forms $S^2(W^*)$ (resp. $\Lambda^2(W^*)$), which is stable under $\mathfrak{S}_{q+2}$. One knows that $S^2(W^*)$ contains the two $\mathfrak{S}_{q+2}$-invariant subspaces:

$$C(\textstyle\sum_i x_i y_i) \simeq \rho_{q+2}, \quad \text{and} \quad \{\sum_i c_i x_i y_i;\ \sum_i c_i = 0\} \simeq \rho_{q+1,1}.$$

Since $S^2(W^*)$ and $\Lambda^2(W^*)$ are not stable under T_1, they must have irreducible $\mathfrak{S}_{q+2}$-subspaces which are mutually isomorphic. Now a simple argument of counting dimensions proves the assertions. Q.E.D.

From Proposition (5.7) and Schur's lemma, it follows that the representation of $\boldsymbol{C}[T_1, T_2]$ induced on M_0 is the $q+1$ copies of a 2-dimensional irreducible one. The characteristic polynomial of T_1T_2 can be computed from the fact that $\mathrm{tr}(\rho^*(T_1T_2) \mid M^1(X^{(2)})) = 0$.

(9.20) Complete multipartite graph $K(s+1, \cdots, s+1)$.

This is a generalization of the complete graph. The vertices are divided into $l+1$ subsets $V_0, V_1, \cdots, V_l$ such that $\sharp(V_i) = s+1$ for $0 \leq i \leq l$. Two vertices $P \in V_i$, $Q \in V_j$ are adjacent if and only if $i \neq j$. If $s=0$, one obtains $K(1, \cdots, 1) = K(l+1)$. Observe that $X = K(s+1, \cdots, s+1)$ is a regular graph of valency $l(s+1)$. To get the spectrum of X, one can apply Proposition (8.34); in fact X can be obtained by the composition $K(l+1)[X_0, \cdots, X_0]$, where X_0 is the (disconnected) regular graph of valency 0 (i.e., $EX_0 = \phi$) with $s+1$ vertices. One obviously has $\mathrm{Spec}(X_0) = \{0, \cdots, 0\ (s+1 \text{ times})\}$, and applying (8.35), one obtains the following results:

(9.21)
$$\mathrm{Spec}(K(s+1, \cdots, s+1)) = \{q+1, 0, \cdots, 0\ (s(l+1) \text{ times}), -s-1, \cdots, -s-1\ (l \text{ times})\},$$

(9.22)
$$Z_{K(s+1,\cdots,s+1)}(u)^{-1} = (1-u)^r(1+u)^{r-1}(1-qu)(1+qu^2)^{s(l+1)}(1+(s+1)u+qu^2)^l,$$

where we put $r = (l+1)(s+1)(q-1)/2+1$, $q = l(s+1)-1$. One can also prove them by a similar computation as in $K(q+2)$.

(9.23) Cube $Q(n)$.

$Q(n)$ is a regular graph of valency n and with 2^n vertices, which represents the cube in $\boldsymbol{R}^n$. Namely $VQ(n)$ is given by

$$VQ(n) = \{P = (p_1, \cdots, p_n) \in \boldsymbol{R}^n;\ p_i = 0, \text{ or } 1\ (1 \leq i \leq n)\},$$

and $P, Q \in VQ(n)$ are adjacent if and only if $|P-Q|=1$. Moreover, it is bipartite:

$$VQ(n)=V_1 \cup V_2, \quad V_i=\{P=(p_1, \cdots, p_n) \in VQ(n); \textstyle\sum_i p_i \equiv i \ (\mathrm{mod}\ 2)\}.$$

To obtain its spectrum, one notes that $Q(1)=K(2)$, and $Q(n)$ is given inductively by the Cartesian product: $Q(n)=Q(n-1)\times K(2)$. Then it is easy to prove by induction and Proposition (8.16), that

$$\text{(9.24)} \qquad \mathrm{Spec}(Q(n))=\left\{n-2i\left(\binom{n}{i}\ \text{times}\right);\ 0\leq i\leq n\right\}.$$

Hence, by (3.7), we have

$$\text{(9.25)} \qquad Z_{Q(n)}(u)^{-1}=(1-u^2)^{(n-1)2^{n-1}}\prod_{i=0}^{n}[1-(n-2i)u+qu^2]^{\binom{n}{i}},$$

with $q=n-1$. Noting that $Q(n)$ is regular bipartite, one can observe that this supplies (infinitely) many examples where the representation ρ^* on $M^1(Q(n))$ has an irreducible component φ of degree 2, of which the characteristic polynomial $p_\varphi(u)=\det(I_2-\varphi(T_1T_2)u)$ decomposes into a product of linear factors over $\boldsymbol{Z}$. In fact, for any positive integer i, put $n=4i+3$, $q=4i+2$. Then from (3.15), one sees that $Z_{Q(n),b}(u)^{-1}$ has the following factor corresponding to $\lambda=n-2i=2i+3 \in \mathrm{Spec}(Q(n))$:

$$\begin{aligned}1-(\lambda^2-2q)u+q^2u^2&=1-(4i^2+4i+5)u+4(2i+1)^2u^2\\&=(1-4u)(1-(2i+1)^2u).\end{aligned}$$

On the other hand, if one assumes that q is a prime number then all $p_\varphi(u)$'s corresponding to the 2-dimensional irreducible factors are irreducible over $\boldsymbol{Q}$.

§ 10. Examples

In the following, we give a list of *all* graphs X, with their characteristic functions $\phi_X(z)$, zeta functions $Z_X(u)$, which have no end point and which satisfy $n=\sharp(VX)\leq 6$, $m=\sharp(EX)\leq 8$. Up to isomorphism, there are 30 such graphs. We added two graphs to complete the list for $n=5$ (cf. [Har]). $\phi_X(z)$ and $Z_X(u)^{-1}$ are decomposed into the products of *irreducible* factors over $\boldsymbol{Q}$. The symbol $X_k(n, m)$ indicates that X is the k-th graph among those satisfying $(\sharp(VX), \sharp(EX))=(n, m)$.

(1) $X_1(3, 3)=\mathrm{Cir}_3: r=1$ (2) $X_1(4, 4)=\mathrm{Cir}_4: r=1$

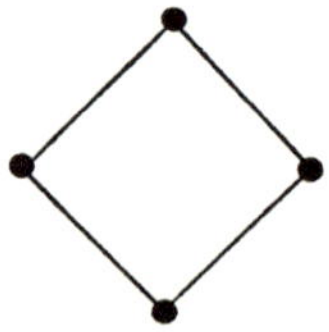

$$\phi_1(z)=(z+1)^2(z-2)$$
$$Z_1(u)^{-1}=(1-u)^2(1+u+u^2)^2$$

$$\phi_2(z)=z^2(z-2)(z+2)$$
$$Z_2(u)^{-1}=(1-u)^2(1+u)^2(1+u^2)^2$$

(3) $X_1(4, 5): r=2$ (4) $X_1(4, 6)=K(4): r=3$

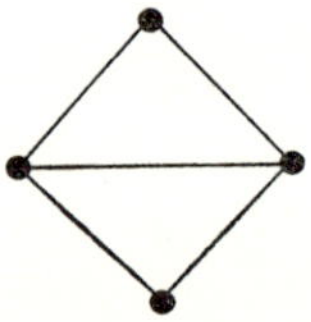

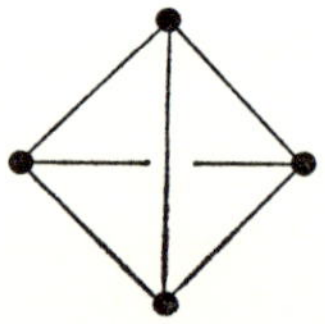

$$\phi_3(z)=z(z+1)(z^2-z-4)$$
$$Z_3(u)^{-1}=(1-u)^2(1+u)(1+u^2)(1+u+2u^2)(1-u^2-2u^3)$$

$$\phi_4(z)=(z+1)^3(z-3)$$
$$Z_4(u)^{-1}=(1-u)^3(1+u)^2(1-2u)(1+u+2u^2)^3$$

(5) $X_1(5, 5)=\mathrm{Cir}_5: r=1$ (6) $X_1(5, 6): r=2$

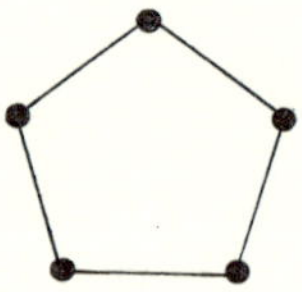

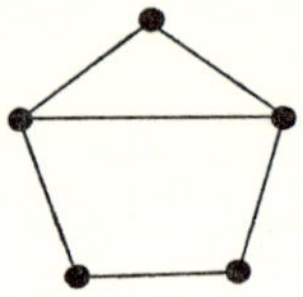

$$\phi_5(z)=(z-2)(z^2+z-1)^2$$
$$Z_5(u)^{-1}=(1-u)^2(1+u+u^2+u^3+u^4)^2$$

$$\phi_6(z)=z(z+2)(z^3-2z^2-2z+2)$$
$$Z_6(u)^{-1}=(1-u)^2(1+u)(1+2u+3u^2+3u^3+2u^4)$$
$$\times(1-u+u^2-2u^3+u^4-2u^5)$$

(7) $X_2(5, 6)$: $r=2$

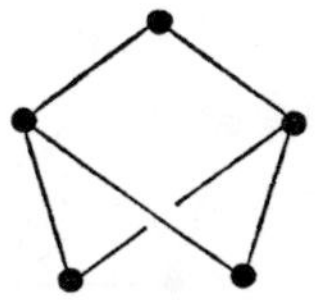

(8) $X_3(5, 6)$: $r=2$

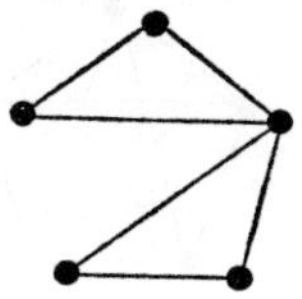

$\phi_7(z)=z^3(z^2-6)$
$Z_7(u)^{-1}=(1-u)^2(1+u)^2(1+u^2)^2(1-2u^2)(1+2u^2)$

$\phi_8(z)=(z-1)(z+1)^2(z^2-z-4)$
$Z_8(u)^{-1}=(1-u)^2(1+u)(1+u+u^2)^2(1-u+u^2)(1-3u^3)$

(9) $X_1(5, 7)$: $r=3$

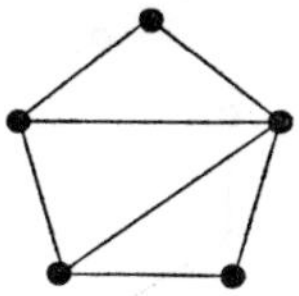

(10) $X_2(5, 7)$: $r=3$

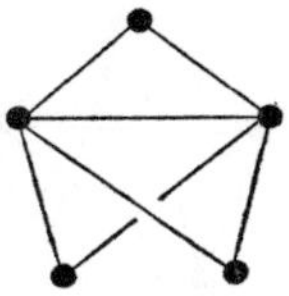

$\phi_9(z)=(z^2+z-1)(z^3-z^2-5z-2)$
$Z_9(u)^{-1}=(1-u)^3(1+u)^2(1+u+2u^2)(1-u-3u^3)$
$\times(1+u+2u^2+u^3+2u^4)$

$\phi_{10}(z)=z^2(z+1)(z+2)(z-3)$
$Z_{10}(u)^{-1}=(1-u)^3(1+u)^2(1+u^2)^2(1+u+3u^2)(1-2u^2-3u^3)$

(11) $X_3(5, 7)$: $r=3$

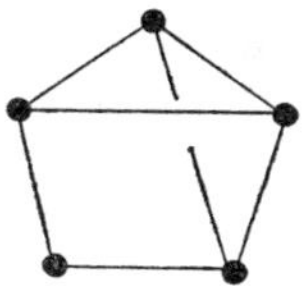

(12) $X_1(5, 8)$: $r=4$

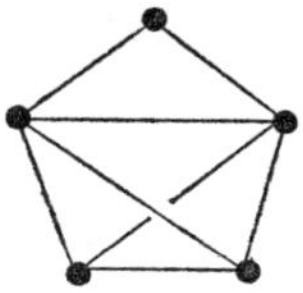

$\phi_{11}(z)=z(z+1)(z^3-z^2-6z+2)$
$Z_{11}(u)^{-1}=(1-u)^3(1+u)^2(1+u+2u^2)(1+2u^2)(1-u^2-2u^3-2u^4-4u^5)$

$\phi_{12}(z)=(z+1)^2(z^3-2z^2-5z+2)$
$Z_{12}(u)^{-1}=(1-u)^4(1+u)^3(1+u+2u^2)(1+u+3u^2)$
$\times(1-u-5u^3-u^4-6u^5)$

(13) $X_2(5, 8)$: $r=4$

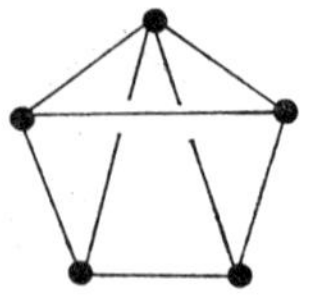

(14) $X_1(5, 9)$: $r=5$

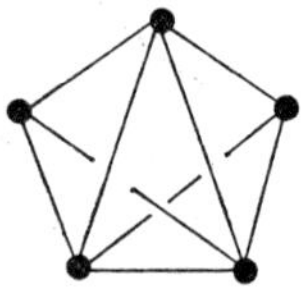

$$\phi_{13}(z)=z^2(z+2)(z^2-2z-4)$$
$$Z_{13}(u)^{-1}=(1-u)^4(1+u)^3(1+2u^2)^2(1+2u+2u^2)(1-u-6u^3)$$

$$\phi_{14}(z)=z(z+1)^2(z^2-2z-6)$$
$$Z_{14}(u)^{-1}=(1-u)^5(1+u)^4(1+2u^2)(1+u+3u^2)^2(1-u-2u^2-6u^3)$$

(15) $X_1(5, 10)=K(5)$: $r=6$

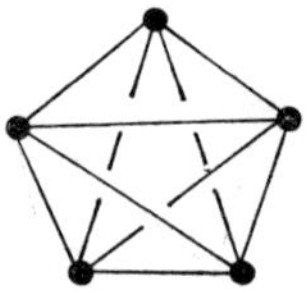

(16) $X_1(6, 6)=\mathrm{Cir}_6$: $r=1$

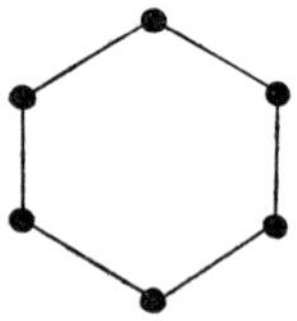

$$\phi_{15}(z)=(z+1)^4(z-4)$$
$$Z_{15}(u)^{-1}=(1-u)^6(1+u)^5(1-3u)(1+u+3u^2)^4$$

$$\phi_{16}(z)=(z-1)^2(z+1)^2(z-2)(z+2)$$
$$Z_{16}(u)^{-1}=(1-u)^2(1+u)^2(1+u+u^2)^2(1-u+u^2)^2$$

(17) $X_1(6, 7)$: $r=2$

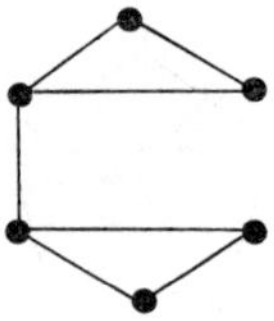

(18) $X_2(6, 7)$: $r=2$

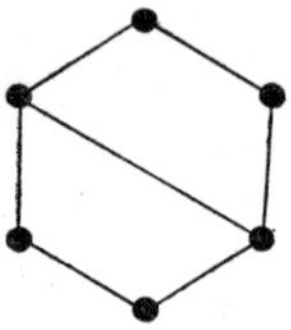

$$\phi_{17}(z)=(z+1)^2(z^2-3)(z^2-2z-1)$$
$$Z_{17}(u)^{-1}=(1-u)^2(1+u)(1+u+u^2)^2(1-u+u^2-2u^3)(1-u^3+2u^4)$$

$$\phi_{18}(z)=(z-1)(z+1)(z^2+2z-1)(z^2-2z-1)$$
$$Z_{18}(u)^{-1}=(1-u)^2(1+u)^2(1+u+u^2)(1-u+u^2)(1+u+u^2+2u^3)$$
$$\times(1-u+u^2-2u^3)$$

(19) $X_3(6, 7)$: $r=2$

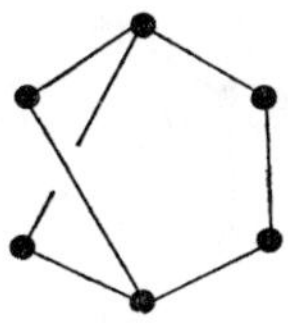

(20) $X_4(6, 7)$: $r=2$

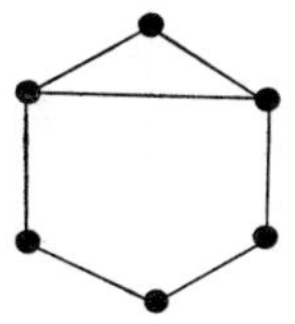

$$\phi_{19}(z)=z(z^2+z-1)(z^3-z^2-5z+4)$$
$$Z_{19}(u)^{-1}=(1-u)^2(1+u)(1+u^2)(1+u+2u^2+2u^3+2u^4)(1-u^2-2u^5)$$

$$\phi_{20}(z)=(z^2+z-1)(z^4-z^3-5z^2+2z+4)$$
$$Z_{20}(u)^{-1}=(1-u)^2(1+u)(1+u+2u^2+u^3+2u^4)(1-u^3-u^5-u^6-2u^7)$$

(21) $X_5(6, 7)$: $r=2$

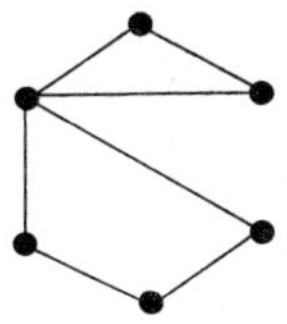

(22) $X_1(6, 8)$: $r=3$

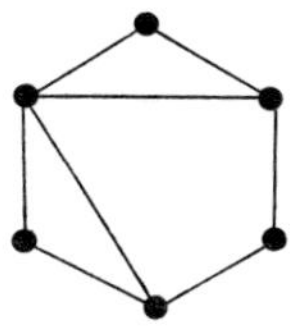

$$\phi_{21}(z)=z(z+1)(z^4-z^3-6z^2+4z+4)$$
$$Z_{21}(u)^{-1}=(1-u)^2(1+u)(1+u^2)(1+u+u^2)(1-u^3-u^4-3u^7)$$

$$\phi_{22}(z)=(z-1)(z+1)(z+2)(z^3-2z^2-3z+2)$$
$$Z_{22}(u)^{-1}=(1-u)^3(1+u)^2(1+2u^2+2u^4)$$
$$\times(1+u+u^2-3u^3-4u^4-8u^5-6u^6-6u^7)$$

(23) $X_2(6, 8)$: $r=3$

(24) $X_3(6, 8)$: $r=3$

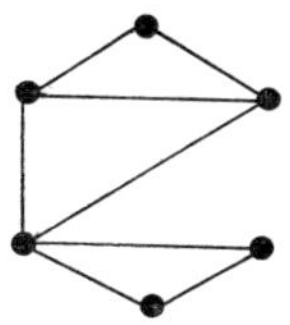

$$\phi_{23}(z)=z^2(z^4-8z^2-2z+7)$$
$$Z_{23}(u)^{-1}=(1-u)^3(1+u)^2(1+u^2)$$
$$\times(1+u+2u^2-4u^4-10u^5-13u^6-17u^7-12u^8-12u^9)$$

$$\phi_{24}(z)=(z+1)^2(z^4-2z^3-5z^2+6z+4)$$
$$Z_{24}(u)^{-1}=(1-u)^3(1+u)^2(1+u+u^2)(1+u+2u^2)$$
$$\times(1-u+u^2-4u^3+3u^4-5u^5+3u^6-6u^7)$$

(25) $X_4(6, 8)=K(2, 4): r=3$ (26) $X_5(6, 8): r=3$

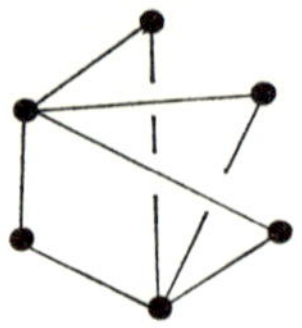

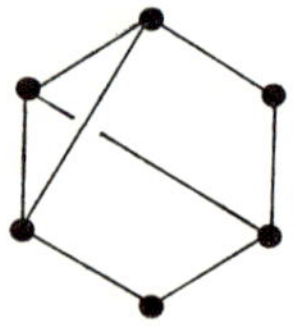

$$\phi_{25}(z)=z^4(z^2-8)$$
$$Z_{25}(u)^{-1}=(1-u)^3(1+u)^3(1-3u^2)(1+3u^2)(1+u^2)^3$$
$$\phi_{26}(z)=(z^2+z-1)(z^4-z^3-6z^2+3z+1)$$
$$Z_{26}(u)^{-1}=(1-u)^3(1+u)^2(1+u+2u^2+u^3+2u^4)\times(1+u^2-u^3-3u^4-6u^5-4u^6-8u^7)$$

(27) $X_6(6, 8): r=3$ (28) $X_7(6, 8): r=3$

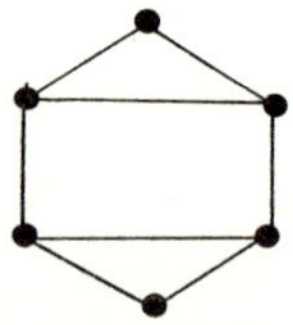

$$\phi_{27}(z)=(z-1)^2(z^4-2z^3-5z^2+8z+1)$$
$$Z_{27}(u)^{-1}=(1-u)^3(1+u)^2(1+u+2u^2)(1+u+2u^2+2u^3+2u^4)\times(1-u-2u^3+2u^4-4u^5)$$
$$\phi_{28}(z)=z(z+2)(z^2-2)(z^2-2z-2)$$
$$Z_{28}(u)^{-1}=(1-u)^3(1+u)^2(1+2u^2)(1+2u+2u^2)(1-u-2u^3)(1+u^2+2u^4)$$

(29) $X_8(6, 8): r=3$ (30) $X_9(6, 8): r=3$

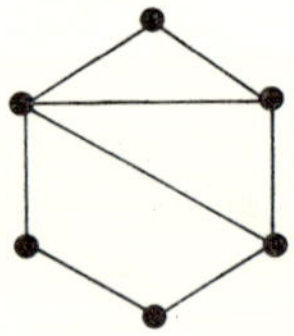

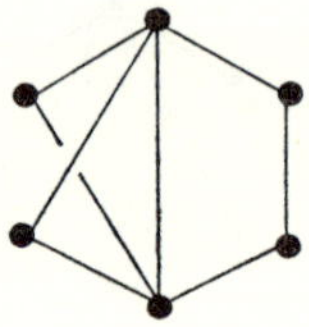

$$\phi_{29}(z)=z^6-8z^4-4z^3+9z^2+4z-1$$
$$Z_{29}(u)^{-1}=(1-u)^3(1+u)^2(1+u+2u^2)\times(1+u^2-2u^3-2u^4-6u^5-4u^6-8u^7-3u^8-6u^9)$$
$$\phi_{30}(z)=z^2(z+2)(z^3-2z^2-4z+4)$$
$$Z_{30}(u)^{-1}=(1-u)^3(1+u)^2(1+u^2)(1+2u+4u^2+4u^3+3u^4)\times(1-u-2u^3+u^4-3u^5)$$

(31) $X_{10}(6, 8): r=3$

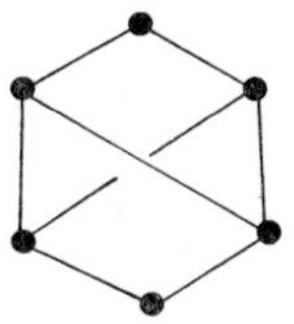

(32) $X_{11}(6, 8): r=3$

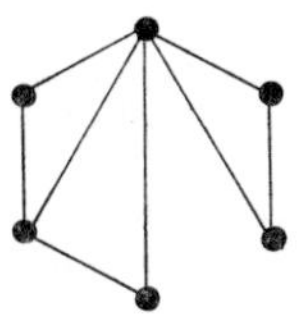

$$\phi_{31}(z)=z^2(z^2+2z-2)(z^2-2z-2)$$
$$Z_{31}(u)^{-1}=(1-u)^3(1+u)^3(1+2u^2)^2(1+u+2u^3)(1-u-2u^3)$$

$$\phi_{32}(z)=z(z+1)(z^4-z^3-7z^2+z+8)$$
$$Z_{32}(u)^{-1}=(1-u)^3(1+u)^2(1+u^2)(1+u+u^2)(1+u^2-5u^3-5u^5-8u^7)$$

References

[Bi] N. Biggs, Algebraic graph theory, Cambridge Tracts in Math., **67** (1974).

[Bo-1] A. Borel, Cohomologie de certains groupes discrets et laplacien p-adique, Sém. Bourbaki 26ème année (1973/74), exp. **437**, 12–35.

[Bo-2] ——, Admissible Representations of a Semi-Simple Group over a Local Field with Vectors Fixed under Iwahori Subgroup, Invent. Math., **35** (1976), 233–259.

[Bou] N. Bourbaki, Groupes et Algèbres de Lie, Chap. IV, Hermann, Paris, 1968.

[B-T] F. Bruhat and J. Tits, Groupes algébriques simples sur un corps local, Proc. Conf. Local Fields, 23–36, Springer-Verlag, 1967.

[Car-1] P. Cartier, Fonctions harmoniques sur un arbre, Sympos. Math., **9** (1972), 203–270.

[Car-2] ——, Representation of p-adic groups: a survey, Proc. Symp. Pure Math., **33** (1979), part I, 111–155.

[Cas] W. Casselman, On a p-adic vanishing theorem of Garland, Bull. Amer. Math. Soc., **80-5** (1974), 1001–1004.

[D-M] K. Doi and T. Miyake, Automorphic forms and the theory of numbers (in Japanese), Kinokuniya, Tokyo (1976).

[Ei] M. Eichler, Quaternäre quadratische Formen und die Riemannsche Vermutung fur die kongruenz Zeta Funktionen, Arch. Math., Vol. **V** (1954), 355–366.

[Gar] H. Garland, p-adic curvature and the cohomology of discrete subgroups of p-adic groups, Ann. of Math., **97** (1973), 375–423.

[Har] F. Harary, Graph Theory, Addison-Wesley, Reading (1969).

[Ha] K. Hashimoto, On Brandt matrices associated with the positive definite quaternion hermitian forms, J. Fac. Sci. Univ. Tokyo, Sect IA Math., **27** (1980), 227–245.

[H-H] K. Hashimoto and A. Hori, Selberg-Ihara's zeta functions for p-adic discrete groups, this volume.

[I-1] Y. Ihara, On discrete subgroups of the two by two projective linear group over p-adic fields, J. Math. Soc. Japan, **18** (1966), 219–235.

[I-2] ——, Discrete subgroups of $PL(2, k_p)$. Proc. Symp. Pure Math. IX, AMS (1966), 272–278.

[I-M] N. Iwahori and H. Matsumoto, On some Bruhat decomposition and the structure of the Hecke ring of p-adic Chevalley groups, Publ. Math.

I.H.E.S., **25** (1965), 5–48.

[L-P-S] A. Lubotzky, R. Phillips and P. Sarnak, Ramanujan Graphs, preprint.

[Mac] I. G. Macdonald, Special functions on a group of p-adic type, Publ. of the Ramanujan Inst., **2**, 1971.

[Ma] H. Hashimoto, Fonctions sphériques sur un groupe semi-simple p-adique, C.R. Acad. Sc. Paris, t. **269** (1969), 829–832.

[Mu] F. D. Murnaghan, The analysis of the Kronecker product of irreducible representations of the symmetric group, Amer. J. Math., **60** (1938), 761–784.

[Sa-1] I. Satake, Theory of spherical functions on reductive algebraic groups over p-adic fields, Publ. Math. I.H.E.S., **18** (1963), 1–69.

[Sa-2] ——, Spherical functions and Ramanujan Conjecture, Proc. Symp. Pure Math. IX. AMS (1966), 258–264.

[Sch] F. K. Schmidt, Die Thorie der Klassenkörper über einem Körper algebraischer Funktionen in einer Unbestimmten und mit endlichem Konstantenbereich, Sitzungsber. phys. med. Sozietät Erlangen, **62** (1930), 267–284.

[Scw] A. J. Schwenk, Computing the characteristic polynomial of a graph, Groups and Combinatorics, Lecture Notes in Math., Springer **406** (1974), 153–172.

[Sel] A. Selberg, Harmonic analysis and discontinuous groups in weakly symmetric Riemannian spaces with applications to Dirichlet series, J. Ind. Math. Soc., **20** (1956), 47–87.

[Ser] J-P. Serre, Arbres, Amalgames, SL_2, Astérisque, no. **46**, Soc. Math. France, 1977.

[Sh] G. Shimura, Introduction to the arithmetic theory of automorphic functions, Publ. Math. Soc. Japan, **11**, 1971.

[Su-1] T. Sunada, L-functions in geometry and some applications, Lecture Notes in Math., Springer, **1201**, 266–284.

[Su-2] ——, Twisted Peron-Frobenius Theorem and L-functions, J. Funct. Anal., **71** (1987), 1–46.

[T-1] J. Tits, Sur le groupe des automorphismes d'un arbre. Essays on Topology and related topics, Mémoires dédiés à Georges de Rham. Springer, 1970, 188–211.

[T-2] ——, Reductive groups over local fields, Proc. Symp. Pure Math., **33-1** (1979), 29–69.

Department of Mathematics
School of Science and Engineering
Waseda University
Tokyo, 160 Japan

Advanced Studies in Pure Mathematics 15, 1989
Automorphic Forms and Geometry of Arithmetic Varieties
pp. 281–286

Any Irreducible Smooth GL_2-Module is Multiplicity Free for any Anisotropic Torus

Hiroaki Hijikata

Dedicated to Prof. Ichiro Satake on his sixtieth birthday

§ 1.

Let k be a non-archimedean local field, B be a quaternion algebra, i.e. a central simple algebra of rank 4 over k. Let L be a separable quadratic subfield of B. The group $G=B^\times$, of the regular elements of B, is a T.D.L.C. (= totally disconnected locally compact) group by the induced topology from B, and $H=L^\times$ is a closed subgroup of G. In other words, G is a k-form of GL_2, and H is a maximal torus anisotropic modulo center. Let (π, E) be a smooth representation of G on the complex vector space E. The purpose of this paper is to prove the following:

Theorem A. *If (π, E) is irreducible as G-module, then it is multiplicity free as H-module. Namely, there is a subset $\hat{H}(\pi)$ of the set $\hat{H}$ of all quasicharacters of H such that*

$$\pi = \bigoplus_{\chi \in \hat{H}(\pi)} \chi \qquad \textit{as H-module.}$$

§ 2.

The irreducible smooth representations of $G=B^\times$ are classified into several series (cf. [J-L], [K] for split G, and [G-G], [Ho] for non-split G). To identify the set $\hat{H}(\pi)$ for all L amounts to get a complete knowledge for the representation π, at least character-theoretically. In this respect, there are no difficulties if k has odd residual characteristic. While, in dyadic case, I have determined $\hat{H}(\pi)$ (for all L) for some series of π's, but not yet for all series.

When G is non-split, i.e. B is a division algebra, there is a close connection between Theorem A and the Basis Problem of modular forms as

Received March 27, 1987.

indicated in Part II Chap. 9 of [H-P-S]. This connection is the motivation of this work.

When G is split, i.e. $B=M_2(k)$, and $G=GL_2(k)$, let K be a maximal compact modulo center subgroup of G. There are two such K's up to conjugacy. The standard one, the normalizer of a maximal compact subgroup of G, contains unramified $L^\times$, while the other one, the normalizer of an Iwahori subgroup of G, contains any ramified $L^\times$. Hence we have the following:

Corollary. *Any irreducible smooth representation* π *of* $GL_2(k)$ *is multiplicity free as* k*-module. In particular,* π *is admissible.* (The last statement is well known, and it is valid for any reductive group G as shown in [B]).

§3.

As for the proof, Theorem A is a formal consequence of the following simple

Proposition B. *For each* L*, there is a topological antiautomorphism* τ *of the algebra* B *satisfying:*

(i) τ *is of order* 2,

(ii) $\tau(a)=a$ *for any* $a \in L$,

(iii) *each coset* Hg *contains a* τ*-fixed element.*

Proof. Let $a \mapsto \bar{a}$ denote the Galois action of L over k. By Skolem-Noether theorem, there exists $y \in B^\times$ such that

$$yay^{-1}=\bar{a} \qquad \text{for any } a \in L.$$

Then it follows that $B=L \oplus yL$, $y^2 \in k^\times$ and

$$i\colon a+yb \longmapsto \bar{a}-yb$$

is the canonical involution of B.

By Hilbert theorem 90, there exists $c \in L^\times$ such that

$$\bar{c}c^{-1}=-1.$$

Define τ as the composite $i \circ I(cy)$ of the canonical involution i and the inner automorphism $I(cy)\colon x \mapsto (cy)x(cy)^{-1}$, i.e.

$$\tau\colon a+yb \longmapsto a+y\bar{b}.$$

Clearly, τ is a topological antiautomorphism of order 2, fixing each element of L. Since $G=B^{\times}=L^{\times}((L+y)\cup\{1\})$, τ also satisfies the last condition (iii).

§ 4.

The formal argument to derive Theorem A from Proposition B can be summarized as Proposition C below after introducing some notation.

We consider a triple (G, Z, ω) consisting of a T.D.L.C. group G, its closed normal subgroup Z, and a locally constant homomorphism $\omega: Z\to C^{\times}$, normalized by G, $\omega(gzg^{-1})=\omega(z)$ for any $z\in Z$, $g\in G$. Let $S(G,\omega)$ denote the vector space of all locally constant complex valued functions f on G, of which supports are compact mod Z, and which are ω-semiinvariant, $f(zg)=\omega(z)f(g)$ for any $z\in Z$. $S(G,\omega)$ is an associative algebra over $\boldsymbol{C}$ by the convolution product,

$$f_1*f_2(g_0)=\int f_1(g)f_2(g^{-1}g_0)d\bar{g},$$

where $d\bar{g}$ is a left invariant Haar measure of $\bar{G}=G/Z$.

Let H be a closed subgroup of G containing Z and having a compact quotient H/Z. Let $\varepsilon: H\to C^{\times}$ be a locally constant homomorphism which coincides with ω on Z. Let $S(G,H,\varepsilon)$ denote the subalgebra of $S(G,\omega)$ consisting of all ε-bi-semiinvariant functions f, $f(hg)=f(gh)=\varepsilon(h)f(g)$ for any $h\in H$. Let (π, E) be a smooth representation of G, on which Z acts as ω^{-1}, $\pi(z)v=\omega(z)^{-1}v$ for $z\in Z$, $v\in V$. Finally let $E(H,\varepsilon^{-1})$ denote the ε^{-1}-eigen subspace under H,

$$E(H,\varepsilon^{-1})=\{v\in E\mid \pi(h)v=\varepsilon(h)^{-1}v \text{ for } h\in H\}.$$

Proposition C. *There are the implications*: (I)$\Rightarrow$(II)$\Rightarrow$(III).

(I) *G has a topological antiautomorphism τ satisfying*:

(1) $\tau(Z)=Z$, $\tau(H)=H$, $\varepsilon\circ\tau=\varepsilon$,

(2) *the automorphism* τ': $g\mapsto\tau(g)^{-1}$ *is of finite order*,

(3) *each double coset HgH contains a τ-fixed element.*

(II) *The algebra $S(G,H,\varepsilon)$ is commutative.*

(III) *If (π, E) is irreducible, then* $\dim E(H,\varepsilon^{-1})\leq 1$.

§ 5.

In the rest of this paper, we retain all the notation of Section 4. The first implication '(I)$\Rightarrow$(II)' is rather obvious. The first assumption (1) implies that the map $f\mapsto\tau f:=f\circ\tau^{-1}$ is a linear isomorphism of $S(G,\omega)$. It also implies that τ' induces an automorphism $\bar{\tau}'$ of $\bar{G}$, hence $d(\bar{\tau}'(\bar{g}))=c\,d\bar{g}$ by some positive constant c. Then the second assumption (2) implies that $c=1$, hence $\tau(f_1 * f_2)=\tau f_2 * \tau f_1$ for $f_1, f_2 \in S(G, \omega)$. The third assumption (3) implies $\tau f=f$ if $f \in S(G, H, \varepsilon)$, hence $f_1 * f_2=f_2 * f_1$ for $f_1, f_2 \in S(G, H, \varepsilon)$.

The next implication '(II)$\Rightarrow$(III)' is more or less known, at least if H is open in G (cf. [C], [B-Z]). In particular, if Z is a trivial subgroup $\{1\}$, hence ω is also trivial, and moreover if H is open and compact, '(II) $\Rightarrow$(III)' is a part of Proposition 2.10 of [B-Z]. Although there is no difficulty to modify their method (of embedding $S(G, \omega)$ into the algebra of distributions) to be capable of covering our case of non-trivial ω and not open H, the points to be checked might not be clear without giving the exact statement at each step. Here, we will give a shorter proof relying on a result of [C], under an extra condition,

(4) Z is a closed subgroup of the center of G.

Note that $(G, Z)=(B^\times, k^\times)$ of Section 1 certainly satisfies (4). Note also, as a general theory, the assumption (4) is not essentially restrictive, since we may work on the quotient by the kernel of ω, of G, Z and everything.

§ 6.

Recall that G is a T.D.L.C. group iff it has a fundamental system of neighbourhoods $\mathcal{U}$ of 1, consisting of open compact subgroups U. Since ε is locally constant, it is trivial on $H \cap U$ for some $U \in \mathcal{U}$. By (4), ZU is an open subgroup normalizing U, and $[H: H \cap ZU]$ is finite, hence the intersection $\bigcap hUh^{-1}$ for $h \in H/(ZU \cap H)$ is an open compact subgroup normalized by H. Thus we may and shall assume that $\mathcal{U}$ consists of open compact subgroups U satisfying

(5) $hUh^{-1}=U$ for $h \in H$, and $U \cap H \subset \ker \omega$.

Hence there is a unique homomorphism $u\colon HU \to \boldsymbol{C}^\times$ satisfying

(6) $u=\varepsilon$ on H, $u=1$ on U.

Let $\mu(HU)$ denote the volume of HU/Z by the Haar measure $d\bar{g}$ of $\bar{G}$ and let $\dot{u}$ denote the function on G which coincides with $\mu(HU)^{-1}u$ on HU, and zero outside. Since HU is open and compact mod Z, $\dot{u}$ is a member of $S(G, \omega)$, and by the definition of convolution, we have:

$$\dot{u} * f=f \text{ iff } f(xg)=u(x)f(g) \quad \text{for any } x \in HU,$$
$$f * \dot{u}=f \text{ iff } f(gx)=u(x)f(g) \quad \text{for any } x \in HU.$$

and

$$S(G, HU, u)=\dot{u} * S(G, \omega) * \dot{u}. \tag{7}$$

Since $S(G, H, \varepsilon)$ is the union of $S(G, HU, u)$, it is commutative iff each $S(G, HU, u)$ is commutative.

By definition, a representation (π, E) is smooth iff E is the union of the U-fixed subspace $E(U, 1)$. Since $E(H, \varepsilon^{-1}) \cap E(U, 1) = E(HU, \varepsilon^{-1})$, $E(H, \varepsilon^{-1})$ is the union of $E(HU, u^{-1})$, and $E(HU, u^{-1}) \subset E(HU', (u')^{-1})$ if $U \supset U'$. Therefore if one knows that $\dim E(HU, u^{-1}) \leq d$ for any $U \in \mathscr{U}$, and $\dim E(hU_0, u_0^{-1}) = d$ for some $U_0 \in \mathscr{U}$, then one can conclude that $E(H, \varepsilon^{-1}) = E(HU_0, u_0^{-1})$.

Since Z acts on E as ω^{-1}, $S(G, \omega)$ acts on E by

$$\pi(f)v = \int f(g)\pi(g)v\, d\bar{g}. \tag{8}$$

In particular, $\pi(\dot{u})$ is the projection operator of E to $E(HU, u^{-1})$, and by (7), $S(G, HU, u)$ acts on $E(HU, u^{-1})$. Also observe

$$\pi(g_0) \circ \pi(f) = \pi(L(g_0)f), \tag{9}$$

where $L(g_0)f = (g \mapsto f(g_0^{-1}g)) \in S(G, \omega)$.

Now '(II) $\Rightarrow$ (III)' is a consequence of the following:

(10) If E is G-irreducible and $E(HU, u^{-1}) \neq 0$, then $E(HU, u^{-1})$ is $S(G, HU, u)$-irreducible. (Hence if $S(G, HU, u)$ is commutative, $\dim E(HU, u^{-1}) = 1$.)

The claim (10) is in [C]. We reproduce its proof. Let v_0 be a non-zero vector in $E(HU, u^{-1})$ and v be an arbitrary vector in $E(HU, u^{-1})$. Since E is G-irreducible, we can find $g_i \in G$, $c_i \in \boldsymbol{C}$ $(i = 1, \cdots, n)$ such that $v = \sum c_i \pi(g_i)v_0$. Since $v_0 = \pi(\dot{u})v_0$, by (9), $\pi(g_i)v_0 = \pi(g_i)\pi(\dot{u})v_0 = \pi(L(g_i)\dot{u})v_0 = \pi(L(g_i)\dot{u})\pi(\dot{u})v_0$. Since $v = \pi(\dot{u})v$, we have $v = \pi(f)v_0$ with

$$f = \sum c_i \dot{u} * L(g_i)\dot{u} * \dot{u}$$

which lies in $S(G, HU, u)$ by (7).

References

[B] I. N. Bernshtein, All reductive p-adic groups are of type I, Funct. Anal. Appl., **8** (1974), 91–93.

[B-Z] I. N. Bernshtein - A. V. Zelevinskii, Representations of the group $GL(n, F)$ where F is a non-archimedean local field, Russian Math. Surveys, **31:3** (1976) 1–68.

[C] W. Casselman, Introduction to the theory of admissible representation of p-adic reductive groups, preprint.

[G-G] I. M. Gelfand and I. M. Graev, Representations of quaternion groups over locally compact and function fields, Funct. Anal. Appl., **2** (1969), 19–33.

[H-P-S] H. Hijikata, A. Pizer and T. Shemanske, The Basis Problem for modular

forms on $\Gamma_0(N)$, preprint.
[Ho] R. Howe, Kirillov theory for compact p-adic groups, Pacific J. Math., **73**, (1977), 365–381.
[J-L] H. Jacquet and R. Langlands, Automorphic forms on $GL(2)$, Lecture Notes in Math., **114** (1970).
[K] P. Kutzko, On the super cuspidal representation of $GL(2)$, I, II, Amer. J. Math., **100** (1978), 43–60 and 705–716.

Department of Mathematics
Kyoto University
Kyoto 606
Japan

Advanced Studies in Pure Mathematics 15, 1989
Automorphic Forms and Geometry of Arithmetic Varieties
pp. 287–300

A Formula for the Dimension of Spaces of Cusp Forms of Weight 1

Toyokazu Hiramatsu

Dedicated to Prof. Ichiro Satake on his sixtieth birthday

Introduction

Let Γ be a fuchsian group of the first kind and denote by d_1 the space of cusp forms of weight 1 on the group Γ. It would be interesting to have a certain formula for d_1. But it is not effective to compute the dimension d_1 by means of the Riemann-Roch theorem. The purpose of this paper is to give some formula of d_1 by making use of the Selberg trace formula ([4], [6], [7]).

§ 1. The Selberg eigenspace

Let S denote the complex upper half-plane and we put $G=SL(2, \boldsymbol{R})$. Consider direct products

$$\tilde{S}=S\times T,\ \tilde{G}=G\times T,$$

where T denotes the real torus. The operation of an element (g, α) of $\tilde{G}$ on $\tilde{S}$ is represented as follows:

$$\tilde{S} \ni (z, \phi) \longrightarrow (g, \alpha)(z, \phi)=\left(\frac{az+b}{cz+d},\ \phi+\arg(cz+d)-\alpha\right) \in \tilde{S},$$

where $g=\begin{pmatrix} a & b \\ c & d \end{pmatrix} \in G$. The space $\tilde{S}$ is a weakly symmetric Riemannian space with the $\tilde{G}$-invariant metric

$$ds^2=\frac{dx^2+dy^2}{y^2}+\left(d\phi-\frac{dx}{2y}\right)^2,$$

and with the isometry μ defined by $\mu(z, \phi)=(-\bar{z}, -\phi)$. The $\tilde{G}$-invariant measure $d(z, \phi)$ associated to the $\tilde{G}$-invariant metric is given by

Received November 15, 1986.

$$d(z, \phi)=d(x, y, \phi)=\frac{dx\wedge dy\wedge d\phi}{y^2}.$$

The ring of $\tilde{G}$-invariant differential operators on $\tilde{S}$ is generated by $\partial/\partial\phi$ and

$$\tilde{\Delta}=y^2\left(\frac{\partial^2}{\partial x^2}+\frac{\partial^2}{\partial y^2}\right)+\frac{5}{4}\frac{\partial^2}{\partial\phi^2}+y\frac{\partial}{\partial\phi}\frac{\partial}{\partial x},$$

which we call the Casimir operator of $\tilde{S}$. By the correspondence

$$G\ni g\longleftrightarrow(g, 0)\in\tilde{G},$$

we identify the group G with a subgroup $G\times\{0\}$ of $\tilde{G}$, and so the subgroup Γ of G with a subgroup $\Gamma\times\{0\}$ of $\tilde{G}$. For an element $(g, \alpha)\in\tilde{G}$, we define a mapping $T_{(g,\alpha)}$ of $L^2(\tilde{S})$ into itself by $(T_{(g,\alpha)}f)(z, \phi)=f((g, \alpha)(z, \phi))$. For an element $g\in G$, we put $T_{(g,0)}=T_g$. Then we have

$$(T_gf)(z, \phi)=f\left(\frac{az+b}{cz+d}, \phi+\arg(cz+d)\right),$$

where $g=\begin{pmatrix}a & b\\ c & d\end{pmatrix}$. Let Γ be a fuchsian group of the first kind not containing the element $\begin{pmatrix}-1 & 0\\ 0 & -1\end{pmatrix}(=-I)$. We denote by $\mathfrak{m}_\Gamma(k, \lambda)=\mathfrak{m}(k, \lambda)$ the set of all functions $f(z, \phi)$ satisfying the following conditions:

(i) $f(z, \phi)\in L^2(\Gamma\backslash\tilde{S})$,

(ii) $\tilde{\Delta}f(z, \phi)=\lambda f(z, \phi), \frac{\partial}{\partial\phi}f(z, \phi)=-\sqrt{-1}\,kf(z, \phi)$.

We call $\mathfrak{m}(k, \lambda)$ the Selberg eigenspace of Γ.

We denote by $S_1(\Gamma)$ the space of cusp forms of weight 1 for the above fuchsian group Γ and put

$$d_1=\dim S_1(\Gamma).$$

Then the following equality holds:

Lemma ([1], [3]). *The notation and the assumption being as above, we have*

$$\mathfrak{m}\left(1, -\frac{3}{2}\right)=\{e^{-\sqrt{-1}\phi}y^{1/2}F(z): F(z)\in S_1(\Gamma)\},$$

and hence

$$(1)\qquad d_1=\dim\mathfrak{m}\left(1, -\frac{3}{2}\right).$$

§ 2. The compact case

In this section we suppose that the group Γ has a compact fundamental domain in the upper half-plane S.

It is well known that every eigenspace $\mathfrak{m}(k, \lambda)$ defined in Section 1 is finite dimensional and orthogonal to each other, and also the eigenspaces span together the space $L^2(\Gamma\backslash\tilde{S})$. We put $\boldsymbol{\lambda}=(k, \lambda)$. For every invariant integral operator with a kernel function $k(z, \phi; z', \phi')$ on $\mathfrak{m}(k, \lambda)$, we have

$$\int_{\tilde{S}} k(z, \phi; z', \phi')f(z', \phi')d(z', \phi')=h(\boldsymbol{\lambda})f(z, \phi),$$

for $f\in\mathfrak{m}(k, \lambda)$. Note that $h(\boldsymbol{\lambda})$ does not depend on f as far as f is in $\mathfrak{m}(k, \lambda)$. We also know that there is a basis $\{f^{(n)}\}_{n=1}^{\infty}$ of the space $L^2(\Gamma\backslash\tilde{S})$ such that each $f^{(n)}$ satisfies the condition (ii) in Section 1. Then we put $\boldsymbol{\lambda}^{(n)}=(k, \lambda)$ for such spectra. We now obtain the following Selberg trace formula for $L^2(\Gamma\backslash\tilde{S})$:

$$(2) \qquad \sum_{n=1}^{\infty} h(\boldsymbol{\lambda}^{(n)})=\sum_{M\in\Gamma}\int_{\tilde{D}} k(z, \phi; M(z, \phi))d(z, \phi),$$

where $\tilde{D}$ denotes a compact fundamental domain of Γ in $\tilde{S}$ and $k(z, \phi; z', \phi')$ is a point-pair invariant kernel of (a)-(b) type in the sense of Selberg such that the series on the left-hand side of (2) is absolutely convergent ([6]). Denote by $\Gamma(M)$ the centralizer of M in Γ and put $\tilde{D}_M=\Gamma(M)\backslash\tilde{S}$. Then

$$(3) \qquad \sum_{M\in\Gamma}\int_{\tilde{D}} k(z, \phi; M(z, \phi))d(z, \phi)=\sum_{l}\int_{\tilde{D}_{M_l}} k(z, \phi; M_l(z, \phi))d(z, \phi),$$

where the sum over $\{M_l\}$ is taken over the distinct conjugacy classes of Γ.

We consider an invariant integral operator on the Selberg eigenspace $\mathfrak{m}(k, \lambda)$ defined by

$$\omega_\delta(z, \phi; z', \phi')=\left|\frac{(yy')^{1/2}}{(z-\bar{z}')/2\sqrt{-1}}\right|^{\delta}\frac{(yy')^{1/2}}{(z-\bar{z}')/2\sqrt{-1}}e^{-\sqrt{-1}(\phi-\phi')}, \ (\delta>1).$$

It is easy to see that our kernel ω_δ is a point-pair invariant kernel of (a)-(b) type under the condition $\delta>1$ and vanishes on $\mathfrak{m}(k, \lambda)$ for all $k\neq 1$. Since $\Gamma\backslash\tilde{G}$ is compact, the distribution of spectra (k, λ) is discrete and we put

$$\mu_1=-\frac{3}{2}, \mu_2, \mu_3, \cdots,$$

$$d_\beta = \dim \mathfrak{m}(1, \mu_\beta), \ (\beta = 1, 2, 3, \cdots).$$

Then the left-hand side of the trace formula (2) equals $\sum_{\beta=1}^{\infty} d_\beta \Lambda_\beta$, where Λ_β denotes the eigenvalue of ω_δ in $\mathfrak{m}(1, \mu_\beta)$. As for the eigenvalue Λ_β, using the special eigenfunction

$$f(z, \phi) = e^{-\sqrt{-1}\phi} y^{v_\beta}, \ \mu_\beta = v_\beta(v_\beta - 1) - \frac{5}{4},$$

for a spectrum $(1, \mu_\beta)$ in $L^2(\tilde{S})$, we obtain

$$\Lambda_\beta = 2^{2+\delta}\pi \frac{\Gamma(1/2)\Gamma((1+\delta)/2)}{\Gamma(\delta)/\Gamma(1+(\delta/2))} \Gamma\left(\frac{\delta-1}{2} + v_\beta\right)\Gamma\left(\frac{\delta+1}{2} - v_\beta\right).$$

If we put $v_\beta = 1/2 + \sqrt{-1}\, r_\beta$, then

$$\Lambda_\beta = 2^{2+\delta}\pi \frac{\Gamma(1/2)\Gamma((1+\delta)/2)}{\Gamma(\delta)\Gamma(1+(\delta/2))} \Gamma\left(\frac{\delta}{2} + \sqrt{-1}\, r_\beta\right)\Gamma\left(\frac{\delta}{2} - \sqrt{-1}\, r_\beta\right). \tag{4}$$

In general, it is known that the series $\sum_{\beta=1}^{\infty} d_\beta \Lambda_\beta$ is absolutely convergent for $\delta > 1$. By the Stirling formula, we see that the above series is also absolutely and uniformly convergent for all bounded δ except $\delta = \pm(2v_\beta - 1)$.

Now we shall calculate the components of trace appearing in the right-hand side of (3) ([2]).

1) Unit class: $M = \begin{pmatrix} 1 & 0 \\ 0 & 1 \end{pmatrix}$.

It is clear that $\omega_\delta(z, \phi; M(z, \phi)) = 1$, and hence

$$J(I) = \int_{\tilde{D}_M} d(z, \phi) = \int_{\tilde{D}} d(z, \phi) < \infty.$$

2) Hyperbolic conjugacy classes.

For the primitive hyperbolic element P, we put

$$g^{-1}Pg = \begin{pmatrix} \lambda_0 & 0 \\ 0 & \lambda_0^{-1} \end{pmatrix} \qquad (g \in G), \ |\lambda_0| > 1$$

and $\Gamma' = g^{-1}\Gamma g$. Then

$$\Gamma'\left(\begin{pmatrix} \lambda_0 & 0 \\ 0 & \lambda_0^{-1} \end{pmatrix}\right) = g^{-1}\Gamma(P)g.$$

The hyperbolic component is calculated as follows:

$$J(P^k)=\int_{\tilde{D}_P}\omega_\delta(z,\phi;P^k(z,\phi))d(z,\phi)$$
$$=\int_{g^{-1}\tilde{D}_P}\omega_\delta(g(z,\phi);P^kg(z,\phi))d(z,\phi)$$
$$=\int_{g^{-1}\tilde{D}_P}\omega_\delta(z,\phi;g^{-1}P^kg(z,\phi))d(z,\phi)$$
$$=(2\pi)(2^{\delta+1}\sqrt{-1})|\lambda_0^k|^{\delta+1}(\operatorname{sgn}\lambda_0)^k\int_{g^{-1}D_P}\frac{y^{\delta-1}}{(z-\lambda_0^{2k}\bar{z})|z-\lambda_0^{2k}\bar{z}|^\delta}dxdy,$$

where $g^{-1}D_P$ is a fundamental domain of $\Gamma'\left(\begin{pmatrix}\lambda_0 & 0\\ 0 & \lambda_0^{-1}\end{pmatrix}\right)$ in S. Thus,

$$J(P^k)=(2^{3+\delta}\pi)\frac{\Gamma(1/2)\Gamma((\delta+1)/2)}{\Gamma((\delta+2)/2)}\frac{(\operatorname{sgn}\lambda_0)^k\log|\lambda_0|}{|\lambda_0^{-k}-\lambda_0^k||\lambda_0^{-k}+\lambda_0^k|^\delta}.$$

Let $\{P_\alpha\}$ be a complete system of representatives of the primitive hyperbolic conjugacy classes in Γ and let $\lambda_{0,\alpha}$ be the eigenvalue ($|\lambda_{0,\alpha}|>1$) of representative P_α. Then, the hyperbolic component $J(P)$ is expressed by the following

$$J(P)=\sum_{\alpha=1}^{\infty}\sum_{k=1}^{\infty}J(P_\alpha^k)$$
$$=\frac{2^{3+\delta}\pi^{3/2}\Gamma((\delta+1)/2)}{\Gamma((\delta+2)/2)}\sum_{\alpha=1}^{\infty}\sum_{k=1}^{\infty}\frac{(\operatorname{sgn}\lambda_{0,\alpha})^k\log|\lambda_{0,\alpha}|}{|\lambda_{0,\alpha}^k-\lambda_{0,\alpha}^{-k}|}|\lambda_{0,\alpha}^k+\lambda_{0,\alpha}^{-k}|^{-\delta}.$$

3) Elliptic conjugacy classes.

Let ρ, $\bar{\rho}$ be the fixed points of an elliptic element $M(\rho\in S)$ and ζ, $\bar{\zeta}$ be the eigenvalues of M. We denote by Φ a linear transformation which maps S into a unit disk:

$$w=\Phi(z)=\frac{z-\rho}{z-\bar{\rho}}.$$

Then we have $\Phi M\Phi^{-1}=\begin{pmatrix}\zeta & 0\\ 0 & \bar{\zeta}\end{pmatrix}$ and

$$\frac{Mz-\rho}{Mz-\bar{\rho}}=\frac{\zeta}{\bar{\zeta}}\frac{z-\rho}{z-\bar{\rho}}.$$

The elliptic component is calculated as follows:

$$J(M)=\int_{\tilde{D}_M}\omega_\delta(z,\phi;M(z,\phi))d(z,\phi)$$
$$=\frac{2^{\delta+1}\sqrt{-1}}{[\Gamma(M):1]}\int_{\tilde{S}}\frac{(yy')^{(\delta+1)/2}}{(z-\bar{z}')|z-\bar{z}'|^\delta}e^{-\sqrt{-1}(\phi-\phi')}d(z,\phi)\quad((z',\phi')=M(z,\phi))$$

$$=\frac{8\pi\bar{\zeta}}{[\Gamma(M):1]}\int_{|w|<1}\frac{(1-w\bar{w})^{\delta-1}}{(1-\bar{\zeta}^2 w\bar{w})|1-\bar{\zeta}^2 w\bar{w}|^{\delta}}dudv \quad (w=u+\sqrt{-1}v)$$

$$=\frac{16\pi^2\bar{\zeta}}{[\Gamma(M):1]}\int_0^1\frac{(1-r^2)^{\delta-1}}{(1-\bar{\zeta}^2 r^2)|1-\bar{\zeta}^2 r^2|^{\delta}}dr.$$

We put

$$I(\delta)=\int_0^1\frac{(1-r^2)^{\delta-1}r}{(1-\bar{\zeta}^2 r^2)|1-\bar{\zeta}^2 r^2|^{\delta}}dr.$$

Then, under the condition $\delta>0$, the function $\frac{\delta(1-r^2)^{\delta-1}r}{1-\bar{\zeta}^2 r^2}$ is Lebesgue-integrable on [0, 1]. Hence

$$\begin{aligned}\lim_{\delta\to 0}\delta I(\delta)&=\lim_{\delta\to 0}\int_0^1\frac{\delta(1-r^2)^{\delta-1}r}{1-\bar{\zeta}^2 r^2}dr\\&=\lim_{\delta\to 0}\left\{\left[-\frac{(1-t)^{\delta}}{2}\frac{1}{1-\bar{\zeta}^2 t}\right]_0^1+\int_0^1(1-t)^{\delta}\left(\frac{1}{1-\bar{\zeta}^2 t}\right)'\frac{dt}{2}\right\}\\&=\frac{1}{2(1-\bar{\zeta}^2)}.\end{aligned}$$

Therefore we obtain

$$\lim_{\delta\to 0}\delta J(M)=\frac{8\pi^2}{[\Gamma(M):1]}\frac{\bar{\zeta}}{1-\bar{\zeta}^2}.$$

Since M and M^{-1} are not conjugate and $\bar{\zeta}/(1-\bar{\zeta}^2)$ is pure imaginary, we have

$$\lim_{\delta\to 0}\delta J(M)+\lim_{\delta\to 0}\delta J(M^{-1})=0.$$

We conclude that the contribution from elliptic classes to d_1 vanishes.

Now we put

$$\zeta_1^*(\delta)=\sum_{\alpha=1}^{\infty}\sum_{k=1}^{\infty}\frac{(\operatorname{sgn}\lambda_{0,\alpha})^k\log|\lambda_{0,\alpha}|}{|\lambda_{0,\alpha}^k-\lambda_{0,\alpha}^{-k}|}|\lambda_{0,\alpha}^k+\lambda_{0,\alpha}^{-k}|^{-\delta}. \tag{5}$$

Then, by the trace formula (2), the Dirichlet series (5) extends to a meromorphic function on the whole δ-plane and has a simple pole at $\delta=0$ whose residue will appear in (6) below. Finally, *multiply the both sides of* (2) *by δ and let δ tend to zero*, then the limit is expressed, by the above 1), 2) and 3), as follows:

$$\dim\mathfrak{m}\left(1,-\frac{3}{2}\right)=\frac{1}{2}\operatorname{Res}_{\delta=0}\zeta_1^*(\delta),$$

namely, by (1) we have

$$d_1 = \frac{1}{2} \operatorname*{Res}_{\delta=0} \zeta_1^*(\delta). \tag{6}$$

Remark 1. Let Γ be a fuchsian group of the first kind which contains the element $-I$, and χ a unitary representation of Γ of degree 1 such that $\chi(-I) = -1$. Let $S_1(\Gamma, \chi)$ be the linear space of cusp forms of weight 1 on the group Γ with character χ, and denote by d_1 the dimension of the linear space $S_1(\Gamma, \chi)$. When the group Γ has a compact fundamental domain in the upper half-plane S, we have the following dimension formula in the same way as in the case $\Gamma \not\ni -I$:

$$d_1 = \frac{1}{2} \sum_{\{M\}} \frac{\chi(M)}{[\Gamma(M): \pm I]} \frac{\bar{\zeta}}{1-\bar{\zeta}^2} + \frac{1}{2} \operatorname*{Res}_{s=0} \zeta_2^*(s), \tag{7}$$

where the sum over $\{M\}$ is taken over the distinct elliptic conjugacy classes of $\Gamma/\{\pm I\}$, $\Gamma(M)$ denotes the centralizer of M in Γ, ζ is one of the eigenvalues of M, and $\zeta_2^*(s)$ denotes the Selberg type zeta-function defined by

$$\zeta_2^*(s) = \sum_{\alpha=1}^{\infty} \sum_{k=1}^{\infty} \frac{\chi(P_\alpha)^k \log \lambda_{0,\alpha}}{\lambda_{0,\alpha}^k - \lambda_{0,\alpha}^{-k}} |\lambda_{0,\alpha}^k + \lambda_{0,\alpha}^{-k}|^{-s}. \tag{8}$$

Here $\lambda_{0,\alpha}$ denotes the eigenvalue ($\lambda_{0,\alpha} > 1$) of representative P_α of the primitive hyperbolic conjugacy classes $\{P_\alpha\}$ in $\Gamma/\{\pm I\}$.

§ 3. The finite case 1 ($:\Gamma \not\ni -I$)

Let Γ be a fuchsian group of the first kind not containing the element $-I$, and suppose that Γ has a non-compact fundamental domain $\tilde{D}$ in the space $\tilde{S}$. Then, we see that the integral

$$\int_{\tilde{D}} \sum_{M \in \Gamma} \omega_\delta(z, \phi; M(z, \phi)) d(z, \phi)$$

is uniformly bounded at a neighborhood of each irregular cusp of Γ, and that by the Riemann-Roch theorem, the number of regular cusps of Γ is even. Therefore we assume for simplicity that $\{\kappa_1, \kappa_2\}$ is a maximal set of cusps of Γ which are regular cusps and not equivalent with respect to Γ. Let Γ_i be the stabilizer in Γ of κ_i, and fix an element $\sigma_i \in SL(2, \boldsymbol{R})$ such that $\sigma_i \infty = \kappa_i$ and such that $\sigma_i^{-1} \Gamma_i \sigma_i$ is equal to the group $\left\{ \begin{pmatrix} 1 & m \\ 0 & 1 \end{pmatrix} : m \in \boldsymbol{Z} \right\}$. Then the Eisenstein series attached to the regular cusp κ_i is defined by

$$E_i(z, \phi; s) = \sum_{\substack{\sigma \in \Gamma_i \backslash \Gamma \\ \sigma_i^{-1} \sigma = \binom{**}{cd}}} \frac{y^s}{|cz+d|^{2s}} e^{-\sqrt{-1}(\phi + \arg(cz+d))} \quad (i=1, 2), \tag{9}$$

where $s=t+\sqrt{-1}r$ with $t>1$. The series (9) has the Fourier expansion at κ_j in the form

$$E_i(\sigma_j(z, \phi); s)=\sum_{m=-\infty}^{\infty} a_{ij,m}(y, \phi; s)e^{2\pi\sqrt{-1}mx}.$$

The constant term $a_{ij,0}(y, \phi; s)$ is given by

$$\begin{aligned} e^{\sqrt{-1}\phi}a_{ij,0}(y, \phi; s) &= a_{ij,0}(y; s) \\ &= \delta_{ij}y^s+\psi_{ij}(s)y^{1-s} \end{aligned}$$

with Kronecker's δ, and

$$\psi_{ij}(s)=-\sqrt{-1}\sqrt{\pi}\frac{\Gamma(s)}{\Gamma(s+(1/2))}\sum_{c\neq 0}\frac{(\operatorname{sgn} c)\cdot N_{ij}(c)}{|c|^{2s}},$$

where $N_{ij}(c)=\#\left\{0\leqq d<|c|:\begin{pmatrix} * & * \\ c & d\end{pmatrix}\in\sigma_i^{-1}\Gamma\sigma_j\right\}$. We put

$$\Phi(s)=(\psi_{ij}(s)).$$

Then it is easy to see that the Eisenstein matrix $\Phi(s)$ is a skew-symmetric matrix.

Since Γ is of finite type, the integral operator defined by ω_δ is not completely continuous on $L^2(\Gamma\backslash\tilde{S})$ in general and the space $L^2(\Gamma\backslash\tilde{S})$ has the following spectral decomposition

$$L^2(\Gamma\backslash\tilde{S})=L^2_0(\Gamma\backslash\tilde{S})\oplus L^2_{Sp}(\Gamma\backslash\tilde{S})\oplus L^2_{\mathrm{cont}}(\Gamma\backslash\tilde{S}),$$

where L^2_0 is the space of cusp forms and is discrete, L^2_{Sp} is the discrete part of the orthogonal complement of L^2_0 and L^2_{cont} is the continuous part of the spectra. We put

$$\begin{aligned} \tilde{H}_\delta(z, \phi; z', \phi') = \frac{1}{8\pi^2}\sum_{i=1}^{2}\int_{-\infty}^{\infty} & h(r)E_i\left(z, \phi; \frac{1}{2}+\sqrt{-1}r\right) \\ & \times\overline{E_i\left(z', \phi'; \frac{1}{2}+\sqrt{-1}r\right)}dr. \end{aligned}$$

Here $h(r)$ denotes the eigenvalue of ω_δ in $\mathfrak{m}(1, \lambda)$ given by (4):

(10) $$h(r)=2^{2+\delta}\pi\frac{\Gamma(1/2)\Gamma((1+\delta)/2)}{\Gamma(\delta)\Gamma(1+(\delta/2))}\Gamma\left(\frac{\delta}{2}+\sqrt{-1}r\right)\Gamma\left(\frac{\delta}{2}-\sqrt{-1}r\right)$$

with $\lambda=s(s-1)-\dfrac{5}{4}$ and $s=\dfrac{1}{2}+\sqrt{-1}r$. We put

$$\kappa_\delta(z, \phi; z', \phi') = \sum_{M \in \Gamma} \omega_\delta(z, \phi; M(z', \phi'))$$

and

$$\tilde{\kappa}_\delta = \kappa_\delta - \tilde{H}_\delta.$$

Then the integral operator $\tilde{\kappa}_\delta$ is now complete continuous on $L^2(\Gamma\backslash\tilde{S})$ and has all discrete spectra of κ_δ. Furthermore, an eigenvalue of $f(z, \phi)$ in $L_0^2(\Gamma\backslash\tilde{S}) \oplus L_{Sp}^2(\Gamma\backslash\tilde{S})$ for $\tilde{\kappa}_\delta$ is equal to that for κ_δ and the image of $\tilde{\kappa}_\delta$ on it is contained in $L_0^2(\Gamma\backslash\tilde{S})$. Considering the trace of $\tilde{\kappa}_\delta$ on $L_0^2(\Gamma\backslash\tilde{S})$, we now obtain the following modified trace formula ([4], [7]):

$$\begin{aligned}\sum_{n=1}^{\infty} h(\lambda^{(n)}) &= \int_{\tilde{D}} \tilde{\kappa}_\delta(z, \phi; z, \phi) d(z, \phi) \\ &= \int_{\tilde{D}} \{\sum_{M \in \Gamma} \omega_\delta(z, \phi; M(z, \phi)) - \tilde{H}_\delta(z, \phi; z, \phi)\} d(z, \phi),\end{aligned}$$

where each of $\lambda^{(n)}$ denotes an eigenvalue corresponding to an orthogonal basis $\{f^{(n)}\}$ for $L_0^2(\Gamma\backslash\tilde{S})$. We put

$$\begin{aligned}&\int_{\tilde{D}} \{\sum_{M \in \Gamma} \omega_\delta(z, \phi; M(z, \phi)) - \tilde{H}_\delta(z, \phi; z, \phi)\} d(z, \phi) \\ &\quad = J(I) + J(P) + J(R) + J(\infty),\end{aligned}$$

where $J(I)$, $J(P)$, $J(R)$ and $J(\infty)$ denote respectively the identity component, the hyperbolic component, the elliptic component and the parabolic component of the traces. Then the components $J(I)$, $J(P)$ and $J(R)$ are as in Section 2 and in the following we shall calculate the component $J(\infty)$ (cf. [9]).

Let $\tilde{D}_i$ be a fundamental domain of the stabilizer Γ_i of cusp κ_i in Γ. Then we have

$$\begin{aligned}J(\infty) = \lim_{Y \to \infty} \Bigg\{ &\sum_{i=1}^{2} \int_{\tilde{D}_i^Y} \sum_{\substack{M \in \Gamma_i \\ M \neq I}} \omega_\delta(z, \phi; M(z, \phi)) d(z, \phi) \\ &- \int_{\tilde{D}_Y} \tilde{H}_\delta(z, \phi; z, \phi) d(z, \phi) \Bigg\},\end{aligned}$$

where $\tilde{D}_i^Y$ denotes the domain consisting of all points (z, ϕ) in $\tilde{D}_i$ such that $\operatorname{Im}(\sigma_i^{-1}z) < Y$, and $\tilde{D}_Y$ the domain consisting of all $(z, \phi) \in \tilde{D}$ such that $\operatorname{Im}(\sigma_i^{-1}z) < Y$ for all $i = 1, 2$. Making use of a summation formula due to Euler-MacLaurin and the Maass-Selberg relation, we have the following (cf. [2], [9]):

$$\int_{\tilde{D}_i^Y} \sum_{\substack{M \in \Gamma_i \\ M \neq I}} \omega_\delta(z, \phi; M(z, \phi)) d(z, \phi) = 2^2 \pi \frac{\Gamma(1/2)\Gamma((\delta+1)/2)}{\Gamma(1+(\delta/2))} \log Y + \varepsilon(\delta) + o(1)$$

as $Y\to\infty$, where $\varepsilon(\delta)$ denotes a function of δ such that $\lim_{\delta\to 0}\delta\varepsilon(\delta)=0$;

$$\frac{1}{8\pi^2}\int_{\tilde{D}_Y}\int_{-\infty}^{\infty} h(r)E_i\Big(z,\phi;\frac{1}{2}+\sqrt{-1}r\Big)\overline{E_i\Big(z,\phi;\frac{1}{2}+\sqrt{-1}r\Big)}dr\,d(z,\phi)$$

$$=2^2\pi\frac{\Gamma(1/2)\Gamma((\delta+1)/2)}{\Gamma(1+(\delta/2))}\log Y$$

$$-\frac{1}{4\pi}\int_{-\infty}^{\infty} h(r)\frac{\psi'_{ij}}{\psi_{ij}}\Big(\frac{1}{2}+\sqrt{-1}r\Big)dr+o(1)$$

as $Y\to\infty$ $(j\neq i)$. By the expression (10) of $h(r)$, we have

$$h(r)=O\Big(\frac{|r|^\delta}{|r|e^{\pi|r|}}\Big);$$

and the operator $\tilde{\kappa}_\delta$ is complete continuous on $L^2(\Gamma\backslash\tilde{S})$. Therefore we have

$$\lim_{\delta\to+0}\delta\int_{-\infty}^{\infty} h(r)\frac{\psi'_{ij}}{\psi_{ij}}\Big(\frac{1}{2}+\sqrt{-1}r\Big)dr=0.$$

It is now clear that the above result, combined with the formula (6), proves the following ([9]):

Theorem 1. *Let Γ be a fuchsian group of the first kind not containing the element $-I$ and suppose that the number of regular cusps of Γ is two. Let d_1 be the dimension for the space consisting of all cusp forms of weight 1 with respect to Γ. Then d_1 is given by*

(11) $$d_1=\frac{1}{2}\operatorname*{Res}_{s=0}\zeta_1^*(s),$$

where $\zeta_1^*(s)$ denotes the Selberg type zeta-function defined by (5) *in Section* 2.

Remark 2. Let Γ be a general discontinuous group of finite type not containing the element $-I$. Then we can prove that in the same way as in the above case, the contribution from parabolic classes to d_1 vanishes.

§ 4. The finite case 2 ($:\Gamma\ni -I$)

Let Γ be a fuchsian group of the first kind and assume that Γ contains the element $-I$ and has a non-compact fundamental domain $\tilde{D}$ in the space $\tilde{S}$. Let χ be a unitary representation of Γ of degree 1 such that $\chi(-I)=-1$. We denote by $S_1(\Gamma,\chi)$ the linear space of cusp forms of weight 1 on the group Γ with the character χ and by d_1 the dimension of the space $S_1(\Gamma, \chi)$. In this section we shall give a similar formula of the number d_1 when the group Γ is of finite type reduced at infinity and $\chi^2 \neq 1$.

Since Γ is of finite type reduced at ∞, ∞ is a cusp of Γ and the stabilizer Γ_∞ of ∞ in Γ is equal to $\pm\Gamma_0$ with $\Gamma_0 = \left\{\begin{pmatrix}1 & m\\ 0 & 1\end{pmatrix} : m \in \boldsymbol{Z}\right\}$. The Eisenstein series $E_\chi(z, \phi; s)$ attached to ∞ and χ is then defined by

$$(12) \qquad E_\chi(z, \phi; s) = \sum_{\substack{M \in \Gamma_\infty \backslash \Gamma \\ M = \binom{*\,*}{c\,d}}} \frac{\bar{\chi}(M) y^s}{|cz+d|^{2s}} e^{-\sqrt{-1}(\phi + \arg(cz+d))},$$

where $s = \sigma + \sqrt{-1}r$ with $\sigma > 1$. The constant term in the Fourier expansion of (12) at ∞ is given by

$$a_0(y, \phi; s) = e^{-\sqrt{-1}\phi}(y^s + \psi_\chi(s) y^{1-s}),$$

$$\psi_\chi(s) = -\sqrt{-1}\sqrt{\pi}\, \frac{\Gamma(s)}{\Gamma\left(s + \dfrac{1}{2}\right)} \sum_{\substack{c>0 \\ d \bmod c \\ \binom{*\,*}{c\,d} \in \Gamma}} \frac{\bar{\chi}(c, d)}{|c|^{2s}}.$$

In the following we only consider the case $\chi\left(\begin{pmatrix}1 & 1\\ 0 & 1\end{pmatrix}\right) = 1$. As shown in [3], the parabolic component $J(\infty)$ in the trace formula is given by

$$J(\infty) = \lim_{Y \to \infty}\left\{2\int_0^Y \int_0^1 \int_0^\pi \sum_{\substack{M \in \Gamma \\ M \neq I}} \omega_\delta(z, \phi; M(z, \phi)) d(z, \phi) - \int_{\tilde{D}_Y} \tilde{H}_\delta(z, \phi; z, \phi) d(z, \phi)\right\}$$

$$= -\frac{1}{4\pi}\int_{-\infty}^{\infty} h(r) \frac{\psi_\chi'((1/2) + \sqrt{-1}r)}{\psi_\chi((1/2) + \sqrt{-1}r)} dr - \frac{1}{4} h(0) \psi_\chi\left(\frac{1}{2}\right) + \varepsilon(\delta)$$

with $\lim_{\delta \to 0} \delta\varepsilon(\delta) = 0$. When we combine this with the formula (7), we are led to the following theorem which is our main purpose in this section.

Theorem 2. *Let Γ be a fuchsian group of the first kind containing the element $-I$ and suppose that Γ is reduced at infinity. Let χ be a one-dimensional unitary representation of Γ such that $\chi(-I) = -1$, $\chi\left(\begin{pmatrix}1 & 1\\ 0 & 1\end{pmatrix}\right) = 1$ and $\chi^2 \neq 1$. We denote by d_1 the dimension of the linear space consisting of cusp forms of weight 1 with respect to Γ with χ. Then the dimension d_1 is given by*

$$(13) \qquad d_1 = \frac{1}{2}\sum_{\{M\}} \frac{\chi(M)}{[\Gamma(M) : \pm I]} \frac{\bar{\zeta}}{1 - \bar{\zeta}^2} + \frac{1}{2} \operatorname*{Res}_{s=0} \zeta_2^*(s) - \frac{1}{4}\psi_\chi\left(\frac{1}{2}\right),$$

where the sum over $\{M\}$ is taken over the distinct elliptic conjugacy classes of $\Gamma/\{\pm I\}$, $\Gamma(M)$ denotes the centralizer of M in Γ, ζ is one of the eigenvalues of M, and $\zeta_2^*(s)$ denotes the Selberg type zeta-function defined by (8) in Section 2.

We may call the formulas (11) and (13) a kind of Riemann-Roch type theorem for automorphic forms of weight 1.

Remark 3. For a general discontinuous group Γ of finite type containing the element $-I$, we obtain the contribution from parabolic classes to d_1 in the same way as in the case of reduced at ∞.

§ 5. The case of $\Gamma_0(p)$

Let p be a prime number such that $p \equiv 3 \bmod 4$, $p \neq 3$ and let $\Phi_0(p)$ be the group generated by the group $\Gamma_0(p)$ and the element $\kappa = \begin{pmatrix} 0 & -\sqrt{p}^{-1} \\ \sqrt{p} & 0 \end{pmatrix}$, namely, $\Phi_0(p) = \Gamma_0(p) + \kappa\Gamma_0(p)$. Let ε be the Legendre symbol on $\Gamma_0(p)$: $\varepsilon(L) = \left(\frac{d}{p}\right)$ for $L = \begin{pmatrix} a & b \\ c & d \end{pmatrix} \in \Gamma_0(p)$. Since $\varepsilon(\kappa^2) = \varepsilon(-I) = -1$, we can define the odd characters $\varepsilon^{\pm}$ on $\Phi_0(p)$ such that $\varepsilon^{\pm}(\kappa) = \pm\sqrt{-1}$. Then we have

$$S_1(\Gamma_0(p), \varepsilon) = S_1(\Phi_0(p), \varepsilon^+) \oplus S_1(\Phi_0(p), \varepsilon^-).$$

We put

$$\mu_1^{\pm} = \dim S_1(\Phi_0(p), \varepsilon^{\pm}).$$

Then

$$\dim S_1(\Gamma_0(p), \varepsilon) = d_1 = \mu_1^+ + \mu_1^-.$$

We denote by $\bar{\Gamma}_0(p)$, $\bar{\Phi}_0(p)$ the inhomogeneous linear transformation group attached to $\Gamma(p)$, $\Phi_0(p)$ respectively. If $\sigma(p)$ is the parabolic class number of $\bar{\Gamma}_0(p)$, then $\sigma(p) = 2$; and if $e_2(p)$, $e_3(p)$ are the number of elliptic classes of order 2, 3 respectively of $\bar{\Gamma}_0(p)$, then

$$e_2(p) = 0, \quad e_3(p) = 1 + \left(\frac{p}{3}\right).$$

Let $\sigma^*(p)$, $e_2^*(p)$, $e_3^*(p)$ denote respectively the number of parabolic classes, the number of elliptic classes of order 2, the number of elliptic classes of order 3 for $\bar{\Phi}_0(p)$. Then we have

$$\sigma^*(p) = \frac{1}{2}\sigma(p) = 1;$$

$$e_3^*(p) = \frac{1}{2}e_3(p) = \frac{1}{2}\left(1 + \left(\frac{p}{3}\right)\right);$$

$$e_2^*(p) = \frac{1}{2}e_2(p) + e_2'(p) = e_2'(p),$$

where $e_2'(p)$ denotes the number of classes of elliptic elements of order 2 of $\kappa\bar{\Gamma}_0(p)$. It is known that

$$e_2'(p)=\left(3-\left(\frac{2}{p}\right)\right)=\begin{cases}4h & \text{if } p\equiv 3 \bmod 8,\\ 2h & \text{if } p\equiv 7 \bmod 8,\end{cases}$$

where h denotes the class number of $Q(\sqrt{-p})$, which is an odd integer. Let ϑ_2 denote the number of the elements L in $\bar{\Gamma}_0(p)$ such that $\varepsilon^-(\kappa L)=+\sqrt{-1}$. Then, by [5], we have the following

$$\vartheta_2=\begin{cases}h & \text{if } p\equiv 3 \bmod 8,\\ 0 & \text{if } p\equiv 7 \bmod 8.\end{cases}$$

In the following, we shall calculate the contribution from elliptic elements to $\mu_1^\pm$. Let $\{M\}$ be a complete system of representatives of the elliptic conjugacy classes of order 2 in $\bar{\Phi}_0(p)$. Then $\{M\}$ is given by $\left\{\kappa\begin{pmatrix} a & b\\ pb & d\end{pmatrix}\right\}$, where $\left\{\begin{pmatrix} a & b\\ pb & d\end{pmatrix}\right\}$ denotes the representatives of positive definite integral quadratic forms $\begin{pmatrix} a & pb\\ pb & pd\end{pmatrix}$ such that $\det\begin{pmatrix} a & pb\\ pb & pd\end{pmatrix}=p$. Then the result of calculation is given in the following table:

p	$\varepsilon(L)$	The number of elliptic classes of order 2	ζ	$\frac{1}{[\Gamma(M):\pm I]}\frac{\zeta}{1-\zeta^2}\varepsilon^\pm(\kappa L)$
$p\equiv 3 \bmod 8$	$\varepsilon(L)=1$	$3h$	$\sqrt{-1}$	$\frac{1}{2}\frac{\sqrt{-1}}{2}(\pm\sqrt{-1})=\mp\frac{1}{4}$
$p\equiv 3 \bmod 8$	$\varepsilon(L)=-1$	h	$\sqrt{-1}$	$\frac{1}{2}\frac{\sqrt{-1}}{2}(\mp\sqrt{-1})=\pm\frac{1}{4}$
$p\equiv 7 \bmod 8$	$\varepsilon(L)=1$	$2h$	$\sqrt{-1}$	$\frac{1}{2}\frac{\sqrt{-1}}{2}(\pm\sqrt{-1})=\mp\frac{1}{4}$

It is clear that there is no contribution from elliptic classes of order 3 to $\mu_1^\pm$. Therefore the contribution from elliptic classes to $\mu_1^\pm$ is given by

$$\frac{1}{2}\sum_{\{M\}}\frac{1}{[\Gamma(M):\pm I]}\frac{\zeta}{1-\zeta^2}\varepsilon^\pm(M)=\mp\frac{1}{4}h.$$

We also have $\psi_{\varepsilon^\pm}(1/2)=\mp 1$. Let $\{P_\alpha\}$ be a complete system of representatives of the primitive hyperbolic conjugacy classes in $\bar{\Gamma}_0(p)$ and let $\lambda_{0,\alpha}$ be the eigenvalue ($\lambda_{0,\alpha}>1$) of representative P_α. We put

$$Z^*(\delta)=\sum_{\alpha=1}^{\infty}\sum_{k=1}^{\infty}\frac{\varepsilon(P_\alpha)^k\log\lambda_{0,\alpha}}{|\lambda_{0,\alpha}^k-\lambda_{0,\alpha}^{-k}|}|\lambda_{0,\alpha}^k+\lambda_{0,\alpha}^{-k}|^{-\delta}.$$

Then, we have consequently the following

$$d_1=\mu_1^+ +\mu_1^- =\frac{1}{2}\operatorname*{Res}_{\delta=0} Z^*(\delta). \tag{14}$$

Remark 4. Combining the above (14) with Serre's result[1], we have the following remarkable equality

$$\operatorname*{Res}_{\delta=0} Z^*(\delta)=(h-1)+4(s+2a).$$

References

[1] Hiramatsu, T., Eichler classes attached to automorphic forms of dimension −1, Osaka J. Math., **3** (1966), 39–48.
[2] ——, On some dimension formula for automorphic forms of weight one I, Nagoya Math. J., **85** (1982), 213–221.
[3] ——, On some dimension formula for automorphic forms of weight one II, Nagoya Math. J., **105** (1987), 169–186.
[4] Kubota, T., Elementary theory of Eisenstein series, Kodansha and Halsted, Tokyo-New York, 1973.
[5] Petersson, H., Über Eisensteinsche Reihen und automorphe Formen von der Demension −1, Comment. Math. Helv., **31** (1956), 111–144.
[6] Selberg, A., Harmonic analysis and discontinuous groups in weakly symmetric Riemannian spaces with applications to Dirichlet series, J. Indian Math. Soc., **20** (1956), 47–87.
[7] ——, Discontinuous groups and harmonic analysis, in Proc. International Math. Congr., Stockholm, 177–189, 1962.
[8] Serre, J.-P., Modular forms of weight one and Galois representations, in Proc. Symposium on Algebraic Number Fields, 193–268, Academic Press, London, 1977.
[9] Hiramatsu, T. and Akiyama, S., On some dimension formula for automorphic forms of weight one III, Nagoya Math. J., **111** (1988), 157–163.

Department of Mathematics
Faculty of Science
Kobe University
Kobe 657
Japan

---

1) cf. Serre [8], p. 253.

Advanced Studies in Pure Mathematics 15, 1989
Automorphic Forms and Geometry of Arithmetic Varieties
pp. 301–349

On Automorphism Groups of Positive Definite Binary Quaternion Hermitian Lattices and New Mass Formula

Tomoyoshi Ibukiyama

Dedicated to Professor Ichiro Satake on his sixtieth birthday

In this paper, we shall give some general method how to calculate the multiplicity of a given finite group which appears as the automorphism groups of the lattices, up to isometry, in a fixed genus in a positive definite metric space, and apply it to the binary quaternion hermitian cases, motivated by the theory of supersingular abelian varieties developed in Katsura-Oort [12]. Our Main Theorems are Theorems 7.1 and 7.2 in § 7. More precisely, we shall consider the following problems. Let B be either the rational number field $\boldsymbol{Q}$, an imaginary quadratic field over $\boldsymbol{Q}$, or a definite quaternion algebra over $\boldsymbol{Q}$. Let (V, h) be a pair of a finite dimensional left B-vector space V over B and a positive definite hermitian metric h on B with respect to the unique positive involution of B. Denote by $G=G(V, h)$ the group of similitudes of (V, h);

$$G=\{g \in GL_B(V);\ h(xg, yg)=n(g)h(x, y)\ (x, y \in V) \text{ for some } n(g) \in \boldsymbol{Q}^{\times}\}.$$

Let $\mathscr{L}$ be a fixed genus of some lattices in V.

Problem 1. Calculate the class number $H=\sharp(\mathscr{L}/G)$ of $\mathscr{L}$.

It is known that Problem 1 can be solved at least in principle by means of the trace formula (cf. Hashimoto [3]), and some explicit calculations have been done by several mathematicians. Now, our main theme in this paper is the following Problem 2. Denote by $L_1, \cdots, L_H$ a complete set of representatives of the classes in $\mathscr{L}$. For each i $(1 \leqq i \leqq H)$, put

$$\Gamma_i=\mathrm{Aut}(L_i)=\{g \in G;\ L_i g=L_i\}.$$

It is easy to see that this is a finite group for each i, because of our assumption that h is positive definite.

Received May 6, 1987.

Problem 2. Let Γ be any finite group. Count the number of classes L_i such that $\Gamma \cong \Gamma_i$.

The usual trace formula(, or its refinement in [3], [5], [6]) gives us some information also on this Problem 2. But, in general, these previously known formulae do not give us enough information to solve the above Problem 2.

In this paper, first, we give a certain new mass formula which gives us enough information to solve Problem 2 (§ 1. Theorem 1.1). This new formula is a generalization of the formula in [3] (cf. also [5], [6]), and the proof itself is obtained in a similar way. But, while the old one in [3] was more or less motivated to explain the trace formula, our new one does not appear as a summand in the usual trace formula in general. And, it does not seem to have been appreciated that such new formula should exist and be useful.

The reason why we need a new mass formula can be roughly explained as follows. We can take a certain big finite group Δ so that it acts on $\mathscr{L}/G$, and that, for each i $(1 \leq i \leq H)$, the subgroup of Δ consisting of all elements which stabilize the class which contains L_i is isomorphic to Γ_i. The usual trace formula is essentially the formula to give the *linear* characters of the permutation representation of Δ on $\mathscr{L}/G$ (although it might sometimes give us slightly more information). To determine each Γ_i, we need the irreducible decomposition of $(\Delta, \mathscr{L}/G)$ as a *permutation* representation. But, in general, a permutation representation *cannot* be determined by the linear representation attached to it. Our new mass formula is quite useful to fill this gap. The usual trace formula expresses some kind of masses by sum of data determined by G-conjugacy classes of elements of G. Our new mass formula expresses some kind of new masses by sum of data determined by G-conjugacy classes of elements of the *direct-product* G^r, where r is a certain natural number. By this formula, we can tell whether several elements of G are contained in a group Γ_i at the same time, or not, and this gives us a general tool in order to solve Problem 2, as we shall explain in § 1.

Secondly, we shall apply this method to some special cases. Hereafter, we shall assume that B is the definite quaternion algebra over Q with fixed prime discriminant p. We shall solve Problem 2 explicitly for arbitrary p for the (unique) non-principal genus in V in the case of $\dim_B V = 2$ (§ 7. Theorem 7.1. As for the case where the discriminant is not necessarily prime, see Theorem 7.2). When $\dim_B V = 1$, the answer to Problems 1, 2 was classically well known by Eichler [2]. When $\dim_B V = 2$, or 3, Problem 1 was solved in Hashimoto-Ibukiyama [5] and Hashimoto [4] by means of the trace formula. Our Problems 1 and 2 have also close relations to the theory of supersingular abelian varieties. For example, when $\dim_B V \geqq 2$, the class number of the principal genus in V is equal to the number of isomorphism classes of principal polarizations on E^n, where E is a supersingular elliptic curve over a field of characteristic p, p is the discriminant of B, and $n = \dim_B V$ (T. Ibukiyama-T. Katsura-F. Oort [11]). In this case, Problem 2 amounts to describe all automorphism groups of all principally polarized abelian varieties (E^n, C), where we denote by C the principal polarizations on E^n. When $n=2$, using this relation, Problem 2 was solved explicitly for all p for the principal genus in V ([11] loc. cit.). When $n=2$ and genus in question is the non-principal genus in V, another geometrical interpretation of Problems 1 and 2 was given in Katsura-Oort [12]: for any natural integer m such that $p \nmid m$, denote by $A_{2,1}(m)$ the coarse moduli space of principally polarized abelian surfaces with level m structure, and by $A_s(m)$ the locus in $A_{2,1}(m)$ of principally polarized supersingular abelian surfaces with level m structure. As shown in [12], the variety $A_s(1)$ is not irreducible in general, and the number of irreducible components of $A_s(1)$ is equal to the class number of the non-principal genus in V. In this case, if $p \neq 2$, Problem 2 amounts to give explicitly the decomposition group of each irreducible component in $A_s(2)$ for the natural covering $A_{2,1}(2) \rightarrow A_{2,1}(1)$ (which is isomorphic to the automorphism group of the Morret-Baily family in [13], [12].) Using this relation and some geometrical methods, Katsura-Oort [12] solved Problem 2 for $p \leqq 31$ in this case. Actually, they used essentially some geometrical alternatives of the usual trace (, or mass) formula, and it worked successfully for such small p. But, for general p, this does not work, and we need a new mass formula which was mentioned above. We shall solve Problem 2 for the non-principal genus in V, $\dim_B V = 2$, completely for all p by a purely *number-theoretical* method. It seems to be an interesting problem to find some geometrical alternatives of our methods, or some direct geometrical interpretation of our results in this paper.

Now, we explain briefly the content of each section. In § 1, after reviewing lattices in hermitian spaces, we give general (but not explicit) new mass formula and explain how to solve Problem 2 in general, by using this formula. From § 2 until the end of this paper, we are devoted into explicit calculations in order to solve Problem 2 in the binary quaternion hermitian case. In § 2, we review the class number formula in [5] II and give some miscellaneous results obtained by that formula. In order to calculate explicit "masses" which we need, in § 3, we classify G-conjugacy classes of some elements of G^2, in § 5, we calculate some "local data", and in § 6, we give explicit "masses". This procedure is more or less similar to that of the explicit calculations of the usual trace formula in [5], but it is more elaborate and complicated in our case. In § 4, we show that some dihedral groups cannot be isomorphic to any Γ_i $(1 \leqq i \leqq H)$, without using mass formula. The argument in this section is special to the case we are considering, but it economizes the calculations. In § 7, we gather the results in the former sections, and solve Problem 2.

The author would like to thank Professors T. Katsura and F. Oort for explaining him their problems in [12] and also for valuable conversations.

Notations. As usual, we denote by $\boldsymbol{Q}$ (resp. $\boldsymbol{Z}$) the field of rational numbers (resp. the ring of rational integers). We denote by $\boldsymbol{Q}_+$ the set of all positive rational numbers. For any algebraic group $\mathfrak{g}$ over $\boldsymbol{Q}$, we denote by $\mathfrak{g}_A$ the adelization of $\mathfrak{g}$, by $\mathfrak{g}_\infty$ (resp. $\mathfrak{g}_q$) the infinite (resp. q-adic) component of $\mathfrak{g}_A$, where q is any prime. For each prime q, we denote by $\boldsymbol{Q}_q$ (resp. $\boldsymbol{Z}_q$) the field of q-adic numbers (resp. the ring of q-adic integers), and by $\boldsymbol{F}_{q^r}$ the finite field of characteristic q with q^r elements. For any ring R, we denote by $R^\times$ the group of units of R, by $M_n(R)$ (n; natural number) the set of n by n matrices, and we put $GL_n(R) = M_n(R)^\times$. We denote by 1_n the unit matrix of $M_n(R)$, which is often denoted by $1 = 1_n$, when n is clear from the context. For any finite set S, we denote by $\#(S)$ the cardinality of S.

§ 1. New mass formula

1.1. First, we review on lattices in metric spaces, and formulate Problems 1, 2 in the adelic language. As in the introduction, let B be either the rational number field $\boldsymbol{Q}$, an imaginary quadratic extension of $\boldsymbol{Q}$, or a positive definite quaternion algebra over $\boldsymbol{Q}$. Let V be a finite dimensional left B-vector space. We denote by $^-$ the unique positive involution of B, which is trivial when $B = \boldsymbol{Q}$, the complex conjugation when B is imaginary quadratic, and the main involution of B when B is quaternionic. We fix a positive definite metric h on V with respect to $^-$: h is a mapping of $V \times V$ to B which satisfies the following conditions (1), (2), (3).

(1) $\quad h(ax+by, z) = ah(x, z) + bh(y, z), \qquad (a, b \in B, x, y, z \in V),$

(2) $\quad h(y, x) = \overline{h(x, y)} \qquad (x, y \in V),$

(3) $\quad h(x, x) \geqq 0, \qquad$ for all $x \in V$, and

$\quad h(x, x) = 0, \qquad$ if and only if $x = 0$.

Denote by $\mathrm{End}_B(V)$ (resp. $GL_B(V)$) the ring (resp. group) of all left B-linear endomorphisms (resp. automorphisms) of V, and by $G = G(V, h)$

the group of similitudes of (V, h):

$$G=G(V, h)=\{g \in GL_B(V);\ h(xg, yg)=n(g)h(x, y),\ x, y \in V\},$$

where $n(g) \in \boldsymbol{Q}_+^\times$ is a scalar depending only on g.

Then, G is an algebraic group over $\boldsymbol{Q}$. We denote by G_A the adelization of G, and for any place v of $\boldsymbol{Q}$, we denote by G_v the v-component of G_A. When v is a finite prime q, we have

$$G_q=\{g \in GL_{B_q}(V_q);\ h_q(xg, yg)=n(g)h_q(x, y),\ x, y \in V_q,\ n(g) \in \boldsymbol{Q}_q^\times\}$$

where $B_q=B\otimes_{\boldsymbol{Q}}\boldsymbol{Q}_q$, $V_q=V\otimes_{\boldsymbol{Q}}\boldsymbol{Q}_q$ and h_q is the continuous prolongation of h to V_q. Let O be a maximal order of B. A left O-module in V is called left O-lattice, when L is a $\boldsymbol{Z}$-lattice in V (, where V is regarded as a vector space over $\boldsymbol{Q}$). Two left O-lattices L and M are said to be isomorphic, if $L=Mg$ for some $g \in G$. For each left O-lattice L and each prime p, put $L_p=L\otimes_{\boldsymbol{Z}}\boldsymbol{Z}_p$. For a fixed left O-lattice L, the genus $\mathscr{L}(L)$ which contains L is defined to be the following set of lattices:

$$\mathscr{L}(L)=\{M\subset V\colon\ M \text{ is a left } O\text{-lattice such that for every prime } p,\ M_p=L_pg_p \text{ for some } g_p \in G_p\}.$$

The number $\sharp(\mathscr{L}(L)/G)$ of isomorphism classes of left O-lattices in $\mathscr{L}(L)$ is known to be finite, and called the class number of $\mathscr{L}(L)$. The class number has the following interpretation in the adelic language. For any prime p, put

$$U_p(L)=\{g \in G_p\colon L_pg=L_p\}, \quad \text{and} \quad \mathfrak{A}(L)=G_\infty \prod_p U_p(L).$$

Decompose G_A into the double cosets as follows:

$$G_A=\coprod_{i=1}^{H} \mathfrak{A}(L)g_jG \qquad \text{(disjoint)}.$$

Then, we have $H=\sharp(\mathscr{L}(L)/G)$. A complete set of representatives of classes in $\mathscr{L}(L)$ is given by $L_1, \cdots, L_H$, where

$$L_i=Lg_i=\bigcap_p (L_pg_{i,p}\cap V) \qquad (i=1, \cdots, H)$$

and $g_{i,p}$ is the p-component of $g_i \in G_A$ for each prime p. Put

$$\Gamma_i=\mathrm{Aut}\,(L_i)=\{g \in G;\ L_ig=L_i\}.$$

Then, we have $\Gamma_i=G\cap g_i^{-1}\mathfrak{A}(L)g_i$, which is a finite group.

Now, we shall explain the essential meaning of Problem 2. We introduce some subgroup $\mathfrak{A}'$ of $\mathfrak{A}=\mathfrak{A}(L)$. Fix a prime p. It is well known that there exists a torsion-free normal subgroup U'_p of $U_p(L)$ of finite index. Put

$$\mathfrak{A}'=G_\infty U'_p \prod_{q\neq p} U_q(L) \quad \text{and} \quad \varDelta=\mathfrak{A}/\mathfrak{A}'=U_p/U'_p.$$

Decompose G_A into double cosets as follows;

$$G_A=\coprod_{j=1}^{H'} \mathfrak{A}'g'_jG \qquad \text{(disjoint).}$$

Denote by $\mathscr{L}'$ the set of H' double cosets in the above decomposition:

$$\mathscr{L}'=\{\mathfrak{A}'g'_jG;\, j=1, \cdots, H'\}.$$

Then, the finite group $\varDelta$ acts on $\mathscr{L}'$ by

$$\bar{\delta}(\mathfrak{A}'g'_jG)=\delta(\mathfrak{A}'g'_jG)=\mathfrak{A}'\delta g'_jG,$$

where $\bar{\delta}\in\varDelta$ and δ is any representative of $\bar{\delta}$ in U_p. It is obvious that $\bar{\delta}(\mathfrak{A}'g'_jG)=\mathfrak{A}'g'_kG$ for some $\bar{\delta}\in\varDelta$, if and only if both $\mathfrak{A}'g'_jG$ and $\mathfrak{A}'g'_kG$ are contained in the same double coset $\mathfrak{A}g_iG$ for some i $(1\leqq i\leqq H)$. In other words, each $\varDelta$-orbit in $\mathscr{L}'$ corresponds one-to-one to each class of $\mathscr{L}(L)$. So, we can calculate H, if the character of the linear representation of $\varDelta$ attached to the permutation representation $(\varDelta, \mathscr{L}')$ is known. On the other hand, for each j $(1\leqq j\leqq H')$, define a subgroup Γ'_j of $\varDelta$ by:

$$\Gamma'_j=\{\bar{\delta}\in\varDelta;\, \bar{\delta}(\mathfrak{A}'g'_jG)=\mathfrak{A}'g'_jG\}.$$

It is trivial that, for each j $(1\leqq j\leqq H')$, there exists the unique i $(1\leqq i\leqq H)$ such that $\mathfrak{A}'g'_jG\subset\mathfrak{A}g_iG$. For that i, we get a group isomorphism $\Gamma'_j\cong\Gamma_i$. In fact, put $g'_j=ug_ia$ $(a\in G,\, u\in\mathfrak{A})$. Then,

$$\begin{aligned}\Gamma'_j&\cong(g'_jGg'^{-1}_j\cap\mathfrak{A})/(g'_jGg'^{-1}_j\cap\mathfrak{A}')\\ &\cong(G\cap g_i^{-1}\mathfrak{A}g_i)/(G\cap g_i^{-1}u^{-1}\mathfrak{A}'ug_i)\\ &\cong\Gamma_i/(G\cap g_i^{-1}u^{-1}\mathfrak{A}'ug_i).\end{aligned}$$

But, $G\cap g_i^{-1}u^{-1}\mathfrak{A}'ug_i=\{1\}$, because G is diagonally embedded in G_A and U'_p is torsion-free. So, $\Gamma'_j\cong\Gamma_i$. In other words, the stabilizer of any point in each $\varDelta$-orbit gives each Γ_i, and the irreducible decomposition of $(\varDelta, \mathscr{L}')$ as *permutation* representation is given by $\oplus_{i=1}^{H}(\varDelta, \varDelta/\Gamma_i)$ through the above isomorphism, where $(\varDelta, \varDelta/\Gamma_i)$ is the permutation representation of $\varDelta$ determined by the action of $\varDelta$ on $\varDelta/\Gamma_i$. This is the essential difference between Problem 1 and Problem 2.

1.2. We shall explain a general method to solve Problem 2 and give a new mass formula. Let Γ be a finite subgroup of G. We would like to know for example how many Γ_i $(1\leqq i\leqq H)$ contain Γ. If Γ is a cyclic group, the usual trace formula gives us a strong tool for this problem: fix a generator γ of Γ and denote by $\{\gamma\}_G$ the set of elements of G which are G-conjugate to γ. Then, our previous trace formula (cf. [2], [3], [5]) expresses the following "mass"

$$\sum_{i=1}^{H} \frac{\#\{\{\gamma\}_G \cap \Gamma_i\}}{\#(\Gamma_i)}$$

by some data on G-conjugacy classes. But, as Γ_i is not necessarily cyclic, we must introduce some new "mass" for non-cyclic Γ. We shall explain this below. For any natural integer r and any group $\mathfrak{g}$, denote by $\mathfrak{g}^r$ the direct product of r copies of $\mathfrak{g}$. The multiplication between elements of $\mathfrak{g}$ and $\mathfrak{g}^r$ is defined through the diagonal embedding of $\mathfrak{g}$ into $\mathfrak{g}^r$:

$$hg=(hg_1, \cdots, hg_r), \qquad gh=(g_1h, \cdots, g_rh),$$
$$(h \in \mathfrak{g},\ g=(g_1, \cdots, g_r) \in \mathfrak{g}^r).$$

Let $\mathfrak{h}$ be a subgroup of $\mathfrak{g}$. Two elements $g, g' \in \mathfrak{g}^r$ are said to be $\mathfrak{h}$-conjugate, if $g'=hgh^{-1}$ for some $h \in \mathfrak{h}$. Now, through the natural embedding $G^r \subset GL_B(V)^r$, we regard elements of G^r also elements of $GL_B(V)^r$. For $\gamma \in G^r$, denote by $\{\gamma\}_G$ (resp. $\{\gamma\}_{GL}$) the set of all elements of G^r which are G (resp. $GL_B(V)$)-conjugate to γ. For each genus $\mathscr{L}$ of lattices in V, and each $\{\gamma\}_G$ (resp. $\{\gamma\}_{GL}$), where $\gamma \in G^r$, we define a new "mass" $m(\mathscr{L}, \{\gamma\}_G)$ (resp. $m(\mathscr{L}, \{\gamma\}_{GL})$) as follows:

$$m(\mathscr{L}, \{\gamma\}_G)=\sum_{i=1}^{H} \frac{\#\{\{\gamma\}_G \cap \Gamma_i^r\}}{\#(\Gamma_i)},$$
$$m(\mathscr{L}, \{\gamma\}_{GL})=\sum_{i=1}^{H} \frac{\#\{\{\gamma\}_{GL} \cap \Gamma_i^r\}}{\#(\Gamma_i)},$$

where H and Γ_i $(1\leqq i\leqq H)$ are defined for each $\mathscr{L}$ as in 1.1.

It is trivial that $m(\mathscr{L}, \{\gamma\}_G)=m(\mathscr{L}, \{\gamma\}_{GL})=0$, unless the group generated by all $\gamma_1, \cdots, \gamma_r$ is of finite order, where $\gamma=(\gamma_1, \cdots, \gamma_r)$. More precisely, take a (finite) subgroup Γ of G, and fix a set of generators $\gamma_1, \cdots, \gamma_r$ of Γ. Then, $h_i^{-1}\Gamma h_i \subset \Gamma_i$ for some i $(1\leqq i\leqq H)$ and some $h_i \in G$ (resp. $h_i \in GL_B(V)$), if and only if $m(\mathscr{L}, \{\gamma\}_G)\neq 0$ (resp. $m(\mathscr{L}, \{\gamma\}_{GL})$ $\neq 0$) for $\gamma=(\gamma_1, \cdots, \gamma_r) \in G^r$. Then above "masses" might be called "masses" of Γ, although they depend on the choice of generators of Γ.

A formula for these masses will be given in Theorem 1.1 later. Here we explain a general method how to solve Problem 2 by using these masses. Assume that $\Delta=U_p'\backslash U_p$ is explicitly known. As each Γ_i $(1\leqq i\leqq H)$ is isomorphic to a subgroup of Δ, we can give explicitly the candidates for the (abstract) group isomorphism classes which contain some Γ_i.

It is more convenient to consider the following refinements (a), (b) of Problem 2. For any finite subgroup Γ of G,

(a) Count the number of i $(1\leqq i\leqq H)$ such that $h_i^{-1}\Gamma h_i=\Gamma_i$ for some $h_i \in GL_B(V)$.

(b) Count the number of i $(1\leqq i\leqq H)$ such that $h_i^{-1}\Gamma h_i=\Gamma_i$ for some $h_i \in G$.

We can solve (a) by the following process (1), (2). (We can solve (b) virtually in the same way.)

(1) Classify $GL_B(V)$-conjugacy classes of all finite subgroups Γ of G which are isomorphic to any subgroup of Δ, satisfying $m(\mathscr{L}, \{\gamma\}_{GL})\neq 0$ for $\gamma=(\gamma_1, \cdots, \gamma_r)$, where $\gamma_1, \cdots, \gamma_r$ are some generators of Γ.

Denote by $\{\mathscr{G}_\varphi; \varphi \in \Phi\}$ a complete set of representatives of the conjugacy classes in (1). For each such class $\mathscr{G}_\varphi$, fix once and for all, a representative $\Gamma_\varphi\subset G$ and a set of generators $\gamma_1^{(\varphi)}, \cdots, \gamma_{r(\varphi)}^{(\varphi)}$ of Γ_φ, and calculate $m(\mathscr{L}, \{\gamma_\varphi\}_{GL})$ for each $\gamma_\varphi=(\gamma_1^{(\varphi)}, \cdots, \gamma_{r(\varphi)}^{(\varphi)}) \in G^{r(\varphi)}$. Now, put $I=\{1, 2, \cdots, H\}$, and for each $\mathscr{G}_\varphi$, define subsets $I(\mathscr{G}_\varphi)$ and $J(\mathscr{G}_\varphi)$ of I as follows:

$$I(\mathscr{G}_\varphi)=\{i \in I;\ \Gamma_i \in \mathscr{G}_\varphi\},$$
$$J(\mathscr{G}_\varphi)=\{j \in I;\ \Gamma_j\supsetneqq\Gamma \text{ for some } \Gamma \in \mathscr{G}_\varphi\}.$$

We want to know $\#(I(\mathscr{G}_\varphi))$ for each $\mathscr{G}_\varphi$. It is obvious that $\{\gamma_\varphi\}_{GL}\cap\Gamma_i^{r(\varphi)}\neq\emptyset$, if and only if $i \in I(\mathscr{G}_\varphi)\cup J(\mathscr{G}_\varphi)$. So we have

$$(*)\qquad m(\mathscr{L}, \{\gamma_\varphi\}_{GL})-\sum_{j\in J(\mathscr{G}_\varphi)}\frac{\#\{\{\gamma_\varphi\}_{GL}\cap\Gamma_j^{r(\varphi)}\}}{\#(\Gamma_j)}=\#(I(\mathscr{G}_\varphi))\times\frac{\#\{\{\gamma_\varphi\}_{GL}\cap\Gamma_\varphi^{r(\varphi)}\}}{\#(\Gamma_\varphi)}.$$

(2) (i) Define the order of Φ as follows: $\varphi<\varphi'(\varphi, \varphi' \in \Phi)$, if and only if $\Gamma\subsetneqq\Gamma'$ for some $\Gamma \in \mathscr{G}_\varphi$ and $\Gamma' \in \mathscr{G}_{\varphi'}$. If φ is maximal with respect to this order among those such that $m(\mathscr{L}, \{\gamma_\varphi\}_{GL})\neq 0$, then $J(\mathscr{G}_\varphi)=\emptyset$, and we have

$$\#(I(\mathscr{G}_\varphi))=m(\mathscr{L}, \{\gamma_\varphi\}_{GL})\times\frac{\#(\Gamma_\varphi)}{\#\{\{\gamma_\varphi\}_{GL}\cap\Gamma_\varphi^{r(\varphi)}\}}.$$

(ii) Fix $\varphi \in \Phi$, and assume that $\#(I(\mathscr{G}_{\varphi'}))$ are known for all φ' such that $\varphi'>\varphi$. As far as we have a good description for $\{\gamma_\varphi\}_{GL}$ and $\Gamma_{\varphi'}^{r(\varphi)}$, we can calculate $\#\{\{\gamma_\varphi\}_{GL}\cap\Gamma_{\varphi'}^{r(\varphi)}\}$, and the left hand side of $(*)$, which is equal to

$$m(\mathscr{L},\{\gamma_\varphi\}_{GL})-\sum_{\substack{\varphi'\in\Phi\\ \varphi'>\varphi}}\#(I(\mathscr{G}_{\varphi'}))\times\frac{\#\{\{\gamma_\varphi\}_{GL}\cap\Gamma_{\varphi'}^{r(\varphi)}\}}{\#(\Gamma_{\varphi'})}.$$

So, we get $\#(I(\mathscr{G}_\varphi))$ also in this case.

1.3. We shall give a mass formula. First, the relation between $m(\mathscr{L},\{\gamma\}_G)$ and $m(\mathscr{L},\{\gamma\}_{GL})$ is given as follows: fix $\gamma\in G^r$, and denote by $\{\gamma_\psi;\ \psi\in\Psi\}$ a complete set of representatives of G-conjugacy classes in $\{\gamma\}_{GL}$. The set Ψ might be infinite, but it is trivial that

$$m(\mathscr{L},\{\gamma\}_{GL})=\sum_{\psi\in\Psi}m(\mathscr{L},\{\gamma_\psi\}_G),$$

and $m(\mathscr{L},\{\gamma_\psi\}_G)=0$, except for finitely many $\psi\in\Psi$. We shall give a formula for $m(\mathscr{L},\{\gamma_\psi\}_G)$.

Theorem 1.1. *Notations and assumptions being as above,*

$$m(\mathscr{L},\{\gamma\}_G)=\sum_{L_G(\Lambda)}M_G(\Lambda)\prod_q c_q(\gamma,U_q,\Lambda),$$

where the product is taken over all primes q and $L_G(\Lambda)$, $M_G(\Lambda)$, and $c_q(\gamma,U_q,\Lambda)$ will be defined below.

Definition of $L_G(\Lambda)$, $M_G(\Lambda)$, **and** $c_q(\gamma,U_q,\Lambda)$**.** These are defined almost in the same way as in [3], or [5] p. 553, except for the small change caused by the fact that $\gamma\notin G$ but $\gamma\in G^r$. We write down their definition here for the sake of completeness. *For any $g=(g_1,\cdots,g_r)\in G^r$, put*

$$Z(g)=\{z\in\mathrm{End}_B(V);\ zg_i=g_iz\ \text{for all}\ i=1,\cdots,r\},$$

and $Z_G(g)=Z(g)\cap G$.

Denote by $Z_G(g)_A$ the adelization of $Z_G(g)$, and by $Z_G(g)_q$ the q-component of $Z_G(g)_A$.

*(1) $L_G(\Lambda)$ runs over the 'G-genera' of **Z**-orders of $Z(\gamma)$: for any **Z**-order $\Lambda\subset Z(\gamma)$,*

$$L_G(\Lambda)=\{\Lambda';\ \Lambda'\text{ is a }\mathbf{Z}\text{-order of }Z(\gamma)\text{ such that, for every prime }q,\ \Lambda'_q=x_q\Lambda_qx_q^{-1}\text{ for some }x_q\in Z_G(\gamma)_q\}.$$

*(2) $M_G(\Lambda)$ is the 'G-mass' of the **Z**-order Λ of $Z(\gamma)$ which is defined as follows: we decompose $Z_G(\gamma)_A$ into disjoint union of double cosets as*

$$Z_G(\gamma)_A=\coprod_{k=1}^{h(\Lambda)} Z_G(\gamma)z_k(\Lambda_A^\times \cap G_A),$$

where $\Lambda_A=Z_G(g)_\infty \prod_q \Lambda_q \subset Z_G(g)_A$. *Put*

$$\Lambda_k=z_k\Lambda z_k^{-1}=\bigcap_q (z_{k,q}\Lambda_q z_{k,q}^{-1}\cap Z(\gamma)).$$

We define

$$M_G(\Lambda)=\sum_{k=1}^{h(\Lambda)}\frac{1}{\#(\Lambda_k^\times \cap G)}.$$

(3) *For any prime q and any* ***Z****-order of* $Z(\gamma)$, *we define the number* $c_q(\gamma, U_q, \Lambda)$ *by*:

$$c_q(\gamma, U_q, \Lambda)=\#(Z_G(\gamma)_q\backslash M_q(\gamma, U_q, \Lambda_q)/U_q),$$

where

$$M_q(\gamma, U_q, \Lambda)=\{x_q \in G_q;\ x_q^{-1}\gamma x_q \in U_q^\gamma \text{ and } Z(\gamma)_q\cap x_qR_qx_q^{-1}=z_q\Lambda_q z_q^{-1} \text{ for some } z_q \in Z_G(\gamma)_q\},$$

and $R_q=\{g_q \in \mathrm{End}_{B_q}(V_q);\ L_qg_q\subset L_q\}$.

Proof of Theorem 1.1. The proof is obtained virtually in the same way as in the proof of Theorem 1 in [3]. We only need to change carefully the objects $Z(g)$, $Z_G(g)$, and $M_q(g, U_q, \Lambda)$ for $g \in G$ in [3] by the new objects $Z(\gamma)$, $Z_G(\gamma)$, and $M_q(\gamma, U_q, \Lambda)$ for $\gamma \in G^r$. The details will be omitted here. q.e.d.

§ 2. Review on quaternion hermitian lattices and some miscellaneous results

From now on unitil the end of this paper, we denote by B the definite quaternion algebra over $\boldsymbol{Q}$ with arbitrary fixed prime discriminant p (except for Theorem 7.2, where the discriminant is not necessarily prime), and by V the two dimensional left B-vector space. Some part of our results is valid also for left B-vector spaces of arbitrary dimension, but we shall not mention on such details.

2.1. First, we review on maximal lattices in V, according to Shimura [16]. The positive definite quaternion hermitian metric on V is unique up to base changes, and we can assume that $V=B^2$ and

$$h(x, y)=x_1\bar{y}_1+x_2\bar{y}_2 \qquad (x=(x_1, x_2),\ y=(y_1, y_2) \in B^2).$$

As in the introduction, we denote by G the group of similitudes of

(B^2, h). Fix a maximal order O of B. For each left O-lattice L, the norm $N(L)$ of L is defined to be the two sided O-ideal spanned by all $h(x, y)$ $(x, y \in L)$. A left O-lattice is called maximal, when L is maximal among those left O-lattices which have the same norm as L. For example, O^2 is a maximal lattice with $N(O^2)=O$. There exists a maximal left O-lattice $M' \subset V$ such that $N(M')=\mathfrak{P}$, where $\mathfrak{P}$ is the unique prime ideal of O dividing p. The set of maximal left O-lattices in V is the disjoint union of $\mathscr{L}(O^2)$ and $\mathscr{L}(M')$, where $\mathscr{L}(O^2)$ (resp. $\mathscr{L}(M')$) is the genus of left O-lattices in V which contains O^2 (resp. M'). The genus $\mathscr{L}(O^2)$ (resp. $\mathscr{L}(M')$) is called the principal (resp. non-principal) genus in V. Our concern in this paper is $\mathscr{L}(M')$, and we describe it more explicitly here. For each prime q, sometimes we use the metric h_q^* defined by:

$$h_q^*(x, y)=x\begin{pmatrix}0 & 1\\ 1 & 0\end{pmatrix}{}^t\bar{y} \qquad (x, y \in B_q^2).$$

We denote by G_q^* the group of similitudes of (B_q^2, h_q^*):

$$G_q^*=\left\{g \in M_2(B_q);\, g\begin{pmatrix}0 & 1\\ 1 & 0\end{pmatrix}{}^t\bar{g}=n(g)\begin{pmatrix}0 & 1\\ 1 & 0\end{pmatrix},\ n(g) \in \boldsymbol{Q}_q^\times\right\}.$$

The above h_q^* is obtained from h_q by a base change of B_q^2. In fact, for each prime q, there exists $\xi_q \in GL_2(O_q)$ such that $\xi_q{}^t\bar{\xi}_q=\begin{pmatrix}0 & 1\\ 1 & 0\end{pmatrix}$. We fix such ξ_q for each q once and for all. It is trivial that $h_q^*(x, y)=h(x\xi_q, y\xi_q)$ and the mapping $G_q \ni g \to \xi_q g \xi_q^{-1} \in G_q^*$ defines an isomorphism $G_q \cong G_q^*$. We often identify G_q with G_q^* by this fixed isomorphism without mentioning it. Now, by Shimura [16], there exists $M \in \mathscr{L}(M')$ such that $M_q=M\otimes_{\boldsymbol{Z}}\boldsymbol{Z}_q=O_q^2$ for every prime $q \neq p$, and that

$$M_p=M\otimes_{\boldsymbol{Z}}\boldsymbol{Z}_p=\{(a\pi, b)\xi_p \in B_p^2;\ a, b \in O_p\},$$

where π is a prime element of O_p. Definite $\mathfrak{U}(M)=G_\infty \prod_q U_q(M)$ as in § 1.1. This $U_q(M)$ is explicitly given as follows: for each prime $q \neq p$, put

$$U_q=G_q \cap GL_2(O_q), \quad \text{and} \quad U_q^*=G_q^* \cap GL_2(O_q)=\xi_q U_q \xi_q^{-1},$$

and for p, put

$$U_p^*=G_q^* \cap \begin{pmatrix}O_p & \pi^{-1}O_p\\ \pi O_p & O_p\end{pmatrix}^\times.$$

Then, $U_q(M)=U_q$ and $U_p(M)=\xi_p^{-1}U_p^*\xi_p$. Denote by H the class number of $\mathscr{L}(M)$, and decompose G_A as in § 1:

$$G_A=\coprod_{i=1}^{H}\mathfrak{A}(M)g_iG \qquad \text{(disjoint)}.$$

For each i $(1\leqq i\leqq H)$, define L_i and Γ_i as in § 1 for this genus $\mathscr{L}(M)$ and the above decomposition.

2.2. In this subsection, we shall give a purely number-theoretical alternative proof of the following Lemma 2.1 which was first obtained by Katsura-Oort [12] by an algebro-geometrical method.

Lemma 2.1 (Katsura-Oort, loc. cit.,). *Assume that $p\geqq 7$. Then, each $\Gamma_i/\{\pm 1\}$ $(1\leqq i\leqq H)$ is isomorphic (as an abstract group) to one of the following groups:*

$$\begin{array}{ll} \boldsymbol{Z}/n\boldsymbol{Z} & (1\leqq n\leqq 6), \\ D_{2n} & (1\leqq n\leqq 6), \\ A_4,\ S_4,\ A_5, & \end{array}$$

where D_{2n} is the dihedral group of order $2n$, S_n (resp. A_n) is the n-th symmetric (resp. alternative) group, and $\pm 1=\pm\begin{pmatrix}1 & 0\\ 0 & 1\end{pmatrix}\in M_2(B)$.

Proof. For any element g of $M_2(B)$, the algebra $\boldsymbol{Q}(g)$ is at most of rank 4 over $\boldsymbol{Q}$. Hence, it is easy to list up all possible characteristic polynomials of elements of $M_2(B)$ which are of finite order, as has been done in [5] p. 590. By this list, we can see that every element of every $\Gamma_i/\{\pm 1\}$ is at most of order 6. Now, put

$$(U_p^*)^1=\left\{u\in U_p^*;\ u\begin{pmatrix}0 & 1\\ 1 & 0\end{pmatrix}{}^t\bar{u}=\begin{pmatrix}0 & 1\\ 1 & 0\end{pmatrix}\right\}.$$

We can embed each Γ_i $(1\leqq i\leqq H)$ into $(U_p^*)^1$ by the mapping:

$$g_i\Gamma_i g_i^{-1}\subset\mathfrak{A}(M)\xrightarrow{\text{pr}}U_p\cong U_p^*,$$

where pr is the natural projection. We denote by Γ_i' the image of Γ_i in $(U_p^*)^1$ by this embedding. Define a subgroup V_p of $(U_p^*)^1$ by:

$$V_p=(U_p^*)^1\cap\begin{pmatrix}1+\pi O_p & O_p\\ \pi^2 O_p & 1+\pi O_p\end{pmatrix}^{\times}.$$

Then, V_p is a normal subgroup of $(U_p^*)^1$, and we get a group isomorphism

$$(U_p^*)^1/V_p\cong SL_2(\boldsymbol{F}_{p^2}).$$

In fact, it is well known that $O_p/\pi O_p \cong \boldsymbol{F}_{p^2}$, and we can show easily that the following map

$$(U_p^*)^1/V_p \ni \begin{pmatrix} a & b\pi^{-1} \\ c\pi & d \end{pmatrix} \longrightarrow \begin{pmatrix} a & b \\ \bar{c} & \bar{d} \end{pmatrix} \bmod \pi \in M_2(\boldsymbol{F}_{p^2})$$

induces the above isomorphism. Now, we want to embed Γ_i into $SL_2(\boldsymbol{F}_{p^2})$. We can show that the order m of any torsion element g of V_p is some multiple of p. In fact, we have

$$g = \begin{pmatrix} 1 & r \\ 0 & 1 \end{pmatrix} + \pi X$$

for some $r \in O_p$ and $X \in M_2(O_p)$, and

$$\begin{pmatrix} 1 & 0 \\ 0 & 1 \end{pmatrix} = g^m \equiv \begin{pmatrix} 1 & mr \\ 0 & 1 \end{pmatrix} \bmod \pi .$$

So, $mr \in \pi O_p$, and if $p \nmid m$, then $r \in \pi O_p$. So, $g \in 1 + \pi M_2(O_p)$. But, it is well known and easy to see that the multiplicative group $1 + \pi M_2(O_p)$ has no torsion element whose order is prime to p. So, we get a contradiction. Next, assume that $p \geqq 7$. Then, by the above considerations, we can show that $\Gamma_i \cap V_p = \{1\}$. In other words, we get an injective group homomorphism $\Gamma_i \hookrightarrow (U_p^*)^1/V_p \cong SL_2(\boldsymbol{F}_{p^2})$. The complete set of subgroups of $PSL_2(\boldsymbol{F}_{p^2})$ is classically wellknown (cf. Huppert [9] p. 213). Besides, the orders of elements $\Gamma_i/\{\pm 1\}$ are at most 6, and not divisible by p. Such subgroups of $PSL_2(\boldsymbol{F}_{p^2})$ are just those listed in Lemma 2.1.

q.e.d.

2.3. Now, we shall review the class number formula for $\mathscr{L}(M)$ and give some miscellaneous results obtained by that formula. The $GL_2(B)$-conjugacy classes $\{g\}_{GL}$ of semi-simple elements of G is parametrized by principal polynomials of g. (The principal polynomial of $g \in M_2(B)$ is defined to be the characteristic polynomial of the image of g by the regular representation $M_2(B) \hookrightarrow M_4(C)$.) More precisely, for a polynomial $f(x) \in \boldsymbol{Q}[x]$ of degree 4, define a subset of G by:

$$C_f = \{g \in G;\ g \text{ is semi-simple and the principal polynomial of } g \text{ is } f(x)\}.$$

It is well known and easy to prove that

$$C_f = \{g\}_{GL}$$

for any $g \in C_f$. So, we sometimes denote $m(\mathscr{L}(M), \{g\}_{GL})$ by $m(\mathscr{L}(M), f(x))$, where $g \in C_f$. We have $m(\mathscr{L}(M), f(x)) = m(\mathscr{L}(M), f(-x))$, because any lattice in V is stable under the action of $\{\pm 1\}$. Define polynomials $f^{(m)}(x)$ $(1\leqq m\leqq 6)$ as follows:

$$f^{(1)}(x)=(x-1)^4,$$
$$f^{(2)}(x)=(x^2+1)^2,$$
$$f^{(3)}(x)=(x^2+x+1)^2,$$
$$f^{(4)}(x)=x^4+1,$$
$$f^{(5)}(x)=x^4+x^3+x^2+x+1,$$
$$f^{(6)}(x)=x^4-x^2+1.$$

For each m $(1\leqq m\leqq 6)$, define $H(m)$ by:

$$H(m)=\begin{cases} m(\mathcal{L}(M), f^{(m)}(x)), & \text{if } f^{(m)}(x)=f^{(m)}(-x) \text{ (i.e., } m=2, 4, 6), \\ 2\times m(\mathcal{L}(M), f^{(m)}(x)), & \text{if } f^{(m)}(x)\neq f^{(m)}(-x) \text{ (i.e., } m=1, 3, 5). \end{cases}$$

We quote the class number formula for $\mathcal{L}(M)$ in [5].

Theorem 2.2 ([5] II). *Assume that* $p\geqq 7$. *Then,* $m(\mathcal{L}(M), f(x))=0$ *except for the case that* $f(x)=f^{(m)}(x)$, *or* $f(x)=f^{(m)}(-x)$ *for some* $m=1, \cdots, 6$, *and*

$$H(1)=\frac{p^2-1}{2880},$$
$$H(2)=\frac{1}{96}\left(4+\left(\frac{-1}{p}\right)\right)\left(p-\left(\frac{-1}{p}\right)\right),$$
$$H(3)=\frac{1}{72}\left(3+\left(\frac{-3}{p}\right)\right)\left(p-\left(\frac{-3}{p}\right)\right),$$
$$H(4)=\begin{cases} \frac{1}{4} & \cdots \text{ if } p\equiv 3 \text{ or } 5 \bmod 8, \\ 0 & \cdots \text{ if } p\equiv 1 \text{ or } 7 \bmod 8, \end{cases}$$
$$H(5)=\begin{cases} \frac{2}{5} & \cdots \text{ if } p\equiv 2 \text{ or } 3 \bmod 5, \\ 0 & \cdots \text{ if } p\equiv 1 \text{ or } 4 \bmod 5, \end{cases}$$
$$H(6)=\begin{cases} \frac{1}{6} & \cdots \text{ if } p\equiv 5 \bmod 12, \\ 0 & \cdots \text{ if } p\equiv 1, 7, \text{ or } 11 \bmod 12, \end{cases}$$

where $\left(\frac{*}{p}\right)$ *is the Legendre symbol.*

The class number H of $\mathscr{L}(M)$ is given by:

$$H=\sum_{m=1}^{6} H(m).$$

Remark. In the notations in [5] II, $H(1)=H_1$, $H(2)=H_6$, $H(3)=H_7$, $H(4)=H_{11}$, $H(5)=H_{10}$, and $H(6)=H_{12}$.

By the above Theorem 2.2, we see that the natural projection on $\Gamma_i/\{\pm 1\}$ of any element γ of any Γ_i $(1\leqq i\leqq H)$ is of order m, if and only if the characteristic polynomial of γ is equal to $f^{(m)}(x)$, or $f^{(m)}(-x)$. For each (abstract) finite group Γ', and each m $(1\leqq m\leqq 6)$, define a rational number $M(\Gamma', m)$ by:

$$M(\Gamma', m)=\frac{\sharp\{\gamma' \in \Gamma';\, \gamma' \text{ is of order } m\}}{\sharp(\Gamma')}.$$

Then, for each $p\geqq 7$, we have

$$H(m)=\sum_{i=1}^{H} M(\Gamma_i/\{\pm 1\}, m)=\sum_{\Gamma'} \sharp(I(\Gamma'))M(\Gamma', m), \tag{2.3}$$

where Γ' runs over a complete set of representatives of all group isomorphism classes of all (abstract) finite groups, and

$$I(\Gamma')=\{i \in I;\, \Gamma_i/\{\pm 1\}\cong\Gamma' \text{ as } \textit{abstract} \text{ groups}\},$$

$I=\{1, \cdots, H\}$. In § 1, we defined the notation $I(\mathscr{G})$ for each $GL_2(B)$-conjugacy class $\mathscr{G}$ of finite subgroups of G. In the case $\mathscr{L}(M)$, the use of the above (new) notation $I(\Gamma')$ is consistent with the old one in § 1 by the following reason: we can show that $\Gamma_i/\{\pm 1\}\cong\Gamma_j/\{\pm 1\}$ for i, j $(1\leqq i, j\leqq H)$, if and only if $h^{-1}\Gamma_i h=\Gamma_j$ for some $h \in GL_2(B)$. In fact, if $\Gamma_i/\{\pm 1\}$ is cyclic, this is obvious by Theorem 2.2 and the Skolem-Noether Theorem. The proof in the case where $\Gamma_i/\{\pm 1\}$ is not cyclic is easily obtained by (3.2) in § 3, the proof of Corollary 4.2 in § 4, and by Lemma 6.7 in § 6, using the Skolem-Noether Theorem. By using Lemma 2.2 and some geometrical alternatives of Theorem 2.2 and (2.3), Katsura-Oort [12] derived some information on Problem 2 in the case of $\mathscr{L}(M)$ and solved it for $p\leqq 31$. For example, we get $M(\boldsymbol{Z}/4\boldsymbol{Z}, 4)=\frac{1}{2}$, $M(\boldsymbol{Z}/5\boldsymbol{Z}, 5)=\frac{4}{5}$, $M(\boldsymbol{Z}/6\boldsymbol{Z}, 6)=\frac{1}{3}$. But, for each prime $p\geqq 7$, we have $H(4)\leqq\frac{1}{4}$, $H(5)\leqq\frac{2}{5}$, and $H(6)\leqq\frac{1}{6}$. So, we see easily that $I(\boldsymbol{Z}/4\boldsymbol{Z})=I(\boldsymbol{Z}/5\boldsymbol{Z})=I(\boldsymbol{Z}/6\boldsymbol{Z})=\varnothing$. In the same way, by Lemma 2.1, Theorem 2.2, and (2.3), we get

$$\sharp(I(D_{12}))=\begin{cases}1 & \cdots \text{ if } p\equiv 5 \bmod 12, \quad (p\neq 5)\\ 0 & \cdots \text{ if } p\equiv 1, 7, \text{ or } 11 \bmod 12.\end{cases}$$

(As for details, see [12] loc. cit.)

If p is very small, Lemma 2.1, Theorem 2.2, and (2.3) give us enough information to solve Problem 2, as was executed in [12]. But, as for general p, it is indispensable to give some new mass formula in order to solve Problem 2. For example, we have $M(D_8, 4)=M(S_4, 4)=\frac{1}{4}$, and $M(D_{10}, 5)=M(A_5, 5)=\frac{2}{5}$, and at least for big prime p, we can not tell which of D_8 or S_4 (, or D_{10} or A_5) appears as $\Gamma_i/\{\pm 1\}$, unless we have some new formula. The rest of this paper will be devoted to a calculation of the new mass formula.

§ 3. Classification of conjugacy classes

In order to calculate the mass $m(\mathscr{L}(M), \{\gamma\}_{GL})$ explicitly for each $\gamma \in G^r$ (r: natural integer) by using Theorem 1.1, first we must classify the G-conjugacy classes in $\{\gamma\}_{GL}=\{h^{-1}\gamma h \in G; h \in GL_2(B)\}$. We shall give a general (but not explicit) parametrization in 3.1, imitating the method in Hijikata [8] (see also [5]), and more explicit theory in the cases we need will be given in 3.2.

3.1. For any natural integer s and any element $g=(g_1, \cdots, g_s) \in M_2(B)$, put ${}^t\bar{g}=({}^t\bar{g}_1, \cdots, {}^t\bar{g}_s)$. Any element $z \in M_2(B)$ is called positive symmetric, $z>0$, if $z={}^t\bar{z}$ and $\operatorname{tr}(xz\,{}^t\bar{x})>0$ for all $x \in M_2(B)$, $x\neq 0$, where tr is the reduced trace of $M_2(B)$. Now, fix a natural number r and an element $\gamma=(\gamma_1, \cdots, \gamma_r) \in G^r$. Denote by $Z(\gamma)\subset M_2(B)$ the "commutor algebra" of γ in $M_2(B)$ as in § 1.2, and denote by $Z(\gamma)^*_+$ the set of all positive symmetric elements in $Z(\gamma)$:

$$Z(\gamma)^*_+=\{z \in Z(\gamma);\ z={}^t\bar{z}>0\}.$$

We define an equivalence relation $\approx$ in $Z(\gamma)^*_+$ by putting $z\approx z'$, if and only if $axz\,{}^t\bar{x}=z'$ for some $x \in Z(\gamma)^\times$ and $a \in \boldsymbol{Q}^\times_+$.

Lemma 3.1. *The mapping $g^{-1}\gamma g\to g\,{}^t\bar{g}$ induces a bijection:*

$$\{\gamma\}_{GL}/\underset{G}{\sim} \approx Z(\gamma)^*_+/\approx,$$

where $\underset{G}{\sim}$ means the equivalence by G-conjugation.

Proof. For $g \in GL_2(B)$, we have $g^{-1}\gamma g \in G^r$, if and only if $(g^{-1}\gamma g)({}^t\bar{g}\,{}^t\bar{\gamma}\,{}^t\bar{g}^{-1})=(1, \cdots, 1) \in G^r$, taking the product in G^r. Namely, $g^{-1}\gamma g \in G^r$, if and only if $g\,{}^t\bar{g} \in Z(\gamma)^*_+\subset Z(\gamma)$. Next, if $g_1^{-1}\gamma g_1, g_2^{-1}\gamma g_2 \in G^r$ and $g_1^{-1}\gamma g_1=h^{-1}g_2^{-1}\gamma g_2 h$ for some $g_1, g_2 \in GL_2(B)$ and $h \in G$, then $g_2 h g_1^{-1} \in Z(\gamma)^\times$. In other words, we have $g_2{}^t\bar{g}_2=cx(g_1{}^t\bar{g}_1){}^t\bar{x}$ for some $c \in \boldsymbol{Q}^\times_+$ and $x \in Z(\gamma)^\times$. So the mapping is well defined. It is well known that any positive symmetric element of $M_2(B)$ is of the form $g{}^t\bar{g}$ for some $g \in GL_2(B)$. Hence we get the surjectivity. For any $c \in \boldsymbol{Q}_+^\times$, there exists $h \in G$ such that $h{}^t\bar{h}=c1_2$. So the injectivity is clear. q.e.d.

3.2. In this subsection, we treat some special cases we need. We assume here that $p \geqq 5$. When $r=1$, the mass formulae were given in the class number formula. So, we treat the case $r \geqq 2$, or those $\gamma=(\gamma_1, \cdots, \gamma_r) \in G^r$ such that the group spanned by $\gamma_1, \cdots, \gamma_r$ is not cyclic. Now, we assume that there exists an injective group homomorphism $D_{2n} \hookrightarrow \Gamma_i/\{\pm 1\}$ for some i $(1 \leqq i \leqq H)$ and some n $(2 \leqq n \leqq 6)$. For each fixed n $(2 \leqq n \leqq 6)$, denote by σ_n', τ_n' the generators of D_{2n} such that $\sigma_n'^2=1$, $\tau_n'^n=1$, and $\sigma_n'\tau_n'=\tau_n'^{-1}\sigma_n'$. Fix a representative σ_n (resp. τ_n) of σ_n' (resp. τ_n') in Γ_i. First, we examine the structure of the subalgebra $\boldsymbol{Q}(\sigma_n, \tau_n)$ of $M_2(B)$ spanned by σ_n, τ_n. As $\sigma_n'^2=1$, we have $\sigma_n^2=\pm 1$.

If $\sigma_n^2=1$, then, by Theorem 2.2, we have $\sigma_n=\pm 1$, and hence $\sigma_n'=1$, which is a contradiction. So, we have $\sigma_n^2=-1$. We get also $\sigma_n\tau_n=\tau_n^{-1}\sigma_n$. In fact, if $\sigma_n\tau_n=-\tau_n^{-1}\sigma_n$, then $(\sigma_n\tau_n)^2=-(\sigma_n\tau_n)(\tau_n^{-1}\sigma_n)=-\sigma_n^2=1$, and hence $\sigma_n'\tau_n'=1$, which is a contradiction. By easy calculation, we can show that $\boldsymbol{Q}(\tau_n+\tau_n^{-1})$ is the center of $\boldsymbol{Q}(\sigma_n, \tau_n)$. Besides, it is easy to see that $\boldsymbol{Q}(\sigma_n, \tau_n)$ is the totally definite quaternion algebra over $\boldsymbol{Q}(\tau_n+\tau_n^{-1})$, and ${}^t\bar{x}$ induces on $\boldsymbol{Q}(\sigma_n, \tau_n)$ the main involution over $\boldsymbol{Q}(\tau_n+\tau_n^{-1})$. More precisely, we get

(3.2) for each $n=2$, or 3, $\boldsymbol{Q}(\sigma_n, \tau_n)$ is the definite quaternion algebra over $\boldsymbol{Q}$, and the discriminant is 2 (resp. 3) when $n=2$ (resp. 3).

(3.3) for each $n=4$, 5, or 6, $\boldsymbol{Q}(\sigma_n, \tau_n)$ is the totally definite quaternion algebra over $\boldsymbol{Q}(\sqrt{2})$, $\boldsymbol{Q}(\sqrt{5})$, or $\boldsymbol{Q}(\sqrt{3})$, respectively, and the discriminant is 1.

The proof consists of standard exercises of the classfield theory, and will be omitted here. Incidentally, by Eichler [2], it can be easily shown that the class number of each $\boldsymbol{Q}(\sigma_n, \tau_n)$ $(2 \leqq n \leqq 5)$ is equal to one. Now, for each n $(2 \leqq n \leqq 6)$, put $\gamma_n=(\sigma_n, \tau_n)$. When $n=6$, the mass $m(\mathcal{L}(M), \{\gamma_6\}_{GL})$ was already known in § 2. When $n=4$, or 5, we have some special method to calculate the masses $m(\mathcal{L}(M), \{\gamma_n\}_{GL})$, as will be shown in the next section. So, we shall give here an explicit parametrization for $\{\gamma_n\}/\underset{G}{\sim}$ only when $n=2$, or 3.

For each $n=2$, or 3, and for each pair (α, β) of elements of $M_2(B)$, or $M_2(B_q)$ (q: prime), we consider the following conditions:

(3.4) (i) $\alpha\beta=\beta^{-1}\alpha$,
(ii) the principal polynomial of α is $f^{(2)}(x)$,
(iii) the principal polynomial of β is $f^{(n)}(x)$.

For each $n=2$, or 3, define a subset C_n of G^2 by:

$$C_n=\{(\alpha, \beta) \in G^2; \text{ the pair } (\alpha, \beta) \text{ satisfies } (3.4)\},$$

and for each prime q, define a subset $C_n(q)$ of G_q^2 by:

$$C_n(q)=\{(\alpha, \beta) \in G_q^2; \text{ the pair } (\alpha, \beta) \text{ satisfies } (3.4)\}.$$

We sometimes regard $C_n(q)$ also as a subset of G_q^* by the fixed identification $G_q^* \cong G_q$. By the Skolem-Noether Theorem and (3.2), it is easy to show that C_n (resp. $C_n(q)$) forms a single $GL_2(B)$ (resp. $GL_2(B_q)$)-conjugacy class for each n and q, if $C_n \neq \emptyset$ (resp. $C_n(q) \neq \emptyset$). Actually, we can show that $C_n \neq \emptyset$ for each $p \geqq 5$. In fact, there exist integers λ, $\mu \in Z$ such that

$$\left(\frac{-(1+\lambda^2)}{p}\right) \neq 1, \quad \text{and} \quad \left(\frac{-3(1+\mu^2)}{p}\right)=1,$$

where $\left(\frac{*}{p}\right)$ is the Legendre symbol. Fix such λ, μ. There exist c, $d \in B$ such that $c^2=-1/(1+\lambda^2)$ and $d^2=-3/(1+\mu^2)$. Define elements α, β_2, $\beta_3 \in G$ by:

$$\alpha=\begin{pmatrix} 0 & -1 \\ 1 & 0 \end{pmatrix}, \quad \beta_2=c\begin{pmatrix} 1 & \lambda \\ \lambda & -1 \end{pmatrix}, \quad \beta_3=\frac{1}{2}\begin{pmatrix} -1+d & d\mu \\ d\mu & -1-d \end{pmatrix}.$$

Then, $(\alpha, \beta_2) \in C_2$ and $(\alpha, \beta_3) \in C_3$. Now, for each $n=2$ or 3, fix $\gamma=(\alpha, \beta) \in C_n$. As we have $\boldsymbol{Q}(\alpha, \beta) \otimes Z(\gamma) \cong M_2(B)$, we can show by (3.2) that $Z(\gamma)$ is the *indefinite* quaternion algebra over $\boldsymbol{Q}$ with discriminant $2p$ (resp. $3p$), when $n=2$ (resp. 3). The algebra $Z(\gamma)$ is stable under the action $x \rightarrow {}^t\bar{x}$, but this does not induce the main involution ρ of $Z(\gamma)$. As ${}^t\bar{x}$ induces a positive involution on $Z(\gamma)$, there exists $a \in Z(\gamma)^\times$ such that ${}^t\bar{x}=a^{-1}x^\rho a$ for any $x \in Z(\gamma)$ and $a^2 \in \boldsymbol{Q}^\times$, $a^2>0$ (Shimura [15]). By the Skolem-Noether Theorem, there exists $b \in Z(\gamma)$ such that $ba=-ab$ and $b^2 \in \boldsymbol{Q}^\times$. As $Z(\gamma)$ is indefinite, we get $b^2>0$. It is obvious that 1, a, b, ab form a basis of $Z(\gamma)$. For an element $z \in Z(\gamma)$, we have $z={}^t\bar{z}$, if and only if $z=x_1+x_2b+x_3ab$ for some x_1, x_2, $x_3 \in \boldsymbol{Q}$. It is easy to see that $z={}^t\bar{z}>0$, if and only if $n(z)>0$ and $\mathrm{tr}(z)>0$, where n (resp. tr) is the reduced norm (resp. trace) of $Z(\gamma)$. Namely, we have

$$Z(\gamma)_+^*=\{z \in Z(\gamma); n(z)>0, \mathrm{tr}(z)>0, \mathrm{tr}(za^{-1})=0\}.$$

So, for each $z \in Z(\gamma)_+^*$ there eixsts the unique square-free positive integer $m(z)$ such that $\boldsymbol{Q}(za^{-1}) \cong \boldsymbol{Q}(\sqrt{-m(z)})$.

Proposition 3.5. *The mapping* $Z(\gamma)_+^* \ni z \rightarrow \boldsymbol{Q}(\sqrt{-m(z)})$ *induces the following bijection*:

$$Z(\Upsilon)^*_+/\approx\ \approx \left\{\begin{array}{l}\textit{the isomorphism classes of imaginary quadratic}\\ \textit{field which can be embedded into } Z(\Upsilon)\end{array}\right\}.$$

Proof. If $z_1, z_2 \in Z(\Upsilon)^*_+$ and $z_1 \approx z_2$, then $z_1a^{-1} = cd(z_2a^{-1})d^\rho$ for some $c \in \boldsymbol{Q}^\times_+$ and $d \in Z(\Upsilon)$. So, we have $\boldsymbol{Q}(z_1a^{-1}) \cong \boldsymbol{Q}(z_2a^{-1})$. So, the mapping factors through $Z(\Upsilon)^*_+/\approx$. If K is an imaginary quadratic field contained in $Z(\Upsilon)$, and if $K \cong \boldsymbol{Q}(\sqrt{-m})$ for some square-free integer $m>0$, then there exists the unique element $y \in K$ such that $y^2 = -m$ and $\mathrm{tr}(ya)>0$. If we put $z=ya$, then ${}^t\bar{z} = {}^t\bar{a}\,{}^t\bar{y} = a^{-1}a^\rho aa^{-1}y^\rho a = a^{-1}(ay)a = ya = z$, and $\mathrm{tr}(z)>0$, $n(z)=n(a)n(y)>0$. So, $z \in Z(\Upsilon)^*_+$, and $m(z)=m$. Hence, the surjectivity is proved. The injectivity is proved as follows: If $\boldsymbol{Q}(z_1a^{-1}) \cong \boldsymbol{Q}(z_2a^{-1})$ for $z_1, z_2 \in Z(\Upsilon)^*_+$, then there exists $c \in \boldsymbol{Q}^\times_+$ such that $(z_1a^{-1})^2 = c^2(z_2a^{-1})^2$. Hence, by the Skolem-Noether Theorem, there exists $z_0 \in Z(\Upsilon)^\times$ such that $z_0^{-1}(z_1a^{-1})z_0 = cz_2a^{-1}$. This means that $z_1 = cn(z_0)^{-1}z_0z_2\,{}^t\bar{z}_0$. q.e.d.

Now, we show that the above parametrization in Proposition 3.5 is "canonical", that is, it does not depend on the choice of initial $\Upsilon \in C_n$. Fix $\Upsilon \in C_n$ as before, and for each $\Upsilon' = g^{-1}\Upsilon g \in \{\Upsilon\}_{GL}$ ($g \in GL_2(B)$), define an imaginary quadratic subfield $K(\Upsilon')$ of $M_2(B)$ by:

$$K(\Upsilon') = \boldsymbol{Q}({}^t\bar{g}a^{-1}g).$$

As $K(\Upsilon') = g^{-1}\boldsymbol{Q}(g\,{}^t\bar{g}a^{-1})g$, the "parameter" of Υ' in Proposition 3.5 is given by the isomorphism class which contains $K(\Upsilon')$. We have $Z(\Upsilon') = g^{-1}Z(\Upsilon)g \supset K(\Upsilon')$, and the involution ${}^t\bar{g}$ induces the complex conjugation on $K(\Upsilon')$. As $Z(\Upsilon')$ is indefinite, by the Skolem-Noether Theorem, there exists $y \in Z(\Upsilon')$ which satisfies the following condition:

(3.6) $\qquad yk = {}^t\bar{k}y$ for any $k \in K(\Upsilon')$, and $0 < y^2 \in \boldsymbol{Q}$.

For any such y, we have $Z(\Upsilon') = K(\Upsilon') + yK(\Upsilon')$.

Proposition 3.7. *Notations and assumptions being as above, $K(\Upsilon')$ is the unique quadratic subfield of $Z(\Upsilon')$ such that its non-trivial automorphism over $\boldsymbol{Q}$ is given by the action $x \to {}^t\bar{x}$. We have also*

$$Z_G(\Upsilon') = Z(\Upsilon') \cap G = K(\Upsilon')^\times \cup yK(\Upsilon')^\times,$$

where y is any element of $Z(\Upsilon')$ which satisfies (3.6).

Proof. First, we show that $y = {}^t\bar{y}$. We have $y = g^{-1}y_0g$ for some $y_0 \in Z(\Upsilon)$, and by definition of a, ${}^t\bar{y}_0 = a^{-1}y_0^\rho a = -a^{-1}y_0a$. As ${}^t\bar{g}ag \in K(\Upsilon')$, we have $y = -({}^t\bar{g}a^{-1}g)y(g^{-1}a\,{}^t\bar{g}^{-1}) = {}^t\bar{g}\,{}^t\bar{y}_0\,{}^t\bar{g}^{-1} = {}^t\bar{y}$. Hence, the action $x \to {}^t\bar{x}$ induces the identity mapping on $yK(\Upsilon')$. This proves the first assertion. Now, it is clear that $K(\gamma')$, $yK(\gamma')\subset Z_G(\gamma')$. On the contrary, if $z\in Z_G(\gamma')$, then $z\,{}^t\bar{z}=c1_2$ for some $c\in \boldsymbol{Q}_+^\times$. By easy calculation, we get ${}^t\bar{z}=({}^t\bar{g}a^{-1}g)z^{\rho'}(g^{-1}a\,{}^t\bar{g})$, where ρ' is the main involution of $Z(\gamma')$. Hence, we get $zk=cn(z)^{-1}kz$ for any $k\in K(\gamma')$, where $n(z)$ is the reduced norm. Taking the norm of both sides, we see that $cn(z)^{-1}=\pm 1$. Hence, $z\in K(\gamma')\cup yK(\gamma')$. q.e.d.

The local version of Propositions 3.5 and 3.7 is obtained in a similar way.

Proposition 3.8. *For each prime q and each $\gamma=(\alpha,\beta)\in C_n(q)$ ($n=2$, or 3), there exists the unique commutative semi-simple subalgebra $K(\gamma)_q$ of $Z(\gamma)_q=\{z\in M_2(B_q);\ z\alpha=\alpha z,\ z\beta=\beta z\}$ of rank 2 over $\boldsymbol{Q}_q$ such that ${}^t\bar{x}$ induces the non-trivial automorphism over $\boldsymbol{Q}_q$ on it. We have*

$$Z_G(\gamma)_q=K(\gamma)_q^\times\cup yK(\gamma)_q^\times,$$

where y is any element of $Z(\gamma)_q$ such that $yk={}^t\bar{k}y$ for any $k\in K(\gamma)_q$. Two elements $\gamma,\gamma'\in C_n(q)$ are G_q-conjugate, if and only if $K(\gamma)\cong K(\gamma')$.

The proof is virtually the same as that of Proposition 3.7, and will be omitted here.

By Propositions 3.7 and 3.8, we get the following "Hasse principle".

Proposition 3.9. *For elements $\gamma,\gamma'\in C_n$ (for fixed $n=2$, or 3), γ is G-conjugate to γ', if and only if γ is G_A-conjugate to γ'.*

Proof. The "only if" part is trivial. Now, we prove the converse. If $\gamma=g^{-1}\gamma' g$ for some $g\in G_A$, then by Proposition 3.8, $K(\gamma)\otimes_{\boldsymbol{Q}}\boldsymbol{Q}_q\cong K(\gamma')\otimes_{\boldsymbol{Q}}\boldsymbol{Q}_q$ for every prime q. So, $K(\gamma)\cong K(\gamma')$. q.e.d.

Remark. As was explained in Hijikata [7], in many cases, the Hasse principle on G-conjugacy classes can be reduced to that on various forms, that is, in our cases, those forms defined by $*$-symmetric positive definite elements of $Z(g)$, where $*$ is the natural involution of $Z(g)$ induced by that of B. For example, the proof of Proposition 3.9 was reduced to the Hasse principle on quaternion anti-hermitian forms of degree one in a "multiplicative" sense (i.e. (H-III) in Hijikata [6]). This method works also in general cases, unless $*$-symmetric elements of $Z(g)$ is equivalent to quaternion anti-hermitian forms (cf. [7]). As the Hasse principle on quaternion anti-hermitian forms is false for general degree (Hijikata [6], Bayer-Fluckiger [1]), we do not know at present whether the Hasse principle on G-conjugacy classes in G^r is valid or not for the remaining cases.

§ 4. Non-existence of D_8 and D_{10}

We assume that $p \geqq 7$. In this section, we shall prove that $I(D_8) = I(D_{10}) = \emptyset$ for $\mathscr{L}(M)$ by showing that, if $\Gamma_i/\{\pm 1\} \hookleftarrow D_8$ (resp. D_{10}), then $\Gamma_i/\{\pm 1\} \cong S_4$ (resp. A_5). We denote by R_i the right order of L_i:

$$R_i = \{g \in M_2(B);\ L_i g \subset L_i\}.$$

Proposition 4.1. *Assume that $\Gamma_i/\{\pm 1\} \hookleftarrow D_{2n}$ for $n=4$, or 5. Then, $R_i \cap \boldsymbol{Q}(\sigma_n, \tau_n)$ is a maximal order of $\boldsymbol{Q}(\sigma_n, \tau_n)$, where σ_n and τ_n are defined for each n as in § 3.2.*

The proof of this proposition will be given later in this section. Here, we shall give Corollaries to this Proposition.

Corollary 4.2. *If $\Gamma_i/\{\pm 1\}$ contains a subgroup which is isomorphic to D_8 (resp. D_{10}), then $\Gamma_i/\{\pm 1\} \cong S_4$ (resp. A_5).*

Proof. For each $n=4$, or 5, define σ_n, τ_n as before, and put $\Omega_n = R_i \cap \boldsymbol{Q}(\sigma_n, \tau_n)$. Denote by W_n the group of units of $\boldsymbol{Z}[\tau_n + \tau_n^{-1}]$, and denote by E_n the subgroup of Ω_n defined by:

$$E_n = \{\gamma \in \Omega_n^\times;\ \gamma\, {}^t\bar{\gamma} = 1\}.$$

Then, $E_n/\{\pm 1\} \cong \Omega_n/W_n$. In fact, if $x \in \Omega_n^\times$, then $x\,{}^t\bar{x} = \varepsilon \in W_n$, but as the norm of the fundamental unit of $\boldsymbol{Q}(\tau_n + \tau_n^{-1})$ is -1, there exists $\varepsilon_0 \in W_n$ such that $\varepsilon_0^2 = \varepsilon$. Hence $(\varepsilon_0 x)\,{}^t\overline{(\varepsilon_0 x)} = 1$. By the usual mass formula (Eichler [1]), we have

$$\frac{1}{\#(\Omega_4/W_4)} \leqq \frac{1}{24} \quad \text{and} \quad \frac{1}{\#(\Omega_5/W_5)} \leqq \frac{1}{60}.$$

As $\Gamma_i \supset E_4$, or E_5, we get $\Gamma_i/\{\pm 1\} \cong S_4$, or A_5, by Lemma 3.1. q.e.d.

Corollary 4.3. *For $\mathscr{L}(M)$ and each $p \geqq 7$,*

$$\#(I(S_4)) = \begin{cases} 1 & \cdots \text{ if } p \equiv 3,\ 5 \bmod 8, \\ 0 & \cdots \text{ if } p \equiv 1,\ 7 \bmod 8, \end{cases}$$

$$\#(I(A_5)) = \begin{cases} 1 & \cdots \text{ if } p \equiv 2,\ 3 \bmod 5, \\ 0 & \cdots \text{ if } p \equiv 1,\ 4 \bmod 5. \end{cases}$$

Proof. By Lemma 3.1, we get

$$H(4) = M(S_4, 4)\,\#(I(S_4)) + M(D_8, 4)\,\#(I(D_8)),$$
$$H(5) = M(A_5, 5)\,\#(I(A_5)) + M(D_{10}, 5)\,\#(I(D_{10})).$$

As $I(D_8)=I(D_{10})=\emptyset$ by Corollary 4.2, we get Corollary 4.3. q.e.d.

Proof of Proposition 4.1. For each $n=4$, or 5, put $\Omega_n=R_i\cap \boldsymbol{Q}(\sigma_n, \tau_n)$ as before, and put $\delta_n=\tau_n+\tau_n^{-1}$. Then, $Z(\delta_n)=\boldsymbol{Q}(\sigma_n, \tau_n)$. It is sufficient to show that $\Omega_n\otimes_{\boldsymbol{Z}}\boldsymbol{Z}_q$ is maximal in $Z(\delta_n)\otimes_{\boldsymbol{Q}}\boldsymbol{Q}_q$ for every prime q. By easy calculation, we can show that the discriminant of the order $\boldsymbol{Z}[\sigma_n, \tau_n]$ of $Z(\delta_n)$ over $\boldsymbol{Z}[\delta_n]$ is equal to $n\boldsymbol{Z}[\delta_n]$ for each $n=4$, or 5. Hence, $\Omega_n\otimes_{\boldsymbol{Z}}\boldsymbol{Z}_q$ is maximal for each prime $q\neq 2$ (resp. $\neq 5$), if $n=4$ (resp. 5). So, we must show that $\Omega_4\otimes_{\boldsymbol{Z}}\boldsymbol{Z}_2$ and $\Omega_5\otimes_{\boldsymbol{Z}}\boldsymbol{Z}_5$ are maximal. Denote by δ_5' the natural projection of $g_i\delta_5 g_i^{-1}\in M_2(O_A)$ in $M_2(O_5)$. Taking $\xi_5\in GL_2(O_5)$ such that $\xi_5{}^t\bar{\xi}_5=\begin{pmatrix}0&1\\1&0\end{pmatrix}$ as in § 3.1, put $\zeta_5=\xi_5\delta_5'\xi_5^{-1}$. It is obvious that $\Omega_5\otimes_{\boldsymbol{Z}}\boldsymbol{Z}_5$ is maximal, if and only if $Z(\zeta_5)_5\cap M_2(O_5)$ is a maximal order of $Z(\zeta_5)_5$. To show that $Z(\zeta_5)\cap M_2(O_5)$ is maximal, it is sufficient to show that $Z(u^{-1}\zeta_5 u)_5\cap M_2(O_5)=u^{-1}Z(\zeta_5)_5 u\cap M_2(O_5)$ is maximal for some $u\in U_5^*$. Now, fix an identification $O_5=M_2(\boldsymbol{Z}_5)$, and define an elements ζ of $M_2(O_5)$ by:

$$\zeta=\begin{pmatrix}h&0\\0&\bar{h}\end{pmatrix},$$

where $h=\begin{pmatrix}0&1\\1&-1\end{pmatrix}$ and $\bar{h}=\begin{pmatrix}-1&-1\\-1&0\end{pmatrix}\in M_2(\boldsymbol{Z}_5)$. We show that $\zeta_5=u^{-1}\zeta u$ for some $u\in U_5^*$. As $2\delta_5+1\in G$, we have $2\zeta_5+1\in G_5^*$. As $2\zeta+1\in G_5^*$, by [4] Theorem 1, we have $x^{-1}(2\zeta+1)x=2\zeta_5+1$ for some $x\in G_5^*$, and hence $x^{-1}\zeta x=\zeta_5$. Now, we show that $x\in Z_G(\zeta)_5\cdot U_5^*$, where $Z_G(\zeta)_5=\{z\in G_5^*;\ z\zeta=\zeta z\}$. For $x, x'\in G_5^*$, we shall write $x\sim x'$, if $x\in Z_G(\zeta)x'U_5^*$. By the Iwasawa decomposition of G_5^* with respect to the compact group U_5^*(cf. [5] p. 579, where 1 is a misprint for 0), and by the fact that $\begin{pmatrix}c&0\\0&1\end{pmatrix}\in Z_G(\zeta)$ for any $c\in \boldsymbol{Q}_5^\times$, we get $x\sim\begin{pmatrix}\bar{\alpha}^{-1}&\beta\\0&\alpha\end{pmatrix}$ for some $\alpha\in B_5^\times$, $\beta\in B_5$ such that $\mathrm{tr}(\bar{\alpha}\beta)=0$. If we put $F=\boldsymbol{Q}_5(h)\subset M_2(\boldsymbol{Q}_5)=B_5$, then by [4] Lemma 10, we have $\alpha\in F^\times\begin{pmatrix}1&0\\0&5^m\end{pmatrix}GL_2(\boldsymbol{Z}_5)$ for some $m\in\boldsymbol{Z}$. As $\begin{pmatrix}\bar{a}^{-1}&0\\0&a\end{pmatrix}\in Z_G(\zeta)$ and $\begin{pmatrix}\bar{b}^{-1}&0\\0&b\end{pmatrix}\in U_5^*$ for any $a\in F^\times$ and $b\in GL_2(\boldsymbol{Z}_5)$, we can also assume that $\alpha=\begin{pmatrix}1&0\\0&5^m\end{pmatrix}$. As $x^{-1}\zeta x\in M_2(O_5)$, we get $\alpha^{-1}\bar{h}\alpha=\begin{pmatrix}-1&5^m\\-5^{-m}&0\end{pmatrix}\in M(\boldsymbol{Z}_5)$. Hence $m=0$, that is, $x\sim\begin{pmatrix}1&\beta\\0&1\end{pmatrix}$ for some $\beta\in B_5$ such that $\mathrm{tr}(\beta)=0$. Now, put $\beta=\begin{pmatrix}a&b\\c&-a\end{pmatrix}\in M_2(Q_5)$. As $\begin{pmatrix}1&-\beta\\0&1\end{pmatrix}\begin{pmatrix}h&0\\0&\bar{h}\end{pmatrix}\begin{pmatrix}1&\beta\\0&1\end{pmatrix}=\begin{pmatrix}h&h\beta-\beta\bar{h}\\0&\bar{h}\end{pmatrix}\in M_2(O_5)$, we have $h\beta-\beta h=(a+b+c)1_2\in M_2(\boldsymbol{Z}_5)$. Put $d=(a+b+c)/3$, and $\beta_0 = d\begin{pmatrix}1 & 1\\ 1 & -1\end{pmatrix}$. As $\begin{pmatrix}1 & \beta_0\\ 0 & 1\end{pmatrix} \in U_5^*$ and $\begin{pmatrix}1 & \beta-\beta_0\\ 0 & 1\end{pmatrix} \in Z_G(\zeta)$, we get $\begin{pmatrix}1 & \beta\\ 0 & 1\end{pmatrix} \sim \begin{pmatrix}1 & 0\\ 0 & 1\end{pmatrix}$. This means that ζ is U_5^*-conjugate to ζ_5. By easy calculation, we see that

$$Z(\zeta)_5 = \begin{pmatrix}1 & 0\\ 0 & y\end{pmatrix} M_2(F) \begin{pmatrix}1 & 0\\ 0 & y\end{pmatrix}^{-1},$$

and

$$Z(\zeta)_5 \cap M_2(O_5) = \begin{pmatrix}1 & 0\\ 0 & y\end{pmatrix} M_2(o_F) \begin{pmatrix}1 & 0\\ 0 & y\end{pmatrix}^{-1}$$

where o_F is the maximal order of F and $y = \begin{pmatrix}0 & 1\\ -1 & 0\end{pmatrix} \in M_2(\boldsymbol{Z}_5)$. Thus, we proved that $\Omega_5 \otimes_{\boldsymbol{Z}} \boldsymbol{Z}_5$ is maximal. The proof of the maximality of $\Omega_4 \otimes_{\boldsymbol{Z}} \boldsymbol{Z}_2$ can be obtained in a similar way, and will be omitted here.

q.e.d.

§ 5. Calculation of local data

In this section, we calculate the "local data" $c_q(\gamma, U_q, \Lambda)$ which are the most important and complicated terms in the mass formula in Theorem 1.1. For each prime q and each $n=2$ or 3, define the subset $C_n(q)$ of G_q^2 as in § 3.2. For each $\gamma = (\alpha, \beta) \in C_n(q)$, put

$$Z(\gamma)_q = \{z \in M_2(B_q);\ z\alpha = \alpha z,\ z\beta = \beta z\},$$

and $Z_G(\gamma)_q = Z(\gamma)_q \cap G_q$. For each order $\Lambda \subset Z(\gamma)_q$, we define $c_q(\gamma, U_q, \Lambda)$ as in Theorem 1.1. For the sake of convenience for calculations, we often identify some elements of $C_n(q) \subset G_q^2$ with some elements of $(G_q^*)^2$ through the fixed isomorphism $G_q \cong G_q^*$ defined by $G_q \ni g \to \xi_q g \xi_q^{-1} \in G_q^*$ as in § 2.1. In such cases, in order to treat everything in G_q^*, we consider G_q^*-conjugacy classes instead of G_q-conjugacy classes, and modify also the definitions of the notations: For each $\gamma \in (G_q^*)^2$, we define $Z(\gamma)_q$ as before, and put $Z_G(\gamma)_q = Z(\gamma)_q \cap G_q^*$. For each order $\Lambda \subset Z(\gamma)_q$, we put

$$c_q(\gamma, U_q^*, \Lambda) = \sharp(Z_G(\gamma)_q \backslash M(\gamma, U_q^*, \Lambda)/U_q^*),$$

where

$$M(\gamma, U_q^*, \Lambda) = \{g \in G_q^*;\ g^{-1}\gamma g \in (U_q^*)^2, \text{ and}$$
$$Z(\gamma)_q \cap gR_q g^{-1} = x\Lambda x^{-1} \text{ for some } x \in Z_G(\gamma)_q\},$$

and $R_q = M_2(O_q)$, if $q \neq p$, and $R_p = \begin{pmatrix}O_p & \pi^{-1}O_p\\ \pi O_p & O_p\end{pmatrix}$, π is a prime element of O_p.

It is easy to see that, for any $\gamma \in (G_q^*)^2$, $Z_G(\xi_q^{-1}\gamma\xi_q)_q = \xi_q^{-1}Z_G(\gamma)_q\xi_q$, and $c_q(\xi_q^{-1}\gamma\xi_q, U_q, \xi_q^{-1}\Lambda\xi_q) = c_q(\gamma, U_q^*, \Lambda)$.

In this section, we calculate $c_q(\gamma, U_q, \Lambda)$ for each representative γ of G_q-conjugacy class in $C_n(q)$ and each order Λ of $Z(\gamma)_q$. We are interested only in those such that $c_q(\gamma, U_q, \Lambda) \neq 0$ for some Λ, or equivalently, $g^{-1}\gamma g \in U_q^2$ for some $g \in G_q$. We say that $\gamma \in G^r$ is q-integral, if $g^{-1}\gamma g \in U_q^r$ for some $g \in G_q$. If $\gamma = (\alpha, \beta) \in C_n(q)$ is q-integral, then α and β are also q-integral. For orders $\Lambda, \Lambda' \subset Z(\gamma)_q (\gamma \in C_n(q))$, we write $\Lambda \sim \Lambda'$, if $\Lambda' = x^{-1}\Lambda x$ for some $x \in Z_G(\gamma)$. We assume that $p \geqq 5$ throughout this section.

5.1. First, we assume that $q \neq 2, n, p$. In this case, we fix, once and for all, an isomorphism $\varphi: B_q \cong M_2(\boldsymbol{Q}_q)$ such that $\varphi(O_q) = M_2(\boldsymbol{Z}_q)$, and identify B_q (resp. O_q) with $M_2(\boldsymbol{Q}_q)$ (resp. $M_2(\boldsymbol{Z}_q)$). There exist $\omega, \eta_2, \eta_3 \in O_q$ such that $\omega^2 = -1$, $\eta_2^2 = -1$, $\eta_3^2 + \eta_3 + 1 = 0$, $\omega\eta_2 = -\eta_2\omega$ and $\omega\eta_3 = \bar{\eta}_3\omega$, where $^-$ denotes the main involution of B_q. For each $n = 2$ or 3, put $\gamma_n = (\alpha, \beta_n) \in C_n(q)$, where $\alpha = \begin{pmatrix} \omega & 0 \\ 0 & \omega \end{pmatrix}$ and $\beta_n = \begin{pmatrix} \eta_n & 0 \\ 0 & \eta_n \end{pmatrix}$. Then, $Z(\gamma_n)_q = M_2(Q_q) \subset M_2(B_q)$. By Proposition 3.8, it is easy to see that a complete set of representatives of G_q-conjugacy classes in $C_n(q)$ are given by $x_i^{-1}\gamma_n x_i$ $(i = 1, \cdots, 4)$, where x_i $(1 \leqq i \leqq 4)$ are any fixed elements of $GL_2(B_q)$ such that $x_i{}^t\bar{x}_i = \begin{pmatrix} 0 & -1 \\ -1 & 0 \end{pmatrix}, \begin{pmatrix} 1 & 0 \\ 0 & -\varepsilon \end{pmatrix}, \begin{pmatrix} 1 & 0 \\ 0 & -q \end{pmatrix}, \begin{pmatrix} 1 & 0 \\ 0 & -\varepsilon q \end{pmatrix}$ for $i = 1, 2, 3$, or 4, respectively, where ε is a fixed element of $\boldsymbol{Z}_q$ such that $\varepsilon \notin (\boldsymbol{Z}_q)^2$. The G_q-conjugacy classes which contains $x_1^{-1}\gamma_n x_1$ and $x_2^{-1}\gamma_n x_2$ are q-integral, because, by [4] Lemma 11, we can take x_1, x_2 so that $x_1, x_2 \in GL_2(O_q)$ and $x_1{}^t\bar{x}_1 = \begin{pmatrix} 0 & -1 \\ -1 & 0 \end{pmatrix}$, $x_2{}^t\bar{x}_2 = \begin{pmatrix} 1 & 0 \\ 0 & -\varepsilon \end{pmatrix}$, and $x_i^{-1}\gamma_n x_i \in U_q^2$ for these x_i $(i = 1, 2)$. We shall show that $x_3^{-1}\gamma_n x_3$ and $x_4^{-1}\gamma_n x_4$ are not q-integral. First, assume that $\left(\frac{-1}{q}\right) = -1$. Fix $g \in GL_2(B_q)$ such that $g{}^t\bar{g} = \begin{pmatrix} 1 & 0 \\ 0 & -q \end{pmatrix}$, or $\begin{pmatrix} 1 & 0 \\ 0 & -\varepsilon q \end{pmatrix}$. If $g^{-1}\gamma_n g$ is q-integral, then $g^{-1}\alpha g$ is also q-integral, and by [4] Proposition 4 and p. 565 (i), there exists $g_1 \in G_q$ such that $g_1^{-1}g^{-1}\alpha g g_1 = \alpha$. Namely, $gg_1 \in Z(\alpha) = M_2(F)$, where $F = \boldsymbol{Q}_q(\alpha)$. So, $g{}^t\bar{g} = cz{}^t\bar{z}$ for some $c \in \boldsymbol{Q}_q^\times$ and $z \in GL_2(F)$. Hence, $\det(g{}^t\bar{g}) \in c^2 N_{F/\boldsymbol{Q}_q}(F^\times)$, which is a contradiction, because F is unramified over $\boldsymbol{Q}_q$. Next, we assume that $\left(\frac{-1}{q}\right) = 1$. In this case, $N_{F/\boldsymbol{Q}_q}(\boldsymbol{Z}_q[\omega]) = Z_q$, because $\boldsymbol{Q}_q(\omega) \cong \boldsymbol{Q}_q \oplus \boldsymbol{Q}_q$. Fix elements $x, y \in \boldsymbol{Z}_q$ such that $x^2 + y^2 = -q$, or $-\varepsilon q$. Put $u = x + y\omega$, and $g_1 = \begin{pmatrix} 1 & 0 \\ 0 & u \end{pmatrix}$. We must show that $\delta = g_1^{-1}\gamma_n g_1$ is not q-integral. Assume that $g^{-1}\delta g \in U_q^2$ for some fixed $g \in G_q$. As $g^{-1}\alpha g \in U_q$, we get $g \in Z_G(\alpha)\cdot U_q$ by [5] Proposition 15 (i). We can assume $g \in Z_G(\alpha)$. By [5] Proposition 3, we have

$$Z_G(\alpha)=F^{\times}\left\{\begin{pmatrix} c & d \\ -\bar{d} & \bar{c} \end{pmatrix};\ c,\ d \in F\right\},$$

where $F=\boldsymbol{Q}_q(\omega)$ and $^-$ is the non-trivial automorphism of F. So, we can write $g=f\begin{pmatrix} c & d \\ -\bar{d} & \bar{c} \end{pmatrix}$, f, c, $d \in F$. Put $\eta'=2\eta_3-1$ when $n=3$, and put $\eta'=\eta$, when $n=2$. Then, $g^{-1}\begin{pmatrix} \eta_n & 0 \\ 0 & u^{-1}\eta_n u \end{pmatrix} g \in U_q$, if and only if

$$g^{-1}\begin{pmatrix} \eta' & 0 \\ 0 & u^{-1}\eta' u \end{pmatrix} g \in U_q.$$

As $\eta'\varphi=\bar{\varphi}\eta'$ for any $\varphi \in F$, we get $u^{-1}\eta' u=u^{-1}\bar{u}\eta'$ and $g^{-1}\begin{pmatrix} \eta' & 0 \\ 0 & u^{-1}\eta' u \end{pmatrix} g \in U_q$, if and only if $g^{-1}\begin{pmatrix} 1 & 0 \\ 0 & u^{-1}\bar{u} \end{pmatrix} \bar{g} \in U_q$, where $\bar{g}=\bar{f}\begin{pmatrix} \bar{c} & \bar{d} \\ -d & c \end{pmatrix}$. Hence, if δ is q-integral, we get $\overline{(u \det g)}/(u \det g) \in \boldsymbol{Z}_q[\omega]^{\times}$. By definition, $\det g= f^2(n(c)+n(d))$, and $\overline{(\det g)}/\det g=(\bar{f}/f)^2$. Namely, $(\bar{f}/f)^2\cdot q/u^2 \in \boldsymbol{Z}_q[\omega]^{\times}$. Hence, $q \in \boldsymbol{Z}_q[\omega]^{\times}(F^{\times})^2$. As $F\cong \boldsymbol{Q}_q\oplus\boldsymbol{Q}_q$ and $\boldsymbol{Z}_q[\omega]^{\times}\cong Z_q^{\times}\oplus Z_q^{\times}$, this is a contradiction. This proves that δ is not q-integral.

Proposition 5.1. *If $q\neq 2, p, n$, then there exist exactly two q-integral G_q-conjugacy classes in $C_n(q)$ for each n and q. A complete set of representatives δ_1, δ_2 of these classes and $K(\delta_i)_q$, $Z_G(\delta_i)_q$ $(i=1, 2)$ are given as follows: We fix $z \in Z_q[\omega]$ such that $z\bar{z}=-\varepsilon$.*

(i) $\delta_1=\left(\begin{pmatrix} \omega & 0 \\ 0 & \omega \end{pmatrix}, \begin{pmatrix} \eta_n & 0 \\ 0 & \eta_n \end{pmatrix}\right) \in (U_q^*)^2$, $K(\delta_1)_q=\left\{\begin{pmatrix} a & 0 \\ 0 & b \end{pmatrix};\ a,\ b \in \boldsymbol{Q}_q\right\}$,

and $Z_G(\delta_1)_q=K(\delta_1)_q^{\times} \cup y_1K(\delta_1)_q^{\times}$,

where $y_1=\begin{pmatrix} 0 & 1 \\ 1 & 0 \end{pmatrix}$,

(ii) $\delta_2=\left(\begin{pmatrix} \omega & 0 \\ 0 & \omega \end{pmatrix}, \begin{pmatrix} \eta_n & 0 \\ 0 & z^{-1}\eta_n z \end{pmatrix}\right) \in U_q^2$, $K(\delta_2)_q=\boldsymbol{Q}_q\left(\begin{pmatrix} 0 & z \\ -\bar{z} & 0 \end{pmatrix}\right)$,

and $Z_G(\delta_2)_q=K(\delta_2)_q^{\times} \cup y_2K(\delta_2)_q^{\times}$,

where $y_2=\begin{pmatrix} 1 & 0 \\ 0 & -1 \end{pmatrix}$.

Proof. Taking γ_n as before, two q-integral G_q-conjugacy classes are represented by $x_i^{-1}\gamma_n x_i$ $(i=1, 2)$, where x_1, x_2 are any elements of $GL_2(O_q)$ such that $x_1{}^t\bar{x}_1=\begin{pmatrix} 0 & -1 \\ -1 & 0 \end{pmatrix}$, $x_2{}^t\bar{x}_2=\begin{pmatrix} 1 & 0 \\ 0 & -\varepsilon \end{pmatrix}$. Put $x_2=\begin{pmatrix} 1 & 0 \\ 0 & z \end{pmatrix}$. Then, $\delta_2=x_2^{-1}\gamma_n x_2$. Next, take $a \in O_q^{\times}=M_2(\boldsymbol{Z}_q)^{\times}$ so that $a^2=-1$, and put $x=\begin{pmatrix} a & 0 \\ 0 & a \end{pmatrix}$. If we put $x_1=x\xi_q$ (where $\xi_q \in GL_2(O_q)$, $\xi_q{}^t\bar{\xi}_q=\begin{pmatrix} 0 & 1 \\ 1 & 0 \end{pmatrix}$), then

$x_1{}^t\bar{x}_1=\begin{pmatrix} 0 & -1 \\ -1 & 0 \end{pmatrix}$, and we get $\xi_q(x_1^{-1}\gamma_n x_1)\xi_q^{-1}=x^{-1}\gamma_n x \in (U_q^*)^2$. As $x \in U_q^*$, identifying $x_1^{-1}\gamma_n x_1$ with $x^{-1}\gamma_n x=\delta_1$ by the isomorphism $U_q \cong U_q^*$, δ_1 represents the class which contains $x_1^{-1}\gamma_n x_1$. The rest is easy by noting that

$$Z(\delta_1)=M_2(\boldsymbol{Q}_q)\subset M_2(B_q), \quad \text{and} \quad Z(\delta_2)=\begin{pmatrix} 1 & 0 \\ 0 & z \end{pmatrix}^{-1} M_2(\boldsymbol{Q}_q)\begin{pmatrix} 1 & 0 \\ 0 & z \end{pmatrix}.$$

q.e.d.

Proposition 5.2. *Assume that $q \neq 2$, p, n, and take δ_1, δ_2 as in Proposition* 5.1.

(i) $$c_q(\delta_1, U_q^*, \Lambda)=\begin{cases} 1 & \cdots \text{ if } \Lambda \sim M_2(\boldsymbol{Z}_q), \\ 0 & \cdots \text{ otherwise,} \end{cases}$$

(ii) $$c_q(\delta_2, U_q, \Lambda)=\begin{cases} 1 & \cdots \text{ if } \Lambda \sim \begin{pmatrix} 1 & 0 \\ 0 & z \end{pmatrix}^{-1} M_2(\boldsymbol{Z}_q)\begin{pmatrix} 1 & 0 \\ 0 & z \end{pmatrix}, \\ 0 & \cdots \text{ otherwise.} \end{cases}$$

Proof. First, we prove (i). Denote the component of δ_1 by α, β: $\delta_1=(\alpha, \beta) \in (G_q^*)^2$. If $g^{-1}\delta_1 g \in (U_q^*)^2$ for some $g \in G_q^*$, then $g^{-1}\alpha g \in U_q^*$. By [4] Proposition 15 (i), (ii), we get $g \in Z_G(\alpha)_q U_q^*$. Now, we write $g_1 \sim g_2$ for $g_1, g_2 \in G_q^*$, if $g_2 \in Z_G(\delta_1)_q g_1 U_q^*$. If we put $F=\boldsymbol{Q}_q(\omega)$, then

$$Z_G(\alpha)_q=F^{\times}\begin{pmatrix} \boldsymbol{Q}_q & \boldsymbol{Q}_q\omega \\ \boldsymbol{Q}_q\omega & \boldsymbol{Q}_q \end{pmatrix}^{\times}.$$

(See [5] p. 579.) First, we treat the case where F is a field. In this case, $F^{\times}=\{q^n; n \in \boldsymbol{Z}\}\cdot \boldsymbol{Z}_q[\omega]^{\times}$, hence we have $g\sim\begin{pmatrix} a & b\omega \\ c\omega & d \end{pmatrix}$ for some $a, b, c, d \in \boldsymbol{Q}_q$ such that $ab+bd \neq 0$. We can also assume that $ad \neq 0$. As $\begin{pmatrix} a & 0 \\ 0 & d \end{pmatrix} \in Z_G(\delta_1)_q$, we get $g\sim\begin{pmatrix} 1 & b'\omega \\ c'\omega & 1 \end{pmatrix}$ ($b', c' \in \boldsymbol{Q}_q$ and $b'c'+1\neq 0$). Now,

$$\begin{pmatrix} 1 & b'\omega \\ c'\omega & 1 \end{pmatrix}^{-1}\beta\begin{pmatrix} 1 & b'\omega \\ c'\omega & 1 \end{pmatrix}=\frac{1}{1+b'c'}\begin{pmatrix} \eta_n+b'c'\bar{\eta}_n & b'\omega(\bar{\eta}_n-\eta_n) \\ c'\omega(\bar{\eta}_n-\eta_n) & b'c'\bar{\eta}_n+\eta_n \end{pmatrix} \in U_q^*.$$

As $q \neq 2$, n, we get $B_q=\boldsymbol{Z}_q+\boldsymbol{Z}_q\omega+\boldsymbol{Z}_q\eta_n+\boldsymbol{Z}_q\omega\eta_n$. Hence, b', c', $1-b'c' \in (1+b'c')\boldsymbol{Z}_q$. Put $m=\mathrm{ord}_q(1+b'c')$ (the q-adic order). If $m>0$, then $b', c' \in q^m\boldsymbol{Z}_q$ and $1+b'c' \in \boldsymbol{Z}_q^{\times}$, which is a contradiction. Now, assume that $m<0$. Then, $m=\mathrm{ord}_q(b'c')$. Put $s=\mathrm{ord}_q(b')$ and $t=\mathrm{ord}_q(c')$. Then, $s+t=m$. If $s>0$, then $\mathrm{ord}_q(c')=t=m-s<m$, and hence $c' \in q^m\boldsymbol{Z}_q$, which is a contradiction. So, $s\leqq 0$, and by the same way, we get $t\leqq 0$. Hence

$$g\sim\begin{pmatrix}q^{-s} & 0\\ 0 & q^{-t}\end{pmatrix}\begin{pmatrix}1 & b'\omega\\ c'\omega & 1\end{pmatrix}\in U_q^*.$$

If $m=0$, then $\begin{pmatrix}1 & b'\omega\\ c'\omega & 1\end{pmatrix}\in U_q^*$. Thus, we get $g\sim 1$ in any case. Next, we treat the case where $F\cong \boldsymbol{Q}_q\oplus\boldsymbol{Q}_q$. If $g^{-1}\delta_1 g\in(U_q^*)^2$, then $g\sim g_1=x\begin{pmatrix}1 & b\omega\\ c\omega & 1\end{pmatrix}$, for some $b, c\in\boldsymbol{Q}_q$, $x\in F^\times$, as before. Put $\beta'=2\beta-1$ when $n=3$, and $\beta'=\beta$, when $n=2$. As we have assumed $q\neq 2$, n, $g^{-1}\beta' g\in U_q^*$, if and only if $g^{-1}\beta g\in U_q^*$. We get $g_1^{-1}\beta' g_1=g_1^{-1}\bar{g}_1\beta'\in U_q^*$, hence $g_1^{-1}\bar{g}_1\in U_q^*$, where $\bar{g}_1=\bar{x}\begin{pmatrix}1 & -b\omega\\ -c\omega & 1\end{pmatrix}$. We get

$$g_1^{-1}\bar{g}_1=\frac{\bar{x}}{x(1+bc)}\begin{pmatrix}1-bc & -2b\omega\\ -2c\omega & 1-bc\end{pmatrix}\in GL_2(\boldsymbol{Z}_q[\omega]).$$

We fix an isomorphism of $\boldsymbol{Z}_q[\omega]$ onto $\boldsymbol{Z}_q\oplus\boldsymbol{Z}_q$ such that ω is mapped to $(1, -1)$. As $N_{F/\boldsymbol{Q}_q}(\bar{x}/x)=1$, we can assume that $\bar{x}/x=(q^e, q^{-e})\in\boldsymbol{Q}_q\oplus\boldsymbol{Q}_q$ for some $e\in\boldsymbol{Z}$. Put $m=\mathrm{ord}_q(1+bc)$. Then, $1-bc$, b, $c\in q^{m+|e|}\boldsymbol{Z}_q$.

If $m>0$, then $2=(1+bc)+(1-bc)\in q^m\boldsymbol{Z}_q$, which is a contradiction, because $q\neq 2$. So, $m\leqq 0$. Put $t=\mathrm{ord}_q(b)$ and $s=\mathrm{ord}_q(c)$. If $m<0$, then $t+s=m$ and $1-bc\in q^m\boldsymbol{Z}_q$. Hence $c=0$, and we can show $g\sim 1$ virtually in the same way as in the case where F is a field. If $m=0$, then $b, c\in q^{|e|}\boldsymbol{Z}_q$ and $1-bc\in q^{|e|}\boldsymbol{Z}$. Hence $e=0$, so $g\sim 1$. Thus, we have proved (i). The proof of (ii) is obtained virtually in the same way as in the proof of (i), and we shall omit it here. q.e.d.

5.2. Next, we treat the case $q=p$. If $\gamma=(\alpha,\beta)\in C_n(p)$ is p-integral, then by [5] (III) Propositions 2.5 and 2.6, we can assume (up to G_q^*-conjugation) that

(5.3) (i) in the case $\left(\dfrac{-1}{p}\right)=-1$,

$$\gamma\in(G_q^*)^2,\qquad \alpha=\begin{pmatrix}\omega & 0\\ 0 & \omega\end{pmatrix}\in G_q^*,$$

where ω is a fixed element of $O_q^\times$ such that $\omega^2=-1$,

(ii) in the case $\left(\dfrac{-1}{p}\right)=1$,

$$\gamma\in(G_q^*)^2,\qquad \alpha=\begin{pmatrix}\omega' & 0\\ 0 & -\omega'\end{pmatrix}\in G_q^*,$$

where $\omega'\in\boldsymbol{Z}_p$, $\omega'^2=-1$.

First, we treat the case $n=2$.

Proposition 5.4. *There are exactly two p-integral G_p^*-conjugacy classes in $C_2(p)$. The representatives δ_1, δ_2 of such classes and $K(\delta_i)_p$, $Z_G(\delta_i)_p$ $(i=1, 2)$ are given as follows:*

(i) *When* $\left(\frac{-1}{p}\right)=-1$,

$$\delta_1=\left(\begin{pmatrix}\omega & 0\\ 0 & \omega\end{pmatrix}, \begin{pmatrix}0 & -\pi^{-1}\\ \pi & 0\end{pmatrix}\right)\in (U_p^*)^2,$$

$$K(\delta_1)_p=\boldsymbol{Q}_p\left(\begin{pmatrix}0 & \omega\\ -p\omega & 0\end{pmatrix}\right),$$

$$Z_G(\delta_1)_p=K(\delta_1)_p^\times \cup yK(\delta_1)_p^\times, \quad \textit{and}$$

$$\delta_2=\left(\begin{pmatrix}\omega & 0\\ 0 & \omega\end{pmatrix}, \begin{pmatrix}0 & \pi^{-1}u\\ \pi u & 0\end{pmatrix}\right)\in (U_p^*)^2,$$

$$K(\delta_2)_p=\boldsymbol{Q}_p\left(\begin{pmatrix}0 & \omega\\ p\omega & 0\end{pmatrix}\right),$$

$$Z_G(\delta_2)_p=K(\delta_2)_p^\times \cup yK(\delta_2)_p^\times,$$

where π is a fixed element of O_q such that $\pi^2=-p$, $\pi\omega=-\omega\pi$, u is a fixed element of $Z_p[\omega]$ such that $u\bar{u}=-1$, and $y=\begin{pmatrix}\omega & 0\\ 0 & -\omega\end{pmatrix}$.

(ii) *When* $\left(\frac{-1}{p}\right)=1$,

$$\delta_1=\left(\begin{pmatrix}\omega' & 0\\ 0 & -\omega'\end{pmatrix}, \begin{pmatrix}0 & -\pi^{-1}\\ \pi & 0\end{pmatrix}\right)\in (U_p^*)^2$$

$$K(\delta_1)_p=\boldsymbol{Q}_p\left(\begin{pmatrix}\pi & 0\\ 0 & \pi\end{pmatrix}\right),$$

$$Z_G(\delta_1)_p=K(\delta_1)_p^\times \cup y'K(\delta_1)_p^\times, \quad \textit{and}$$

$$\delta_2=\left(\begin{pmatrix}\omega' & 0\\ 0 & -\omega'\end{pmatrix}, \begin{pmatrix}0 & \pi^{-1}u\\ \pi u & 0\end{pmatrix}\right)\in (U_q^*)^2,$$

$$K(\delta_2)_p=Q_p\left(\begin{pmatrix}\pi u & 0\\ 0 & \pi u\end{pmatrix}\right),$$

$$Z_G(\delta_2)_p=K(\delta_2)_p^\times \cup y'K(\delta_2)_p^\times,$$

where π is as in (i), *u is an element of $O_q^\times$ such that $u\pi=\pi\bar{u}$ and $n(u)=-1$, and* $y'=\begin{pmatrix}u & 0\\ 0 & -u\end{pmatrix}$.

Proof. First, we assume that $\left(\frac{-1}{p}\right)=-1$. Take δ_1 as in (i). By easy calculation, we get

$$Z(\delta_1)_p=\left\{\begin{pmatrix} a & b \\ p\bar{b} & \bar{a}\end{pmatrix};\ a,\ b\in \boldsymbol{Q}_p(\omega)\right\}.$$

The main involution ρ of $Z(\delta_1)$ is given by:

$$\begin{pmatrix} a & b \\ p\bar{b} & \bar{a}\end{pmatrix}^{\rho}=\begin{pmatrix} \bar{a} & -b \\ -p\bar{b} & a\end{pmatrix}.$$

Put $c=\begin{pmatrix} 0 & -\omega \\ p\omega & 0\end{pmatrix}$. As we are dealing with elements of G_p^*, we put $z^*=\begin{pmatrix} 0 & 1 \\ 1 & 0\end{pmatrix}{}^t\bar{z}\begin{pmatrix} 0 & 1 \\ 1 & 0\end{pmatrix}$ for $z\in M_2(B_p)$. Then, $z^*=c^{-1}z^{\rho}c$ for any $z\in Z(\delta_1)_p$, and $z^*=z$, if and only if $z=\begin{pmatrix} a & b \\ pb & \bar{a}\end{pmatrix}$ for some $a\in \boldsymbol{Q}_p(\omega)$, $b\in \boldsymbol{Q}_p$. A quadratic semisimple algebra K over $\boldsymbol{Q}_p$ can be embedded into $Z(\delta_1)$, if and only if $K=\boldsymbol{Q}_p(\sqrt{-1})$, $\boldsymbol{Q}_p(\sqrt{p})$, or $\boldsymbol{Q}_p(\sqrt{-p})$. Hence, by Proposition 3.5, representatives of G_p^*-conjugacy classes in $C_2(p)$ is given by $x_i^{-1}\delta_1 x_i$ $(i=1, 2, 3)$, where

$$x_1x_1^*=\begin{pmatrix} 1 & 0 \\ 0 & 1\end{pmatrix}=\begin{pmatrix} 0 & -\omega \\ p\omega & 0\end{pmatrix}^{-1}c,\qquad x_2x_2^*=\begin{pmatrix} u & 0 \\ 0 & \bar{u}\end{pmatrix}=\begin{pmatrix} 0 & b' \\ p\bar{b}' & 0\end{pmatrix}^{-1}c,$$

and $x_3x_3^*=\begin{pmatrix} \omega & 0 \\ 0 & -\omega\end{pmatrix}c=\begin{pmatrix} 0 & 1 \\ p & 0\end{pmatrix}$, where b' is a fixed element of $\boldsymbol{Z}_p[\omega]$ such that $b'\bar{b}'=-1$ and $u=\bar{b}'^{-1}\omega$. If we put $x_2=\begin{pmatrix} u & 0 \\ 0 & 1\end{pmatrix}$, then

$$x_2x_2^*=x_2\begin{pmatrix} 0 & 1 \\ 1 & 0\end{pmatrix}{}^t\bar{x}_2\begin{pmatrix} 0 & 1 \\ 1 & 0\end{pmatrix}=\begin{pmatrix} u & 0 \\ 0 & \bar{u}\end{pmatrix}.$$

Hence, δ_1 or $\delta_2=x_2^{-1}\delta_1x_2$ is p-integral, and corresponds with $\boldsymbol{Q}_p(\sqrt{p})$, or $\boldsymbol{Q}_p(\sqrt{-p})$, respectively. Now, we show that $x_3^{-1}\delta_1x_3$ is not p-integral. If this is p-integral, then, by [5] (iii) Proposition 2.5 (i), we have

$$g^{-1}x_3^{-1}\begin{pmatrix} \omega & 0 \\ 0 & \omega\end{pmatrix}x_3g=\begin{pmatrix} \omega & 0 \\ 0 & \omega\end{pmatrix}$$

for some $g\in G_p^*$. Hence, $x_3g\in M_2(F)$, where $F=\boldsymbol{Q}_p(\omega)$. This means that $dx_3x_3^*=ff^*$ for some $f\in GL_2(F)$ and $d\in \boldsymbol{Q}_p^{\times}$. Hence, $-d^2p=(\det f)(\det f^*)=(\det f)\ (\det {}^t\bar{f})\in N_{F/\boldsymbol{Q}_p}(F)$. As F is unramified over $\boldsymbol{Q}_p$, this is a contradiction. This proves (i).

Next, we prove the case (ii). Taking δ_1 as in (ii), we get

$$Z\left(\begin{pmatrix} \omega' & 0 \\ 0 & -\omega'\end{pmatrix}\right)_p=\left\{\begin{pmatrix} a & 0 \\ 0 & b\end{pmatrix};\ a,\ b\in B_p\right\},\quad \text{and}$$

$$Z(\delta_1)_p=\left\{\begin{pmatrix} a & 0 \\ 0 & \pi a\pi^{-1}\end{pmatrix};\ a\in B_p\right\}.$$

A complete set of representatives of G_p^*-conjugacy classes in $C_2(p)$ is given by $x_i^{-1}\delta_1 x_i$ $(i=1, 2, 3)$, where

$$x_1x_1^*=\begin{pmatrix}1&0\\0&1\end{pmatrix},\quad x_2x_2^*=\begin{pmatrix}u&0\\0&u\end{pmatrix},\quad\text{and}\quad x_3x_3^*=\begin{pmatrix}u\pi&0\\0&u\pi\end{pmatrix},$$

taking u as in (ii). For $i=1, 2$, $x_i^{-1}\delta_1 x_i$ is G_p-conjugate to δ_i. Now, we show that $x_3^{-1}\delta_1 x_3$ is not p-integral. By [4] (III) Proposition 2.6 (i), if $g^{-1}x_3^{-1}\delta_1 x_3 g \in U_p^*$ for some $g \in G_p^*$, then $g^{-1}x_3^{-1}\begin{pmatrix}\omega'&0\\0&-\omega'\end{pmatrix}x_3 g$ is U_p^*-conjugate to $\begin{pmatrix}\omega'&0\\0&-\omega'\end{pmatrix}$. Namely, we can assume $x_3 g \in Z\left(\begin{pmatrix}\omega'&0\\0&-\omega'\end{pmatrix}\right)$, and hence $x_3 g=\begin{pmatrix}a&0\\0&b\end{pmatrix}$ for some $a, b \in B_p^\times$. In other words,

$$\begin{pmatrix}a&0\\0&b\end{pmatrix}^{-1}\begin{pmatrix}0&-\pi^{-1}\\\pi&0\end{pmatrix}\begin{pmatrix}a&0\\0&b\end{pmatrix}=\begin{pmatrix}0&-a^{-1}\pi^{-1}b\\b^{-1}\pi a&0\end{pmatrix}\in U_p^*.$$

So, we get $a^{-1}b \in O_p^\times$. On the other hand, we get $dx_3x_3^*=d\begin{pmatrix}u\pi&0\\0&u\pi\end{pmatrix}=\begin{pmatrix}a\bar{b}&0\\0&b\bar{a}\end{pmatrix}$ for some $d \in \boldsymbol{Q}_p^\times$. That is, $\mathrm{ord}_\pi(a\bar{b})$ is odd, which is a contradiction. q.e.d.

Proposition 5.5. *Let notations and assumptions be as in Proposition* 5.4.

(i) *If* $\left(\frac{-1}{p}\right)=-1$, *then for each* $i=1, 2$,

$$c_p(\delta_i, U_p^*, \Lambda)=\begin{cases}1 & \cdots \text{ if } \Lambda\sim\Lambda(i),\\ 0 & \cdots \text{ otherwise,}\end{cases}$$

where $\Lambda(i)$ *is the unique maximal order of* $Z(\delta_i)_p$.

(ii) *If* $\left(\frac{-1}{p}\right)$, *then for each* $i=1, 2$,

$$c_p(\delta_i, U_p^*, \Lambda)=\begin{cases}1 & \cdots \text{ if } \Lambda\sim\Lambda(i),\\ 0 & \cdots \text{ otherwise,}\end{cases}$$

where $\Lambda(i)$ *is the unique maximal order of* $Z(\delta_i)_p$.

Proof. We assume $\left(\frac{-1}{p}\right)=-1$, and give the proof for δ_1. If $g^{-1}\delta_1 g \in (U_p^*)^2$ for some $g \in G_p^*$, then by [5] III Proposition 2.5 (i), $g \in$

$(GL_2(F)\cap G_p^*)\cdot U_p^*$, where $F=\boldsymbol{Q}_p(\omega)$. By a similar argument as in the proof of Proposition 5.4, we can assume $g=\begin{pmatrix}1 & b\omega \\ c\omega & d\end{pmatrix}$ for some $b, c, d\in \boldsymbol{Q}_p$. As $\begin{pmatrix}1 & c\omega/p \\ -c\omega & 1\end{pmatrix}\in K(\delta_1)_p$, we get $g\sim\begin{pmatrix}1 & c\omega/p \\ -c\omega & 1\end{pmatrix}g\sim\begin{pmatrix}1 & x\omega \\ 0 & y\end{pmatrix}$ for some $x, y\in \boldsymbol{Q}_p$. Now,

$$\begin{pmatrix}1 & x\omega \\ 0 & y\end{pmatrix}^{-1}\begin{pmatrix}0 & -\pi^{-1} \\ \pi & 0\end{pmatrix}\begin{pmatrix}1 & x\omega \\ 0 & y\end{pmatrix}=\begin{pmatrix}-xy^{-1}\omega\pi, & -x^2y^{-1}\pi-y\pi^{-1} \\ y^{-1}\pi, & xy^{-1}\pi\omega\end{pmatrix}\in U_p^*.$$

Hence, $y^{-1}\in \boldsymbol{Z}_p$, $xy^{-1}\in \boldsymbol{Z}_p$, and $x^2y^{-1}p-y\in \boldsymbol{Z}_p$. Then, $-(xy^{-1})^2p+1\in y^{-1}\boldsymbol{Z}_p$. So, $y^{-1}\notin p\boldsymbol{Z}_p$. This means that $y\in \boldsymbol{Z}_p^\times$ and $x\in \boldsymbol{Z}_p$. Hence, $g\sim 1$. This proves the assertion. The proof for δ_2 is virtually the same, and will be omitted here. Next, assume that $\left(\frac{-1}{p}\right)=1$. If $g^{-1}\delta_1 g\in U_p^*$ for some $g\in G_p^*$, then $g\sim\begin{pmatrix}a & 0 \\ 0 & da\end{pmatrix}$ ($d\in \boldsymbol{Q}_p^\times$, $a\in B_p^\times$), by [5] III Proposition 2.5 (i). As

$$\begin{pmatrix}a & 0 \\ 0 & da\end{pmatrix}^{-1}\begin{pmatrix}0 & -\pi^{-1} \\ \pi & 0\end{pmatrix}\begin{pmatrix}a & 0 \\ 0 & da\end{pmatrix}=\begin{pmatrix}0 & -da^{-1}\pi^{-1}a \\ d^{-1}a^{-1}\pi a & 0\end{pmatrix}\in U_p^*,$$

we get $d\in \boldsymbol{Z}_p^\times$. Besides, $B_p=\{\pi^r;\, r\in \boldsymbol{Z}\}\times O_p^\times$. Hence, $g\sim 1$. The proof for δ_2 is virtually the same, and will be omitted here. q.e.d.

Next, we treat the case $n=3$ and $q=p$. In this case, the representatives of p-integral G_p^*-conjugacy classes have not so simple form. Here, it is more convenient to proceed as follows. Take $\alpha\in G_p^*$ as in (5.3) in each case. We shall find all $\beta\in G_p^*$ such that $(\alpha, \beta)\in C_3(p)$ for this α, and investigate whether (α, β) is p-integral, or not. For this purpose, it is more convenient to replace each element $(\alpha, \beta)\in C_3(p)$ by $(\alpha, 2\beta-1)$. Namely, for each q, put

$$C_3'(q)=\{(\alpha, \beta')\in (G_q^*)^2;\ \alpha\beta'=-\beta'\alpha,\ (\beta')^2=-3\}.$$

Then, $C_3'(q)$ corresponds bijectively to $C_3(q)$ by the mapping $(\alpha, \beta)\in C_3(q)$ to $(\alpha, 2\beta-1)\in C_3'(q)$. Besides, if $q\neq 2, 3$, (α, β) is q-integral, if and only if $(\alpha, 2\beta-1)$ is q-integral.

Proposition 5.6. *There are exactly two p-integral G_p^*-conjugacy classes in $C_3'(p)$, and the representatives δ_1, δ_2 of such classes and $K(\delta_i)_p$, $Z_G(\delta_i)_p$ ($i=1, 2$) are given as follows:*

(i) *When* $\left(\frac{-1}{p}\right)=-1$, *for each* $i=1, 2$,

$$\delta_i=\left(\begin{pmatrix}\omega & 0\\ 0 & \omega\end{pmatrix}, \begin{pmatrix}0 & -\pi^{-1}\\ \varepsilon_i\pi & 0\end{pmatrix}b_i\right) \in (U_p^*)^2,$$

$$K(\delta_i)_p=\boldsymbol{Q}_p\left(\begin{pmatrix}0 & \omega\\ -p\varepsilon_i\omega & 0\end{pmatrix}\right),$$

$$Z_G(\delta_i)_p=K(\delta_i)_p^\times \cup yK(\delta_i)_p^\times,$$

where $\varepsilon_1=1$, $\varepsilon_2=-1$, b_1, b_2 *are fixed elements of* $\boldsymbol{Z}_p[\omega]$ *[such* $N(b_1)=3$, $N(b_2)=-3$, π *is a prime element of* O_p *such that* $\pi^2=-p$, $\pi\omega=-\omega\pi$, *and* $y=\begin{pmatrix}\omega & 0\\ 0 & -\omega\end{pmatrix}$.

(ii) *When* $\left(\frac{-1}{p}\right)=1$, *for each* $i=1, 2$,

$$\delta_i=\left(\begin{pmatrix}\omega' & 0\\ 0 & -\omega'\end{pmatrix}, \begin{pmatrix}0 & \pi_i^{-1}\\ -3\pi_i & 0\end{pmatrix}\right) \in (U_p^*)^2,$$

$$K(\delta_i)_p=\boldsymbol{Q}_p\left(\begin{pmatrix}\pi_i & 0\\ 0 & \pi_i\end{pmatrix}\right),$$

$$Z_G(\delta_i)_p=K(\delta_i)_p^\times \cup y'K(\delta_i)_p^\times,$$

where π_1, π_2 *and* y' *are defined as follows: Fix an element* $\varepsilon \in \boldsymbol{Z}_p^\times \backslash (\boldsymbol{Z}_p^\times)^2$, *and* $c \in O_p^\times$ *such that* $c^2=\varepsilon$. *We denote by* π_1, π_2 *any fixed elements of* O_p *such that* $\pi_i c=-c\pi_i$ $(i=1, 2)$ *and* $\pi_1^2=p$, $\pi_2^2=p\varepsilon$. *We put* $y'=\begin{pmatrix}c & 0\\ 0 & -c\end{pmatrix}$.

Proof. First, assume that $\left(\frac{-1}{p}\right)=-1$. Put $\alpha=\begin{pmatrix}\omega & 0\\ 0 & \omega\end{pmatrix}$. If $(\alpha, \beta') \in C_3'(p)$, then by easy calculation, we can show that $\beta'=\pi b\begin{pmatrix}\mu\omega & 1\\ \lambda & \mu\omega\end{pmatrix}$ for some λ, $\mu \in \boldsymbol{Q}_p$ and $b \in \boldsymbol{Q}_p(\omega)$ such that $N(b)(\mu^2+\lambda)=3/p$. If we put $\gamma=(\alpha, \beta')$, then

$$K(\gamma)_p=\boldsymbol{Q}_p+\boldsymbol{Q}_p\begin{pmatrix}-\mu & \omega\\ -\lambda\omega & \mu\end{pmatrix}\cong \boldsymbol{Q}_p(\sqrt{m}),$$

where $m=\mu^2+\lambda$. As we have assumed $p\geq 5$, and $\boldsymbol{Q}_p(\omega)$ is unramified over $\boldsymbol{Q}_p$, we get $mp \in N_{F/\boldsymbol{Q}_p}(\boldsymbol{Q}_p(\omega))$, and hence $m \in p(\boldsymbol{Q}_p)^2 \cup (-p)(\boldsymbol{Q}_p)^2$. Putting $\mu=0$ and $\lambda=\pm p$, we get the required results. The proof for the case $\left(\frac{-1}{p}\right)=1$ is similarly obtained, and the details will be omitted here.

q.e.d.

Proposition 5.7. *Let notations and assumptions be as in Proposition* 5.6.

(i) *If* $\left(\frac{-1}{p}\right)=-1$, *then for each* $i=1, 2$

$$c_p(\delta_i, U_p^*, \Lambda)=\begin{cases}1 & \cdots \text{ if } \Lambda\sim\Lambda(i),\\ 0 & \cdots \text{ otherwise,}\end{cases}$$

where $\Lambda(i)=\left\{\begin{pmatrix} a & b \\ \varepsilon_i p\bar{b} & \bar{a}\end{pmatrix}; a, b\in \boldsymbol{Z}_p[\omega]\right\}$ *for each* $i=1, 2$.

(ii) *If* $\left(\frac{-1}{p}\right)=1$, *then, for each* $i=1, 2$.

$$c_p(\delta_i, U_p^*, \Lambda)=\begin{cases}1 & \cdots \text{ if } \Lambda\sim\Lambda(i),\\ 0 & \cdots \text{ otherwise}\end{cases}$$

where $\Lambda(i)=\boldsymbol{Z}_p+\boldsymbol{Z}_p\pi_i 1_2+\boldsymbol{Z}_p y+\boldsymbol{Z}_p y\pi_i 1_2$, *for each* $i=1, 2$.

Proof. The proof is more or less similar as the proof of Proposition 5.5, and will be omitted here. q.e.d.

5.3. Now, we treat the case $q=n=3$.

Proposition 5.8. *There exist exactly two* 3-*integral* G_q^*-*conjugacy classes in* $C_3(3)$. *The representatives* δ_1, δ_2 *of such classes, and* $K(\delta_i)_3$, $Z_G(\delta_i)_p$ $(i=1, 2)$ *are given as follows*:

$$\delta_1=\left(\begin{pmatrix}\omega & 0\\ 0 & \omega\end{pmatrix}, \frac{1}{2}\begin{pmatrix}1 & \varepsilon\\ -3\varepsilon & 1\end{pmatrix}\right)\in (U_3^*)^2,$$

$$K(\delta_1)_3=\boldsymbol{Q}_3\left(\begin{pmatrix}0 & \omega\\ 3\omega & 0\end{pmatrix}\right),$$

$$Z_G(\delta_1)_3=K(\delta_1)_3^{\times}\cup yK(\delta_1)^{\times}, \quad \textit{and}$$

$$\delta_2=\left(\begin{pmatrix}\omega & 0\\ 0 & \omega\end{pmatrix}, \frac{b}{3}\begin{pmatrix}1 & \varepsilon\\ 3\varepsilon & 1\end{pmatrix}\right)\in (U_3^*)^2,$$

$$K(\delta_2)_3=\boldsymbol{Q}_3\left(\begin{pmatrix}0 & \omega\\ -3\omega & 0\end{pmatrix}\right),$$

$$Z_G(\delta_2)_3=K(\delta_2)_3^{\times}\cup yK(\delta_2)_3^{\times},$$

where

$$\omega=\begin{pmatrix}0 & 1\\ -1 & 0\end{pmatrix},\quad \varepsilon=\begin{pmatrix}1 & 0\\ 0 & -1\end{pmatrix}\in M_2(\boldsymbol{Q}_3)=B_3,\quad y=\begin{pmatrix}\omega & 0\\ 0 & -\omega\end{pmatrix}\in M_2(B_3),$$

and b *is a fixed element of* $\boldsymbol{Q}_3(\omega)$ *such that* $N(b)=-1$.

The local data are given as follows: *for each* $i=1, 2$,

$$c_3(\delta_i,\ U_3^*,\ \Lambda)=\begin{cases}1 & \cdots\ \text{if } \Lambda\sim\Lambda(i),\\ 0 & \cdots\ \text{otherwise,}\end{cases}$$

where

$$\Lambda(1)=\left\{\begin{pmatrix} x & y \\ 3\bar{y} & \bar{x}\end{pmatrix};\ x,\ y\in \boldsymbol{Z}_3[\omega]\right\},\quad \text{and}$$

$$\Lambda(2)=\left\{\begin{pmatrix} x & y \\ -3\bar{y} & \bar{x}\end{pmatrix};\ x,\ y\in \boldsymbol{Z}_3[\omega]\right\}.$$

Proof. The proof is more or less similar to the proof of Proposition 5.5, and will be omitted here. q.e.d.

5.4. Finally, we treat the case $q=2$ (and $n=2$ or 3). This is the most elaborate case, but here we shall omit the proofs, because it seems too lengthy to write them down. First, we treat $C_3(2)$. There exist elements $\omega,\ \eta\in O_2$ such that $\eta^2+\eta+1=0$, $\omega^2=-1$, and $\omega\eta=\eta^{-1}\omega$. We fix such pair $\eta,\ \omega$ once and for all, and put $\eta'=2\eta+1$.

Proposition 5.9. *There exists exactly four 2-integral G_2^*-conjugacy classes in $C_3(2)$. The representatives $\delta_1,\ \delta_2,\ \delta_3,\ \delta_4$ of such classes and $K(\delta_i)_2$, $Z_G(\delta_i)_2$ $(i=1,\cdots,4)$ are given as follows:*

$$\delta_i=\left(\omega b_i\begin{pmatrix}0 & 1\\ \lambda_i & 0\end{pmatrix},\ \begin{pmatrix}\eta & 0\\ 0 & \eta\end{pmatrix}\right)\in (U_2^*)^2\quad (i=1,\cdots,4),$$

$$K(\delta_i)_2=\boldsymbol{Q}_2\left(\begin{pmatrix}0 & \eta'\\ -\lambda_i\eta' & 0\end{pmatrix}\right),$$

$$Z_G(\delta_i)_2=K(\delta_i)_2^\times\cup yK(\delta_i)_2^\times\quad (i=1,\cdots,4),$$

where $\lambda_1=1$, $\lambda_2=-1$, $\lambda_3=\dfrac{1}{3}$, $\lambda_4=\dfrac{-1}{3}$, each b_i $(1\leqq i\leqq 4)$ is a fixed element of $\boldsymbol{Z}_2[\eta]$ such that $N(b_i)=\lambda$, and $y=\begin{pmatrix}\eta' & 0\\ 0 & -\eta'\end{pmatrix}$.

The local data are given as follows:

$$c_2(\delta_1,\ U_2^*,\ \Lambda)=\begin{cases}1 & \cdots\ \text{if } \Lambda\sim\Lambda(1),\\ 0 & \cdots\ \text{otherwise,}\end{cases}$$

$$c_2(\delta_2,\ U_2^*,\ \Lambda)=\begin{cases}1 & \cdots\ \text{if } \Lambda\sim\Lambda(2),\\ 1 & \cdots\ \text{if } \Lambda\sim\Lambda'(2),\\ 0 & \cdots\ \text{otherwise,}\end{cases}$$

$$c_2(\delta_3,\ U_2^*,\ \Lambda)=\begin{cases}1 & \cdots\ \text{if } \Lambda\sim\Lambda(3),\\ 1 & \cdots\ \text{if } \Lambda\sim\Lambda'(3),\\ 0 & \cdots\ \text{otherwise,}\end{cases}$$

$$c_4(\delta_4, U_2^*, \Lambda)=\begin{cases}1 & \cdots \text{ if } \Lambda\sim\Lambda(4),\\ 0 & \cdots \text{ otherwise,}\end{cases}$$

where

$$\Lambda(i)=\left\{\begin{pmatrix} x & y \\ \lambda_i\bar{y} & \bar{x}\end{pmatrix}; x, y\in \boldsymbol{Z}_2[\eta]\right\} \quad (i=1, \cdots, 4),$$

and

$$\Lambda'(i)=\begin{pmatrix}1 & -\eta' \\ 0 & 2\end{pmatrix}^{-1} M_2(O_2)\begin{pmatrix}1 & -\eta' \\ 0 & 2\end{pmatrix}\cap(K(\delta_i)_2+yK(\delta_i)_2) \quad (i=2, 3).$$

Next, we consider $C_2(2)$. In this case, $Z(\Upsilon)_2$ is the division quaternion algebra over $\boldsymbol{Q}_2$ for any $C_2(2)$. So, by Proposition 3.6, $C_2(2)$ consists of seven G_2-conjugacy classes. Fix an isomorphism O_2 with $M_2(\boldsymbol{Z}_2)$, and identify O_2 with $M_2(\boldsymbol{Z}_2)$. Put $\omega=\begin{pmatrix}0 & 1\\ -1 & 0\end{pmatrix}$, $\varepsilon=\begin{pmatrix}1 & 0\\ 0 & -1\end{pmatrix}\in O_2$.

Proposition 5.10. *Any G_2-conjugacy classes in $C_2(2)$ are 2-integral. The representatives $\delta_1, \cdots, \delta_7$ of conjugacy classes in $C_2(2)$, and $K(\delta_i)_2$, $Z_G(\delta_i)_2$ $(i=1, \cdots, 7)$ are given as follows (, identifying G_2 with G_2^* for some of them)*:

$$\delta_i=\left(\begin{pmatrix}\omega & 0\\ 0 & \omega\end{pmatrix}, \varepsilon\begin{pmatrix}a_i & b_i\\ \bar{b}_i & d_i\end{pmatrix}\right)\in(U_2^*)^2 \quad (i=1, 2, 3),$$

$$K(\delta_i)_2=\boldsymbol{Q}_2\left(\begin{pmatrix}1 & -\bar{a}_i/\bar{b}_i\\ -a_i/b_i & -1\end{pmatrix}\right),$$

and

$$Z_G(\delta_i)_2=K(\delta_i)_2^\times\cup y_iK(\delta_i)_2^\times \quad (i=1, 2, 3),$$

where a_i, b_i $(i=1, 2, 3)$ are fixed elements of $\boldsymbol{Z}_2[\omega]$ such that $N(a_i)+N(b_i)=-1$, $N(b_1)=2$, $N(b_2)=10$, $N(b_3)=5$, $d_i=-\bar{b}_i^{-1}\bar{b}_i\bar{a}_i$ $(i=1, 2, 3)$, and $y_i=\begin{pmatrix}0 & \bar{a}_i/\bar{b}_i\\ a_i/b_i & 0\end{pmatrix}$

$$\delta_j=\left(\begin{pmatrix}\omega & 0\\ 0 & \omega\end{pmatrix}, \begin{pmatrix}0 & 1\\ \lambda_j & 0\end{pmatrix}\right)\in(U_2^*)^2 \quad (j=4, 5),$$

$$K(\delta_j)_2=\boldsymbol{Q}_2\left(\begin{pmatrix}0 & \omega\\ -\lambda_j\omega & 0\end{pmatrix}\right),$$

$$Z_G(\delta_j)_2=K(\delta_j)_2^\times\cup\begin{pmatrix}\omega & 0\\ 0 & -\omega\end{pmatrix}K(\delta_j)^\times \quad (j=4, 5),$$

where $\lambda_4=-1$, $\lambda_5=-5$, and

$$\delta_k=\left(\begin{pmatrix}\omega & \bar{\pi}\omega\varepsilon\\ 0 & \omega\end{pmatrix}, \begin{pmatrix}-b_k\lambda_k & \pi^{-1}(\pi\bar{b}_k-\pi^{-1}\bar{b}_k^{-1})\varepsilon\\ \varepsilon\pi\lambda_k b_k & \bar{b}_k\lambda_k\end{pmatrix}\right)\in(U_2^*)^2,\quad (k=6,7),$$

$$K(\delta_k)_2=\left\{\begin{pmatrix}x+y\lambda_k\pi^{-1}\omega\varepsilon, & y\omega(2^{-1}\lambda_k-1)\\ y\lambda_k\omega, & x+y\lambda_k\omega\pi^{-1}\varepsilon\end{pmatrix};\ x,y\in \boldsymbol{Q}_2\right\},$$

$$Z_G(\delta_k)_2=K(\delta_k)^\times\cup\begin{pmatrix}\omega & 0\\ 0 & -\omega\end{pmatrix}K(\delta_k)^\times\quad (k=6,7),$$

where $\lambda_6=-2$, $\lambda_7=-10$, $\pi=1+\omega$, *and* b_6, b_7 *are fixed elements of* $\boldsymbol{Q}_2[\omega]$ *such that* $N(b_6)=\dfrac{1}{2}$, $N(b_7)=\dfrac{1}{10}$.

Proposition 5.11. *Notations being as in Proposition* 5.10, *local data for* δ_i $(i=1,\cdots,7)$ *are given as follows*:

For each $i=1$, 2, *or* 3,

$$c_2(\delta_i, U_2, \Lambda)=\begin{cases}1 & \cdots\ \text{if } \Lambda\sim\Lambda(i),\\ 0 & \cdots\ \text{otherwise}.\end{cases}$$

where $\Lambda(i)=M_2(O_2)\cap(K(\delta_i)+y_iK(\delta_i))$ $(i=1,2,3)$.

For each $i=4$, 5,

$$c_2(\delta_i, U_2^*, \Lambda)=\begin{cases}3 & \cdots\ \text{if } \Lambda\sim\Lambda(i),\\ 0 & \cdots\ \text{otherwise},\end{cases}$$

where $\Lambda(i)=\left\{\begin{pmatrix}a & b\\ -\lambda_i\bar{b} & \bar{a}\end{pmatrix};\ a,b\in \boldsymbol{Z}_2[\omega]\right\}$ $(i=4,5)$.

For each $i=6$, 7,

$$c_2(\delta_i, U_2^*, \Lambda)=\begin{cases}1 & \cdots\ \text{if } \Lambda\sim\Lambda(i),\\ 0 & \cdots\ \text{otherwise},\end{cases}$$

where $\Lambda(i)=\left(K(\delta_i)_2+\begin{pmatrix}\omega & 0\\ 0 & -\omega\end{pmatrix}K(\delta_i)_2\right)\cap M_2(O_2)$ $(i=6,7)$).

For each $i=1,\cdots,7$, put $\delta_i=(\alpha_i,\beta_i)\in C_2(2)$, where δ_i are as in Proposition 10. Then, $\boldsymbol{Q}_2(\alpha_i,\beta_i)$ is the division quaternion algebra over $\boldsymbol{Q}_2$. In the next section, we shall use the following Lemma.

Lemma 5.12. *The order* $\boldsymbol{Q}_2(\alpha_i,\beta_i)\cap M_2(O_2)$ *is the (unique) maximal order of* $\boldsymbol{Q}_2(\alpha_i,\beta_i)$ *for each* $i=1$, 2, 6, *or* 7, *and not maximal for each* $i=3$, 4, 5.

Proof. The unique maximal order of $\boldsymbol{Q}_2(\alpha_i,\beta_i)$ is explicitly given by

$$\boldsymbol{Z}_2+\boldsymbol{Z}_2\alpha_i+\boldsymbol{Z}_2\beta_i+\boldsymbol{Z}_2\frac{1+\alpha_i+\beta_i+\alpha_i\beta_i}{2}.$$

By direct calculation, we can show that $(1+\alpha_i+\beta_i+\alpha_i\beta_i)/2 \in M_2(O_2)$, if and only if $i=1, 2, 6$, or 7. q.e.d.

§ 6. Explicit mass formulae

In this section, we give the explicit mass formulae for $m(\mathscr{L}(M), \{\gamma\}_{GL})$ for any $\gamma \in C_2$ or C_3, gathering local data in § 5 together. Any element $g \in G^r$, or the G-conjugacy class which contains g is called locally integral, if g is q-integral for every prime q.

Proposition 6.1. *For any $g \in C_2$, g is locally integral, if and only if $K(g) \cong \boldsymbol{Q}(\sqrt{-p})$ and $p \not\equiv 7 \bmod 8$, or $K(g) \cong \boldsymbol{Q}(\sqrt{-2p})$. For any $g \in C_3$, g is locally integral, if and only if $K(g) \cong \boldsymbol{Q}(\sqrt{-3p})$.*

Proof. If $g \in C_2$ is locally integral, then by Proposition 5.4, $K(g) \otimes_{\boldsymbol{Q}} \boldsymbol{Q}_p$ is ramified over $\boldsymbol{Q}_p$, and by Proposition 5.1, $K(g) \otimes_{\boldsymbol{Q}} \boldsymbol{Q}_q$ is not ramified over $\boldsymbol{Q}_q$ for any $q \neq 2$. As $K(g)$ is imaginary quadratic, and $K(g)_2 \not\cong \boldsymbol{Q}_2 \oplus \boldsymbol{Q}_2$, we get the assertion for C_2. The assertion for $g \in C_3$ is obtained similarly by Propositions 5.1, 5.6, and 5.8. q.e.d.

Now, we shall calculate $M_G(\Lambda)$ for each locally integral $g \in C_2$, or C_3, and $\Lambda \subset Z(g) = K(g) + yK(g)$, where $y \in G$ is taken as in Proposition 3.7 for each g. We have

$$\Lambda^\times \cap G = \Lambda^\times \cap Z_G(g) = (\Lambda^\times \cap K(g)^\times) \cup (\Lambda^\times \cap yK(g)^\times).$$

For any order Λ, we get $\Lambda^\times \cap G = \{\pm 1\}$. In fact, as g is locally integral, $K(g) \cong \boldsymbol{Q}(\sqrt{-p})$, $\boldsymbol{Q}(\sqrt{-2p})$, or $\boldsymbol{Q}(\sqrt{-3p})$. Denote by o_K the maximal order of $K(g)$. Then, as we assumed $p \geqq 5$, $o_K^\times = \{\pm 1\} = K(g)^\times \cap \Lambda^\times$. On the other hand, any element $y_1 \in yK(g)^\times$ satisfies $0 < y_1^2 \in \boldsymbol{Q}^\times$, because $Z(g)$ is indefinite and $K(g)$ is imaginary. As $Z(g)$ is division, $y_1^2 \neq 1$ and $y_1 y_1^\rho \neq -1$, where ρ is the main involution of $Z(g)$. If y_1 is integral, then $(y_1 y_1^\rho)^{-1} \notin \boldsymbol{Z}$, so y_1^{-1} is not integral. Hence we get $\Lambda^\times \cap yK(g)^\times = \varnothing$. By definition of $M_G(\Lambda)$, we get

$$M_G(\Lambda) = \frac{1}{2} h(\Lambda),$$

where

$$h(\Lambda) = \sharp(Z_G(g) \backslash Z_G(g)_A / (\Lambda_A^\times \cap G_A)).$$

Now, we calculate $h(\Lambda)$ for each locally integral $g \in C_2$, or C_3, and each $\Lambda \subset Z(g)$ such that $c_q(g, U_q, \Lambda) \neq 0$ for all primes q. We denote by $h(-m)$ the class number of the imaginary quadratic field $\boldsymbol{Q}(\sqrt{-m})$.

Proposition 6.2. *Assume that* $g \in C_2$ *and* $K(g) \cong \boldsymbol{Q}(\sqrt{-p})$, $p \not\equiv 7 \bmod 8$. *Then, the G-genus which contains* Λ *such that* $c_q(g, U_q, \Lambda) \neq 0$ *for all primes q is unique, and for that* Λ, *we get*

$$h(\Lambda)=\begin{cases} \dfrac{h(-p)}{2} & \cdots \text{ if } p\equiv 1, \text{ or } 5 \bmod 8, \\ 3h(-p) & \cdots \text{ if } p\equiv 3 \bmod 8. \end{cases}$$

Proof. The first assertion is obvious by Propositions 5.2, 5.5, and 5.11. To show the rest, first, assume that $p\equiv 1$, or $5 \bmod 8$.

Then, by Propositions 5.2, 5.5, 5.11, there exists $y_q \in yK(g)_q^\times \cap \Lambda_q^\times$ for each prime q. Besides, it is easy to see that

$$\Lambda_q^\times \cap K(g)_q^\times = (o_K \otimes_{\boldsymbol{Z}} \boldsymbol{Z}_q)^\times \quad \text{and} \quad \Lambda_q^\times \cap yK(g)_q^\times = y_q(o_K \otimes_{\boldsymbol{Z}} \boldsymbol{Z}_q)^\times$$

for any prime q, where o_K is the maximal order of $K(g)$. Hence, it is easy to see that the representatives of $Z_G(g)\backslash Z_G(g)_A/(\Lambda_A^\times \cap G_A)$ can be taken in $K(g)_A^\times$. Two elements $a, b \in K(g)_A^\times$ belong to the same double coset, if and only if $kau=b$ for some $k \in K(g)^\times$, $u \in \Lambda_A^\times \cap K(g)^\times$, or $kya(u_v')=b$ for some $k \in K(g)^\times$ and $(u_v') \in \Lambda_A^\times \cap G_A$ such that $u_q'=y_q u_q \in y_q(\Lambda_q^\times \cap K(g)_q^\times)$ for all primes q. It is easy to see that $kya(u_v')=k\bar{a}(yu_v')=k\bar{a}(yy_q)(u_v)$, where $^-$ denotes the non-trivial automorphism of $K(g)_A^\times$ which is induced from the complex conjugation of $K(g)$. Deote by $\mathfrak{y}$ the fractional ideal of $K(g)$ defined by:

$$\mathfrak{y}=\bigcap_q (yy_q(o_K \otimes_{\boldsymbol{Z}} \boldsymbol{Z}_q) \cap K(g)).$$

Denote by $C(K(g))$ the ideal class group of $K(g)$. For any classes c_1, $c_2 \in C(K(g))$, write $c_1 \sim c_2$, if $c_1=\bar{c}_2 c(\mathfrak{y})$, where $c(\mathfrak{y})$ is the ideal class which contains $\mathfrak{y}$. It is obvious that $h(\Lambda)=\sharp(C(K(g))/\sim)$. Now, we shall show that, if $c_1 \sim c_2$, then $c_1 \neq c_2$. Take a prime p' such that $\left(\dfrac{-p}{p'}\right)=1$ and $p'\equiv 3 \bmod 4$. Then, it is easy to see that the Hilbert symbol $(-p, p')_v = -1$ for $v=2$ and p, and $(-p, p')_v=1$ for all the other places v of $\boldsymbol{Q}$. As the discriminant of $Z(g)$ is $2p$, we can assume $y^2=p'$ and $y_q=y$ for any prime $q \neq p'$. So, $c(\mathfrak{y})$ is the class which contains the prime ideal $\mathfrak{p}'$ such that $\mathfrak{p}'\bar{\mathfrak{p}}'=(p')$. For $c \in C(K(g))$, assume that $c \sim c$. Take an ideal $\mathfrak{a} \in c$. Then, $\mathfrak{a}(k)=\bar{\mathfrak{a}}\mathfrak{p}'$ for some $k \in K(g)^\times$. Taking the norm of both sides, we get $N(k)=N(\mathfrak{p}')=p'$. This contradicts to the fact that $\left(\dfrac{p'}{p}\right)=-1$. Thus, we get $h(\Lambda)=h(-p)/2$. Next, assume that $p\equiv 3 \bmod 8$. In this case, by Proposition 5.11, we can show that $\Lambda_2 \cap yK(g_2)=\varnothing$.

Besides, we can assume that $y^2=2$. So, $y \in \Lambda_q^\times \cap yK(g)_q = y(o_K \otimes_Z Z_q)^\times$ for any prime $q \neq 2$. Hence, it is obvious that the representatives of $Z_G(g)\backslash Z_G(g)_A/(\Lambda_A^\times \cap G_A)$ can be taken in $K(g)_A^\times$. As $\Lambda_2 \cap yK(g)_2 = \emptyset$, we can see that for any $a, b \in K(g)_A^\times$, $b \in Z_G(g)a(\Lambda_A^\times \cap G_A)$, if and only if $b \in K(g)a(\Lambda_A^\times \cap K(g)_A^\times)$. By Propositions 5.2, 5.5, 5.11, we get $\Lambda_q^\times \cap K(g)_q^\times = (o_K \otimes_Z Z_q)^\times$ for any prime $q \neq 2$, and

$$[(o_K \otimes_Z Z_2)^\times : \Lambda_2^\times \cap K(g)_2^\times] = \left[Z_2\left[\frac{1+\sqrt{-3}}{2}\right]^\times : Z_2[\sqrt{-3}]^\times \right] = 3.$$

Hence, $h = 3h(-p)$. q.e.d.

Proposition 6.3. *Assume that $g \in C_2$ and $K(g) \cong Q(\sqrt{-2p})$. Then, the G-genus which contains Λ such that $c_q(g, U_q, \Lambda) \neq 0$ for all primes q is unique, and for that Λ, we get*

$$h(\Lambda) = \frac{h(-2p)}{2}.$$

Proof. Take a prime p' such that $p' \equiv 5 \bmod 8$ and $\left(\frac{p'}{p}\right) = -1$. We can assume that $y^2 = p'$, where y is as before. As

$$\Lambda_q^\times \cap K(g)_q^\times = (o_K \otimes_Z Z_q)^\times \quad \text{and} \quad \Lambda_q^\times \cap yK(g)_q = y_q(o_K \otimes_Z Z_q)^\times$$

for some $y_q \in \Lambda_q^\times$ for every prime q, the proof is obtained virtually in the same way as in the case $K(g) \cong Q(\sqrt{-p})$ and $p \equiv 1 \bmod 4$. The details will be omitted here. q.e.d.

Proposition 6.4. *Assume that $g \in C_3$ and $K(g) \cong Q(\sqrt{-3p})$. If* $p \equiv 3 \bmod 4$, *the G-genus which contains Λ such that $c_q(g, U_q, \Lambda) \neq 0$ for all primes q is unique, and for that Λ, we get*

$$h(\Lambda) = \frac{h(-3p)}{2}.$$

If $p \equiv 1 \bmod 4$, *there are exactly two G-genera $L_G(\Lambda)$, $L_G(\Lambda')$ such that $c_q(g, U_q, \Lambda) \neq 0$ and $c_q(g, U_q, \Lambda') = 0$ for all primes q. We can assume $\Lambda_2 \sim \Lambda(i)$ and $\Lambda_2' \sim \Lambda'(i)$, where $i = 2$ (resp. 3), if $p \equiv 5$ (resp. 1) mod 8 and $\Lambda(i)$ and $\Lambda'(i)$ are as in Proposition 5.9. For these orders, we get*

$$h(\Lambda) = \begin{cases} \dfrac{h(-3p)}{2} & \cdots \ \text{if } p \equiv 5 \bmod 8, \\[2ex] \dfrac{3h(-3p)}{2} & \cdots \ \text{if } p \equiv 1 \bmod 8, \end{cases}$$

$$h(\Lambda')=\frac{h(-3p)}{2} \quad \cdots \text{ if } p\equiv 1 \bmod 4.$$

Proof. Take a prime p' such that $p'\equiv 5 \bmod 12$ and $\left(\frac{p'}{p}\right)=-1$. We can assume $y^2=p'$. If $p\not\equiv 1 \bmod 8$, or $q\neq 2$, we get

$$\Lambda_q^\times \cap K(g)_q^\times=(\Lambda_q')^\times \cap K(g)_q^\times=(o_K\otimes_{\boldsymbol{Z}} \boldsymbol{Z}_q)^\times,$$

and

$$\Lambda_q^\times \cap yK(g)_q^\times=(\Lambda_q')^\times \cap yK(g)_q^\times=y_q(o_K\otimes_{\boldsymbol{Z}} \boldsymbol{Z}_q)^\times$$

for some $y_q \in \Lambda_q$. When $p\equiv 1 \bmod 8$, we get $(\Lambda_2')^\times \cap K(g)_2^\times=(o_K\otimes_{\boldsymbol{Z}} \boldsymbol{Z}_2)^\times$,

$$(\Lambda_2')^\times \cap yK(g)_2^\times=y(o_K\otimes_{\boldsymbol{Z}} \boldsymbol{Z}_2)^\times, \quad \text{and} \quad \Lambda_2^\times \cap K(g)_2\cong \boldsymbol{Z}_2[\sqrt{-3}\,]^\times.$$

So, the proof is obtained virtually in the same way as in the proof of Proposition 6.2. q.e.d.

Theorem 6.5. *Assume that $p\geqq 5$. Then,*

$$m(\mathscr{L}(M), C_2)=\begin{cases} \dfrac{3h(-p)}{4}+\dfrac{h(-2p)}{4} & \cdots \text{ if } p\equiv 1, \text{ or } 5 \bmod 8, \\ \dfrac{3h(-p)}{2}+\dfrac{h(-2p)}{4} & \cdots \text{ if } p\equiv 3 \bmod 8, \\ \dfrac{h(-2p)}{4} & \cdots \text{ if } p\equiv 7 \bmod 8, \end{cases}$$

and

$$m(\mathscr{L}(M), C_3)=\begin{cases} h(-3p) & \cdots \text{ if } p\equiv 1 \bmod 8, \\ \dfrac{h(-3p)}{4} & \cdots \text{ if } p\equiv 3, \text{ or } 7 \bmod 8, \\ \dfrac{h(-3p)}{2} & \cdots \text{ if } p\equiv 5 \bmod 8. \end{cases}$$

Proof. This is easily obtained by Theorem 1.1 and Propositions in § 5 and Propositions 6.2, 6.3, 6.4. q.e.d.

Finally, we shall give some refinement of Proposition 6.1. Assume that $\Gamma_i \ni \alpha, \beta$ for some $\gamma=(\alpha, \beta) \in C_2$ and some i $(1\leqq i\leqq H)$. By Proposition 6.1, we have $K(\gamma)=\boldsymbol{Q}(\sqrt{-p})$, or $\boldsymbol{Q}(\sqrt{-2p})$. More precisely, we have

Proposition 6.6. *Assumptions being as above, Γ_i has a subgroup Γ such that $\Gamma \ni \alpha, \beta$ and $\Gamma/\{\pm 1\} \cong A_4$, if and only if $K(\gamma) \cong \boldsymbol{Q}(\sqrt{-2p})$.*

Proof. Put $R_i = \{g \in M_2(B);\ L_i g \subset L_i\}$ as before. By Lemma 5.12, $\Lambda = \boldsymbol{Q}(\alpha, \beta) \cap R_i$ is a maximal order of $\boldsymbol{Q}(\alpha, \beta)$, if and only if $K(\gamma) \cong \boldsymbol{Q}(\sqrt{-2p})$. If Λ is maximal, then, noting that $\Lambda \supset Z[\alpha, \beta]$, we can see easily that

$$\Lambda = Z + Z\alpha + Z\beta + Z\frac{1+\alpha+\beta+\alpha\beta}{2},$$

and

$$\Lambda^\times \cap G = \left\{\pm 1,\ \pm\alpha,\ \pm\beta,\ \frac{\pm 1 \pm \alpha \pm \beta \pm \alpha\beta}{2}\right\},$$

and $(\Lambda^\times \cap G)/\{\pm 1\} \cong A_4$. This proves that the condition $K(\gamma) \cong \boldsymbol{Q}(\sqrt{-2p})$ is sufficient. Now, we show that the condition is necessary. Assume that there exists a subgroup $\Gamma \subset \Gamma_i$ such that $\Gamma \ni \alpha, \beta$, and $\Gamma/\{\pm 1\} \cong A_4$. As we shall see in the next Lemma 6.7, we get $\Gamma \subset \boldsymbol{Q}(\alpha, \beta)$. If $K(\gamma) \cong \boldsymbol{Q}(\sqrt{-p})$, then $\Lambda = \boldsymbol{Z}[\alpha, \beta]$, and $\Lambda^\times \cap G = \{\pm 1, \pm\alpha, \pm\beta, \pm\alpha\beta\}$. As $(\Lambda^\times \cap G)/\{\pm 1\} = (\boldsymbol{Z}/2\boldsymbol{Z})^2$ in this case, we get a contradiction. Hence, $K(\gamma) \cong \boldsymbol{Q}(\sqrt{-2p})$. q.e.d.

Lemma 6.7. *Assume that there exists a subgroup $\Gamma \subset \Gamma_i$ such that $\Gamma/\{\pm 1\} \cong A_4$. Then, there exists $\gamma = (\alpha, \beta) \in C_2$ such that $\alpha, \beta \in \Gamma$. For any such $\gamma = (\alpha, \beta) \in C_2$, we get $\Gamma \subset \boldsymbol{Q}(\alpha, \beta)$.*

Proof. Fix an isomorphism $\Gamma/\{\pm 1\} \cong A_4$ and fix any representatives σ (resp. τ) in Γ of (12) (34) (resp. (123)) $\in A_4$. Replacing σ by $-\sigma$, or τ by $-\tau$, if necessary, we can assume that $\tau^3 = 1$ and $(\sigma\tau)^3 = 1$. Then, by Theorem 3.1, we have $\sigma^2 = -1$, $\tau^2 + \tau + 1 = 0$, and $(\sigma\tau)^2 + (\sigma\tau) + 1 = 0$. So, we get $\tau\sigma\tau^{-1} = -\tau^2 + \sigma\tau$ and $\sigma\tau\sigma\tau^{-1} = -\sigma\tau^2 - \tau$. Now, put $\alpha = \sigma$, $\beta = \tau\sigma\tau^{-1}$. Then $(\alpha, \beta) \in C_2$, and $(-1+\alpha+\beta-\alpha\beta)/2 = (-1+\tau-\tau^2+\sigma(1+\tau+\tau^2)) = \tau$. Hence, $\tau \in \boldsymbol{Q}(\alpha, \beta)$. As A_4 is generated by (12) (34), (13) (24), and (123), we get $\Gamma \subset \boldsymbol{Q}(\alpha, \beta)$. q.e.d.

§ 7. Main Theorems

Here, we give our main results. For the readers' convenience, we recall here the notations and assumptions in the previous sections. We denote by B the definite quaternion algebra with prime discriminant p. We denote by $L_1, \cdots, L_H$ a complete set of representatives of classes of lattices in the *non-principal* genus $\mathscr{L}(M)$ of the positive definite binary quaternion hermitian space V over B. For each i $(1 \leqq i \leqq H)$, define a subgroup Γ_i of G by:

$$\Gamma_i = \{g \in G;\ L_i g = L_i\},$$

where G is the group of similitudes of V. For any (abstract) finite group Γ, we put

$$I(\Gamma) = \{i \in \{1, \cdots, H\};\ \Gamma_i/\{\pm 1\} \cong \Gamma\}.$$

Theorem 7.1. *Assume that $p \geqq 7$. Then, for each prime p,*

$$\begin{aligned}
\sharp(I(\{1\})) &= \frac{p^2-1}{2880} - \frac{1}{96}\left(4+\left(\frac{-1}{p}\right)\right)\left(p-\left(\frac{-1}{p}\right)\right) \\
&\quad - \frac{1}{144}\left(3+\left(\frac{-3}{p}\right)\right)\left(p-\left(\frac{-3}{p}\right)\right) + \frac{1}{8}(1-\chi(2))h(-p) \\
&\quad + \frac{1}{8}h(-2p) + \begin{cases} \dfrac{1}{8}(3-\chi'(2))h(-3p) & \cdots \text{ if } p \equiv 1 \bmod 4, \\[2mm] \dfrac{1}{8}h(-3p) & \cdots \text{ if } p \equiv 3 \bmod 4, \end{cases} \\
&\quad - \frac{5}{12}\sharp(I(D_{12})) - \frac{1}{2}\sharp(I(S_4)) - \frac{3}{5}\sharp(I(A_5)),
\end{aligned}$$

$$\begin{aligned}
\sharp(I(\boldsymbol{Z}/2\boldsymbol{Z})) &= \frac{1}{48}\left(4+\left(\frac{-1}{p}\right)\right)\left(p-\left(\frac{-1}{p}\right)\right) - \frac{3}{8}(1-\chi(2))h(-p) \\
&\quad - \frac{h(-2p)}{8} - \begin{cases} \dfrac{1}{4}(3-\chi'(2))h(-3p) & \cdots \text{ if } p \equiv 1 \bmod 4, \\[2mm] \dfrac{h(-3p)}{4} & \cdots \text{ if } p \equiv 3 \bmod 4, \end{cases} \\
&\quad + \frac{5}{4}\sharp(I(S_4)) + \frac{4}{3}\sharp(I(D_{12})) + \sharp(I(A_5)),
\end{aligned}$$

$$\begin{aligned}
\sharp(I(\boldsymbol{Z}/3\boldsymbol{Z})) &= \frac{1}{48}\left(3+\left(\frac{-3}{p}\right)\right)\left(p-\left(\frac{-3}{p}\right)\right) - \frac{h(-2p)}{4} \\
&\quad - \begin{cases} \dfrac{1}{8}(3-\chi'(2))h(-3p) & \cdots \text{ if } p \equiv 1 \bmod 4, \\[2mm] \dfrac{h(-3p)}{8} & \cdots \text{ if } p \equiv 3 \bmod 4. \end{cases} \\
&\quad + \frac{1}{4}\sharp(I(D_{12})) + \frac{1}{2}\sharp(I(S_4)) + \sharp(I(A_5)),
\end{aligned}$$

$$\#(I((Z/2Z)^2))=\frac{1}{4}(1-\chi(2))h(-p)-\#(I(D_{12}))-\frac{1}{2}\#(I(S_4)),$$

$$\#(I(S_3))=\begin{cases}\frac{1}{4}(3-\chi'(2))h(-3p) & \cdots \text{ if } p\equiv 1 \bmod 4,\\ \frac{h(-3p)}{4} & \cdots \text{ if } p\equiv 3 \bmod 4,\end{cases}$$
$$-\#(I(A_5))-\#(I(D_{12}))-\#(I(S_4)),$$

$$\#(I(A_4))=\frac{h(-2p)}{4}-\#(I(A_5))-\frac{1}{2}\#(I(S_4)),$$

$$\#(I(D_{12}))=\begin{cases}1 & \cdots \text{ if } p\equiv 5 \bmod 12,\\ 0 & \cdots \text{ if } p\equiv 1,\ 7,\ \text{or } 11 \bmod 12,\end{cases}$$

$$\#(I(S_4))=\begin{cases}1 & \cdots \text{ if } p\equiv 3,\ \text{or } 5 \bmod 8,\\ 0 & \cdots \text{ if } p\equiv 1,\ \text{or } 7 \bmod 8,\end{cases}$$

$$\#(I(A_5))=\begin{cases}1 & \cdots \text{ if } p\equiv 2,\ \text{or } 3 \bmod 5,\\ 0 & \cdots \text{ if } p\equiv 1,\ \text{or } 4 \bmod 5,\end{cases}$$

and $\#(I(\Gamma))=0$, unless Γ is isomorphic to one of the above groups, where $h(-m)$ is the class number of the field $\boldsymbol{Q}(\sqrt{-m})$, and $\left(\frac{*}{p}\right)$ is the Legendre symbol, χ (resp. χ') is the Kronecker symbol of $\boldsymbol{Q}(\sqrt{-p})$ (resp. $\boldsymbol{Q}(\sqrt{-3p})$), that is,

$$\chi(2)=\begin{cases}1 & \cdots \text{ if } p\equiv 7 \bmod 8,\\ -1 & \cdots \text{ if } p\equiv 3 \bmod 8,\\ 0 & \cdots \text{ if } p\equiv 1 \bmod 4,\end{cases} \qquad \chi'(2)=\begin{cases}1 & \cdots \text{ if } p\equiv 5 \bmod 8,\\ -1 & \cdots \text{ if } p\equiv 1 \bmod 8,\\ 0 & \cdots \text{ if } p\equiv 3 \bmod 4.\end{cases}$$

Remark 1. If $p=2$, 3, or 5, then $H=1$ ([5]). If $p=2$, then $\#(\Gamma_1)=1920$ and Γ_1 was explicitly given in [10] p. 592. It was shown in Katsura-Oort [12] that, if $p=3$, then $\Gamma_1/\{\pm 1\}\cong A_6$, and if $p=5$, then $\Gamma_1/\{\pm 1\}\cong PGL(2, \boldsymbol{F}_5)$.

Remark 2. By Theorem 7.1 and the fact that $\#(I(A_4))$ and $\#(I((Z/2Z)^2)$ are integers, we can show that $h(-p)\equiv 0$ (resp. 2) mod 4, if $p\equiv 1$ (resp. 5) mod 8, and that $h(-2p)\equiv 0$ (resp. 2) mod 4, if $p\equiv \pm 1$ (resp. ± 3) mod 8. These results are actually well known. (cf. Redei-Reichart [14])

Proof. The results on D_{12}, S_4, and A_5 have been already given in § 3 and § 4. Now, we show the rest. For each (abstract) finite group Γ and for each $n=2$, or 3, define a number $c_n(\Gamma)$ as follows:

$$c_n(\Gamma)=\#\{(\alpha,\beta)\in\Gamma^2;\ \alpha^2=\beta^n=1,\ \alpha\beta=\beta^{-1}\alpha,\ \alpha\neq\beta\}/\#(\Gamma)\times\begin{cases}2 & \cdots \text{ if } n=2,\\ 1 & \cdots \text{ if } n=3.\end{cases}$$

By the argument in § 3.2, we can easily show that

$$\begin{aligned}c_n(\Gamma_i/\{\pm1\})&=\#(C_n\cap\Gamma_i^2)/\#(\Gamma_i)\\ &=\#\{\gamma=(\alpha,\beta)\in C_n;\ \alpha,\beta\in\Gamma_i\}/\#(\Gamma_i)\quad(i=1,\cdots,H).\end{aligned}$$

We have $c_3(S_3)=c_3(D_{12})=c_3(S_4)=c_3(A_5)=1$. Hence, by Lemma 2.1 and by definition of the "mass", we get

$$m(\mathscr{L}(M),C_3)=\#(I(S_3))+\#(I(D_{12}))+\#(I(S_4))+\#(I(A_5)).$$

Hence, by Theorem 6.5, we get the assertion for $I(S_3)$. We get $c_2((\boldsymbol{Z}/2\boldsymbol{Z})^2)=c_2(D_{12})=3$, $c_2(A_4)=c_2(A_5)=1$, and $c_2(S_4)=2$. Now, define a number $c_2'(\Gamma)$ for each Γ by

$$c_2'(\Gamma)=\frac{2}{\#(\Gamma)}\times\#\left\{(\alpha,\beta)\in\Gamma^2;\ \begin{array}{c}\alpha^2=\beta^2=1,\ \alpha\beta=\beta\alpha,\ \alpha\neq\beta\\ \alpha,\beta\text{ are contained in a subgroup}\\ \text{of }\Gamma\text{ which is isomorphic to }A_4\end{array}\right\}.$$

Then, $c_2'((\boldsymbol{Z}/2\boldsymbol{Z})^2)=c_2'(D_{12})=0$, $c_2'(A_4)=c_2'(A_5)=1$, and $c_2'(S_4)=\frac{1}{2}$. By Lemma 2.1, Theorem 6.5, and Proposition 6.6, we get

$$3\#(I((\boldsymbol{Z}/2\boldsymbol{Z})^2)+\frac{3}{2}\#(I(S_4))+3\#(I(D_{12}))=\frac{1}{4}(1-\chi(2))h(-p),$$

and

$$\#(I(A_4))+\frac{1}{2}\#(I(S_4))+\#(I(A_5))=\begin{cases}\frac{1}{4}(3-\chi'(2))h(-3p), & \text{if } p\equiv1 \bmod 4,\\ \frac{h(-3p)}{4} & \text{if } p\equiv3 \bmod 4.\end{cases}$$

Hence, we get the assertion for $(\boldsymbol{Z}/2\boldsymbol{Z})^2$ and A_4. The rest is easily obtained by Theorem 2.2 and (2.3), noting that $M(\Gamma,1)=\frac{1}{\#(\Gamma)}$ for any finite group Γ, and

$$M(Z/2Z, 2)=\frac{1}{2},\quad M((Z/2Z)^2, 2)=\frac{3}{4},\quad M(S_3, 2)=\frac{1}{2},$$

$$M(A_4, 2)=\frac{1}{4},\quad M(D_{12}, 2)=\frac{7}{12},\quad M(S_4, 2)=\frac{3}{8},\quad M(A_5, 2)=\frac{1}{4},$$

$$M(Z/3Z, 3)=\frac{2}{3},\quad M(S_3, 3)=\frac{1}{3},\quad M(A_4, 3)=\frac{2}{3},\quad M(D_{12}, 3)=\frac{1}{6},$$

$$M(S_4, 3)=\frac{1}{3},\quad \text{and}\quad M(A_5, 3)=\frac{1}{3}.$$ q.e.d.

Numerical examples. For small primes p, the explicit values of $\sharp(I(\Gamma))$ for each finite group Γ are given in the following table. When $p\leqq 31$ and $p\neq 23$, such table has been already given in Katsura-Oort [12], but we include also these cases below for the readers' convenience. In order to calculate the following table by Theorem 7.1, we used the table of class numbers of imaginary quadratic fields by H. Wada [17].

p \ Γ	$\{1\}$	$Z/2Z$	$Z/3Z$	$(Z/2Z)^2$	S_3	A_4	S_4	D_{12}	A_5
7	0	0	0	0	0	0	0	0	1
11	0	0	0	0	0	0	1	0	0
13	0	0	0	0	0	0	1	0	1
17	0	0	0	0	0	0	0	1	1
19	0	0	0	0	0	1	1	0	0
23	0	0	0	0	1	0	0	0	1
29	0	0	0	0	1	0	1	1	0
31	0	0	0	0	1	2	0	0	0
37	0	0	0	0	2	1	1	0	1
41	0	0	0	1	1	1	0	1	0
43	0	0	1	0	1	1	1	0	1
47	0	1	0	0	1	1	0	0	1
53	0	1	0	0	2	0	1	1	1
59	0	1	1	1	0	1	1	0	0
61	0	0	1	1	3	2	1	0	0
67	0	1	2	0	1	2	1	0	1
71	0	2	1	0	2	1	0	0	0
73	0	1	1	1	3	3	0	0	1
79	0	1	3	0	3	2	0	0	0
83	1	1	1	1	1	1	1	0	1
89	0	3	1	2	1	2	0	1	0
97	0	3	2	1	3	4	0	0	1
101	1	2	1	2	3	1	1	1	0
103	1	2	3	0	2	4	0	0	1
107	2	1	2	1	3	0	1	0	1
109	0	3	4	1	5	2	1	0	0
113	1	4	1	1	4	1	0	1	1
127	2	2	5	0	4	3	0	0	1

Now, we give here explicit answers to Problem 2 also in some other cases. Consider the definite quaternion algebra B over $\boldsymbol{Q}$ with discriminant D, where D is not necessarily prime. Let O be a maximal order of B. Take any positive divisor D_2 of D and put $D_1=D/D_2$. Let $\mathscr{L}(D_1, D_2)$ be the set of all maximal O-lattices $L\subset B^2$ such that $L\otimes_{\boldsymbol{Z}}\boldsymbol{Z}_g\cong O_q^2$, if and only if $p\nmid D_2$. This $\mathscr{L}(D_1, D_2)$ forms a single genus (Shimura [16]). Take a complete set of representatives $L_1, \cdots, L_H$ of the classes in $\mathscr{L}(D_1, D_2)$, and put $\Gamma_i=\mathrm{Aut}\,(L_i)$ $(1\leqq i\leqq H)$. For any finite group Γ, put $I(\Gamma)=\sharp\{i;\ \Gamma_i/\{\pm 1\}\cong\Gamma,\ 1\leqq i\leqq H\}$.

Theorem 7.2. *Assume that* 2, 3, $5\nmid D$ *and* $D_2\neq 1$. *Put* $t=\sharp\{p; p\,|\,D\}$ *and* $r=\sharp\{p; p\,|\,D_1\}$. *Then,*

$$\sharp(I(\boldsymbol{Z}/5\boldsymbol{Z}))=\begin{cases}2^{t+r-2}, & \text{if } p\equiv -1 \bmod 5 \text{ for all } p\,|\,D_1\ (D_1\neq 1)\\ & \text{and } p\equiv \pm 2 \bmod 5 \text{ for all } p\,|\,D_2,\\ 0 & \text{otherwise,}\end{cases}$$

$$\sharp(I(A_5))=\begin{cases}2^{t-1}, & \text{if } D_1=1 \text{ and } p\equiv \pm 2 \bmod 5 \text{ for all } p\,|\,D,\\ 0 & \text{otherwise,}\end{cases}$$

$$\sharp(I(\boldsymbol{Z}/4\boldsymbol{Z}))=\begin{cases}2^{t+s-1}, & \text{if } D_1\neq 1,\ p\not\equiv 1 \bmod 8 \text{ for all } p\,|\,D_1\\ & \text{and } p\equiv \pm 3 \bmod 8 \text{ for all } p\,|\,D_2,\\ 0 & \text{otherwise,}\end{cases}$$

where $s=\sharp\{p\,|\,D_1;\ p\equiv -1 \bmod 8\}$,

$$\sharp(I(S_4))=\begin{cases}2^{t-1}, & \text{if } D_1=1 \text{ and } p\equiv \pm 3 \bmod 8, \text{ for all } p\,|\,D,\\ 0 & \text{otherwise,}\end{cases}$$

$$\sharp(I(\boldsymbol{Z}/6\boldsymbol{Z}))=\begin{cases}2^{t+u-3}, & \text{if } D_1(-1;\,12)\neq\varnothing \text{ and}\\ & D(1;\,12)=D_2(-1:\,12)=\varnothing,\\ 2^{t-2}, & \text{if } D_1\neq 1,\ D(1;\,12)=D(-1;\,12)=\varnothing \text{ and}\\ & \sharp(D(5;\,12))=odd,\\ 0 & \text{otherwise,}\end{cases}$$

where for any integers i, j, $D(i;j)=\{p\,|\,D;\ p\equiv i \bmod j\}$, $D_2(i;j)=\{p\,|\,D_2;\ p\equiv i \bmod j\}$, *and* $u=\sharp\{p\,|\,D_1;\ p\equiv -1 \bmod 12\}$,

$$\sharp(I(D_{12}))=\begin{cases}2^{t-1}, & \text{if } D_1=1,\ D(1;\,12)=D(-1;\,12)=\varnothing, \text{ and}\\ & \sharp(D(5;\,12))=odd,\\ 0 & \text{otherwise,}\end{cases}$$

$$\sharp(I(A_4))=\begin{cases}2^{r-2}h(-2D_2)-\sharp(I(A_5))-2^{-1}\sharp(I(S_4)), \\ \qquad\qquad \text{if } \left(\dfrac{-2D_2}{p}\right)=-1 \text{ for all } p\mid D_1, \\ 0 \qquad\quad \text{otherwise,}\end{cases}$$

$$\sharp(I((\boldsymbol{Z}/2\boldsymbol{Z})^2)=\begin{cases}2^{r-2}h(-D_2), & \text{if } \left(\dfrac{-D_2}{p}\right)=-1 \text{ for all } p\mid D_1, \\ & D_2\not\equiv -1 \bmod 8, \text{ and } D_1\neq 1, \\ \dfrac{h(-D_2)}{4}, & \text{if } D_1=1 \text{ and } D_2\equiv 1 \bmod 4, \\ \dfrac{h(-D_2)}{2}, & \text{if } D_1=1 \text{ and } D_2\equiv 3 \bmod 8,\end{cases}$$

$$-\sharp(I(D_{12}))-2^{-1}\sharp(I(S_4)),$$

$\sharp(I(S_3))=0$, *if* $\left(\dfrac{-3D_2}{p}\right)=1$ *for some* $p\mid D_1$. *In other cases,*

$$\sharp(I(S_3))=\begin{cases}2^r h(-3D_2), & \text{if } -3D_2\equiv 5 \bmod 8, \\ 2^{r-1}h(-3D_2), & \text{if } -3D_2\equiv 1 \bmod 8, \\ 2^{r-2}h(-3D_2), & \text{if } -3D_2\equiv 3 \bmod 4,\end{cases}$$

$$-\sharp(I(S_4))-\sharp(I(A_5))-\sharp(I(D_{12})),$$

$$\sharp(I(\boldsymbol{Z}/3\boldsymbol{Z}))=\frac{3}{2}H_6-\frac{1}{2}\sharp(I(S_3))-\sharp(I(A_4))-\frac{1}{2}\sharp(I(S_4))$$

$$-\frac{1}{4}\sharp(I(D_{12}))-\frac{1}{2}\sharp(I(A_5))-\frac{1}{2}\sharp(I(\boldsymbol{Z}/6\boldsymbol{Z})),$$

where H_6 *is given in* [5] (II) *Theorem* (p. 696) *for each* D_1, D_2,

$$\sharp(I(\boldsymbol{Z}/2\boldsymbol{Z}))=2H_7-\frac{3}{2}\sharp(I((\boldsymbol{Z}/2\boldsymbol{Z})^2)-\sharp(I(S_3))-\sharp(I(A_4))$$

$$-\frac{3}{4}\sharp(I(S_4))-\frac{7}{6}\sharp(I(D_{12}))-\frac{1}{2}\sharp(I(A_5))$$

$$-\sharp(I(\boldsymbol{Z}/4\boldsymbol{Z}))-\frac{1}{3}\sharp(I(\boldsymbol{Z}/6\boldsymbol{Z})),$$

where H_7 *is given in* [5] (II) *Theorem for each* D_1, D_2,

$$\sharp(I(\{1\}))=\frac{1}{2^6 3^2 5}\prod_{q\mid D_1}(p-1)(p^2+1)\prod_{p\mid D_2}(p^2-1)$$

$$-\sum_{k=2}^{6}\frac{1}{k}\,\sharp(I(Z/kZ))-\frac{1}{6}\,\sharp(I(S_3))-\frac{1}{12}\,\sharp(I(A_4))$$

$$-\frac{1}{24}\,\sharp(I(S_4))-\frac{1}{12}\,\sharp(I(D_{12}))-\frac{1}{60}\,\sharp(I(A_5)).$$

Proof. The proof is similar to that of Theorem 7.1. So, we only sketch it and omit the details here. By the assumption that $D_2 \neq 1$ and $2, 3, 5 \nmid D$, Lemma 2.1 is also valid, and almost all the results in § 2, 3, and 4 are valid also in this case. Now, the local data at $q \nmid D_1$ in § 5 can be used without any change. The calculation of local data at $q \mid D_1$ can be done virtually in the same way as in § 5. Finally, we obtain mass formulae and $\sharp(I(\Gamma))$ for each Γ. q.e.d.

References

[1] E. Bayer-Fluckiger, Principe de Hasse faible pour les systèmes des formes quadratiques, J. reine angew. Math., **378** (1987), 53–59.
[2] M. Eichler, Über die Idealklassenzahl totaler definiter Quaternionen Algebren, Math. Z., **43** (1938), 127–151.
[3] K. Hashimoto, On Brandt matrices associated with the positive definite quaternion hermitian forms, J. Fac. Sci. Univ. Tokyo Sect. IA, **27** (1980), 227–245.
[4] ——, Class numbers of positive definite ternary quaternion hermitian forms, Proc. Japan Acad., **59** Ser. A no. 10 (1983), 490–493.
[5] K. Hashimoto and T. Ibukiyama, On class numbers of positive definite binary quaternion hermitian forms (I): J. Fac. Sci. Univ. Tokyo Sect. IA, **27** (1980), 549–601; (II) J. Fac. Sci. Univ. Tokyo Sect. IA, **28** (1983), 695–699; (III) J. Fac. Sci. Univ. Tokyo Sect. IA, **30** (1983), 393–401.
[6] ——, On relations of dimensions of automorphic forms of $Sp(2, R)$ and its compact twist $Sp(2)$ (II), Advanced Studies in Pure Math., **7** (1985), 31–102.
[7] H. Hijikata, Hasse principle on quaternionic anti-hermitian forms, J. Math. Soc. Japan, **15** (1963), 165–175.
[8] ——, Hasse principle for the conjugacy classes of the orthogonal group (in Japanese), Reports of the Symposium on algebraic groups held at Yamanaka-Kyodo-Kenshujo, 1973.
[9] B. Huppert, Endliche Gruppen I, Springer-Verlag, Berlin, Heidelberg, New York, 1979.
[10] T. Ibukiyama, On symplectic Euler factors of genus two, J. Fac. Sci. Univ. Tokyo Sect. IA, **30** (1984), 587–614.
[11] T. Ibukiyama, T. Katsura and F. Oort, Supersingular curves of genus two and class numbers, Compositio Math., **57** (1986), 127–152.
[12] T. Katsura and F. Oort, Families of supersingular abelian surfaces, Compositio Math., **62** (1987), 107–167.
[13] L. Moret-Bailly, Familles de courbes et de variétés abéliennes sur P^1, Astérisque, **86** (1981), 109–140.
[14] L. Redei and H. Reichardt, Die Anzahl der durch 4 teilbaren Invarianten der Klassengruppe eines beliebigen quadratischen Zahlkörpers, J. reine angew. Math., **170** (1933), 69–74.
[15] G. Shimura, On the theory of automorphic functions, Ann. Math., **70** (1959), 101–144.

[16] ——, Arithmetic of alternating forms and quaternion hermitian forms, J. Math. Soc. Japan, **15** (1963), 33–65.
[17] H. Wada, The table of the class numbers of the quadratic field $Q(\sqrt{-m})$, $1 \leqq m < 24{,}000$, Sūriken-Kōkyūroku.

Department of Mathematics
College of General Education
Kyushu University
Ropponmatsu, Fukuoka
810, Japan

Advanced Studies in Pure Mathematics 15, 1989
Automorphic Forms and Geometry of Arithmetic Varieties
pp. 351–364

T-Complexes and Ogata's Zeta Zero Values

Masa-Nori Ishida

Dedicated to Prof. Ichiro Satake and Prof. Friedrich Hirzebruch on their sixtieth birthdays

Introduction

In [T1], Tsuchihashi defined the notion of cusp singularities in arbitrary dimension. They include the Hilbert modular cusp singularities as a special case. In this paper, we will show the rationality of the zeta zero value $Z(C, \Gamma; 0)$ of the zeta function associated to a Tsuchihashi cusp singularity which was defined by Ogata [Og]. He gave a formula for the zero value as a sum of integrals of C^∞-functions described by the characteristic function of the convex cone C. By this formula, he showed that the value is a half-integer in odd-dimensional case [Og, Theorem 2.3]. In two dimensional case, the singularity is a Hilbert modular cusp and 12 times the zeta zero value is an integer by [Z].

By the construction of Tsuchihashi cusp singularities, they have toroidal resolutions and the exceptional sets are toric divisors in the sense of [S2]. In order to describe toric divisors, we introduce the notion of *T-complexes* which is essentially equal to that of the weighted dual graphs which appear in [T1]. A T-complex Σ is a category with a finite number of objects. We define a functor $D^0_{\boldsymbol{Q}}$ from Σ to the category of $\boldsymbol{Q}$-vector spaces. We show that the rational number field $\boldsymbol{Q}$ has a natural injection into the inductive limit $\operatorname{ind\,lim}_\Sigma D^0_{\boldsymbol{Q}}$. We define a special element ω_Σ of $\operatorname{ind\,lim}_\Sigma D^0_{\boldsymbol{Q}}$. When Σ is the T-complex associated to a toroidal resolution of a Tsuchihashi cusp singularity (C, Γ), Ogata's formula means that there exists an explicit retraction $\operatorname{ind\,lim}_\Sigma D^0_{\boldsymbol{Q}} \otimes \boldsymbol{R} \to \boldsymbol{R}$ and the zero value $Z(C, \Gamma; 0)$ is the image of ω_Σ in $\boldsymbol{R}$. By using an equality in Section 1 for a nonsingular complete fan, we show that ω_Σ is in the image of $\boldsymbol{Q}$ in $\operatorname{ind\,lim}_\Sigma D^0_{\boldsymbol{Q}}$ for any T-complex Σ. This implies that the image of ω_Σ in $\boldsymbol{R}$ is independent of the retraction and is a rational number.

Received December 4, 1986.

§ 1. An equality for a nonsingular complete fan

Let N be a free $\boldsymbol{Z}$-module of finite rank and let $N_{\boldsymbol{R}}=N\otimes_{\boldsymbol{Z}}\boldsymbol{R}$. A nonempty subset σ of $N_{\boldsymbol{R}}$ is said to be a *strongly convex rational polyhedral cone* (*s.c.r.p. cone* for short) if there exists a finite subset $\{n_1, \cdots, n_s\}\subset N$ such that $\sigma=\boldsymbol{R}_0 n_1+\cdots+\boldsymbol{R}_0 n_s$ with $\sigma\cap(-\sigma)=\{0\}$, where $\boldsymbol{R}_0=\{c \in \boldsymbol{R};\ c\geq 0\}$. An s.c.r.p. cone σ is said to be *nonsingular* if there exists a $\boldsymbol{Z}$-basis $\{n_1, \cdots, n_r\}$ of N such that $\sigma=\boldsymbol{R}_0 n_1+\cdots+\boldsymbol{R}_0 n_s$ for an integer $0\leq s\leq r$. We call $\{n_1, \cdots, n_s\}$ *the canonical set of generators for* σ and denote it by gen σ when σ is nonsingular. This set is uniquely determined by σ.

A nonempty collection Δ of s.c.r.p. cones is said to be a *fan* if (1) $\tau \in \Delta$ and $\sigma\prec\tau$, which means that σ is a face of τ, imply $\sigma \in \Delta$, and (2) if $\sigma, \tau \in \Delta$ then $\sigma\cap\tau$ is a common face of σ and τ. For an s.c.r.p. cone π in $N_{\boldsymbol{R}}$, we denote by $\Gamma(\pi)$ the set of faces of π. Clearly, $\mathbf{0}:=\{0\}$ and π are in $\Gamma(\pi)$, and $\Gamma(\pi)$ is a fan in $N_{\boldsymbol{R}}$. A fan Δ in $N_{\boldsymbol{R}}$ is said to be *complete* if it is finite and the union $\bigcup_{\sigma\in\Delta}\sigma$ is equal to $N_{\boldsymbol{R}}$. It is said to be *nonsingular* if it consists of nonsingular cones.

For an s.c.r.p. cone ρ, we denote $N[\rho]:=N/(N\cap(\rho+(-\rho)))$. $N[\rho]$ is a free $\boldsymbol{Z}$-module with its rank equal to codim ρ. For a cone σ with $\rho\prec\sigma$, we denote by $\sigma[\rho]$ the image of σ in the quotient $N[\rho]_{\boldsymbol{R}}=N_{\boldsymbol{R}}/(\rho+(-\rho))$. If ρ is an element of a fan Δ, then $\Delta[\rho]=\{\sigma[\rho];\ \sigma \in \Delta \text{ and } \rho\prec\sigma\}$ is a fan in $N[\rho]_{\boldsymbol{R}}$.

For each nonzero element x in N, we denote $\gamma(x):=\boldsymbol{R}_0 x$. The cone $\gamma(x)$ is nonsingular of dimension one. If x is primitive, then gen $\gamma(x)=\{x\}$.

Let Δ be a nonsingular fan in $N_{\boldsymbol{R}}$. For a subset Φ of Δ, we set

$$f(\Phi):=\sum_{\sigma\in\Phi}\prod_{x\in\operatorname{gen}\sigma}\frac{1}{\exp(x)-1},$$

where we understand $\prod_{x\in\operatorname{gen}\sigma}(1/(\exp(x)-1))=1$ for $\sigma=\{0\}$. Let $M:=\operatorname{Hom}_{\boldsymbol{Z}}(N, \boldsymbol{Z})$ be the $\boldsymbol{Z}$-module dual to N. Since $x \in N$ is a linear function on the complex space $M_{\boldsymbol{C}}=M\otimes_{\boldsymbol{Z}}\boldsymbol{C}$, $\exp(x)$ is understood to be a holomorphic function on $M_{\boldsymbol{C}}$. Hence $f(\Phi)$ is a meromorphic function on $M_{\boldsymbol{C}}$. Note that the meromorphic function $1/(\exp(z)-1)$ of a complex variable z has poles of order one at each point of $2\pi i\boldsymbol{Z}$ and it has no pole elsewhere. Since no gen σ contains both x and $-x$ for any $x \in N$, $f(\Phi)$ may have poles of order at most one along the hyperplanes $H_{x,d}=(x=2\pi id)\subset M_{\boldsymbol{C}}$ for $x \in G(\Delta)$ and $d \in \boldsymbol{Z}$, where $G(\Delta):=\bigcup_{\sigma\in\Delta}\operatorname{gen}\sigma$. Note that $G(\Delta)$ is in one-to-one correspondence with the set $\Delta(1)$ of cones of dimension one in Δ by $x\mapsto\gamma(x)$.

In the rest of this section, we devote ourselves to proving the following theorem.

Theorem 1.1. *Let n be the rank of N. If $n \geq 1$ and Δ is a nonsingular complete fan, then $f(\Delta)$ is equal to zero.*

We prove the theorem by induction on n.

If $n=1$, then there is only one complete fan and the theorem is true since

$$1/(\exp(x)-1)+1/(\exp(-x)-1)+1=0.$$

Let $r \geq 2$ be an integer. We assume that the theorem is true for $1 \leq n \leq r-1$. We now suppose $n=r$.

Lemma 1.2. *In the above situation, $f(\Delta)$ is an entire function on $M_{\boldsymbol{C}}$.*

Proof. Since $f(\Delta)$ is periodic with respect to the subgroup $2\pi i M \subset M_{\boldsymbol{C}}$, it is sufficient to show that $f(\Delta)$ has no pole along the hyperplane $H_{x,0}$ for each $x \in G(\Delta)$. Let $\gamma := \gamma(x)$ and $\Delta(\gamma \prec) := \{\sigma \in \Delta; \gamma \prec \sigma\}$. By definition, we have $f(\Delta) = f(\Delta(\gamma \prec)) + f(\Delta \backslash \Delta(\gamma \prec))$. It is easy to see that the restriction $xf(\Delta(\gamma \prec))|_{H_{x,0}}$ is equal to $f(\Delta[\gamma])$, where $\Delta[\gamma]$ is the fan in $N[\gamma]_{\boldsymbol{R}} = N_{\boldsymbol{R}}/(\gamma+(-\gamma))$ induced by Δ. Since $\Delta[\gamma]$ is also a nonsingular complete fan and $N[\gamma] = N/(N_{\boldsymbol{R}} \cap (\gamma+(-\gamma)))$ is of rank $r-1$, $f(\Delta[\gamma])$ is equal to zero by the induction assumption. Hence $f(\Delta(\gamma \prec))$ has no pole along $H_{x,0}$. If $-\gamma$ is not an element of Δ, then $f(\Delta \backslash \Delta(\gamma \prec))$ has no pole at $H_{x,0}$ by definition. Hence so is $f(\Delta)$. Suppose $-\gamma \in \Delta$. Then $f(\Delta \backslash \Delta(\gamma \prec)) = f(\Delta(-\gamma \prec)) + f(\Delta \backslash (\Delta(\gamma \prec) \cup \Delta(-\gamma \prec)))$ has no pole at $H_{x,0}$, since $(-x)f(\Delta(-\gamma \prec))$ is zero on $H_{x,0}$ similarly as above. q.e.d.

Let m be an element of M. We define the $\boldsymbol{C}$-linear mapping $\varphi_m: \boldsymbol{C} \to M_{\boldsymbol{C}}$ by $\varphi_m(t) = tm$ for $t \in \boldsymbol{C}$. We denote by $g_m(t)$ the pull-back $\varphi_m^*(f(\Delta))$. By the above lemma, $g_m(t)$ is an entire function on $\boldsymbol{C}$.

Lemma 1.3. *Suppose m is not in $x^{\perp} := \{v \in M_{\boldsymbol{R}}; \langle v, x\rangle = 0\}$ for any $x \in G(\Delta)$. Then $g_m(t)=0$.*

Proof. By the definition of $f(\Delta)$, we have

$$g_m(t) = \textstyle\sum_{\sigma \in \Delta} \prod_{x \in \operatorname{gen} \sigma} \{(1/(\exp(a_x t)-1)\},$$

where $a_x = \langle m, x\rangle$ for each $x \in G(\Delta)$. The integers a_x's are not zero by assumption. By Lemma 1.2, $g_m(t)$ is an entire function on $\boldsymbol{C}$. Since a_x's are integers, $g_m(t)$ has the periodicity $g_m(t+2\pi i) = g_m(t)$. Hence, in order to prove $g_m(t)=0$, it is sufficient to show that

$$|g_m(t)| \longrightarrow 0, \qquad \text{as } |\operatorname{Re} t| \longrightarrow \infty.$$

Let $H(m)$ be the hyperplane $\{u \in N_R; \langle m, u\rangle = 0\}$ in N_R, and let $H^+(m) := \{u \in N_R; \langle m, u\rangle \geq 0\}$ and $H^-(m) := \{u \in N_R; \langle m, u\rangle \leq 0\}$. It is clear that

$$1/(\exp(a_x t)-1) \longrightarrow 0, \quad \text{as } \operatorname{Re} a_x t \longrightarrow \infty, \quad \text{and}$$
$$1/(\exp(a_x t)-1) \longrightarrow -1, \quad \text{as } \operatorname{Re} a_x t \longrightarrow -\infty.$$

Since $\sigma \in \Delta$ is contained in $H^+(m)$(resp. $H^-(m)$) if and only if $a_x > 0$ (resp. $a_x < 0$) for every $x \in \operatorname{gen} \sigma$, we have

$$g_m(t) \longrightarrow \textstyle\sum_{\sigma \in \Delta^-}(-1)^{\dim \sigma}, \quad \text{as } \operatorname{Re} t \longrightarrow \infty, \quad \text{and}$$
$$g_m(t) \longrightarrow \textstyle\sum_{\sigma \in \Delta^+}(-1)^{\dim \sigma}, \quad \text{as } \operatorname{Re} t \longrightarrow -\infty,$$

where $\Delta^- := \{\sigma \in \Delta; \sigma \subset H^-(m)\}$ and $\Delta^+ := \{\sigma \in \Delta; \sigma \subset H^+(m)\}$. We set $\Phi := \{\sigma \cap H^+(m); \sigma \in \Delta \backslash \Delta^-\}$ and $\Phi_0 := \{\sigma \cap H(m); \sigma \in \Delta\}$. Then $\Phi \cap \Phi_0 = \emptyset$ and $\Phi \cup \Phi_0$ is a finite polyhedral decomposition of $H^+(m)$. By [1, Lemma 1.6], for example, we have $\sum_{\tau \in \Phi}(-1)^{\dim \tau} = (-1)^r$. The same lemma also implies that $\sum_{\sigma \in \Delta}(-1)^{\dim \sigma} = (-1)^r$. Since the map $\Delta \backslash \Delta^- \rightarrow \Phi$ which sends σ to $\sigma \cap H^+(m)$ is bijective and preserves the dimension of cones, the limit $\sum_{\sigma \in \Delta^-}(-1)^{\dim \sigma}$ in the first case is equal to zero. The second limit is also zero since the equality $\sum_{\sigma \in \Delta^+}(-1)^{\dim \sigma} = 0$ is proved similarly. q.e.d.

By Lemma 1.3, the entire function $\varphi(\Delta)$ is zero on rational lines $\boldsymbol{R}m$ in M_R which are not contained in $\bigcup_{x \in G(\Delta)} x^\perp$. Clearly, the union of such lines is dense in M_R. Thus $\varphi(\Delta)$ is zero on M_R. Since $\varphi(\Delta)$ is an analytic function, it is also zero on M_C. q.e.d.

Remark 1.4. Let $\sum_{k=0}^{\infty} B_k t^{k-1}/k!$ be the power series expansion of $1/(\exp(t)-1)$. The coefficients B_k's are known as the Bernoulli numbers. The above theorem for $n=1$ means the wellknown fact that $B_1 = -1/2$ and $B_{2k+1} = 0$ for $k > 0$.

§ 2. *T*-complexes

We denote by $\mathscr{C}$ the category of pairs (N, σ) of a free $\boldsymbol{Z}$-module N of finite rank and an s.c.r.p. cone σ in N_R. For an object $\alpha \in \mathscr{C}$, we denote $\alpha = (N(\alpha), \sigma(\alpha))$. For two objects $\alpha, \beta \in \mathscr{C}$, a morphism $u : \alpha \rightarrow \beta$ consists of an isomorphism $u_Z : N(\alpha) \rightarrow N(\beta)$ such that $u_R(\sigma(\alpha))$ is a face of $\sigma(\beta)$, where $u_R = u_Z \otimes 1_R : N(\alpha)_R \xrightarrow{\sim} N(\beta)_R$. Since any morphism u is determined by the isomorphism u_Z, the following lemma is obvious.

Lemma 2.1. *Every morphism $u : \alpha \rightarrow \beta$ in $\mathscr{C}$ is epimorphic and monomorphic.*

For an object $\alpha \in \mathscr{C}$, we denote $r(\alpha) := \operatorname{rank} N(\alpha)$ and $d(\alpha) := \dim \sigma(\alpha)$. Clearly, $0 \leq d(\alpha) \leq r(\alpha)$ for any α. For each nonnegative integer r, we denote by $\mathscr{C}_r$ the subcategory of $\mathscr{C}$ consisting of $\alpha \in \mathscr{C}$ with $r(\alpha) = r$. It is obvious that the category $\mathscr{C}$ is the disjoint union of $\mathscr{C}_r$'s.

Definition 2.2. A subcategory Σ of $\mathscr{C}$ is said to be a *graph of cones* if the class $\operatorname{mor} \Sigma$ of morphisms in Σ is a finite set.

If Σ is a graph of cones, then Σ consists of finite objects, since $1_\alpha \in \operatorname{mor} \Sigma$ for each $\alpha \in \Sigma$.

Example 2.3. Let Δ be a finite fan of $N_{\boldsymbol{R}}$. Then we regard Δ as a graph of cones by identifying Δ with $\{(N, \sigma); \sigma \in \Delta\}$ and by defining morphism $u : (N, \sigma) \to (N, \tau)$ to be in $\operatorname{mor} \Delta$ if and only if $u_{\boldsymbol{Z}}$ is the identity. Furthermore, any subset Φ of Δ is similarly considered to be a graph of cones.

Let Σ be a graph of cones and let ρ be an element of Σ. We denote by $\Sigma(\rho\prec)$ the comma category consisting of the pairs $\beta' = (\beta, v)$ of an element $\beta \in \Sigma$ and a morphism $v : \rho \to \beta$ in $\operatorname{mor} \Sigma$. A morphism $u' : \beta' = (\beta, v) \to \gamma' = (\gamma, w)$ in the category $\Sigma(\rho\prec)$ consists of a morphism $u : \beta \to \gamma$ such that $u \circ v = w$. By defining $N(\beta') = N(\beta)$, $\sigma(\beta') = \sigma(\beta)$ and $u'_{\boldsymbol{Z}} = u_{\boldsymbol{Z}}$, we see that $\Sigma(\rho\prec)$ is also a graph of cones and is contained in $\mathscr{C}_{r(\rho)}$. Similarly, the comma category $\Sigma(\prec\rho)$ consists of pairs $\beta' = (\beta, v)$ of $\beta \in \Sigma$ and $(v : \beta \to \rho) \in \operatorname{mor} \Sigma$. $\Sigma(\prec\rho)$ is also a graph of cones contained in $\mathscr{C}_{r(\rho)}$ if we define $N(\beta')$ and $\sigma(\beta')$ in the same way as above.

Let $\beta' = (\beta, v)$ be an element of $\Sigma(\rho\prec)$. Then $v_{\boldsymbol{R}}(\sigma(\rho)) \subset N(\beta)_{\boldsymbol{R}}$ is an s.c.r.p. cone of dimension $d(\rho)$. We denote by $\beta'[\rho]$ the object of $\mathscr{C}_{r(\rho)-d(\rho)}$ with $N(\beta'[\rho]) := N(\beta)[v_{\boldsymbol{R}}(\sigma(\rho))]$ and $\sigma(\beta'[\rho]) := \sigma(\beta)[v_{\boldsymbol{R}}(\sigma(\rho))]$. Let $u' : \beta' \to \gamma' = (\gamma, w)$ be a morphism in $\Sigma(\rho\prec)$. Then the isomorphism $u'_{\boldsymbol{Z}} : N(\beta) \to N(\gamma)$ induces a morphism $\beta'[\rho] \to \gamma'[\rho]$ in $\mathscr{C}_{r(\rho)-d(\rho)}$ which we denote by $u'[\rho]$. Hence if we set $\Sigma[\rho] := \{\beta'[\rho] : \beta' \in \Sigma(\rho\prec)\}$ and $\operatorname{mor} \Sigma[\rho] := \{u'[\rho];\ u' \in \operatorname{mor} \Sigma(\rho\prec)\}$, $\Sigma[\rho]$ is a graph of cones contained in $\mathscr{C}_{r(\rho)-d(\rho)}$ which is naturally isomorphic to $\Sigma(\rho\prec)$ as categories. We call $\Sigma[\rho]$ *the link of* Σ *at* ρ.

A subcategory Φ of Σ is said to be *star closed* if $\Phi(\rho\prec) = \Sigma(\rho\prec)$ for every $\rho \in \Phi$, and *star open* if $\Phi(\prec\rho) = \Sigma(\prec\rho)$ for every $\rho \in \Phi$. Since star closed or star open subcategories are full subcategories, we also call them star closed or star open *subsets* of Σ, respectively.

Definition 2.4. A *homomorphism* $\varphi : \Sigma \to \Sigma'$ of graphs of cones consists of a functor $\bar{\varphi} : \Sigma \to \Sigma'$ and a collection $\{\varphi_\alpha; \alpha \in \Sigma\}$ of morphisms $\varphi_\alpha : \alpha \to \bar{\varphi}(\alpha)$ such that the diagram

$$\begin{array}{ccc} \alpha & \xrightarrow{\quad u \quad} & \beta \\ {\scriptstyle \varphi_\alpha}\downarrow & & \downarrow{\scriptstyle \varphi_\beta} \\ \bar{\varphi}(\alpha) & \xrightarrow{\quad \bar{\varphi}(u) \quad} & \bar{\varphi}(\beta) \end{array}$$

is commutative for every $u:\alpha\to\beta$ in mor Σ. φ is said to be an *isomorphism* if $\bar{\varphi}$ is isomorphic, i.e., $\bar{\varphi}$ induces a bijection mor $\Sigma\to$mor Σ', and all φ_α are isomorphisms.

For a morphism $u:\alpha\to\beta$ in $\mathscr{C}$, we denote $i(u):=\alpha$ and $f(u):=\beta$. The connectedness of a graph of cones is defined in the same way as that of usual graphs. Namely, Σ is connected if and only if the equivalence relation generated by $i(u)\sim f(u)$ for $u\in$ mor Σ has at most one equivalence class.

Now we are ready to define the notion of T-complexes.

Definition 2.5. A graph of cones Σ is said to be a *T-complex* if

(1) Σ is nonempty and connected,

(2) for any $\rho\in\Sigma$, the comma category $\Sigma(\prec\rho)$ is isomorphic to $\Gamma(\sigma(\rho))\backslash\{\mathbf{0}\}$ as graphs of cones, and

(3) for any $\rho\in\Sigma$ the link $\Sigma[\rho]$ is isomorphic to a complete fan. Furthermore, Σ is said to be *nonsingular* if $\sigma(\alpha)\subset N(\alpha)_{\boldsymbol{R}}$ is a nonsingular cone for every $\alpha\in\Sigma$.

All the known examples of T-complexes are essentially written as follows:

There exist a fan $\tilde{\Sigma}$ of $N_{\boldsymbol{R}}$ and a subgroup $\Gamma\subset\mathrm{Aut}_{\boldsymbol{Z}}(N)$ such that

(1) $U=(\bigcup_{\sigma\in\tilde{\Sigma}}\sigma)\backslash\{0\}$ is a nonempty connected open cone of $N_{\boldsymbol{R}}$, and

(2) Γ induces a free action on $\tilde{\Sigma}\backslash\{\mathbf{0}\}$ and $\sharp(\tilde{\Sigma}\backslash\{\mathbf{0}\})$ is finite modulo Γ.

Let Σ be a set of representatives of $\tilde{\Sigma}\backslash\{\mathbf{0}\}$ modulo Γ. For each element $\alpha\in\Sigma$, we set $N(\alpha):=N$ and $\sigma(\alpha):=\alpha$. For elements $\alpha,\beta\in\Sigma$, a morphism $u:\alpha\to\beta$ consists of an element $u_{\boldsymbol{Z}}\in\Gamma$ such that $u_{\boldsymbol{R}}(\alpha)\subset N_{\boldsymbol{R}}$ is a face of the cone β. Then we see that Σ is a T-complex. Σ is nonsingular if and only if so is $\tilde{\Sigma}$.

Examples 2.6. We will give examples of T-complexes of the above form.

(1) *Toric variety type.* Let $\tilde{\Sigma}$ be a complete fan of $N_{\boldsymbol{R}}$ and $\Gamma=\{1_N\}$. Then $\Sigma=\tilde{\Sigma}\backslash\{\mathbf{0}\}$ is a T-complex.

(2) *Degenerate Abelian variety type.* Let $N=\boldsymbol{Z}^{n+1}$, $C=\{(x_1,\cdots,x_{n+1})\in\boldsymbol{R}^{n+1}; x_{n+1}>0\}$ and Γ be a subgroup of finite index of the group of the matrices of the form

$$\begin{pmatrix} 1 & & 0 & b_1 \\ & \ddots & & \vdots \\ 0 & & 1 & b_n \\ 0 & \cdots & 0 & 1 \end{pmatrix} \quad (b_i \in \boldsymbol{Z}).$$

Then, for a Γ-invariant polyhedral decomposition $\tilde{\Sigma}$ of $C \cup \{0\}$, we get a T-complex Σ.

(3) *Tsuchihashi cusp singularity type.* When C is an open convex cone which contains no lines in $N_{\boldsymbol{R}}$, such a pair (C, Γ) induces an isolated singularity which is independent of the choice of $\tilde{\Sigma}$. A Hilbert modular cusp singularity is a special case of this type of singularities. Other cases and some explicit examples were studied by Tsuchihashi [T1].

(4) *Inoue-Kato manifold type.* Let A be an $n \times n$-matrix of positive integers with the determinant ± 1, and let $N = \boldsymbol{Z}^n$ and $\pi = \{(x_1, \cdots, x_n) \in N_{\boldsymbol{R}}; x_1, \cdots, x_n \geq 0\}$. Then $\bigcup_{m \in \boldsymbol{Z}} A^m(\pi) \backslash \{0\}$ is an open half-space and $\bigcap_{m \in \boldsymbol{Z}} A^m(\pi)$ is a closed half-line. Let $C := (\bigcup_{m \in \boldsymbol{Z}} A^m(\pi)) \backslash (\bigcap_{m \in \boldsymbol{Z}} A^m(\pi))$ and $\Gamma = \{A^m; m \in \boldsymbol{Z}\}$. Then there exists a nonsingular Γ-invariant polyhedral decomposition $\tilde{\Sigma}$ of $C \cup \{0\}$. By these data, we can construct a compact non-Kähler manifold of dimension n with the fundamental group $\boldsymbol{Z}$ [T2]. When $n = 2$, this is known as a hyperbolic Inoue surface (see [MO, Sec. 15]). C is connected if $n \geq 3$. The associated T-complex corresponds to an anti-canonical divisor of the manifold if $\det A = 1$.

§ 3. Functors on a graph of cones

We denote by $\mathscr{C}^{n.s.}$ the full subcategory of $\mathscr{C}$ consisting of $\alpha \in \mathscr{C}$ such that the cone $\sigma(\alpha)$ is nonsingular. We denote the canonical set of generators $\operatorname{gen} \sigma(\alpha) \subset N(\alpha)$ simply by $\operatorname{gen} \alpha$. For a morphism $u: \alpha \to \beta$ in $\mathscr{C}^{n.s.}$, we have $u_{\boldsymbol{Z}}(\operatorname{gen} \alpha) \subset \operatorname{gen} \beta$. For each $\alpha \in \mathscr{C}^{n.s.}$, we denote $x(\alpha) = \prod_{x \in \operatorname{gen} \alpha} x$ which is an element of the symmetric power $S^{d(\alpha)} N(\alpha)$ over $\boldsymbol{Z}$. For a morphism $u: \alpha \to \beta$ in $\mathscr{C}^{n.s.}$, we set $x(u) := \prod_{x \in \operatorname{gen} \beta \backslash u_{\boldsymbol{Z}}(\operatorname{gen} \alpha)} x \in S^{d(\beta) - d(\alpha)} N(\beta)$.

Let k be a commutative ring with unity. For each nonnegative integer m, we define the functor $D_k^m : \mathscr{C}^{n.s.} \to (k\text{-modules})$ as follows. For each $\alpha \in \mathscr{C}^{n.s.}$, we set $D_k^m(\alpha) := S^{d(\alpha)+m} N(\alpha)_k$ where $N(\alpha)_k := N(\alpha) \otimes_{\boldsymbol{Z}} k$ and the symmetric power is taken over the ring k. For a morphism $u: \alpha \to \beta$, we define the homomorphism $D_k^m(u)(z) := x(u) \cdot S^{d(\alpha)+m} u_k(z)$, where $S^d u_k : S^d N(\alpha)_k \to S^d N(\beta)_k$ is the symmetric power of $u_k = u_{\boldsymbol{Z}} \otimes 1_k$. It is easy to see that D_k^m satisfies the axiom of functors. We denote by $k^{\sim}$ the constant functor defined by $k^{\sim}(\alpha) := k$ and $k^{\sim}(u) := 1_k$ for all $\alpha \in \mathscr{C}^{n.s.}$ and $u \in \operatorname{mor} \mathscr{C}^{n.s.}$. We define the morphism of functors $\varepsilon : k^{\sim} \to D_k^0$ by $\varepsilon(\alpha)(a) := a x(\alpha) \in D_k^0(\alpha)$ for $\alpha \in \mathscr{C}^{n.s.}$ and $a \in k$. Since $x(u) \cdot S^{d(\alpha)} u_k(x(\alpha)) = x(\beta)$, for $u: \alpha \to \beta$, this is indeed a morphism of functors.

Let Φ be a graph of cones. Then, for a functor $V: \mathscr{C}^{n.s.} \to (k\text{-modules})$, the inductive limit ind $\lim_\Phi V$ is described as the cokernel

$$\bigoplus_{u \in \operatorname{mor} \Phi} V(i(u)) \overset{p}{\underset{q}{\rightrightarrows}} \bigoplus_{\alpha \in \Phi} V(\alpha) \longrightarrow \operatorname{ind} \lim_\Phi V$$

where p consists of the identities $1_{V(i(u))} : V(i(u)) \to V(i(u)) \subset \bigoplus_{\alpha \in \Phi} V(\alpha)$ and q consists of the homomorphisms $V(u) : V(i(u)) \to V(f(u)) \subset \bigoplus_{\alpha \in \Phi} V(\alpha)$. For a graph of cones Φ, we get a homomorphism ind $\lim_\Phi \varepsilon$: ind $\lim_\Phi k^{\sim} \to$ ind $\lim_\Phi D_k^0$. Note that ind $\lim_\Phi k^{\sim} = k$ if Φ is nonempty and connected.

Lemma 3.1. *Let Σ be a nonsingular T-complex. Then, there exists a morphism of functors $\nu : D_k^0|_\Sigma \to k^{\sim}|_\Sigma$ such that $\nu \circ \varepsilon$ is the identity on Σ. In particular,* ind $\lim_\Sigma \varepsilon$ *defines an injection $k \hookrightarrow$* ind $\lim_\Sigma D_k^0$ *and the image is a direct summand.*

Proof. Let $\Sigma_1 = \{\gamma \in \Sigma ; d(\gamma) = 1\}$. By the condition (2) in Definition 2.5, Σ does not contain zero-dimensional cone. Hence Σ_1 is a star open subset of Σ. Let γ be an element of Σ_1 and let gen $\gamma = \{x\}$. Since x is a primitive element of $N(\gamma)$, kx is a direct summand of $N(\gamma)_k$. Hence, there exists a k-homomorphism $\nu(\gamma) : N(\gamma)_k \to k$ such that $\nu(\gamma)(x) = 1$. By the condition (2) in Definition 2.5, there is no morphism $u : \gamma \to \gamma'$ if $\gamma, \gamma' \in \Sigma_1$ and $\gamma \neq \gamma'$. Hence we get a morphism of functors $\nu : D_k^0|_{\Sigma_1} \to k^{\sim}|_{\Sigma_1}$ which satisfies $\nu \circ \varepsilon = \mathrm{id}$ on Σ_1. Let Φ be a maximal star open subset of Σ such that $\Sigma_1 \subset \Phi$ and that there exists a morphism of functors $\nu : D_k^0|_\Phi \to k^{\sim}|_\Phi$ with $\nu \circ \varepsilon = \mathrm{id}$ on Φ. Assume $\Phi \neq \Sigma$ and let ρ be an element of $\Sigma \backslash \Phi$ with the smallest $d(\rho) =: d$. By Definition 2.5, (2), we have an isomorphism $\Sigma(\prec \rho) \simeq \Gamma(\rho) \backslash \{\mathbf{0}\}$. By the minimality of $d(\rho)$, ν induces a morphism of functors $\nu' : D_k^0|_{\Gamma(\rho)\backslash\{0,\rho\}} \to k^{\sim}|_{\Gamma(\rho)\backslash\{0,\rho\}}$. Let $N = N(\rho)$ and let $\{x_1, \cdots, x_r\}$ be a basis of N such that gen $\rho = \{x_1, \cdots, x_d\}$. The free k-module $S^d N_k$ has $\{x_1^{a_1} \cdots x_r^{a_r} : a_1, \cdots, a_r \geq 0, a_1 + \cdots + a_r = d\}$ as a basis. For each face α of ρ, gen α is a subset of $\{x_1, \cdots, x_d\}$ and the image of $S^{d(\alpha)} N_k$ in $S^d N_k$ is generated by monomials of degree d which is divisible by $x(\rho/\alpha) = \prod_{x \in \operatorname{gen} \rho \backslash \operatorname{gen} \alpha} x$. For each monomial z of degree d, we define $\nu'(\rho)(z) := \nu'(\alpha)(y)$ if $z = x(\rho/\alpha) y$ for some $\alpha \in \Gamma(\rho) \backslash \{\mathbf{0}, \rho\}$ and for a monomial y of degree $d(\alpha)$, and we define $\nu'(\rho)(z) := 0$ otherwise. Since $d = d(\rho) \geq 2$, $\Gamma(\rho) \backslash \{\mathbf{0}, \rho\}$ is nonempty. Since $x(\rho) = x(\rho/\alpha) x(\alpha)$, we have $\nu'(\rho)(x(\rho)) = 1$. We see easily that the definition does not depend on the choice of α and hence $\nu'(\rho) \circ D_k^0(u) = \nu'(\alpha)$ for every $u : \alpha \to \rho$. Thus the morphism of functors ν' is extended to $\Gamma(\rho) \backslash \{\mathbf{0}\}$. Since there is no morphism $\rho \to \alpha$ in Σ with $\alpha \in \Phi$, we can combine this extended ν' with ν, and we get an extension of ν to $\Phi \cup \{\rho\}$. This contradicts the maximality of Φ and we have $\Phi = \Sigma$. q.e.d.

We call ν in the above lemma a *retraction* of $D_k^0|_\Sigma$ to $k^\sim|_\Sigma$. In the proof of the above lemma, the extension of ν' to $\Gamma(\rho)\backslash\{\mathbf{0}\}$ depends on the choice of the basis $\{x_1, \cdots, x_r\}$ of N. Hence the retraction ν is neither unique nor canonical. We will see in Section 5 that there exists an explicit retraction for $k=\boldsymbol{R}$ in the case of Thuchihashi cusp singularities.

Let Σ be a graph of nonsingular cones, i.e., a graph of cones contained in $\mathscr{C}^{n.s.}$, and let ρ be an element of Σ. We are going to define the restriction

$$\text{ind lim}_\Sigma D_k^0 \longrightarrow \text{ind lim}_{\Sigma[\rho]} D_k^{d(\rho)}$$

of the inductive limit of D_k^0 to the link of ρ.

We define the homomorphism

$$h'_\rho : \oplus_{\alpha\in\Sigma} D_k^0(\alpha) \longrightarrow \oplus_{\alpha'[\rho]\in\Sigma[\rho]} D_k^{d(\rho)}(\alpha'[\rho])$$

by $h'_\rho((y_\alpha)):=(\bar{y}_{\alpha'})$ where $y_\alpha \in S^{d(\alpha)}N(\alpha)_k$ and $\bar{y}_{\alpha'}$ is the image of y_α by the natural homomorphism $S^{d(\alpha)}N(\alpha)_k \to S^{d(\alpha)}N(\alpha'[\rho])_k$ if $\alpha'=(\alpha, u)\in\Sigma(\rho\prec)$ for a morphism $u:\rho\to\alpha$. Note that $d(\alpha)=d(\alpha'[\rho])+d(\rho)$ in this case. Similarly, we define the homomorphism

$$h''_\rho : \oplus_{u\in\text{mor}\,\Sigma} D_k^0(i(u)) \longrightarrow \oplus_{u'[\rho]\in\text{mor}\,\Sigma[\rho]} D_k^{d(\rho)}(i(u'[\rho]))$$

by $h''_\rho((z_u)):=(\bar{z}_{u'})$ for $z_u \in S^{d(i(u))}N(i(u))_k$ and $\bar{z}_{u'}$ is the image of z_u in $S^{d(i(u'))}N(i(u'[\rho]))_k$ if $u'\in\text{mor}\,\Sigma(\rho\prec)$ is defined by $u\in\text{mor}\,\Sigma$.

Proposition 3.2. *Let Σ be a nonsingular T-complex and let ρ be an element of Σ. Then the diagram*

$$\begin{array}{ccc}
\oplus_{u\in\text{mor}\,\Sigma} D_k^0(i(u)) & \overset{p}{\underset{q}{\rightrightarrows}} & \oplus_{\alpha\in\Sigma} D_k^0(\alpha) \\
\downarrow h''_\rho & & \downarrow h'_\rho \\
\oplus_{u'[\rho]\in\text{mor}\,\Sigma[\rho]} D_k^{d(\rho)}(i(u'[\rho])) & \overset{p}{\underset{q}{\rightrightarrows}} & \oplus_{\alpha'[\rho]\in\Sigma[\rho]} D_k^{d(\rho)}(\alpha'[\rho])
\end{array}$$

commutes for p and q, respectively.

Proof. Let v be in mor Σ and let z be an element of the direct summand $D_k^0(i(v))$ of $\oplus_{u\in\text{mor}\,\Sigma} D_k^0(i(u))$. We have $h'_\rho(p(z))=p(h''_\rho(z))$, since their components for $\alpha'[\rho]\in\Sigma[\rho]$ are both equal to the image z_u of z in $S^{d(i(v))}N(\alpha'[\rho])_k$ if $\alpha'=(i(v), u)$ for some $u:\rho\to i(v)$ and are both zero otherwise. Hence the diagram is commutative for p.

Now we prove the commutativity for q. Let $\beta=f(v)$. The component for $\beta'[\rho]$ of $h'_\rho(q(z))$ is equal to the image $v(z)_w$ of $v(z):=D_k^0(v)(z)$ in $S^{d(\beta)}N(\beta'[\rho])_k$ if $\beta'=(\beta, w)$ for some $w:\rho\to\beta$ and zero otherwise. On the other hand, the same component of $q(h''_\rho(z))$ is equal to $v_t[\rho](z_t)$ if $\beta' =(\beta, v\circ t)$ for some $t:\rho\to i(v)$ and is zero otherwise, where $v_t:(i(u), t)\to(\beta, v\circ t)$ is defined by v and z_t is the image of z in $S^{d(i(v_t))}N(i(v_t)[\rho])_k$. Here note that such t is unique by Lemma 2.1. Clearly, these components $v(z)_w$ and $v_t[\rho](z_t)$ for $\beta'[\rho]$ are equal if there exists such a t. By the condition (2) in Definition 2.5, such a t exists if and only if $w_{\boldsymbol{Z}}(\text{gen }\rho)\subset v_{\boldsymbol{Z}}(\text{gen }i(v))$. If t does not exist, there exists $x\in\text{gen }\rho$ such that $w_{\boldsymbol{Z}}(x)\notin v_{\boldsymbol{Z}}(\text{gen }i(v))$. Hence, $x(v)$ is divisible by $w_{\boldsymbol{Z}}(x)$ and $v(z)_w\in S^{d(\beta)}N(\beta'[\rho])_k$ is also zero since $N(\beta'[\rho])=N(\beta)/\boldsymbol{Z}w_{\boldsymbol{Z}}(\text{gen }\rho)$. q.e.d.

We denote by $h_{\rho,\Sigma}$, or simply h_ρ, the homomorphism $\text{ind lim}_\Sigma D_k^0\to\text{ind lim}_{\Sigma[\rho]}D_k^{d(\rho)}$ induced by the diagram in the above proposition.

Let N be a free $\boldsymbol{Z}$-module of rank $r\geq 0$. We denote by $B(N_k)$ the total ring of homogeneous quotients of the symmetric algebra S^*N_k. $B(N_k)$ is written as the direct sum $\bigoplus_{m\in\boldsymbol{Z}}B(N_k)_m$ of the k-vector spaces consisting of the homogeneous elements of degree m.

Let Δ be a nonsingular fan in $N_{\boldsymbol{R}}$, and let m be a nonnegative integer. For each $\alpha\in\Delta$, we define the homomorphism $\lambda_\Delta^\alpha: D_k^m(\alpha)\to B(N_k)_m$ by $\lambda_\Delta^\alpha(z):=z/x(\alpha)$ for $z\in S^{d(\alpha)+m}N_k$. It is easy to see that these homomorphisms commute with $D_k^m(u)$ for every morphism $u:\alpha\to\beta$ in Δ. Hence we get the limit homomorphism $\lambda_\Delta:\text{ind lim}_\Delta D_k^m\to B(N_k)_m$.

For a nonsingular T-complex Σ and for an element $\rho\in\Sigma$, we denote by $\bar{h}_\rho$ the composite $\lambda_{\Sigma[\rho]}\circ h_\rho:\text{ind lim}_\Sigma D_k^0\to B(N[\rho]_k)_{d(\rho)}$, where $N[\rho]=N(\rho)[\rho]$.

Lemma 3.3. *Let Σ be a nonsingular T-complex. Then an element z in* ind $\lim_\Sigma D_k^0$ *is in the image of* ind $\lim_\Sigma \varepsilon$ *if and only if $\bar{h}_\rho(z)=0$ for every $\rho\in\Sigma$.*

Proof. The image of the morphism $\varepsilon(\alpha):k^\sim(\alpha)\to D_k^0(\alpha)$ is equal to $kx(\alpha)\subset S^{d(\alpha)}N(\alpha)_k$. Since the image of $x(\alpha)$ in $S^{d(\alpha)}N[\alpha]_k$ is zero for every α, the necessity of the condition is obvious.

Now we suppose $z\in\text{ind lim}_\Sigma D_k^0$ satisfies $\bar{h}_\rho(z)=0$ for every $\rho\in\Sigma$. We may assume $z\neq 0$ because otherwise the assertion is obvious. Let $(z_\alpha)\in\bigoplus_{\alpha\in\Sigma}D_k^0(\alpha)$ be a representative of z such that $d=\max\{d(\alpha);\, z_\alpha\neq 0\}$ is minimal. We will show $d=1$. Assume $d>1$ and take (z_α) so that the cardinality of $\{\alpha\in\Sigma;\, d(\alpha)=d \text{ and } z_\alpha\neq 0\}$ is the smallest. Let ρ be an element of Σ such that $d(\rho)=d$ and $z_\rho\neq 0$. By the definition of d, we have $z_\alpha=0$ for any $\alpha\in\Sigma(\rho\prec)$ with $\alpha\neq\rho$. Hence the condition $\bar{h}_\rho(z)=0$ implies that the image of $z_\rho\in S^dN(\rho)_k$ in $S^dN[\rho]_k$ is zero. Let gen $\rho=\{x_1,\cdots,x_d\}$. Since $N[\rho]=N(\rho)/(\boldsymbol{Z}x_1+\cdots+\boldsymbol{Z}x_d)$, the kernel of the homomorphism $S^dN(\rho)_k\to S^dN[\rho]_k$ is equal to $\sum_{i=1}^d x_iS^{d-1}N(\rho)_k$. Hence z_ρ is of the form $\sum_{i=1}^{d} x_i y_i$ for $y_i \in S^{d-1}N(\rho)_k$. By the condition (2) in Definition 2.5, there exists $u^i : \mu_i \to \rho$ in Σ such that u^i (gen μ_i) = gen $\rho \setminus \{x_i\}$, i.e., $x(u^i) = x_i$, for each $i = 1, \cdots, d$. Let $y'_i \in S^{d-1}N(\mu_i)_k$ be the element which satisfies $S^{d-1}(u^i)(y'_i) = y_i$, for each an i. Let $y' = (y'_u)$ be the element of $\bigoplus_{u \in \mathrm{mor}\, \Sigma} D_k^0(i(u))$ defined by $y'_u = y'_i$ if $u = u^i$ for some i and $y'_u = 0$ otherwise. Then we have $q(y') = \sum x_i y_i = z_\rho$, while the components of $p(y')$ are zero for every α with $d(\alpha) \geq d$. Let $(z'_\alpha) = (z_\alpha) + p(y') - q(y') \in \bigoplus_{\alpha \in \Sigma} D_k^0(\alpha)$. Then clearly, $\max\{d(\alpha) : z'_\alpha \neq 0\} \leq d$ and $z'_\alpha = z_\alpha$ for $\alpha \in \Sigma$ with $d(\alpha) = d$ except when $\alpha = \rho$. Since $z'_\rho = 0$ and (z'_α) is also a representative of z, this contradicts the minimality of $\{\alpha \in \Sigma; d(\alpha) = d$ and $z_\alpha \neq 0\}$. Hence we have $d = 1$.

Let $\rho \in \Sigma$ be an element with $z_\rho \neq 0$. Then since $d(\rho) = 1$, we have gen $\rho = \{x(\rho)\}$. By the condition $\bar{h}_\rho(z) = 0$, we see that z_ρ is in $kx(\rho) = \ker(N(\rho)_k \to N[\rho]_k)$ which is equal to the image of $\varepsilon(\rho) : k^{\sim}(\rho) \to D_k^0(\rho)$. Hence every z_α is in the image of $k^{\sim}(\alpha)$. q.e.d.

§ 4. ω-invariant of a T-complex

Let Δ be a nonsingular fan in N_R and let m be a nonnegative integer. For each $\alpha \in \Delta$, we set

$$\omega_\alpha^m := \left[\prod_{x \in \mathrm{gen}\, \alpha} \frac{x}{\exp(x) - 1} \right]_{d(\alpha)+m},$$

where $[f]_d$ denotes the homogeneous part of degree d of a power series f. Note that $x/(\exp(x) - 1)$ is an element of the completion of the symmetric algebra S^*N_Q with respect to the natural grading. Hence $(\omega_\alpha^m)_{\alpha \in \Delta}$ is an element of $\bigoplus_{\alpha \in \Delta} D_Q^m(\alpha)$.

Lemma 4.1. *Let ω_Δ^m be the image of (ω_α^m) in* ind $\lim_\Delta D_Q^m$. *If Δ is a nonsingular complete fan, then $\lambda_\Delta(\omega_\Delta^m) \in B(N_Q)_m$ is equal to zero.*

Proof. Since $x(\alpha) = \prod_{x \in \mathrm{gen}\, \alpha} x$, we see that $\lambda_\Delta^\alpha(\omega_\alpha^m)$ is equal to the homogeneous part of degree m of $\prod_{x \in \mathrm{gen}\, \alpha} 1/(\exp(x) - 1)$. Hence $\lambda_\Delta(\omega_\Delta^m)$ is equal to that of $\sum_{\alpha \in \Delta} \prod_{x \in \mathrm{gen}\, \alpha} 1/(\exp(x) - 1)$, which is zero by Theorem 1.1. q.e.d.

Let Σ be a nonsingular T-complex. For each $\alpha \in \Sigma$, we set $\omega_\alpha := [\prod_{x \in \mathrm{gen}\, \alpha} x/(\exp(x) - 1)]_{d(\alpha)} \in D_Q^0(\alpha)$.

Proposition 4.2. *In the above situation, let ω_Σ be the class of $(\omega_\alpha)_{\alpha \in \Sigma}$ in* ind $\lim_\Sigma D_Q^0$. *Then ω_Σ is in the image of* ind $\lim_\Sigma \varepsilon$.

Proof. By Lemma 3.3, it is sufficient to show $\bar{h}_\rho(\omega_\Sigma) = 0$ for every

$\rho \in \Sigma$. Let $N = N(\rho)$. Since ρ is the initial object of $\Sigma(\rho \prec)$, we may regard $N(\alpha) = N$ for every $\alpha \in \Sigma(\rho \prec)$ and $u_Z = 1_N$ for every $u \in \mathrm{mor}\,\Sigma(\rho \prec)$. For each $\alpha \in \Sigma(\rho \prec)$, $\alpha[\rho]$ is the nonsingular cone of $N[\rho]_R$ with $\mathrm{gen}\,\alpha[\rho] = \{\bar{x};\, x \in \mathrm{gen}\,\alpha \backslash \mathrm{gen}\,\rho)$ where $\bar{x}$ denotes the image of $x \in N$ in $N[\rho]$. Since $x/(\exp(x)-1) = 1$ on $M[\rho]_C$ for $x \in \mathrm{gen}\,\rho$, the restriction of $\prod_{x \in \mathrm{gen}\,\alpha} x/(\exp(x)-1)$ to $M[\rho]_C$ is equal to $\prod_{\bar{x} \in \mathrm{gen}\,\alpha[\rho]} \bar{x}/(\exp(\bar{x})-1)$. Since $d(\alpha[\rho]) = d(\alpha) - d(\rho)$, we have $\omega_\alpha|_{M[\rho]_C} = \omega^{d(\rho)}_{\alpha[\rho]}$. Hence $h_\rho(\omega_\Sigma) = \omega^{d(\rho)}_{\Sigma[\rho]} \in \mathrm{ind}\,\lim_{\Sigma[\rho]} D^{d(\rho)}_k$. Since $\Sigma[\rho]$ is a nonsingular complete fan by Definition 2.5, (3), $\bar{h}_\rho(\omega_\Sigma) = \lambda_\Delta \circ h_\rho(\omega_\Sigma)$ is zero by Lemma 4.1. q.e.d.

Lemma 3.1 and Proposition 4.2 imply that there exists a unique rational number a for each nonsingular T-complex Σ such that $(\mathrm{ind}\,\lim_\Sigma \varepsilon)(a) = \omega_\Sigma$. We also denote $\omega_\Sigma := a$ and call it *the ω-invariant* of the T-complex Σ.

Proposition 4.2 implies obviously the following:

Corollary 4.3. *Let $\nu : D^0_R|_\Sigma \to R^\sim|_\Sigma$ be an arbitrary retraction. Then* $\mathrm{ind}\,\lim_\Sigma \nu((\omega_\alpha)) \in \boldsymbol{R}$ *is equal to the rational number ω_Σ.*

Remark 4.4. Let Σ be a nonsingular T-complex, and let d be a positive integer such that

$$(1) \qquad d\omega_\alpha \in D^0_Z(\alpha) \qquad \text{for every } \alpha \in \Sigma.$$

Then $(d\omega_\alpha) \in \oplus_{\alpha \in \Sigma} D^0_Z(\alpha)$ satisfies the condition of Lemma 3.3 for $k = \boldsymbol{Z}$. Hence $d\omega_\Sigma$ is an integer. The minimal number satisfying (1) depends only the dimension r of Σ. For example, $d = 12$ for $r = 2$ and $d = 720$ for $r = 4$.

When r is odd, we can show that ω_Σ is a half-integer in the same way as Ogata's [Og, Theorem 2.3].

§ 5. Ogata's zeta zero value

Let C, Γ, $\tilde{\Sigma}$ and Σ be as in Example 2.6, (3). The characteristic function ϕ_C on the open convex cone C is given by

$$\phi_C(x) := \int_{C^*} \exp(-\langle x, x^* \rangle)\, dx^*,$$

where $C^* \subset M_R$ is the dual cone of C and dx^* is a Euclidean metric. Ogata [Og] defined the zeta function of the pair (C, Γ) by

$$Z(C, \Gamma; s) := \sum_{x \in (N \cap C)/\Gamma} \phi_C(x)^s$$

which converges for complex numbers s with $\mathrm{Re}\, s > 1$ and can be extended meromorphically to the whole complex plane. He proved in [Og, Proposition 3.10] that this zeta function is regular at $s=0$ and the value is equal to

$$\sum_{\alpha\in\Sigma}\int_{\alpha}[\prod_{x\in\mathrm{gen}\,\alpha}(\partial_x/(1-\exp(-\partial_x)))]_{d(\alpha)}G_2(t)dt_\alpha$$

where ∂_k is the first order derivation defined by

$$\partial_x f(t)=\lim_{h\to 0}\{(f(t+hx)-f(t))/h\},$$

dt_α is the Lebesgue measure on α normalized with respect to the basis gen α and $G_2(t)=\exp(-\phi_C(t)^{-2})$.

Let α be an element of Σ. By extending the correspondence $x\mapsto -\partial_x$ to their products, we get an isomorphism $z\mapsto D_z$ from $D^0_{\boldsymbol{R}}(\alpha)=S^{d(\alpha)}N_{\boldsymbol{R}}$ to the $\boldsymbol{R}$-module of derivations of order $d(\alpha)$ with constant coefficients.

Proposition 5.1. *For each $\alpha\in\Sigma$, we define the homomorphism $F(\alpha)$: $S^{d(\alpha)}N_{\boldsymbol{R}}\to\boldsymbol{R}$ by*

$$F(\alpha)(z):=\int_\alpha D_z G_2(t)dt_\alpha .$$

Then F is a retraction of the morphism of functors $\varepsilon|_\Sigma:\boldsymbol{R}^{\sim}|_\Sigma\to D^0_{\boldsymbol{R}}|_\Sigma$, i.e., F is a morphism of functors and $F\circ\varepsilon|_\Sigma$ is identity.

Proof. Let α be an element of Σ and let gen $\alpha=\{x_1,\cdots,x_d\}$. We take a coordinate $(t_1,\cdots,t_r)$ of $N_{\boldsymbol{R}}$ such that $t_i(x_j)=\delta_{i,j}$, where $\delta_{i,j}$ is Kronecker's delta. Since $D_{\varepsilon(\alpha)(1)}=\prod_{i=1}^{d}(-\partial/\partial t_i)$, we have

$$(F\circ\varepsilon)(\alpha)(1)=\int_0^\infty\cdots\int_0^\infty\prod_{i=1}^{d}(-\partial/\partial t_i)G_2(t)dt_1\cdots dt_d .$$

By [Og, Lemma 3.5], the partial derivatives of $G_2(t)$ goes to zero at infinity. Hence this integral is equal to $G_2(0)=1$. Hence it is sufficient to show that F is a morphism of functors. Let $u:\beta\to\alpha$ be a homomorphism in Σ. We may regard gen $\beta=\{x_1,\cdots,x_{d'}\}\subset$ gen α for an integer $0<d'\leq d$. Furthermore, it is sufficient to show the commutativity in the case $d'=d-1$. Then, for an element $z\in D^0_{\boldsymbol{R}}(\beta)$, we have $D_{z'}=(-\partial/\partial t_d)D_z$ for $z'=D^0_{\boldsymbol{R}}(u)(z)$. Hence

$$\begin{aligned}F(\alpha)(D^0_{\boldsymbol{R}}(u)(z))&=\int_0^\infty\cdots\int_0^\infty\left(\int_0^\infty(-\partial/\partial t_d)D_zG_2(t)dt_d\right)dt_1\cdots dt_{d-1}\\&=\int_0^\infty\cdots\int_0^\infty D_zG_2(t)dt_1\cdots dt_{d-}\\&=F(\beta)(z).\end{aligned}$$

q.e.d.

Let α be in Σ. Then, for ω_α in Section 4, we have

$$D_{\omega_\alpha}=[\prod_{x\in \mathrm{gen}\,\alpha}(\partial_x/(1-\exp(-\partial_x)))]_{d(\alpha)}.$$

Hence by Ogata's formula, Corollary 4.3 and Proposition 5.1, we have the following:

Theorem 5.2. *The zeta zero value* $Z(C, \Gamma; 0)$ *is equal to the* ω*-invariant* ω_Σ *of the T-complex* Σ. *In particular, it is a rational number.*

References

[I] M.-N. Ishida, Torus embeddings and de Rham complexes, in Commutative Algebra and Combinatorics, Proceedings of USA-Japan Conf., Kyoto 1985, (M. Nagata and H. Matsumura, eds.), Advanced Studies in Pure Mathematics, **11** (1987), 111–145.

[MO] T. Oda, Lectures on Torus Embeddings and Applications (Based on joint work with K. Miyake), Tata Inst. of Fund. Res., Bombay, Springer-Verlag, Berlin, Heidelberg, New York, 1978.

[Od] T. Oda, Convex Bodies and Algebraic Geometry, An Introduction to the Theory of Toric varieties, Ergebnisse der Math. (3) **15**, Springer-Verlag, Berlin, Heidelberg, New York, Tokyo, 1988.

[Og] S. Ogata, Special values of zeta functions associated to cusp singularities, Tôhoku Math. J., **37** (1985), 367–384.

[S1] I. Satake, Special values of zeta functions associated with self-dual homogeneous cones, in Manifolds and Lie groups (Notre Dame Ind., 1980), Progress in Math., **14**, Birkhäuser, Boston, 1981, 359–384.

[S2] I. Satake, On numerical invariants of arithmetic varieties of Q-rank one, Automorphic forms of several variables, Taniguchi Symp., Katata, 1983 (I. Satake and Y. Morita, eds.), Progress in Math., **46**, Birkhäuser, Basel, Boston, Stuttgart, 1984, 353–369.

[T1] H. Tsuchihashi, Higher dimensional analogues of periodic continued fractions and cusp singularities, Tôhoku Math. J., **35** (1983), 607–639.

[T2] H. Tsuchihashi, Certain compact complex manifolds with infinite cyclic fundamental groups, Tôhoku Math. J., **39** (1987), 519–532.

[Z] D. Zagier, Valeur des fonctions zêta des corps quadratiques réels aux entiers négatifs, Journées Arithmétiques de Caen (Univ. Caen, Caen, 1976), Astérisque, Nos. **41–42**, Soc. Math. France, Paris, 1977, 135–151.

Mathematical Institute
Tohoku University
Sendai, 980
Japan

Advanced Studies in Pure Mathematics 15, 1989
Automorphic Forms and Geometry of Arithmetic Varieties
pp. 365–372

The Structure of the Icosahedral Modular Group

Ryoichi Kobayashi and Isao Naruki

Dedicated to Prof. Friedrich Hirzebruch and Prof. Ichiro Satake on their sixtieth birthdays

Introduction

By the group in the title we mean the Hilbert modular group PSL$(2, \mathcal{O})$ where $\mathcal{O}$ is the ring of integers in the number field $\boldsymbol{Q}(\sqrt{5})$. This naming here comes from the following fact: Hirzebruch [H] studied the irreducible action over $H\times H$ ($H:=\{z\in \boldsymbol{C};\, \mathrm{Im}(z)>0\}$) of the principal congruence subgroup $\Gamma=\Gamma(2)$ of $SL(2, \mathcal{O})$ associated with the prime ideal $(2)\subseteq\mathcal{O}$ and showed that thecom pactified quotient $\overline{H\times H/\Gamma}$ is equivariantly birational to the projective plane $P_2(\boldsymbol{C})$ acted nontrivially by the icosahedral group $\mathfrak{A}_5\cong PSL(2, \mathcal{O}/(2))$ ($\mathcal{O}/(2)\cong F_4$; $\mathfrak{A}_5$: the alternating group of degree five). Γ can be regarded as a subgroup of $PSL(2, \mathcal{O})$ since $-1\notin\Gamma$. It acts even freely on $H\times H$ and the description in [H] is so explicit that one can describe the arithmetic group Γ as some quotient group of π_1 of the complement of the icosahedral arrangement. This arrangement consists of the reflecting lines of fifteen involutions in $\mathfrak{A}_5$ and is therefore mapped into itself by the group. Thus, if one calculates π_1 of the quotient by $\mathfrak{A}_5$ of the complement, then one succeeds in determining the structure of $PSL(2, \mathcal{O})$. Since we have only a few Hilbert modular groups for which the group-structure is combinatorially described, this is naturally of interest and is the purpose of this note. The main result is stated as follows:

Theorem. *The icosahedral modular group $PSL(2, \mathcal{O})$ is isomorphic to the group generated by three elements α, β, γ with the fundamental relation;*

$$\begin{cases}\alpha\beta=\beta\alpha \quad \alpha\gamma\alpha=\gamma\alpha\gamma \quad \beta\gamma\beta\gamma\beta=\gamma\beta\gamma\beta\gamma\\ (\beta\gamma)^2=1 \quad (\alpha\gamma)^3=1.\end{cases}$$

Moreover, if we define the homomorphism $h: PSL(2, \mathcal{O})\to\mathfrak{A}_5$ by setting $h(\alpha)=(1, 4)(2, 5)$, $h(\beta)=(1, 2)(4, 5)$, $h(\gamma)=(1, 4)(2, 3)$, then we obtain the exact sequence:

Received April 24, 1987.
Revised March 2, 1988.

$$1 \to \Gamma \to PSL(2, \mathcal{O}) \xrightarrow{h} \mathfrak{A}_5 \to 1.$$

The motivation to this work was the hope that this description will serve the purpose of giving a fundamental region for the action over $H \times H$, which is a much harder problem.

§ 1. Quotient of $P_2(C)$ by $\mathfrak{A}_5$.

As the icosahedral group acts on $P_1(C) \cong S^2$, it acts also over $P_1(C) \times P_1(C)$, commuting the transposition τ of product components. It thus acts on $P_2(C) \cong P_1(C) \times P_1(C)/\tau$. This action was thoroughly investigated by F. Klein [K]. It is induced by the three-dimensional irreducible representation of $\mathfrak{A}_5$ which is unique up to automorphisms, and the most reasonable generators for the ring of invariants are given by him. They are denoted by A, B, C, D and are of degree 2, 6, 10, 15. For the present section we are mainly interested in A and D. In this note we denote the curves $A=0$, $D=0$ etc. by $[A]$, $[D]$, etc. $[A]$ is the image of the diagonal of $P_1(C) \times P_1(C)$ and $[D]$ is the union of the fifteen reflecting lines mentioned in the introduction. We sometimes call this set of lines the icosahedral arrangement. Outside $[D]$ there are exactly two orbits of points, at which $\mathfrak{A}_5$ has a non-trivial isotropy subgroup: they are isolated points. They all lie on the invariant conic $[A]$. At one of the orbits the isotropy subgroup is of order 5 and it is of order 3 at the other orbit. Thus the quotient $S := P_2(C)/\mathfrak{A}_5$ has exactly two quotient singular points which are both cyclic. One is resolved by the exceptional set of the graph:

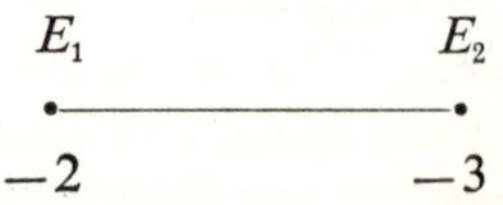

and the graph for the other is:

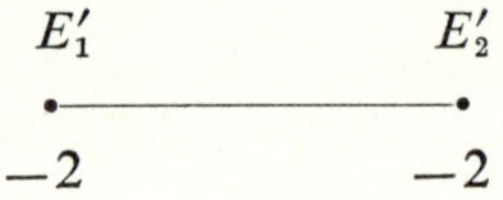

E_i, E_i' $(i=1, 2)$ are all rational curves on the minimal desingularization of S, which we denote by $\hat{S}$. We denote the images on S of $[A]$, $[D]$ etc. by $[\![A]\!]$, $[\![D]\!]$ etc. and let $[\![\hat{A}]\!]$, $[\![\hat{D}]\!]$ stand for their strict transforms on $\hat{S}$. $[\![\hat{A}]\!]$ is nonsingular; $[\![\hat{D}]\!]$ has three singular points which are rational double of Types A_1, A_2, A_4; they correspond to the intersection points of the icosahedral arrangement with multiplicity 2, 3, 5. $[\![\hat{D}]\!]$ is disjoint from E_i, E_i'; but $[\![\hat{A}]\!]$ cuts once the exceptional sets at E_2, E_2' transversally (E_i'

are so indexed). One sees moreover that $[\![\hat{A}]\!]$ is an exceptional curve of the first kind. This implies that one can successively blow down the curves $[\![\hat{A}]\!]$, E_2', E_2 and E_1, and that, by this, E_1' is mapped to a rational curve of self-intersection number 1. We denote this blowing down of $\hat{S}$ by $\check{S}$ and the image of E_1' by L. By computing directly we see further that the Euler number of $\check{S}$ is equal to 3. Thus $\check{S}$ is also isomorphic to the plane $P_2(C)$ and L is a line on it. Now we denote the image to $\check{S}$ of the curve $[\![\hat{D}]\!]$ by $[\![\check{D}]\!]$; it has four singular points of Types A_1, A_2, A_3, A_4 in which the last one arises from the blowing down. As the image of a line in $[D]$, $[\![\check{D}]\!]$ is rational; so it is a quintic curve on $\check{S}\cong P_2(C)$. $[\![\check{D}]\!]$ and L intersect only at the last A_4-singular point with multiplicity 5.

§ 2. Computation of the fundamental group

In this section we preserve the notation in the previous section. Our object here is to compute the fundamental group of $P_2(C)-[D]$. Note that $[A]$ intersects $[D]$ transversally, Thus, by using [O-S] and by putting a line l in $[D]$ at infinity, we have the isomorphism:

$$(2.1) \qquad \pi_1(P_2(C)-([A]\cup[D]),\, *)\cong\pi_1(P_2(C)-[D],\, *)\times \boldsymbol{Z}$$

where $*$ denotes the reference point chosen. Here the product is direct and we have used $\pi_1(P_2(C)-([A]\cup l),\, *)\cong \boldsymbol{Z}$. Thus it suffices that we compute the left hand side of (2.1). Now recall that $\mathfrak{A}_5$ acts freely on the complement $P_2(C)-([A]\cup[D])$. On the other hand we have seen in Section 1 that

$$\begin{aligned}\{P_2(C)-([A]\cup[D])\}/\mathfrak{A}_5 &\cong P_2(C)-(L\cup[\![\check{D}]\!])\\ &\cong \boldsymbol{C}^2-([\![\check{D}]\!]\cap \boldsymbol{C}^2)\end{aligned}$$

where we have introduced the identifications $\check{S}=P_2(C)$, $P_2(C)-L=\boldsymbol{C}^2$, Denoting for brevity the affine quintic curve $[\![\check{D}]\!]\cap\boldsymbol{C}^2$ by $\tilde{D}$, we thus obtain the exact sequence:

$$(2.2) \qquad 1\rightarrow \pi_1(P_2(C)-([A]\cup[D]),\, *)\rightarrow \pi_1(\boldsymbol{C}^2-\tilde{D},\, *)\xrightarrow{\ h\ }\mathfrak{A}_5\rightarrow 1.$$

By this we see that it suffices to compute $\pi_1(\boldsymbol{C}^2-\tilde{D},\, *)$. Now it is necessary to write down the equation of $\tilde{D}$ explicitly i.e. we have to recall the construction of Section 1 with explicit expressions. It is summarized in the following:

$$\begin{array}{ccc} P_2(C) & \overset{\delta}{\dashrightarrow} & \check{S}\cong P_2(C)\\ {\scriptstyle q}\downarrow & & \uparrow{\scriptstyle \beta}\,\wr\\ S=P_2(C)/\mathfrak{A}_5 & \underset{\pi}{\overset{\sim}{\longleftarrow}} & \hat{S}\end{array}$$

where $\rightrightarrows$ means a birational morphism, q is the natural map, π is the minimal desingularization of S, β is the blowing-down mentioned in Section 1 and δ is the rational mapping defined by the diagarm and to be written down explicitly. Now, using the properties of Klein's invariants A, B, C, D, we can see that the mapping δ is given by assigning the proportionality $(A^5 : A^2B : C) \in P_2(\boldsymbol{C})$ to the generic point of the original $P_2(\boldsymbol{C})$ on which the curves $[A]$, $[D]$ sit. The line L is obviously given by putting the first coordinate to be zero. If we introduce the affine coordinates x, y for $\check{S}-L$ by identifying (x, y) with $(1 : x : y)$ in the target plane of δ, then δ is given by $x=B/A^3$, $y=C/A^5$. Now, by [K], we have the following identity:

$$144D^2 = -1728B^2 + 720ACB^3 - 80A^2C^2B + 64A^3(5B^2 - AC)^2 + C^3.$$

This implies that the equation of the affine curve $\tilde{D}$ is:

$$\text{(2.3)} \qquad -1728x^5 + 720x^3y - 80xy^2 + 64(5x^2 - y)^2 + y^3 = 0.$$

The line-family $x=$const. are exaclty the lines passing through the A_4-singular point of $[\![\check{D}]\!]$ at infinity. They intersect $[\![\check{D}]\!]$ at this double point with multiplicity 2, since L, the unique line with higher contact, is not included in the family. This explains why the left hand side is of degree 3 with respect to y. Note also that all the singular points of $\tilde{D}$ are in the real part $\boldsymbol{R}^2 \subseteq \boldsymbol{C}$. This suggests that the real configuration is essential for the computation of $\pi_1(\boldsymbol{C}^2 - \tilde{D}, *)$: It looks like the following figure:

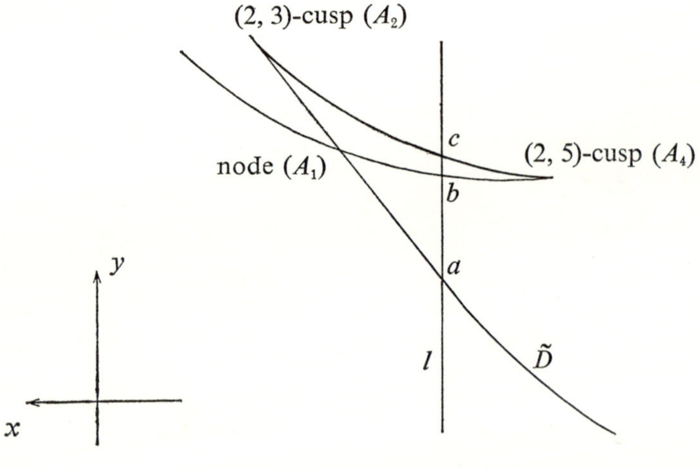

The vertical line, denoted by l, is the standard line which we fix for the computation of $\pi_1(\boldsymbol{C}^2 - \tilde{D}, *)$, and a, b, c are the intersection points of l

and $\tilde{D}$, ordered from below. Now we take three paths α, β, γ (issuing from the base point $*$ in the upper half-plane which one can for example take to be i) in the complex feature of the line l as follows:

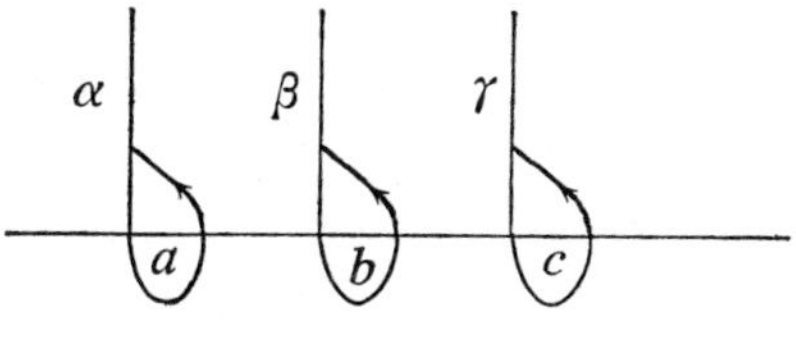

Then we apply the well known method, which is attributed to Zariski and Van Kampen, to the computation of the required fundamental group. The first fact to be mentioned is that α, β, γ generate $\pi_1(C^2-\tilde{D}, *)$; so we have only to get adequate relations among them. For this, the method above suggests us to move the standard line l within the family x=const. along the real axis in the x-plane so far as it does not meet any singular point of $\tilde{D}$, to surround or avoid the singular point by escaping in a sufficient small size into the complex part, and to turn back to the original position of l; the paths α, β, γ are by this kind of processes deformed to other paths say $\tilde{\alpha}$, $\tilde{\beta}$, $\tilde{\gamma}$, which can of course be written as some products of α, β, γ. Since $\tilde{\alpha}$, $\tilde{\beta}$, $\tilde{\gamma}$ are obviously identical with α, β, γ in $\pi_1(C^2-\tilde{D}, *)$, we get the corresponding relations among α, β, γ. Skipping detailed discussions, we simply assert that the relations:

$$(2.4)\qquad \begin{cases} \alpha\beta=\beta\alpha \qquad \alpha\gamma\alpha=\gamma\alpha\gamma \\ \beta\gamma\beta\gamma\beta=\gamma\beta\gamma\beta\gamma, \end{cases}$$

are obtained by the above procedure, corresponding to the node, the (2, 3)-cusp and the (2, 5)-cusp of the curve $\tilde{D}$ respectively. There is no essential relation other than (2.4) since we do not have any line x=const. which is tangent to $\tilde{D}$. Now we note that the composite path $\alpha\beta\gamma$ surrounds once the (2, 5)-cusp at infinity. According to the description of the blowing down β in Section 1, this implies that $\alpha\beta\gamma$ surrounds only the exceptional curve E_1 in $\hat{S}$. Since $E_1\cup E_2$ is the exceptional set of the cyclic quotient singularity on $[\![A]\!]$ of order 5, the power $(\alpha\beta\gamma)^5$ is the image of a path which surrounds the conic $[A]$ once. By (2.1), (2.2) this should lie in the center of $\pi_1(C^2-\tilde{D})$, which one can in fact check directly by using the relations (2.4). We have also seen by this argument that $\alpha\beta\gamma$ is mapped to an element of order five by the homomorphism $h:\pi_1(C^2-\tilde{D})\to\mathfrak{A}^5$ of (2.2). We may of course assume that $h(\alpha\beta\gamma)=(1, 2, 3, 4, 5)$. Then $h(\alpha)$, $h(\beta)$, $h(\gamma)$ are almost uniquely determined:

$$(2.5)\qquad \begin{cases} h(\alpha)=(1,4)(2,5) \\ h(\beta)=(1,2)(4,5) \\ h(\gamma)=(1,4)(2,3). \end{cases}$$

(From the definition of the arrangement D, the elements α, β, γ should be mapped onto some involutions of $\mathfrak{A}_5$.) We have thus completely described $\pi_1(C^2-\tilde{D})$, and hence also $\pi_1(P_2(C)-([A]\cup[D]), *)$, etc. In particular we obtain the exact sequence:

$$1 \longrightarrow \pi_1(P_2(C)-[D], *) \longrightarrow G_1 \xrightarrow{h} \mathfrak{A}_5 \longrightarrow 1$$

where G_1 is the group generated by α, β, γ with the fundamental relations (2.4) and

$$(2.6)\qquad (\alpha\beta\gamma)^5=1.$$

§ 3. The icosahedral modular group

As in the introduction we let $\mathcal{O}$ be the ring of integers of the number field $Q(\sqrt{5})$, $\mathfrak{p}=(2)$ the prime ideal generated by 2 and Γ the principal congruence subgroup of $SL(2,\mathcal{O})$ for the level $\mathfrak{p}$. We begin by recalling Hirzebruch's description [H] of the compactified quotient $\check{Y}=\overline{H\times H/\Gamma}$; it has exactly five cusps, each of which is resolved minimally by a triangle of rational curves with self-intersection number -3. He shows that the closure of the image of diagonal $\{(z,z); z\in H\}$ is lifted to an exceptional curves of the first kind on the minimal desingularization Y of $\check{Y}$ and that there are exactly ten transforms of it by $\mathfrak{A}_5\cong PSL(2,\mathcal{O}/\mathfrak{p})$ which are disjoint from each other. The separate blowing down of these ten curves is the cubic diagonal surface of Clebsch and Klein (X: $\sum_i y_i=\sum_i y_i^3=0$ in $P_4(C)$: $(y_0, y_1, \cdots, y_4)$) with their image being the Eckardt points of the surface. (See [S] for this terminaology.) The five exceptional sets give rise to fifteen lines on the cubic surface and the remaining twelve lines on the surface separate themselves into two orbits under the action of $\mathfrak{A}_5$ which form a "double six" [S]. We take one of the two and blow down the lines belonging to it in order to get the plane $P_2(C)$ and the action over it. This last action is identical with the one discussed in Section 1 and the fifteen lines on X are mapped onto the icosahedral arrangement. The inverse procedure of getting quotient $H\times H/\Gamma$ from the arrangement is also clear for us: We blow up the six fivefold intersection points of the arrangement to get the diagonal surface and then we blow up the ten threefold intersection points of the arrangement to get Y. To obtain $H\times H/\Gamma$ one needs only delete the strict transforms on Y of the lines in the arrange ment. Since we also know by [H] that Γ acts freely on $H\times H$, the group

Γ is described to be π_1 of this complement in Y. By the first blowing up we should add to (2.4), (2.6) the following relation:

$$(\beta\gamma)^5=1 \tag{3.1}$$

and by the second blowing up the relation:

$$(\alpha\gamma)^3=1. \tag{3.2}$$

The paths $(\beta\gamma)^5$ and $(\alpha\gamma)^3$ correspond to paths surrounding the exceptional curves for the two kinds of blowings-up, which therefore become upstairs homotope to zero. (Check that $(\beta\gamma)^5$, $(\alpha\gamma)^3$ both lie in the kernel of the homomorphism h in (2.2).) Thus we have arrived at the exact sequence

$$1 \longrightarrow \Gamma \longrightarrow G_2 \xrightarrow{h} \mathfrak{A}_5 \longrightarrow 1 \tag{3.3}$$

where G_2 is the group generated by α, β, γ with the fundamental relations (2.4), (2.6), (3.1), (3.2) and h is again defined by (2.5). It is obvious that we have also prove the isomorphism

$$G_2 \cong PSL(2, \mathcal{O}). \tag{3.4}$$

This and the sequence (3.3) should be regarded as the main result of the note. It is now an easy work to find the matrix elements of the group in the right hand side corresponding to α, β, γ:

$$\alpha=\begin{pmatrix}1 & 1\\ 0 & 1\end{pmatrix}$$

$$\beta=\begin{pmatrix}1 & u^2\\ 0 & 1\end{pmatrix} \qquad u=\frac{\sqrt{5}-1}{2}$$

$$\gamma=\begin{pmatrix}0 & 1\\ -1 & 2\end{pmatrix}$$

We finally remark that these are all in $SL(2, \mathcal{O})$, that the relations (2.4), (2.6) are satisfied in $SL(2, \mathcal{O})$-level while we have $(\beta\gamma)^5=(\alpha\gamma)^3=-1$.

References

[H] Hirzebruch, F., Hilbert modular group of the field $\mathbf{Q}(\sqrt{5})$ and the cubic diagonal surface of Clebsch and Klein, Russian Math. Surveys, **31**(5) (1976), 96–110.

[K] Klein, F., Weitere Untersuchungen über das Ikosaeder, Gesam. Math. Abh. Bd. 2, 321–384, Springer (1973).

[K-N] Kobayashi, R. and Naruki, I., Holomorphic conformal structures and uniformization of complex surfaces, Math. Ann., **279** (1988), 485–500.

[O–S] Oka, M. and Sakamoto, K., Product theorem of the fundamental group of a reducible curve, J. Math. Soc. Japan, **30** (1978), 599–602.
[S] Segre, B., The nonsingular cubic surfaces, Oxford University Press, 1942.

R. Kobayashi
College of arts and sciences
University of Tokyo
Tokyo, 153 Japan

I. Naruki
Res. Inst. for Math. Sci.
Kyoto University
Kyoto, 606 Japan

Advanced Studies in Pure Mathematics 15, 1989
Automorphic Forms and Geometry of Arithmetic Varieties
pp. 373–413

Invariants and Hodge Cycles

Michio Kuga*

Dedications

I would like to dedicate this paper to Professor I. Satake and Professor F. Hirzebruch on their birthdays, whose mathematics greatly influenced me.

I would like to thank Professor Parry, who pointed out an initial miscalculation in this work, and the decomposition of $P=P_0+P_1$ (which looks trivial now, but initially was not), which made later calculations easier.

I would also like to thank Professor Goldman, Mrs. Goldman, Professor Iwasawa, Professor Kra, Professor Kuranishi, and Professor Sah, without which this paper would not be possible.

§ 1. Introduction

Let F be an even $2N$ dimensional vector space over $\boldsymbol{Q}$, and β be a non degenerate alternating bilinear form on F. We put

$$F_{\boldsymbol{R}}=F\otimes\boldsymbol{R}\cong\boldsymbol{R}^{2N},$$
$$Sp(F,\beta)=\{g\in \operatorname{aut}(F)\,|\,\beta(gu,gv)=\beta(u,v)\},$$

and

$$\mathfrak{H}(F,\beta)=\left\{J\in\operatorname{aut}(F_{\boldsymbol{R}})\,\middle|\,\begin{array}{l}J^2=-1\\ \beta(u,Jv)=\text{symmetric on } u, v\\ \beta(u,Ju)>0 \text{ for } 0\neq{}^{\forall}u\in F\end{array}\right\};$$

and we call $\mathfrak{H}(F,\beta)$ the Siegel half space.

An element J of $\mathfrak{H}(F,\beta)$ is a complex structure of the real vector space $F_{\boldsymbol{R}}$, therefore it defines a complex vector space $(F_{\boldsymbol{R}},J)$ of N dimension, which we denote by E or E_J. The group $Sp(F,\beta)$ operates on $\mathfrak{H}(F,\beta)$ by

Received April 8, 1987.
Revised April 20, 1988.
* Supported by The Paul and Gabriella Rosenbaum Foundation.

$$\left.\begin{array}{r} Sp(F, \beta) \ni g \\ \mathfrak{H}(F, \beta) \ni J \end{array}\right\} \longrightarrow g(J)=g^{-1}Jg$$

and $\mathfrak{H}(F, \beta)$ is actually a symmetric space of the simple Lie group $Sp(F, \beta)(\boldsymbol{R})$ of R-points of $Sp(F, \beta)$ modulo a maximal compact subgroup.

A lattice $L \subset F$ is said to be integral, iff $\beta(L, L) \subset Z$. If an integral lattice L is given, $(F_{\boldsymbol{R}}, L, J, \beta)$ is a polarized abelian variety for all J in $\mathfrak{H}(F, \beta)$. Put

$$Sp(L, \beta)=\{\gamma \in Sp(F, \beta) \mid \gamma(L)=L\}.$$

$Sp(L, \beta)$ acts on $\mathfrak{H}(F, \beta)$ properly discontinuously and two polarized abelian varieties $(F_{\boldsymbol{R}}, L, J, \beta)$ and $(F_{\boldsymbol{R}}, L, J', \beta)$ with two elements J, J' of $\mathfrak{H}(F, \beta)$, are isomorphic iff J', J' are $Sp(L, \beta)$ equivalent.

The symmetric space $\mathfrak{H}(F, \beta)$ is actually a symmetric domain. Namely $\mathfrak{H}(F, \beta)$ has complex structure, which is invariant under the action of g for every $g \in Sp(F, \beta)(\boldsymbol{R})$. There are exactly 2 such complex structures of $\mathfrak{H}(F, \beta)$; one is complex conjugate of the other. Among these two, one satisfies the following condition.

A system of complex manifolds

$$A \xrightarrow{\pi} D$$

is called 1 dimensional family of abelian varieties iff

1) A, D are complex manifolds; π is smooth surjective holomorphic map;
2) D is a disc and π is proper;
3) $\pi^{-1}(\lambda)$ $(\lambda \in D)$ are all abelian varieties.

When such a 1-dimensional family $A \xrightarrow{\pi} D$ of abelian varieties is given, there exist a polarization of $\pi^{-1}(\lambda)$, (F, β, L), and a map $t: D \rightarrow \mathfrak{H}(F, \beta)$ such that

$$\pi^{-1}(\lambda) \cong (F_{\boldsymbol{R}}, L, \beta, t(\lambda)) \quad \text{(as polarized abelian varieties)}$$

for all $\lambda \in$ small neighborhood of origin of D. Such t is called a moduli-map of $A \xrightarrow{\pi} D$.

$\mathfrak{H}(F, \beta)$ is a complex manifold. The complex structure in $\mathfrak{H}(F, \beta)$ is characterized as the unique one such that the moduli-map t is holomorphic in a small neighborhood of origin of D for every 1-dimensional family $A \xrightarrow{\pi} D$ of abelian varieties (belonging to β).

If a symplectic representation $\rho: G \rightarrow Sp(F, \beta)$ of a semi-simple connected hermitian algebraic group $G/\boldsymbol{Q}$, not containing trivial representation, together with an holomorphic map $\tau: X \rightarrow \mathfrak{H}(F, \beta)$ of the corresponding hermitian symmetric domain $X=G(\boldsymbol{R})^0$/maximal compact into the Siegel half space $\mathfrak{H}(F, \beta)$ of (F, β) commuting with symmetries, satisfies the compatibility condition:

$$\rho(g)[\tau(x)]=\tau[g(x)] \qquad \forall g \in G \quad \forall x \in X$$

the pair (ρ, τ) defines so-called GTAS A over an arithmetic variety $V=\Gamma\backslash X$, if $\rho(\Gamma)\subset Sp(L, \beta)$ for a cocompact subgroup Γ of G without torsion. Here L is an integral lattice.

A is defined as follows; The semi direct product $\Gamma\ltimes_\rho L$, with respect to the representation $\rho: \Gamma\rightarrow \mathrm{aut}(L)$, operates on the product manifold $X\times F_{\boldsymbol{R}}$ properly discontinuously, so that it produces the quotient space $\Gamma\ltimes L\backslash X\times F_{\boldsymbol{R}}$, which is denoted by A. And the natural map of division by $\Gamma\ltimes_\rho L$, from $X\times F_{\boldsymbol{R}}$ to A, is denoted by p. From A to $V=\Gamma\backslash X$ taking the left factor, we have a natural map π. $A\xrightarrow{\pi}V$ is obviously a fibre bundle over the arithmetic variety V with torus as fibres A_λ $(\lambda \in V)$ (associated to the fundamental group).

Take a point $\lambda_0 \in X$, we use the same symbol λ_0 for the point in V which is covered by $\lambda_0 \in X$. Identify $\pi_1(V, \lambda_0)$ with Γ, then the action of $\pi_1(V, \lambda_0)=\Gamma$ on the homology group $H_1(A_{\lambda_0}, \boldsymbol{Q})=F$ of the fibre is nothing else than ρ.

Therefore the representation of $\pi_1(V, \lambda_0)=\Gamma$ on the cohomology group $H^1(A_{\lambda_0}, \boldsymbol{Q})={}^tF$ (dual space of F) of the fibre is the dual representation ρ^* of ρ.

The representation of $\pi_1(V, \lambda_0)=\Gamma$ on the higher cohomology group $H^r(A_{\lambda_0}, \boldsymbol{Q})=\Lambda^r({}^tF)$ is therefore $\Lambda^r(\rho^*)$.

Since the representation ρ is always self-dual ($\because$ existence of β), the difference between ρ or ρ^* is not essential.

In the manifold A, we can define a complex structure; and A becomes a complex manifold. The complex structure of A is canonical by the following properties.

(1) $A\xrightarrow{\pi}V$ is holomorphic

(2) The map of $E_{\tau(x)}$ to A defined by $p\circ\varphi_x$ is holomorphic for every $x \in X$.

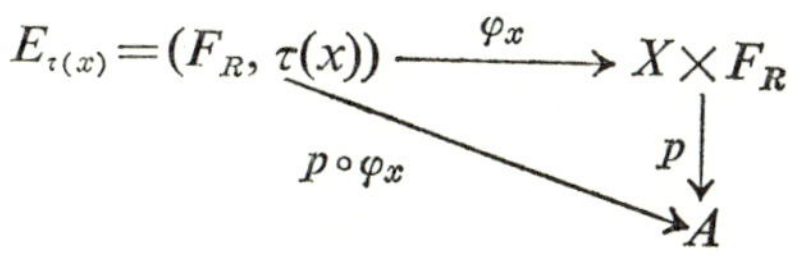

where φ_x is defined by

$$\varphi_x(u)=x\times u \qquad \text{for all } u \in F_{\boldsymbol{R}}.$$

(3) The complex manifold $\tilde{A}=\Gamma\backslash X\times F_{\boldsymbol{R}}$, (which is a covering of $A=\Gamma\ltimes L\backslash X\times F_{\boldsymbol{R}}$) is a complex vector bundle $\boldsymbol{E}$ over V, where Γ is identified with the subgroup $\Gamma\ltimes\{0\}$ of $\Gamma\ltimes L$.

A, as complex manifold with this complex structure, is actually a projective algebraic variety, and π is a rational map. Moreover addition $A\underset{V}{\times}A\to A$, and inverse $A\to A$ are also algebraic.

Therefore, fibres $A_\lambda=\pi^{-1}(\lambda)$ are Abelian varieties. Thus such variety A is called GTAS over V. (Group theoretical abelian scheme over V). It has the following properties:

(1) $A\xrightarrow{\pi}V$ are projective algebraic varieties and π is everywhere defined surjective smooth rational map.

(2) $A\underset{V}{\times}A\xrightarrow{+}A$, $A\xrightarrow{-1}A$ are also algebraic.

(3) As C^∞-manifold $A\xrightarrow{\pi}V$ is a fibre bundle, whose fibres are tori, and associated to the fundamental group.

(4) Fibres $A_\lambda=\pi^{-1}(\lambda)$ $(\lambda\in V)$ are abelian varieties.

(5) The base V is an arithmetic variety, i.e., $V=\Gamma\backslash X$; $X=G(\boldsymbol{R})^0$/maximal compact to $\Gamma\subset G$, cocompact and no torsion.

(6) The action ρ of $\pi_1(V,\lambda)=\Gamma$ on $F=H_1(A_\lambda,\boldsymbol{Q})$ is extendable to an algebraic representation $\rho: G\to\mathrm{aut}(F)$ of G defined over $\boldsymbol{Q}$.

Conversely, does the properties (1) ~ (6) characterize GTAS? No, perhaps not. Actually from (1) ~ (6), we can prove that there exist (F, L, β), and holomorphic map $\tau: X\to\mathfrak{H}(F,\beta)$ and $\rho: G\to S_P(F,\beta)$, such that $\rho(\gamma)[\tau(x)]=\tau[\gamma(x)]$ for all $\gamma\in\Gamma$.

But we cannot prove

$$(*)\quad \rho(g)[\tau(x)]=\tau[g(x)] \qquad \forall g\in G \quad \forall x\in X$$

and

(**) the commutativity with symmetries.

If this (*) (**) be true, our A must be GTAS. In order to have (*), perhaps we must discuss Hecke-operators, densely distributed, beside Γ. Anyway, in this note, we use only those properties (1) ~ (6) of GTAS.

Satake classified all G and most ρ which admits some GTAS [6]. Susan Addington, inherited Satake, (roughly) classified all ρ of G which admits some GTAS when G is the quaternion type. [2], [3]. For that purpose, she invented some combinatorix in a finite group $\mathscr{G}$, called "*chemistry*". For that, see [2] or [3] [4].

When $A\xrightarrow{\pi}V$ is a rigid GTAS over an arithmetic variety $V=\Gamma\backslash X$, the space of Hodge cycles in a generic fibre $A_\lambda=\pi^{-1}(\lambda)$ $(\lambda\in V)$ is denoted by

$$HH^r(A_\lambda,\boldsymbol{Q}) \qquad (HH^r=0 \text{ if } r=\text{odd})$$

and is equal to

$$H^r(A_\lambda, Q)^G = \wedge^r({}^tF)^G$$

where, $F = H_1(A_\lambda, Q)$ on which $\Gamma = \pi_1(V, \lambda)$ operates; and the operation is extendable to an action of G on F. Here, $\wedge^r({}^tF)$ is identified canonically with $H^r(A_\lambda, Q)$: in which Γ-invariant part is equal to the G-invariant part, and coincides with the space of Hodge cycles. (See [1], [4]).

Thus the problem of determining the space of Hodge cycles is reducible to the problem of determining the space of group invariants. But unlike the ordinary invariants-theory, this invariants-theory is a bit complicated, because it relates also to a combinatrix of the Galois-group ("*chemistry*").

In this paper, the author tries to calculate $\dim HH^r(A_\lambda, Q)$ in some examples with simple "*chemistries*". We put $\dim HH^r(A_\lambda, Q) = \dim \wedge^r({}^tF)^G = b_r$ also we put the polynomial $\sum_{r=0}^N b_r t^r = F(t)$ where $N = 2\dim_C A_\lambda = \dim \rho_{A_\lambda} = \dim P$. See [5].

In the most of the following cases, G is of the quaternion type. In this case, the assumption that ρ which admits a rigid GTAS does not contain 1 implies that $\wedge^{2r+1} \circ \rho^*$ is also trivial part free; i.e. $b_{2r+1} = 0$ for odd $2r+1$.

Let k be a totally real number field with $[k : Q] = m$; K be a Galois extension of Q such that $K \supset k$ with $\mathscr{G} = \mathrm{Gal}(K/Q)$, $S =$ the set of ∞-places of k on which $\mathscr{G}$ acts transitively. Let B be a quaternion-algebra over k; and $S_0 (\subset S)$ be the set of ∞-places of k which splits in B. We assume that $S_0 \neq \phi$. Then $(\mathscr{G}, S, S_0)$ is a *chemistry* ([2] [4]). Take a (maximal) order $\mathfrak{o}$ of B, and put

$$\Gamma(1) = \{\gamma \in \mathfrak{o}^\times \mid \nu(\gamma) = 1\}$$

where $\nu : B^\times \to k^\times$ is the reduced norm. If $\Gamma \subset \Gamma(1)$, $\Gamma \sim \Gamma(1)$, (commensurable), $V = \Gamma \backslash \mathfrak{h}^{|S_0|}$ is an arithmetic variety, where $\mathfrak{h} =$ the upper half plain, of which product $\mathfrak{h}^{|S_0|}$, Γ acts as usual. Susan Addington found a functor

$$\begin{array}{l} A \\ \Big| \rightsquigarrow P = \sum X_i \\ V \end{array}$$

associating a $\mathscr{G}$-invariant stable polymer P in the *chemistry* $(\mathscr{G}, S, S_0)$ to each group-theoretical abelian scheme A over V. Moreover if this P is furthermore rigid, then A can not have deformations [1]. Conversely, for any $\mathscr{G}$-invariant stable polymer P in $(\mathscr{G}, S, S_0)$ there exists a group theoretical scheme A over V and a positive integer μ such that

$$\begin{array}{l} A \\ \Big| \rightsquigarrow \mu P. \\ V \end{array}$$

Moreover Salmon Abdulali [1] showed that $H^r(A_\lambda, \boldsymbol{Q})^G = \wedge^r({}^tF)^G$ coincides with the space $HH^r(A_\lambda, \boldsymbol{Q})$ of Hodge cycles in the generic fibre A_λ if the polymer is rigid.

So, in this paper, we are going to calculate

$$\dim \wedge^{2r}({}^tF)^G$$

for our rigid polymer representation space ${}^tF$ of $\mathscr{G}$, in some special cases. As notation and notions, we use the same notation and notions as in [2], [4] except group theoretical abelian scheme.

We called group theoretical abelian scheme a GTFabV there and denoted it by the notation $V \xrightarrow{\pi} U$ there instead of the notation $A \xrightarrow{\pi} V$ here.

§ 2. The character-algebra of $SL(2, C)$

The trivial representation of $SL(2, \boldsymbol{C})$ is denoted by 1. The class of the identity representation

$$x \longmapsto x \in SL(2, C)$$

of $SL(2, \boldsymbol{C})$ is denoted by X. The class of the symmetric tensor representation of $SL(2, \boldsymbol{C})$ of the degree ν is denoted by X_ν. Hence $X = X_1$, $1 = X_0$.

For 2 classes P, Q of representation of $SL(2, \boldsymbol{C})$, the class $P \oplus Q$ of representation is also denoted by $P+Q$ the class $P \otimes Q$ of representation is also denoted by $P \cdot Q$. Thus the all virtual classes of representation of $SL(2, \boldsymbol{C})$ generates a ring $\mathscr{R}$ which is generated by X, and actually isomorphic to the polynomial ring $Z[t]$ of 1 variable over Z. The "multiplicity" of 1 in $P \in \mathscr{R}$ is denoted $(P, 1)$. $\mathscr{R}$ contains $1, X_1, X_2, X_3, \cdots$, and all the elements P of $\mathscr{R}$ are expressable uniquely as linear combinations of them with Z as coefficients; and the coefficient of 1 in that linear expression of P is nothing else than $(P, 1)$.

The multiplicative structure of $\mathscr{R}$ with respect to the above additive structure is given by the Clebsch-Gordan formula: $X_\nu \cdot X_\mu = X_{\nu+\mu} + X_{\nu+\mu-2} + X_{\nu+\mu-4} + \cdots + X_{|\nu-\mu|}$.

For example:

$$X^2 = X \cdot X = X_2 + 1$$

$$X^3 = X^2 \cdot X = (X_2+1)X = X_2X + X = X_3 + X + X = X_3 + 2X$$

$$X^4 = X^3 \cdot X = (X_3+2X)X = X_3X + 2XX = X_4 + X_2 + 2(X_2+1)$$
$$= X_4 + 3X_2 + 2$$

$$X^5 = X^4 \cdot X = (X_4 + 3X_2 + 2)X = X_4X + 3X_2X + 2X$$
$$= X_5 + X_3 + 3(X_3 + X) + 2X = X_5 + 4X_3 + 5X$$
$$\vdots$$
$$X_2^2 = X_2 \cdot X_2 = X_4 + X_2 + 1$$
$$X_2^3 = X_2^2 \cdot X_2 = (X_4 + X_2 + 1)X_2 = X_4X_2 + X_2X_2 + X_2$$
$$= X_6 + X_4 + X_2 + X_4 + X_2 + 1 + X_2 = X_6 + 2X_4 + 3X_2 + 1$$
$$X_2^4 = X_2^3 \cdot X_2 = (X_6 + 2X_4 + 3X_2 + 1)X_2 = X_8 + X_6 + X_4 + 2(X_6 + X_4 + X_2)$$
$$+ 3(X_4 + X_2 + 1) + X_2 = X_8 + 3X_6 + 6X_4 + 6X_2 + 3$$
$$\vdots$$
$$XX_2 = X_3 + X$$
$$X^2X_2 = X(X_3 + X) = XX_3 + XX = X_4 + X_2 + X_2 + 1 = X_4 + 2X_2 + 1$$
$$\vdots$$
$$XX_2^2 = X(X_4 + X_2 + 1) = X_5 + X_3 + X_3 + X + X = X_5 + 2X_3 + 2X$$
$$X^2X_2^2 = X(X_5 + 2X_3 + 2X) = X_6 + X_4 + 2(X_4 + X_2) + 2(X_2 + 1)$$
$$= X_6 + 3X_4 + 4X_2 + 1$$
$$\vdots$$

Lemma. *If $n>0$ is an even integer X^n is a linear combination of $1, X_2, X_4, \cdots, X_{2k}, \cdots, X_n$ of even suffixes $2k$, with positive integers a_{2k} as coefficients, and $a_n = 1$; i.e.*

$$X^n = a_0 1 + a_2 X_2 + \cdots + a_{2k} X_{2k} + \cdots + X_n \qquad Z \ni a_{2k} > 0,\ a_n = 1$$

If $n>0$ is an odd integer X^n is a linear combination of $X, X_3, \cdots, X_{2k+1}, \cdots, X_n$ of odd suffixes $2k+1$, with positive integers a_{2k+1} as coefficients, and $a_n = 1$; i.e.

$$X^n = a_1 X + a_3 X_3 + \cdots + a_{2k+1} X_{2k+1} + \cdots + X_n \quad Z \ni a_{2k+1} > 0,\ a_n = 1$$

Proof. By the induction on n. □

Definition. If a representation P of $SL(2, C)$ is a linear combination of $1, X_2, X_4, \cdots$ with non-negative coefficients, P is said to be *even.*

Lemma. *For $m, n \in Z$, $n>0, m>0$,*

$$X^n \text{ is even} \rightleftarrows n \text{ even}$$
$$X^n X_2^m \text{ is even} \rightleftarrows n \text{ even.}$$

Lemma. *For $n, m \in Z$, $n>0, m>0$,*

$$(X^n, 1) > 0 \rightleftarrows n \text{ even}$$
$$(X^n X_2^m, 1) > 0 \rightleftarrows n \text{ even}.$$

Easy.

Lemma. *For integers* $\nu \geqq 0$, $\mu \geqq 0$,

$$(X_\nu \cdot X_\mu; 1) = \begin{cases} 1 & \nu = \mu \\ 0 & \nu \neq \mu. \end{cases}$$

§ 3. The character-algebra of (a product of) $SL_2(C)$

For each atom $\alpha \in S$, we prepare a Lie group $G_\alpha \cong SL_2(C)$. The product $G_S = \prod_{\alpha \in S} G_\alpha$, is isomorphic to the complex form $G(C)$ of G. By this isomorphism, G is considered as a subgroup of G_S. i.e.

$$G \xrightarrow{\iota} G_S$$

For an atom $\alpha \in S$, the projection to the α-th component G_α of $G_S = \prod G_\alpha$ is a 2 dimensional representation of G_S. The class of this representation, we denote that by ρ_α or α. The corresponding representation space, we denote by W_α or by α. For a molecule $X = \langle \alpha, \beta, \cdots, \omega \rangle$ the representation (class) $\alpha \otimes \beta \otimes \cdots \otimes \omega$ of G_S is denoted by ρ_X or by X. The corresponding representation space, is denoted by W_X or by X. For a polymer $P = \sum_{i=1} X_i$ the representation class $X_1 \oplus \cdots \oplus X_k$ is denoted by ρ_P or P. The corresponding representation space we denote by W_P or P. Finally symbol $\oplus$ is often abbreviated as $+$; and symbol $\otimes$ is often abbreviated as $\cdot$. For each non-negative integer ν, and for each atom $\alpha \in S$, we put $\alpha_\nu = X_\nu \circ \alpha$: $G \xrightarrow{\alpha} G_\alpha = SL_2(2, C) \xrightarrow{X_\nu} SL(\nu+1, C)$. In particular, $\alpha_1 = \alpha$, $\alpha_0 = 1$ for all α. Then $1, \alpha_1, \alpha_2, \alpha_3, \cdots, \beta_1, \beta_2, \beta_3, \cdots, \gamma_1, \gamma_2, \gamma_3, \cdots$ are in the commutative ring $\mathscr{R}(S)$ of the all virtual representation-classes of G_S; $\mathscr{R}(S)$ is generated by them, in fact, they are spanned by the product: $\alpha_{\nu_\alpha} \beta_{\nu_\beta} \gamma_{\nu_\gamma} \cdots$ of them: i.e.

$$\mathscr{R}(S) \ni P;\ P = \sum_{(\nu)} a_{\nu_\alpha \nu_\beta \cdots \nu_\omega} \alpha_{\nu_\alpha} \beta_{\nu_\beta} \cdots \omega_{\nu_\omega} \qquad a_{(\nu)} \in Z$$

This linear expression is unique, and $(P, 1) = a_{0,\ldots,0}$ is obviously the multiplicty of 1 in P.

The function:

$$P \rightsquigarrow (P, 1)$$

is an additive homomorphism of $\mathscr{R}(S)$ to Z, but unfortunately, it is not multiplicative, but we have

Lemma. *If* $P(\alpha, \beta, \cdots, \omega)$, $Q(\xi, \eta, \cdots, \zeta)$ *has no letter in common, then*

$$(P(\alpha, \cdots, \omega)Q(\xi, \cdots, \zeta), 1)=(P(\alpha, \cdots, \omega), 1)(Q(\xi, \cdots, \zeta), 1).$$

Easy.

The multiplicity $(P, 1)$ is also denoted by

$$(P, 1)=\int\cdots\int P(\alpha, \cdots, \omega)d\alpha\cdots d\omega.$$

If we take the maximal compact subgroup $G_u=SU(2)^m$ of $G=SL(2, \boldsymbol{C})^m$ and if we write the Haar measure of G_u as $d\alpha\cdots d\omega$, normalized with $\int\cdots\int_{G_u} d\alpha\cdots d\omega=1$, the above symbol of the multiplicity acquaires reality. I.e., identifying the representation P with its character functions $\operatorname{tr} P$,

$$(P, 1)=\int\cdots\int P(\alpha, \cdots, \omega)d\alpha\cdots d\omega$$

is just the orthogonality relation via the unitary trick.

If P involves variables $\alpha, \beta, \cdots, \omega; \xi, \cdots, \eta$, obviously we have

Lemma.

$$\begin{aligned}\int\cdots\int\Big(\int\cdots\int P(\alpha, \beta\cdots\omega, \xi\cdots\eta)d\alpha\cdots d\omega\Big)d\xi\cdots d\eta \\ =\int\cdots\int\Big(\int\cdots\int P(\alpha, \beta\cdots\omega, \xi\cdots\eta)d\xi\cdots d\eta\Big)d\alpha\cdots d\omega \\ =\int\cdots\int P(\alpha\cdots\eta)d\alpha\cdots d\eta.\end{aligned}$$

This is because $\operatorname{tr}(P\otimes Q)=\operatorname{tr} P\cdot\operatorname{tr} Q$, $\operatorname{tr}(P+Q)=\operatorname{tr} P+\operatorname{tr} Q$ and Fubini's theorem.

§ 4. Chemistry and abelian scheme

The fact that a G.T. abelian scheme $A\xrightarrow{\pi}V=\Gamma\backslash\mathfrak{H}^{|S_0|}$ over V, corresponds to a polymer P implies the fact that the action of $\Gamma=\pi_1(V, \lambda)$, (λ is a generic point of V) on the cohomology group $H^{2r}(A_\lambda, \boldsymbol{Q})=\Lambda^{2r}({}^tF)$, (${}^tF=H_0^1(A_\lambda, \boldsymbol{Q})$), is extendable to a $\boldsymbol{Q}$-representation $\rho_A^{(r)}$ of G, which is $\boldsymbol{C}$-isomorphic to $\Lambda^r(\rho_P)$.

Now, therefore, if

$$P=\sum_{i=0}^{k} X_i \qquad X_i=\{\alpha_i, \beta_i, \cdots, \omega_i\}$$

then ${}^tF\underset{Q}{\otimes}C$ has a unique subspaces $F_{X_1}, F_{X_2}, \cdots, F_{X_k}$ such that

$$ {}^tF\otimes C=\bigoplus_{i=0}^{k} F_{X_i} \qquad F_{X_i}\cong W_{X_i}\ (=X_i) $$

Therefore:

$$ \begin{aligned} \wedge^r({}^tF)\otimes C&=\wedge^r({}^tF\otimes C)=\wedge^r(F_{X_i}\oplus\cdots\oplus F_{X_k})\\ &=\bigoplus_{r_1+r_2+\cdots+r_k=r}\wedge^{r_1}(F_{X_1})\otimes\wedge^{r_2}(F_{X_2})\otimes\cdots\otimes\wedge^{r_k}(F_{X_k}) \end{aligned} $$

Now if $X=\{\alpha, \beta, \cdots, \omega\}$, there exist two-dimensional C-linear G_S-spaces $F_{X,\alpha}, F_{X,\beta}, \cdots, F_{X,\omega}$ such that

$$ F_X=F_{X,\alpha}\otimes F_{X,\beta}\otimes\cdots\otimes F_{X,\omega};\quad F_{X,\alpha}\cong W_\alpha, \cdots, F_{X,\omega}\cong W_\omega $$

as G_S-spaces. Now, in order to compute

$$ \begin{aligned} F(t)&=\sum_{r=0}^{N} b_r t^r=\sum \dim_Q[\wedge^r({}^tF)^G]t^r=\sum \dim_C[\wedge^r({}^tF)^G\otimes C]t^r\\ &=\sum \dim_C[\wedge^r({}^tF\otimes C)^{G_S}]t^r, \end{aligned} $$

we define:

$$ \begin{aligned} G(t)&=\sum_r \dim_C \wedge^r({}^tF\otimes C)t^r\\ &=\sum \dim \wedge^{r_1}(F_{X_1})t^{r_1} \dim \wedge^{r_2}(F_{X_2})t^{r_2}\cdots\dim \wedge^{r_k}(F_{X_k})t^{r_k} \end{aligned} $$

and

$$ H(t)=\bigoplus_{r_1\cdots r_k}\wedge^{r_1}(F_{X_1})t^{r_1}\otimes\cdots\otimes\wedge^{r_k}(F_{X_k})t^k=f(F_{X_1})\otimes\cdots\otimes f(F_{X_k}) $$

where $f(F_X)=\sum_{r=0}^{\dim F_X}\wedge^r(F_X)t^r$ is a formal polynomial of t with the G vector spaces as coefficients.

§ 5. $|X|=2$

Lemma. *For two linear spaces A, B over a same field k.*

$$ \wedge^2(A\otimes B)\cong[\wedge^2(A)\otimes S^2(B)]\oplus[S^2(A)\otimes\wedge^2(B)] $$

where $S^2(V)$, $\wedge^2(V)$ are the spaces of symmetric and alternating 2-tensors in V.

∵ Well known. □

Putting $X=\{\alpha, \beta\}$, since $F_X\cong W_X=\alpha\beta$, we have

Lemma. $\wedge^0(F_X)\cong \boldsymbol{C}$

$$\wedge^1(F_X)\cong F_X\cong \alpha\beta$$

$$\wedge^2(F_X)\cong \alpha_2+\beta_2$$

$$\wedge^3(F_X)\cong \alpha\beta$$

$$\wedge^4(F_X)\cong \boldsymbol{C}$$

The first, second, fourth, fifth formulas are obvious. Because of the last lemma,

$$\wedge^2(F_X)\cong \wedge^2(\alpha\beta)=\wedge^2(\alpha\otimes\beta)\cong(\wedge^2(\alpha)\otimes S^2(\beta))\oplus(S^2(\alpha)\otimes\wedge^2(\beta))=\beta_2+\alpha_2.$$

Therefore

$$\begin{aligned} f(F_X)&=\sum_{r=0}^{4}\wedge^r(F_X)t^r\cong \boldsymbol{C}+\alpha\beta t+(\alpha_2+\beta_2)t^2+\alpha\beta t^3+\boldsymbol{C}t^4\\ &=(1+t^4)1+(t+t^3)\alpha\beta+t^2\alpha_2+t^2\beta_2\end{aligned}$$

where 1 means the trivial vector space $\boldsymbol{C}$ over $\boldsymbol{C}$. We write this by $f(\alpha, \beta)$: i.e.

$$\begin{aligned} f(F_X)=f(\alpha,\beta)&=(1+t^4)1+(t+t^3)\alpha\beta+t^2\alpha_2+t^2\beta_2\\ &=(1,\alpha,\alpha_2)\begin{pmatrix}1+t^4 & 0 & t^2\\ 0 & t+t^3 & 0\\ t^2 & 0 & 0\end{pmatrix}\begin{pmatrix}1\\ \beta\\ \beta_2\end{pmatrix}={}^t\hat{\alpha}P\hat{\beta},\end{aligned}$$

where

$$\hat{\alpha}=\begin{pmatrix}1\\ \alpha\\ \alpha_2\end{pmatrix}\qquad \hat{\beta}=\begin{pmatrix}1\\ \beta\\ \beta_2\end{pmatrix}\quad \text{etc. and}$$

$$P=\begin{pmatrix}1+t^4 & 0 & t^2\\ 0 & t+t^3 & 0\\ t^2 & 0 & 0\end{pmatrix}.$$

For the later purpose we put

$$M(\alpha)=\begin{pmatrix}1\\ \alpha\\ \alpha_2\end{pmatrix}(1,\alpha,\alpha_2)=\begin{pmatrix}1 & \alpha & \alpha_2\\ \alpha & \alpha\alpha & \alpha\alpha_2\\ \alpha_2 & \alpha_2\alpha & \alpha_2\alpha_2\end{pmatrix}.$$

Because of the Lemma we have

$$\int M(\alpha)d\alpha=E \qquad (3\times 3 \text{ identity matrix}).$$

We put for a non-negative integer ν

$$f(\alpha, \beta)^{\nu}=[(1+t^4)1+(t+t^3)\alpha\beta+t^2\alpha_2+t^2\beta_2]^{\nu}={}^t\hat{\alpha}_{\nu}P_{\nu}\hat{\beta}_{\nu}$$

where

$$\hat{\alpha}_{\nu}=\begin{pmatrix}1\\ \alpha\\ \alpha_2\\ \vdots\\ \alpha_{2\nu}\end{pmatrix} \qquad \hat{\beta}_{\nu}=\begin{pmatrix}1\\ \beta\\ \vdots\\ \beta_{2\nu}\end{pmatrix}$$

and P_{ν} is $(2\nu+1)\times(2\nu+1)$ matrix, of which entries are polynomials in t, of degree 4ν. Similarly to $M(\alpha)$, we define for non-negative ν

$$M_{\nu}(\alpha)={}^t\hat{\alpha}_{\nu}\cdot\hat{\alpha}_{\nu}=\begin{pmatrix}1 & \alpha & \alpha_2\cdots & \alpha_{2\nu}\\ \alpha & & & \\ \vdots & & \alpha_1\alpha_j & \\ \vdots & & & \ddots \\ \alpha_{2\nu} & & & \alpha_{2\nu}\alpha_{2\nu}\end{pmatrix}.$$

This is a $(2\nu+1)\times(2\nu+1)$ matrix, and

$$\int M_{\nu}(\alpha)d\alpha=E \qquad ((2\nu+1)\times(2\nu+1) \text{ unit matrix}).$$

Also we define $P_{\nu,0}$, the $(\nu+1)\times(\nu+1)$ matrix which consists of even-th rows and even-th columns of P_{ν}, and define $P_{\nu,1}$ the $\nu\times\nu$ matrix which consists of odd-th rows and odd-th columns of P_{ν}.

Then

$$P_{\nu}\sim P_{\nu,0}+P_{\nu,1}$$

and

$$P_{\nu}^n\sim P_{\nu,0}^n+P_{\nu,1}^n \qquad \operatorname{tr} P_{\nu}^n=\operatorname{tr} P_{\nu,0}^n+\operatorname{tr} P_{\nu,1}^n.$$

For small $\mu'S$, P_{μ}, $P_{\mu,0}$, $P_{\mu,1}$ are

$$P_1=P=\begin{pmatrix}1+t^4 & 0 & t^2\\ 0 & t+t^3 & 0\\ t^2 & 0 & 0\end{pmatrix}$$

$$P_{1,0}=\begin{pmatrix}1+t^4 & t^4\\ t^2 & 0\end{pmatrix}, \quad P_{1,1}=(t+t^3)$$

$$P_2=\begin{bmatrix} 1+t^2+6t^4+t^6+t^8 & 0 & 3t^2+3t^4+3t^6 & 0 & t^4 \\ 0 & 2t+6t^3+6t^5+2t^7 & 0 & 2t^3+2t^5 & 0 \\ 3t^2+3t^4+3t^6 & 0 & t^2+4t^4+t^6 & 0 & 0 \\ 0 & 2t^3+2t^5 & 0 & 0 & 0 \\ t^4 & 0 & 0 & 0 & 0 \end{bmatrix}$$

$$P_{2,0}=\begin{pmatrix} 1+t^2+6t^4+t^6+t^8 & 3t^2+3t^4+3t^6 & t^4 \\ 3t^2+3t^4+3t^6 & t^2+4t^4+t^6 & 0 \\ t^4 & 0 & 0 \end{pmatrix}$$

$$P_{2,1}=\begin{pmatrix} 2t+6t^3+6t^5+2t^7 & 2t^3+2t^5 \\ 2t^3+2t^5 & 0 \end{pmatrix}.$$

§ 6. Cyclic case, $\mu=1$

Let $\mathscr{G}=Z_n=\{0, 1, \cdots, n-1\}$, $P=\sum gX$, $X=\langle 0, 1\rangle$. Put $0=\alpha$, $1=\beta$, $2=\gamma$, $3=\delta$, $\cdots, n-2=\eta$, $n-1=\omega$, $X_i=\langle i, i+1\rangle$, $n=0$: then

$$\begin{aligned} P&=\sum_g gX=X_1+\cdots+X_{n-1}=\{0, 1\}+\{1, 2\}+\{2, 3\}+\cdots+\{n-1, 0\} \\ &=\{\alpha, \beta\}+\{\beta, \gamma\}+\cdots+\{\eta, \omega\}+\{\omega, \alpha\} \\ H(t)&=f(F_{X_1})f(F_{X_2})\cdots f(F_{X_{n-1}})=f(\alpha, \beta)f(\beta, \gamma)f(\gamma, \delta)\cdots f(\eta, \omega)f(\omega, \alpha) \\ &={}^t\hat{\alpha}P\hat{\beta}{}^t\hat{\beta}P\hat{\gamma}{}^t\hat{\gamma}P\hat{\delta}\cdots P\hat{\omega}{}^t\hat{\omega}P=\mathrm{tr}({}^t\hat{\alpha}P\hat{\beta}{}^t\hat{\beta}P\cdots P\hat{\omega}{}^t\hat{\omega}P\hat{\alpha}) \\ &=\mathrm{tr}(P\hat{\beta}{}^t\hat{\beta}P\hat{\gamma}{}^t\hat{\gamma}\cdots P\hat{\omega}{}^t\hat{\omega}P\hat{\alpha}{}^t\hat{\alpha}). \end{aligned}$$

Now put $\hat{\beta}{}^t\hat{\beta}=M(\beta)$, $\cdots$, $\hat{\alpha}{}^t\hat{\alpha}=M(\alpha)$, then

$$M(\alpha)=\hat{\alpha}{}^t\hat{\alpha}=\begin{pmatrix}1\\ \alpha\\ \alpha_2\end{pmatrix}(1, \alpha, \alpha_2)=\begin{pmatrix}1 & \alpha & \alpha_2\\ \alpha & \alpha\alpha & \alpha\alpha_2\\ \alpha_2 & \alpha_2\alpha & \alpha_2\alpha_2\end{pmatrix}$$

and

$$H(t)=\mathrm{tr}(PM(\beta)PM(\gamma)\cdots PM(\omega)PM(\alpha)),$$

therefore

$$\begin{aligned} F(t)&=(H(t), 1)=\int\cdots\int H(t)d\alpha\cdots d\omega \\ &=\mathrm{tr}\Big(P\Big(\int M(\beta)d\beta\Big)P\int M(\gamma)d\gamma\cdots P\int M(\omega)d\omega P\int M(\alpha)d\alpha\Big) \\ &=\mathrm{tr}(PE\ PE\cdots PE\ PE)=\mathrm{tr}(P^n). \end{aligned}$$

Now since: $$P\sim\begin{pmatrix}1+t^4 & t^2\\ t^2 & 0\end{pmatrix}\oplus(t+t^3),$$

the eigen polynomial of $P=\{Z^2-(1+t^4)Z-t^4\}\{Z-(t+t^3)\}$. Eigenvalues of P are

$$\lambda=\frac{(1+t^4)+\sqrt{1+6t^4+t^8}}{2},\qquad \mu=\frac{(1+t^4)-\sqrt{1+6t^4+t^8}}{2},\qquad \rho=t+t^3.$$

Therefore for even $n=2m$,

$$\begin{aligned}\operatorname{tr}(P^n)&=\lambda^n+\mu^n+\rho^n\\&=\left(\frac{(1+t^4)+\sqrt{1+6t^4+t^8}}{2}\right)^n+\left(\frac{(1+t^4)-\sqrt{1+6t^4+t^8}}{2}\right)^n+(t+t^3)^n\\&=\frac{2}{2^n}\left[\sum_{k=0}^{m}\binom{n}{2k}(1+t^4)^{2k}(1+6t^4+t^8)^{m-k}\right]+(t+t^3)^n.\end{aligned}$$

It is also determinable recursively by:

$$f^{(n)}+\sigma_1 f^{(n-1)}+\sigma_2 f^{(n-2)}+\sigma_3 f^{(n-3)}=0\quad(n=3,4,\cdots).$$

from $f^{(2)}, f^{(1)}, f^{(0)}$, where

$$\begin{aligned}f^{(n)}\quad&=\operatorname{tr}(P^n)\in \mathbf{Z}[t]\\f^{(2)}(t)&=1+t^2+10t^4+t^5+t^8\\f^{(1)}(t)&=1+t^4\\f^{(0)}(t)&=3\end{aligned}$$

where

$$\sigma_i=\sigma_i(t)\in \mathbf{Z}[t]\qquad(i=1,2,3)$$

are coefficients of the characteristic polynomial

$$\begin{aligned}Z^3+\sigma_1(t)Z^2+\sigma_2(t)Z+\sigma_3(t)&=\det(ZI_3-P)\\&=Z^3-(1+t+t^3+t^4)Z^4+(t+t^3-t^4+t^5+t^7)Z+(t^5+t^7).\end{aligned}$$

Therefore

$$\begin{aligned}\sigma_1(t)&=-1-t-t^3-t^4,\\\sigma_2(t)&=t+t^3-t^4+t^5+t^7,\\\sigma_3(t)&=t^5+t^7.\end{aligned}$$

For example:

$$\begin{aligned}
f^{(3)}(t) &= -\sigma_1(t)f^{(2)}(t)-\sigma_2(t)f^{(1)}(t)-\sigma_3(t)f^{(0)}(t)\\
&=(1+t+t^3+t^4)(1+t^2+10t^4+t^6+t^8)\\
&\quad -(t+t^3-t^4+t^5+t^7)(1+t^4)-(t^5+t^7)3\\
&=1+t^2+t^3+12t^4+6t^5+2t^6+6t^7+12t^8+t^9+t^{10}+t^{12}\\
f^{(4)}(t) &= -\sigma_1\cdot f^{(3)}-\sigma_2\cdot f^{(2)}-\sigma_3\cdot f^{(1)}\\
&=(1+t+t^3+t^4)(1+t^2+t^3+\cdots+t^{12})\\
&\quad -(t+t^3-t^4+t^5+t^7)(1+t^2+10t^4+t^6+t^8)-(t^5+t^7)(1+t^4)\\
&=1+t^2+t^3+15t^4+6t^5+11t^6+7t^7+41t^8+7t^9+11t^{10}\\
&\quad +6t^{11}+15t^{12}+t^{13}+t^{14}+t^{16}.
\end{aligned}$$

§ 7. Cyclic case, $\mu=2$ or $\mu\geqq 2$

If $\mathscr{G}=Z_n=\{0, 1, \cdots, n-1\}$, $X=\{0, 1\}$, $P=2\sum_g gX=2\{0, 1\}+2\{1, 2\}+\cdots+2\{n-1, 0\}$, it is treatable as if $\mu=1$ if we count the same molecule twice in P i.e.; put $X_0=X_1=\{0, 1\}$, $X_2=X_3=\{1, 2\}$, $X_4=X_5=\{2, 3\}$, $\cdots$, $X_{2n-2}=X_{2n-1}=\{n-1, 0\}$

$$P=X_0+X_1+X_2+\cdots+X_{2n-2}+X_{2n-1}.$$

Therefore: by putting $\alpha=0$, $\beta=1$, $\cdots$, $\omega=n-1$,

$$\begin{aligned}
H(t) &= f(\alpha, \beta)f(\alpha, \beta)f(\beta, \gamma)f(\beta, \gamma)f(\gamma, \delta)f(\gamma, \delta)\\
&\qquad \cdots f(\eta, \omega)f(\eta, \omega)f(\omega, \alpha)f(\omega, \alpha)\\
&= f(\alpha, \beta)^2 f(\beta, \gamma)^2 f(\gamma, \delta)^2\cdots f(\eta, \omega)^2 f(\omega, \alpha)^2\\
&= {}^t\hat{\alpha}_2P_2\hat{\beta}_2{}^t\hat{\beta}_2P_2\hat{\gamma}_2{}^t\hat{\gamma}_2P_2\hat{\delta}_2\cdots {}^t\hat{\eta}_2P_2\hat{\omega}_2{}^t\hat{\omega}_2P_2\hat{\alpha}_2\\
&= {}^t\hat{\alpha}_2P_2M_2(\beta)P_2M_2(\gamma)P_2M_2(\delta)\cdots P_2M_2(\omega)P_2\hat{\alpha}_2\\
&= \mathrm{tr}(P_2M_2(\beta)P_2M_2(\gamma)\cdots P_2M_2(\omega)P_2M_2(\alpha))
\end{aligned}$$

Therefore:

$$\begin{aligned}
F(t) &= (H(t), 1)\\
&= \int\cdots\int \mathrm{tr}(P_2M_2(\beta)P_2M_2(\gamma)\cdots P_2M_2(\alpha)d\alpha d\beta\cdots d\omega\\
&= \mathrm{tr}\Big(P_2\Big(\int M_2(\beta)d\beta\Big)P_2\Big(\int M(\gamma)d\gamma\Big)\cdots P_2\Big(\int M(\alpha)d\alpha\Big)\Big)\\
&= \mathrm{tr}(P_2EP_2E\cdots P_2E)=\mathrm{tr}(P_2^n)
\end{aligned}$$

Similarly for general μ;

$$F(t)=\mathrm{tr}(P_\mu^n).$$

Therefore, they are determinable recursively by

$$f_\mu^{(n)}+\sigma_1 f_\mu^{(n-1)}+\sigma_2 f_\mu^{(n-2)}+\cdots+\sigma_m f_\mu^{(n-m)}=0$$

for $n\geqq m$ from $f_\mu^{(0)}, \cdots, f_\mu^{(m-1)}$ where

$$f_\mu^{(n)}=f_\mu^{(n)}(t)=\mathrm{tr}(P_\mu^n)\in \boldsymbol{Z}[t]$$
$$f_\mu^{(0)}=m\in \boldsymbol{Z}$$

and

$$\sigma_1=\sigma_1(t), \cdots, \sigma_m=\sigma_m(t)\in \boldsymbol{Z}[t]$$

are coefficients of the characteristic polynomial

$$Z^m+\sigma_1(t)Z^{m-1}+\sigma_2(t)Z^{m-2}+\cdots+\sigma_m(t)$$

of the matrix P_μ, where $m=2\mu+1$.

For small μ and n, the values of $\mathrm{tr}\, P_\mu^n$ are

$$\mathrm{tr}\, P_2^1=1+2t+2t^2+6t^3+10t^4+6t^5+2t^6+2t^7+t^8$$
$$\mathrm{tr}\, P_2^2=1+6t^2+56t^4+126t^6+210t^8+126t^{10}+56t^{12}+6t^{14}+t^{16}$$
$$\mathrm{tr}\, P_2^4=1+4t^2+82t^4+452t^6+2600t^8+8208t^{10}+20574t^{12}+33224t^{14}$$
$$+40790t^{16}+33224t^{18}+20574t^{20}+8208t^{22}+2600t^{24}+452t^{26}$$
$$+82t^{28}+4t^{30}+t^{32}.$$

We write here the characteristic polynomial of P_2:

$$P_2=\begin{pmatrix} 1+t^2+6t^4+t^6+t^8 & 0 & 3t^2+3t^4+3t^6 & 0 & t^4 \\ 0 & 2t+6t^3+6t^5+2t^7 & 0 & 2t^3+2t^5 & 0 \\ 3t^2+3t^4+3t^6 & 0 & t^2+4t^4+t^6 & 0 & 0 \\ 0 & 2t^3+2t^5 & 0 & 0 & 0 \\ t^4 & 0 & 0 & 0 & 0 \end{pmatrix}$$
$$\sim P_{2,0}\oplus P_{2,1}$$

where

$$P_{2,0}=\begin{pmatrix} 1+t^2+6t^4+t^6+t^8 & 3t^2+3t^4+3t^6 & t^4 \\ 3t^2+3t^4+3t^6 & t^2+4t^4+t^6 & 0 \\ t^4 & 0 & 0 \end{pmatrix}$$

$$P_{1,0}=\begin{pmatrix} 2+6t^3+6t^5+2t^7 & 2t^3+2t^5 \\ 2t^3+2t^5 & 0 \end{pmatrix}.$$

Therefore the characteristic polynomial of P_2=the characteristic polynomial of $P_{2,0}\times$the characteristic polynomial of $P_{2,1}$.

The characteristic polynomial of

$$\begin{aligned} P_{2,0}&=\begin{vmatrix} Z-(1+t^2+6t^4+t^6+t^8) & -(3t^2+3t^4+3t^6) & -t^4 \\ -(3t^2+3t^4+3t^6) & Z-(t^2+4t^4+t^6) & 0 \\ -t^4 & 0 & Z \end{vmatrix} \\ &=[Z-(1+t^2+6t^4+t^6+t^8)][Z-(t^2+4t^4+t^6)]Z \\ &\quad -t^8(Z-(t^2+4t^4+t^6))-(3t^2+3t^4+3t^6)^2Z \\ &=Z^3-(1+2t^2+10t^4+2t^6+t^8)Z^2 \\ &\quad +(t^2-4t^4-7t^6-2t^8-7t^{10}-4t^{12}+t^{14})Z+(t^{10}+4t^{12}+t^{14}) \end{aligned}$$

The characteristic polynomial of

$$\begin{aligned} P_{2,1}&=\begin{vmatrix} Z-(2t+6t^3+6t^5+2t^7) & -(2t^3+2t^5) \\ -(2t^3+2t^5) & Z \end{vmatrix} \\ &=Z^2-(2t+6t^3+6t^5+2t^7)Z-(4t^6+8t^8+4t^{10}) \end{aligned}$$

Therefore the characteristic polynomial of

$$\begin{aligned} P_2&=[Z^3-(1+2t^2+10t^4+2t^6+t^8)Z^2 \\ &\quad +(t^2-4t^4-7t^6-2t^8-7t^{10}-4t^{12}+t^{14})Z+(t^{10}+4t^{12}+t^{14})] \\ &\quad \times[Z^2-(2t+6t^3+6t^5+2t^7)Z-(4t^6+8t^8+4t^{10})] \\ &=Z^5-(1+2t+2t^2+6t^3+10t^4+6t^5+2t^6+2t^7+t^8)Z^4 \\ &\quad +(2t+t^2+10t^3-4t^4+38t^5-11t^6+78t^7-10t^8+78t^9-11t^{10} \\ &\quad +38t^{11}-4t^{12}+10t^{13}+t^{14}+2t^{15})Z^3 \\ &\quad +(-2t^3+2t^5+4t^6+32t^7+16t^8+68t^9+61t^{10}+76t^{11}+100t^{12} \\ &\quad +76t^{13}+61t^{14}+68t^{15}+16t^{16}+32t^{17}+4t^{18}+2t^{19}-2t^{21})Z^2 \\ &\quad +(-4t^8+8t^{10}-2t^{11}+56t^{12}-14t^{13}+80t^{14}-32t^{17}+72t^{16} \\ &\quad -32t^{17}+80t^{18}-14t^{19}+56t^{20}-2t^{21}+8t^{22}-4t^{24})Z \\ &\quad -(4t^{16}+24t^{18}+40t^{20}+24t^{22}+4t^{24}). \end{aligned}$$

Let $\mathscr{G}$ be an arbitral finite group. Assume in this section, that $S=\mathscr{G}$ and the action of $\mathscr{G}$ on $S(=\mathscr{G})$ is the left multiplication. We are still assuming that $|X|=2$, and put

$$X=\{\alpha, \xi\}.$$

Take the unique element $h \in \mathscr{G}$ such that $\xi=h\alpha$. Then it is easy

$$\begin{aligned}P=\sum gX=&\{\alpha, h\alpha\}+\{h\alpha, h^2\alpha\}+\cdots+\{h^{n-1}\alpha, \alpha\}+\{\beta, h_2\beta\}\\&+\{h_2\beta, h_2^2\beta\}+\cdots+\{h^{n-1}\beta, \beta\}+\{\gamma, h_3\gamma\}+\cdots\\&+\{h_3^{n-1}\gamma, \gamma\}+\cdots+\{\omega, h_k\omega\}+\cdots\{h_k^{n-1}\omega, \omega\}\end{aligned}$$

where:

n=the order of h in $\mathscr{G}$ (=the smallest positive integer n such that $h^n=1$

$k=|\mathscr{G}/H|$

$g_1=1, g_2, g_3, \cdots, g_k$=representatives of $\mathscr{G}/H$

$\beta=g_2\alpha, \gamma=g_3\alpha, \cdots, \omega=g_k\alpha, h_2=g_2hg_2^{-1}, h_3=g_3hg_3^{-1}, \cdots, h_k=g_khg_k^{-1}$

$H=\{1, h, h^2, \cdots, h^{n-1}\}$=the cyclic subgroup generated by h

And, in this case, by the same calculation as before,

$$F(t)=\sum_{j=0}^{2\dim A_X} \dim HH^{\gamma}(A_\lambda, \mathbf{Q})t^{\gamma}=(\operatorname{tr} P^n)^k.$$

Where $A\xrightarrow{\pi}V$ is an abelian scheme belonging to the polymer $P=\sum gX$ and A_λ is a generic fibre in that. In the same way, if we taken an abelian scheme belonging to $P=\mu\sum g_X$ with higher $\mu>0$, $F(t)$ becomes:

$$F(t)=(\operatorname{tr}(P_\mu^n))^k$$

§ 8.

If $\mathscr{G}\neq S$, the calculation will become difficult, and only for some examples we can calculates the dimension of G-invariant cycles.

I) Edges of a tetrahedron.

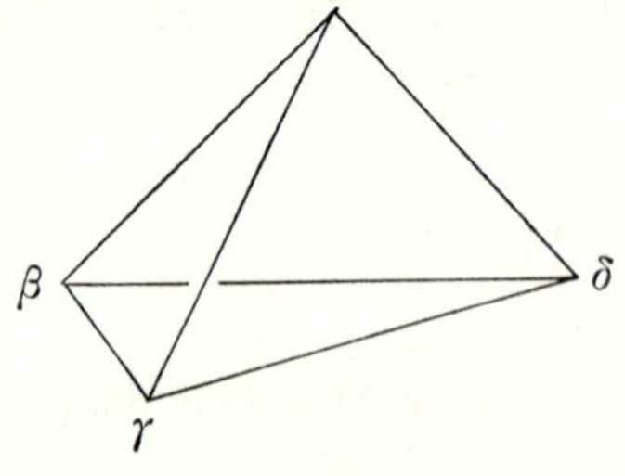

$\mathscr{G}$ = tetrahedral group $\cong A_4$

$$P=\{\alpha,\beta\}+\{\alpha,\gamma\}+\{\alpha,\delta\}+\{\beta,\gamma\}+\{\beta,\delta\}+\{\gamma,\delta\}$$
$$F(t)=(f(\alpha,\beta)f(\alpha,\gamma)f(\alpha,\delta)f(\beta,\gamma)f(\beta,\delta)f(\gamma,\delta),1).$$

Lemma. $((a+bX+cX_2)(a'+b'X+c'X_2)(a''+b''X+c''X_2),1)$
$$=aa'a''+(ab'b''+ba'b''+bb'a'')+(ac'c''+ca'c''+cc'a'')+(bc'c''+cb'c''+cc'b'')+cc'c''$$

$\because$ Calculation □

We put this 3-linear form

$$\begin{pmatrix} a & b & c \\ a' & b' & c' \\ a'' & b'' & c'' \end{pmatrix}=aa'a''+ab'b''+ba'b''+bb'a''+ac'c''+ca'c''$$
$$+cc'a''+bc'c''+cb'c''+cb'c''+cc'b''+cc'c''.$$

We put furthermore,

Definition.

$$\{A,B,C\}=\mathrm{tr}(PAPAPA)+\mathrm{tr}(PAPBPB)\times 3+\mathrm{tr}(PAPCPC)\times 3+\mathrm{tr}(PBPCPC)\times 3+\mathrm{tr}(PCPCPC)$$

for three 3×3 matrices A, B, C.

$$F(t)=\iiiint f(\beta,\gamma)f(\gamma,\delta)f(\delta,\beta)\cdot f(\alpha,\beta)f(\alpha,\gamma)f(\alpha,\delta)\,d\alpha d\beta d\gamma d\delta$$
$$=\iiint \mathrm{tr}(PM(\gamma)PM(\delta)PM(\beta))\left[\int\!\!\int\{(a+c\beta_2)+b\beta\alpha+c\alpha_2\}\times\{(a+c\gamma_2)+b\gamma\alpha+c\alpha_2\}\{(a+c\delta_2)+b\delta\alpha+c\alpha_2\}d\alpha\right]d\beta d\gamma d\delta$$
$$=\iiint \mathrm{tr}(PM(\gamma)PM(\delta)PM(\beta)\begin{pmatrix} a+c\beta_2 & b\beta & c \\ a+c\gamma_2 & b\gamma & c \\ a+c\delta_2 & b\delta & c\end{pmatrix}d\beta d\gamma d\gamma,$$

where $a=1+t^4$
$b=t+t^3$
$c=t^2$

$$=\iiint \mathrm{tr}(PM(\beta)PM(\gamma)PM(\delta))\times[(a+c\beta_2)(a+c\gamma_2)(a+c\delta_2)$$

$$+\{(x+c\beta_2)b\gamma b\delta+b\beta(a+c\gamma_2)b\delta+b\beta b\gamma(a+c\delta_2)\}$$
$$+\{(a+c\beta_2)cc+c(a+c\gamma_2)c+cc(a+c\delta_2)\}$$
$$+\{b\beta cc+cb\gamma c+ccb\delta\}+ccc]d\beta d\gamma d\delta$$
$$=\iiint \operatorname{tr}(P(a1+c\beta_2M(\beta))P(a1+c\gamma_2M(\gamma))P(a1+c\delta_2M(\delta))d\beta d\gamma d\delta$$
$$+\iiint \operatorname{tr}(P(a1+c\beta_2M\beta))P(\gamma M(\gamma))P(\delta M(\delta))d\beta d\gamma d\delta$$
$$+\iiint \operatorname{tr}(P(b\beta M(\beta))P(a+c\gamma_2M(\gamma))P(b\delta M(\delta))d\beta d\gamma d\delta$$
$$+\iiint \operatorname{tr}(P(b\beta M(\beta))P(b\gamma M(\gamma))P(a1+c\delta_2M(\delta))d\beta d\gamma d\delta$$
$$+\iiint \operatorname{tr}P(a1+c\beta_2M(\beta))P(cM(\gamma))P(cM(\delta))d\beta d\gamma d\delta$$
$$+\iiint \operatorname{tr}P(cM(\beta))P(a1+c\gamma_2M(\gamma))P(cM(\delta))d\beta d\gamma d\delta$$
$$+\iiint \operatorname{tr}P(cM(\beta))P(cM(\gamma))P(a1+c\delta_2M(\delta))d\beta d\gamma d\delta$$
$$+\iiint \operatorname{tr}P(b\beta M(\beta))P(cM(\gamma))P(cM(\delta))d\beta d\gamma d\delta$$
$$+\iiint \operatorname{tr}P(cM(\beta))P(b\gamma M(\gamma))P(cM(\delta))d\beta d\gamma d\delta$$
$$+\iiint \operatorname{tr}P(cM(\beta))P(cM(\gamma))P(b\delta M(\delta))d\beta d\gamma d\delta$$
$$+\iiint \operatorname{tr}P(cM(\beta))P(cM(\gamma))P(cM(\delta))d\beta d\gamma d\delta.$$

$\therefore$ Defining matrices A, B, C, by

$$\int (a1+cX_2M(X))dX=A$$
$$\int bXM(X)dX=B$$
$$\int cM(X)dX=C,$$

we have

$$F(t)=\operatorname{tr}(PAPAPA)+\operatorname{tr}(PAPBPB)\times 3+\operatorname{tr}(PAPCPC)\times 3$$
$$+\operatorname{tr}(PBPCPC)\times 3+\operatorname{tr}(PCPCPC)$$
$$=\{ABC\}$$

In order to compute these matrices A, B, C, we put:

Lemma.

$$\int XM(X)dX=\begin{pmatrix}0&1&0\\1&0&1\\0&1&0\end{pmatrix}=K$$

$$\int X_2M(X)dX=\begin{pmatrix}0&0&1\\0&1&0\\1&0&1\end{pmatrix}=J$$

then obviously

$$A=a1+cJ=\begin{pmatrix}a&0&c\\0&a+c&0\\c&0&a+c\end{pmatrix}$$

$$B=bk=\begin{pmatrix}0&b&0\\b&0&b\\0&b&0\end{pmatrix}$$

$$C=c1=\begin{pmatrix}c&0&0\\0&c&0\\0&0&c\end{pmatrix},$$

where $a=1+t^4$, $b=t+t^3$, $c=t^2$.

Now

$$\operatorname{tr}(PAPAPA)=\operatorname{tr}(((1+t^4)P+t^2PJ)^3),\ (A=(1+t^4)I_3+t^2J)$$

$$=\operatorname{tr}\left(\left((1+t^4)\begin{pmatrix}1+t^4&0&t^2\\0&t+t^3&0\\t^2&0&0\end{pmatrix}+t^2\begin{pmatrix}1+t^4&0&t^2\\0&t+t^3&0\\t^2&0&0\end{pmatrix}\begin{pmatrix}0&0&1\\0&1&0\\1&0&1\end{pmatrix}\right)^3\right)$$

$$=\operatorname{tr}\left|\left(\begin{pmatrix}1+3t^2+t^4&2t^2+t^4+2t^6\\t^2+t^6&t^4\end{pmatrix}\oplus(t+2t^3+2t^5+t^7)\right)^3\right|$$

$$=\operatorname{tr}\begin{bmatrix}1+13t^4+2t^6+52t^8+9t^{10}+81t^{12}+9t^{14}+52t^{16}+2t^{18}+13t^{20}+t^{24} & 2t^2+t^4+20t^6+11t^8+57t^{10}+23t^{12}+57t^{14}+11t^{16}+20t^{18}+t^{20}+2t^{22}\\ t^2+10t^6+t^8+28t^{10}+2t^{12}+28t^{14}+t^{16}+10t^{18}+t^{22} & 2t^4+t^6+14t^8+6t^{10}+25t^{12}+6t^{14}+14t^{16}+t^{18}+2t^{24}\end{bmatrix}+(t+2t^3+2t^5+t^7)^3$$

$$=1+t^3+15t^4+6t^5+3t^6+18t^7+66t^8+35t^9+15t^{10}+48t^{11}+106t^{12}$$
$$+48t^{13}+15t^{14}+35t^{15}+66t^{16}+18t^{17}+3t^{18}+6t^{19}+15t^{20}+t^{21}+t^{24}.$$

On the other hand

Lemma.

$$PK=\begin{pmatrix}1+t^4 & 0 & t^2\\ 0 & t+t^3 & 0\\ t^2 & 0 & 0\end{pmatrix}\begin{pmatrix}0 & 1 & 0\\ 1 & 0 & 1\\ 0 & 1 & 0\end{pmatrix}=\begin{pmatrix}0 & 1+t^2+t^4 & 0\\ t+t^3 & 0 & t+t^3\\ 0 & t^2 & 0\end{pmatrix}$$

$$PJ=\begin{pmatrix}1+t^4 & 0 & t^2\\ 0 & t+t^3 & 0\\ t^2 & 0 & 0\end{pmatrix}\begin{pmatrix}0 & 0 & 1\\ 0 & 1 & 0\\ 1 & 0 & 1\end{pmatrix}=\begin{pmatrix}t^2 & 0 & 1+t^2+t^4\\ 0 & t+t^3 & 0\\ 0 & 0 & t^2\end{pmatrix}$$

$\because$ Calculation.

Now, since $A=aI+bJ$, $B=bK$,

$$\begin{aligned}\operatorname{tr}(PAPBPB)&=a\operatorname{tr}(PPBPB)+c\operatorname{tr}(PJPBPB)\\ &=ab^2\operatorname{tr}(PPKPK)+cb^2\operatorname{tr}(PJPKPK)\\ &=ab^2\operatorname{tr}(PKPKP)+cb^2\operatorname{tr}(JPKPKP)\\ &=\operatorname{tr}((ab^2I_3+cb^2J)(PKPKP))\end{aligned}$$

and

$$\begin{aligned}ab^2I_3+cb^2J&=(1+t^4)(t+t^3)^2I_3+t^2(t+t^3)^2J\\ &=\begin{bmatrix}(1+t^4)(t+t^3)^2 & 0 & t^2(t+t^3)^2\\ 0 & (1+t^2+t^4)(t+t^3)^2 & 0\\ t^2(t+t^3)^2 & 0 & (1+t^2+t^4)(t+t^3)^2\end{bmatrix}\\ &=\begin{bmatrix}t^2+2t^4+2t^6+2t^8+t^{10} & 0 & t^4+2t^6+t^8\\ 0 & t^2+3t^4+4t^6+3t^8+t^{10} & 0\\ t^4+2t^6+t^8 & 0 & t^2+3t^4+4t^6+3t^8+t^{10}\end{bmatrix}=Q\end{aligned}$$

and

$$\begin{aligned}PKPKP&=\begin{bmatrix}0 & 1+t^2+t^4 & 0\\ t+t^3 & 0 & t+t^3\\ 0 & t^2 & 0\end{bmatrix}\begin{bmatrix}0 & 1+t^2+t^4 & 0\\ t+t^3 & 0 & t+t^3\\ 0 & t^2 & 0\end{bmatrix}\begin{bmatrix}1+t^4 & 0 & t^2\\ 0 & t+t^3 & 0\\ t^2 & 0 & 0\end{bmatrix}\\ &=\begin{bmatrix}(1+t^2+t^4)\times(t+t^3) & 0 & (1+t^2+t^4)\times(t+t^3)\\ 0 & (t+t^3)(1+t^2+t^4)+(t+t^3)t^2 & 0\\ t^2(t+t^3) & 0 & t^2(t+t^3)\end{bmatrix}\begin{bmatrix}1+t^4 & 0 & t^2\\ 0 & t+t^3 & 0\\ t^2 & 0 & 0\end{bmatrix}\end{aligned}$$

$$=\begin{bmatrix} t+3t^3+5t^5+5t^7 \\ +3t^9+t^{11} & 0 & t^3+2t^5+2t^7+t^9 \\ 0 & t^2+4t^4+6t^6+4t^8 \\ & +t^{11} & 0 \\ t^3+2t^5+2t^7+t^9 & 0 & t^5+t^7 \end{bmatrix}$$

hence

$$Q(PKPKP)=\begin{bmatrix} t^2+2t^4+2t^6+2t^8 \\ +t^{10} & 0 & t^4+2t^6+t^8 \\ 0 & t^2+3t^4+4t^6+3t^8 \\ & +t^{10} & 0 \\ t^4+2t^6+t^8 & 0 & t^2+3t^4+4t^6+3t^8 \\ & & +t^{10} \end{bmatrix}$$

$$\times\begin{bmatrix} t+3t^2+5t^5+5t^7 \\ +3t^9+t^{11} & 0 & t^3+2t^5+2t^7+t^9 \\ 0 & t^2+4t^4+6t^6+4t^8 \\ & +t^{10} & 0 \\ t^3+2t^5+2t^7+t^9 & 0 & t^5+t^7 \end{bmatrix}$$

$$=\begin{bmatrix} t^3+5t^5+14t^7+27t^9 & & t^5+4t^7+9t^9+14t^{11} \\ +37t^{11}+37t^{13}+27t^{15} & 0 & +14t^{13}+9t^{15}+4t^{17} \\ +14t^{17}+5t^{19}+21t^{21} & & +t^{19} \\ & t^4+7t^6+22t^8 \\ 0 & +41t^{10}+50t^{12} & 0 \\ & +41t^{14}+22t^{16} \\ & +7t^{18}+t^{20} \\ 2t^5+10t^7+24t^9 \\ +36t^{11}+36t^{13}+24t^{15} & 0 & 2t^7+8t^9+14t^{11} \\ +10t^{17}+2t^{19} & & +14t^{13}+8t^{15}+2t^{17} \end{bmatrix}$$

Therefore

$$\begin{aligned}\operatorname{tr}(QPKPKP)=&\,t^3+t^4+5t^5+7t^6+16t^7+22t^8+35t^9+41t^{10}+51t^{11}\\ &+50t^{12}+51t^{13}+41t^{14}+35t^{15}+22t^{16}+16t^{17}+7t^{18}\\ &+5t^{19}+t^{20}+t^{21}.\end{aligned}$$

Therefore

$$\begin{aligned}\operatorname{tr}(PAPBPB)\times 3=\operatorname{tr}(QPKPKP)\times 3=&\,3t^3+3t^4+15t^5+21t^6+48t^7\\ &+66t^8+105t^9+123t^{10}+153t^{11}+150t^{12}+153t^{13}\cdots.\end{aligned}$$

Lastly

$$\begin{aligned}&\operatorname{tr}(PAPCPC)\times 3+\operatorname{tr}(PBPCPC)\times 3+\operatorname{tr}(PCPCPC)\\ &=\operatorname{tr}(3APCPCP)+\operatorname{tr}(3BPCPCP)+\operatorname{tr}(PCPCPC)\\ &=\operatorname{tr}(3t^4AP^3)+\operatorname{tr}(3t^4BP^3)+\operatorname{tr}(t^6P^3)\\ &=\operatorname{tr}((3t^4A+3t^4D+t^6I_3)P^3).\end{aligned}$$

Now we put

$$3t^4A+3t^3B+t^3I_3=R,$$

then our 3-ple sum is equal to $\operatorname{tr}(RP^3)$, and since

$$R=\begin{bmatrix} 3t^4+t^6+3t^8 & 3t^5+3t^7 & 3t^6 \\ 3t^5+3t^7 & 3t^5+4t^6+3t^8 & 3t^5+3t^7 \\ 3t^6 & 3t^5+3t^7 & 3t^4+4t^6+3t^8 \end{bmatrix}$$

$$P^3=\begin{bmatrix} 1+t^4 & 0 & t^2 \\ 0 & t+t^3 & 0 \\ t^2 & 0 & 0 \end{bmatrix}^3$$

$$=\begin{bmatrix} 1+5t^4+5t^8+t^{12} & 0 & t^2+3t^6+t^{10} \\ 0 & t^3+3t^5+3t^7+t^9 & 0 \\ t^2+3t^6+t^{10} & 0 & t^4+t^8 \end{bmatrix},$$

$\therefore$ We have

$$\operatorname{tr}(RP^3)=\operatorname{tr}\begin{bmatrix} 3t^4+t^6+3t^8 & 3t^5+3t^7 & 3t^6 \\ 3t^5+3t^7 & 3t^4+4t^6+3t^8 & 3t^5+3t^7 \\ 3t^6 & 3t^5+3t^7 & 3t^4+4t^6+3t^8 \end{bmatrix}$$

$$\times\begin{bmatrix} 1+5t^4+5t^8+t^{12} & 0 & t^2+3t^6+t^{10} \\ 0 & t^3+3t^5+3t^7+t^9 & 0 \\ t^2+3t^6+t^{10} & 0 & t^4+t^8 \end{bmatrix}$$

$$=\operatorname{tr}\begin{bmatrix} \begin{array}{l} 3t^4+t^6+21t^8+5t^{10} \\ +39t^{12}+5t^{14}+21t^{16} \\ +t^{18}+3t^{20} \end{array} & \begin{array}{l} 3t^6+12t^{10}+18t^{12} \\ +12t^{14}+3t^{16} \end{array} & \begin{array}{l} 3t^6+t^8+15t^{10} \\ +3t^{12}+15t^{14} \\ +t^{16}+3t^{18} \end{array} \\ ? & \begin{array}{l} 3t^7+13t^9+24t^{11} \\ +24t^{13}+13t^{15} \\ +3t^{17} \end{array} & ? \\ ? & ? & \begin{array}{l} 6t^8+4t^{10}+15t^{12} \\ +4t^{14}+6t^{16} \end{array} \end{bmatrix},$$

$$=3t^4+t^6+3t^7+27t^8+13t^9+9t^{10}+24t^{11}+54t^{12}+24t^{13}+9t^{14}$$
$$+13t^{15}+27t^{16}+3t^{17}+t^{18}+3t^{20}.$$

Therefore

$$\operatorname{tr}(PA)^3+\operatorname{tr}(PAPBPB)\times 3+\operatorname{tr}(RP^3)=(1+4t^3+21t^4+21t^5+25t^6$$
$$+69t^7+159t^8+153t^9+147t^{10}+225t^{11}+310t^{12}+225t^{13}+147t^{14}$$
$$+153t^{15}+159t^{16}+69t^{17}+25t^{18}+21t^{19}+21t^{20}+4t^{21}+t^{24}).$$

This is the $F(t)$. This polymer is not rigid for any S.

II) Hexahedron

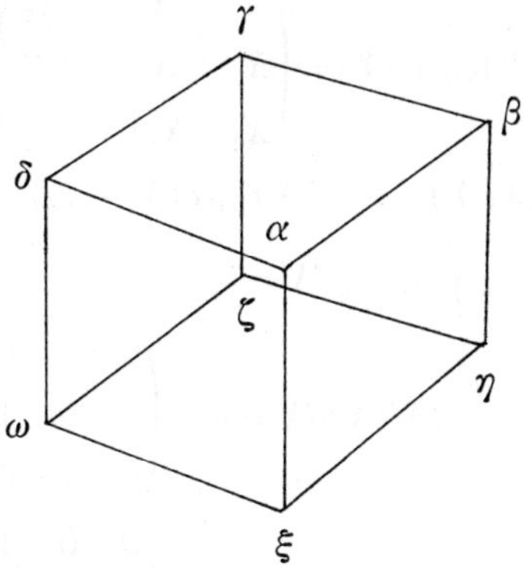

$\mathscr{G}$ = hexahedral group $\cong S_4$

$$P=\{\alpha, \beta\}+\{\beta, \gamma\}+\{\gamma, \delta\}+\{\delta, \alpha\}+\{\alpha, \xi\}+\{\beta, \eta\}+\{\gamma, \zeta\}+\{\delta, \omega\} +\{\xi, \eta\}+\{\eta, \zeta\}+\{\zeta, \omega\}+\{\omega, \xi\}$$

$$\begin{aligned} F(t)&=(f(\alpha, \beta)f(\beta, \gamma)f(\gamma, \delta)f(\delta, \alpha)f(\alpha, \xi)f(\beta, \eta)f(\gamma, \zeta)f(\delta, \omega)f(\xi, \eta) \\ &\quad\times f(\eta, \zeta)f(\zeta, \omega)f(\omega, \xi), 1) \\ &=([\operatorname{tr}(PM(\alpha)PM(\beta)PM(\gamma)PM(\delta))][\operatorname{tr}(PM(\xi)PM(\eta)PM(\zeta)PM(\omega))] \\ &\quad\times f(\alpha, \xi)f(\beta, \eta)f(\gamma, \zeta)f(\delta, \omega), 1) \end{aligned}$$

Lemma.

$$\begin{aligned} \operatorname{tr}(ACB)\operatorname{tr}(A'C'B')&=\operatorname{tr}[(ACB)\otimes(A'C'B')] \\ &=\operatorname{tr}[(A\otimes A')(C\otimes C')(B\otimes B')]. \end{aligned}$$

∵ Obvious □

Therefore

$$F(t)=(\operatorname{tr}(P^{(2)}M(\alpha, \xi)P^{(2)}M(\beta, \eta)P^{(2)}M(\gamma, \zeta)P^{(2)}M(\delta, \omega)) \times f(\alpha, \xi)f(\beta, \eta)f(\gamma, \zeta)f(\delta, \omega), 1)$$

where:

$$P^{(2)}=P\otimes P$$
$$M(X, Y)=M(X)\otimes M(Y).$$

Now

Lemma.

$$\iint AM(X, Y)Bf(X, Y)dXdY=A[a1\otimes 1+bK\otimes K+c(J\otimes 1+1\otimes J)]B$$

where $a=1+t^4$, $b=t+t^3$, $c=t^2$ *and* A, B *are any* 1×9 *or* 9×1 *constant matrices and*

$$M(X, Y)=M(X)\otimes M(Y)=\begin{pmatrix}1 & X & X_2\\ X & X^2 & XX_2\\ X_2 & X_2X & X_2^2\end{pmatrix}\otimes\begin{pmatrix}1 & Y & Y_2\\ Y & Y^2 & YY^2\\ Y_2 & Y_2Y & Y_2^2\end{pmatrix}.$$

$\because$ Since $f(X, Y)=a+bXY+c(X_2+Y_2)$,

$$\int M(X)dX=1_3$$

$$\int XM(X)dX=K=\begin{pmatrix}0 & 1 & 0\\ 1 & 0 & 1\\ 0 & 1 & 0\end{pmatrix}$$

$$\int X_2M(X)dX=J=\begin{pmatrix}0 & 0 & 1\\ 0 & 1 & 0\\ 1 & 0 & 1\end{pmatrix}$$

$$\iint AM(X, Y)Bf(X, Y)dXdY$$

$$=A\left(\iint (a+bXY+dX_2+Y_2)M(X)\otimes M(Y)dXdY\right)B$$

$$=A\Big[a\int M(X)dX\otimes\int M(Y)dY+b\int XM(X)dX\otimes\int YM(Y)dY$$

$$+c\int X_2M(X)dX\otimes\int M(Y)dY+c\int M(X)dX\otimes\int Y_2M(Y)dY\Big]B$$

$$=A[a1_3\otimes 1_3+bK\otimes K+c(J\otimes 1_3+1_3\otimes J)]B.$$

Q.E.D.

Therefore

$$F(t)=\mathrm{tr}(P^{(2)}QP^{(2)}QP^{(2)}QP^{(2)}Q)$$

where

$$Q=a1\otimes 1+bK\otimes K+c(J\otimes 1+1\otimes J)$$

$$=\begin{bmatrix}1+t^4 & & & & & & & & \\ & 1+t^4 & & & & & & 0 & \\ & & 1+t^4 & & & & & & \\ & & & 1+t^4 & & & & & \\ & & & & 1+t^4 & & & & \\ & & & & & 1+t^4 & & & \\ & & 0 & & & & 1+t^4 & & \\ & & & & & & & 1+t^4 & \\ & & & & & & & & 1+t^4\end{bmatrix}$$

$$
+(t+t^{3})\left[\begin{array}{ccc|ccc|ccc}
 & & & 0 & 1 & 0 & & & \\
 & & & 1 & 0 & 1 & & & \\
 & & & 0 & 1 & 0 & & & \\
\hline
0 & 1 & 0 & & & & 0 & 1 & 0 \\
1 & 0 & 1 & & & & 1 & 0 & 1 \\
0 & 1 & 0 & & & & 0 & 1 & 0 \\
\hline
 & & & 0 & 1 & 0 & & & \\
 & & & 1 & 0 & 1 & & & \\
 & & & 0 & 1 & 0 & & &
\end{array}\right]
$$

$$
+t^{2}\left[\left[\begin{array}{ccc|ccc|ccc}
0 & 0 & 1 & & & & & & \\
0 & 1 & 0 & & & & & & \\
1 & 0 & 1 & & & & & & \\
\hline
 & & & 0 & 0 & 1 & & & \\
 & & & 0 & 1 & 0 & & & \\
 & & & 1 & 0 & 1 & & & \\
\hline
 & & & & & & 0 & 0 & 1 \\
 & & & & & & 0 & 1 & 0 \\
 & & & & & & 1 & 0 & 1
\end{array}\right]+\left[\begin{array}{ccc|ccc|ccc}
 & & & & & & 1 & 0 & 0 \\
 & 0 & & & 0 & & 0 & 1 & 0 \\
 & & & & & & 0 & 0 & 1 \\
\hline
 & & & 1 & 0 & 0 & & & \\
 & & & 0 & 1 & 0 & & & \\
 & & & 0 & 0 & 1 & & & \\
\hline
1 & 0 & 0 & & & & 1 & 0 & 0 \\
0 & 1 & 0 & & 0 & & 0 & 1 & 0 \\
0 & 0 & 1 & & & & 0 & 0 & 1
\end{array}\right]\right]
$$

$$
=\left[\begin{array}{ccc|ccc|ccc}
1+t^{4} & 0 & t^{2} & 0 & t+t^{3} & 0 & t^{2} & 0 & 0 \\
0 & 1+t^{2}+t^{4} & 0 & t+t^{3} & 0 & t+t^{3} & 0 & t^{3} & 0 \\
t^{2} & 0 & 1+t^{2}+t^{4} & 0 & t+t^{3} & 0 & 0 & 0 & t^{2} \\
\hline
0 & t+t^{3} & 0 & 1+t^{2}+t^{4} & 0 & t^{2} & 0 & t+t^{2} & 0 \\
t+t^{3} & 0 & t+t^{3} & 0 & 1+2t^{2}+t^{4} & 0 & t+t^{3} & 0 & t+t^{3} \\
0 & t+t^{3} & 0 & t^{2} & 0 & 1+2t^{2}+t^{4} & 0 & t+t^{3} & 0 \\
\hline
t^{2} & 0 & 0 & 0 & t+t^{3} & 0 & 1+t^{2}+t^{4} & 0 & t^{2} \\
0 & t^{2} & 0 & t+t^{3} & 0 & t+t^{3} & 0 & 1+2t^{2}+t^{4} & 0 \\
0 & 0 & t^{2} & 0 & t+t^{3} & 0 & t^{2} & 0 & 1+2t^{2}+t^{4}
\end{array}\right].
$$

Now

$$P^{(2)}=P\otimes P=\begin{bmatrix}1+t^4 & 0 & t^2\\ 0 & t+t^3 & 0\\ t^2 & 0 & 0\end{bmatrix}\otimes\begin{bmatrix}1+t^4 & 0 & t^2\\ 0 & t+t^3 & 0\\ t^2 & 0 & 0\end{bmatrix}$$

$$=\left[\begin{array}{ccc|ccc|ccc}
(1+t^4)^2 & 0 & (1+t^4)t^2 & & & & t^2(1+t^4) & 0 & t^4\\
0 & (1+t^4)\times(t+t^3) & 0 & & 0 & & 0 & t^2(t+t^3) & 0\\
(1+t^4)t^2 & 0 & 0 & & & & t^4 & 0 & 0\\
\hline
 & & & (t+t^3)\times(1+t^4) & 0 & (t+t^3)t^2 & & & \\
 & 0 & & 0 & (t+t^3)^2 & 0 & & 0 & \\
 & & & (t+t^3)t^2 & 0 & 0 & & & \\
\hline
t^2(1+t^4) & 0 & t^4 & & & & & & \\
0 & t^2(t+t^3) & 0 & & 0 & & & 0 & \\
t^4 & 0 & 0 & & & & & &
\end{array}\right]$$

$$=\left[\begin{array}{ccc|ccc|ccc}
1+2t^4+t^8 & 0 & t^2+t^6 & & & & t^2+t^6 & 0 & t^4\\
0 & t+t^3+t^5+t^7 & 0 & & 0 & & 0 & t^3+t^5 & 0\\
t^2+t^6 & 0 & 0 & & & & t^4 & 0 & 0\\
\hline
 & & & t+t^3+t^5+t^7 & 0 & t^3+t^5 & & & \\
 & 0 & & 0 & t^2+2t^4+t^6 & 0 & & 0 & \\
 & & & t^3+t^5 & 0 & 0 & & & \\
\hline
t^2+t^6 & 0 & t^4 & & & & & & \\
0 & t^3+t^5 & 0 & & 0 & & & 0 & \\
t^4 & 0 & 0 & & & & & &
\end{array}\right]$$

$$
P^{(2)}Q=\left(\begin{array}{ccc|ccc|ccc}
1+5t^4+5t^8+t^{12} & & 2t^2+t^4+5t^6+t^8+2t^{10} & & t+3t^3+5t^5+5t^7+3t^9+t^{11} & & 2t^2+t^4+5t^6+t^8+2t^{10} & & 3t^4+4t^6+3t^8 \\
 & t+2t^3+4t^5+4t^7+2t^9+t^{11} & & t^2+3t^4+4t^6+3t^8+t^{10} & & t^2+3t^4+4t^6+3t^8+t^{10} & & 2t^3+4t^5+4t^7+2t^9 & \\
t^2+3t^6+t^{10} & & t^4+t^8 & & t^3+2t^5+2t^7+t^9 & & 2t^4+t^6+2t^8 & & t^6 \\
\hline
 & t^2+3t^4+4t^6+3t^8+t^{10} & & t+2t^3+4t^5+4t^7+2t^9+t^{11} & & 2t^3+4t^5+4t^7+2t^9 & & t^2+3t^4+4t^6+3t^8+t^{10} & \\
t^3+3t^5+3t^7+t^9 & & t^3+3t^5+3t^7+t^9 & & t^2+4t^4+6t^6+4t^8+t^{10} & & t^3+3t^5+3t^7+t^9 & & t^3+3t^5+3t^7+t^9 \\
 & t^4+2t^6+t^8 & & t^3+2t^5+2t^7+t^9 & & t^5+t^7 & & t^4+2t^6+t^8 & \\
\hline
t^2+3t^6+t^{10} & 0 & 2t^4+t^6+2t^8 & 0 & t^3+2t^5+2t^7+t^9 & 0 & t^4+t^8 & 0 & t^6 \\
0 & t^3+2t^5+2t^7+t^9 & 0 & t^4+2t^6+t^8 & 0 & t^3+2t^5+t^8 & 0 & t^4+t^8 & 0 \\
t^4+t^8 & 0 & t^6 & 0 & t^5+t^7 & 0 & t^6 & 0 & 0
\end{array}\right)
$$

We put this $P^{(2)}Q=A$, then

$$F(t)=\mathrm{tr}(A^4).$$

Since $A\sim A_0+A_1$, where

$$A_0=\begin{pmatrix} 1+5t^4+5t^8+t^{12} & 2t^2+t^4+5t^6+t^8+2t^{10} & t+3t^3+5t^5+5t^7+3t^9+t^{11} & 2t^2+t^4+5t^6+t^8+2t^{10} & 3t^4+4t^6+3t^8 \\ t^2+3t^6+t^{10} & t^4+t^8 & t^3+2t^5+2t^7+t^9 & 2t^4+t^6+2t^8 & t^6 \\ t^3+3t^5+3t^7+t^9 & t^3+3t^5+3t^7+t^9 & t^2+4t^4+6t^6+4t^8+t^{10} & t^3+3t^5+3t^7+t^9 & t^3+3t^5+3t^7+t^9 \\ t^2+3t^6+t^{10} & 2t^4+t^6+2t^8 & t^3+2t^5+2t^7+t^9 & t^4+t^8 & t^6 \\ t^4+t^8 & t^6 & t^5+t^7 & t^6 & 0 \end{pmatrix}$$

$$A_1=\begin{pmatrix} t+2t^3+4t^5+4t^7+2t^9+t^{11} & t^2+3t^4+4t^6+3t^8+t^{10} & t^2+3t^4+4t^6+3t^8+t^{10} & 2t^3+4t^5+4t^7+2t^9 \\ t^2+3t^4+4t^6+3t^8+t^{10} & t+2t^3+4t^5+4t^7+2t^9+t^{11} & 2t^3+4t^5+4t^7+2t^9 & t^2+3t^4+4t^6+3t^8+t^{10} \\ t^4+2t^6+t^8 & t^3+2t^5+2t^7+t^9 & t^5+t^7 & t^4+2t^6+t^8 \\ t^3+2t^5+2t^7+t^9 & t^4+2t^6+t^8 & t^4+2t^6+t^8 & t^5+t^7 \end{pmatrix}$$

Therefore

$$F(t)=\mathrm{tr}(A_0^4)+\mathrm{tr}(A_1^4).$$

We know that for odd power t^ν the coefficient C_ν in $F(t)=\mathrm{tr}(A^4)=\sum C_\nu t^\nu$ is $0\leqq C_\nu=\dim\wedge^\nu(F)^G\leqq\dim\wedge^\nu(\mu F)^G=\dim H^\nu(A_\lambda Q)=0$, for a large μ. So we calculate $\mathrm{tr}\, A^4=\sum C_\nu t^\nu$ only for even power.

We put $A_0^+=$ the even power polynomial part of A_0.

$$=\begin{pmatrix} 1+5t^4+5t^8+t^{10} & 2t^2+t^4+5t^6+t^8+2t^{10} & 0 & 2t^2+t^4+5t^6+t^8+2t^{10} & 3t^4+4t^6+3t^8 \\ t^2+3t^6+t^{10} & t^4+t^8 & 0 & 2t^4+t^6+2t^8 & t^6 \\ 0 & 0 & t^2+4t^4+6t^6+4t^8+t^{10} & 0 & 0 \\ t^2+3t^6+t^{10} & 2t^4+t^6+2t^8 & 0 & t^4+t^8 & t^6 \\ t^4+t^8 & t^6 & 0 & t^6 & 0 \end{pmatrix}$$

We put $A_0^- =$ the odd power polynomial part of A_0

$$= \begin{pmatrix} 0 & 0 & \begin{matrix} t+3t^3+5t^5 \\ 5t^7+3t^9+t^{11} \end{matrix} & 0 & 0 \\ 0 & 0 & \begin{matrix} t^3+2t^5+2t^7 \\ +t^9 \end{matrix} & 0 & 0 \\ \begin{matrix} t^3+3t^5+3t^7 \\ +t^9 \end{matrix} & \begin{matrix} t^3+3t^5+3t^7 \\ +t^9 \end{matrix} & 0 & \begin{matrix} t^3+3t^5+3t^7 \\ +t^9 \end{matrix} & \begin{matrix} t^3+3t^5+3t^7 \\ +t^9 \end{matrix} \\ 0 & 0 & \begin{matrix} t^3+2t^5+2t^7 \\ +t^9 \end{matrix} & 0 & 0 \\ 0 & 0 & t^5+t^7 & 0 & 0 \end{pmatrix}$$

$$A_0 = A_0^+ + A_0^-$$

$$A_0^2 = A_0^{+2} + A_0^+A_0^- + A_0^-A_0^+ + A_0^{-2}$$

$$\begin{aligned} A_0^4 = {} & A_0^{+4} + A_0^{+3}A_0^- + A_0^{+2}A_0^-A_0^+ + A_0^{+2}A_0^{-2} + A_0^+A_0^-A_0^{+2} + A_0^+A_0^-A_0^+A_0^- \\ & + A_0^+A_0^-A_0^-A_0^+ + A_0^+A_0^-A_0^-A_0^- + A_0^-A_0^+A_0^+A_0^+ + A_0^-A_0^+A_0^+A_0^- \\ & + A_0^-A_0^+A_0^-A_0^+ + A_0^-A_0^+A_0^-A_0^- + A_0^-A_0^-A_0^+A_0^+ + A_0^-A_0^-A_0^+A_0^- \\ & + A_0^-A_0^-A_0^-A_0^+ + A_0^-A_0^-A_0^-A_0^- \\ \equiv {} & A_0^{+4} + A_0^{-4} + A_0^+A_0^-A_0^+A_0^- + A_0^+A_0^-A_0^-A_0^+ + A_0^-A_0^+A_0^+A_0^- \\ & + A_0^-A_0^+A_0^-A_0^+ + A_0^-A_0^-A_0^+A_0^+ + A_0^+A_0^+A_0^-A_0^- \\ & \text{(mod odd power of } t\text{)}, \end{aligned}$$

$$\operatorname{tr} A_0^4 \equiv \operatorname{tr} A_0^{+4} + \operatorname{tr} A_0^{-4} + \operatorname{tr}(A_0^+A_0^-A_0^+A_0^-)\times 2 + \operatorname{tr}(A_0^{+2}A_0^{-2})\times 4.$$

Now since

$$A_0^+ = \begin{pmatrix} a(1,1) & a(1,2) & 0 & a(1,4) & a(1,5) \\ a(2,1) & a(2,2) & 0 & a(2,4) & a(2,5) \\ 0 & 0 & a(3,3) & 0 & 0 \\ a(4,1) & a(4,2) & 0 & a(4,4) & a(4,5) \\ a(5,1) & a(5,2) & 0 & a(5,4) & a(5,5) \end{pmatrix},$$

$$A_0^- = \begin{pmatrix} 0 & 0 & b(1,3) & 0 & 0 \\ 0 & 0 & b(2,3) & 0 & 0 \\ b & b & 0 & b & b \\ 0 & 0 & b(4,3) & 0 & 0 \\ 0 & 0 & b(5,3) & 0 & 0 \end{pmatrix}$$

$$a(1,1)=1+5t^4+5t^8+t^{12} \qquad a(1,2)=2t^2+t^4+5t^6+t^8+2t^{10}$$
$$a(2,1)=t^2+3t^6+t^{10} \qquad a(2,2)=t^4+t^8$$
$$a(4,1)=t^2+3t^6+t^{10} \qquad a(4,2)=2t^4+t^6+2t^8$$
$$a(5,1)=t^4+t^8 \qquad a(5,2)=t^6$$
$$a(3,3)=t^2+4t^4+6t^6+4t^8+t^{10}$$
$$b(1,3)=t+3t^3+5t^5+5t^7+3t^9+t^{11}$$
$$b(2,3)=t^3+2t^5+2t^7+t^9$$
$$b(4,3)=t^3+2t^5+2t^7+t^9$$
$$b(5,3)=t^5+t^7$$
$$b \qquad =t^3+3t^5+3t^7+t^9.$$

$$\therefore \quad A_0^+A_0^-=\begin{pmatrix} 0 & 0 & c(1,3) & 0 & 0 \\ 0 & 0 & c(2,3) & 0 & 0 \\ a(3,3)b & a(3,3)b & 0 & a(3,3)b & a(3,3)b \\ 0 & 0 & c(4,3) & 0 & 0 \\ 0 & 0 & c(5,3) & 0 & 0 \end{pmatrix}=B$$

where
$$c(1,3)=\sum a(1,k)b(k,3)$$
$$c(2,3)=\sum a(2,k)b(k,3)$$
$$c(4,3)=\sum a(4,k)b(k,3)$$
$$c(5,3)=\sum a(5,k)b(k,3).$$

We put this matrix B.

$$\begin{aligned}
\therefore \quad \operatorname{tr}(B^2) &= \operatorname{tr} A_0^+A_0^-A_0^+A_0^- \\
&= c(1,3)a(3,3)b+c(2,3)a(3,3)b+a(3,3)c(1,3)b \\
&\quad +a(3,3)c(2,3)b+a(3,3)c(4,3)b+a(3,3)c(5,3)b \\
&\quad +c(4,3)a(3,3)b+c(5,3)a(3,3)b \\
&= 2[c(1,3)+c(2,3)+c(4,3)+c(5,3)]a(3,3)b \\
&= 2[\textstyle\sum_k a(1,k)b(k,3)+\sum a(2,k)b(k,3) \\
&\quad +\textstyle\sum a(4,k)b(k,3)+\sum a(5,k)b(k,3)]a(3,3)b \\
&= 2[\textstyle\sum_k (a(1,k)+a(2,k)+a(4,k)+a(5,k))b(k,3)]a(3,3)b \\
&= 2[(1+2t^2+6t^4+6t^6+6t^8+2t^{10}+t^{12}) \\
&\quad \times(t+3t^3+5t^5+5t^7+3t^9+t^{11}) \\
&\quad +(2t^2+4t^4+7t^6+4t^8+2t^{10})(t^3+2t^5+2t^7+t^9) \\
&\quad +(2t^2+4t^4+7t^6+4t^8+2t^{10})(t^3+2t^5+2t^7+t^9)
\end{aligned}$$

$$+(3t^4+6t^6+3t^8)(t^5+t^7)](t^2+4t^4+6t^6+4t^8+t^{10})$$
$$\times(t^3+3t^5+3t^7+t^9)$$
$$=2\times(t^6+12t^8+77t^{10}+342t^{12}+1144t^{14}+2994t^{16}+6256t^{18}$$
$$+10546t^{20}+14410t^{22}+15983t^{24}+14410t^{26}+10546t^{28}$$
$$+6256t^{30}+2994t^{32}+1144t^{34}+342t^{36}+77t^{38}+12t^{40}+t^{42}).$$

Now

$$A_0^{-2}=\begin{bmatrix} b(1,3)b & b(1,3)b & 0 & b(1,3)b & b(1,3)b \\ b(2,3)b & b(2,3)b & 0 & b(2,3)b & b(2,3)b \\ 0 & 0 & \triangle & 0 & 0 \\ b(4,3)b & b(4,3)b & 0 & b(4,3)b & b(4,3)b \\ b(5,3)b & b(5,3)b & 0 & b(5,3)b & b(5,3)b \end{bmatrix}$$

where $\triangle=b\cdot b(1,3)+b\cdot b(2,3)+b\cdot b(4,3)+b\cdot b(5,3)$,

$$\therefore\quad A_0^{-2}=\begin{bmatrix} b(1,3) & b(1,3) & 0 & b(1,3) & b(1,3) \\ b(2,3) & b(2,3) & 0 & b(2,3) & b(2,3) \\ 0 & 0 & \square & 0 & 0 \\ b(4,3) & b(4,3) & 0 & b(4,3) & b(4,3) \\ b(5,3) & b(5,3) & 0 & b(5,3) & b(5,3) \end{bmatrix}\times b$$

where $\square=b(1,3)+b(2,3)+b(4,3)+b(5,3)$.

$$\therefore\quad A_0^{-4}=\begin{bmatrix} \begin{matrix} b_1^2+b_1b_2 \\ +b_1b_4+b_1b_5 \end{matrix} & // & 0 & // & // \\ // & \begin{matrix} b_2b_1+b_2^2 \\ +b_2b_4+b_2b_5 \end{matrix} & 0 & // & // \\ 0 & 0 & \square^2 & 0 & 0 \\ // & // & 0 & \begin{matrix} b_4b_1+b_4b_2 \\ +b_4^2+b_4b_5 \end{matrix} & // \\ // & // & 0 & // & \begin{matrix} b_5b_1+b_5b_2 \\ +b_5b_4+b_5^2 \end{matrix} \end{bmatrix} b^2$$

where

$$b_i=b(1,3)\quad(i=1,2,4,5).$$

$$\therefore\quad \mathrm{tr}(A_0^{-4})=(b_1^2+b_1b_2+\cdots+\square^2+\cdots+b_5b_1+\cdots+b_5^2)b^2$$
$$=\{(b_1+b_2+b_4+b_5)^2+\square^2\}b^2=2\times(b_1+b_2+b_4+b_5)^2\times b^2.$$

Therefore

$$\begin{aligned}\operatorname{tr}((A_0^-)^4)&=2\times(t+5t^3+10t^5+10t^7+5t^9+t^{11})^2+(t^3+3t^5+3t^7+t^9)^2\\&=2\times(t^8+16t^{10}+120t^{12}+560t^{14}+1820t^{16}+4368t^{18}+8008t^{20}\\&\quad+11440t^{22}+12878t^{24}+11440t^{26}+8008t^{28}+4368t^{30}\\&\quad+1820t^{32}+560t^{34}+120t^{36}+16t^{38}+t^{40}).\end{aligned}$$

Next put

$$A_0^{+2}=4(B(ij))\oplus(b)$$
$$A^{-2}\sim(C(ij))\oplus(\Delta).$$

We know $B(i,j)$, $C(i,j)$; b, Δ are polynomial of t. Now

$$(C(i,j))=\left[\begin{array}{cc|cc} b(1,3) & b(1,3) & b(1,3) & b(1,3)\\ b(2,3) & b(2,3) & b(2,3) & b(2,3)\\ \hline b(4,3) & b(4,3) & b(4,3) & b(4,3)\\ b(5,3) & b(5,3) & b(5,3) & b(5,3)\end{array}\right]b$$

and $(B(i,j))$ is known by a computer.

Namely, if we put,

$$(A_0^+)^2\sim\left[\begin{array}{cc|cc} f_{11}(t) & f_{12}(t) & f_{14}(t) & f_{15}(t)\\ f_{21}(t) & f_{22}(t) & f_{24}(t) & f_{25}(t)\\ \hline f_{41}(t) & f_{42}(t) & f_{44}(t) & f_{45}(t)\\ f_{51}(t) & f_{52}(t) & f_{54}(t) & f_{55}(t)\end{array}\right]\oplus(f(t))$$

and put

$$\sum_i f_{ij}(t)=F_j(t).$$

Then

$$\begin{aligned}F_1(t)&=4+0t^2+56t^4+8t^6+240t^8+48t^{10}+144t^{12}+48t^{14}+240t^{16}\\&\quad+8t^{16}+56t^{20}+0t^{22}+4t^{24}\\F_2(t)&=0+8t^2+4t^4+84t^6+33t^8+248t^{10}+100t^{12}+248t^{14}+44t^{16}\\&\quad+84t^8+4t^{20}+8t^{22}+0t^{24}\\F_4(t)&=0+8t^2+4t^4+84t^6+44t^8+248t^{10}+100t^{12}+248t^{14}+44t^{16}\\&\quad+84t^{18}+4t^{20}+8t^{22}+0t^{24}=F_2(t)\end{aligned}$$

$$F_5(t)=0+0t^2+12t^4+16t^6+88t^8+88t^{10}+160t^{12}+88t^{14}+88t^{16}$$
$$+16t^{18}+12t^{20}+0t^{22}+0t^{24}.$$

Since

$$A_0^{-2}=(C(ij))+(\Delta)=\begin{bmatrix} b(1,3) & b(1,3) & b(1,3) & b(1,3) \\ b(2,3) & b(2,3) & b(2,3) & b(2,3) \\ b(4,3) & b(4,3) & b(4,3) & b(4,3) \\ b(5,3) & b(5,3) & b(5,3) & b(5,3) \end{bmatrix} b\oplus(\Delta),$$

$$\mathrm{tr}[(A_0^{+2})(A_0^{-2})]-\Delta f$$
$$=\mathrm{tr}(f(ij))(C(ij))=\sum_{i,j} f(i,j)b(j,3)b=\sum_j F_j\cdot b(j,3)\cdot b$$
$$=[F_1(t)b(1,3)+F_2(t)b(2,3)+F_4b(4,3)+F_5b(5,3)]\times b$$
$$=(4t+12t^3+92t^5+236t^7+784t^9+1376t^{11}+2816t^{13}+3712t^{15}$$
$$+4296t^{17}+4296t^{19}+3712t^{21}+2816t^{23}+1376t^{25}+784t^{27}$$
$$+236t^{29}+92t^{31}+12t^{33}+4t^{35})(t^3+3t^5+3t^7+t^9)$$
$$=4t^4+24t^6+140t^8+552t^{10}+1780t^{12}+4528t^{14}+11532t^{16}$$
$$+17072t^{18}+25256t^{20}+31136t^{22}+33200t^{24}+31136t^{26}$$
$$+25256t^{28}+17072t^{30}+11532t^{32}+4528t^{34}+1780t^{36}$$
$$+552t^{38}+140t^{40}+24t^{42}+4t^{44}$$
$$=X.$$
$$\therefore\quad \mathrm{tr}(A_0^{+2})(A_0^{-2})=X+\Delta f.$$

$$\Delta=\Box b=b(b(1,3)+b(2,3)+b(4,3)+b(5,3))$$
$$=(t^3+3t^5+3t^7+t^9)(t+5t^3+10t^5+10t^7+5t^9+t^{11})$$
$$=(t^2+4t^4+6t^6+4t^8+t^{10})^2=\{(t^2(1+t^2)^2)\}$$
$$\Delta=t^4+8t^6+28t^8+56t^{10}+70t^{12}+56t^{14}+28t^{16}+8t^{18}+t^{20}$$
$$f=t^4+8t^6+28t^8+56t^{10}+70t^{12}+56t^{14}+28t^{16}+8t^{18}+t^{20}=\Delta$$
$$\Delta f=(t^2(1+t^2)^4)^4=t^8(1+t^2)^{16}$$
$$=t^8+16t^{10}+120t^{12}+560t^{14}+1820t^{16}+4368t^{18}+8008t^{20}$$
$$+11440t^{22}+12870t^{24}+11440t^{26}+8008t^{28}+4368t^{30}$$
$$+1820t^{32}+560t^{34}+120t^{36}+16t^{38}+t^{40}.$$

Now by a computer,

$$\mathrm{tr}(A_0^{+})^4-g^4=1+36t^4+8t^6+510t^8+200t^{10}+3692t^{12}+1816t^{14}$$
$$+14899t^{16}+7576t^{18}+34452t^{20}+15400t^{22}+45622t^{24}$$

$$+15400t^{26}+34452t^{28}+7576t^{30}+14899t^{32}+1816t^{34}$$
$$+3692t^{36}+200t^{38}+510t^{40}+8t^{42}+36t^{44}+t^{46}$$

with $g=t^2(1+t^2)4=t^2+4t^4+6t^6+4t^8+t^{10}$

$$g^4=t^8(1+t^2)^{16}=f^2$$
$$=t^8+16t^{10}+120t^{12}+560t^{14}+1820t^{16}+4368t^{18}+8008t^{20}$$
$$+11440t^{22}+12870t^{24}+11440t^{26}+8008t^{28}+4368t^{30}+1820t^{32}$$
$$+560t^{34}+120t^{36}+16t^{38}+t^{40}.$$

Now

$$\operatorname{tr} A_0^{+4}=1+36t^4+8t^6+510t^8+200t^{10}+3692t^{12}+1816t^{14}$$
$$+14899t^{16}+7976t^{18}+34452t^{20}+15400t^{22}+45622t^{24}$$
$$+15400t^{26}+34452t^{28}+7976t^{30}+14899t^{32}+1816t^{34}$$
$$+3692t^{36}+200t^{38}+510t^{40}+8t^{42}+36t^{44}+t^{48}+g^4$$

$$\operatorname{tr} A_0^{-4}=2t^8+32t^{10}+240t^{12}+1120t^{14}+3640t^{16}+8736t^{18}$$
$$+16016t^{20}+22880t^{22}+25756t^{24}+22880t^{26}+16016t^{28}$$
$$+8736t^{30}+3640t^{32}+1120t^{34}+240t^{36}+32t^{38}+2t^{40}$$

$$\operatorname{tr}(A_0^{+2}A_0^{-2})\times 4=16t^4+96t^6+560t^8+2208t^{10}+7120t^{16}+18112t^{18}$$
$$+46128t^{16}+68288t^{18}+101024t^{20}+124544t^{22}+132800t^{24}$$
$$+124544t^{26}+101024t^{28}+68288t^{30}+46128t^{32}+18112t^{34}$$
$$+7120t^{36}+2208t^{38}+560t^{40}+96t^{42}+16t^{44}+4\Delta f$$

$$\operatorname{tr}(A_0^+A_0^-)^2\times 2=4t^6+48t^8+308t^{10}+1368t^{12}+4576t^{14}+11976t^{16}+25024t^{18}$$
$$+42184t^{20}+57640t^{22}+63952t^{24}+57640t^{26}+42184t^{28}$$
$$+25024t^{30}+11976t^{32}+4576t^{34}+1368t^{36}+308t^{38}+48t^{40}$$
$$+4t^{42}$$

$$g^4+4\Delta f=5g^4=5t^8+80t^{10}+600t^{12}+2800t^{14}+9100t^{16}+21840t^{18}$$
$$+40040t^{20}+57200t^{22}+64350t^{24}+57200t^{26}+40040t^{28}$$
$$+21840t^{30}+9100t^{32}+2800t^{34}+600t^{36}+80t^{38}+5t^{40}$$

$$\operatorname{tr} A_0^4\equiv\operatorname{tr}(A_0^+)^4+\operatorname{tr}(A_0^-)^4+[\operatorname{tr}(A_0^{+2}A_0^{-2})]\times 4+[\operatorname{tr}(A_0^+A_0^-A_0^+A_0^-)]\times 2$$

(mod odd powers of t)

$$=1+52t^4+108t^6+1125t^8+2828t^{10}+13020t^{12}+28424t^{14}$$
$$+85743t^{16}+131864t^{18}+233720t^{20}+277664t^{22}+332480t^{24}$$
$$+277664t^{26}+233720t^{28}+131864t^{30}+85743t^{32}+28424t^{34}$$
$$+13020t^{36}+2828t^{38}+1125t^{40}+108t^{42}+52t^{44}+t^{48}.$$

Since

$$A_1^+ = \begin{bmatrix} 0 & \begin{matrix} t^2+3t^4+4t^6 \\ +3t^8+t^{10} \end{matrix} & \begin{matrix} t^2+3t^4+4t^6 \\ +3t^8+t^{10} \end{matrix} & 0 \\ \begin{matrix} t^2+3t^4+4t^6 \\ +3t^8+t^{10} \end{matrix} & 0 & 0 & \begin{matrix} t^2+3t^4+4t^6 \\ 3t^8+t^{10} \end{matrix} \\ t^4+2t^6+t^8 & 0 & 0 & t^4+2t^6+t^8 \\ 0 & t^4+2t^6+t^8 & t^4+2t^6+t^8 & 0 \end{bmatrix}$$

$$A_1^+ \sim \begin{pmatrix} a & a \\ b & b \end{pmatrix} \oplus \begin{pmatrix} a & a \\ b & b \end{pmatrix}$$

with $a = t^2+3t^4+4t^6+3t^8+t^{10}$, $b = t^4+2t^6+t^8$.

$$\therefore \quad \operatorname{tr}(A_1^+)^4 = 2\operatorname{tr}\begin{pmatrix} a & a \\ b & b \end{pmatrix}^4 = 2\operatorname{tr}\begin{pmatrix} a^2+ab & a^2+ab \\ ba+b^2 & ba+b^2 \end{pmatrix}^2 = 2\operatorname{tr}\begin{pmatrix} c & c \\ d & d \end{pmatrix}\begin{pmatrix} c & c \\ d & d \end{pmatrix}$$

$$\begin{aligned}
&= 2[(c^2+cd)+(cd+d^2)] = 2(c+d)^2 \\
&= 2(a+b)^4 && c = a^2+ab \\
&= 2(t^2+4t^4+6t^6+4t^8+t^{10})^4 && d = b^2+ab \\
&= 2[t^2(1+t^2)^4]^4 && c+d = (a+b)^2 \\
&= 2g^4 \\
&= 2(t^8+16t^{10}+120t^{12}+560t^{14}+1820t^{16}+4368t^{18} \\
&\quad +8008t^{20}+11440t^{22}+12870t^{24}+11440t^{26}+8008t^{28} \\
&\quad +4368t^{30}+1820t^{32}+560t^{34}+120t^{36}+16t^{38}+t^{40}).
\end{aligned}$$

Similarly

$$A_1^- \sim \begin{pmatrix} \alpha & \beta \\ \gamma & \delta \end{pmatrix} + \begin{pmatrix} \alpha & \beta \\ \gamma & \delta \end{pmatrix}$$

with

$$\begin{aligned}
\alpha &= t+2t^3+4t^5+4t^7+2t^9+t^{11} \\
\beta &= t^3+2t^5+2t^7+t^9 \\
\gamma &= 2t^3+4t^5+4t^7+2t^9 \\
\delta &= t^5+t^7.
\end{aligned}$$

Putting λ_1, λ_2 eigenvalues of $\begin{pmatrix} \alpha & \beta \\ \gamma & \delta \end{pmatrix}$,

$$\begin{aligned}
\operatorname{tr}[(A_1^-)^4] &= 2\operatorname{tr}\left[\begin{pmatrix} \alpha & \beta \\ \gamma & \delta \end{pmatrix}^4\right] \\
&= 2(\lambda_1^4+\lambda_2^4) = 2[(\lambda_1+\lambda_2)^4 - (4\lambda_1^3\lambda_2+6\lambda_1^2\lambda_2^2+4\lambda_1\lambda_2^3)] \\
&= 2[(\alpha+\delta)^4 - \lambda_1\lambda_2(4\lambda_1^2+8\lambda_1\lambda_2+\lambda_2^2-\lambda_1\lambda_2)]
\end{aligned}$$

$$
\begin{aligned}
&=2[(\alpha+\delta)^4-(\alpha\cdot\delta-\beta\cdot\gamma)[4(\lambda_1+\lambda_2)^2-2(\alpha\cdot\delta-\beta\cdot\gamma)]]\\
&=2[(\alpha+\delta)^4-(\alpha\cdot\delta-\beta\cdot\gamma)[4(\alpha+\delta)^2-2(\alpha\cdot\delta-\beta\cdot\gamma)]]\\
&\qquad \alpha+\delta=t+2t^3+5t^5+5t^7+2t^9+t^{11}\\
&\qquad \alpha\cdot\delta-\beta\cdot\gamma=-t^6-5t^8-10t^{10}-12t^{12}-10t^{14}-5t^{16}-t^{18}\\
&=2[(t+2t^3+5t^5+5t^7+2t^9+t^{11})^4\\
&\quad +(t^6+5t^8+10t^{10}+12t^{12}+10t^{14}+5t^{16}+t^{18})\\
&\quad \times(4t^2+16t^4+57t^6+125t^8+206t^{10}+252t^{12}+206t^{14}\\
&\quad +125t^{16}+57t^{18}+16t^{20}+4t^{22})]\\
&=2(t^4+8t^6+48t^8+208t^{10}+711t^{12}+1970t^{14}+4483t^{16}\\
&\quad +8468t^{18}+13271t^{20}+17478t^{22}+19136t^{24}+17478t^{26}\\
&\quad +13271t^{28}+8468t^{30}+4483t^{32}+1970t^{34}+711t^{36}\\
&\quad +208t^{38}+48t^{40}+8t^{42}+t^{44}).
\end{aligned}
$$

Now

$$
A^+A^-=\begin{pmatrix}0&a&a&0\\a&0&0&a\\b&0&0&b\\0&b&b&0\end{pmatrix}\begin{pmatrix}\alpha&0&0&\gamma\\0&\alpha&\gamma&0\\0&\beta&\delta&0\\\beta&0&0&\delta\end{pmatrix}
$$

$$
=\begin{pmatrix}0&a\alpha+a\beta&a\gamma+a\delta&0\\a\alpha+a\beta&0&0&a\gamma+a\delta\\b\alpha+b\beta&0&0&b\gamma+b\delta\\0&b\alpha+b\beta&b\gamma+b\delta&0\end{pmatrix}
$$

$$
\begin{aligned}
\operatorname{tr}(A_1^+A_1^-)^2&=(a\alpha+a\beta)^2+(a\gamma+a\delta)(b\alpha+b\beta)\\
&\quad +(a\alpha+a\beta)^2+(a\gamma+a\delta)(b\alpha+b\beta)\\
&\quad +(b\alpha+b\beta)(a\gamma+a\delta)+(b\gamma+b\delta)^2\\
&\quad +(b\alpha+b\beta)(a\gamma+a\delta)+(b\gamma+b\delta)^2\\
&=2a^2(\alpha+\beta)^2+2ab(\gamma+\delta)(\alpha+\beta)\\
&\quad +2ab(\alpha+\beta)(\gamma+\delta)+2b^2(\gamma+\delta)^2\\
&=2a^2(\alpha+\beta)^2+2b^2(\gamma+\delta)^2+4ab(\alpha+\beta)(\gamma+\delta)\\
&=2(aX)^2t^2+2(bY)^2t^2+4abXYt^2,\\
&\qquad a=t^2+3t^4+4t^6+3t^8+t^{10}\\
&\qquad b=t^4+2t^6+t^8\\
&\qquad tX=\alpha+\beta=t+3t^3+6t^5+6t^7+3t^9+t^{11}
\end{aligned}
$$

$$tY = \gamma + \delta = 2t^3 + 5t^5 + 5t^7 + 2t^9$$
$$X = 1 + 3t^2 + 6t^4 + 6t^6 + 3t^8 + t^{10}$$
$$Y = 2t^2 + 5t^4 + 5t^6 + 2t^9.$$

By a computer,

$$\begin{aligned}\operatorname{tr}(A_1^+ A_1^-)^2 = 2(&t^6 + 12t^8 + 78t^{10} + 348t^{12} + 1161t^{14} + 3024t^{16} + 6288t^{18} \\ &+ 10554t^{20} + 14376t^{22} + 15932t^{24} + 14376t^{26} + 10554t^{28} \\ &+ 6288t^{30} + 3024t^{32} + 1161t^{34} + 348t^{38} + 78t^{38} + 12t^{40} + t^{42}).\end{aligned}$$

Now

$$A_1^{+2} = \begin{pmatrix} 0 & a & a & 0 \\ a & 0 & 0 & a \\ b & 0 & 0 & b \\ 0 & b & b & 0 \end{pmatrix} \begin{pmatrix} 0 & a & a & 0 \\ a & 0 & 0 & a \\ b & 0 & 0 & b \\ 0 & b & b & 0 \end{pmatrix}$$

$$= \begin{pmatrix} a^2 + ab & 0 & 0 & a^2 + ab \\ 0 & a^2 + ab & a^2 + ab & 0 \\ 0 & ab + b^2 & ba + b^2 & 0 \\ ba + b^2 & 0 & 0 & ba + b^2 \end{pmatrix}$$

$$A_1^{-2} = \begin{pmatrix} \alpha & 0 & 0 & \gamma \\ 0 & \alpha & \gamma & 0 \\ 0 & \beta & \delta & 0 \\ \beta & 0 & 0 & \delta \end{pmatrix} \begin{pmatrix} \alpha & 0 & 0 & \gamma \\ 0 & \alpha & \gamma & 0 \\ 0 & \beta & \delta & 0 \\ \beta & 0 & 0 & \delta \end{pmatrix}$$

$$= \begin{pmatrix} \alpha^2 + \gamma\beta & 0 & 0 & \alpha\gamma + \gamma\delta \\ 0 & \alpha^2 + \gamma\beta & \alpha\gamma + \gamma\delta & 0 \\ 0 & \alpha\beta + \delta\beta & \beta\gamma + \delta^2 & 0 \\ \beta\alpha + \delta\beta & 0 & 0 & \beta\gamma + \delta^2 \end{pmatrix}$$

$$A_1^{+2} = \begin{pmatrix} A & 0 & 0 & A \\ 0 & A & A & 0 \\ 0 & B & B & 0 \\ B & 0 & 0 & B \end{pmatrix} \qquad \begin{aligned} A &= a^2 + ab \\ B &= ab + b^2 \\ C &= \alpha^2 + \alpha\beta \end{aligned}$$

$$A_1^{-2} = \begin{pmatrix} C & 0 & 0 & D \\ 0 & C & D & 0 \\ 0 & E & F & 0 \\ E & 0 & 0 & F \end{pmatrix} \qquad \begin{aligned} D &= \alpha\gamma + \gamma\delta \\ E &= \alpha\beta + \beta\delta \\ F &= \beta\gamma + \delta^2 \end{aligned}$$

$$A_1^{+2}A_1^{-2}=\begin{pmatrix} A & 0 & 0 & A \\ 0 & A & A & 0 \\ 0 & B & B & 0 \\ B & 0 & 0 & B \end{pmatrix}\begin{pmatrix} C & 0 & 0 & D \\ 0 & C & D & 0 \\ 0 & E & F & 0 \\ E & 0 & 0 & F \end{pmatrix}$$

$$=\begin{pmatrix} AC+AE & & \multicolumn{2}{c}{X} \\ & AC+AE & & \\ \multicolumn{2}{c}{X} & BD+BF & \\ & & & BD+BF \end{pmatrix}$$

$$\begin{aligned}
\operatorname{tr}(A_1^{+2}A_1^{-2}) &= 2A(C+E)+2B(D+F) \\
&= 2(a^2+ab)(\alpha^2+\gamma\beta+\alpha\beta+\delta\beta)+2(ab+b^2)(\alpha\gamma+\gamma\delta+\beta\gamma+\delta^2) \\
&= 2a(a+b)(\alpha(\alpha+\beta)+\beta(\gamma+\delta))+2b(2+b)(\delta(\gamma+\delta)+\gamma(\alpha+\beta)) \\
&= 2a(a+b)(\alpha_0X+\beta_0Y)t^2+2b(a+b)(\gamma_0X+\delta_0Y)t^2 \\
&= 2a(a+b)\xi+2b(a+b)\eta
\end{aligned}$$

$$a+b=t^2+4t^4+6t^6+4t^8+t^{10}=t^2(1+t^2)^4$$

$$\alpha+\beta=t+3t^3+6t^5+6t^7+3t^9+t^{11}=tX$$

$$\gamma+\delta=2t^3+5t^5+5t^7+2t^9=tY,$$

$$a=t^2+3t^4+4t^6+3t^8+t^{10}$$

$$b=t^4+2t^6+t^8$$

$$\alpha=t+2t^3+4t^5+4t^7+2t^9+t^{11}=\alpha_0t$$

$$\beta=t^3+2t^5+2t^7+t^9=\beta_0t$$

$$\gamma=2t^3+4t^5+4t^7+2t^9=\gamma_0t$$

$$\delta=t^5+t^7=\delta_0t$$

where

$$\begin{aligned}
\xi &= (\alpha_0X+\beta_0Y)t^2 \\
&= t^2+5t^4+18t^6+43t^8+72t^{10}+86t^{12}+72t^{14}+43t^{16}+18t^{18} \\
&\quad +5t^{20}+t^{22} \\
\eta &= (\gamma_0X+\delta_0Y)t^2 \\
&= 2t^4+10t^6+30t^8+57t^{10}+70t^{12}+57t^{14}+30t^{16}+10t^{16}+2t^{20}.
\end{aligned}$$

By a computer,

$$\begin{aligned}
\operatorname{tr}(A^{+2}A_1^{-2}) = 2(&t^6+12t^8+77t^{10}+342t^{12}+1144t^{14}+2992t^{16}+6240t^{18} \\
&+10490t^{20}+14298t^{22}+15848t^{24}+14298t^{26}+10490t^{28} \\
&+6240t^{30}+2992t^{32}+1144t^{34}+342t^{36}+77t^{38}+12t^{40}+t^{44})
\end{aligned}$$

Now

$$\operatorname{tr} A_1^4 \equiv \operatorname{tr}(A_1^{+4})^4 + \operatorname{tr}(\Delta_1^{-4}) + [\operatorname{tr}(A_1^+ A_1^- A_1^+ A_1^-)] \times 2 + [\operatorname{tr}(A_1^{+2} A_1^{-2})] \times 4 \quad \text{(mod odd powers of } t\text{)}$$

$$= 2 \times (t^4 + 14t^6 + 121t^8 + 688t^{10} + 2955t^{12} + 9428t^{14} + 24319t^{16} + 25412t^{18} + 84347t^{20} + 114862t^{22} + 127262t^{24} + 114862t^{26} + 84347t^{28} + 25412t^{30} + 24319t^{32} + 9428t^{34} + 2955t^{36} + 688t^{38} + 121t^{40} + 14t^{42} + t^{44})$$

$$\operatorname{tr} A^4 = \operatorname{tr} A_0^4 + \operatorname{tr} A_1^4$$

$$= 1 + 54t^4 + 136t^6 + 1367t^8 + 4204t^{10} + 18930t^{12} + 47280t^{14} + 134381t^{16} + 182688t^{18} + 402414t^{20} + 507388t^{22} + 587004t^{24} + 507388t^{26} + 402414t^{28} + 182688t^{30} + 134381t^{32} + 47280t^{34} + 18930t^{36} + 4204t^{38} + 1367t^{40} + 136t^{42} + 54t^{44} + t^{48}.$$

This is the wanted $F(t)$. This polymer is rigid for $S_0 = \{\alpha, \beta, \xi, \zeta\}$.

References

[1] Abdulali, Salman, Thesis, SUNY at Stony Brook, 1984–5.
[2] Addington, Susan, Thesis, SUNY at Stony Brook, 1981.
[3] ——, Equivariant Holomorphic Maps of Symmetric Domains, Duke Math. J., 1987.
[4] Kuga, Michio, "Chemistry and GTFabV's", Progress in Math., **46**, Birkhauser (1984), 269–281.
[5] ——, "Algebraic cycles in GTFabV", J. Fac. Sci. Univ. Tokyo 1A, **29 : 1** (1982), 13–29.
[6] Satake, Ichiro, Algebraic Structures of Symmetric Domains, Princeton Univ. Press, 1980.

Department of Mathematics
State University of New York at Stony Brook
Stony Brook, NY 11794
U.S.A.

Advanced Studies in Pure Mathematics 15, 1989
Automorphic Forms and Geometry of Arithmetic Varieties
pp. 415–428

A Note on Zeta Functions Associated with Certain Prehomogeneous Affine Spaces

Atsushi Murase and Takashi Sugano

Dedicated to Prof. Ichiro Satake on his sixtieth birthday

§ 0. Introduction

The theory of zeta functions associated with prehomogeneous vector spaces (briefly P.V.) was founded by M. Sato and studied by several authors. In [5], Sato and Shintani established the analytic continuation and the functional equations of zeta functions associated with irreducible P.V.'s, and F. Sato [4] extended their results to regular P.V.'s under some mild assumptions. (See also Shintani [6, 7] and Suzuki [8].)

In this paper we shall study some zeta functions associated with *non-regular* P.V. in a special case and prove their analytic continuation and functional equations. Recall that the dual of a regular P.V. is also a regular P.V. and that functional equations hold between zeta functions associated with a regular P.V. and its dual. In our case, however, the dual of our non-regular P.V. is not even a P.V. Thus, instead of the dual, we are led to introduce some prehomogeneous *affine* spaces, the precise definition of which is given below.

Let G be a complex connected linear algebraic group, V a finite dimensional vector space and ρ a rational homomorphism from G into the group of affine transformations of V. We call a triple (G, ρ, V) a *prehomogeneous affine space* (briefly P.A.) if there exists a proper algebraic subset S of V such that $V-S$ is a single G-orbit. The set S is called the *singular set* of (G, ρ, V). In particular, when $\mathrm{Im}\,\rho$ is contained in $GL(V)$, such a triple is called a *prehomogeneous vector space* (briefly P.V.).

In § 1, we define a non-reductive algebraic group $\boldsymbol{G}$ and introduce a pair of non-regular P.V. and P.A. with $\boldsymbol{G}$-action. Zeta functions associated with them are defined and studied in § 2. Though our P.V. and P.A. are not dual to each other, we can prove functional equations between these two types of zeta functions. The next section (§ 3) is devoted to the preparation for the last one. Finally we prove that some contribution to the

Received April 7, 1987.

dimension formula for Jacobi cusp forms of degree n is expressed in terms of special values of our zeta functions (associated with P.A.).

We note here that our result is an analogue of Shintani's work on zeta functions associated with the space of quadratic forms ([7]). In fact, most of our results are obtained by a slight modification of Shintani's argument.

The authors express their profound gratitude to Professor F. Sato for his kind advice and warm encouragement. They are also deeply grateful to Professor T. Arakawa for his helpful information and for giving the authors the opportunity to learn his manuscript ([1]) before publication.

Notation. As usual, $\boldsymbol{Z}, \boldsymbol{Q}, \boldsymbol{R}$ and $\boldsymbol{C}$ are the ring of rational integers, the rational number field, the real number field and the complex number field. For any complex number x, we put $\boldsymbol{e}[x]=\exp(2\pi\sqrt{-1}x)$. We denote by $M_{m,n}$ and Sym_n the space of matrices with m rows and n columns and the space of symmetric matrices of degree n respectively. For any finite dimensional vector space V over $\boldsymbol{R}$, $\mathscr{S}(V)$ is the space of rapidly decreasing functions on V. When X is a smooth manifold, we denote by $C_0^\infty(X)$ the space of smooth functions with compact support. For any complex number λ we define a holomorphic function $\det(Z)^\lambda$ on $\{Z \in \mathrm{Sym}_n(\boldsymbol{C}) \mid \mathrm{Re}\, Z>0\}$ so that it coincides with the usual one on $\{X \in \mathrm{Sym}_n(\boldsymbol{R}) \mid X>0\}$.

§ 1. Prehomogeneous affine spaces

Fix once for all positive integers n, m and put $V=\mathrm{Sym}_n$, $W=M_{m,n}$ and $\boldsymbol{V}=V\times W$. Let $\boldsymbol{G}$ be the semi-direct product of W and $G=GL_n$ with composition rule

$$(\xi, g)\cdot(\xi', g')=(\xi+\xi' g^{-1}, gg') \qquad (\xi, \xi' \in W, g, g' \in G).$$

Note that $\boldsymbol{G}$ is isomorphic to the subgroup

$$\left\{\begin{bmatrix} 1_m & \xi \\ & 1_n \end{bmatrix}\begin{bmatrix} 1_m & \\ & g \end{bmatrix} \middle| \xi \in W, g \in G\right\}$$

of GL_{m+n}.

As is well-known, the triple $(G, \cdot, V)$ is a P.V. with G-action $\cdot$ on V given by $x\cdot g={}^t gxg$ $(x \in V, g \in G)$, whose singular set is $\{x \in V \mid \det x=0\}$. We now introduce two types of prehomogeneous affine structure to $(\boldsymbol{G}, \boldsymbol{V})$. We fix once for all a semi-integral positive definite symmetric matrix S of degree m. For simplicity we write $S(u, v)$ and $S[u]$ for ${}^t uSv$ and ${}^t uSu$ respectively $(u, v \in W)$. For $X=[x, u] \in \boldsymbol{V}$ and $\boldsymbol{g}=(\xi, g) \in \boldsymbol{G}$, set

$$X\cdot\boldsymbol{g}=[g^{-1}x\,{}^t g^{-1}, (u-\xi x)\,{}^t g^{-1}] \qquad \text{(dot action)}$$

and

$$X * \boldsymbol{g} = [{}^t g(x + S[\xi] + S(u, \xi) + S(\xi, u))g, (u+\xi)g] \qquad \text{(star action)}.$$

It is easily verified that $(X, \boldsymbol{g}) \mapsto X \cdot \boldsymbol{g}$, $X * \boldsymbol{g}$ give rise to right affine actions of $\boldsymbol{G}$ on $\boldsymbol{V}$. We put $P(X) = \det x$ and $P^*(X) = \det(x - S[u])$ $(X = [x, u] \in \boldsymbol{V})$. Then

$$P(X \cdot \boldsymbol{g}) = (\det g)^{-2} P(X)$$

and

$$P^*(X * \boldsymbol{g}) = (\det g)^2 P^*(X) \qquad (\boldsymbol{g} = (\xi, g) \in \boldsymbol{G}, X \in \boldsymbol{V}).$$

Lemma 1.1. *The triple* $(\boldsymbol{G}, \cdot, \boldsymbol{V})$ *(resp.* $(\boldsymbol{G}, *, \boldsymbol{V})$*) is a prehomogeneous affine space with the singular set* $\boldsymbol{S} = \{X \in \boldsymbol{V} \mid P(X) = 0\}$ *(resp.* $\boldsymbol{S}^* = \{X \in \boldsymbol{V} \mid P^*(X) = 0\}$*).*

Remark 1.2. In fact $(\boldsymbol{G}, \cdot, \boldsymbol{V})$ is a prehomogeneous *vector* space and its relative invariants are $P(X)^n$ $(n \in \boldsymbol{Z})$. Thus $(\boldsymbol{G}, \cdot, \boldsymbol{V})$ is not regular and the general theory of F. Sato is not applicable to our case.

Let $G_{\boldsymbol{R}}^+ = \{g \in G_{\boldsymbol{R}} \mid \det g > 0\}$ and $\boldsymbol{G}_{\boldsymbol{R}}^+ = W_{\boldsymbol{R}} G_{\boldsymbol{R}}^+$. We now define two representations r and R of $\boldsymbol{G}_{\boldsymbol{R}}^+$ on $\mathscr{S}(\boldsymbol{V}_{\boldsymbol{R}})$ in the following manner:

$$\begin{aligned} &r(\boldsymbol{g})f(X) = f(X \cdot \boldsymbol{g})\boldsymbol{e}[\operatorname{tr}(xS[\xi] - 2S(u, \xi))], \\ &R(\boldsymbol{g})f(X) = (\det g)^{m+n+1} f(X * \boldsymbol{g}), \\ &\qquad (\boldsymbol{g} = (\xi, g) \in \boldsymbol{G}_{\boldsymbol{R}}^+, X = [x, u] \in \boldsymbol{V}_{\boldsymbol{R}}, f \in \mathscr{S}(\boldsymbol{V}_{\boldsymbol{R}})). \end{aligned}$$

The Fourier transform $\mathscr{F}f$ of $f \in \mathscr{S}(\boldsymbol{V}_{\boldsymbol{R}})$ is defined by

$$\mathscr{F}f(Y) = \int_{\boldsymbol{V}_{\boldsymbol{R}}} f(X)\boldsymbol{e}[\langle X, Y \rangle] dX.$$

Here the inner product $\langle\ ,\ \rangle$ on $\boldsymbol{V}_{\boldsymbol{R}}$ is given by

$$\langle [x, u], [y, v] \rangle = \operatorname{tr}(xy + 2S(u, v))$$

and dX is the usual Lebesgue measure on $\boldsymbol{V}_{\boldsymbol{R}}$; for $X = [x, u] \in \boldsymbol{V}_{\boldsymbol{R}}$

$$dX = dx du, \quad dx = \prod_{1 \le i \le j \le n} dx_{ij}, \quad du = \prod_{\substack{1 \le k \le m \\ 1 \le l \le n}} du_{kl}.$$

The following formula is easily verified.

Lemma 1.3. $r(\boldsymbol{g})\mathscr{F} = \mathscr{F}R(\boldsymbol{g})$ $(\boldsymbol{g} \in \boldsymbol{G}_{\boldsymbol{R}}^+)$.

For $0 \le i \le n$, let $V_i = \{x \in V_{\boldsymbol{R}} \mid \operatorname{sgn} x = (i, n-i)\}$, $\boldsymbol{V}_i = V_i \times W_{\boldsymbol{R}}$ and $\boldsymbol{V}_i^* = \{[x, u] \in \boldsymbol{V}_{\boldsymbol{R}} \mid x - S[u] \in V_i\}$. Then we have

$$\boldsymbol{V_R}-\boldsymbol{S_R}=\coprod_{i=0}^{n} V_i, \qquad \boldsymbol{V_R}-\boldsymbol{S_R^*}=\coprod_{i=0}^{n} V_i^* \quad \text{(disjoint union)}.$$

For $f \in \mathcal{S}(\boldsymbol{V_R})$ and $s \in \boldsymbol{C}$, we put

$$\Phi_i(s,f)=\int_{V_i} |P(X)|^s \, \boldsymbol{e}[\mathrm{tr}\,(x^{-1}S[u])]f(X)dX$$

and

$$\Phi_i^*(s,f)=\int_{V_i^*} |P^*(X)|^s f(X)dX$$

The integrals are absolutely convergent for $\mathrm{Re}\, s>0$. It is clear that

$$\Phi_i(s, r(\boldsymbol{g})f)=(\det g)^{2s+m+n+1}\,\Phi_i(s,f)$$

and

$$\Phi_i^*(s, R(\boldsymbol{g})f)=(\det g)^{-2s}\Phi_i^*(s,f) \qquad (\boldsymbol{g} \in \boldsymbol{G_R^+}, f \in \mathcal{S}(\boldsymbol{V_R})).$$

Proposition 1.4 (cf. Lemma 15 in [7]). *Let $f \in \mathcal{S}(\boldsymbol{V_R})$.*

(i) *$\Phi_i(s,f)$ and $\Phi_i^*(s,f)$ are continued to meromorphic functions on the whole s-plane.*

(ii) $$\Phi_i(s-(m+n+1)/2, \mathcal{F}f)=C_{n,S}^{-1/2}(2\pi)^{n(n-1)/4-ns}\boldsymbol{e}[m(2i-n)/8+ns/4] \times \Gamma_n(s)\sum_{j=0}^{n} u_{ij}(s)\Phi_j^*(-s,f),$$

where $C_{n,S}=(\det 2S)^n 2^{n(n-1)/2}$, $\Gamma_n(s)=\prod_{k=0}^{n-1}\Gamma(s-k/2)$ and $u_{ij}(s)$ are polynomials in $\boldsymbol{e}[-s/2]$ given by (2.4) *in* [7].

Proof. Let T be the linear operator of $\mathcal{S}(\boldsymbol{V_R})$ given by $Tf([x,u])=f([x-S[u],u])$. For $F \in \mathcal{S}(V_R)$, we define its Fourier transform $\hat{F}$ by

$$\hat{F}(x)=\int_{V_R} F(y)\boldsymbol{e}[\mathrm{tr}\,(xy)]dy$$

and put

$$\phi_i(s, F)=\int_{V_i} |\det x|^s F(x)dx.$$

In view of Lemma 15 in [7], our proof is reduced to the following formula.

Lemma 1.5. *Assume that* $\mathrm{Re}\, s>0$ *and* $f \in \mathcal{S}(\boldsymbol{V_R})$. *Then*

$$\Phi_j^*(s, Tf) = \phi_j(s, F_f),$$
$$\Phi_j(s, \mathcal{F}Tf) = (\det 2S)^{-n/2} \boldsymbol{e}[m(2j-n)/8]\phi_j(s+m/2, \hat{F}_f),$$

where

$$F_f(x) = \int_{W_{\boldsymbol{R}}} f([(x, u)]du \in \mathcal{S}(V_{\boldsymbol{R}}).$$

Proof. The first part is obvious from the definitions. The Lebesgue convergence theorem implies

$$\Phi_j(s, \mathcal{F}Tf)$$
$$= \int_{V_j} |\det x|^s \, dx \left\{ \lim_{\varepsilon \downarrow 0} \int_{W_{\boldsymbol{R}}} \boldsymbol{e}[\mathrm{tr}\,(x^{-1} + \sqrt{-1}\varepsilon)S[u]]\mathcal{F}Tf([x, u])du \right\}.$$

Changing the order of integration, we have

$$\int_{W_{\boldsymbol{R}}} \boldsymbol{e}[\mathrm{tr}\,(x^{-1} + \sqrt{-1}\varepsilon)S[u]]\mathcal{F}Tf([x, u])du$$
$$= \int_{W_{\boldsymbol{R}}} du \int_{V_{\boldsymbol{R}}} dydv f([y, v])\boldsymbol{e}[\mathrm{tr}\,\{(x^{-1} + \sqrt{-1}\varepsilon)S[u + v(x^{-1} + \sqrt{-1}\varepsilon)^{-1}]$$
$$+ \sqrt{-1}\varepsilon x(x^{-1} + \sqrt{-1}\varepsilon)^{-1}S[v] + xy\}]$$
$$= \det\,(\varepsilon - \sqrt{-1}x^{-1})^{-m/2}\,(\det 2S)^{-n/2}$$
$$\int_{V_{\boldsymbol{R}}} \boldsymbol{e}[\mathrm{tr}\,\{xy + \sqrt{-1}\varepsilon x(x^{-1} + \sqrt{-1}\varepsilon)^{-1}S[v]\}]f([y, v])dydv.$$

Taking the limit, we obtain

$$\lim_{\varepsilon \downarrow 0} \int_{W_{\boldsymbol{R}}} \boldsymbol{e}[\mathrm{tr}\,(x^{-1} + \sqrt{-1}\varepsilon)S[u]]\mathcal{F}Tf([x, u])dxdu$$
$$= \boldsymbol{e}[m(2j-n)/8]\,|\det x|^{m/2}\,(\det 2S)^{-n/2}\hat{F}_f(x),$$

which completes the proof of the lemma. Q.E.D.

§ 2. Zeta functions associated with prehomogeneous affine spaces

In this section we study zeta functions associated with P.A. introduced in the previous section.

For $x \in V_{\boldsymbol{R}}$ ($\det x \neq 0$) put $G_x = \{g \in G_{\boldsymbol{R}}^+ \mid x \cdot g = x\} = SO(x)$. Similarly for $X = [x, u] \in \boldsymbol{V}_{\boldsymbol{R}} - \boldsymbol{S}_{\boldsymbol{R}}$ (resp. $\boldsymbol{V}_{\boldsymbol{R}} - \boldsymbol{S}_{\boldsymbol{R}}^*$), we put $\boldsymbol{G}_X = \{\boldsymbol{g} \in \boldsymbol{G}_{\boldsymbol{R}}^+ \mid X \cdot \boldsymbol{g} = X\}$ (resp. $\boldsymbol{G}_{\boldsymbol{R}}^* = \{\boldsymbol{g} \in \boldsymbol{G}_{\boldsymbol{R}}^+ \mid X * \boldsymbol{g} = X\}$). Then G_x (resp. $G_{x-S[u]}$) is isomorphic to $\boldsymbol{G}_X$ (resp. $\boldsymbol{G}_X^*$) through the mapping ι (resp. ι^*), where $\iota(h) = (u(1-h^{-1})x^{-1}, {}^t h^{-1})$ (resp. $\iota^*(h) = (u(h^{-1} - 1), h)$). We normalize a Haar measure dg (resp. $d\xi$) on $G_{\boldsymbol{R}}^+$ (resp. $W_{\boldsymbol{R}}$) by

$$dg=(\det g)^{-n}\prod dg_{ij} \quad (\text{resp. } d\xi=\prod d\xi_{ij}).$$

Then $d_r\boldsymbol{g}=d\xi dg$ ($\boldsymbol{g}=(\xi, g)\in \boldsymbol{G}_{\boldsymbol{R}}^+$) gives a right invariant measure on $\boldsymbol{G}_{\boldsymbol{R}}^+$. We let $x\in V_i$ (resp. $X\in \boldsymbol{V}_i$, $X\in \boldsymbol{V}_i^*$) and normalize a Haar measure $d\nu_x$ (resp. $d\nu_X$, $d\nu_X^*$) on G_x (resp. $\boldsymbol{G}_X$, $\boldsymbol{G}_X^*$) by the following formula:

$$\int_{G_{\boldsymbol{R}}^+}\phi(g)dg=\int_{G_x\backslash G_{\boldsymbol{R}}^+}|\det x\cdot g|^{-(n+1)/2}d(x\cdot g)\int_{G_x}\phi(hg)d\nu_x(h)$$

$$(\text{resp. } \int_{\boldsymbol{G}_{\boldsymbol{R}}^+}\phi(g)d_r\boldsymbol{g}=\int_{\boldsymbol{G}_X\backslash \boldsymbol{G}_{\boldsymbol{R}}^+}|P(X\cdot\boldsymbol{g})|^{-(m+n+1)/2}d(X\cdot\boldsymbol{g})\int_{\boldsymbol{G}_X}\phi(\boldsymbol{hg})d\nu_X(\boldsymbol{h}),$$

$$\int_{\boldsymbol{G}_X^*\backslash \boldsymbol{G}_{\boldsymbol{R}}^+}|P^*(X*\boldsymbol{g})|^{-(m+n+1)/2}d(X*\boldsymbol{g})\int_{\boldsymbol{G}_X^*}\phi(\boldsymbol{hg})d\nu_X^*(\boldsymbol{h})),$$

where $d(x\cdot g)$ (resp. $d(X\cdot\boldsymbol{g})$, $d(X*\boldsymbol{g})$) is the usual Lebesgue measure on $V_i=x\cdot G_{\boldsymbol{R}}^+$ (resp. $\boldsymbol{V}_i=X\cdot\boldsymbol{G}_{\boldsymbol{R}}^+$, $\boldsymbol{V}_i^*=X*\boldsymbol{G}_{\boldsymbol{R}}^+$). It is easy to see that

$$d\nu_X(\iota(h))=|\det x|^{-m/2}d\nu_x(h) \qquad (X=[x, u],\ h\in G_x),$$

and

$$d\nu_X^*(\iota^*(h))=|\det y|^{m/2}d\nu_y(h) \qquad (X=[x, u],\ y=x-S[u],\ h\in G_y).$$

Let $L=V\cap M_n(\boldsymbol{Z})$, $L^*=\{x=(x_{ij})\in V_{\boldsymbol{R}}\,|\,x_{ij}\in 2^{-1}\boldsymbol{Z},\ x_{ii}\in \boldsymbol{Z}\}$, $M=M_{m,n}(\boldsymbol{Z})$, $M^*=(2S)^{-1}M$, $\boldsymbol{L}=L\times M$, $\boldsymbol{L}^*=L^*\times M^*$, $\Gamma=SL_n(\boldsymbol{Z})$ and $\boldsymbol{\Gamma}=\{(\xi, g)\in \boldsymbol{G}_{\boldsymbol{R}}\,|\,\xi\in M,\ g\in\Gamma\}$. Then $\boldsymbol{L}$ and $\boldsymbol{L}^*$ are dual to each other with respect to the inner product $\langle\ ,\ \rangle$ and $\boldsymbol{L}$ (resp. $\boldsymbol{L}^*$) is $\boldsymbol{\Gamma}$-invariant under the dot action (resp. the star action). For $x\in V_{\boldsymbol{Q}}$ ($\det x\neq 0$) (resp. $X\in \boldsymbol{V}_{\boldsymbol{Q}}-\boldsymbol{S}_{\boldsymbol{Q}}$, $X\in \boldsymbol{V}_{\boldsymbol{Q}}-\boldsymbol{S}_{\boldsymbol{Q}}^*$) we put $\Gamma_x=\Gamma\cap G_x$ (resp. $\boldsymbol{\Gamma}_X=\boldsymbol{\Gamma}\cap\boldsymbol{G}_X$, $\boldsymbol{\Gamma}_X^*=\boldsymbol{\Gamma}\cap\boldsymbol{G}_X^*$). It is easy to see that $\Gamma_x\backslash G_x$ (resp. $\boldsymbol{\Gamma}_X\backslash\boldsymbol{G}_X$, $\boldsymbol{\Gamma}_X^*\backslash\boldsymbol{G}_X^*$) has a finite volume and we define

$$\mu(x)=\int_{\Gamma_x\backslash G_x}d\nu_x$$

$$(\text{resp. } \mu(X)=\int_{\boldsymbol{\Gamma}_X\backslash\boldsymbol{G}_X}d\nu_X,\ \mu^*(X)=\int_{\boldsymbol{\Gamma}_X^*\backslash\boldsymbol{G}_X^*}d\nu_X^*)$$

unless $n=2$ and $-(\det x)$ (resp. $-P(X)$, $-P^*(X)$) has a square root in $\boldsymbol{Q}$. In this exceptional case we put $\mu(x)=0$ (resp. $\mu(X)=0$, $\mu^*(X)=0$).

For $0\leq i\leq n$, we put $\boldsymbol{L}_i=\boldsymbol{L}\cap\boldsymbol{V}_i$ and $\boldsymbol{L}_i^*=\boldsymbol{L}^*\cap\boldsymbol{V}_i^*$. Let $\boldsymbol{L}_i/\cdot$ (resp. $\boldsymbol{L}_i^*/*$) be a complete set of representatives of $\boldsymbol{L}_i$ (resp. $\boldsymbol{L}_i^*$) under the dot action (resp. the star action) of $\boldsymbol{\Gamma}$. We now define Dirichlet series $\xi_i(s, \boldsymbol{L})$ and $\xi_i^*(s, \boldsymbol{L}^*)$ by

$$\xi_i(s, \boldsymbol{L})=\sum_{X=[x,u]\in \boldsymbol{L}_i/\cdot}\mu(X)\boldsymbol{e}[-\mathrm{tr}\,(x^{-1}S[u])]\,|P(X)|^{-s}$$

and

$$\xi_i^*(s, \boldsymbol{L}^*) = \sum_{X \in \boldsymbol{L}_i^*/*} \mu^*(X)\,|P^*(X)|^{-s}.$$

Lemma 2.1. *The Dirichlet series* $\xi_i(s, \boldsymbol{L})$ *and* $\xi_i^*(s, \boldsymbol{L}^*)$ *converge absolutely if* $\operatorname{Re} s > (m+n+1)/2$.

Proof. On accounting the relation between $d\nu_X^*$ and $d\nu_y$ we have

$$\mu^*([x, u]) = |\det y|^{m/2}\,[\Gamma_y : \iota^{*-1}(\boldsymbol{\Gamma}_X^*)]\,\mu(y),$$

where $y = x - S[u]$. For each fixed $y \in V_Q$, we say that two elements u and u' in M^* are equivalent if there exists a $\gamma \in \boldsymbol{\Gamma}_y$ such that $u' \equiv u\gamma \pmod{M}$. Take a complete set T_y^* of representatives of such equivalence classes in M^*. Let $\mathscr{R}_i$ be a complete set of representatives of V_i under the action of Γ. Then we can take

$$\{[y+S[u], u]\,|\, y \in \mathscr{R}_i,\ y+S[u] \in L^*,\ u \in T_y^*\}$$

as $\boldsymbol{L}_i^*/*$. Since

$$\sum_{\substack{u \in T_y^* \\ S[u]+y \in L^*}} [\Gamma_y : \iota^{*-1}(\boldsymbol{\Gamma}_{[y+S[u],u]})] = \sum_{\substack{u \in M^*/M \\ S[u]+y \in L^*}} 1,$$

we have

$$\sum_{X \in \boldsymbol{L}_i^*/*} \mu(X)\,|P^*(X)|^{-s} = \sum_{y \in \mathscr{R}_i} \sum_{\substack{u \in M^*/M \\ S[u]+y \in L^*}} \mu(y)\,|\det y|^{-(s-m/2)}.$$

Noting that $\{y \in V_i \,|\, S[u]+y \in L^* \text{ for some } u \in M^*\}$ is contained in a lattice of $V_{\boldsymbol{R}}$, we see that $\xi_i^*(s, \boldsymbol{L}^*)$ converges absolutely if $\operatorname{Re}(s - m/2) > (n+1)/2$ (cf. (2.6) in [7]). The convergence of $\xi_i(s, \boldsymbol{L})$ is similarly proved.

Q.E.D.

Remark 2.2. The above proof implies that

$$\xi_n^*(s, \boldsymbol{L}^*) = 2^{-(n+1)} C_n \xi_n^*(s - m/2; S),$$

where

$$\xi_n^*(s; S) = \sum_{u \in M^*/M} \sum_{\substack{y \in \mathscr{R}_n \\ y+S[u] \in L^*}} |\Gamma_y|^{-1} (\det y)^{-s}$$

and

$$C_n = \prod_{k=1}^{n} \frac{2\pi^{k/2}}{\Gamma(k/2)}.$$

To show the analytic continuation and the functional equation of $\xi_i(s, \boldsymbol{L})$ and $\xi_i^*(s, \boldsymbol{L})$, consider the integrals

$$Z(f, \boldsymbol{L}, s)=\int_{\Gamma\backslash G_{\boldsymbol{R}}^{+}} \det g^{-2s} \sum_{X\in \boldsymbol{L}'} r(\boldsymbol{g})f(X)d_r\boldsymbol{g},$$

$$Z^*(f, \boldsymbol{L}^*, s)=\int_{\Gamma\backslash G_{\boldsymbol{R}}^{+}} \det g^{2s-(m+n+1)} \sum_{X\in \boldsymbol{L}^{*\prime}} R(\boldsymbol{g})f(X)d_r\boldsymbol{g}.$$

Here f is a function on $\boldsymbol{V}_{\boldsymbol{R}}$, $\boldsymbol{L}'=\boldsymbol{L}\cap(\boldsymbol{V}_{\boldsymbol{R}}-\boldsymbol{S}_{\boldsymbol{R}})$ and $\boldsymbol{L}^{*\prime}=\boldsymbol{L}^*\cap(\boldsymbol{V}_{\boldsymbol{R}}-\boldsymbol{S}_{\boldsymbol{R}}^*)$.

Lemma 2.3 (cf. Lemma 16 in [7]). *When $n\neq 2$, the integral $Z(f, \boldsymbol{L}, s)$ and $Z^*(f, \boldsymbol{L}^*, s)$ $(f\in \mathscr{S}(\boldsymbol{V}_{\boldsymbol{R}}))$ are absolutely convergent if* $\operatorname{Re} s>(m+n+1)/2$. *Furthermore we have*

$$Z(f, \boldsymbol{L}, s)=\sum_{0\le i\le n} \xi_i(s, \boldsymbol{L})\Phi_i(s-(m+n+1)/2, f),$$

$$Z^*(f, \boldsymbol{L}^*, s)=\sum_{0\le i\le n} \xi_i^*(s, \boldsymbol{L}^*)\Phi_i^*(s-(m+n+1)/2, f).$$

If $f\in C_0^\infty(V_n)$ (resp. $f\in C_0^\infty(V_n^*)$) this is true even for $n=2$.

Proof. It is easy to see that for real s

$$\int_{\Gamma\backslash G_{\boldsymbol{R}}^{+}} |\det g|^{-s} \sum_{X\in \boldsymbol{L}'} |r(\boldsymbol{g})f(X)|d_r\boldsymbol{g}$$

$$\le\int_{\Gamma\backslash G_{\boldsymbol{R}}^{+}} \det g^{2s-m} \sum_{x\in L'} F_{|f|}({}^t gxg)dg,$$

where $L'=\{x\in L\,|\det x\neq 0\}$. Thus the convergence of $Z(f, \boldsymbol{L}, s)$ is an immediate consequence of Lemma 16 in [7]. For $X\in \boldsymbol{V}_{\boldsymbol{R}}-\boldsymbol{S}_{\boldsymbol{R}}$, put

$$P_s(X)=|\det x|^s\, \boldsymbol{e}[\operatorname{tr}(x^{-1}S[u])].$$

Then

$$P_s(X\cdot\boldsymbol{g})/P_s(X)=(\det g)^{-2s}\, \boldsymbol{e}[\operatorname{tr}(xS[\xi]-2S(u, \xi))]$$

and we have

$$Z(f, \boldsymbol{L}, s)=\int_{\Gamma\backslash G_{\boldsymbol{R}}^{+}} \sum_{X\in \boldsymbol{L}'} P_s(X\cdot\boldsymbol{g})/P_s(X)f(X\cdot\boldsymbol{g})d_r\boldsymbol{g}$$

$$=\sum_{0\le i\le n}\sum_{X\in \boldsymbol{L}_i/\cdot} \mu(X)P_s(X)^{-1}\int_{G_X\backslash G_{\boldsymbol{R}}^{+}} P_{s-(m+n+1)/2}(X\cdot\boldsymbol{g})d(X\cdot\boldsymbol{g}),$$

which proves the assertion on $Z(f, \boldsymbol{L}, s)$. The remaining part is checked in quite a similar manner. Q.E.D.

Set

$$Z_+(f, \boldsymbol{L}, s)=\int_{\substack{\Gamma\backslash G_{\boldsymbol{R}}^{+}\\ \det g\le 1}} (\det g)^{-2s} \sum_{X\in \boldsymbol{L}'} r(\boldsymbol{g})f(X)d_r\boldsymbol{g}$$

and

$$Z_+^*(f, \boldsymbol{L}^*, s)=\int_{\substack{\Gamma\backslash G_{\boldsymbol{R}}^+ \\ \det g\geq 1}} (\det g)^{2s-(m+n+1)} \sum_{X\in \boldsymbol{L}^{*\prime}} R(\boldsymbol{g})f(X)d_r\boldsymbol{g}.$$

These integrals are absolutely convergent for any $s \in \boldsymbol{C}$ and hence define entire functions if $f \in \mathscr{S}(\boldsymbol{V_R})$ and $n\neq 2$.

Theorem 2.4 (cf. Theorem 5, in [7]). *Assume* $n\neq 2$.
(i) *Dirichlet series* $\xi_i(s, \boldsymbol{L})$ *and* $\xi_i^*(s, \boldsymbol{L}^*)$ $(0\leq i\leq n)$ *are continued to meromorphic functions in the whole s-plane.*
(ii) *They are holomorphic except for possible simple poles at* $s=(m+k+1)/2$ $(1\leq k\leq n)$.
(iii) *They satisfy the following functional equations*:

$$\xi_i^*((m+n+1)/2-s, \boldsymbol{L}^*)=C_{n,S}^{1/2}\,\boldsymbol{e}\left[\frac{ns}{4}\right](2\pi)^{n(n-1)/4-ns}\Gamma_n(s)$$
$$\times\sum_{j=0}^{n} u_{ji}(s)\boldsymbol{e}\left[\frac{m(2j-n)}{8}\right]\xi_j(s, \boldsymbol{L}).$$

Proof. Let $\mathscr{Q}=\det((1/(2-\delta_{ij}))(\partial/\partial x_{ij}))$ be a differential operator on $\boldsymbol{V_R}$. For $f_i^{(0)} \in C_0^\infty(\boldsymbol{V}_i^*)$ put $f_i=\mathscr{Q}f_i^{(0)}$. Note that

$$\Phi_i^*(s,f_i)=(-1)^i\prod_{k=0}^{n-1}(s+k/2)\Phi_i^*(s-1,f_i^{(0)})$$

and that for any $s \in \boldsymbol{C}$ there exists an $f_i^{(0)}$ such that $\Phi_i^*(s,f_i^{(0)})$ is not zero. It is easy to see that

$$\mathscr{F}f_i(X)=(-2\pi\sqrt{-1})^n P(X)\mathscr{F}f_i^{(0)}(X).$$

Thus f_i (resp. $\mathscr{F}f_i$) vanishes on $\boldsymbol{S}_{\boldsymbol{R}}^*$ (resp. $\boldsymbol{S_R}$). The Poisson summation formula and Lemma 1.2 now imply

$$Z(\mathscr{F}f_i, \boldsymbol{L}, s)=Z_+(\mathscr{F}f_i, \boldsymbol{L}, s)+C_{n,S}^{-1}Z_+^*(f_i, \boldsymbol{L}^*, (m+n+1)/2-s),$$
$$Z^*(f_i, \boldsymbol{L}^*, s)=Z_+^*(f_i, \boldsymbol{L}^*, s)+C_{n,S}Z_+(\mathscr{F}f_i, \boldsymbol{L}, (m+n+1)/2-s).$$

Hence $Z(\mathscr{F}f_i, \boldsymbol{L}, s)$ and $Z^*(f_i, \boldsymbol{L}^*, s)$ are entire functions of s and satisfy the functional equation

$$Z^*(f_i, \boldsymbol{L}^*, (m+n+1)/2-s)=C_{n,S}Z(\mathscr{F}f_i, \boldsymbol{L}, s).$$

Applying Lemma 2.3 and Proposition 1.4 to both sides, we have the analytic continuation of ξ_i^* and the functional equation (iii) in $\operatorname{Re} s>0$. Taking account of the fact

$$(2\pi)^{-n}\boldsymbol{e}[n(n+1)/8]\Gamma_n(s)\Gamma_n((n+1)/2-s)$$
$$\sum_{0\leq l\leq n} u_{i,l}(s)u_{n-l,j}((n+1)/2-s) \qquad =\delta_{i,j},$$

we have the analytic continuation of ξ_i. Therefore (iii) holds for any $s \in \boldsymbol{C}$. Q.E.D.

Remark 2.5. As we can see from Lemma 3.1 and Theorem 3.2 below, even when $n=2$, $\xi_2^*(s, \boldsymbol{L}^*)$ has an analytic continuation to a meromorphic function on $\boldsymbol{C}$ and is holomorphic except for possible simple poles at $s=(m+2)/2$ and $(m+3)/2$.

§ 3. Integrals $Z(f_n(X; \lambda), \boldsymbol{L}, s)$ and $Z^*(f_n^*(X; \lambda), \boldsymbol{L}^*, s)$

For $\lambda \in \boldsymbol{C}$, we define functions $f_n(X; \lambda)$ and $f_n^*(X; \lambda)$ on $\boldsymbol{V_R}$ by

$$f_n([x, u]; \lambda)=\det\left(1+\frac{1}{2\sqrt{-1}}x\right)^{-\lambda} \boldsymbol{e}[-\operatorname{tr}(2\sqrt{-1}+x)^{-1}S[u]]$$

and

$$f_n^*([x, u]; \lambda)=\begin{cases}\det(x-S[u])^{\lambda-(m+n+1)/2}\exp(-4\pi \operatorname{tr}(x)) & \text{if } x>S[u] \\ 0 & \text{otherwise.}\end{cases}$$

Put

$$d_{n,S}(\lambda)=C_{n,S}^{1/2}(2\pi)^{n(n-1)/4-n(\lambda-m/2)}2^{-n\lambda}\Gamma_n(\lambda-m/2).$$

A direct calculation shows

Lemma 3.1. *The integral* $Z^*(f_n^*(X; \lambda), \boldsymbol{L}^*, s)$ *is absolutely convergent and is equal to*

$$C_{n,S}^{-1}d_{n,S}(s+\lambda-(m+n+1)/2)\xi_n^*(s, \boldsymbol{L}^*),$$

if $\operatorname{Re} s>(m+n+1)/2$ *and* $\operatorname{Re}(\lambda+s)>m+n$.

Thus $Z^*(f_n^*(X; \lambda), \boldsymbol{L}^*, s)$ is continued to a meromorphic function of (λ, s) on the whole $\boldsymbol{C}^2$. It is easily verified that the Fourier transform $\mathscr{F}f_n^*(X; \lambda)$ exists if $\operatorname{Re}\lambda>(m+n+1)/2$ and is equal to $C_{n,S}^{-1}d_{n,S}(\lambda)f_n(X; \lambda)$. Furthermore the identity

$$\sum_{X \in \boldsymbol{L}^*} f_n^*(X; \lambda)=d_{n,S}(\lambda)\sum_{X \in \boldsymbol{L}} f_n(X; \lambda)$$

holds if $\operatorname{Re} s>m+n+1$. By a similar argument to that in the proof of Lemma 21 in [7], we obtain the following theorem.

Theorem 3.2 (cf. Lemma 21 in [7]).
(i) *Assume that* $\operatorname{Re} s>(n-1)/2$ *and that the pair* (λ, s) *satisfies the following inequalities*:

$\operatorname{Re}\lambda > m+1,\ \operatorname{Re} s < \operatorname{Re}\lambda - m/2$ *if* $n=1$,

$\operatorname{Re}\lambda > \operatorname{Max}(13/2,\ 2\operatorname{Re} s + 7/2) + m$ *if* $n=2$,

$\operatorname{Re}\lambda > m+n+7/2,\ \operatorname{Re} s < \operatorname{Re}\lambda - (m+n-1)/2$ *if* $n \geq 3$.

Then the integral $Z(f_n(X;\lambda), \boldsymbol{L}, s)$ *is absolutely convergent.*

(ii) *Both* $Z(f_n(X;\lambda), \boldsymbol{L}, s)$ *and* $Z^*(f_n(X;\lambda), \boldsymbol{L}^*, s)$ *are continued to meromorphic functions of* (λ, s) *on* $\boldsymbol{C}^2$ *and satisfy*

$$Z^*(f_n^*(X;\lambda), \boldsymbol{L}^*, s) = d_{n,S}(\lambda) Z(f_n(X;\lambda), \boldsymbol{L}, (m+n+1)/2 - s).$$

§ 4. The contribution of purely parabolic conjugacy classes to the dimension formula for the space of Jacobi cusp forms

In this section, we show that special values of $\xi_n^*(s, \boldsymbol{L}^*)$ appear in the calculation of the dimension formula for the space of Jacobi cusp forms.

We recall the definition of Jacobi cusp forms from Shintani's unpublished work (for more precise treatment refer to [2]; see also Yamazaki [9]). Let $\tilde{G} = \tilde{G}_{n,m}$ be the subgroup of Sp_{m+n} consisting of

$$(\xi, \eta, \kappa)g = \left(\begin{array}{c|cc} 1_{m+n} & \begin{matrix}\kappa & \eta \\ {}^t\eta & 0\end{matrix} \\ \hline & 1_{m+n} \end{array}\right)\left(\begin{array}{cc|cc} 1_m & \xi & & \\ & 1_n & & \\ \hline & & 1_m & \\ & & -{}^t\xi & 1_n \end{array}\right)\left(\begin{array}{cc|cc} 1_m & & & \\ & a & & b \\ \hline & & 1_m & \\ & c & & d \end{array}\right),$$

where $\xi, \eta \in M_{m,n}$, $\kappa \in \mathrm{Sym}_m$ and $g = \begin{bmatrix} a & b \\ c & d \end{bmatrix} \in Sp_n$. Then $\tilde{G}$ is the semi-direct product of $H = \{(\xi, \eta, \kappa)\}$ and $G = Sp_n$. The center Z of $\tilde{G}$ is $\{(0, 0, \kappa)\}$ and hence $\tilde{G}$ is not reductive if $m \geq 1$. We let $\tilde{G}_{\boldsymbol{R}}$ act on $\mathscr{D} = \mathscr{D}_{n,m} = \mathfrak{H}_n \times M_{m,n}(\boldsymbol{C})$ ($\mathfrak{H}_n$ = the upper half plane of degree n) as follows:

$$\tilde{g}\langle Z\rangle = (g\langle z\rangle, wj(g, z)^{-1} + \xi \cdot g\langle z\rangle + \eta)$$
$$(\tilde{g} = (\xi, \eta, \kappa)g \in \tilde{G}_{\boldsymbol{R}},\ Z = (z, w) \in \mathscr{D}),$$

where $g\langle z\rangle = (az+b)(cz+d)^{-1}$ and $j(g, z) = cz+d$. We define an automorphy factor $J_{S,l}(\tilde{g}, Z)$ by

$$J_{S,l}(\tilde{g}, Z) = \det j(g, z)^l \boldsymbol{e}[\mathrm{tr}\{-S\kappa + S[w]j(g, z)^{-1}c$$
$$-2S(\xi, w)j(g, z)^{-1} - S[\xi]g\langle z\rangle\}],$$

where S is a positive definite semi-integral symmetric matrix of degree m and l is a positive integer. Then the space $\mathfrak{S}(S, l)$ of Jacobi cusp forms of index S and weight l with respect to $\tilde{\Gamma} = \tilde{G}(\boldsymbol{Z})$ is defined to be the $\boldsymbol{C}$-vector space of holomorphic functions f on $\mathscr{D}$ which satisfy

(i) $f(\tilde{\gamma}\langle Z\rangle)=J_{S,l}(\tilde{\gamma}, Z)f(Z)$ for $\forall\tilde{\gamma}\in\tilde{\Gamma}$, $\forall Z\in\mathscr{D}$,
(ii) as a function on $\tilde{G}_{\boldsymbol{R}}$, $J_{S,l}(\tilde{g}, Z_0)^{-1}f(\tilde{g}\langle Z_0\rangle)$ is bounded $(Z_0=(\sqrt{-1}1_n, 0)\in\mathscr{D})$.

For $Z=(z, w)$ and $Z'=(z', w')\in\mathscr{D}$, we put

$$K_{S,l}(Z, Z')=A_{S,l}\det\left(\frac{z-\bar{z}'}{2\sqrt{-1}}\right)^{-l}\boldsymbol{e}[-\operatorname{tr}(z-\bar{z}')^{-1}S[w-\bar{w}']],$$

where

$$A_{S,l}=(\det 2S)^n 2^{-n(n+3)/2}\pi^{-n(n+1)/2}\prod_{i=0}^{n-1}\prod_{j=1}^{n-i}\left(l-\frac{m+i}{2}-j\right).$$

Let $d\mu(Z)$ be the $\tilde{G}_{\boldsymbol{R}}$-invariant measure on $\mathscr{D}$ given by

$$d\mu(Z)=(\det y)^{-m-n-1}\prod_{1\le i\le j\le n}dx_{ij}dy_{ij}\prod_{\substack{1\le i\le m\\1\le j\le n}}du_{ij}dv_{ij}$$
$$(Z=(z, w),\ z=x+\sqrt{-1}y,\ w=u+\sqrt{-1}v).$$

In view of Satake [3] and applying the Selberg trace formula, we obtain the following dimension formula for $\mathfrak{S}(S, l)$.

Lemma 4.1. *Assume* $l>m+2n$. *Then*

$$\dim_{\boldsymbol{C}}\mathfrak{S}(S, l)=\int_{\tilde{\Gamma}\backslash\mathscr{D}}\sum_{\tilde{\gamma}\in\tilde{\Gamma}/Z(\tilde{\Gamma})}K_{S,l}(\tilde{\gamma}\langle Z\rangle, Z)J_{S,l}(\tilde{\gamma}, Z)^{-1}$$
$$|J_{S,l}(\tilde{g}_Z, Z_0)|^{-2}\,d\mu(Z).$$

Here $\tilde{g}_Z$ *is an element of* $\tilde{G}_{\boldsymbol{R}}$ *such that* $\tilde{g}_Z\langle Z_0\rangle=Z$ *and* $Z(\tilde{\Gamma})=\{(0, 0, \kappa)\,|\,\kappa\in\mathrm{Sym}_m(\boldsymbol{Z})\}$ *is the center of* $\tilde{\Gamma}$.

We set $\tilde{\Gamma}'=\{\tilde{\gamma}\in\tilde{\Gamma}\,|\,\boldsymbol{e}[\operatorname{tr}(S\psi_{\tilde{\gamma}}(h))]=1$ for $\forall h\in H(\tilde{\gamma})_{\boldsymbol{R}}\}$, where $H(\tilde{g})=\{h\in H\,|\,h^{-1}\tilde{g}h\tilde{g}^{-1}=(0, 0, \psi_{\tilde{g}}((h))\in Z\}$ for $\tilde{g}\in\tilde{G}$. The following is easily verified.

Lemma 4.2. *Assume* $l>m+2n$. *Then*

$$\dim_{\boldsymbol{C}}\mathfrak{S}(S, l)=\int_{\tilde{\Gamma}\backslash\mathscr{D}}\sum_{\tilde{\gamma}\in\tilde{\Gamma}'/Z(\tilde{\Gamma})}K_{S,l}(\tilde{\gamma}\langle Z\rangle, Z)J_{S,l}(\tilde{\gamma}, Z)^{-1}$$
$$|J_{S,l}(\tilde{g}_Z, Z_0)|^{-2}\,d\mu(Z).$$

For an integer r $(0\le r\le n)$, let $\tilde{\Pi}_r$ be the set consisting of $\tilde{\gamma}\in\tilde{\Gamma}$ which is $\tilde{\Gamma}$-conjugate to $h\begin{bmatrix}1 & x\\ 0 & 1\end{bmatrix}$ $(h=(\xi, \eta, \kappa)\in\tilde{\Gamma}$, $x\in\mathrm{Sym}_n(\boldsymbol{Z})$, rank $x=r)$. We set

$$I_{S,l}(\tilde{\Pi}_r)=\int_{\Gamma\backslash\mathfrak{D}} \sum_{\tilde{\gamma}\in(\tilde{\Pi}_r\cap\tilde{\Gamma}')/Z(\Gamma)} K_{S,l}(\tilde{\gamma}\langle Z\rangle, Z)J_{S,l}(\tilde{\gamma}, Z)^{-1} |J_{S,l}(\tilde{g}_Z, Z_0)|^{-2}\,d\mu(Z).$$

The sum $\sum_{r=0}^{n} I_{S,l}(\tilde{\Pi}_r)$ is called "the contribution of purely parabolic conjugacy classes to the dimension formula for $\mathfrak{S}(S, l)$". Observe that $\tilde{\Pi}_r\cap\tilde{\Gamma}'$ is

$$\left\{\tilde{\gamma}\in\tilde{\Pi}_r \,\middle|\, \tilde{\gamma} \text{ is } \tilde{\Gamma}\text{-conjugate to } (0, (\eta', 0), \kappa)\left(\begin{array}{c|cc} 1_n & \begin{matrix} x' & 0 \\ 0 & 0 \end{matrix} \\ \hline & 1_n \end{array}\right),\ \eta'\in M_{m,r}(\mathbf{Z}),\ x'\in \mathrm{Sym}_r(\mathbf{Z})\right\}.$$

Shintani's argument in the proof of Proposition 8 in [7] works also for our case (see also Theorem 3.2 (i)) and we have the following theorem.

Theorem 4.3. *The integral $I_{S,l}(\tilde{\Pi}_r)$ is absolutely convergent if $l\geq 2n+m+3$ and equals*

$$\frac{2^n\omega_{n-r}A_{S,l}}{U_{n-r}C_r}Z\left(f_r(X; l), \mathbf{L}^{(r)}, n-\frac{r-1}{2}\right).$$

Here

$$U_r=\prod_{k=1}^{r}\frac{2\pi^k}{\Gamma(k)},\quad C_r=\prod_{k=1}^{r}\frac{2\pi^{k/2}}{\Gamma(k/2)},\quad \omega_l=\zeta(2)\zeta(4)\cdots\zeta(2l)$$

and $\mathbf{L}^{(r)}=\mathrm{Sym}_r(\mathbf{Z})\times M_{m,r}(\mathbf{Z})$.

Combining Theorem 3.2 (ii), Lemma 3.1 and Remark 2.2 we have

Corollary 4.4.

$$I_{S,l}(\tilde{\Pi}_r)=(\det 2S)^{n-r}2^{r(n-r)-1}(2\pi)^{-(n-r)(n-r+1)/2}\,\omega_{n-r}U_{n-r}^{-1}\xi_r^*(r-n; S)\prod_{i=1}^{n-r}\prod_{j=1}^{i}\left(l-\frac{m+n-i}{2}-j\right),$$

where we put $\xi_0^*(s; S)\equiv 2$.

Remark 4.5. Special values of $\xi_n^*(s; S)$ at non-positive integers are known only for $n\leq 2$. If $n=1$, as is well-known, we have

$$\xi_1^*(1-k; S)=-\sum_{u\in M^*/M} B_k(\langle -S[u]\rangle)/k,$$

where $B_k(x)$ is the k-th Bernoulli polynomial and for $x\in\mathbf{R}$ we define $\langle x\rangle$

by the condition: $x \equiv \langle x \rangle$ (mod $\boldsymbol{Z}$) and $0 < \langle x \rangle \leq 1$. On the other hand, when $n=2$, $\xi_2^*(s; S)$ is a linear combination of partial zeta functions studied in Arakawa [1]. Applying his results to this case we have $\xi_2^*(1-k; S) \in \boldsymbol{Q}$. (We can also evaluate $\xi_2^*(1-k; S)$ explicitly using formulas in [1]. For example $\xi_2^*(0; 1) = 2^{-6}$, if $m=1$ and $S=1$.)

References

[1] Arakawa, T., Special values of L-functions associated with the space of quadratic forms and the representation of $Sp(2n, F_p)$ in the space of Siegel cusp forms, in this volume.

[2] Murase, A., L-functions attached to Jacobi forms of degree, n, Part I. The basic identity, to appear in J. Reine Angew. Math.

[3] Satake, I., A group extension and its unitary representation (in Japanese), Sugaku, **21** (1969), 241–253.

[4] Sato, F., Zeta functions in several variables associated with prehomogeneous vector spaces I: Functional equation, Tôhoku Math. J., **34** (1982), 437–483.

[5] Sato, M. and Shintani, T., On zeta functions associated with prehomogeneous vector spaces, Ann. of Math., **100** (1974), 131–170.

[6] Shintani, T., On Dirichlet series whose coefficients are class-numbers of integral binary cubic forms, J. Math. Soc. Japan, **24** (1972), 132–188.

[7] ——, On zeta functions associated with the vector space of quadratic forms, J. Fac. Sci. Univ. Tokyo, **22** (1975), 25–65.

[8] Suzuki, T., On zeta functions associated with quadratic forms of variable coefficients, Nagoya Math. J., **73** (1979), 117–147.

[9] Yamazaki, T., Jacobi forms and a Maass relation for Eisenstein series, J. Fac. Sci. Univ. Tokyo, **33** (1986), 295–310.

A. Murase
Department of Mathematics
Kyoto Sangyo University
Kyoto
603 Japan

T. Sugano
Department of Mathematics
Faculty of Education
Mie University
Tsu
514 Japan

Advanced Studies in Pure Mathematics 15, 1989
Automorphic Forms and Geometry of Arithmetic Varieties
pp. 429–463

On Zeta Functions Associated with the Exceptional Lie Group of Type E_6

Masakazu Muro

Dedicated to Prof. Ichiro Satake on his sixtieth birthday

Abstract. We define a zeta function associated with the exceptional Lie group of type E_6 and compute their functional equations and residues as an application of microlocal analysis.

Contents

Received April 13, 1987.

Introduction

In 1974, Sato and Shintani [Sm-Sh] proposed a method to construct a new type of zeta functions associated with a "prehomogeneous" action of a linear algebraic group on a finite dimensional vector space. Such zeta functions are called *zeta functions associated with prehomogeneous vector spaces* and some interesting examples have been calculated by Sato-Shintani [Sm-Sh] and by Shintani [Sh].

In particular, Shintani [Sh] analyzed precisely the zeta functions associated with the vector space of symmetric matrices as a typical example of such zeta functions. They are called *Shintani's zeta function.* The reason why Shintani's zeta function has called attention is that some special values of the zeta function defined for positive definite symmetric matrices has close connection with the dimension of Siegel cusp forms through Selberg's trace formula. Though Shintani did not give the special values of his zeta functions explicitly, he calculated the functional equations completely and computed some residues in terms of some special values of the zeta functions of low rank. Similar zeta functions may be defined for other spaces of the same type, for example, zeta functions associated with complex Hermitian matrices or quaternion Hermitian matrices. See [Sm-Sh].

In our paper we shall construct a zeta function of the same type for modular forms on an exceptional symmetric domain. It is a zeta function associated with the prehomogeneous vector space $(\boldsymbol{G}_{\mathbf{C}}, \boldsymbol{X}_{\mathbf{C}})$. In our case, the reductive group $\boldsymbol{G}_{\mathbf{C}} := \boldsymbol{GL}_1 \times \boldsymbol{E}_6$ acts on the space $\boldsymbol{X}_{\mathbf{C}}$ of 3×3 Hermitian matrices over the complex Cayley algebra $\mathfrak{C}_{\mathbf{C}}$. Our purpose of this paper is to analyze invariant tempered distributions on $\boldsymbol{X}_{\mathbf{C}}$ by microlocal analysis. As its consequence, we shall calculate the functional equations and residues of the zeta functions explicitly. The calculation of functional equations is reduced to the computations of the Fourier transforms of tempered distributions obtained as a complex power of a relatively invariant polynomial and the calculation of the residues are reduced to the computation of Fourier transforms of tempered distributions which are given as invariant measures on low dimensional orbits. These facts have already been pointed out in [Sm-Sh]. After all, the key problem is explicit calculations of Fourier transforms of relatively invariant distributions. Microlocal analysis presents us a strong tool to solve this problem.

In Chapter I, we will sketch the procedure of the calculations, which we explained in [Mr1] precisely. In Chapter II, the reader will find how to reduce the calculations of functional equations and residues to the explicit computations of Fourier transforms under suitable convergence conditions. More precise calculations for some specific examples will appear in a future paper.

We close the introduction by giving some comments to relevant papers. Satake and Faraut [Sk-Fa] carried out the Fourier transform of the complex powers of relative invariants on some prehomogeneous vector spaces including our case by making use of the analysis on Jordan algebra. In [Mr 1] the author did similar calculations by using microlocal analysis in the cases of the space of symmetric matrices, complex Hermitian matrices and quaternion Hermitian matrices. On the other hand, F. Sato [Sf 1–3] has obtained a generalization of Sato-Shintani's construction of zeta functions, but it is unknown that microlocal analysis works well for F. Sato's examples.

Notations. We denote by $\mathbb{N}, \mathbb{Z}, \mathbb{Q}, \mathbb{R}$ and $\mathbb{C}$, the set of natural numbers, the ring of rational integers, the rational number field, the real number field and the complex number field, respectively. For a vector space X, $GL(X)$ means the group of invertible linear endomorphisms of X.

Chapter I. Microlocal Analysis of Invariant Hyperfunctions with respect to the Group E_6

In order to introduce zeta functions associated with the exceptional Lie group of type E_6 and to carry out calculations of their functional equations and residues, we need to analyze invariant hyperfunctions under the group action of the exceptional Lie group of type E_6. In particular, computing Fourier transforms of invariant distributions is the most important. Our final goal of this chapter is to give explicit formulas of Fourier transforms of some type of invariant distributions with respect to the group of type E_6, which are given in § 7 and § 8. Our tool is microlocal analysis mainly developed by Sato and Kashiwara.

§ 1. A realization of the exceptional Lie group of type E_6

We begin with the definition of complex Cayley algebra. Let $\mathbb{H}_C$ be the complex quaternion algebra, which is uniquely determined up to isomorphisms. We define the *complex Cayley algebra* $\mathfrak{C}_C$ to be an $\mathbb{H}_C$-free module of rank two whose basis is $\{1, e\}$; the multiplication low of the algebra $\mathfrak{C}_C$ is given by:

$$(1.1) \qquad (q+re)\cdot(s+te) := (qs-\bar{t}r)+(tq+r\bar{s})e,$$

where q, r, s and t are elements in $\mathbb{H}_C$ and $\bar{t}$ and $\bar{s}$ stand for the conjugate quaternions of t and s, respectively. Then $\mathfrak{C}_C$ is an 8-dimensional non-associative algebra defined over $\mathbb{C}$. For an element $a := q+re \in \mathfrak{C}_C$, we define the *conjugate* Cayley number $\bar{a}$ by $\bar{a} := \bar{q} - re$. We may easily prove that $\overline{a \cdot b} = \bar{b} \cdot \bar{a}$ for any $a, b \in \mathfrak{C}_{\mathrm{C}}$. We put:

$$(1.2) \qquad N(a) := a \cdot \bar{a} \quad \text{and} \quad T(a) := a + \bar{a},$$

and call them the *norm* and the *trace* of $a \in \mathfrak{C}_{\mathrm{C}}$, respectively. The norm and the trace of a are complex numbers.

We put $\boldsymbol{X}_{\mathrm{C}} := \mathbf{Her}_3(\mathfrak{C}_{\mathrm{C}})$, where $\mathbf{Her}_3(\mathfrak{C}_{\mathrm{C}})$ is the set of 3×3 Cayley Hermitian matrices, i.e., the set of 3×3 matrices x whose entries are in $\mathfrak{C}_{\mathrm{C}}$ and which satisfy ${}^t\bar{x} = x$. Then $\boldsymbol{X}_{\mathrm{C}}$ is a 27-dimensional complex vector space. For an element

$$x := \begin{bmatrix} \xi_1, & x_3, & \bar{x}_2 \\ \bar{x}_3, & \xi_2, & x_1 \\ x_2, & \bar{x}_1, & \xi_3 \end{bmatrix} \qquad \text{in } \boldsymbol{X}_{\mathrm{C}},$$

we put:

$$(1.3) \qquad \begin{aligned} \det(x) &:= \xi_1\xi_2\xi_3 - \xi_1 N(x_1) - \xi_2 N(x_2) - \xi_3 N(x_3) + T((x_3 \cdot x_1) \cdot x_2), \\ \operatorname{tr}(x) &:= \xi_1 + \xi_2 + \xi_3, \end{aligned}$$

and call them the *determinant* and the *trace* of x, respectively. Let $\boldsymbol{G}_{\mathrm{C}}^1$ be the subgroup of the complex general linear group $\boldsymbol{GL}(\boldsymbol{X}_{\mathrm{C}})$ consisting of elements which keep the polynomial $\det(x)$ invariant. Then $\boldsymbol{G}_{\mathrm{C}}^1$ is a closed linear algebraic subgroup of $\boldsymbol{GL}(\boldsymbol{X}_{\mathrm{C}})$, which is isomorphic to the complex exceptional Lie group of type $\boldsymbol{E}_6$. We denote it by $\boldsymbol{E}_{6\mathrm{C}}$. The group $\boldsymbol{G}_{\mathrm{C}}^1$ is a 78-dimensional complex Lie group.

We put $\boldsymbol{G}_{\mathrm{C}} := \boldsymbol{GL}_1(\mathbf{C}) \times \boldsymbol{E}_{6\mathrm{C}}$. We define a representation ρ of $\boldsymbol{G}_{\mathrm{C}}$ into $\boldsymbol{GL}(\boldsymbol{X}_{\mathrm{C}})$ by:

$$(1.4) \qquad \rho(g) : x \longmapsto g_1(g_2 \cdot x),$$

with $g := (g_1, g_2) \in \boldsymbol{G}_{\mathrm{C}}$ and $x \in \boldsymbol{X}_{\mathrm{C}}$. We may naturally regard $\boldsymbol{G}_{\mathrm{C}}$ as a subgroup of $\boldsymbol{GL}(\boldsymbol{X}_{\mathrm{C}})$ by the inclusion map ρ. On the other hand, when we put:

$$(1.5) \qquad \langle x, y \rangle := \operatorname{tr}(x \cdot y),$$

it gives a non-degenerate complex bilinear form on $\boldsymbol{X}_{\mathrm{C}}$ and hence it is an inner product on $\boldsymbol{X}_{\mathrm{C}}$. We may identify $\boldsymbol{X}_{\mathrm{C}}$ with its dual vector space $\boldsymbol{X}_{\mathrm{C}}^*$ through the inner product (1.5). The contragredient representation ρ^* on $\boldsymbol{X}_{\mathrm{C}}$ with respect to (1.5) is defined naturally and $\boldsymbol{G}_{\mathrm{C}}$ is embedded in $\boldsymbol{GL}(\boldsymbol{X}_{\mathrm{C}})$ by ρ^*. The images of $\boldsymbol{G}_{\mathrm{C}}^1$ by ρ and ρ^* are the same. The orbit structure of $\boldsymbol{X}_{\mathrm{C}}$ with respect to the group action of $\rho(\boldsymbol{G}_{\mathrm{C}})$ and $\rho^*(\boldsymbol{G}_{\mathrm{C}})$ is the same, that is to say, the orbital decompositions of $\boldsymbol{X}_{\mathbf{C}}$ by the group $\rho(\boldsymbol{G}_{\mathbf{C}})$ and $\rho^*(\boldsymbol{G}_{\mathbf{C}})$ are the same.

We put $P(X):=\det(x)$. Then $P(X)$ is an irreducible polynomial of degree 3 on $\boldsymbol{X}_{\mathbf{C}}$. Apparently we have:

$$P(\rho(g)x)=\chi(g)P(x),$$

for all $g:=(g_1, g_2)\in \boldsymbol{G}$ where $\chi(g)=g_1^3$. Namely $P(x)$ is an irreducible relatively invariant polynomial corresponding to the character $\chi(g)$ on $\boldsymbol{X}_{\mathbf{C}}$ under the action $\rho(g)$. For the contragredient action ρ^*, we have $P(\rho^*(g)\cdot x)=\chi(g)^{-1}P(x)$, which means that $P(x)$ is a relatively invariant polynomial corresponding to the character $\chi(g)^{-1}$ under the contragredient action $\rho^*(g)$. We put:

$$\boldsymbol{S}_{\mathbf{C}}:=\{x\in \boldsymbol{X}_{\mathbf{C}};\, P(x)=0\}, \tag{1.6}$$

and call it the *singular set* of $\boldsymbol{X}_{\mathbf{C}}$.

Proposition 1.1. 1) *The triplets* $(\boldsymbol{G}_{\mathbf{C}}, \rho, \boldsymbol{X}_{\mathbf{C}})$ *and* $(\boldsymbol{G}_{\mathbf{C}}, \rho^*, \boldsymbol{X}_{\mathbf{C}})$ *are regular irreducible prehomogeneous vector spaces with an irreducible relatively invariant polynomial* $P(x)=\det(x)$.

2) *The vector space* $\boldsymbol{X}_{\mathbf{C}}$ *decomposes into four orbits*; *the open orbit is unique and it is* $\boldsymbol{X}_{\mathbf{C}}-\boldsymbol{S}_{\mathbf{C}}$; *the singular set* $\boldsymbol{S}_{\mathbf{C}}$ *decomposes into the following three orbits*:

$$\begin{aligned}
&\boldsymbol{S}_{1\mathbf{C}}:=\rho(\boldsymbol{G}_{\mathbf{C}})\cdot\begin{bmatrix}1&&\\&1&\\&&0\end{bmatrix}=\rho^*(\boldsymbol{G}_{\mathbf{C}})\cdot\begin{bmatrix}1&&\\&1&\\&&0\end{bmatrix},\\
&\boldsymbol{S}_{2\mathbf{C}}:=\rho(\boldsymbol{G}_{\mathbf{C}})\cdot\begin{bmatrix}1&&\\&0&\\&&0\end{bmatrix}=\rho^*(\boldsymbol{G}_{\mathbf{C}})\cdot\begin{bmatrix}1&&\\&0&\\&&0\end{bmatrix},\\
&\boldsymbol{S}_{3\mathbf{C}}:=\{0\}.
\end{aligned} \tag{1.7}$$

3) *The orbits in* (1.7) *are not only* $\boldsymbol{G}_{\mathbf{C}}$*-orbits but also* $\boldsymbol{G}^1_{\mathbf{C}}$*-orbits.*

Proof. The claims 1) and 2) were proved in Sato-Kimura [Sm-Ki] Proposition 47 and Kimura [Ki] Proposition 6-2, respectively. For 3), we may easily check the validity by computing the dimension of the $\boldsymbol{G}^1_{\mathbf{C}}$-orbits generated by

$$\begin{bmatrix}1&&\\&1&\\&&0\end{bmatrix},\ \begin{bmatrix}1&&\\&0&\\&&0\end{bmatrix}\quad\text{and}\quad\begin{bmatrix}0&&\\&0&\\&&0\end{bmatrix}.$$

q.e.d.

§ 2. An invariant holonomic system $\mathfrak{M}_s$ and its holonomy diagram

In this section we shall give a holonomic system $\mathfrak{M}_s$ having the solution $P(x)^s$ and give the complex holonomy diagram of it. For the details about theoretical aspect of microlocal analysis, see [Sm-Ka-Ki-Os], and for the technical aspect of it, see [Ki] Proposition 6-2.

Let $\mathscr{G}_{\mathrm{C}}$ be the Lie algebra of $\boldsymbol{G}_{\mathrm{C}}$ and let $d\rho$ and $\delta\chi$ be infinitesimal representations of ρ and χ, respectively. Consider the following system of linear differential equations with one unknown function $u(x)$:

$$(2.1) \qquad \mathfrak{M}_s\colon \left(\left\langle d\rho(A)x, \frac{\partial}{\partial x}\right\rangle - s\delta\chi(A)\right)u(x)=0, \qquad \text{for all } A \in \mathscr{G}_{\mathrm{C}}.$$

on $\boldsymbol{X}_{\mathrm{C}}$. Here, s is a complex number, $\langle\ ,\ \rangle$ is the canonical bilinear form on $\boldsymbol{X}_{\mathrm{C}}\times\boldsymbol{X}_{\mathrm{C}}^*$, and $\partial/\partial x$ is the first order homogeneous differential operator corresponding to the dual basis.

We see that $\mathfrak{M}_s$ is a holonomic system, i.e., the characteristic variety $\mathrm{ch}(\mathfrak{M}_s)$ is a Lagrangian subvariety in $T^*\boldsymbol{X}_{\mathrm{C}}$, as a consequence of the following proposition:

Proposition 2.1. *The characteristic variety* $\mathrm{ch}(\mathfrak{M}_s)$ *is contained in* $\bigcup_{i=1}^n \overline{T^*_{\boldsymbol{S}_{i\mathrm{C}}}\boldsymbol{X}_{\mathrm{C}}}$ *where* $T^*_{\boldsymbol{S}_{i\mathrm{C}}}\boldsymbol{X}_{\mathrm{C}}$ *is the conormal bundle of* $\boldsymbol{S}_{i\mathrm{C}}$ *in* $\boldsymbol{X}_{\mathrm{C}}$. *Here the upper bar means the Zariski closure of the variety.*

For the proof, see the argument in Kashiwara [Ka 1] p. 116 Example 1 or Muro [Mr 1].

From the definition, each conormal bundles $T^*_{\boldsymbol{S}_{i\mathrm{C}}}\boldsymbol{X}_{\mathrm{C}}$ is an irreducible Lagrangian subvariety in $T^*\boldsymbol{X}_{\mathrm{C}}$. Therefore $\mathfrak{M}_s$ is a holonomic system on $\boldsymbol{X}_{\mathrm{C}}$. We denote by $\Lambda_{i\mathrm{C}}$ the Zariski closure $\overline{T^*_{\boldsymbol{S}_{i\mathrm{C}}}\boldsymbol{X}_{\mathrm{C}}}$. From the argument in Kimura [Ki] § 6, $\mathfrak{M}_s$ is a simple holonomic system on each $\Lambda_{i\mathrm{C}}$ and the order of $\mathfrak{M}_s$ on $\Lambda_{i\mathrm{C}}$ is $-is-i(4i-3)/2$. We may look upon the complex cotangent bundle $T^*\boldsymbol{X}_{\mathrm{C}}$ of $\boldsymbol{X}_{\mathrm{C}}$ as the product $\boldsymbol{X}_{\mathrm{C}}\times\boldsymbol{X}_{\mathrm{C}}^*$. The group $\boldsymbol{G}_{\mathrm{C}}$ acts on $T^*\boldsymbol{X}_{\mathrm{C}}\simeq\boldsymbol{X}_{\mathrm{C}}\times\boldsymbol{X}_{\mathrm{C}}^*\colon g\cdot(x, y)=(\rho(g)\cdot x, \rho^*(g)\cdot y)$ with $g\in\boldsymbol{G}_{\mathrm{C}}$ and $(x, y)\in\boldsymbol{X}_{\mathrm{C}}\times\boldsymbol{X}_{\mathrm{C}}^*$. It is easily proved that $T^*_{\boldsymbol{S}_{i\mathrm{C}}}\boldsymbol{X}_{\mathrm{C}}$ is a $\boldsymbol{G}_{\mathrm{C}}$-stable subset and so is the closure $\Lambda_{i\mathrm{C}}=\overline{T^*_{\boldsymbol{S}_{i\mathrm{C}}}\boldsymbol{X}_{\mathrm{C}}}$. Each $\Lambda_{i\mathrm{C}}$ decomposes into finitely many $\boldsymbol{G}_{\mathrm{C}}$-orbits. Namely:

Proposition 2.2. *The* $\boldsymbol{G}_{\mathrm{C}}$*-stable Lagrangian subvariety* $\Lambda_{i\mathrm{C}}$ *has the* $\boldsymbol{G}_{\mathrm{C}}$*-orbital decomposition*:

$$(2.2) \qquad \coprod_{\substack{p\geq i\\ q\geq 3-i}} \boldsymbol{G}_{\mathrm{C}}\cdot\left(\begin{bmatrix} I_{3-p} & \\ & 0_p\end{bmatrix}, \begin{bmatrix} 0_q & \\ & I_{3-q}\end{bmatrix}\right)=\Lambda_{i\mathrm{C}}.$$

Here I_i *means* $i\times i$ *identity matrix.*

This proposition can be easily proved by the direct calculation so we omit the proof. We denote by $\Sigma_{p,q\mathbf{C}}$ the $\boldsymbol{G}_{\mathbf{C}}$-orbit

$$\boldsymbol{G}_{\mathbf{C}}\cdot\left(\begin{bmatrix} I_{3-p} & \\ & 0_p \end{bmatrix}, \begin{bmatrix} 0_q & \\ & I_{3-q} \end{bmatrix}\right) \qquad \text{for each } 0\leq p,\, q\leq 3.$$

Finally we give the complex holonomy diagram of $\mathfrak{M}_s$.

Proposition 2.3. 1) *The Lagrangian subvarieties $\Lambda_{i\mathbf{C}}$ and $\Lambda_{j\mathbf{C}}$ have an intersection of codimension one if $|i-j|=1$. Otherwise there are no intersections of codimension one. The $\boldsymbol{G}_{\mathbf{C}}$-orbit $\Sigma_{i+1,3-i\mathbf{C}}$ is an open dense orbit in $\Lambda_{i\mathbf{C}}\cap\Lambda_{i+1\mathbf{C}}$ $(i=0, 1, 2)$.*

2) *For each $p\in\Sigma_{i+1,3-i}$, we have*

$$(T(\Lambda_{i\mathbf{C}}\cap\Lambda_{i+1\mathbf{C}}))_p=(T\Lambda_{i\mathbf{C}})_p\cap(T\Lambda_{i+1\mathbf{C}})_p$$

where $(TA)_p$ means the tangent space of A at p. Namely $\Lambda_{i\mathbf{C}}$ and $\Lambda_{i+1\mathbf{C}}$ have a regular intersection at each $p\in\Sigma_{i+1,3-i\mathbf{C}}$.

3) *The complex holonomy diagram of $\mathfrak{M}_s$ is*

$$(2.3)\qquad \begin{array}{ccccccc} 0 & & -s-\frac{1}{2} & & -2s-5 & & -3s-\frac{27}{2} \\ \circ & \longrightarrow & \circ & \longrightarrow & \circ & \longrightarrow & \circ \\ \Lambda_{0\mathbf{C}} & (s+1) & \Lambda_{1\mathbf{C}} & (s+5) & \Lambda_{2\mathbf{C}} & (s+9) & \Lambda_{3\mathbf{C}} \end{array}\ .$$

Each $\circ$ stands for an irreducible Lagrangian subvariety and the line $\longrightarrow$ means an intersection of codimension one. This implies that the b-function is:

$$(2.4)\qquad b(s):=(s+1)(s+5)(s+9).$$

Namely we have $P(\partial/\partial x)P(x)^{s+1}=\text{const. } b(s)\cdot P(x)^s$.

For the proof see [Ki].

§ 3. Real forms of the prehomogeneous vector spaces and real holonomy diagrams of $\mathfrak{M}_s$

In this section, we shall determine all the real forms of the prehomogeneous vector space $(\boldsymbol{G}_{\mathbf{C}}, \rho, \boldsymbol{X}_{\mathbf{C}}):=(\boldsymbol{GL}_1(\mathbb{C})\times\boldsymbol{E}_{6\mathbf{C}}, \rho, \mathbf{Her}_3(\mathfrak{C}_{\mathbf{C}}))$ and give their orbit structure. Let $(\boldsymbol{G}_{\mathbf{C}}, \rho, \boldsymbol{X}_{\mathbf{C}})$ be a given complex prehomogeneous vector space. For a real form $\boldsymbol{G}_{\mathbf{R}}$ of the complex Lie group $\boldsymbol{G}_{\mathbf{C}}$ and a real form $\boldsymbol{X}_{\mathbf{R}}$ of the complex vector space $\boldsymbol{X}_{\mathbf{C}}$, we say that $(\boldsymbol{G}_{\mathbf{R}}, \rho, \boldsymbol{X}_{\mathbf{R}})$ is a *real form* of the complex prehomogeneous vector space $(\boldsymbol{G}_{\mathbf{C}}, \rho, \boldsymbol{X}_{\mathbf{C}})$ if $\rho(\boldsymbol{G}_{\mathbf{R}})\subset \boldsymbol{GL}(\boldsymbol{X}_{\mathbf{R}})\cap\rho(\boldsymbol{G}_{\mathbf{C}})$.

We begin with determining all the possible real forms of the complex quaternion algebra. As well known, there are following two real forms of the complex quaternion algebras $\mathbb{H}_C$ and they are all:

(3.1) 1) (quaternion division algebra)

$\mathbb{H}_R^d :=$ a four dimensional real vector space generated by $\{1, u, v, w\}$ endowed with the structure of algebra defined over $\mathbb{R}$; the multiplication low is given by $w=uv=-vu$ and $u^2=v^2=-1$.

2) (quaternion split algebra)

$\mathbb{H}_R^s :=$ a four dimensional real vector space generated by $\{1, u, v, w\}$ endowed with the structure of algebra defined over $\mathbb{R}$; the multiplication low is given by $w=uv=-vu$ and $u^2=v^2=+1$.

For any element $p=\alpha+\beta u+\gamma v+\delta w$ in $\mathbb{H}_R^d$ or in $\mathbb{H}_R^s$, the quaternion conjugate $\bar{p}$ is defined to be $\bar{p}=\alpha-\beta u-\gamma v-\delta w$. The complexifications of $\mathbb{H}_R^d$ and $\mathbb{H}_R^s$ are naturally isomorphic to $\mathbb{H}_C$.

According to the real form of $\mathbb{H}_C$, there are two types of real forms of the complex Cayley algebra: one is the division Cayley algebra $\mathfrak{C}_R^d$ and the other is the split Cayley algebra $\mathfrak{C}_R^s$, which are defined to be a $\mathbb{H}_R^d$- and $\mathbb{H}_R^s$- free modules of rank two generated by $\{1, e\}$, respectively. The multiplication law is given by $(q+re)(s+te) := (qs-\bar{t}r)+(tq+r\bar{s})e$ for two elements $q+re$ and $s+te$ in $\mathfrak{C}_R^d$ or $\mathfrak{C}_R^s$. The trace $T(a)$ and the norm $N(a)$ of $a \in \mathfrak{C}_R^d$ or $\mathfrak{C}_R^s$ are defined in the same manner as in the case of the complex Cayley algebra $\mathfrak{C}_C$. The complexifications $\mathfrak{C}_R^d \otimes_R \mathbb{C}$ and $\mathfrak{C}_R^s \otimes_R \mathbb{C}$ are naturally isomorphic to $\mathfrak{C}_C$.

According to the real forms of $\mathfrak{C}_C$, there are two real forms of the complex vector space $\boldsymbol{X}_C = \mathbf{Her}_3(\mathfrak{C}_C)$. They are

(3.2) 1) the division case: $\boldsymbol{X}_R := \mathbf{Her}_3(\mathfrak{C}_R^d)$

and

2) the split case: $\boldsymbol{X}_R := \mathbf{Her}_3(\mathfrak{C}_R^s)$.

The *determinant* $\det(x)$ and the *trace* $\operatorname{tr}(x)$ of an element x in $\boldsymbol{X}_R$ are defined in the same way as the complex case. By definition, they coincide with the restrictions of the polynomials $\det(x)$ and $\operatorname{tr}(x)$ on $\boldsymbol{X}_C$ to $\boldsymbol{X}_R$. Putting $\boldsymbol{G}_R^1 := \boldsymbol{G}_C^1 \cap \boldsymbol{GL}(\boldsymbol{X}_R)$, $\boldsymbol{G}_R^1$ is a real form of the complex Lie group $\boldsymbol{G}_C^1$ from the definition. Then $\boldsymbol{G}_R^+ := \boldsymbol{GL}_1(\mathbb{R})^+ \times \boldsymbol{G}_R^1$ is a real form of $\boldsymbol{G}_C$ and $(\boldsymbol{G}_R^+, \rho, \boldsymbol{X}_R)$ is a real form of the complex prehomogeneous vector space $(\boldsymbol{G}_C, \rho, \boldsymbol{X}_C)$. Here $\boldsymbol{GL}_1(\mathbb{R})^+$ is the connected component of $\boldsymbol{GL}_1(\mathbb{R})$ containing the identity. Thus we have two real forms of $(\boldsymbol{G}_C, \rho, \boldsymbol{X}_C)$.

Proposition 3.1. *All possible real forms of $(\boldsymbol{G}_C, \rho, \boldsymbol{X}_C)$ are the above two.*

Proof. As proved in Rubenthaler [Ru], there are only two possible real forms of the complex prehomogeneous vector space $(\boldsymbol{G}_{\mathrm{C}}, \rho, \boldsymbol{X}_{\mathrm{C}})$. The two real forms given in (3.2) are actually real forms. When $\boldsymbol{X}_{\mathrm{R}}=\mathbf{Her}_3(\mathfrak{C}^d_{\mathrm{R}})$, the group $\boldsymbol{G}^1_{\mathrm{R}}$ is the real exceptional Lie group of type $\boldsymbol{E}_6$ whose signature of the killing form is -26, which is denoted by $\boldsymbol{E}_6^{(-26)}$. When $\boldsymbol{X}_{\mathrm{R}}=\mathbf{Her}_3(\mathfrak{C}^s_{\mathrm{R}})$, $\boldsymbol{G}^1_{\mathrm{R}}$ is the real form of signature 6, which is denoted by $\boldsymbol{E}_6^{(6)}$. Thus the above two real forms in (3.2) are different to each other and they are all possible real forms. q.e.d.

After all, the following two real forms are obtained.

(3.3) 1) The division case:

$$\boldsymbol{G}^+_{\mathrm{R}} := \boldsymbol{GL}_1(\mathbb{R})^+ \times \boldsymbol{E}_6^{(-26)} \quad \text{and} \quad \boldsymbol{X}_{\mathrm{R}} := \mathbf{Her}_3(\mathfrak{C}^d_{\mathrm{R}}).$$

2) The split case:

$$\boldsymbol{G}^+_{\mathrm{R}} := \boldsymbol{GL}_1(\mathbb{R})^+ \times \boldsymbol{E}_6^{(6)} \quad \text{and} \quad \boldsymbol{X}_{\mathrm{R}} := \mathbf{Her}_3(\mathfrak{C}^s_{\mathrm{R}}).$$

The restriction of the inner product in (1.5) to $\boldsymbol{X}_{\mathrm{R}}$ is a real valued inner product on the real vector space $\boldsymbol{X}_{\mathrm{R}}$. The contragredient representation ρ^* of $\boldsymbol{G}_{\mathrm{R}}$ gives a real form $(\boldsymbol{G}_{\mathrm{R}}, \rho^*, \boldsymbol{X}_{\mathrm{R}})$ of $(\boldsymbol{G}_{\mathrm{C}}, \rho^*, \boldsymbol{X}_{\mathrm{C}})$. We denote by $\boldsymbol{G}^+_{\mathrm{R}}$ the connected component of $\boldsymbol{G}_{\mathrm{R}}$ containing the identity element. The character $\chi(g)$ is real-valued and positive on $\boldsymbol{G}^+_{\mathrm{R}}$ because we see by a direct computation that the restriction of $P(x)$ on $\boldsymbol{X}_{\mathrm{R}}$ is a polynomial with real coefficients in each real form.

For a fixed real form $(\boldsymbol{G}^+_{\mathrm{R}}, \rho, \boldsymbol{X}_{\mathrm{R}})$, we denote $\boldsymbol{S}_{\mathrm{R}} := \boldsymbol{S}_{\mathrm{C}} \cap \boldsymbol{X}_{\mathrm{R}}$ and call it the *real singular set*. The set $\boldsymbol{X}_{\mathrm{R}}$ has the same orbital decomposition under the action $\rho(\boldsymbol{G}^+_{\mathrm{R}})$ and $\rho^*(\boldsymbol{G}^+_{\mathrm{R}})$. Namely:

Proposition 3.2. 1) (*The division case*): *When $\boldsymbol{X}_{\mathrm{R}}=\mathbf{Her}_3(\mathfrak{C}^d_{\mathrm{R}})$, each real locus of the $\boldsymbol{G}_{\mathrm{C}}$-orbits in $\boldsymbol{X}_{\mathrm{C}}$ decomposes into the following connected components which are $\boldsymbol{G}^+_{\mathrm{R}}$-orbits*:

i) $\boldsymbol{X}_{\mathrm{R}}-\boldsymbol{S}_{\mathrm{R}}=\coprod_{i=0}^{3}\boldsymbol{O}_i$ (*disjoint union*), *where* $\boldsymbol{O}_i := \rho(\boldsymbol{G}^+_{\mathrm{R}})\cdot I_3^{(i)} = \rho^*(\boldsymbol{G}^+_{\mathrm{R}})\cdot I_3^{(i)}$ *with* $I_3^{(i)} := \begin{bmatrix} -I_i & \\ & I_{3-i}\end{bmatrix}$ $(i=0, 1, 2, 3)$.

ii) $\boldsymbol{S}_{1\mathrm{R}} := \boldsymbol{S}_{1\mathrm{C}} \cap \boldsymbol{X}_{\mathrm{R}} = \boldsymbol{S}_1^0 \amalg \boldsymbol{S}_1^1 \amalg \boldsymbol{S}_1^2$ (*disjoint union*), *where* $\boldsymbol{S}_1^i := \rho(\boldsymbol{G}^+_{\mathrm{R}})\cdot I_2^{(i)} = \rho^*(\boldsymbol{G}^+_{\mathrm{R}})\cdot I_2^{(i)}$ *with* $I_2^{(i)} := \begin{bmatrix} -I_i & & \\ & I_{2-i} & \\ & & 0\end{bmatrix}$ $(i=0, 1, 2)$.

iii) $\boldsymbol{S}_{2\mathrm{R}} := \boldsymbol{S}_{2\mathrm{C}} \cap \boldsymbol{X}_{\mathrm{R}} = \boldsymbol{S}_2^0 \amalg \boldsymbol{S}_2^1$ (*disjoint union*), *where* $\boldsymbol{S}_2^i := \rho(\boldsymbol{G}^+_{\mathrm{R}})\cdot I_1^{(i)} = \rho^*(\boldsymbol{G}^+_{\mathrm{R}})\cdot I_1^{(i)}$ *with* $I_1^{(i)} := \begin{bmatrix} (-1)^i & \\ & 0_2\end{bmatrix}$ $(i=0, 1)$.

vi) $\boldsymbol{S}_{3\mathrm{R}} := \boldsymbol{S}_{3\mathrm{C}} \cap \boldsymbol{X}_{\mathrm{R}} = \boldsymbol{S}_3^0 := \{0\}$ (*the origin*).

2) *The split case*: *When $\boldsymbol{X}_{\mathrm{C}}=\mathbf{Her}_3(\mathfrak{C}^s_{\mathrm{R}})$, each real locus of the $\boldsymbol{G}_{\mathrm{C}}$-orbits in $\boldsymbol{X}_{\mathrm{C}}$ decomposes into the following connected components which are $\boldsymbol{G}^+_{\mathrm{R}}$-orbits.*

i) $X_{\mathbb{R}}-S_{\mathbb{R}}=O_+\amalg O_-$ *with* $O_+ :=\{x\in X_{\mathbb{R}};\det(x)>0\}$ *and* $O_- :=\{x\in X_{\mathbb{R}};\det(x)<0\}$.

ii) $S_{i\mathbb{R}}:=S_{i\mathbb{C}}\cap X_{\mathbb{R}}\ (i=1,2,3)$ *is a single* $G^+_{\mathbb{R}}$*-orbit. We denote it by* S^0_i.

3) *In both of the cases* 1) *and* 2), *the* $G^+_{\mathbb{R}}$*-orbits in the real singular set* $S_{\mathbb{R}}:=S_{\mathbb{C}}\cap X_{\mathbb{R}}$ *are* $G^1_{\mathbb{R}}$*-orbits.*

For the proof of this proposition we have only to carry out similar computation as we did in the complex case (see [Ki]). We omit the proof because it is complicated.

We conclude this section by calculating the real holonomy diagrams of $\mathfrak{M}_s$, which indicates the configuration of the real Lagrangian subvarieties appearing in $\mathrm{ch}(\mathfrak{M}_s)_{\mathbb{R}}:=\mathrm{ch}(\mathfrak{M}_s)\cap T^*X_{\mathbb{R}}$. For the precise definition of a real holonomy diagram, see [Ka-Mi], [Mr 1], p. 414 or [Mr 3]. Each real locus $\Lambda_{i\mathbb{R}}:=\Lambda_{i\mathbb{C}}\cap T^*X_{\mathbb{R}}$ of $\Lambda_{i\mathbb{C}}$ is a real Lagrangian subvariety. Indeed, we may easily check that $\Lambda_{i\mathbb{R}}:=\overline{T^*_{S_{i\mathbb{R}}}X_{\mathbb{R}}}$ where $S_{i\mathbb{R}}:=S_{i\mathbb{C}}\cap X_{\mathbb{R}}$. For each irreducible component $\Lambda_{i\mathbb{C}}$ of $\mathrm{ch}(\mathfrak{M}_s)$, we denote by $\Lambda^\circ_{i\mathbb{C}}$ the subset of $\Lambda_{i\mathbb{C}}$ consisting of points at which $\mathrm{ch}(\mathfrak{M}_s)$ is non-singular, i.e., if $\xi\in\Lambda_{i\mathbb{C}}$, then there exists a neighborhood U of ξ such that $U\cap\mathrm{ch}(\mathfrak{M}_s)=U\cap\Lambda_{i\mathbb{C}}$ and it is a non-singular variety. Then $\Lambda^\circ_{i\mathbb{C}}$ is an open dense non-singular subset of $\Lambda_{i\mathbb{C}}$. In our case, we see that $\Lambda^\circ_{i\mathbb{C}}=\Sigma_{i,3-i}$. The real locus $\Lambda^\circ_{i\mathbb{R}}:=\Lambda^\circ_{i\mathbb{C}}\cap T^*X_{\mathbb{R}}$ decomposes into finitely many connected components, each of which is a $G^+_{\mathbb{R}}$-orbit. Explicit calculation is easy and the result is the following.

(3.5) 1) The division case: $X_{\mathbb{R}}=\mathbf{Her}_3(\mathfrak{C}^d_{\mathbb{R}})$.

$$\Lambda^\circ_{i\mathbb{R}}:=\coprod_{\substack{0\le p\le 3-i\\ 0\le q\le i}}\Lambda^{p,q}_i \qquad (i=0,1,2,3)$$

where $$\Lambda^{p,q}_i:=G^+_{\mathbb{R}}\cdot\left(\begin{bmatrix} I^{(p)}_{3-i} & \\ & 0_i\end{bmatrix},\begin{bmatrix}0_{3-i} & \\ & I^{(q)}_i\end{bmatrix}\right),$$

with $$I^{(k)}_j:=\begin{bmatrix}I_k & \\ & -I_{j-k}\end{bmatrix}\in M_j(\mathbb{R}).$$

2) The split case: $X_{\mathbb{R}}=\mathbf{Her}_3(\mathfrak{C}^s_{\mathbb{R}})$.

$$\Lambda^\circ_{i\mathbb{R}}:=\coprod_{\varepsilon=\pm}\Lambda^\varepsilon_i \qquad (i=0,1,2,3)$$

where $$\Lambda^\varepsilon_i:=G^+_{\mathbb{R}}\cdot\left(\begin{bmatrix}I_2 & \\ & \varepsilon\end{bmatrix},\begin{bmatrix}0_3\end{bmatrix}\right) \qquad (i=0) \quad \text{or}$$

$$\Lambda^\varepsilon_i:=G^+_{\mathbb{R}}\cdot\left(\begin{bmatrix}I_{3-i} & \\ & 0_i\end{bmatrix},\begin{bmatrix}0_{3-i} & & \\ & \varepsilon & \\ & & I_{i-1}\end{bmatrix}\right) \qquad (i\ge 1).$$

Next we consider the real locus of the codimension one intersection $\Sigma_{i+1,3-i\mathbb{C}}$ between $\Lambda_{i\mathbb{C}}$ and $\Lambda_{i+1\mathbb{C}}$: $\Sigma_{i+1,3-i\mathbb{C}}\cap T^*X_{\mathbb{R}}$ $(i=0,1,2)$. It is a regular intersection of real codimension one between $\Lambda_{i\mathrm{R}}$ and $\Lambda_{i+1\mathrm{R}}$. Each connected component of $\Sigma_{i+1,3-i\mathrm{R}}$ is a G_{R}^{+}-orbit. We have the following decomposition of $\Sigma_{i+1,3-i\mathrm{R}}$ into G_{R}^{+}-orbits in each case.

(3.6) 1) The division case: $X_{\mathrm{R}}=\mathbf{Her}_3(\mathfrak{C}_{\mathrm{R}}^{d})$

$$\Sigma_{i+1,3-i\mathrm{R}}=\coprod_{\substack{0\le p\le 2-i\\ 0\le q\le i}}\Sigma_i^{p,q}\qquad (i=0,1,2)$$

where $\Sigma_i^{p,q}$ is a G_{R}^{+}-orbit generated by

$$\left(\begin{bmatrix} I_{2-i}^{(p)} & \\ & 0_{i+1}\end{bmatrix},\begin{bmatrix} 0_{3-i} & \\ & I_i^{(q)}\end{bmatrix}\right).$$

2) The split case: $X_{\mathrm{R}}=\mathbf{Her}_3(\mathfrak{C}_{\mathrm{R}}^{s})$.

$\Sigma_{i+1,3-i\mathrm{R}}$ is a single G_{R}^{+}-orbit generated by

$$\left(\begin{bmatrix} I_{2-i} & \\ & 0_{i+1}\end{bmatrix},\begin{bmatrix} 0_{3-i} & \\ & I_i\end{bmatrix}\right),$$

which is denoted by Σ_i $(i=0, 1, 2)$.

Let Σ be a connected component of $\Sigma_{i+1,3-i\mathrm{R}}$. Take a point (x, y) in Σ. Then there exist two connected components Λ_i^e and Λ_i^f in $\Lambda_{i\mathrm{R}}^{\circ}$, and, Λ_{i+1}^g and Λ_{i+1}^h in $\Lambda_{i+1\mathrm{R}}^{\circ}$ such that $\Lambda_{i\mathrm{R}}-\Sigma=\Lambda_i^e\amalg\Lambda_i^f$ and $\Lambda_{i+1\mathrm{R}}-\Sigma=\Lambda_{i+1}^g\amalg\Lambda_{i+1}^h$. Such Λ_i^e, Λ_i^f, Λ_{i+1}^g and Λ_{i+1}^h do not depend on the choice of the point (x, ξ) in Σ. In other words, $\Lambda_i^e\cup\Lambda_i^f\cup\Sigma$ and $\Lambda_{i+1}^g\cup\Lambda_{i+1}^h\cup\Sigma$ coincide with $\Lambda_{i\mathrm{R}}$ and $\Lambda_{i+1\mathrm{R}}$ in a neighborhood of (x, y), respectively and they have the intersection Σ. We denote this situation by the following diagram.

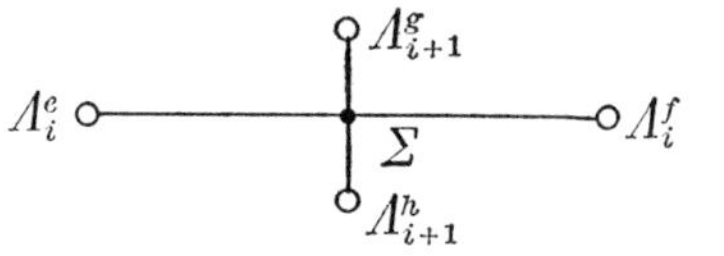

The real holonomy diagram between $\Lambda_{i\mathrm{R}}$ and $\Lambda_{i+1\mathrm{R}}$ is a diagram obtained by drawing all the intersections of codimension one between $\Lambda_{i\mathrm{R}}$ and $\Lambda_{i+1\mathrm{R}}$. In our case we have the following real holonomy diagrams.

(3.7) The division case: $X_{\mathrm{R}}=\mathbf{Her}_3(\mathfrak{C}_{\mathrm{R}}^{d})$,

1) The intersection of $\Lambda_{0\mathrm{R}}$ and $\Lambda_{1\mathrm{R}}$.

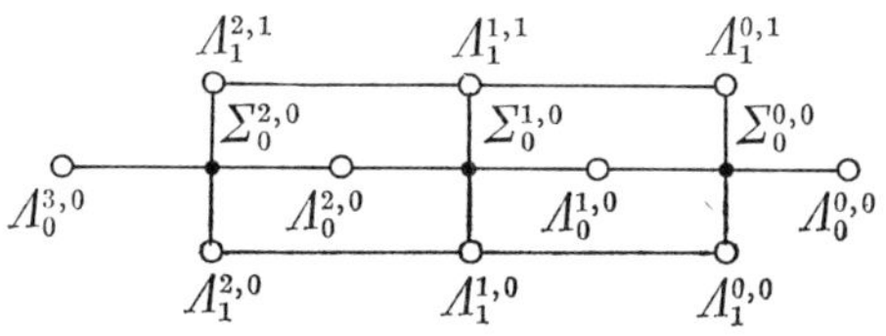

2) The intersection of $\Lambda_{1\mathrm{R}}$ and $\Lambda_{2\mathrm{R}}$.

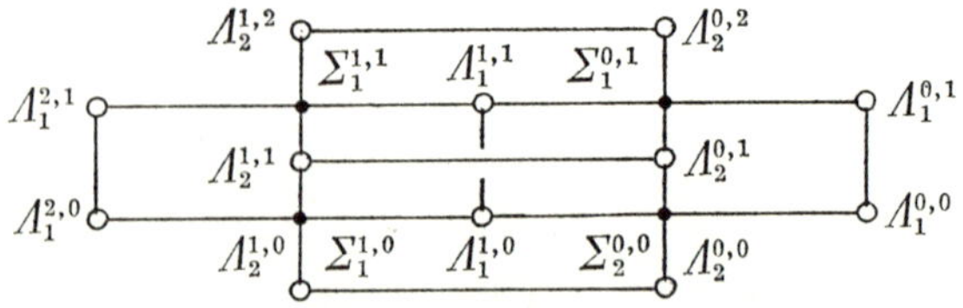

3) The intersection of $\Lambda_{2\mathrm{R}}$ and $\Lambda_{3\mathrm{R}}$.

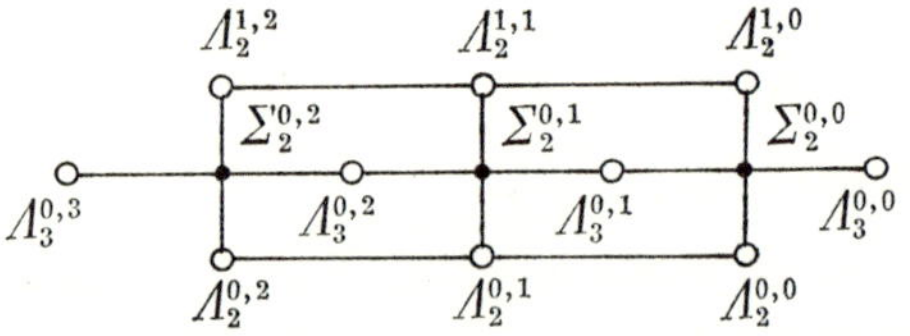

(3.8) The split case: $\boldsymbol{X}_{\mathrm{R}}=\mathbf{Her}_3(\mathfrak{C}_{\mathrm{R}}^s)$,
The intersection of $\Lambda_{i\mathrm{R}}$ and $\Lambda_{i+1\mathrm{R}}$:

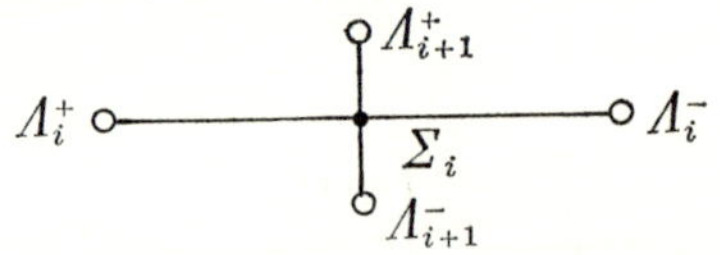

§ 4. Local zeta functions as a solution of the holonomic system $\mathfrak{M}_s$

Let $\mathfrak{M}_s$ be the holonomic system defined in (2.1) and let $(\boldsymbol{G}_{\mathrm{R}}^+, \rho, \boldsymbol{X}_{\mathrm{R}})$ be a real form of the complex prehomogeneous vector space $(\boldsymbol{G}_{\mathrm{C}}, \rho, \boldsymbol{X}_{\mathrm{C}})$. For a hyperfunction $f(x)$ on $\boldsymbol{X}_{\mathrm{R}}$, we say that $f(x)$ is a *χ^s-invariant* hyperfunction if it satisfies $f(\rho(g)\cdot x)=\chi(g)^s\cdot f(x)$ for all $g\in \boldsymbol{G}_{\mathrm{R}}^+$. Note the following proposition.

Proposition 4.1. *If $f(x)$ is a hyperfunction solution of $\mathfrak{M}_s$, then it is a χ^s-invariant tempered distribution. Conversely, if $f(x)$ is a χ^s-invariant hyperfunction, it is a tempered distribution and is a solution to $\mathfrak{M}_s$.*

Proof. Since $\mathfrak{M}_s$ is a simple holonomic system, $\mathfrak{M}_s$ is a completely regular holonomic system, hence any solution of $\mathfrak{M}_s$ is a tempered distribution on $X_{\mathbf{R}}$. It is easily checked that a solution $f(x)$ of $\mathfrak{M}_s$ is χ^s-invariant if it is a solution to $\mathfrak{M}_s$ in a more general situation. See for example Proposition 2.3 in [Mr 2]. q.e.d.

This proposition enables us to treat any χ^s-invariant hyperfunction as a solution of the holonomic system $\mathfrak{M}_s$.

Let $\coprod_{i\in \boldsymbol{I}_0} \boldsymbol{O}_i = \boldsymbol{X}_{\mathbf{R}} - \boldsymbol{S}_{\mathbf{R}}$ be the connected component decomposition of $\boldsymbol{X}_{\mathbf{R}} - \boldsymbol{S}_{\mathbf{R}}$ obtained in Proposition 3.2. Here the index set $\boldsymbol{I}_0$ is $\{0, 1, 2, 3\}$ in the division case and $\{+, -\}$ in the split case. We put:

$$|P(x)|_i^s := \begin{cases} |P(x)|^s & x \in \boldsymbol{O}_i \\ 0 & x \notin \boldsymbol{O}_i \end{cases} \tag{4.1}$$

with $s \in \mathbb{C}$. When the real part of s is sufficiently large, $|P(x)|_i^s$ is a continuous function. It can be continued with respect to s to the whole complex plane $\mathbb{C}$ as a tempered distribution with a meromorphic parameter $s \in \mathbb{C}$. The function defined by (4.1) stands for the extended one as a distribution with a meromorphic parameter in the whole complex plane. The possible poles of $|P(x)|_i^s$ is contained in the set of *critical points*: $\mathrm{Cri}\,(P(x)^s) := \{s \in ; b(s+k)=0 \text{ with } k \in \mathbb{N} \cup \{0\}\}$. Following Sato-Shintani [Sm-Sh], we call $|P(x)|_i^s$ a *local zeta function*. For a fixed complex number $\lambda \in \mathbb{C}$, $|P(x)|_i^\lambda$ is a χ^λ-invariant tempered distribution if $\lambda \notin \mathrm{Cri}\,(P(x)^s)$. Moreover let λ be a point in $\mathbb{C}$ and let $a_i(s)$ $(i \in \boldsymbol{I}_0)$ be holomorphic functions defined near $s=\lambda$ such that $\sum_{i\in \boldsymbol{I}_0} a_i(s)\cdot|P(x)|_i^s$ is holomorphic at $s=\lambda$. Then

$$\sum_{i\in \boldsymbol{I}_0} a_i(s)\cdot|P(x)|_i^s|_{s=\lambda}, \tag{4.2}$$

is a χ^λ-invariant tempered distribution. Conversely we have the following proposition.

Proposition 4.2. *Every χ^λ-invariant tempered distribution is written in the form in* (4.2).

The proof of this proposition has been given as a special case of the main theorem in [Mr 3]. A more general result related to Proposition 4.2 is announced in [Mr 5].

§ 5. Real principal symbols of the solutions of $\mathfrak{M}_s$ and their explicit expressions

Let $\mathscr{B}_{X_{\mathbf{R}}}$ be the sheaf of hyperfunctions on $X_{\mathbf{R}}$ and let $\mathscr{C}_{X_{\mathbf{R}}}$ be the sheaf of microfunctions on $T^*X_{\mathbf{R}}$ (and not on $\mathring{T}^*X_{\mathbf{R}} := T^*X_{\mathbf{R}} - X_{\mathbf{R}}$ as usually used). Then there is an isomorphism:

$$sp: \mathscr{B}_{X_{\mathrm{R}}} \xrightarrow{\sim} \pi_*(\mathscr{C}_{X_{\mathrm{R}}}), \tag{5.1}$$

where π is the projection map from $T^*\boldsymbol{X}_{\mathrm{R}}$ to $\boldsymbol{X}_{\mathrm{R}}$. Especially, for the restriction sheaf of $\mathscr{C}_{X_{\mathrm{R}}}$ to $T^*\boldsymbol{X}_{\mathrm{R}}-\boldsymbol{X}_{\mathrm{R}}$, we have Sato's fundamental exact sequence:

$$0 \longrightarrow \mathscr{A}_{X_{\mathrm{R}}} \longrightarrow \mathscr{B}_{X_{\mathrm{R}}} \longrightarrow \pi_*(\mathscr{C}_{X_{\mathrm{R}}}|_{T^*X_{\mathrm{R}}-X_{\mathrm{R}}}) \longrightarrow 0, \tag{5.2}$$

where $\mathscr{A}_{X_{\mathrm{R}}}$ is the sheaf of real analytic functions on $\boldsymbol{X}_{\mathrm{R}}$.

Let $f(x)$ be a hyperfunction solution to $\mathfrak{M}_s$ on $\boldsymbol{X}_{\mathrm{R}}$. It is a section of $\mathscr{B}_{X_{\mathrm{R}}}$ on $\boldsymbol{X}_{\mathrm{R}}$. Through the map sp in (5.1), $f(x)$ is regarded as a section of $\mathscr{C}_{X_{\mathrm{R}}}$ on $T^*\boldsymbol{X}_{\mathrm{R}}$, which is denoted by $sp(f(x))$. From Sato's fundamental theorem (Theorem 3.4.3 in [Ka-Kw-Ki]), we see that:

$$\boldsymbol{SS}(f(x)) := \mathrm{supp}\,(sp(f(x))) \subset \mathrm{ch}\,(\mathfrak{M}_s) \cap T^*\boldsymbol{X}_{\mathrm{R}}. \tag{5.3}$$

Let $\Lambda_{i\mathrm{C}}$ be an irreducible component of $\mathrm{ch}\,(\mathfrak{M}_s)$ and $\Lambda_{i\mathrm{R}} := \Lambda_{i\mathrm{C}} \cap T^*\boldsymbol{X}_{\mathrm{R}}$. The open dense subset $\Lambda_{i\mathrm{R}}^{\circ}$ in $\Lambda_{i\mathrm{R}}$ defined in § 3 decomposes into a finite number of connected components. We denote it by $\Lambda_{i\mathrm{R}}^{\circ} := \coprod_{j=1}^{j_i} \Lambda_i^j$: the connected component decomposition of $\Lambda_{i\mathrm{R}}^{\circ}$. We see that two solutions which coincide with each other on every Λ_i^j are the same. Namely:

Proposition 5.1. *Let $f(x)$ and $g(x)$ be two hyperfunction solutions of the holonomic system $\mathfrak{M}_s$ on $\boldsymbol{X}_{\mathrm{R}}$. If $sp(f(x))$ and $sp(g(x))$ coincide on every Λ_i^j ($i=0, 1, 2, 3$ and $j=1, \cdots, j_i$), then $f(x)=g(x)$ as hyperfunctions on $\boldsymbol{X}_{\mathrm{R}}$.*

Proof. The proof of this proposition is almost the same as the proof of Lemma 2.15 and Corollary 2.16 in [Mr 1]. A generalization of this theorem has been given in [Mr 4]. We shall give an outline of the proof.

We may show this proposition by applying an induction on the orbits and the Holmgren's uniqueness theorem. We have seen that $f(x)-g(x)$ is zero on $\boldsymbol{X}_{\mathrm{R}}-\boldsymbol{S}_{\mathrm{R}}$. Let $x \in \boldsymbol{S}_1$. There exists a neighborhood U of x such that $U \cap \boldsymbol{S}_{\mathrm{R}} = U \cap \boldsymbol{S}_{1\mathrm{R}}$. From the Holmgren's uniqueness theorem, we see that $(f-g)(x)=0$ on U by estimating the singular spectrum of $(f-g)$ on $\mathring{T}^*U$. Next we assume that $f-g$ is zero on $\boldsymbol{X}_{\mathrm{R}}-(\boldsymbol{S}_{k\mathrm{R}} \cup \cdots \cup \boldsymbol{S}_{3\mathrm{R}})$ $(2 \leq k \leq 3)$. Then, for any $x \in \boldsymbol{S}_{k\mathrm{R}}$, there exists a neighborhood U of x such that $U \cap (\boldsymbol{S}_{k\mathrm{R}} \cup \cdots \cup \boldsymbol{S}_{3\mathrm{R}}) = U \cap \boldsymbol{S}_{k\mathrm{R}}$. By applying Holmgren's uniqueness theorem, $f-g$ is 0 on U, which means $f-g=0$ on $X_{\mathrm{R}}-(\boldsymbol{S}_{k+1\mathrm{R}} \cup \cdots \cup \boldsymbol{S}_{3\mathrm{R}})$. Therefore by induction we obtain that $f-g$ is zero on $\boldsymbol{X}_{\mathrm{R}}$.

q.e.d.

With a microfunction solution $u(x)$ on Λ_i^j, we may associate a real analytic section of $|\Omega_{\Lambda_{i\mathbb{R}}}|^{\otimes 1/2}\otimes|\Omega_{X_{\mathbb{R}}}|^{\otimes -1/2}$ on Λ_i^j. Here $|\Omega_{\Lambda_{i\mathbb{R}}}|$ and $|\Omega_{X_{\mathbb{R}}}|$ are the sheaf of volume forms on $\Lambda_{i\mathbb{R}}$ and $\boldsymbol{X}_{\mathbb{R}}$, respectively. It is called the *real principal symbol* of the solution $u(x)$ and is denoted by $\sigma_{\Lambda_i^j}(u)$. For the precise definition of real principal symbols, see [Ka-Mi] or [Mr 1], pp. 416–419. Since the holonomic system $\mathfrak{M}_s$ is simple on each $\Lambda_{i\mathbb{C}}$, the dimension of the microfunction solution on each Λ_i^j is one. Any microfunction solution on Λ_i^j is written as a constant multiple of a non-trivial section of the solution space on Λ_i^j. When we denote by $\mathcal{S}ol\,(\mathfrak{M}_s)(\Lambda_i^j)$ (resp. $\mathcal{S}ymbol\,(\mathfrak{M}_s)(\Lambda_i^j)$) the space of microfunction solutions of $\mathfrak{M}_s$ on Λ_i^j (resp. the space of real principal symbols of the solutions on Λ_i^j), $\sigma_{\Lambda_i^j}$ gives a linear isomorphism:

$$
(5.4)\qquad
\begin{array}{ccc}
\sigma_{\Lambda_i^j}\colon \mathcal{S}ol\,(\mathfrak{M}_s)(\Lambda_i^j) & \longrightarrow & \mathcal{S}ymbol\,(\mathfrak{M}_s)(\Lambda_i^j) \\
\text{\rotatebox{90}{$\in$}} & & \text{\rotatebox{90}{$\in$}} \\
u(x) & \longmapsto & \sigma_{\Lambda_i^j}(u).
\end{array}
$$

Thus we may consider real principal symbols on Λ_i^j instead of microfunction solutions on Λ_i^j. Proposition 5.1 means that two hyperfunction solutions whose real principal symbols on each Λ_i^j coincide with each other are the same.

Next we will see that we may take a canonical basis for the space $\mathcal{S}ymbol\,(\mathfrak{M}_s)(\Lambda_i^j)$. Namely, for a microfunction solution $u(x)$ on Λ_i^j, we may express its principal symbol as:

$$
(5.5)\qquad \sigma_{\Lambda_i^j}(u(x))=c_{\Lambda_i^j}(u)\cdot|P_{\Lambda_{i\mathbb{R}}}|^s\cdot\sqrt{|\omega_{\Lambda_{i\mathbb{R}}}|}/\sqrt{|dx|}
$$

with a constant $c_{\Lambda_i^j}(u)$ depending only on u. Here:

$$
(5.6)\quad 1)\quad P_{\Lambda_{i\mathbb{R}}}:=P\circ\pi/\sigma^{m_{\Lambda_i}}|_{\Lambda_{i\mathbb{R}}^{\circ}},
$$

$$
\qquad\quad 2)\quad \omega_{\Lambda_{i\mathbb{R}}}:=\frac{\pi^{-1}(dx)\wedge d\sigma}{\sigma^{\mu_{\Lambda_i}}}\Big/ d\sigma\Big|_{\Lambda_{i\mathbb{R}}^{\circ}},
$$

and; 1) π is the projection map from $\boldsymbol{W}$ to $\boldsymbol{X}_{\mathbb{R}}$ where $\boldsymbol{W}$ is the Zariski closure of the set $\{(x,y)\in\boldsymbol{X}_{\mathbb{R}}\times\boldsymbol{X}_{\mathbb{R}}^*;\,x=s\cdot\operatorname{grad}(\log P(y)),\,y\in\boldsymbol{X}_{\mathbb{R}}^*-\boldsymbol{S}_{\mathbb{R}},\,s\in\mathbb{C}\}$; 2) $\sigma:=\langle x,y\rangle/3$, (a function on $\boldsymbol{W}$); 3) $dx:=d\xi_1\wedge d\xi_2\wedge d\xi_3\wedge dx_1\wedge dx_2\wedge dx_3$ where $dx_i=\bigwedge_{j=1}^{8}dx_i^j$ with $x_i:=(x_i^1+x_i^2u+x_i^3v+x_i^4w)+(x_i^5+x_i^6u+x_i^7v+x_i^8w)e\in\mathfrak{C}_{\mathbb{R}}^d$ or $\mathfrak{C}_{\mathbb{R}}^s$ ($x_i^j\in\mathbb{R}$ and $\{1,u,v,w\}$ is the basis of $\mathbb{H}^d$ or $\mathbb{H}^s$ defined in (3.1)); 4) the numbers m_{Λ_i} and μ_{Λ_i} are constants appearing in the expression of the order $-m_{\Lambda_i}s-(\mu_{\Lambda_i}/2)$ on $\Lambda_{i\mathbb{C}}$, namely $m_{\Lambda_i}=i$ and $\mu_{\Lambda_i}=4i^2-3i$. As proved in [Sm-Ka-Ki-Os], $P_{\Lambda_{i\mathbb{R}}}(x,y)$ (resp. $\omega_{\Lambda_{i\mathbb{R}}}(x,y)$) is a real analytic function (resp. a real analytic volume form) on $\Lambda_{i\mathbb{R}}^{\circ}$ which does not vanish on $\Lambda_{i\mathbb{R}}^{\circ}$. The above argument is a consequence of fundamental calculations developed in [Sm-Ka-Ki-Os]. For details see [Sm-Ka-Ki-Os].

Let $f(x)$ be a hyperfunction solution of $\mathfrak{M}_s$ on $\boldsymbol{X}_{\mathbb{R}}$. Then the real principal symbol of $sp(f(x))$ on Λ_i^j is written as:

$$\sigma_{\Lambda_i^j}(sp(f))=c_i^j(f)\cdot|P_{\Lambda i\mathrm{R}}|^s\cdot\sqrt{|\omega_{\Lambda i\mathrm{R}}|}/\sqrt{|dx|}, \tag{5.7}$$

with a constant $c_i^j(f)$ depending on f. We call the constant $c_i^j(f)$ the coefficient of $f(x)$ on Λ_i^j. Proposition 5.1 is rewritten in terms of the coefficients of microfunctions.

Proposition 5.2. *For two hyperfunction solutions $f(x)$ and $g(x)$, if the coefficient $c_i^j(sp(f))$ coincides with $c_i^j(sp(g))$ for each Λ_i^j in* $\mathrm{ch}(\mathfrak{M})_{\mathrm{R}}$, *then $f(x)=g(x)$ as hyperfunctions on X_{R}.*

As a result of this section, a hyperfunction solution f of $\mathfrak{M}_s$ is completely determined by its coefficients $\{c_i^j(f)\}$ over all the connected components of $\mathrm{ch}(\mathfrak{M}_s)_{\mathrm{reg}}:=$the regular locus of $\mathrm{ch}(\mathfrak{M}_s)$. In other words, a hyperfunction solution with a given set of coefficients is unique. However there may exist sets of coefficients to which no hyperfunction solutions correspond. For a given set of coefficients, there are some linear relations between the coefficients in order that it is a set of coefficients of a hyperfunction solution of $\mathfrak{M}_s$. In the next section we shall give their relations.

§ 6. A relation between real principal symbols on two Lagrangian varieties having an intersection of codimension one

We assume the same situation as in § 5. Let $f(x)$ be a hyperfunction solution. Since $\Lambda_{i\mathrm{C}}$ and $\Lambda_{i+1\mathrm{C}}$ have an intersection of codimension one, the microfunction solutions on $\Lambda_{i\mathrm{R}}^{\circ}$ and $\Lambda_{i+1\mathrm{R}}^{\circ}$ have linear relations. It was one of the main results in [Ka-Mi]. In [Mr 1], the author has given such solutions for relatively invariant microfunctions on other prehomogeneous vector spaces closely related to our case. It was presented as linear relation between the coefficients on $\Lambda_{i\mathrm{R}}^{\circ}$ and $\Lambda_{i+1\mathrm{R}}^{\circ}$. In our case, we have the following relations of the coefficients.

Proposition 6.1. *Let $f(x)$ be a hyperfunction solution of $\mathfrak{M}_s$. Then there are following linear relations among the coefficients of f on $\Lambda_{i\mathrm{R}}^{\circ}$ and $\Lambda_{i+1\mathrm{R}}^{\circ}$.*

i) *In the division case*:

$$\begin{bmatrix} c_{\Lambda_{i+1}^{j-1,k+1}}(f) \\ c_{\Lambda_{i+1}^{j-1,k}}(f)\end{bmatrix}=\frac{\Gamma(s+4i+1)}{\sqrt{2\pi}} \times\begin{bmatrix} \exp\left(-\frac{\pi}{2}\sqrt{-1}\,(s+1)\right), & \exp\left(\frac{\pi}{2}\sqrt{-1}\,(s+1)\right) \\ \exp\left(\frac{\pi}{2}\sqrt{-1}\,(s+1)\right), & \exp\left(-\frac{\pi}{2}\sqrt{-1}\,(s+1)\right)\end{bmatrix}\begin{bmatrix} c_{\Lambda_i^{j,k}}(f) \\ c_{\Lambda_i^{j-1,k}}(f)\end{bmatrix}, \tag{6.1}$$

provided that $s \neq -(4i+1), -(4i+1)-1, \cdots$. *For* $s=-(4i+1), -(4i+1)-1, \cdots$, *the inverse matrix of*

$$\frac{\Gamma(s+4i+1)}{\sqrt{2\pi}} \times \begin{bmatrix} \exp\left(-\frac{\pi}{2}\sqrt{-1}(s+1)\right), & \exp\left(\frac{\pi}{2}\sqrt{-1}(s+1)\right) \\ \exp\left(\frac{\pi}{2}\sqrt{-1}(s+1)\right), & \exp\left(-\frac{\pi}{2}\sqrt{-1}(s+1)\right) \end{bmatrix}$$

is *well defined and the relation is given by this inverse matrix.*

ii) *In the split case*:

$$\begin{bmatrix} c_{A^+_{i+1}}(f) \\ c_{A^-_{i+1}}(f) \end{bmatrix} = \frac{\Gamma(s+4i+1)}{\sqrt{2\pi}} \times \begin{bmatrix} \exp\left(-\frac{\pi}{2}\sqrt{-1}(s+1)\right), & \exp\left(\frac{\pi}{2}\sqrt{-1}(s+1)\right) \\ \exp\left(\frac{\pi}{2}\sqrt{-1}(s+1)\right), & \exp\left(-\frac{\pi}{2}\sqrt{-1}(s+1)\right) \end{bmatrix} \begin{bmatrix} c_{A^+_i}(f) \\ c_{A^-_i}(f) \end{bmatrix}, \tag{6.2}$$

provided that $s \neq -(4i+1), -(4i+1)-1, \cdots$. *When* $s=-(4i+1), -(4i+1)-1, \cdots$, *the inverse matrix of*

$$\frac{\Gamma(s+4i+1)}{\sqrt{2\pi}} \times \begin{bmatrix} \exp\left(-\frac{\pi}{2}\sqrt{-1}(s+1)\right), & \exp\left(\frac{\pi}{2}\sqrt{-1}(s+1)\right) \\ \exp\left(\frac{\pi}{2}\sqrt{-1}(s+1)\right), & \exp\left(-\frac{\pi}{2}\sqrt{-1}(s+1)\right) \end{bmatrix}$$

is well defined and the relation of the coefficients are given by this inverse matrix.

Proof. As proved in [Ka-Mi], the relations between the coefficients are obtained from the normal relation matrix corrected by the Maslov indices of Lagrangian subvarieties and their intersections. See [Ka-Mi] p. 139, formula 3.5. In our case there is no effect of correction by the Maslov indices. Following the method presented by [Ka-Mi], we have:

i) In the division case.

$$\begin{bmatrix} c_{A^{j-1,k+1}_{i+1}}(f) \\ c_{A^{j-1,k}_{i+1}}(f) \end{bmatrix} = \frac{\Gamma(s+4i-3)}{\sqrt{2\pi}} \times \begin{bmatrix} \exp\left(-\frac{\pi}{2}\sqrt{-1}(s+4i+1)\right), & \exp\left(\frac{\pi}{2}\sqrt{-1}(s+4i+1)\right) \\ \exp\left(\frac{\pi}{2}\sqrt{-1}(s+4i+1)\right), & \exp\left(-\frac{\pi}{2}\sqrt{-1}(s+4i+1)\right) \end{bmatrix} \tag{6.3}$$

$$\times\begin{bmatrix}\exp(2\pi\sqrt{-1}(i-2k)), & 0 \\ 0 & , \exp(-2\pi\sqrt{-1}(i-2k))\end{bmatrix}$$

$$\times\begin{bmatrix}c_{\Lambda_i^{j,k}}(f) \\ c_{\Lambda_i^{j-1,k}}(f)\end{bmatrix}.$$

ii) In the split case:

$$\begin{bmatrix}c_{\Lambda_{i+1}^+}(f) \\ c_{\Lambda_{i+1}^-}(f)\end{bmatrix}=\frac{\Gamma(s+4i-3)}{\sqrt{2\pi}}$$

$$(6.4)\qquad \times\begin{bmatrix}\exp\left(-\frac{\pi}{2}\sqrt{-1}(s+4i+1)\right), & \exp\left(\frac{\pi}{2}\sqrt{-1}(s+4i+1)\right) \\ \exp\left(\frac{\pi}{2}\sqrt{-1}(s+4i+1)\right), & \exp\left(-\frac{\pi}{2}\sqrt{-1}(s+4i+1)\right)\end{bmatrix}$$

$$\times\begin{bmatrix}\exp\left(\frac{\pi}{4}\sqrt{-1}\,0\right), & 0 \\ 0 & , \exp\left(\frac{\pi}{4}\sqrt{-1}\,0\right)\end{bmatrix}\begin{bmatrix}c_{\Lambda_i^+}(f) \\ c_{\Lambda_i^-}(f)\end{bmatrix}.$$

These relations (6.3) and (6.4) imply (6.1) and (6.2), respectively. q.e.d.

Remark. Conversely for a given set of the coefficients $\{c_{\Lambda_i^*}\}$ satisfying the relations (6.1) and (6.2), there exists a unique hyperfunction solution to $\mathfrak{M}_s$ whose coefficients $c_{\Lambda_i^*}(f)$ are the given ones. This fact is proved as a consequence derived from the main theorem in [Mr 3].

§ 7. Real principal symbols on the zero section and on the conormal bundle of the origin

In this section, we shall prove that the real principal symbol of a hyperfunction solution of $\mathfrak{M}_s$ has a specific expression especially on the zero-section and the conormal bundle of the origin. Moreover the real principal symbol on the conormal bundle is expressed in terms of the Fourier transform of $f(x)$. This enables us to reduce the calculation of the Fourier transform of $f(x)$ to the computation of the coefficients of $f(x)$ on the conormal bundle of the origin.

For a hyperfunction solution $f(x)$ of $\mathfrak{M}_s$, the real principal symbols of $f(x)$ on the zero section: $\Lambda_{i\mathrm{R}}^{\circ}:=(\boldsymbol{X}_{\mathrm{R}}-\boldsymbol{S}_{\mathrm{R}})\times\{0\}\subset\boldsymbol{X}_{\mathrm{R}}\times\boldsymbol{X}_{\mathrm{R}}^*$ (resp. the conormal bundle of the origin: $\Lambda_{3\mathrm{R}}^{\circ}:=\{0\}\times(\boldsymbol{X}_{\mathrm{R}}^*-\boldsymbol{S}_{\mathrm{R}})\subset\boldsymbol{X}_{\mathrm{R}}\times\boldsymbol{X}_{\mathrm{R}}^*$) have expression in terms of the coordinate system on $\boldsymbol{X}_{\mathrm{R}}$ (resp. $\boldsymbol{X}_{\mathrm{R}}^*$). In fact, let $\Lambda_0^{\boldsymbol{j}}$ ($\boldsymbol{j}\in\{(p,0); p=0,1,2,3\}$ in the division case and $\boldsymbol{j}\in\{+,-\}$ in the split case) be a connected component of Λ_{0R}° and let Λ_3^k ($k \in \{(0, q); q = 0, 1, 2, 3\}$ in the division case and $k \in \{+, -\}$ in the split case) be the connected component decomposition of Λ_{3R}°. Then we have:

$$(7.1) \quad 1) \quad \sigma_{\Lambda_0^j}(sp(f(x)) = c_0^j(f) \cdot |P(x)|_j^s$$

$$2) \quad \sigma_{\Lambda_3^k}(sp(f(x)) = c_3^k(f) \cdot |K_0|^s \cdot |K_1|^{1/2} \cdot |P(y)|_k^{-s-9} \sqrt{|dy|}/\sqrt{|dx|}$$

where $c_0^j(f)$ (resp. $c_3^k(f)$) are the coefficients of $f(x)$ on Λ_0^j (resp. Λ_3^k) with respect to $|P_{\Lambda 0R}|^s \sqrt{|\omega_{\Lambda 0R}|}/\sqrt{|dx|}$ (resp. $|P_{\Lambda 3R}|^s \sqrt{|\omega_{\Lambda 3R}|}/\sqrt{|dx|}$) and the constants $|K_0|$ and $|K_1|$ are defined by:

$$|K_0| := |P(y) \cdot P(\text{grad}\,(\log P(y)))| = 1,$$
$$|K_1| := |P(y)|^{-18} |\text{Hessian}\,(\log P(y))| = 2^{24}.$$

The equalities (7.1) 1) and 2) are proved by a direct calculation applying the equation (5.6). Namely substituting $|P_{\Lambda 0R}(x, y)| := |P(x)|$ and $|\omega_{\Lambda 0R}| := |dx|$, we have (7.1), 1). For $P_{\Lambda 3R}$ and $\omega_{\Lambda 3R}$, we see that

$$\begin{aligned} |P_{\Lambda 3R}(x, y)| &= |P(s \cdot \text{glad}\,(\log P(y)))/s^3| \\ &= |P(y)P(\text{grad}\,(\log P(y))) \cdot P(y)^{-1}| = |K_0| \cdot |P(y)^{-1}| \\ &= |P(y)^{-1}|. \\ |\omega_{\Lambda 3R}(x, y)| &= d(s \cdot \text{grad}\,(\log P(y)))/s^{27} \\ &= |P(y)|^{18} |\text{Hessian}\,(\log P(y))| \cdot |P(y)|^{-18} \cdot |dy| \\ &= |K_1| \cdot |P(y)|^{-18} |dy| = 2^{24} \cdot |P(y)|^{-18} |dy|. \end{aligned}$$

Thus by using (5.6), we get (7.1), 1) and 2).

Next we shall see that the real principal symbols of a solution $u(x)$ on the conormal bundle of the origin are written in terms of the Fourier transform of $u(x)$. For a tempered distribution $u(x)$ on X_R we put:

$$(7.2) \quad \begin{aligned} u^{\vee}(y) &:= \int u(x) \cdot \exp\,(-2\pi\sqrt{-1}\langle x, y\rangle) dx \\ u^{\wedge}(x) &:= \int u(y) \cdot \exp\,(2\pi\sqrt{-1}\langle x, y\rangle) dy. \end{aligned}$$

Then we have:

$$(7.3) \quad (u^{\vee})^{\wedge}(x) = (u^{\wedge})^{\vee}(x) = 2^{24} \cdot u(x).$$

We can write the real principal symbol on the conormal bundle of the origin by using the Fourier transform. Namely we have:

Proposition 7.1. *Let $f(x)$ be a solution of a holonomic system $\mathfrak{M}_s$.*

Then the Fourier transform $f^{\vee}(y)$ *is a real analytic function on* $\boldsymbol{X}_{\mathrm{R}}^{*}-\boldsymbol{S}_{\mathrm{R}}$ *and the real principal symbol on the conormal bundle of the origin is given by*:

$$\sigma_{\Lambda_3^k}(sp(f))=(2\pi)^{3s+(27/2)}\cdot 2^{24}\cdot f^{\vee}(y)|_{\Lambda_3^k}\sqrt{|dy|}/\sqrt{|dx|}. \tag{7.4}$$

Here we regard Λ_3^k *as a subset of* $\boldsymbol{X}_{\mathrm{R}}^{*}$.

Proof. By utilizing the same method as Proposition 3.5 in [Mr 3], we see that $f^{\vee}(y)$ is a solution to the holonomic system $\mathfrak{M}_{-s-9}$ when we identify $\boldsymbol{X}_{\mathrm{R}}$ and $\boldsymbol{X}_{\mathrm{R}}^{*}$. Thus $f^{\vee}(y)$ is real analytic on $\boldsymbol{X}_{\mathrm{R}}-\boldsymbol{S}_{\mathrm{R}}$. The distribution $f(x)$ is χ^s-invariant. As stated in Proposition 4.2, any χ^s-invariant hyperfunction is written as

$$\sum_{i\in \boldsymbol{I}_0} a_i(r)\cdot|P(x)|_i^r|_{r=s} \tag{7.5}$$

where $a_i(r)$'s are holomorphic functions in $r\in\mathbb{C}$ defined near $r=s$. For the χ^s-invariant hyperfunction $|P(x)|_i^s$ ($i\in \boldsymbol{I}_0$) with $s\notin \mathrm{Cri}\,(P(x)^s)$, we may prove (7.4) in the same way as the proof of Proposition 3.6, Formula (3.35) in [Mr 3]. Thus for any function $f(x)$ written in the form (7.5), we may prove by analytic continuation that the principal symbol of $f(x)$ on Λ_3^k is expressed as (7.4). q.e.d.

Comparing (7.1), 2) and (7.4), we have:

Proposition 7.2. *Let* $f(x)$ *be a solution of the holonomic system* $\mathfrak{M}_s$. *Then the Fourier transform is given by*:

$$f^{\vee}(y)|_{X_{\mathrm{R}}-S_{\mathrm{R}}}=(2\pi)^{-3s-(27/2)}\cdot 2^{-12}\cdot\sum_{k} c_3^k(f)\cdot|P(x)|^{-s-9}|_{\Lambda_3^k}, \tag{7.6}$$

where $\boldsymbol{k}$ *runs through the set* $\{(0, q);\ q=0, 1, 2, 3\}$ *in the division case or the set* $\{+, -\}$ *in the split case.*

§ 8. The Fourier transforms of local zeta functions

In this section, we shall give the formula of the Fourier transform of the relatively invariant tempered distributions $|P(x)|_i^s$ explicitly. As proved in the preceding section, we have to compute the real principal symbol on the conormal bundle of the origin. More precisely, our problem is to compute the coefficients of $|P(x)|_i^s$ on the conormal bundle of the origin (Proposition 7.2).

Theorem 8.1. (The formula of the Fourier transform of $|P(x)|_i^s$)

1) (*In the division case*): *We have*:

$$\int |P(x)|_i^s \cdot \exp(-2\pi\sqrt{-1}\langle x, y\rangle)dx$$
$$= (2\pi)^{-3s-(27/2)} \cdot 2^{-12} \cdot \sum_{j=0}^{3} c_{ij}(s) \cdot |P(y)|_j^{-s-9}, \qquad (i=0, 1, 2, 3) \tag{8.1}$$

where $c_{ij}(s) := A_{i+1,j+1}$ *with the* 4×4 *matrix* $A := (A_{p,q})_{1\le p,q\le 4}$ *defined by*:

$$A = (2\pi)^{-(3/2)} \cdot \Gamma(s+1) \cdot \Gamma(s+5) \cdot \Gamma(s+9) \times (-\sqrt{-1})$$
$$\times \begin{bmatrix} -\exp\left(-\frac{3\pi}{2}\sqrt{-1}s\right), & 3\exp\left(\frac{\pi}{2}\sqrt{-1}s\right), & -3\exp\left(-\frac{\pi}{2}\sqrt{-1}s\right), & \exp\left(\frac{3\pi}{2}\sqrt{-1}s\right) \\ \exp\left(-\frac{\pi}{2}\sqrt{-1}s\right), & -\exp\left(-\frac{3\pi}{2}\sqrt{-1}s\right) - 2\exp\left(\frac{\pi}{2}\sqrt{-1}s\right), & \exp\left(\frac{3\pi}{2}\sqrt{-1}s\right) + 2\exp\left(-\frac{\pi}{2}\sqrt{-1}s\right), & -\exp\left(\frac{\pi}{2}\sqrt{-1}s\right) \\ -\exp\left(\frac{\pi}{2}\sqrt{-1}s\right), & \exp\left(\frac{3\pi}{2}\sqrt{-1}s\right) + 2\exp\left(-\frac{\pi}{2}\sqrt{-1}s\right), & -\exp\left(-\frac{3\pi}{2}\sqrt{-1}s\right) - 2\exp\left(\frac{\pi}{2}\sqrt{-1}s\right), & \exp\left(-\frac{\pi}{2}\sqrt{-1}s\right) \\ \exp\left(\frac{3\pi}{2}\sqrt{-1}s\right), & -3\exp\left(-\frac{\pi}{2}\sqrt{-1}s\right), & 3\exp\left(\frac{\pi}{2}\sqrt{-1}s\right), & -\exp\left(-\frac{3\pi}{2}\sqrt{-1}s\right) \end{bmatrix}. \tag{8.2}$$

2) (*In the split case*); *We have*:

$$\int |P(x)|_i^s \cdot \exp(-2\pi\sqrt{-1}\langle x, y\rangle)dx$$
$$= (2\pi)^{-3s-(27/2)} \cdot 2^{-12} \cdot \sum_{j\in\{+,-\}} c_{ij}(s) \cdot |P(y)|_j^{-s-9}, \qquad (i=+,-) \tag{8.3}$$

where

$$\begin{bmatrix} c_{++}(s), & c_{+-}(s) \\ c_{-+}(s), & c_{--}(s) \end{bmatrix} = (2\pi)^{-(3/2)} \cdot \Gamma(s+1) \cdot \Gamma(s+5) \cdot \Gamma(s+9) \times -\sqrt{-1}$$
$$\times \begin{bmatrix} -3\exp\left(\frac{\pi}{2}\sqrt{-1}s\right) - \exp\left(-\frac{3\pi}{2}\sqrt{-1}s\right), \\ 3\exp\left(-\frac{\pi}{2}\sqrt{-1}s\right) + \exp\left(\frac{3\pi}{2}\sqrt{-1}s\right), \end{bmatrix} \tag{8.4}$$

$$\left.\begin{array}{c} 3\exp\left(-\frac{\pi}{2}\sqrt{-1}s\right)+\exp\left(\frac{3\pi}{2}\sqrt{-1}s\right) \\ -3\exp\left(\frac{\pi}{2}\sqrt{-1}s\right)-\exp\left(-\frac{3\pi}{2}\sqrt{-1}s\right)\end{array}\right].$$

Proof. 1) First we suppose that s is not a negative integer. As we have seen in (7.1), the coefficients of $|P(x)|_i^s$ $(i=0, 1, 2, 3)$ on $\Lambda_{0\mathrm{R}}^{\circ}$ are given by:

$$c_{\Lambda_0^{p,\circ}}(|P(x)|_i^s)=\begin{cases}1 & \text{if } p=i,\\ 0 & \text{if } p\neq i.\end{cases}$$

Applying the relation (6.1), we can compute the coefficients of $|P(x)|_i^s$ on $\Lambda_{1\mathrm{R}}^{\circ}$, $\Lambda_{2\mathrm{R}}^{\circ}$ and $\Lambda_{3\mathrm{R}}^{\circ}$ explicitly. Especially the coefficient on $\Lambda_3^{0,j}$ is the coefficient $c_3^{(0,j)}(|P(x)|_i^s)$ appearing in (7.6). By Proposition 7.2, we get the formula of the Fourier transform, namely the coefficient $c_3^{(0,j)}(|P(x)|_i^s)$ is $c_{ij}(s)$ in (8.1). Thus, by computing the coefficients on $\Lambda_{0\mathrm{R}}^{\circ}$ explicitly, we have the Fourier transform (8.1) for every $s\in\mathbb{C}$ except when it is a negative integer. When s is a negative integer, we also get (8.1) by analytic continuation. After all we obtain (8.1) for all $s\in\mathbb{C}$.

2) In the case of 2), we may prove (8.3) in the same way. q.e.d.

§ 9. Invariant measures on singular orbits and their Fourier transforms

As proved in Proposition 3.2, the singular set $\boldsymbol{S}_{\mathrm{R}}$ consists of three real loci $\boldsymbol{S}_{1\mathrm{R}}$, $\boldsymbol{S}_{2\mathrm{R}}$ and $\boldsymbol{S}_{3\mathrm{R}}$. Each $\boldsymbol{S}_{i\mathrm{R}}$ decomposes into a finite number of $\boldsymbol{G}_{\mathrm{R}}^{+}$-orbits denoted by $\coprod_{j\in\boldsymbol{I}_i}\boldsymbol{S}_i^j$ $(i=1, 2, 3)$. Here, in the division case the index set is $\boldsymbol{I}_i:=\{0, 1, \cdots, 3-i\}$ and in the split case it is $\boldsymbol{I}_i:=\{0\}$. We have seen that each $\boldsymbol{G}_{\mathrm{R}}^{+}$-orbit $\boldsymbol{S}_i^j$ is a $\boldsymbol{G}_{\mathrm{R}}^{1}$-orbit in Proposition 3.2 3). By investigating the action of Lie algebra of $\boldsymbol{G}_{\mathrm{R}}^{1}$ on $\boldsymbol{S}_i^j$, we see that each $\boldsymbol{S}_i^j$ admit a non-trivial $\boldsymbol{G}_{\mathrm{R}}^{1}$-invariant measure, which is uniquely determined up to a constant multiple. We denote it by $d\nu_i^j$.

Proposition 9.1. *Let $\mathscr{S}(\boldsymbol{X}_{\mathrm{R}})$ be the space of rapidly decreasing functions on $\boldsymbol{X}_{\mathrm{R}}$ and let $d\nu_i^j$ $(i=1, 2, 3$, and $j\in\boldsymbol{I}_i)$ be a non-trivial $\boldsymbol{G}_{\mathrm{R}}^{1}$-invariant measure on $\boldsymbol{S}_i^j$. For any element $f(x)$ in $\mathscr{S}(\boldsymbol{X}_{\mathrm{R}})$, the integral $\int f(x)d\nu_i^j(x)$ is absolutely convergent and the functional,*

$$f\longmapsto\int f(x)d\nu_i^j(x), \tag{9.1}$$

defines a tempered distribution on $\boldsymbol{X}_{\mathrm{R}}$ supported on $\boldsymbol{S}_i^j$.

Proof. If $i=3$, $d\nu_i^j(x)$ is a measure supported on the origin $\{0\}$. Then (9.1) is apparently defines a tempered distribution on $\boldsymbol{X}_{\mathbb{R}}$. So we confine ourselves to the cases $i=1$ and $i=2$. Note that $d\nu_i^j(x)$ is a $\boldsymbol{G}_{\mathbb{R}}^1$-invariant measure on the orbit $\boldsymbol{S}_i^j$ and $\boldsymbol{S}_i^j$ is stable under the multiplication operation of positive numbers. From the definition of $d\nu_i^j(x)$, we see that it is homogeneous of degree $36-12i$, i.e., $d\nu_i^j(r\cdot x)=r^{36-12i}d\nu_i^j(x)$ for any $r>0$. Thus the integral $\int f(x)d\nu_i^j(x)$ is written in terms of the polar coordinate $(r, \omega)\in \mathbb{R}_+\times \boldsymbol{S}^{(26)}$ where $\mathbb{R}_+ :=\{r\in \mathbb{R};\, r\geq 0\}$ and $\boldsymbol{S}^{(26)}$ is a 26-dimensional sphere. The integral $\int f(x)d\nu_i^j(x)$ is written as

$$\int_{\boldsymbol{X}_{\mathbb{R}}} f(x)d\nu_i^j(x)=\int_{\mathbb{R}_+\times \boldsymbol{S}^{(26)}} f(r\cdot\omega)\cdot r^{35-12i}dr\cdot d\tilde{\nu}_i^j(\omega), \tag{9.2}$$

where $d\tilde{\nu}_i^j(\omega)$ is the restriction of $d\nu_i^j(x)$ on the unit sphere in $\boldsymbol{X}_{\mathbb{R}}$ and dr is an Euclidean measure on $\mathbb{R}_+$. When $f(x)\in \mathscr{S}(\boldsymbol{X}_{\mathbb{R}})$, the integral

$$\int_{\boldsymbol{S}^{(26)}} f(r\cdot\omega)d\tilde{\nu}_i^j(\omega)$$

is a continuous bounded function on $\mathbb{R}_+$, hence the left hand side is absolutely integrable. It is a routine work to show that (9.1) is a continuous linear functional on $\mathscr{S}(\boldsymbol{X}_{\mathbb{R}})$. q.e.d.

As proved in Proposition 9.1, (9.1) defines a homogeneous distribution of degree $36-12i$ supported on $\boldsymbol{S}_i^j$. We may denote it by $T_i^j(x)$, i.e.,

$$\int f(x)d\nu_i^j(x)=\int T_i^j(x)f(x)dx \qquad \text{for any } f(x)\in \mathscr{S}(\boldsymbol{X}_{\mathbb{R}}).$$

Since $d\nu_i^j(x)$ is $\boldsymbol{G}_{\mathbb{R}}^1$-invariant and homogeneous of degree $36-12i$, we have

$$\int f(\rho(g)\cdot x)\cdot T_i^j(x)dx=\chi(g)^{4i-12}\int f(x)T_i^j(x)dx,$$

which means that

$$T_i^j(\rho(g)\cdot x)=\chi(g)^{-(4i-3)}\cdot T_i^j(x), \tag{9.3}$$

for all $g\in \boldsymbol{G}_{\mathbb{R}}^+$. Thus the distribution defined by (9.1) is a $\chi(g)^{-(4i-3)}$-invariant tempered distribution $T_i^j(x)$ supported on $\boldsymbol{S}_i^j$.

The problem we want to solve in this section is to calculate explicitly the image of the Fourier transform of $T_i^j(x)$ on $\boldsymbol{X}_{\mathbb{R}}^*-\boldsymbol{S}_{\mathbb{R}}$. We normalize $d\nu_i^j$ suitably and compute the Fourier transform on $\boldsymbol{X}_{\mathbb{R}}^*-\boldsymbol{S}_{\mathbb{R}}$ in the same manner as in § 8.

(In the division case)

Let $d\nu_i^j$ be a G_{R}^1-invariant measure on S_i^j. We shall normalize $d\nu_i^j$ in the following way. We put

$$x_i^j := \begin{bmatrix} I_{3-i}^{(j)} & \\ & 0_i \end{bmatrix},$$

a point in S_i^j. We identify the tangent space $(TX_{\mathrm{R}})_{x_i^j}$ of X_{R} at x_i^j with X_{R}. We introduce a coordinate on X_{R} by

$$X = \begin{bmatrix} \xi_1, & x_3, & \bar{x}_2 \\ \bar{x}_3, & \xi_2, & x_1 \\ x_2, & \bar{x}_1, & \xi_3 \end{bmatrix}, \tag{9.4}$$

with $\xi_p \in \mathbb{R}$ and $x_q := (x_p^1 + x_p^2 u + x_p^3 v + x_p^4 w) + (x_p^5 + x_p^6 u + x_p^7 v + x_p^8 w)e \in \mathfrak{C}_{\mathrm{R}}^d$. Then the tangent space of S_i^j at x_i^j is a subspace of $(TX_{\mathrm{R}})_{x_i^j}$, which is given by:

(9.5) 1) $(TS_1^j)_{x_1^j} = \{X \in X_{\mathrm{R}};\ \xi_3 = 0\}$ $(j=0, 1, 2)$,

2) $(TS_2^j)_{x_2^j} = \{X \in X_{\mathrm{R}};\ \xi_2 = \xi_3 = 0,\ x_1 = 0\}$ $(j=0, 1)$,

3) $(TS_3^j)_{x_3^j} = \{X \in X_{\mathrm{R}};\ \xi_1 = \xi_2 = \xi_3 = 0,\ x_1 = x_2 = x_3 = 0\}$ $(j=0)$.

Now we normalize $d\nu_i^j$ so that

$$d\nu_i^j|_{x_i^j} = \begin{cases} |d\xi_1 \wedge d\xi_2 \wedge dx_1 \wedge dx_2 \wedge dx_3| & i=1 \\ |d\xi_1 \wedge dx_2 \wedge dx_3| & i=2 \\ 1 & i=3 \end{cases} \tag{9.6}$$

where $dx_p := \bigwedge_{q=1}^{8} dx_p^q$.

Definition (In the division case). Let $T_i^j(x)$ be the tempered distribution defined by the linear functional (9.1) where $d\nu_i^j$ is normalized by (9.6).

Theorem 9.2 (In the division case). *The Fourier transform of the tempered distribution $T_i^j(x)$ is given by*:

$$\int T_i^j(x) \exp(-2\pi\sqrt{-1}\langle x, y\rangle) dx|_{X_{\mathrm{R}}^* - S_{\mathrm{R}}} = \sum_{k=0}^{3} b_{ik}^j \cdot |P(x)|^{-4i-12}|_{O_k}. \tag{9.8}$$

where b_{ik}^j's are:

$$b_{1k}^2 := (2\pi)^{-12} \cdot 2^{-12} \cdot \Gamma(4)\Gamma(8),$$
$$b_{1k}^1 := (2\pi)^{-12} \cdot 2^{-11} \cdot \Gamma(4)\Gamma(8),$$
$$b_{1k}^0 := (2\pi)^{-12} \cdot 2^{-12} \cdot \Gamma(4)\Gamma(8),$$
$$b_{2k}^1 := (2\pi)^{-4} \cdot 2^{-12} \cdot \Gamma(4),$$
$$b_{2k}^0 := (2\pi)^{-4} \cdot 2^{-12} \cdot \Gamma(4),$$
$$b_{3k}^0 := 2^{-12}.$$

Proof. The coefficients of the solution $T_1^j(x)$ on Λ_{0R}° and on Λ_{1R}° are from the definition

$$c_{\Lambda_0^{p},0}(T_1^j(x)) = 0 \qquad \text{for all } p=0, 1, 2, 3.$$
$$c_{\Lambda_1^{j},q}(T_1^j(x)) = (2\pi)^{-(1/2)} \qquad \text{for all } q=0, 1.$$
$$c_{\Lambda_1^{k},q}(T_1^j(x)) = 0 \qquad \text{for } k \neq j \text{ and for all } q=0, 1.$$

By using the relation of the coefficients (6.1) for $s=-1$, we can easily compute the coefficients of $T_1^j(x)$ on Λ_{3R}° from the data of coefficients on Λ_{1R}°. Namely we have

$$c_{\Lambda_3^0,k}(T_1^j(x)) = (2\pi)^{-1}\Gamma(4)\Gamma(8)(2\pi)^{-(1/2)} \qquad \text{for all } k.$$

Similarly, for $T_2^j(x)$ ($j=0, 1$), the coefficients of $T_2^j(x)$ on Λ_{0R}° and on Λ_{1R}° are zero, $c_{\Lambda_2^j,q}(T_2^j(x)) = (2\pi)^{-5}$ ($q=0, 1, 2$) and $c_{\Lambda_2^k,q}(T_2^j(x)) = 0$ for $k \neq j$ ($q=0, 1, 2$). Calculating the coefficients by using the relation matrix of the coefficients (6.1) for $s=-5$, we have

$$c_{\Lambda_3^0,k}(T_2^j(x)) = (2\pi)^{-(1/2)}\Gamma(4)(2\pi)^{-5} \qquad \text{for all } k.$$

Thus, by Proposition 7.2, we have the above formula (9.8) of the Fourier transform.

For $T_3^0(x)$, the coefficients of $T_3^0(x)$ on Λ_{0R}°, on Λ_{1R}° and on Λ_{2R}° are zero. On Λ_{3R}°, we see that

$$c_{\Lambda_3^0,k}(T_3^0(x)) = (2\pi)^{-(27/2)} \qquad \text{for all } k.$$

Thus, by Proposition 7.2, we have the above formula (9.8) of the Fourier transform of $T_3^0(x)$. q.e.d.

(In the split case)

Let $d\nu_i^0$ be a $\boldsymbol{G}_{\mathrm{R}}^1$-invariant measure on $\boldsymbol{S}_i^0$. We shall normalize $d\nu_i^0$ in the following way. We put

$$x_i^0 := \begin{bmatrix} I_{3-i} & \\ & 0_i \end{bmatrix},$$

a point in $\boldsymbol{S}_i^0$. We identify the tangent space $(T\boldsymbol{X}_{\mathrm{R}})_{x_i^0}$ of $\boldsymbol{X}_{\mathrm{R}}$ at x_i^0 with $\boldsymbol{X}_{\mathrm{R}}$. We introduce a coordinate on $\boldsymbol{X}_{\mathrm{R}}$ on by (9.4) where each x_p is an element in the split Cayley algebra $\mathfrak{C}_{\mathrm{R}}^s$. Then the tangent space of $\boldsymbol{S}_i^0$ at x_i^0 is a subspace of $(T\boldsymbol{X}_{\mathrm{R}})_{x_i^0}$, which is given by

(9.9) 1) $(T\boldsymbol{S}_1^0)_{x_1^0}=\{X\in\boldsymbol{X}_{\mathrm{R}};\ \xi_3=0\}$.

2) $(T\boldsymbol{S}_2^0)_{x_2^0}=\{X\in\boldsymbol{X}_{\mathrm{R}};\ \xi_2=\xi_3=0,\ x_1=0\}$.

3) $(T\boldsymbol{S}_3^0)_{x_3^0}=\{X\in\boldsymbol{X}_{\mathrm{R}};\ \xi_1=\xi_2=\xi_3=0,\ x_1=x_2=x_3=0\}$.

We normalize $d\nu_i^0$ so that

$$(9.10)\qquad d\nu_i^0|_{x_i^0}=\begin{cases}|d\xi_1\wedge d\xi_2\wedge dx_1\wedge dx_2\wedge dx_3| & i=1\\ |d\xi_1\wedge dx_2\wedge dx_3| & i=2\\ 1 & i=3\end{cases}$$

where $dx_p := \bigwedge_{q=1}^{8} dx_p^q$.

Definition (In the split case). Let $T_i^0(x)$ be the tempered distribution defined by the linear functional (9.1) where $d\nu_i^0$ is normalized by (9.10).

Theorem 9.3 (In the split case). *The Fourier transform of the tempered distribution $T_i^0(x)$ is given by*:

$$(9.8)\qquad \begin{aligned}&\int T_i^0(x)\exp(-2\pi\sqrt{-1}\langle x, y\rangle)dx|_{X_{\mathrm{R}}^*-S_{\mathrm{R}}}\\ &=\sum_{\varepsilon=\pm} b_{ik}^0\cdot|P(x)|^{-4i-12}|_{O_k}.\end{aligned}$$

where b_{ik}^0's are:

$$b_{1k}^0 :=(2\pi)^{-12}\cdot 2^{-10}\cdot\Gamma(4)\Gamma(8),$$
$$b_{2k}^0 :=(2\pi)^{-4}\cdot 2^{-11}\cdot\Gamma(4),$$
$$b_{3k}^0 :=2^{-12}.$$

The proof of this theorem is the same as that of Theorem 9.2.

Chapter II. Zeta functions associated with the exceptional group of type E_6

In this chapter we shall give a definition of zeta functions $\xi_i(\Gamma, L, s)$ and $\xi_j^*(\Gamma, L^*, s)$ $(i, j\in \boldsymbol{I}_0)$ which are defined for a given discrete subgroup in $\boldsymbol{G}_{\mathrm{R}}^+$ and a $\mathbb{Z}$-lattice in $\boldsymbol{X}_{\mathrm{R}}$. They are combined with one another through a functional equation. They have a finite number of simple poles. Our purpose is to give them explicitly. After defining the zeta functions as Dirichlet series, we will prove the possibility of their analytic continuation and compute the functional equation explicitly in § 10. In § 11, we will show that computation of their residues is reduced to calculation of the Fourier transforms of invariant measures on singular orbits and we will give their values.

§ 10. Zeta functions and functional equations

Let $(G_{\mathbb{R}}^+, \rho, X_{\mathbb{R}})$ be one of the real forms of the complex prehomogeneous vector space $(G_{\mathbb{C}}, \rho, X_{\mathbb{C}})=(GL_1(\mathbb{C})\times E_{6\mathbb{C}}, \rho, \mathbf{Her}_3(\mathfrak{C}_{\mathbb{C}}))$. One is 1) the division case and the other is 2) the split case in (3.3). Let $\coprod_{i\in I_0} O_i$ be the connected component decomposition of $X_{\mathbb{R}}-S_{\mathbb{R}}$, where the index set $I_0:=\{0, 1, 2, 3\}$ in the division case and $I_0:=\{+, -\}$ in the split case (Proposition 3.2). Let $X_{\mathbb{Q}}$ be a $\mathbb{Q}$-vector space in $X_{\mathbb{R}}$ such that $\mathbb{R}\otimes_{\mathbb{Q}} X_{\mathbb{Q}} = X_{\mathbb{R}}$. We put $G_{\mathbb{Q}}^1:=\{g \in G_{\mathbb{R}}^1; \rho(g) \in GL(X_{\mathbb{Q}})\}$. It is easily checked that $\rho^*(G_{\mathbb{Q}}^1)=\rho(G_{\mathbb{Q}}^1)$ where ρ^* is the contragredient representation of ρ with respect to the inner product $\langle x, y\rangle=\mathrm{tr}\,(x\cdot y)$.

Let L be a $\mathbb{Z}$-lattice in $X_{\mathbb{Q}}$ and let L^* be its dual lattice. Take a subgroup Γ in $G_{\mathbb{Q}}^1$ such that L is *Γ-stable*, i.e. $\rho(\Gamma)\cdot L=L$. Then L^* is also a Γ-stable lattice, i.e. $\rho^*(\Gamma)\cdot L^*=L^*$. For a function $f \in \mathcal{S}(X_{\mathbb{R}})$ and for a Γ-stable subset M in L (resp. L^*), we put:

$$(10.1)\qquad Z(f, M, s):=\int_{G_{\mathbb{R}}^+/\Gamma} |\chi(g)|^s\cdot \sum_{x\in M} f(\rho(g)\cdot x)dg,$$

$$\left(\text{resp. } Z^*(f, M, s):=\int_{G_{\mathbb{R}}^+/\Gamma} |\chi(g)|^{-s}\cdot \sum_{y\in M} f(\rho^*(g)\cdot y)dg,\right)$$

where dg is a non-trivial Haar measure on $G_{\mathbb{R}}^+$. In particular, when we put $L_i:=L\cap O_i$ and $L_i^*:=L^*\cap O_i$, they are Γ-stable subsets. We suppose that:

(10.2) *If the real part* $\mathrm{Re}\,(s)$ *is sufficiently large, then* $Z(f, L_i, s)$ *and* $Z^*(f, L_i^*, s)$ *are absolutely convergent for any* $f \in \mathcal{S}(X_{\mathbb{R}})$ *and for any* $i \in I_0$.

Then we may express $Z(f, L_i, s)$ and $Z^*(f, L_i^*, s)$ by:

$$(10.3)\qquad Z(f, L_i, s):=\xi_i(\Gamma, L, s)\cdot \int f(x)\cdot |P(x)|_i^{s-9}\, dx,$$

$$\left(\text{resp. } Z^*(f, L_i^*, s):=\xi_i^*(\Gamma, L^*, s)\cdot \int f(x)\cdot |P(x)|_i^{s-9}\, dx\right)$$

with

$$(10.4)\qquad \xi_i(\Gamma, L, s) := \sum_{[x]\in\Gamma\backslash L_i} \left(\int_{G^1_{\mathbf{R}x}/\Gamma_x} d\nu_x\right)\cdot|P(x)|^{-s}.$$

$$\left(\text{resp. } \xi^*_i(\Gamma, L^*, s) := \sum_{[x]\in\Gamma\backslash L^*_i} \left(\int_{G^1_{\mathbf{R}x}/\Gamma_x} d\nu^*_x\right)\cdot|P(x)|^{-s}\right).$$

Here $G^+_{\mathbf{R}x}$ is the isotropy subgroup of $G^+_{\mathbf{R}}$ at x; Γ_x is the isotropy subgroup of Γ at x; $d\nu_x$ (resp. $d\nu^*_x$) is the left invariant measure on $G^+_{\mathbf{R}x}$ normalized so that (10.3) is valid.

The absolute convergence of $Z(f, L_i, s)$ (resp, $Z^*(f, L^*_i, s)$) guarantees that $\xi_i(\Gamma, L, s)$ (resp. $\xi_i(\Gamma, L^*, s)$) is absolutely convergent when $\mathrm{Re}(s)$ is sufficiently large. We call $\xi_i(\Gamma, L, s)$ and $\xi^*_i(\Gamma, L^*, s)$ zeta functions associated with the exceptional Lie group of type E_6.

Under the convergence condition (10.2), we may prove that $\xi_i(\Gamma, L, s)$ and $\xi^*_i(\Gamma, L^*, s)$ are continued to the whole complex plane as a meromorphic function with finitely many poles and they are combined with one another by functional equations. We shall explain it briefly following Sato-Shintani [Sm-Sh].

Recall Poisson's summation formula:

$$(10.5)\qquad \sum_{x\in L} f(x) = v(L)^{-1}\cdot \sum_{y\in L^*} f^\vee(y),$$

for any $f(x)\in\mathscr{S}(X_{\mathbf{R}})$ with

$$v(L) := \int_{L\backslash X_{\mathbf{R}}} dx.$$

Here

$$f^\vee(y) = \int_{X_{\mathbf{R}}} f(x)\exp(-2\pi\sqrt{-1}\langle x, y\rangle)dx.$$

We may put $f(x)$ belonging to $C^\infty_0(O_i)$:= the space of compactly supported functions on O_i. Then for $g\in G^+_{\mathbf{R}}$, $f(\rho(g)\cdot x)$ is also in $C^\infty_0(O_i)$. Substituting $f(\rho(g)\cdot x)$ in (10.5), we have:

$$(10.6)\qquad \sum_{x\in L} f(\rho(g)\cdot x) = v(L)^{-1}\cdot \sum_{y\in L^*} f^\vee(\rho^*(g)\cdot y)\cdot \chi(g)^{-9}.$$

Integrating the left hand side of (10.6) on $G^+_{\mathbf{R}}/\Gamma$ with respect to $\chi(g)^s dg$, we see that the integral is absolutely convergent and coincides with $Z(f, L, s)$ if $\mathrm{Re}(s)$ is sufficiently large. In particular, since $f(x)\in C^\infty_0(O_i)$, $Z(f, L, s)$ is actually

$$\xi_i(\Gamma, L, s)\int f(x)\cdot|P(x)|_i^{s-9}\,dx.$$

We divide the integral $Z(f, L, s)$ into the sum of the two integrals $Z^+(f, L, s)$ and $Z^-(f, L, s)$, i.e., $Z(f, L, s)=Z^+(f, L, s)+Z^-(f, L, s)$, with

$$Z^+(f, L, s) := \int_{G_{\mathbf{R}}/\Gamma, \chi(g)\geq 1} \sum_{x\in L} f(\rho(g)\cdot x)\cdot \chi(g)^s dg$$

and

$$Z^-(f, L, s) := \int_{G_{\mathbf{R}}/\Gamma, \chi(g)<1} \sum_{x\in L} f(\rho(g)\cdot x)\cdot \chi(g)^s dg.$$

Then $Z^+(f, L, s)$ is absolutely convergent for any $s\in\mathbf{C}$ and defines an entire function. The next part $Z^-(f, L_i, s)$ is written by using (10.6) as

$$\begin{aligned} Z^-(f, L, s) &= v(L)^{-1}\cdot\int_{G_{\mathbf{R}}^+/\Gamma, \chi(g)<1} \sum_{y\in L^*} f^\vee(\rho^*(g)\cdot y)\cdot\chi(g)^{s-9}dg \\ \text{(10.7)} \qquad &= v(L)^{-1}\cdot\int_{G_{\mathbf{R}}^+/\Gamma, \chi(g)<1} \sum_{y\in L^*-S_{\mathbf{R}}} f^\vee(\rho^*(g)\cdot y)\cdot\chi(g)^{s-9}dg \quad (1) \\ &\quad + v(L)^{-1}\cdot\int_{G_{\mathbf{R}}^+/\Gamma, \chi(g)<1} \sum_{y\in L^*\cap S_{\mathbf{R}}} f^\vee(\rho^*(g)\cdot y)\cdot\chi(g)^{s-9}dg \quad (2) \end{aligned}$$

It is easily checked that the first term (1) is absolutely convergent for any $s\in\mathbf{C}$ and defines an entire function.

We substitute $P(\partial/\partial x)f(x)$ for $f(x)$. Then the Fourier transform $(P(\partial/\partial x)f(x))^\vee = P(y)f^\vee(y)$ vanishes on $\boldsymbol{S}_{\mathbf{R}}$. Thus the second term (2) in the calculation of $Z^-(P(\partial/\partial x)\cdot f(x), L, s)$ of (10.7) vanishes and hence $Z^-(P(\partial/\partial x)\cdot f(x), L, s)$ is an entire function. This yields that $Z(P(\partial/\partial x)\cdot f(x), L, s)$ is continued to an entire function. Then

$$\begin{aligned} Z\left(P\left(\frac{\partial}{\partial x}\right)f(x), L, s\right) &= \xi_i(\Gamma, L, s)\cdot\int P\left(\frac{\partial}{\partial x}\right)f(x)\cdot|P(x)|_i^{s-9}\,dx \\ &= \xi_i(\Gamma, L, s)\cdot\int(-1)^3\cdot f(x)\cdot P\left(\frac{\partial}{\partial x}\right)|P(x)|_i^{s-9}\,dx \\ &= \text{const.}\cdot\xi_i(\Gamma, L, s)\cdot b(s-10)\cdot\int f(x)\cdot|P(x)|_i^{s-10}\,dx, \end{aligned}$$

with $b(s)=(s+1)(s+5)(s+9)$. Since $\int f(x)\cdot|P(x)|_i^{s-10}\,dx$ is entire for any $f(x)\in C_0^\infty(\boldsymbol{O}_i)$, $\xi_i(\Gamma, L, s)$ is continued to the whole complex plane as a meromorphic function. Furthermore, for any $s\in\mathbf{C}$, we may choose $f(x)$ so that $\int f(x)\cdot|P(x)|_i^{s-10}\,dx$ does not vanish. Thus $\xi_i(\Gamma, L, s)\cdot b(s-10)$ is an entire function. For the integral $Z^*(f, L^*, s)$, the same argument is possible and we can prove that $\xi_i^*(\Gamma, L^*, s)\cdot b(s-10)$ is entire.

Next we shall compute the functional equation combining $\xi_i(\Gamma, L, s)$ with $\xi_i^*(\Gamma, L^*, s)$. For $f(x) \in C_0^\infty(\boldsymbol{O}_i)$,

$$\begin{aligned}\xi_i(\Gamma, L, s)\cdot\int P\left(\frac{\partial}{\partial x}\right)f(x)\cdot|P(x)|_i^{s-9}\,dx \\ =Z\left(P\left(\frac{\partial}{\partial x}\right)f, L, s\right)=Z^+\left(P\left(\frac{\partial}{\partial x}\right)f, L, s\right) \\ +Z^-\left(P\left(\frac{\partial}{\partial x}\right)f, L, s\right).\end{aligned} \tag{10.8}$$

We have seen that both $Z^+(P(\partial/\partial x)f, L, s)$ and $Z^-(P(\partial/\partial x)f, L, s)$ are entire functions. By using Poisson's summation formula (10.6), we have:

(10.9) 1) $$Z^+\left(P\left(\frac{\partial}{\partial x}\right)f, L, s\right) = v(L)^{-1}\cdot\int_{G_{\mathbf{R}}^+/\Gamma,\chi(g)\geq 1}\sum_{y\in L^*}(P\cdot f^\vee)(\rho^*(g)\cdot y)\cdot\chi(g)^{s-9}dg$$

2) $$Z^-\left(P\left(\frac{\partial}{\partial x}\right)f, L, s\right) = v(L)^{-1}\cdot\int_{G_{\mathbf{R}}^+/\Gamma,\chi(g)<1}\sum_{y\in L^*}(P\cdot f^\vee)(\rho^*(g)\cdot y)\cdot\chi(g)^{s-9}dg.$$

The integrals in the right hand sides of (10.9), 1) and 2) are not absolutely convergent for $\mathrm{Re}(s)\gg 0$. However if $\mathrm{Re}(s)$ is sufficiently small, both of them are absolutely convergent and the sum of the two is

$$\begin{aligned}&v(L)^{-1}\cdot\int_{G_{\mathbf{R}}^+/\Gamma}\sum_{y\in L^*}(P\cdot f^\vee)(\rho^*(g)\cdot y)\cdot\chi(g)^{s-9}dg \\ &=v(L)^{-1}\cdot Z^*(P\cdot f^\vee, L^*, 9-s) \\ &=v(L)^{-1}\cdot\sum_{j\in \boldsymbol{I}_0}\xi_j^*(\Gamma, L^*, 9-s)\cdot\int(P\cdot f^\vee)(x)\cdot|P(x)|_j^{-s}\,dx \\ &=v(L)^{-1}\cdot\sum_{j\in \boldsymbol{I}_0}\xi_j^*(\Gamma, L^*, 9-s)\cdot\int P\left(\frac{\partial}{\partial y}\right)f(y)\cdot(|P(x)|_j^{-s})^\vee dy.\end{aligned} \tag{10.10}$$

By using the formula of the Fourier transform of $|P(x)|_j^{-s}$:

$$\begin{aligned}(|P(x)|_j^{-s})^\vee &:= \int\exp(-2\pi\sqrt{-1}\langle x, y\rangle)\cdot|P(y)|_j^{-s}\,dy \\ &=\sum_{i\in \boldsymbol{I}_0}d_{ji}(-s)\cdot|P(y)|_i^{s-9},\end{aligned}$$

we have:

$$(10.10) = v(L)^{-1} \cdot \sum_{j \in I_0} \xi_j^*(\Gamma, L^*, 9-s) \cdot d_{ji}(-s) \int P\left(\frac{\partial}{\partial y}\right) f(y) \cdot |P(y)|_i^{s-9} \, dy.$$

Comparing this formula with (10.8), we have

$$(10.11) \qquad \xi_i(\Gamma, L, s) = v(L)^{-1} \cdot \sum_{j \in I_0} \xi_j^*(\Gamma, L^*, 9-s) \cdot d_{ji}(-s),$$

which is the functional equation we want to compute. The explicit form of $d_{ji}(-s)$ is given in § 8, Theorem 8.1:

$$d_{ij}(s) = (2\pi)^{-3s-(27/2)} \cdot 2^{-12} \cdot c_{ij}(s).$$

After all we have obtained the following theorem.

Theorem 10.1. *For a lattice L in $X_{\mathbf{Q}}$ and for a subgroup Γ in $G_{\mathbf{Q}}$ satisfying $\rho(\Gamma) \cdot L \subset L$, we suppose the convergence condition* (10.2). *Then the Dirichlet series $\xi_i(\Gamma, L, s)$ and $\xi_i^*(\Gamma, L^*, s)$ ($i \in I_0$) are absolutely convergent if the real part* Re(s) *is sufficiently large. They are continued to the whole complex plane as meromorphic functions with a finite number of poles. They have the functional equation* (10.11).

Remark. The convergence condition (10.2) is always satisfied in the division case. In fact, Weil [We] gave a sufficient condition for absolute convergence of integrals of the form $Z(f, L, s)$ and Igusa [Ig] actually proved the convergence of $Z(f, L_i, s)$ in the division case by checking the Weil's condition.

§ 11. Computations of residues

In this section we shall explain how we calculate the residues of $\xi_k(\Gamma, L, s)$. The residues of $\xi_k^*(\Gamma, L^*, s)$ are computed in the same way. As proved in the preceding section,

$$\xi_k(\Gamma, L, s) \cdot \int f(x) \cdot |P(x)|_k^{s-9} \, dx$$

$$(11.1) \qquad = Z^+(f, L, s) \qquad (1)$$

$$+ v(L)^{-1} \cdot \int_{G_{\mathbf{R}}^+/\Gamma, \chi(g)<1} \sum_{y \in L^* - S_{\mathbf{R}}} f^\vee(\rho^*(g) \cdot y) \cdot \chi(g)^{s-9} dg \qquad (2)$$

$$+ v(L)^{-1} \cdot \int_{G_{\mathbf{R}}^+/\Gamma, \chi(g)<1} \sum_{y \in L^* \cap S_{\mathbf{R}}} f^\vee(\rho^*(g) \cdot y) \cdot \chi(g)^{s-9} dg \qquad (3)$$

for $f(x) \in C_0^\infty(\boldsymbol{O}_k)$ and $s \in \mathbb{C}$ with sufficiently large real part. We have seen that the terms (1) and (2) in (11.1) are entire functions for any $f(x) \in C_0^\infty(\boldsymbol{O}_k)$. Then the poles of $\xi_k(\Gamma, L, s)$ appear in the third term (3) if they exist.

We shall compute the term (3) more precisely. Since Γ is a subgroup of $\boldsymbol{G}_{\mathbb{R}}^1$, we may divide the integral on $\boldsymbol{G}_{\mathbb{R}}^+/\Gamma$ into the product space $\boldsymbol{GL}_1(\mathbb{R})^+ \times (\boldsymbol{G}_{\mathbb{R}}^1/\Gamma)$. Thus the term (3) in (11.1) is written as:

$$(11.2) \quad v(L)^{-1} \cdot \int_1^\infty \chi(g_1)^{s-9} \frac{d\chi(g_1)}{\chi(g_1)} \int_{G_{\mathbb{R}}^1/\Gamma} \sum_{y \in L^* \cap S_{\mathbb{R}}} f^\vee(\rho^*(g_1 \cdot g_2) \cdot y) dg_2,$$

where $g = g_1 \cdot g_2$ with $(g_1, g_2) \in \boldsymbol{GL}_1(\mathbb{R})^+ \times \boldsymbol{G}_{\mathbb{R}}^1/\Gamma$ and dg_2 is a Haar measure on $\boldsymbol{G}_{\mathbb{R}}^1$ satisfying $dg = d\chi(g_1)/\chi(g_1) \times dg_2$. In order to simplify the argument, we suppose that:

$$(11.3) \quad \int_{G_{\mathbb{R}}^1/\Gamma} \sum_{y \in L^* \cap S_{\mathbb{R}}} f^\vee(\rho^*(g_2) \cdot y) dg_2 \textit{ is absolutely convergent.}$$

Let $\coprod_j \boldsymbol{S}_j = \boldsymbol{S}_{\mathbb{R}}$ be the $\boldsymbol{G}_{\mathbb{R}}^1$-orbital decomposition of $\boldsymbol{S}_{\mathbb{R}}$, which coincides with the $\boldsymbol{G}_{\mathbb{R}}^1$-orbital decomposition. The integral in (11.3) is calculated as:

$$(11.4) \quad \begin{aligned}
&\int_{G_{\mathbb{R}}^1/\Gamma} \sum_{y \in L^* \cap S_{\mathbb{R}}} f^\vee(\rho^*(g_2) \cdot y) dg_2 \\
&= \int_{G_{\mathbb{R}}^1/\Gamma} \sum_{i=1}^3 \sum_{j \in I_i} \sum_{y \in L^* \cap S_i^j} f^\vee(\rho^*(g_2) \cdot y) dg_2 \\
&= \sum_{i=1}^3 \sum_{j \in I_i} \sum_{[y] \in \Gamma \backslash L^* \cap S_i^j} \sum_{\gamma \in \Gamma/\Gamma_y} \int_{G_{\mathbb{R}}^1/\Gamma} f^\vee(\rho^*(g_2 \cdot \gamma) \cdot y) dg_2 \\
&= \sum_{i=1}^3 \sum_{j \in I_i} \sum_{[y] \in \Gamma \backslash L^* \cap S_i^j} \int_{G_{\mathbb{R}}^1/\Gamma_y} f^\vee(\rho^*(g_2) \cdot y) dg_2 \\
&= \sum_{i=1}^3 \sum_{j \in I_i} \sum_{[y] \in \Gamma \backslash L^* \cap S_i^j} \int_{S_i^j} f^\vee(x) d\nu_i^j(x) \cdot \int_{G_{\mathbb{R}_y}^1/\Gamma_y} d\nu_y \\
&= \sum_{i=1}^3 \sum_{j \in I_i} \left(\sum_{[y] \in \Gamma \backslash L^* \cap S_i^j} \left(\int_{G_{\mathbb{R}_y}^1/\Gamma_y} d\nu_y \right) \right) \cdot \int_{S_i^j} f^\vee(x) d\nu_i^j(x),
\end{aligned}$$

where $d\nu_i^j(x)$ is a $\boldsymbol{G}_{\mathbb{R}}^1$-invariant measure on $\boldsymbol{S}_i^j$ and $d\nu_y$ is a left invariant measure on the isotropy group $\boldsymbol{G}_{\mathbb{R}_y}^1$ of $\boldsymbol{G}_{\mathbb{R}}^1$ at y normalized according to the normalization of $d\nu_i^j$.

As we have seen in § 8, the integral

$$(11.5) \quad h(x) \longmapsto \int_{S_i^j} h(x) d\nu_i^j(x), \qquad (h(x) \in \mathcal{S}(\boldsymbol{X}_{\mathbb{R}})).$$

is absolutely convergent and defines a $\chi^{-(4i-3)}$-invariant tempered distribution $T_i^j(x)$ on $\boldsymbol{X}_{\mathbb{R}}$. Since $f \in C_0^\infty(\boldsymbol{O}_k)$, we have

$$\int_{S_i^j} f^\vee(\rho^*(g)\cdot y)d\nu_i^j(y)$$
$$=\int f^\vee(\rho^*(g)\cdot y)\cdot T_i^j(y)dy$$
$$=\chi(g)^{9-(4i-3)}\int f(y)\cdot T_i^{j\vee}(y)dy$$
$$=\chi(g)^{9-(4i-3)}\int f(y)\cdot b_{ik}^j\cdot|P(x)|_k^{4i-12}\,dy,$$

where b_{ik}^j are the constants calculated in § 9. See Theorem 9.2 and Theorem 9.3. Thus the value of the integral (11.2) is,

$$v(L)^{-1}\cdot\int_1^\infty \chi(g_1)^{s-(4i-3)}\frac{d\chi(g_1)}{\chi(g_1)}$$
$$\times\sum_{i=1}^3\sum_{j\in I_i}\Bigl(\sum_{[y]\in\Gamma\backslash L^*\cap S_i^j}\Bigl(\int_{G^1_{\mathbf{R}_y}/\Gamma_y} d\nu_y\Bigr)\Bigr)$$
$$\cdot\int_{S_i^j} f(x)\cdot b_{ik}^j\cdot|P(x)|_k^{4i-12}\,dx \tag{11.6}$$
$$=v(L)^{-1}\cdot(-(s-4i+3)^{-1})$$
$$\times\sum_{i=1}^3\sum_{j\in I_i}\Bigl(\sum_{[y]\in\Gamma\backslash L^*\cap S_i^j}\Bigl(\int_{G^1_{\mathbf{R}_y}/\Gamma_y} d\nu_y\Bigr)\Bigr)$$
$$\cdot\int_{S_i^j} f(x)\cdot b_{ik}^j\cdot|P(x)|_k^{4i-12}\,dx.$$

Thus we have:

$$\xi_k(\Gamma, L, s)\cdot\int f(x)\cdot|P(x)|_k^{s-9}\,dx$$
$$=((1)\text{ in }(11.1))+((2)\text{ in }(11.1))$$
$$-\sum_{i=1}^3(s-4i+3)^{-1}\cdot v(L)^{-1}$$
$$\times\sum_{j\in I_i}\Bigl(\sum_{[y]\in\Gamma\backslash L^*\cap S_i^j}\Bigl(\int_{G^1_{\mathbf{R}_y}/\Gamma_y} d\nu_y\Bigr)\Bigr)\cdot\int_{S_i^j} f(x)\cdot b_{ik}^j\cdot|P(x)|_k^{4i-12}\,dx.$$

Hence the residues of the pole of $\xi_k(\Gamma, L, s)$ at $s=4i-3$ $(i=1, 2, 3)$ is

$$-v(L)^{-1}\cdot\sum_{j\in I_i}\Bigl(\sum_{[y]\in\Gamma\backslash L^*\cap S_i^j}\Bigl(\int_{G^1_{\mathbf{R}_y}/\Gamma_y} d\nu_y\Bigr)\Bigr)b_{ik}^j. \tag{11.7}$$

Thus the explicit computation is reduced to the computation of the sum of the absolutely convergent series:

$$(11.8)\qquad \Big(\sum_{[y]\in\Gamma\backslash L^*\cap S^i_j}\Big(\int_{G^1_{\mathbf{R}_y}/\Gamma_y} d\nu_y\Big)\Big),$$

and the computation of the Fourier transform:

$$(11.9)\qquad \begin{aligned}&\int T^j_i(x)\exp(-2\pi\sqrt{-1}\langle x,y\rangle)dx|_{X^*_{\mathbf{R}}-S_{\mathbf{R}}}\\ &=\sum_{k\in I_0} b^j_{ik}\cdot|P(y)|^{4i-12}|_{O_k}.\end{aligned}$$

Theorem 11.1 (Calculation of the residues). *Under the convergence condition* (11.3), *the residue of the simple pole of* $\xi_k(\Gamma, L, s)$ *at* $s=4i-3$ $(i=1, 2, 3)$ *is given by* (11.7). *The values of* b^j_{ik}*'s have been computed in Theorems* 9.2 *and* 9.3.

Remark. The evaluation of the series (11.8) is another problem. It highly depends on the choice of the lattice L. It may be expressed as a special value of the Riemann's zeta function or a zeta function of quadratic forms (so-called Siegel's zeta function), but we can say nothing about it.

References

[Ba 1] Baily, W. L. Jr., An exceptional arithmetic group and its Eisenstein series, Ann. of Math., **91** (1970), 512–549.

[Ba 2] ——, Introductory lectures on automorphic forms, Tokyo, Iwanami Shoten Publishers, (1973).

[Ig] Igusa, J-i., On certain representations of semisimple algebraic groups and their arithmetic of the corresponding invariants(I), Invent. Math., **12** (1971), 62–94.

[Ja] Jacobson, N., Exceptional Lie algebras, (Marcel Dekker Lecture Note in Math. 1), Marcel Dekker, (1971).

[Ka 1] Kashiwara, M., Systems of microdifferential equations, (Progress in Math. 34), Boston, Birkhäuser, (1983).

[Ka 2] ——, Microlocal calculus of simple microfunctions, in Analysis and Algebraic Geometry, Papers in honor of Professor K. Kodaira, Baily, W. L. and Shioda, T., ed., Tokyo, Iwanami Shoten Publishers., (1977), 369–374.

[Ka-Mi] Kashiwara, M. and Miwa, T., Microlocal calculus and Fourier transforms of relative invariants of prehomogeneous vector spaces, (in Japanese), Surikaiseki Kenkyusho Kokyuroku, **283** (1974), 60–147.

[Ka-Kw-Ki] Kashiwara, M., Kawai, T. and Kimura, T., Foundations of Algebraic Analysis (in Japanese), Tokyo, Kinokuniya, (1980), English translation (Princeton Mathematical Series No. 37), Princeton U.P., (1986).

[Ki] Kimura, T., The b-functions and holonomy diagrams of irreducible regular prehomogeneous vector spaces, Nagoya Math. J., **85** (1982), 1–80.

[Mr 1] Muro, M., Microlocal analysis and calculations on some relatively invariant hyperfunctions related to zeta functions associated with the vector spaces of quadratic forms, Publ. RIMS, Kyoto Univ., **22** (1986), 395–463.

[Mr 2] ——, Singular invariant tempered distributions on regular prehomogeneous vector spaces, J. Funct. Anal., **76** (1988), 317–345.

[Mr 3] ——, Invariant hyperfunctions on regular prehomogeneous vector spaces of commutative parabolic type, (preprint), (1988).

[Mr 4] ——, On uniqueness of hyperfunction solutions of holonomic systems, Ark. Mat., **26** (1988), 305–314.

[Mr 5] ——, The dimension of the space of relatively invariant hyperfunctions on regular prehomogeneous vector spaces, Proc. Japan Acad., **63**, Ser. A, (1987), 66–68.

[Mu-Ru-Sc] Muller, I., Rubenthaler, H. and Schiffmann, G., Structure des espaces préhomogènes associés à certaines algèbres de Lie graduées, Math. Ann., **274** (1986), 95–123.

[Ru] Rubenthaler, H., Forms réeles infinitésimales préhomogènes de type parabolique, Annales de l'Inst. de Fourier, Grenoble, **36-1**, (1986), 1–38.

[Sk-Fa] Satake, I. and Faraut, J., The functional equations of zeta-distributions associated with formally real Jordan algebras, Tôhoku Math. J., **36**, No. 3 (1984), 469–482.

[Sf 1] Sato, F., Zeta functions in several variables associated with prehomogeneous vector spaces I, Functional equations, Tôhoku Math. J., **34**, No. 3 (1982), 437–483.

[Sf 2] ——, Zeta functions in several variables associated with prehomogeneous vector spaces II, A convergence criterion, Tôhoku Math. J., **35** (1983), 77–99.

[Sf 3] ——, Zeta functions in several variables associated with prehomogeneous vector spaces III, Eisenstein series for indefinite quadratic forms, Ann. of Math., **116** (1982), 177–212.

[Sm-Ki] Sato, M. and Kimura, T., A classification of irreducible prehomogeneous vector spaces and their relative invariants, Nagoya Math. J., **65** (1977), 1–155.

[Sm-Sh] Sato, M. and Shintani, T., On zeta functions associated with prehomogeneous vector spaces, Ann. of Math., **100** (1974), 131–170.

[Sm-Ka-Ki-Os] Sato, M., Kashiwara, M., Kimura, T. and Oshima, T., Micro-local analysis of prehomogeneous vector spaces, Invent. math., **62** (1980), 117–179.

[Sh] Shintani, T., On zeta functions associated with the vector spaces of quadratic forms, J. Fac. Sci. Univ. Tokyo Sect. IA, **22** (1975), 25–65.

[We] Weil, A., Sur la formule de Siegel dan la théorie des groupes classiques, Acta Math., **113** (1965), 1–87.

Department of Mathematics
Kochi University
Kochi 780, Japan

Current address:
Faculty of General Education
Gifu University
Gifu 501–11, Japan

Advanced Studies in Pure Mathematics 15, 1989
Automorphic Forms and Geometry of Arithmetic Varieties
pp. 465–508

On Functional Equations of Zeta Distributions

Fumihiro Sato

Dedicated to Prof. Ichiro Satake on his sixtieth birthday

Introduction

Recent development in the theory of prehomogeneous vector spaces (in particular the works of Gyoja-Kawanaka [10] on prehomogeneous vector spaces defined over finite fields and of Igusa [17] on prehomogeneous vector spaces defined over $\mathfrak{p}$-adic number fields) has revealed a striking resemblance between the theories over finite fields, $\mathfrak{p}$-adic number fields, real and complex number fields and algebraic number fields, as is common in the theory of representations of algebraic groups.

Now we give a brief sketch of the fundamental theorem in the theory of prehomogeneous vector spaces. Let K be one of the fields mentioned above and $(\boldsymbol{G}, \rho, \boldsymbol{V})$ be a K-regular prehomogeneous vector space (satisfying some additional conditions, if necessary). Take K-irreducible polynomials $P_1, \cdots, P_n$ defining the K-irreducible hypersurfaces contained in the singular set $\boldsymbol{S}$. Let $\Omega(K^\times)$ be the set of quasi-characters of the multiplicative group $K^\times$ and $\mathcal{S}(\boldsymbol{V}(K))$ the space of Schwartz-Bruhat functions on $\boldsymbol{V}(K)$. For an $\omega \in \Omega(K^\times)^n$ we can define a tempered distribution (zeta distribution) $Z(\omega)$ on $\boldsymbol{V}(K)$ by analytic continuation of the integral

$$Z(\omega)(\phi)=\int_{V(K)-S(K)} \prod_{i=1}^{n} \omega_i(P_i(x))\phi(x)d_V^\times(x) \qquad (\phi \in \mathcal{S}(\boldsymbol{V}(K))),$$

where $d_V^\times(x)$ is a certain relatively $\boldsymbol{G}(K)$-invariant measure on $\boldsymbol{V}(K)-\boldsymbol{S}(K)$. Starting from the prehomogeneous vector space $(\boldsymbol{G}, \rho^*, \boldsymbol{V}^*)$ dual to $(\boldsymbol{G}, \rho, \boldsymbol{V})$, we can obtain a tempered distribution $Z^*(\omega)$ $(\omega \in \Omega(K^\times)^n)$ on $\boldsymbol{V}^*(K)$.

Roughly speaking, the fundamental theorem states that the Fourier transform of the tempered distribution $Z(\omega)$ coincides with $Z^*(\omega^*)$ for certain ω^* up to a constant multiple $\gamma(\omega)$ depending meromorphically on ω: $\hat{Z}(\omega)=\gamma(\omega)Z^*(\omega^*)$.

Received March 3, 1987.

The aim of this paper, which is expository in part, is to explain in detail the analogy between the theories over various fields. However our attention will be focused mainly on the cases of $\mathfrak{p}$-adic number fields and the rational number field $\boldsymbol{Q}$. The case of algebraic number fields can be reduced to the case of $\boldsymbol{Q}$ by using Weil's functor $R_{K/\boldsymbol{Q}}$.

When the base field K is a $\mathfrak{p}$-adic number field $k_{\mathfrak{p}}$, the fundamental theorem has been proved by Igusa [17] under the assumptions that

1) $\boldsymbol{G}$ is reductive and self-adjoint over $k_{\mathfrak{p}}$,
2) $\boldsymbol{S}$ is absolutely irreducible,
3) $\boldsymbol{S}$ contains only a finite number of $\boldsymbol{G}^1$-orbits, where $\boldsymbol{G}^1$ is the kernel of rational characters corresponding to relative invariants.

After recalling some basic properties of regular prehomogeneous vector spaces and the fundamental theorem for $K=\boldsymbol{R}$ in Section 1, the fundamental theorem for $k_{\mathfrak{p}}$ is proved in Section 2 under certain weaker conditions so as to include the case of several independent relative invariants. Namely the assumptions 1) and 2) will be replaced by the assumption that $(\boldsymbol{G}, \rho, \boldsymbol{V})$ is $k_{\mathfrak{p}}$-regular; however some modified form of the assumption 3) on the finiteness of singular orbit is inevitable at present.

Compared with the previous result for $\boldsymbol{R}$ ([37], [40], [35]), the result for $k_{\mathfrak{p}}$ is still unsatisfactory in some respects. Firstly the finiteness assumption on singular orbits should be removed in future. Secondly the nature of the so-called Γ-matrices $\gamma(\omega)$, in particular a possible relation of Γ-matrices with b-functions, remains to be clarified. Hence it seems profitable at the present stage to compare the Γ-matrix over a $\mathfrak{p}$-adic number field with the Γ-matrix over $\boldsymbol{R}$ for a prehomogeneous vector space defined over $\boldsymbol{Q}$. In Section 3 we make such a comparison of Γ-matrices for some concrete examples. Our calculation of Γ-matrices along with the recent results of Muller [25] and Igusa [18] suggests that Γ-matrices are under the control of b-functions also in the case of $\mathfrak{p}$-adic number fields (see § 3.5).

For the case $K=\boldsymbol{Q}$, by introducing an analogue $\mathscr{S}(\boldsymbol{V}(\boldsymbol{Q}))$ of the Schwartz-Bruhat space and a relatively $\boldsymbol{G}(\boldsymbol{Q})$-invariant "measure" on $\boldsymbol{V}(\boldsymbol{Q})-\boldsymbol{S}(\boldsymbol{Q})$, we reformulate in Section 4 the functional equations of zeta functions associated with prehomogeneous vector spaces in order to make it easier to see the analogy to the theories over the other fields. Moreover, as a result of the introduction of $\mathscr{S}(\boldsymbol{V}(\boldsymbol{Q}))$, we are able to discuss L-functions and zeta functions of Hurwitz type associated with prehomogeneous vector spaces as well (§ 4.6). The necessity of this kind of generalization of zeta functions, which has been recognized independently by Hoffmann [16], results from the fact that certain special values of L-functions and zeta functions of Hurwitz type appear in the contribution of parabolic conjugacy classes to the Selberg trace formula. This kind of application of prehomogeneous vector spaces will be reviewed in Section 4.7. Another by-product of our formulation is that the arithmetic meaning of zeta functions is clarified by considering the "measure" on $\boldsymbol{V}(\boldsymbol{Q})-\boldsymbol{S}(\boldsymbol{Q})$ (§ 4.4).

The fundamental theorem over finite fields due to Gyoja-Kawanaka is explained in Section 4 in connection with L-functions associated with prehomogeneous vector spaces.

In the present paper we do not formulate the fundamental theorems in their full generality, namely we do not discuss partial Fourier transforms with respect to regular subspaces, since it is easy to extend most of the results in this paper to the case of partial Fourier transforms as treated in [35].

The author would like to express his sincere gratitude to Professors Hashimoto, Arakawa, Murase and Sugano for their helpful information. The present paper was written during author's stay in Göttingen. He is deeply grateful to the Sonderforschungsbereich 170 "Geometrie und Analysis", especially to Prof. U. Christian for his warm hospitality.

§ 1. Preliminaries

1.1. Let K be a field of characteristic 0. We denote by $\bar{K}$ the algebraic closure of K. Let $\boldsymbol{G}$ be a connected linear algebraic group defined over K, $\boldsymbol{V}$ a finite dimensional $\bar{K}$-vector space with K-structure and $\rho: \boldsymbol{G}\to \boldsymbol{GL}(\boldsymbol{V})$ a rational representation of $\boldsymbol{G}$ on $\boldsymbol{V}$ defined over K. We assume that the triple $(\boldsymbol{G}, \rho, \boldsymbol{V})$ is a prehomogeneous vector space defined over K. Then by definition there exists a proper algebraic subset $\boldsymbol{S}$ of $\boldsymbol{V}$ such that $\boldsymbol{V}(\bar{K})-\boldsymbol{S}(\bar{K})$ is a single $\boldsymbol{G}(\bar{K})$-orbit. The set $\boldsymbol{S}$ is called the singular set of $(\boldsymbol{G}, \rho, \boldsymbol{V})$ and is also defined over K. Denote by $K[\boldsymbol{V}]$ (resp. $K(\boldsymbol{V})$) the ring of polynomial functions (resp. the field of rational functions) defined over K. Let $\boldsymbol{S}_1, \cdots, \boldsymbol{S}_n$ be the K-irreducible hypersurfaces in $\boldsymbol{V}$ contained in $\boldsymbol{S}$. Take a K-irreducible polynomial function $P_i \in K[\boldsymbol{V}]$ defining $\boldsymbol{S}_i$ for each $i=1, \cdots, n$. Then $P_1, \cdots, P_n$ are relative invariants of $(\boldsymbol{G}, \rho, \boldsymbol{V})$ and any relative invariant in $K(\boldsymbol{V})$ can be written uniquely as $cP_1^{\nu_1}\cdots P_n^{\nu_n}$ with $c\in K^\times$, $\nu_1, \cdots, \nu_n \in \boldsymbol{Z}$ ([35, Lemma 1.3]). Let $\chi_1 \cdots \chi_n$ be the rational characters of $\boldsymbol{G}$ corresponding to $P_1, \cdots, P_n$, respectively and $X_\rho(\boldsymbol{G})_K$ the group of K-rational characters of $\boldsymbol{G}$ corresponding to relative invariants. The group $X_\rho(\boldsymbol{G})_K$ is a free abelian group of rank n generated by $\chi_1, \cdots, \chi_n$ ([35, Lemma 1.4]).

1.2. Let $\boldsymbol{V}^*$ be the $\bar{K}$-vector space dual to $\boldsymbol{V}$. The vector space $\boldsymbol{V}^*$ has a natural K-structure determined by the K-structure of $\boldsymbol{V}$. Denote by $\rho^*: \boldsymbol{G}\to\boldsymbol{GL}(\boldsymbol{V}^*)$ the rational representation of $\boldsymbol{G}$ on $\boldsymbol{V}^*$ contragredient to ρ. Fix a K-basis of $\boldsymbol{V}(K)$ and identify $\boldsymbol{V}$ with $\bar{K}^N$ ($N=\dim_K \boldsymbol{V}$). Then

$K[V]$ can be identified with the ring $K[x_1, \cdots, x_N]$ of polynomials in N variables with coefficients in K. We also identify V^* with $\bar{K}^N$ via the K-basis of $V^*(K)$ dual to the fixed basis of $V(K)$.

For a relative invariant $P(x) \in K[V]=K[x_1, \cdots, x_N]$, we define a mapping $\phi_P: V-S \to V^*$ by

$$\phi_P(x)=\left(\frac{1}{P(x)}\frac{\partial P}{\partial x_1}(x), \cdots, \frac{1}{P(x)}\frac{\partial P}{\partial x_N}(x)\right).$$

The mapping ϕ_P is independent of the choice of K-basis of $V(K)$ and defines a rational mapping defined over K. Moreover it is known that

$$\phi_P(\rho(g)x)=\rho^*(g)\phi_P(x) \qquad (x \in V-S,\ g \in G).$$

Definition. If there exists a relative invariant $P(x) \in K[V]$ such that the mapping $\phi_P: V-S \to V^*$ is dominant, then (G, ρ, V) is called *K-regular* and such a P is called *nondegenerate*.

Lemma 1.1 ([37, Chap. 1 § 1], [39, § 4 Prop. 10, Remark 11], [35, Lemmas 2.4, 2.5]). *Let (G, ρ, V) be a K-regular prehomogeneous vector space and $P(\in K[V])$ be a nondegenerate relative invariant. Then*

(1) *(G, ρ^*, V^*) is also a K-regular prehomogeneous vector space.*

(2) *Let S^* be the singular set of (G, ρ^*, V^*). The mapping ϕ_P induces a G-equivariant biregular rational mapping defined over K of $V-S$ onto V^*-S^*.*

(3) *For an $x \in V-S$, put $x^*=\phi_P(x)$, Then $G_x=G_{x^*}$, where $G_x=\{g \in G;\ \rho(g)x=x\}$ and $G_{x^*}=\{g \in G;\ \rho^*(g)x^*=x^*\}$.*

(4) *$X_\rho(G)_K=X_{\rho^*}(G)_K$.*

(5) *The rational character of G defined by $g \mapsto \det \rho(g)^2$ is in $X_\rho(G)_K$.*

(6) *S is a hypersurface in V if and only if S^* is a hypersurface in V^*.*

Let $S_1^*, \cdots, S_n^*$ be the K-irreducible hypersurfaces contained in S^*. Note that by Lemma 1.1 (4) the number of K-irreducible hypersurfaces contained in S^* is equal to the number n of K-irreducible hypersurfaces contained in S. Take a K-irreducible relative invariant $P_i^* \in K[V^*]$ defining S_i^* for each $i=1, \cdots, n$ and denote by χ_i^* the rational character of G corresponding to P_i^*. Since $\chi_1, \cdots, \chi_n$ and $\chi_1^*, \cdots, \chi_n^*$ form two systems of generators of the free abelian group $X_\rho(G)_K=X_{\rho^*}(G)_K$, there exists a unimodular matrix $U=(u_{ij}) \in GL(n, Z)$ satisfying

$$\chi_i=\prod_{j=1}^{n} \chi_j^{*u_{ij}} \qquad (1 \leqq i \leqq n).$$

Let λ and λ^* be the elements of $(\frac{1}{2}\boldsymbol{Z})^n$ defined by

$$\det \rho(g)^2 = \prod_{i=1}^{n} \chi_i(g)^{2\lambda_i}, \qquad \det \rho^*(g)^2 = \prod_{i=1}^{n} \chi_i^*(g)^{2\lambda_i^*}.$$

Since $\det \rho(g) = \det \rho^*(g)^{-1}$, we have $\lambda^* = -\lambda U$.

Another important fact derived immediately from Lemma 1.1 is the one-to-one correspondence between $\boldsymbol{G}(K)$-orbits in $\boldsymbol{V}(K)-\boldsymbol{S}(K)$ and $\boldsymbol{G}(K)$-orbits in $\boldsymbol{V}^*(K)-\boldsymbol{S}^*(K)$. In particular, if the number of $\boldsymbol{G}(K)$-orbits in $\boldsymbol{V}(K)-\boldsymbol{S}(K)$ is finite, then so is the number of $\boldsymbol{G}(K)$-orbits in $\boldsymbol{V}^*(K)-\boldsymbol{S}^*(K)$ and they coincide with each other. Note that this is the case if K is a local field ([42, III 4.2 Example d), 4.4 Theorem 5]).

1.3. Now we recall the definition of b-functions (for details, see [37, Chap. 1] and [38]). In the following we assume that $(\boldsymbol{G}, \rho, \boldsymbol{V})$ is K-regular.

For $\chi \in X_\rho(\boldsymbol{G})_K = X_{\rho^*}(\boldsymbol{G})_K$, let $\delta(\chi)$ and $\delta^*(\chi)$ be the elements in $\boldsymbol{Z}^n$ such that

$$\chi = \prod_{i=1}^{n} \chi_i^{\delta(\chi)_i} = \prod_{i=1}^{n} \chi_i^{*\delta^*(\chi)_i}.$$

Then $\delta(\chi)U = \delta^*(\chi)$.

We put

$$P^\chi(x) = \prod_{i=1}^{n} P_i(x)^{\delta(\chi)_i}, \qquad P^{*\chi}(y) = \prod_{i=1}^{n} P_i^*(y)^{\delta^*(\chi)_i} \qquad (\chi \in X_\rho(\boldsymbol{G})_K).$$

Define a partial differential operator $P^\chi(\mathrm{grad}_y)$ (resp. $P^{*\chi}(\mathrm{grad}_x)$) with constant coefficients by

$$P^\chi(\mathrm{grad}_y) e^{\langle x, y\rangle} = P^\chi(x) e^{\langle x, y\rangle}$$
$$(\text{resp. } P^{*\chi}(\mathrm{grad}_x) e^{\langle x, y\rangle} = P^{*\chi}(y) e^{\langle x, y\rangle}).$$

If $\delta^*(\chi)_i \geqq 0$ for all i, then there exists a polynomial $b_\chi(s)$ in $s_1, \cdots, s_n$ of degree $\deg P^{*\chi}(y)$ satisfying

$$P^{*\chi}(\mathrm{grad}_y) P^s(x) = b_\chi(s) P^{s+\delta(\chi)}(x),$$

where $P^s(x) = \prod_{i=1}^{n} P_i(x)^{s_i}$. By the cocycle property

$$b_{\chi\psi}(s) = b_\chi(s) b_\varphi(s+\delta(\chi)),$$

we can define $b_\chi(s)$ for all $\chi \in X_\rho(\boldsymbol{G})$. The polynomial $b_\chi(s)$ is called the *b-function* of $(\boldsymbol{G}, \rho, \boldsymbol{V})$.

In the case of $\bar{K} = \boldsymbol{C}$ the b-function $b_\chi(s)$ has the following expression in terms of the gamma function:

$$b_\chi(s)=c(\chi)\gamma(s)/\gamma(s+\delta(\chi))$$

$$\gamma(s)=\prod_{i=1}^{m}\left\{\prod_{j=1}^{\alpha_i}\Gamma\Big(\sum_{l=1}^{n}a_l^{(i)}s_l-p_{ij}\Big)\right\}\Big/\left\{\prod_{j=1}^{\beta_i}\Gamma\Big(\sum_{l=1}^{n}a_l^{(i)}s_l-q_{ij}\Big)\right\},$$

where c is a homomorphism of $X_\rho(\boldsymbol{G})_K$ into $\boldsymbol{C}^\times$ and $a_l^{(i)}\in\boldsymbol{Z}$, $a_l^{(i)}\geqq 0$, p_{ij}, $q_{ij}\in\boldsymbol{C}$.

1.4. In this paragraph we assume that K is a subfield of $\boldsymbol{R}$ and we describe briefly the main result in the theory over $\boldsymbol{R}$.

Put $V=\boldsymbol{V}(\boldsymbol{R})$, $V^*=\boldsymbol{V}^*(\boldsymbol{R})$, $S=\boldsymbol{S}(\boldsymbol{R})$ and $S^*=\boldsymbol{S}^*(\boldsymbol{R})$.

Let $V-S=V_1\cup\cdots\cup V_\nu$ and $V^*-S^*=V_1^*\cup\cdots\cup V_\nu^*$ be the decompositions into connected components. Each connected component is an orbit of the identity component G^+ of $\boldsymbol{G}(\boldsymbol{R})$.

We denote by $\mathscr{S}(V)$ and $\mathscr{S}(V^*)$ the spaces of rapidly decreasing functions on V and V^*, respectively.

Consider the integrals

$$Z_j(s)(f)=\int_{V_j}\prod_{i=1}^{n}|P_i(x)|^{s_i-\lambda_i}f(x)dx \qquad (f\in\mathscr{S}(V),\ 1\leqq j\leqq\nu)),$$

$$Z_j^*(s)(f^*)=\int_{V_j^*}\prod_{i=1}^{n}|P_i(y)|^{s_i-\lambda_i^*}f^*(y)dy \qquad (f^*\in\mathscr{S}(V^*),\ 1\leqq j\leqq\nu),$$

where dx and dy are Euclidean measures on V and V^*, respectively. The integrals $Z_j(s)(f)$ (resp. $Z_j^*(s)(f^*)$) are absolutely convergent at least for $\operatorname{Re}s_i\geqq\lambda_i$ (resp. $\operatorname{Re}s_i\geqq\lambda_i^*$). The analytic continuation of $Z_j(s)(f)$ (resp. $Z_j^*(s)(f^*)$) to a meromorphic function of s in $\boldsymbol{C}^n$ always exists and defines a tempered distribution

$$Z_j(s): f\longmapsto Z_j(s)(f)$$

$$(\text{resp. } Z_j^*(s): f^*\longmapsto Z_j^*(s)(f^*))$$

on V (resp. V^*) depending on s meromorphically, which we call the *zeta distribution* associated with $(\boldsymbol{G},\rho,\boldsymbol{V})$ over $\boldsymbol{R}$.

The Fourier transform $\hat{Z}_j(s)$ of the zeta distribution $Z_j(s)$, which is a tempered distribution on V^*, is defined by

$$\hat{Z}_j(s)(f^*)=Z_j(s)(\hat{f}^*),$$

$$\hat{f}^*(x)=\int_{V^*}f^*(y)e^{2\pi i\langle x,y\rangle}dy.$$

Now the fundamental theorem over $\boldsymbol{R}$ can be stated as follows:

Theorem *R* (cf. [37, Chap. 2 § 1], [40, § 1], [35, § 5]). *Assume that* $(\boldsymbol{G}, \rho, \boldsymbol{V})$ *is* $\boldsymbol{R}$*-regular and* $\boldsymbol{S}$ *is a hypersurface. Then*

$$\hat{Z}_i(s)=\sum_{j=1}^{\nu}\Gamma_{ij}(s-\lambda)Z_j^*(\lambda^*+s^*) \qquad (1\leqq i\leqq \nu)$$

with

$$\Gamma_{ij}(s)=c(-s)(-2\pi\sqrt{-1})^{\Sigma_{l=1}^{n} s_l \deg P_l^*}\gamma(s)\cdot\prod_{l=1}^{n}\varepsilon_l(i)^{-s_l}\cdot\prod_{l=1}^{n}\varepsilon_l^*(j)^{s_l^*}\cdot t_{ij}(s),$$

where $\varepsilon_l(i)=\operatorname{sgn} P_l(x)$ $(x\in V_i)$, $\varepsilon_l^*(j)=\operatorname{sgn} P_l^*(y)$ $(y\in V_j^*)$, $s^*=(s_1^*, \cdots, s_n^*)$ $=sU$ *and* $t_{ij}(s)$ *are polynomials in* $e^{\pm 2\pi\sqrt{-1}s_1}, \ldots, e^{\pm 2\pi\sqrt{-1}s_n}$.

We call the matrix $(\Gamma_{ij}(s))_{i,j=1}^{\nu}$ the *Γ-matrix* of $(\boldsymbol{G}, \rho, \boldsymbol{V})$.

Remark. If we take $K=\boldsymbol{C}$ and assume that $(\boldsymbol{G}, \rho, \boldsymbol{V})$ is ($\boldsymbol{C}$-) regular, then $(\tilde{\boldsymbol{G}}, \tilde{\rho}, \tilde{\boldsymbol{V}})=R_{\boldsymbol{C}/\boldsymbol{R}}(\boldsymbol{G}, \rho, \boldsymbol{V})$ is $\boldsymbol{R}$-regular and $\tilde{\boldsymbol{V}}(\boldsymbol{R})-\tilde{\boldsymbol{S}}(\boldsymbol{R})=\boldsymbol{V}(\boldsymbol{C})-\boldsymbol{S}(\boldsymbol{C})$. Hence Theorem $\boldsymbol{R}$ for $(\tilde{\boldsymbol{G}}, \tilde{\rho}, \tilde{\boldsymbol{V}})$ immediately yields "Theorem $\boldsymbol{C}$" (the fundamental theorem for prehomogeneous vector spaces defined over $\boldsymbol{C}$) for $(\boldsymbol{G}, \rho, \boldsymbol{V})$. Note that $\nu=1$ in this case and hence the Γ-matrix is a scalar. However more is known for $K=\boldsymbol{C}$. Namely the Γ-matrix is explicitly calculated (up to sign) at least when $\boldsymbol{G}$ is reductive ([37, § 3]). If we further assume that $\boldsymbol{S}$ is absolutely irreducible, then the sign left to be determined was settled by Igusa ([18, Theorem 2]). A remarkable fact in the case of $K=\boldsymbol{C}$ is that the Γ-matrix is completely determined by the b-function.

§ 2. Fourier transforms of *p*-adic complex powers

2.1. Let K be a $\mathfrak{p}$-adic number field, namely a finite extension of the p-adic number field $\boldsymbol{Q}_p$. Denote by $\mathfrak{o}$ the ring of integers of K and by $\mathfrak{p}$ the unique maximal ideal of $\mathfrak{o}$. Let π be a (fixed) prime element of K and $\mathfrak{o}^\times$ the unit group of $\mathfrak{o}$. We normalize the absolute value of $\alpha\in K^\times$ by

$$|\alpha|=q^{-\operatorname{ord}_{\mathfrak{p}}(\alpha)}, \qquad q=N(\mathfrak{p})=\sharp(\mathfrak{o}/\mathfrak{p}).$$

Let $\Omega(K^\times)$ be the set of quasi-characters of $K^\times$. For a complex number $s\in\boldsymbol{C}$, we define an $\omega_s\in\Omega(K^\times)$ by

$$\omega_s(\alpha)=|\alpha|^s \qquad (\alpha\in K^\times).$$

We denote the dual group of $\mathfrak{o}^\times$ by $\widehat{\mathfrak{o}^\times}$. We identify an element $\phi\in\widehat{\mathfrak{o}^\times}$ with the character of $K^\times$ obtained by extending ϕ to $K^\times$ so that $\phi(\pi)=1$.

Every $\omega \in \Omega(K^{\times})$ can be expressed as $\omega=\omega_s\phi$ with

$$s \in \boldsymbol{C}\Big/\Big(\frac{2\pi i}{\log q}\Big)\boldsymbol{Z} \quad \text{and} \quad \phi \in \widehat{\mathfrak{o}^{\times}}.$$

Then we put $\operatorname{Re}(\omega)=\operatorname{Re} s$ ($s=\log \omega(\pi)/\log q$).

Put

$$\Omega(K^{\times})_{\phi}=\Big\{\omega_s\phi;\ s \in \boldsymbol{C}\Big/\Big(\frac{2\pi i}{\log q}\Big)\boldsymbol{Z}\Big\} \qquad (\phi \in \widehat{\mathfrak{o}^{\times}}).$$

Since

$$\Omega(K^{\times})=\coprod_{\phi \in \widehat{\mathfrak{o}^{\times}}} \Omega(K^{\times})_{\phi} \quad \text{and} \quad \Omega(K^{\times})_{\phi}\simeq \boldsymbol{C}\Big/\Big(\frac{2\pi i}{\log q}\Big)\boldsymbol{Z}\simeq \boldsymbol{C}^{\times},$$

we may regard $\Omega(K^{\times})$ as a 1-dimensional complex Lie group. We call a function F: $\Omega(K^{\times})^n \to \boldsymbol{C}$ a *rational function on* $\Omega(K^{\times})^n$, if the function $F_{\phi}(s_1, \cdots, s_n)=F(\omega_{s_1}\phi_1, \cdots, \omega_{s_n}\phi_n)$ is a rational function of $q^{-s_1}, \cdots, q^{-s_n}$ for each $\phi=(\phi_1, \cdots, \phi_n) \in (\widehat{\mathfrak{o}^{\times}})^n$.

2.2. We now consider a prehomogeneous vector space $(\boldsymbol{G}, \rho, \boldsymbol{V})$ defined over a $\mathfrak{p}$-adic number field K and assume that

(A.1) *$(\boldsymbol{G}, \rho, \boldsymbol{V})$ is K-regular.*

Retain the notation in Section 1.1 and Section 1.2 and put $G=\boldsymbol{G}(K)$, $V=\boldsymbol{V}(K)$, $S=\boldsymbol{S}(K)$ and $S^*=\boldsymbol{S}^*(K)$. As remarked at the end of Section 1.2, the number of $\rho(G)$-orbits in $V-S$ is finite and is equal to the number of $\rho^*(G)$-orbits in V^*-S^*. Let

$$V-S=V_1 \cup \cdots \cup V_{\nu} \quad \text{and} \quad V^*-S^*=V_1^* \cup \cdots \cup V_{\nu}^*$$

be the $\rho(G)$-orbit decomposition of $V-S$ and the $\rho^*(G)$-orbit decomposition of V^*-S^*, respectively.

Denote by $\mathcal{S}(V)$ and $\mathcal{S}(V^*)$ the spaces of Schwartz-Bruhat functions on V and V^*, respectively. We fix a nontrivial additive character ψ of K and define the Fourier transform of $f^* \in \mathcal{S}(V)$ by

$$\hat{f}^*(x)=\int_{V^*} f^*(y)\psi(\langle x, y\rangle)dy,$$

where dy is a (fixed) Haar measure on V^*. Denote by dx the Haar measure on V dual to dy.

For $\omega=(\omega_1, \cdots, \omega_n) \in \Omega(K^{\times})^n$, put

$$\omega(P(x))=\prod_{i=1}^{n}\omega_i(P_i(x)) \quad \text{and} \quad \omega(P^*(y))=\prod_{i=1}^{n}\omega_i(P_i^*(y)).$$

Further put

$$\omega_\lambda=(\omega_{\lambda_1}, \cdots, \omega_{\lambda_n}) \quad \text{and} \quad \omega_{\lambda^*}=(\omega_{\lambda_1^*}, \cdots, \omega_{\lambda_n^*})$$

(for the definitions of λ and λ^*, see § 1.2).

Consider the integrals

$$Z_i(\omega)(f)=\int_{V_i}(\omega/\omega_\lambda)(P(x))f(x)dx \qquad (f \in \mathcal{S}(V),\ 1\leqq i\leqq \nu)$$

and

$$Z_j^*(\omega)(f^*)=\int_{V_j^*}(\omega/\omega_{\lambda^*})(P^*(y))f^*(y)dy \qquad (f^* \in \mathcal{S}(V^*),\ 1\leqq j\leqq \nu).$$

It is clear that $Z_i(\omega)(f)$ (resp. $Z_j^*(\omega)(f^*)$) are absolutely convergent for $\operatorname{Re}\omega_i>\lambda_i$ $(1\leqq i\leqq n)$ (resp. $\operatorname{Re}\omega_i>\lambda_i^*$ $(1\leqq i\leqq n)$) and represent holomorphic functions in $\{\omega \in \Omega(K^\times)^n;\ \operatorname{Re}\omega_i>\lambda_i\ (1\leqq i\leqq n)\}$ (resp. $\{\omega \in \Omega(K^\times)^n;\ \operatorname{Re}\omega_i>\lambda_i^*\ (1\leqq i\leqq n)\}$).

Since the theorem of Denef (see [8, Theorems 3.2 and 7.4] and [9, Theorem 3.1]) on the rationality of p-adic integrals, which Igusa used in his proof of [17, Lemma 2], can be easily generalized to the case of several complex variables (cf. [8, Proof of Theorem 5.1], [2, § 2 Theorem A]), we can prove the following lemma in the same manner as in the proof of [17, Lemma 2].

Lemma 2.1. *The functions $Z_i(\omega)(f)$ $(1\leqq i\leqq \nu, f\in \mathcal{S}(V))$ have analytic continuations to rational functions in $\Omega(K^\times)^n$ in the sense of Section 2.1. Moreover for each $\phi \in \widehat{(\mathfrak{o}^\times)^n}$, there exists a collection of integers $\{a_1^{(l)}, \cdots, a_n^{(l)}, b^{(l)}, m_l;\ 1\leqq l\leqq h,\ 0\leqq m_l\leqq [K\colon \boldsymbol{Q}_p]\dim V\}$ independent of $f \in \mathcal{S}(V)$ such that*

$$\prod_{l=1}^{h}(1-q^{-\Sigma_{j=1}^{n}a_j^{(l)}s_j-b^{(l)}})^{m_l}\cdot Z_i(\omega_s\phi)(f)$$

are holomorphic functions and hence polynomials in $q^{\pm s_1}, \cdots, q^{\pm s_n}$.

The same statement holds also for $Z_j^*(\omega)(f^*)$.

The tempered distributions $Z_i(\omega)$ and $Z_j^*(\omega)$ defined by

$$Z_i(\omega)\colon \mathcal{S}(V) \ni f \longmapsto Z_i(\omega)(f) \in \boldsymbol{C}$$

and

$$Z_j^*(\omega)\colon \mathcal{S}(V^*) \ni f^* \longrightarrow Z_j^*(\omega)(f^*) \in \boldsymbol{C}$$

are called the *zeta distributions associated with* $(\boldsymbol{G}, \rho, \boldsymbol{V})$ and $(\boldsymbol{G}, \rho^*, \boldsymbol{V}^*)$, respectively. They depend on ω meromorphically.

2.3. In the following we further assume that

(A.2) *$\boldsymbol{S}$ decomposes into a finite number of $\boldsymbol{G}$-orbits. Moreover for any $\boldsymbol{G}$-orbit $\boldsymbol{S}'$ in $\boldsymbol{S}$, there exists a $\chi \in X_\rho(\boldsymbol{G})_K$ $(\chi \neq 1)$ such that $\boldsymbol{S}'$ is a $\boldsymbol{G}^{(\chi)}$-orbit, where $\boldsymbol{G}^{(\chi)}$ denotes the kernel of χ.*

The assumption (A.2) enables us to analyze relatively invariant distributions on $S = \boldsymbol{S}(K)$ and singularities of zeta distributions.

The condition is equivalent to the following:

(A.2)′ *$\boldsymbol{S}$ decomposes into a finite number of $\boldsymbol{G}$-orbits. Moreover for any $x \in \boldsymbol{S}$, there exists a $\chi \in X_\rho(\boldsymbol{G})_K$ $(\chi \neq 1)$ such that the restriction of χ to the identity component of $\boldsymbol{G}_x = \{g \in \boldsymbol{G};\ \rho(g)x = x\}$ is nontrivial.*

Define a homomorphism $\nu\colon G \to \boldsymbol{Z}^n$ by

$$\nu(g) = (\mathrm{ord}_\mathfrak{p}\, \chi_1(g), \cdots, \mathrm{ord}_\mathfrak{p}\, \chi_n(g)) \qquad (g \in G).$$

Lemma 2.2. *For any $x \in S$, $\nu(G_x)$ is a non-zero submodule of $\boldsymbol{Z}^n$, where $G_x = \boldsymbol{G}_x \cap G$.*

Proof. Let $S^{(1)}$ be the $\rho(G)$-orbit containing x and $\boldsymbol{S}_1$ be the $\rho(\boldsymbol{G})$-orbit in $\boldsymbol{S}$ such that $\boldsymbol{S}_1(K) \supset S^{(1)}$. By (A.2) there exists a $\chi \in X_\rho(\boldsymbol{G})_K$ $(\chi \neq 1)$ such that $\boldsymbol{S}_1$ is a $\boldsymbol{G}^{(\chi)}$-orbit. Put $G^{(\chi)} = \boldsymbol{G}^{(\chi)}(K)$. Then $S^{(1)}$ is $G^{(\chi)}$-stable and decomposes into a finite number of $G^{(\chi)}$-orbits. Since $G^{(\chi)}$ is a normal subgroup of G, every element in G induces a permutation of $G^{(\chi)}$-orbits in $S^{(1)}$. Let G' be the subgroup of G consisting of all elements in G which stabilize all $G^{(\chi)}$-orbits. It is clear that the index $[G\colon G']$ is finite. Put $G'_x = G' \cap G_x$. Then we have $G' = G'_x \cdot G^{(\chi)}$. Recall that χ can be written as $\chi = \prod_{i=1}^n \chi_i^{r_i}$ with $r_1, \cdots, r_n \in \boldsymbol{Z}$. For an $\boldsymbol{m} = (m_1, \cdots, m_n) \in \boldsymbol{Z}^n$, put $\langle \boldsymbol{r}, \boldsymbol{m} \rangle = \sum_{i=1}^n r_i m_i$. If $\nu(G_x) = \{0\}$, then $\langle \boldsymbol{r}, \nu(G') \rangle = \langle \boldsymbol{r}, \nu(G'_x \cdot G^{(\chi)}) \rangle = \langle \boldsymbol{r}, \nu(G'_x) \rangle = \{0\}$. Since G' is of finite index in G, we have $\langle \boldsymbol{r}, \nu(G) \rangle = \{0\}$. On the other hand $\chi_1, \cdots, \chi_n$ are multiplicatively independent K-rational characters of $\boldsymbol{G}$ and hence $\boldsymbol{G}$ contains a K-split algebraic torus $\boldsymbol{T}$ of dimension n such that the restictions of $\chi_1, \cdots, \chi_n$ to $\boldsymbol{T}$ generate a subgroup of the character group of $\boldsymbol{T}$ of finite index. This implies that $\nu(G)$ is a subgroup of $\boldsymbol{Z}^n$ of finite index. This contradicts $\langle \boldsymbol{r}, \nu(G) \rangle = \{0\}$, since $\boldsymbol{r} \neq 0$. Therefore $\nu(G) \neq \{0\}$.

Let

$$S=S^{(1)}\cup\cdots\cup S^{(m)}$$

be the $\rho(G)$-orbit decomposition of S. The number m of $\rho(G)$-orbits is finite by the assumption (A.2) and [42, III 4.4 Theorem 5]. We fix a $\rho(G)$-orbit in S, say $S^{(1)}$. Let $\mathscr{S}(S^{(1)})$ be the $\boldsymbol{C}$-vector space of locally constant functions on $S^{(1)}$ with compact support and $\mathscr{S}(S^{(1)})^*$ the space of distributions on $S^{(1)}$. By definition we have $\mathscr{S}(S^{(1)})^*=\mathrm{Hom}_{\boldsymbol{C}}(\mathscr{S}(S^{(1)}),\boldsymbol{C})$. The group G acts on $\mathscr{S}(S^{(1)})$ and $\mathscr{S}(S^{(1)})^*$ as follows:

$$f^g(x)=f(\rho(g)x)\qquad (f\in\mathscr{S}(S^{(1)}),\ g\in G),$$
$$(gT)(f)=T(f^g)\qquad (T\in\mathscr{S}(S^{(1)})^*,\ f\in\mathscr{S}(S^{(1)}),\ g\in G).$$

For an $\omega\in\Omega(K^\times)^n$, we define a continuous homomorphism $\omega\circ\chi\colon G\to\boldsymbol{C}^\times$ by

$$\omega\circ\chi(g)=\prod_{i=1}^n\omega_i(\chi_i(g))\qquad (g\in G).$$

Put

$$\mathscr{E}_{S^{(1)}}(\omega)=\{T\in\mathscr{S}(S^{(1)})^*;\ gT=\omega\circ\chi(g)^{-1}T,\ g\in G\}\qquad (\omega\in\Omega(K^\times)^n)$$

and

$$\Omega(S^{(1)})=\{\omega\in\Omega(K^\times)^n;\ \mathscr{E}_{S^{(1)}}(\omega)\neq\{0\}\}.$$

Also put

$$\Omega(S^{(1)})_\phi=\Omega(S^{(1)})\cap(\Omega(K^\times)_{\phi_1}\times\cdots\times\Omega(K^\times)_{\phi_n})$$

for $\phi=(\phi_1,\cdots,\phi_n)\in(\widehat{\mathfrak{o}^\times})^n$.

We say that a subset E of

$$\Omega(K^\times)^n_\phi=\Omega(K^\times)_{\phi_1}\times\cdots\times\Omega(K^\times)_{\phi_n}\simeq\boldsymbol{C}^n\Big/\left\{\left(\frac{2\pi i}{\log q}\right)\boldsymbol{Z}\right\}^n$$

a *linear subspace*, if there exists a finite number of inhomogeneous linear functions $L_1,\cdots,L_t$ on $\boldsymbol{C}^n$ such that the coefficients of $s_1,\cdots,s_n$ in $L_1,\cdots,L_t$ are integral and

$$E=\left\{(\omega_{s_1}\phi_1,\cdots,\omega_{s_n}\phi_n);\ L_i(s_1,\cdots,s_n)\in\left(\frac{2\pi i}{\log q}\right)\boldsymbol{Z}\ (1\leq i\leq t)\right\}.$$

Lemma 2.3. *If $\Omega(S^{(1)})_\phi$ is not empty, then $\Omega(S^{(1)})_\phi$ is a linear subspace of $\Omega(K^\times)^n_\phi$. In fact let $\boldsymbol{m}^{(i)}=(m_1^{(i)},\cdots,m_n^{(i)})$ $(1\leq i\leq r)$ be a set of* *generators of the free abelian group* $\nu(G_x)$ $(x \in S^{(1)})$. *Then for some* $\alpha = (\alpha_1, \cdots, \alpha_n) \in \boldsymbol{C}^n$, *we have*

$$\Omega(S^{(1)})_\phi = \left\{(\omega_{s_1+\alpha_1}\phi_1, \cdots, \omega_{s_n+\alpha_n}\phi_n);\ \sum_{i=1}^{n} m_i^{(j)} s_i \in \left(\frac{2\pi i}{\log q}\right)\boldsymbol{Z}\quad (1 \leq j \leq r)\right\}.$$

Proof. Assume that $\Omega(S^{(1)})_\phi$ is not empty. For $\omega = \omega_s\phi \in \Omega(K^\times)^n_\phi$, ω belongs to $\Omega(S^{(1)})_\phi$ if and only if $(\omega \circ \chi) \cdot \Delta_G|_{G_x} = \Delta_{G_x}$, where $x \in S^{(1)}$ and Δ_G (resp. Δ_{G_x}) are the modules of G (resp. G_x). Fix an $\alpha \in \boldsymbol{C}^n$ such that $\omega_\alpha\phi \in \Omega(S^{(1)})_\phi$. Then $\omega_{s+\alpha}\phi$ belongs to $\Omega(S^{(1)})_\phi$ if and only if $\omega_s \circ \chi|_{G_x} = 1$. This is equivalent to the condition that

$$\sum_{i=1}^{n} m_i s_i \in \left(\frac{2\pi i}{\log q}\right)\boldsymbol{Z} \qquad \text{for any} \quad (m_1, \cdots, m_n) \in \nu(G_x).$$

This proves the lemma.

We define $\mathscr{E}_S(\omega)$ $(\omega \in \Omega(K^\times)^n)$, $\Omega(S)$ and $\Omega(S)_\phi$ in the same way as in the definitions of $\mathscr{E}_{S^{(1)}}(\omega)$, $\Omega(S^{(1)})_\phi$ and $\Omega(S^{(1)})$. Then by (A.2) and [17, p. 1016] we have $\Omega(S)_\phi = \bigcup_{i=1}^m \Omega(S^{(i)})_\phi$ and $\Omega(S) = \bigcup_{i=1}^m \Omega(S^{(i)})$, where $S = S^{(1)} \cup \cdots \cup S^{(m)}$ is the $\rho(G)$-orbit decomposition of S. Therefore Lemma 2.3 immediately implies the following

Corollary. *$\Omega(S)_\phi$ is a finite union of linear subspaces of $\Omega(K^\times)^n_\phi$.*

Proposition 2.4. *The poles of $Z_i(\omega)(f)$ are contained in $\Omega(S)$.*

Proof. Fix a $\phi \in \widehat{(\mathfrak{o}^\times)}^n$. Then by Lemma 2.1 we have

$$Z_i(\omega_s\phi)(f) = \frac{P(f, \omega_s\phi)}{\prod_{l=1}^{h} (1 - q^{-\sum_{j=1}^n a_j^{(l)} s_j - b^{(l)}})^{m_l}},$$

where $P(f, \omega_s\phi)$ is a polynomial in $q^{\pm s_1}, \cdots, q^{\pm s_n}$, $a_j^{(l)}$, $b^{(l)}$, m_l are integers independent of f and $1 \leq m_l \leq [k: \boldsymbol{Q}_p] \dim \boldsymbol{V}$. Assume that $s \in \boldsymbol{C}^n$ satisfies the linear equation

$$\sum_{j=1}^{n} a_j^{(l)} s_j + b^{(l)} \in \left(\frac{2\pi i}{\log q}\right)\boldsymbol{Z}$$

for some l. Then the distribution T on V defined by

$$T(f) = (1 - q^{-\sum_{j=1}^n a_j^{(l)} s_j - b^{(l)}})^{m_l} Z_i(\omega_s\phi)(f) \qquad (f \in \mathscr{S}(V))$$

satisfies the identity $gT = (\omega_s\phi) \circ \chi(g)^{-1} T$. We may assume that the integer m_l is chosen so that $T(f) \neq 0$ for some $f \in \mathscr{S}(V)$. Since $Z_i(\omega_s)(f)$ is entire for every $f \in \mathscr{S}(V-S)$, we have $T(f)=0$ for all $f \in \mathscr{S}(V-S)$. In other words, T is in $\mathscr{E}_S(\omega_s\phi)$. Hence $\omega=\omega_s\phi \in \Omega(S)$.

2.4. For $\omega \in \Omega(K^\times)^n$, put

$$\omega^*=\left(\prod_{i=1}^n \omega_i^{u_{i1}}, \cdots, \prod_{i=1}^n \omega_i^{u_{in}}\right) \in \Omega(K^\times)^n,$$

where $U=(u_{ij})$ is the unimodular matrix defined in Section 1.2.

The Fourier transform $\hat{Z}_i(\omega)$ of the zeta distribution $Z_i(\omega)$ is defined by $\hat{Z}_i(\omega)\,(f^*)=Z_i(\omega)(\hat{f}^*)$ $(f^* \in \mathscr{S}(V^*))$. Then we have the following

Theorem $k_\mathfrak{p}$. *If* (G, ρ, V) *and* (G, ρ^*, V^*) *satisfy the assumptions* (A.1) *and* (A.2), *then the following functional equations hold*:

$$\hat{Z}_i(\omega)=\sum_{j=1}^{\nu} \Gamma_{ij}(\omega) Z_j^*(\omega^*\omega_{\lambda^*}) \qquad (1 \leqq i \leqq \nu),$$

where $\Gamma_{ij}(\omega)$ $(1 \leqq i, j \leqq \nu)$ *are rational functions in* $\Omega(K^\times)^n$.

Proof. For $f \in \mathscr{S}(V)$ and $f^* \in \mathscr{S}(V^*)$, we put

$$f^g(x)=f(\rho(g)x), \qquad f^{*g}(y)=f^*(\rho^*(g)y) \qquad (g \in G).$$

Then we have

$$(f^{*g})^\wedge(x)=\omega_\lambda \circ \chi(g)(\hat{f}^*)^g(x) \qquad (f^* \in \mathscr{S}(V^*)).$$

Moreover it is easy to check the following identities:

$$Z_i(\omega)((f^{*g})^\wedge)=(\omega_\lambda/\omega) \circ \chi(g) Z_i(\omega)(\hat{f}^*),$$
$$Z_j^*(\omega^*\omega_{\lambda^*})(f^{*g})=(\omega_\lambda/\omega) \circ \chi(g) Z_j^*(\omega^*\omega_{\lambda^*})(f^*).$$

Therefore by the uniqueness of relatively invariant distributions on homogeneous spaces (cf. [17, Lemma 1]), we obtain

$$Z_i(\omega)(\hat{f}^*)=\Gamma_{ij}(\omega) Z_j^*(\omega^*\omega_{\lambda^*})(f^*) \qquad (f^* \in \mathscr{S}(V_j^*))$$

and hence

$$(*) \qquad Z_i(\omega)(\hat{f}^*)=\sum_{j=1}^{\nu} \Gamma_{ij}(\omega) Z_j^*(\omega^*\omega_{\lambda^*})(f^*) \qquad (f^* \in \mathscr{S}(V^*-S^*)),$$

where $\Gamma_{ij}(\omega)$ is independent of f^*. It is obvious that $\Gamma_{ij}(\omega)$ is a rational function on $\Omega(K^\times)^n$. Consider the distribution T_ω on V^* defined by

$$T_\omega(f^*)=Z_i(\omega)(\hat{f}^*)-\sum_{j=1}^{\nu}\Gamma_{ij}(\omega)Z_j^*(\omega^*\omega_{\lambda^*})(f^*) \qquad (f^*\in\mathcal{S}(V^*)).$$

By the identity (*) and Proposition 2.4, T_ω defines a distribution with support in S^* for ω in $U=\{\omega\in\Omega(K^\times)^n;\ \omega\notin\Omega(S),\ \omega^*\omega_{\lambda^*}\notin\Omega(S^*)\}$. Since $gT_\omega=(\omega_\lambda/\omega)\circ\chi(g)T$, we have $T_\omega=0$ unless $\omega^*\omega_{\lambda^*}\in\Omega(S^*)$. Therefore $T_\omega=0$ for ω in U. It follows from the assumption (A.2) and Corollary to Lemma 2.3 that U is a dense open subset of $\Omega(K^\times)^n$. Since T_ω is a distribution depending on ω meromorphically, we have $T_\omega=0$; namely the identity (*) holds for all $f^*\in\mathcal{S}(V^*)$.

Remark 1. As remarked in the introduction, Theorem $k_\mathfrak{p}$ has been proved by Igusa ([17]) under the following three conditions:

1) $\boldsymbol{G}$ is a connected reductive algebraic group defined over a $\mathfrak{p}$-adic number field and there exists an involution of End($\boldsymbol{V}$) defined over the base field under which $\rho(\boldsymbol{G})$ is stable.

2) $\boldsymbol{S}$ is an absolutely irreducible hypersurface.

By the condition 2) there exists a unique (up to a constant factor) irreducible relative invariant. Let $\boldsymbol{G}^1$ be the kernel of its corresponding rational character of $\boldsymbol{G}$.

3) $\boldsymbol{S}$ decomposes into a finite number of $\boldsymbol{G}^1$-orbits.

By [39, § 4 Remark 26] the conditions 1) and 2) imply our condition (A.1). It is obvious that (A.2) is a natural generalization of the condition 3) above.

Remark 2. One can find many examples of prehomogeneous vector spaces satisfying the conditions (A.1) and (A.2) among reductive prehomogeneous vector spaces which admit only a finite number of orbits classified completely by Kimura, Kasai and Yasukura ([23]). Another examples can be obtained in the following manner. Let $(\boldsymbol{G}, \rho, \boldsymbol{V})$ be a prehomogeneous vector space of commutative parabolic type defined over a $\mathfrak{p}$-adic number field K. Then the restriction of ρ to a K-parabolic subgroup remains to be prehomogeneous ([26, Prop. 3.21]) and supplies us an example with the properties (A.1) and (A.2). In the following section we shall examine three examples of this kind.

Remark 3. It is very likely that the condition (A.2) for a regular $(\boldsymbol{G}, \rho, \boldsymbol{V})$ implies the same condition for $(\boldsymbol{G}, \rho^*, \boldsymbol{V}^*)$. Note that $(\boldsymbol{G}, \rho^*, \boldsymbol{V}^*)$ inherits the finiteness of $\boldsymbol{G}$-orbits from $(\boldsymbol{G}, \rho, \boldsymbol{V})$ (Pyasetskii [31]).

§ 3. Calculation of Γ-matrices

3.1. In this section K is a local field of characteristic 0. We denote by ψ the additive character of K defined by

$$\psi(\alpha)=\begin{cases}\exp(2\pi i\,\alpha), & \text{if } K=\boldsymbol{R},\\ \exp(4\pi i\,\mathrm{Re}\,\alpha), & \text{if } K=\boldsymbol{C},\\ \exp(2\pi i\,\lambda_0(\mathrm{tr}_{K/\boldsymbol{Q}_p}(\alpha))), & \text{if } K \text{ is a finite extension of } \boldsymbol{Q}_p.\end{cases}$$

Here λ_0 stands for the canonical mapping

$$\boldsymbol{Q}_p \longrightarrow \boldsymbol{Q}_p/\boldsymbol{Z}_p \longrightarrow \boldsymbol{Q}/\boldsymbol{Z} \longrightarrow \boldsymbol{R}/\boldsymbol{Z}.$$

A Haar measure $|d\alpha|_K$ on K is always normalized such that $|d\alpha|_K$ is auto dual with respect to ψ.

For a quasi-character $\omega \in \Omega(K^\times)$, the Tate Γ-factor $\Gamma_K(\omega)$ is defined by the functional equation

$$(3.1) \qquad (\omega(\alpha)/|\alpha|_K)^\wedge = \Gamma_K(\omega)\omega(\alpha)^{-1},$$

where $\wedge$ means the Fourier transformation. The identity (3.1) is nothing but the fundamental theorems $\boldsymbol{R}$, $\boldsymbol{C}$ and $k_{\mathfrak{p}}$ for the simplest prehomogeneous vector space obtained from the standard-one dimensional representation of $\boldsymbol{GL}(1)$.

If K is an archimedean local field, then $\Gamma_K(\omega)$ is given explicitly by

$$\Gamma_{\boldsymbol{R}}(|\cdot|_{\boldsymbol{R}}^s\cdot\mathrm{sgn}(\cdot)^\delta)=i^\delta\pi^{(1-2s)/2}\Gamma\left(\frac{s+\delta}{2}\right)\Big/\Gamma\left(\frac{1-s+\delta}{2}\right) \qquad (\delta=0,1),$$

$$\Gamma_{\boldsymbol{C}}\left(|\cdot|_{\boldsymbol{C}}^s\cdot\left(\frac{\cdot}{|\cdot|_{\boldsymbol{C}}^{1/2}}\right)^p\right)=i^{|p|}(2\pi)^{1-2s}\Gamma\left(s+\frac{|p|}{2}\right)\Big/\Gamma\left(1-s+\frac{|p|}{2}\right) \qquad (p\in\boldsymbol{Z}).$$

If K is a finite extension of $\boldsymbol{Q}_p$, then

$$\Gamma_K(|\cdot|_K^s\cdot\phi(\cdot))=N(\mathfrak{d})^{s-1/2}\times\begin{cases}(1-q^{-(1-s)})/(1-q^{-s}) & (\phi=1),\\ g_\phi\cdot q^{fs} & (\phi\neq 1),\end{cases}$$

where $q=N(\mathfrak{p})$, $\phi\in\widehat{\mathfrak{o}^\times}$, f is the conductor of ϕ, $\mathfrak{d}$ is the absolute different of K and

$$g_\phi=q^{-f}\sum_{u\in\mathfrak{o}^\times/1+\mathfrak{p}^f}\phi(u)\psi(u/\pi^{f+\mathrm{ord}_{\mathfrak{p}}(\mathfrak{d})}).$$

For $\varepsilon\in K^\times/(K^\times)^2$, we put

$$(3.2) \qquad \Gamma_K(\omega;\varepsilon)=[K^\times:(K^\times)^2]^{-1}\sum_\phi\phi(\varepsilon)\Gamma_K(\omega\phi) \qquad (\omega\in\Omega(K^\times)),$$

where ϕ runs through all unitary characters of $K^\times$ of order 2.

Next we recall Weil's result on Fourier transforms of quadratic characters.

Let Y be a nondegenerate m by m symmetric matrix with entries in K. The matrix defines a quadratic character $\psi({}^t uYu)$ $(u \in K^m)$ on K^m. Let $|du|_K = |du_1|_K \cdots |du_m|_K$ be the autodual Haar measure on K^m with respect to the pairing $(u, v) = \psi(\langle u, v\rangle)$ with $\langle u, v\rangle = u_1 v_1 + \cdots + u_m v_m$.

The following formula for the Fourier transforms of quadratic characters was obtained by Weil:

$$\int_{K^m} \psi({}^t uYu)\psi(\langle u, v\rangle)|du|_K = |2|_K^{-m/2} |\det Y|_K^{-1/2} \gamma(Y)\psi(-{}^t vY^{-1}v/4), \tag{3.3}$$

for some constant $\gamma(Y)$ with absolute value 1 ([50, n° 14, Theorem 2]). It is known that

$$\gamma(Y_1 \perp Y_2) = \gamma(Y_1)\gamma(Y_2), \quad \gamma({}^t gYg) = \gamma(Y) \quad (g \in \boldsymbol{GL}(m, K)).$$

By considering an element $\alpha \in K^\times$ as a nondegenerate symmetric matrix of size 1, we can define a constant $\gamma(\alpha)$. The constant $\gamma(\alpha)$ depends only on $\mathrm{sgn}(\alpha)$, the residue class of α in $K^\times/(K^\times)^2$. As for the explicit values of $\gamma(\alpha)$, we quote the following results from [50, n° 26] and [32]:

(1) if $K = \boldsymbol{C}$, then $\gamma(\alpha) = 1$ for all $\alpha \in \boldsymbol{C}^\times$,
(2) if $K = \boldsymbol{R}$, then $\gamma(\alpha) = \exp(\mathrm{sgn}(\alpha) \cdot \pi i/4)$ for $\alpha \in \boldsymbol{R}^\times$,
(3) if $K = \boldsymbol{Q}_p$, then for $\alpha \in \boldsymbol{Z}_p^\times$

$$\gamma(\alpha) = \begin{cases} 1 & (p > 2) \\ \exp(\pi i/4) & (p = 2,\ \alpha \equiv 1 \pmod{4\boldsymbol{Z}_2}) \\ \exp(-\pi i/4) & (p = 2,\ \alpha \equiv -1 \pmod{4\boldsymbol{Z}_2}) \end{cases}$$

and

$$\gamma(p\alpha) = \begin{cases} \left(\dfrac{\alpha}{p}\right) \exp((1-p)\pi i/4) & (p > 2) \\ \psi(\alpha/8) & (p = 2). \end{cases}$$

3.2. **$(\mathrm{Trig}(n), \mathrm{Sym}(n))$:** Let $\boldsymbol{G} = \boldsymbol{G}^{(n)} = \mathbf{Trig}(n)$ be the group of all nondegenerate lower triangular matrices of size n and $\boldsymbol{V} = \boldsymbol{V}^{(n)} = \mathbf{Sym}(n)$ the vector space of all symmetric matrices of size n. Let ρ be the rational representation of $\boldsymbol{G}$ on $\boldsymbol{V}$ defined by $\rho(g)x = gx\,{}^t g$ $(g \in \boldsymbol{G}, x \in \boldsymbol{V})$. With respect to the natural K-structure, the triple $(\boldsymbol{G}, \rho, \boldsymbol{V})$ is a prehomogeneous vector space defined over K with the singular set

$$\boldsymbol{S} = \bigcup_{i=1}^{n} \boldsymbol{S}_i, \qquad \boldsymbol{S}_i = \{x \in \boldsymbol{V};\ d_i(x) = 0\},$$

where $d_i(x)$ is the determinant of the upper left i by i block of x. For

$i=n$, $d_n(x)=\det x$. The character χ_i corresponding to $d_i(x)$ is given by

$$\chi_i\left(\begin{pmatrix} a_1 & & 0 \\ & \ddots & \\ * & & a_n \end{pmatrix}\right)=(a_1\cdots a_i)^2 \qquad (1\leqq i\leqq n).$$

We identify V with its dual vector space via the symmetric bilinear form $\langle x, y\rangle=\operatorname{tr} xy$ $(x, y \in V)$. Then the contragredient representation ρ^* is given by $\rho^*(g)y={}^tg^{-1}yg^{-1}$ $(g \in G, x \in V)$. Since $d_n(x)$ is a nondegenerate relative invariant in $K[V]$, (G, ρ, V) satisfies the condition (A.1). Hence the triple (G, ρ^*, V) is also a K-regular prehomogeneous vector space and the singular set S^* is given by

$$S^*=\bigcup_{i=1}^{n} S_i^*, \qquad S_i^*=\{y \in V;\ d_i^*(y)=0\},$$

where $d_i^*(y)$ is the determinant of lower right i by i block of y. The character χ_i^* corresponding to $d_i^*(y)$ is given by

$$\chi_i^*\left(\begin{pmatrix} a_1 & & 0 \\ & \ddots & \\ * & & a_n \end{pmatrix}\right)=(a_{n-i+1}\cdots a_n)^{-2}=\chi_n^{-1}\chi_{n-i}.$$

Therefore we have

$$U=\left[\begin{array}{ccc|c} & & 1 & -1 \\ & \cdot^{\cdot^{\cdot}} & & \vdots \\ 1 & & & -1 \\ \hline & & & -1 \end{array}\right]\begin{matrix}\left.\vphantom{\begin{matrix}1\\ \vdots\\ 1\end{matrix}}\right\} n-1 \\ \\ \end{matrix}, \qquad \lambda=\lambda^*=\left(0, \cdots, 0, \frac{n+1}{2}\right).$$

Lemma 3.1. (G, ρ, V) *and* (G, ρ^*, V) *satisfy the condition* (A.2) *in Section* 2.

Proof. We prove the condition (A.2)$'$ by induction on n. For $n=1$, the condition (A.2)$'$ is clearly satisfied. Now assume that the condition (A.2)$'$ is satisfied by $(G^{(m)}, \rho, V^{(m)})$ $(m\leqq n-1)$. It is easy to see that under the action of $\rho(G^{(n)})$ every point x in S is equivalent to one of points of the following forms:

$$(\mathrm{I}) \qquad \left[\begin{array}{c|c} 1 & 0\cdots 0 \\ \hline 0 & \\ \vdots & x_0 \\ 0 & \end{array}\right] \qquad (x_0 \in S^{(n-1)}),$$

(II)
$$\begin{pmatrix} 0 & 0\cdots 0 \\ 0 & \\ \vdots & x_0 \\ 0 & \end{pmatrix} \qquad (x_0 \in V^{(n-1)}),$$

(III)
$$\left(\begin{array}{c|c|c|c} 0 & 0\cdots 0 & 1 & 0\cdots 0 \\ \hline 0 & & 0 & \\ \vdots & x_1 & \vdots & x_2 \\ 0 & & 0 & \\ \hline 1 & 0\cdots 0 & 0 & 0\cdots 0 \\ \hline 0 & & 0 & \\ \vdots & {}^t x_2 & \vdots & x_3 \\ 0 & & 0 & \end{array}\right) \begin{array}{l} \\ \left.\vphantom{\begin{matrix}0\\0\\0\end{matrix}}\right\} i-2 \\ \\ \left.\vphantom{\begin{matrix}0\\0\\0\end{matrix}}\right\} n-i \end{array} \qquad \left(\begin{pmatrix} x_1 & x_2 \\ {}^t x_2 & x_3 \end{pmatrix} \in V^{(n-2)}\right).$$

By the inductive assumption we see that $\boldsymbol{S}$ is decomposed into a finite number of $\rho(\boldsymbol{G}^{(n)})$-orbits. Therefore it is sufficient to prove the second condition in (A.2)′ for all points of the form (I), (II) or (III). If x is of type (I), then

$$\boldsymbol{G}_x = \left\{ \begin{pmatrix} \pm 1 & 0\cdots 0 \\ 0 & \\ \vdots & g_1 \\ 0 & \end{pmatrix} \Big\} n-1;\ g_1 x_0 {}^t g_1 = x_0 \right\}.$$

Hence by the induction hypothesis there exists a character χ of the form $\chi = \prod_{i=2}^{n} \chi_i^{m_i}$ $(m_2, \cdots, m_n \in \boldsymbol{Z})$ such that $\chi|_{G_x}$ is nontrivial. If x is of type (II), then $\boldsymbol{G}_x$ includes the group $\left\{\begin{pmatrix} a_1 & \\ & I_{n-1} \end{pmatrix};\ a_1 \in \boldsymbol{GL}(1)\right\}$; hence $\chi_1|\boldsymbol{G}_x$ is nontrivial. If x is of type (III), then $\boldsymbol{G}_x$ includes the group

$$\left\{\begin{pmatrix} t & & & \\ & I_{i-2} & & \\ & & t^{-1} & \\ & & & I_{n-i} \end{pmatrix};\ t \in \boldsymbol{GL}(1)\right\};\ \text{hence } \chi_1|_{G_x} \text{ is nontrivial.}$$

This proves the lemma for $(\boldsymbol{G}, \rho, \boldsymbol{V})$. The proof for $(\boldsymbol{G}, \rho^*, \boldsymbol{V})$ is quite the same; so we omit it.

Since $(\boldsymbol{G}, \rho^*, \boldsymbol{V})$ and $(\boldsymbol{G}, \rho^*, \boldsymbol{V})$ satisfy the conditions (A.1), (A.2) and the singular sets $\boldsymbol{S}$ and $\boldsymbol{S}^*$ are hypersurfaces, the fundamental theorem holds over an arbitrary local field of characteristic 0.

For an $\varepsilon=(\varepsilon_1, \cdots, \varepsilon_n) \in (K^\times/(K^\times)^2)^n$, put

$$V_\varepsilon=\{x \in \boldsymbol{V}(K)-\boldsymbol{S}(K);\ \operatorname{sgn}(d_i(x))=\varepsilon_1\cdots\varepsilon_i\ (1\leqq i\leqq n)\}$$

and

$$V_\varepsilon^*=\{y \in \boldsymbol{V}(K)-\boldsymbol{S}^*(K);\ \operatorname{sgn}(d_i^*(y))=\varepsilon_{n-i+1}\cdots\varepsilon_n\ (1\leqq i\leqq n)\}.$$

Then the $\boldsymbol{G}(K)$-orbit decompositions of $\boldsymbol{V}(K)-\boldsymbol{S}(K)$ and $\boldsymbol{V}(K)-\boldsymbol{S}^*(K)$ are given by

$$\boldsymbol{V}(K)-\boldsymbol{S}(K)=\bigcup_{\varepsilon\in\{K^\times/(K^\times)^2\}^n} V_\varepsilon, \qquad \boldsymbol{V}(K)-\boldsymbol{S}^*(K)=\bigcup_{\varepsilon\in\{K^\times/(K^\times)^2\}^n} V_\varepsilon^*.$$

The zeta distributions are defined by analytic continuations of the following integrals:

$$Z_\omega(\omega)(\phi)=\int_{V_\varepsilon}\prod_{i=1}^n \omega^{(i)}(d_i(x))\phi(x)\frac{|dx|_K}{|\det x|_K^{(n+1)/2}},$$

$$Z_\varepsilon^*(\omega)(\phi)=\int_{V_\varepsilon^*}\prod_{i=1}^n \omega^{(i)}(d_i^*(x))\phi(x)\frac{|dx|_K}{|\det x|_K^{(n+1)/2}}$$

$$(\varepsilon \in \{K^\times/(K^\times)^2\}^n, \quad \omega=(\omega^{(1)}, \cdots, \omega^{(n)}) \in \Omega(K^\times)^n, \quad \phi \in \mathscr{S}(\boldsymbol{V}(K))),$$

where $|dx|_K=\prod_{1\leqq i\leqq j\leqq n}|dx_{ij}|_K$ and $|dx_{ij}|_K$ is the Haar measure on K normalized as in Section 3.1.

Theorem 3.2.

$$\hat{Z}_\varepsilon(\omega)=\sum_{\eta\in\{K^\times/(K^\times)^2\}^n}\Gamma_{\varepsilon\eta}(\omega)Z_\eta^*(\omega^*\omega_\lambda),$$

where

$$\Gamma_{\varepsilon\eta}(\omega)=|2|_K^{-n(n-1)/4}\prod_{1\leqq i<j\leqq n}\gamma(\varepsilon_i\eta_j)\prod_{i=1}^n \Gamma_K(\omega^{(i)}\cdots\omega^{(n)}\omega_{(1-i)/2};\ \varepsilon_i\eta_i)$$

and $\omega^*=(\omega^{(n-1)}, \cdots, \omega^{(1)}, (\omega^{(1)}\cdots\omega^{(n)})^{-1})$. *(For the definitions of* $\gamma(\alpha)$ *and* $\Gamma_K(\omega; \varepsilon)$*, see* § 3.1.)

Proof. By Theorem $\boldsymbol{R}$, Theorem $\boldsymbol{C}$ and the combination of Theorem $k_\mathfrak{p}$ with Lemma 3.1, the functional equation holds for some meromorphic functions $\Gamma_{\varepsilon\eta}(\omega)$. Therefore it is enough to calculate $\Gamma_{\varepsilon\eta}(\omega)$ for $\phi \in \mathscr{S}(\boldsymbol{V}(K))$ with support in $\boldsymbol{V}(K)-\boldsymbol{S}^*(K)$ and for ω with sufficiently large real part. This restriction on ω and ϕ assures the convergence of all integrals appearing in the following calculation. First we note that the formula for $\Gamma_{\varepsilon\eta}(\omega)$ is an immediate consequence of (3.1) and the definition (3.2) of $\Gamma_K(\omega; \varepsilon)$, if $n=1$. Let us prove the formula for general n by induction on n.

Since any $x \in V_\varepsilon$ can be decomposed as

$$x=\left(\begin{array}{c|c} x_1 & y \\ \hline {}^t y & a+{}^t y x_1^{-1} y \end{array}\right)$$

with $x_1 \in V_{\varepsilon'}^{(n-1)}$ ($\varepsilon'=(\varepsilon_1, \cdots, \varepsilon_{n-1})$), $y \in K^{n-1}$, $a \in \varepsilon_n \cdot (K^\times)^2$), we have

$$\begin{aligned}\hat{Z}_\varepsilon(\omega)(\phi)&=Z_\varepsilon(\omega)(\hat{\phi})\\ &=\int_{V_{\varepsilon'}^{(n-1)}} \prod_{i=1}^{n-2} \omega^{(i)}(d_i(x_1)) \cdot (\omega^{(n-1)}\omega^{(n)}\omega_{-1/2})(d_{n-1}(x_1)) \frac{|dx_1|_K}{|\det x_1|_K^{n/2}}\\ &\quad\times \int_{K^{n-1}} |dy|_K \int_{V_{\varepsilon_n}^{(1)}} (\omega^{(n)}\omega_{(1-n)/2})(a)\hat{\phi}\left(\left(\begin{array}{c|c} x_1 & y \\ \hline {}^t y & a+{}^t y x_1^{-1} y \end{array}\right)\right)\frac{|da|_K}{|a|_K}.\end{aligned}$$

Applying the theorem for $n=1$ to the integral with respect to a, we get

$$\begin{aligned}\hat{Z}_\varepsilon(\omega)(\phi)&=\sum_{\eta_n \in K^\times/(K^\times)^2} \Gamma_K(\omega^{(n)}\omega_{(1-n)/2};\ \varepsilon_n\eta_n)\\ &\quad\times \int_{V_{\varepsilon'}^{(n-1)}} \prod_{i=1}^{n-2} \omega^{(i)}(d_i(x_1)) \cdot (\omega^{(n-1)}\omega^{(n)}\omega_{-1/2})(d_{n-1}(x_1)) \frac{|dx_1|_K}{|\det x_1|_K^{n/2}}\\ &\quad\times \int_{V_{\eta_n}^{(1)}} (\omega_{(1+n)/2}/\omega^{(n)})(a^*) \frac{|da^*|_K}{|a^*|_K} \int_{K^{n-1}} \psi(a^*\, {}^t y x_1^{-1} y)\hat{\phi}_1\left(\left(\begin{array}{c|c} x_1 & y \\ \hline {}^t y & a^* \end{array}\right)\right)|dy|_K,\end{aligned}$$

where

$$\hat{\phi}_1\left(\begin{pmatrix} x_1 & y \\ {}^t y & a^* \end{pmatrix}\right)=\int_{V^{(n-1)}(K)\times K^{n-1}} \phi\left(\begin{pmatrix} x_1^* & y^* \\ {}^t y^* & a^* \end{pmatrix}\right) \times \psi(\operatorname{tr} x_1 x_1^* + 2\, {}^t y y^*)|dx_1^*|_K |dy^*|_K.$$

The integral with respect to y can be calculated by using (3.3) and the formula

$$\gamma(a^* x_1^{-1})=\gamma(\langle a^*\varepsilon_1^{-1}\rangle \perp \cdots \perp \langle a^*\varepsilon_{n-1}^{-1}\rangle)=\prod_{i=1}^{n-1} \gamma(\varepsilon_i \eta_n).$$

Thus we have

$$\begin{aligned}\hat{Z}_\varepsilon(\omega)(\phi)&=|2|_K^{-(n-1)/2} \sum_{\eta_n \in K^\times/(K^\times)^2} \Gamma_K(\omega^{(n)}\omega_{(1-n)/2};\ \varepsilon_n\eta_n) \prod_{i=1}^{n-1} \gamma(\varepsilon_i\eta_n)\\ &\quad\times \int_{V_{\eta_n}^{(1)}} (\omega_1/\omega^{(n)})(a^*) \frac{|da^*|_K}{|a^*|_K} \int_{K^{n-1}} |dy^*|_K\\ &\quad\times \int_{V_{\varepsilon'}^{(n-1)}} \prod_{i=1}^{n-2} \omega^{(i)}(d_i(x_1)) \cdot (\omega^{(n-1)}\omega^{(n)})(d_{n-1}(x_1))\\ &\quad\cdot \psi(-a^{*-1}\, {}^t y^* x_1 y^*)\hat{\phi}_2\left(\begin{pmatrix} x_1 & y^* \\ {}^t y^* & a^* \end{pmatrix}\right)\frac{|dx_1|_K}{|\det x_1|_K^{n/2}},\end{aligned}$$

where

$$\hat{\phi}_2\left(\begin{pmatrix} x_1 & y^* \\ {}^t y^* & a^* \end{pmatrix}\right)=\int_{V^{(n-1)}(K)} \phi\left(\begin{pmatrix} x_1^* & y^* \\ {}^t y^* & a^* \end{pmatrix}\right)\psi(\operatorname{tr} x_1x_1^*)|dx_1^*|_K.$$

Since

$$\begin{aligned}&\psi(-a^{*-1}\,{}^t y^* x_1 y^*)\hat{\phi}_2\left(\begin{pmatrix} x_1 & y^* \\ {}^t y^* & a^* \end{pmatrix}\right)\\ &\quad=\int_{V^{(n-1)}(K)} \phi\left(\left(\begin{array}{c|c} x_1^*+a^{*-1}y^*\,{}^t y^* & y^* \\ \hline {}^t y^* & a^* \end{array}\right)\right)\psi(\operatorname{tr} x_1x_1^*)|dx_1^*|_K,\end{aligned}$$

it follows from the induction hypothesis that

$$\begin{aligned}\hat{Z}_\varepsilon(\omega)(\phi)=&\,|2|_K^{-n(n-1)/4}\sum_{\eta_n\in K^\times/(K^\times)^2}\prod_{i=1}^{n-1}\gamma(\varepsilon_i\eta_n)\Gamma_K(\omega^{(n)}\omega_{(1-n)/2};\,\varepsilon_n\eta_n)\\ &\times\sum_{\eta'=(\eta_1,\cdots,\eta_{n-1})}\prod_{1\leq i<j\leq n-1}\gamma(\varepsilon_i\eta_j)\prod_{i=1}^{n-1}\Gamma_K(\omega^{(i)}\cdots\omega^{(n)}\omega_{(1-i)/2};\,\varepsilon_i\eta_i)\\ &\times\int_{K^{n-1}}|dy^*|_K\int_{V^{(1)}_{\eta_n}}(\omega_1/\omega^{(n)})(a^*)\frac{|da^*|_K}{|a^*|_K}\\ &\times\int_{V^{*(n-1)}_{\eta'}}\prod_{i=1}^{n-2}\omega^{(n-i-1)}(d_i^*(x_1^*))(\omega_{n/2}/\omega^{(1)}\cdots\omega^{(n)})(d_{n-1}^*(x_1^*))\\ &\quad\cdot\phi\left(\left(\begin{array}{c|c} x_1^*+a^{*-1}y^*\,{}^t y^* & y^* \\ \hline {}^t y^* & a^* \end{array}\right)\right)\frac{|dx_1^*|_K}{|d_{n-1}^*(x_1^*)|_K^{n/2}}\\ =&\,|2|_K^{-n(n-1)/4}\sum_{\eta\in\{K^\times/(K^\times)^2\}^n}\prod_{1\leq i<j\leq n}\gamma(\varepsilon_i\eta_j)\\ &\times\prod_{i=1}^{n}\Gamma_K(\omega^{(i)}\cdots\omega^{(n)}\omega_{(1-i)/2};\,\varepsilon_i\eta_i)Z_\eta^*(\omega^*\omega_\lambda)(\phi).\end{aligned}$$

Remark. Theorem 3.2 for $K=\boldsymbol{R}$ was used previously in proving [36, Theorem 8 (3)], where the proof was omitted. Micro-local calculus provides us another method to prove Theorem 3.2 for $K=\boldsymbol{R}$. The calculation of $\Gamma_{\varepsilon\eta}(\omega)$ for $K=\boldsymbol{R}$ based on the method was carried out by T. Miwa several years ago (unpublished, see also [48]).

3.3. (Trig $(n)\times$ Trig (n), $M(n)$): We now consider the group $\boldsymbol{G}=$ **Trig** $(n)\times$ **Trig** (n) and the vector space $\boldsymbol{V}=M(n)$ of n-rowed square matrices. We define a rational representation ρ of $\boldsymbol{G}$ on $\boldsymbol{V}$ by $\rho(g_1, g_2)x=g_1x{}^tg_2$. With respect to the natural K-structure, the triple $(\boldsymbol{G}, \rho, \boldsymbol{V})$ is a prehomogeneous vector space defined over K with the singular set

$$\boldsymbol{S}=\bigcup_{i=1}^{n}\{x\in\boldsymbol{V}\colon d_i(x)=0\},$$

where $d_i(x)$ is the same as in Section 3.2. The character χ_i corresponding to $d_i(x)$ is given by

$$\chi_i\left(\begin{pmatrix} a_1 & & 0 \\ & \ddots & \\ * & & a_n \end{pmatrix}, \begin{pmatrix} b_1 & & 0 \\ & \ddots & \\ * & & b_n \end{pmatrix}\right)=(a_1\cdots a_i)\cdot(b_1\cdots b_i) \qquad (1\leqq i\leqq n).$$

Since $d_n(x)=\det x$ is a nondegenerate relative invariant, $(\boldsymbol{G}, \rho, \boldsymbol{V})$ is K-regular.

We identify $\boldsymbol{V}$ with its dual vector space via the symmetric bilinear form $\langle x, y\rangle=\mathrm{tr}\,{}^t xy$ $(x, y \in \boldsymbol{V})$. Then the representation ρ^* contragredient to ρ is given by $\rho^*(g_1, g_2)y={}^t g_1^{-1} y g_2^{-1}$ and the triple $(\boldsymbol{G}, \rho^*, \boldsymbol{V})$ is a K-regular prehomogeneous vector space with the singular set

$$\boldsymbol{S}^*=\bigcup_{i=1}^{n}\{y \in \boldsymbol{V};\ d_i^*(y)=0\}.$$

The character corresponding to $d_i^*(y)$ is given by

$$\chi_i^*\left(\begin{pmatrix} a_1 & & 0 \\ & \ddots & \\ * & & a_n \end{pmatrix}, \begin{pmatrix} b_1 & & 0 \\ & \ddots & \\ * & & b_n \end{pmatrix}\right)=(a_{n-i+1}\cdots a_n)^{-1}\cdot(b_{n-i+1}\cdots b_n)^{-1}.$$

From the formulas for χ_i and χ_i^* we have

$$U=\left[\begin{array}{ccc|c} & & 1 & -1 \\ & \cdot^{\cdot^{\cdot}} & & \vdots \\ 1 & & & -1 \\ \hline & & & -1 \end{array}\right]\Bigg\}{\scriptstyle n-1}, \qquad \lambda=\lambda^*=(0, \cdots, 0, n).$$

Lemma 3.3. *$(\boldsymbol{G}, \rho, \boldsymbol{V})$ and $(\boldsymbol{G}, \rho^*, \boldsymbol{V})$ satisfy the condition* (A.2) *in* Section 2.

Since the proof is quite similar to that of Lemma 3.1, we omit it.

It is easy to see that $\boldsymbol{V}(K)-\boldsymbol{S}(K)$ (resp. $\boldsymbol{V}(K)-\boldsymbol{S}^*(K)$) is a single $\rho(\boldsymbol{G}(K))$- (resp. $\rho^*(\boldsymbol{G}(K))$-) orbit. The zeta distributions are defined by the analytic continuations of the integrals

$$Z(\omega)(\phi)=\int_{\boldsymbol{V}(K)-\boldsymbol{S}(K)}\prod_{i=1}^{n}\omega^{(i)}(d_i(x))\phi(x)\frac{|dx|_K}{|\det x|_K^n},$$

$$Z^*(\omega)(\phi)=\int_{\boldsymbol{V}(K)-\boldsymbol{S}^*(K)}\prod_{i=1}^{n}\omega^{(i)}(d_i^*(y))\phi(y)\frac{|dy|_K}{|\det y|_K^n}$$

$$(\omega=(\omega^{(1)}, \cdots, \omega^{(n)}) \in \Omega(K^\times)^n,\ \phi \in \mathcal{S}(V(K))),$$

where $|dx|_K=\prod_{1\leq i,j\leq n}|dx_{ij}|_K$. The normalization of the Haar measure $|dx_{ij}|_K$ on K is the same as in Sections 3.1, 3.2.

Theorem 3.4.

$$\hat{Z}(\omega)=\prod_{i=1}^{n}\Gamma_K(\omega^{(i)}\cdots\omega^{(n)}\omega_{1-i})Z^*(\omega^*\omega_\lambda),$$

where $\omega^*=(\omega^{(n-1)}, \cdots, \omega^{(1)}, (\omega^{(1)}\cdots\omega^{(n)})^{-1})$.

Theorem 3.4 can be proved by an inductive calculation similar to that in the proof of Theorem 3.2. In the present case the functional equation (3.1) is sufficient for the calculation and we need not appeal to the formula for Fourier transforms of quadratic characters.

Remark. One can get an explicit formula for the Γ-matrix for (**Trig** $(2n)$), ρ, **Alt** $(2n)$), where **Alt** $(2n)=\{x \in M(2n);\ x=-{}^t x\}$ and $\rho(g)x=gx\,{}^t g$, by a similar calculation. In [§ 14, § 4], by using the theory of spherical functions on **Alt** $(2n, k_\mathfrak{p})$, Hironaka and the author gave an explicit formula for the zeta function $Z(\omega)(\phi)$ for $\omega=(\omega_{s_1}, \cdots, \omega_{s_n})$ and for $\mathrm{ch}_{\mathbf{Alt}(2n,\mathfrak{o})}$, the characteristic function of **Alt** $(2n, \mathfrak{o})$, which gives another method to calculate the Γ-matrix. The formula for the Γ-matrix for (**Trig** $(2n)$, ρ, **Alt** $(2n)$) has an application to calculation of local densities of alternating forms ([15]). The spherical functions introduced in [14] are closely related to the zeta functions associated with ($\boldsymbol{Sp}(n)\times$ **Trig** $(2n-2)$, $\Lambda_1\otimes\Lambda_1$, $M(2n, 2n-2)$); Theorem 6 in [14] combined with Lemma 3.1 in [14] can be viewed as an explicit formula for $Z(\omega)(\phi)$ with $\omega=(\omega_{s_1}, \cdots, \omega_{s_{n-1}})$ and $\phi=\mathrm{ch}_{M(2n,2n-2;\mathfrak{o})}$ for various $\mathfrak{o}$-forms of the prehomogeneous vector space.

3.4. *The vector representation of a maximal parabolic subgroup of an orthogonal group*: Let $Q(x)$ be a nondegenerate isotropic quadratic form in $m+2$ variables with coefficients in K. We assume that $Q(x)$ is of the form

$$Q(x)=x_0x_{m+1}+\sum_{1\leq i,j\leq m}a_{ij}x_ix_j \qquad (a_{ij}=a_{ji}).$$

The matrix of Q is given by

$$\begin{pmatrix} & & 1/2 \\ & A & \\ 1/2 & & \end{pmatrix} \qquad \text{with } A=(a_{ij}).$$

Let $\boldsymbol{P}$ be the maximal parabolic subgroup of the orthogonal group $\boldsymbol{O}(Q)=\{g\in \boldsymbol{GL}(m+2);\ Q(gx)=Q(x)\}$ defined by

$$\boldsymbol{P}(K)=\left\{\begin{pmatrix} a & -2a^{t}uAh & -a^{t}uAu \\ 0 & h & u \\ 0 & 0 & a^{-1}\end{pmatrix} : \begin{array}{l} a\in K^{\times}, \\ u\in K^{m}, \\ h\in \boldsymbol{O}(A)(K)\end{array}\right\}.$$

We put $\boldsymbol{G}=\boldsymbol{GL}(1)\times\boldsymbol{P}$ and denote by ρ the standard $m+2$-dimensional representation of $\boldsymbol{G}$. The representation space $\boldsymbol{V}$ is identified with the vector space of column vectors with $m+2$ components. The $\boldsymbol{GL}(1)$-part of $\boldsymbol{G}$ acts on $\boldsymbol{V}$ as scalar multiplication. Then the triple $(\boldsymbol{G}, \rho, \boldsymbol{V})$ is a prehomogeneous vector space defined over K with the singular set

$$\bigcup_{i=1}^{2}\{x\in \boldsymbol{V};\ P_i(x)=0\},\quad P_1(x)=x_{m+1},\quad P_2(x)=Q(x).$$

The character χ_i corresponding to P_i is given by

$$\chi_i\left(t, \begin{pmatrix} a & * & * \\ 0 & h & * \\ 0 & 0 & a^{-1}\end{pmatrix}\right)=\begin{cases} ta^{-1} & (i=1), \\ t^{2} & (i=2).\end{cases}$$

Since $P_2(x)=Q(x)$ is a nondegenerate relative invariant, $(\boldsymbol{G}, \rho, \boldsymbol{V})$ is K-regular.

Now we identify $\boldsymbol{V}$ with its dual vector space via the symmetric bilinear form $\langle x, y\rangle=x_0y_0+\cdots+x_{m+1}y_{m+1}$. Then the representation ρ^* contragredient to ρ is given by $\rho^*(t, p)y=t^{-1}\,{}^{t}p^{-1}y$ $(t\in \boldsymbol{GL}(1),\ p\in \boldsymbol{P},\ y\in \boldsymbol{V})$. The triple $(\boldsymbol{G}, \rho^*, \boldsymbol{V})$ is a K-regular prehomogeneous vector space with the singular set

$$\boldsymbol{S}^*=\bigcup_{i=1}^{2}\{y\in \boldsymbol{V};\ P_i^*(y)=0\},\quad P_1^*(y)=y_0,\quad P_2^*(y)=Q^*(y),$$

where Q^* is the quadratic form defined by

$$Q^*(y)=y_0y_{m+1}+4^{-1}\sum_{1\le i,j\le m} a_{ij}^*y_iy_j \qquad \text{with } A^{-1}=(a_{ij}^*).$$

The character χ_i^* corresponding to $P_i^*(y)$ is given by

$$\chi_i^*\left(t, \begin{pmatrix} a & * & * \\ 0 & h & * \\ 0 & 0 & a^{-1}\end{pmatrix}\right)=\begin{cases} (ta)^{-1} & (i=1) \\ t^{-2} & (i=2).\end{cases}$$

From the formulas for χ_i and χ_i^* we have

$$U=\begin{pmatrix}1 & -1\\ 0 & -1\end{pmatrix}, \qquad \lambda=\lambda^*=\left(0,\ \frac{m+2}{2}\right).$$

Lemma 3.5. $(\boldsymbol{G}, \rho, \boldsymbol{V})$ and $(\boldsymbol{G}, \rho^*, \boldsymbol{V})$ *satisfies the condition* (A.2) *in* Section 2.

Proof. We give a proof only for $(\boldsymbol{G}, \rho, \boldsymbol{V})$. It is easy to check that $\boldsymbol{V}$ admits the following 6 $\boldsymbol{G}$-orbits:

(1) $\boldsymbol{V}-\boldsymbol{S}$,
(2) $\{x \in \boldsymbol{V};\ x_{m+1}\neq 0,\ P_2(x)=0\}$,
(3) $\{x \in \boldsymbol{V};\ x_{m+1}=0,\ P_2(x)\neq 0\}$,
(4) $\{x \in \boldsymbol{V};\ x_{m+1}=0,\ P_2(x)=0,\ (x_1, \cdots, x_m)\neq 0\}$,
(5) $\{x \in \boldsymbol{V};\ x_1=\cdots=x_{m+1}=0,\ x_0\neq 0\}$,
(6) $\{0\}$.

The group $\boldsymbol{G}^{(\chi_2)}$ (resp. $\boldsymbol{G}^{(\chi_1)}$) acts transitively on the orbits (2), (4), (5), (6) (resp. (3), (6)). This proves the lemma.

For $\varepsilon \in K^\times/(K^\times)^2$, we put

$$V_\varepsilon=\{x \in \boldsymbol{V}(K);\ \mathrm{sgn}(P_2(x))=\varepsilon\},\ V_\varepsilon^*=\{y \in \boldsymbol{V}(K);\ \mathrm{sgn}(P_2^*(y))=\varepsilon\}.$$

Then

$$\boldsymbol{V}(K)-\boldsymbol{S}(K)=\bigcup_{\varepsilon\in K^\times/(K^\times)^2} V_\varepsilon$$

and

$$\boldsymbol{V}(K)-\boldsymbol{S}^*(K)=\bigcup_{\varepsilon\in K^\times/(K^\times)^2} V_\varepsilon^*$$

give the $\boldsymbol{G}(K)$-orbit decompositions of $\boldsymbol{V}(K)-\boldsymbol{S}(K)$ and $\boldsymbol{V}(K)-\boldsymbol{S}^*(K)$, respectively.

The zeta distributions are defined by analytic continuations of the following integrals:

$$Z_\varepsilon(\omega)(\phi)=\int_{V_\varepsilon}\omega^{(1)}(P_1(x))\omega^{(2)}(P_2(x))\phi(x)\frac{|dx|_K}{|P_2(x)|_K^{(m+2)/2}},$$

$$Z_\varepsilon^*(\omega)(\phi)=\int_{V_\varepsilon^*}\omega^{(1)}(P_1^*(x))\omega^{(2)}(P_2^*(x))\phi(x)\frac{|dx|_K}{|P_2^*(x)|_K^{(m+2)/2}},$$

$(\omega=(\omega^{(1)}, \omega^{(2)}) \in \Omega(K^\times)^2,\ \phi \in \mathcal{S}(\boldsymbol{V}(K)))$, where $|dx|_K=\prod_{i=0}^{m+1}|dx_i|_K$ is the normalized Haar measure on $\boldsymbol{V}(K)$.

Theorem 3.6.

$$\hat{Z}_\varepsilon(\omega)=\sum_{\eta\in K^\times/(K^\times)^2}\Gamma_{\varepsilon\eta}(\omega)Z^*_\eta(\omega^*\omega_\lambda),$$

where

$$\Gamma_{\varepsilon\eta}(\omega)=|2|_K^{-m/2}|\det A|_K^{-1/2}\times\sum_{\mu\in K^\times/(K^\times)^2}\gamma(-\mu A)\Gamma_K(\omega^{(2)}\omega_{-m/2};\,\mu\varepsilon)\Gamma_K(\omega^{(1)}\omega^{(2)};\,\mu\eta)$$

and $\omega^*=(\omega^{(1)},(\omega^{(1)}\omega^{(2)})^{-1})$.

Proof. We shall calculate $\Gamma_{\varepsilon\eta}(\omega)$ for $\phi\in\mathcal{S}(V(K))$ with support in $V(K)-S^*(K)$ and ω with sufficiently large real part. Changing the variable x to y defined by

$$\begin{cases} y_0=x_0+{}^ty'Ay'/y_{m+1} & (y'={}^t(y_1,\cdots,y_m)) \\ y_i=x_i & (1\leqq i\leqq m+1),\end{cases}$$

we have from Theorem 3.2 for $n=1$ the identity

$$\begin{aligned}\hat{Z}_\varepsilon(\omega)(\phi)=&\int_{\substack{y_{m+1}\neq 0\\ \mathrm{sgn}(y_0y_{m+1})=\varepsilon}}(\omega^{(1)}\omega^{(2)}\omega_{-(m+2)/2})(y_{m+1})(\omega^{(2)}\omega_{-(m+2)/2})(y_0)\\ &\times\hat{\phi}(y_0-{}^ty'Ay'/y_{m+1},\,y',\,y_{m+1})|dy_0|_K|dy'|_K|dy_{m+1}|_K\\ =&\sum_{\varepsilon_1\in K^\times/(K^\times)^2}\sum_{\eta_1\in K^\times/(K^\times)^2}\Gamma_K(\omega^{(2)}\omega_{-m/2};\,\varepsilon_1\eta_1)\\ &\times\int_{\mathrm{sgn}\,y_0^*=\eta_1}(\omega_{m/2}/\omega^{(2)})(y_0^*)|dy_0^*|_K\\ &\times\int_{\mathrm{sgn}\,y_{m+1}=\varepsilon_1\varepsilon}(\omega^{(1)}\omega^{(2)}\omega_{-(m+2)/2})(y_{m+1})|dy_{m+1}|_K\\ &\times\int_{K^m}\psi(-(y_0^*/y_{m+1})\,{}^ty'Ay')\hat{\phi}_1(y_0^*,\,y',\,y_{m+1})|dy'|_K,\end{aligned}$$

where

$$\hat{\phi}_1(y_0^*,y',y_{m+1})=\int_{K^{m+1}}\phi(y_0^*,y_1^*,\cdots,y_{m+1}^*)\psi\Big(\sum_{i=1}^{m+1}y_iy_i^*\Big)\prod_{i=1}^{m+1}|dy_i^*|_K.$$

From (3.3) it follows that

$$\begin{aligned}\hat{Z}_\varepsilon(\omega)(\phi)=&|2|_K^{-m/2}|\det A|_K^{-1/2}\\ &\times\sum_{\varepsilon_1\in K^\times/(K^\times)^2}\sum_{\eta_1\in K^\times/(K^\times)^2}\Gamma_K(\omega^{(2)}\omega_{-m/2};\,\varepsilon_1\eta_1)\gamma(-\varepsilon_1\eta_1\varepsilon A)\\ &\times\int_{\mathrm{sgn}\,y_0^*=\eta_1}\omega^{(2)}(y_0^*)^{-1}|dy_0^*|_K\int_{K^m}|dy^{*\prime}|_K\end{aligned}$$

$$\times\int_{\operatorname{sgn} y_{m+1}=\varepsilon_1\varepsilon}(\omega^{(1)}\omega^{(2)}\omega_{-1})(y_{m+1})\psi\left(\frac{y_{m+1}}{4y_0^*}{}^t y^{*\prime}A^{-1}y^{*\prime}\right)$$
$$\cdot\hat{\phi}_2(y_0^*, y^{*\prime}, y_{m+1})|dy_{m+1}|_K,$$

where

$$\hat{\phi}_2(y_0^*, y^{*\prime}, y_{m+1})=\int_K \phi_2(y_0^*, \cdots, y_{m+1}^*)\psi(y_{m+1}y_{m+1}^*)|dy_{m+1}^*|_K.$$

Applying Theorem 3.2 for $n=1$ to the integral with respect to y_{m+1}, we obtain

$$\hat{Z}_\varepsilon(\omega)(\phi)=|2|_K^{-m/2}|\det A|_K^{-1/2}\sum_{\varepsilon_1,\eta_1}\Gamma_K(\omega^{(2)}\omega_{-m/2};\varepsilon_1\eta_1)\gamma(-\varepsilon_1\eta_1\varepsilon A)$$
$$\times\int_{\eta_2\in K^\times/(K^\times)^2}\Gamma_K(\omega^{(1)}\omega^{(2)};\varepsilon_1\eta_2\varepsilon)\int_{\operatorname{sgn} y_0^*=\eta_1}\omega^{(2)}(y_0^*)^{-1}|dy_0^*|_K$$
$$\times\int_{K^m}|dy^{*\prime}|_K\int_{\operatorname{sgn} y_{m+1}^*=\eta_2}(\omega^{(1)}\omega^{(2)})^{-1}(y_{m+1}^*)$$
$$\cdot\phi(y_0^*, y^{*\prime}, y_{m+1}^*-{}^t y^{*\prime}A^{-1}y'/4y_0^*)|dy_{m+1}^*|_K$$
$$=|2|_K^{-m/2}|\det A|_K^{-1/2}$$
$$\times\sum_{\varepsilon_1,\eta_1,\eta_2}\gamma(-\varepsilon_1\eta_1\varepsilon A)\Gamma_K(\omega^{(2)}\omega_{-m/2};\varepsilon_1\eta_1)\Gamma_K(\omega^{(1)}\omega^{(2)};\varepsilon_1\eta_2\varepsilon)$$
$$\times\int_{\substack{\operatorname{sgn} P_1^*(y^*)=\eta_1\\ \operatorname{sgn} P_2^*(y^*)=\eta_1\eta_2}}\omega^{(1)}(P_1^*(y^*))(\omega_{(m+2)/2}/\omega^{(1)}\omega^{(2)})(P_2^*(y^*))\phi(y^*)\cdot\frac{|dy^*|_K}{|P_2^*(y^*)|_K^{(m+2)/2}}.$$

In the right hand side of the identity above, we put $\mu=\varepsilon_1\eta_1\varepsilon$ and $\eta=\eta_1\eta_2$. Then it is easy to see that the right hand side is equal to

$$|2|_K^{-m/2}|\det A|_K^{-1/2}\sum_\eta Z_\eta^*(\omega^*\omega_\lambda)(\phi)$$
$$\times\sum_\mu\gamma(-\mu A)\Gamma_K(\omega^{(2)}\omega_{-m/2};\mu\varepsilon)\Gamma_K(\omega^{(1)}\omega^{(2)};\mu\eta).$$

This completes the proof.

Remark. Theorem 3.6 for $K=\boldsymbol{R}$ has been already obtained by Muro (private communication). His calculation is based on the method of micro-local calculus.

3.5. *Observation.* It is remarkable that the entries of the Γ-matrices calculated in Theorems 3.2, 3.4 and 3.6 have the following form in common:

$$a_0 \sum_i c_i \sum_j \Gamma_K(e_j(\omega)\chi_{ij}),$$

where a_0 is a completely elementary constant, the coefficients c_i involve constants of rather delicate nature such as $\gamma(Y)$, the arguments $e_j(\omega)$ of Tate Γ-factors are determined by the b-function of a prehomogeneous vector space under consideration and χ_{ij} are unitary characters of K. Only c_i and χ_{ij} depend on open orbits in $\boldsymbol{V}(K)-\boldsymbol{S}(K)$. It seems to be a fairly general phenomenon that the entries of Γ-matrices can be written in the form above, if independent relative invariants $P_1, \cdots, P_n$ are suitably normalized. The recent results of Muller [25] and Igusa [18] as well as the formula for Γ-matrices over $\boldsymbol{R}$ obtained by the method of micro-local calculus (cf. [19], [22, Theorem 7.10], [28]) support this expectation. In particular it is expected that the exponential polynomials $t_{ij}(s)$ in Theorem $\boldsymbol{R}$ are also under the control of b-functions. In the results of Muller the coefficients c_i are expressed in terms of the coefficients of the Fourier transforms of quadratic characters, on the other hand in the formula based on the micro-local calculus they are expressed in terms of the Maslov indices; it is noteworthy that they are determined by signatures of certain quadratic forms in both cases.

§ 4. Zeta functions as distribution

4.1. Let V be a finite dimensional $\boldsymbol{Q}$-vector space. A complex valued function ϕ on V is called a *Schwartz-Bruhat function*, if there exist two lattices L_1 and L_2 in V such that the support of ϕ is contained in L_1 and $\phi(x)=\phi(x')$ whenever $x-x' \in L_2$. We denote by $\mathscr{S}(V)$ the vector space of all Schwartz-Bruhat functions on V. Let V^* be the vector space dual to V.

Now we fix a lattice L_0 in V and let L_0^* be the lattice in V^* dual to L_0. Define a Fourier transform $\hat{\phi}$ of $\phi \in \mathscr{S}(V)$ by setting

$$\hat{\phi}(y)=\frac{1}{[L_0: L]} \sum_{x \in V/L} \phi(x) e^{2\pi i \langle x, y \rangle} \qquad (y \in V^*),$$

where L is a sublattice of L_0 such that y is in the lattice L^* dual to L and $\phi(x)=\phi(x')$ if $x-x' \in L$. It is easy to see that $\hat{\phi}(y)$ is independent of the choice of L and $\hat{\phi}$ defines a Schwartz-Bruhat function on V^*. The Fourier transformation $\phi \mapsto \hat{\phi}$ induces a linear isomorphism of $\mathscr{S}(V)$ onto $\mathscr{S}(V^*)$.

Let $\mathscr{S}(V)'$ be the dual space of $\mathscr{S}(V)$. We call an element in $\mathscr{S}(V)'$ a *distribution on* V. For a distribution $T \in \mathscr{S}(V^*)'$, the Fourier transform $\hat{T}$ ($\in \mathscr{S}(V)'$) is defined by $\hat{T}(\phi)=T(\hat{\phi})$ ($\phi \in \mathscr{S}(V)$).

Put $V(\boldsymbol{R})=V\otimes_{\boldsymbol{Q}}\boldsymbol{R}$ and denote by $\mathscr{S}(V(\boldsymbol{R}))$ the vector space of all rapidly decreasing functions. Let dx be the Euclidean measure on $V(\boldsymbol{R})$ normalized by $\int_{V(\boldsymbol{R})/L_0} dx=1$. We define a Fourier transform $\hat{f}$ of $f \in \mathscr{S}(V(\boldsymbol{R}))$ by

$$\hat{f}(y)=\int_{V(\boldsymbol{R})} f(x)e^{-2\pi i(x,y)}dx.$$

The function $\hat{f}$ is a rapidly decreasing function on $V^*(\boldsymbol{R})=V^*\otimes_{\boldsymbol{Q}}\boldsymbol{R}$. For an $f \in \mathscr{S}(V(\boldsymbol{R}))$ and a $\phi \in \mathscr{S}(V)$, the following variant of the Poisson summation formula holds:

$$\sum_{m\in V} f(m)\phi(m)=\sum_{m^*\in V^*} \hat{f}(m)\hat{\phi}(m^*).$$

Remark. The Poisson summation formula above is essentially an interpretation of the Poisson summation formula on the adelized vector space $V(\boldsymbol{A})$.

4.2. Let $(\boldsymbol{G}, \rho, \boldsymbol{V})$ be a prehomogeneous vector space defined over the rational number field $\boldsymbol{Q}$ and $\boldsymbol{S}$ its singular set. Let $\boldsymbol{S}_1, \cdots, \boldsymbol{S}_n$ be the $\boldsymbol{Q}$-irreducible hypersurfaces contained in $\boldsymbol{S}$ and fix a $\boldsymbol{Q}$-irreducible polynomial P_i defining $\boldsymbol{S}_i$ for each $i=1, \cdots, n$. We assume that

(4.1) *for any* $x \in \boldsymbol{V}(\boldsymbol{Q})-\boldsymbol{S}(\boldsymbol{Q})$, *the identity component of* $\boldsymbol{G}_x=\{g \in \boldsymbol{G};\ \rho(g)x=x\}$ *has no nontrivial* $\boldsymbol{Q}$-*rational character.*

Under this assumption we shall define a relatively $\boldsymbol{G}(\boldsymbol{Q})$-invariant "measure" on $\boldsymbol{V}(\boldsymbol{Q})-\boldsymbol{S}(\boldsymbol{Q})$.

Let G^+ be the connected component of the identity element of $\boldsymbol{G}(\boldsymbol{R})$. Let $\boldsymbol{V}(\boldsymbol{R})-\boldsymbol{S}(\boldsymbol{R})=V_1\cup\cdots\cup V_\nu$ be the G^+-orbit decomposition. Let dg be a right invariant measure on G^+ and $d_V^\times(x)$ be a relatively G^+-invariant measure on $\boldsymbol{V}(\boldsymbol{R})-\boldsymbol{S}(\boldsymbol{R})$ with multiplier $\Delta(h)=d(hg)/dg$. The existence of such a measure is assured by (4.1) and can be normalized as follows:

$$d_V^\times(x)=\prod_{i=1}^{n} |P_i(x)|^{-\delta_i}dx,$$

where $\delta_1, \cdots, \delta_n$ are some rational numbers (see [35, § 4]).

For any $x \in V_i$, we normalize a Haar measure $d\mu_x$ on $G_x^+=\boldsymbol{G}_x\cap G^+$ by the formula

$$\int_{G^+} f(g)dg=\int_{G^+/G_x^+} d_V^\times(\rho(\dot{g})x)\int_{G_x^+} f(\dot{g}h)d\mu_x(h).$$

Put $x'=\rho(h)x$ $(h \in G^+)$. Then the mapping $g \mapsto hgh^{-1}$ induces an isomorphism of G_x^+ onto $G_{x'}^+$ and the pull back of the measure $d\mu_{x'}$ by the isomorphism coincides with $d\mu_x$.

Let $C^\infty(\boldsymbol{V}(\boldsymbol{Q})-\boldsymbol{S}(\boldsymbol{Q}))$ be the space of all complex valued functions on $\boldsymbol{V}(\boldsymbol{Q})-\boldsymbol{S}(\boldsymbol{Q})$ invariant under some arithmetic subgroup of $\boldsymbol{G}(\boldsymbol{Q})$ and $\mathscr{S}(\boldsymbol{V}(\boldsymbol{Q})-\boldsymbol{S}(\boldsymbol{Q}))$ the subspace of $C^\infty(\boldsymbol{V}(\boldsymbol{Q})-\boldsymbol{S}(\boldsymbol{Q}))$ consisting of all functions whose supports are contained in a finite union of orbits of some arithmetic subgroup of $\boldsymbol{G}(\boldsymbol{Q})$.

We fix an arithmetic subgroup Γ_0 of $\boldsymbol{G}(\boldsymbol{Q})$. For an $f \in \mathscr{S}(\boldsymbol{V}(\boldsymbol{Q})-\boldsymbol{S}(\boldsymbol{Q}))$, one can find an arithmetic subgroup Γ of $\boldsymbol{G}(\boldsymbol{Q})$ contained in G^+ and rational points $x_1, \cdots, x_m \in \boldsymbol{V}(\boldsymbol{Q})-\boldsymbol{S}(\boldsymbol{Q})$ such that

$$f(x)=\sum_{i=1}^{m} c_i ch_{\Gamma\cdot x_i}(x) \qquad (c_i \in \boldsymbol{C}), \tag{4.2}$$

where $ch_{\Gamma\cdot x_i}$ stands for the characteristic function of $\rho(\Gamma)x_i$.

Define a linear form μ on $\mathscr{S}(\boldsymbol{V}(\boldsymbol{Q})-\boldsymbol{S}(\boldsymbol{Q}))$ by

$$\mu(f)=[\Gamma:\Gamma_0]\sum_{i=1}^{m} c_i \int_{G_{x_i}^+/\Gamma_{x_i}} d\mu_{x_i}, \tag{4.3}$$

where $[\Gamma:\Gamma_0]=[\Gamma:\Gamma\cap\Gamma_0]/[\Gamma_0:\Gamma\cap\Gamma_0]$ and $\Gamma_{x_i}=\Gamma\cap G_{x_i}^+$. Note that the right hand side of (4.3) is finite by the assumption (4.1).

In order to see that $\mu(f)$ is independent of the decomposition (4.2) of f, we need the following lemma.

Lemma 4.1. *Let Γ_1 and Γ_2 be two arithmetic subgroups of $\boldsymbol{G}(\boldsymbol{Q})\cap G^+$ with $\Gamma_1\supset\Gamma_2$. For an $x \in \boldsymbol{V}(\boldsymbol{Q})-\boldsymbol{S}(\boldsymbol{Q})$, let*

$$\rho(\Gamma_1)x=\bigcup_{i=1}^{t} \rho(\Gamma_2)x_i$$

be the Γ_2-orbit decomposition. Then

$$[\Gamma_1:\Gamma_0]\int_{G_x^+/\Gamma_{1,x}} d\mu_x=[\Gamma_2:\Gamma_0]\sum_{i=1}^{t}\int_{G_{x_i}^+/\Gamma_{2,x_i}} d\mu_{x_i}.$$

Proof. Fox a $\gamma_i \in \Gamma_1$ such that $x_i=\rho(\gamma_i)x$ for each $i=1, \cdots, t$. Then we have the double coset decomposition

$$\Gamma_1=\bigcup_{i=1}^{t} \Gamma_2\gamma_i\Gamma_{1,x}.$$

Hence

$$[\Gamma_1\colon \Gamma_2]=\sum_{i=1}^{t} \sharp(\Gamma_2\backslash\Gamma_2\gamma_i\Gamma_{1,x})$$
$$=\sum_{i=1}^{t} [\gamma_i\Gamma_{1,x}\gamma_i^{-1}\colon \Gamma_2\cap\gamma_i\Gamma_{1,x}\gamma_i^{-1}]$$
$$=\sum_{i=1}^{t} [\Gamma_{1,x_i}\colon \Gamma_{2,x_i}].$$

Since x and x_i are equivalent under Γ_1, we have

$$\int_{G_x^+/\Gamma_{1,x}} d\mu_x=\int_{G_{x_i}^+/\Gamma_{1,x_i}} d\mu_{x_i}=[\Gamma_{1,x_i}\colon \Gamma_{2,x_i}]^{-1}\cdot\int_{G_{x_i}^+/\Gamma_{2,x_i}} d\mu_{x_i}.$$

Therefore

$$[\Gamma_2\colon \Gamma_0]\cdot\sum_{i=1}^{t}\int_{G_{x_i}^+/\Gamma_{2,x_i}} d\mu_{x_i}=[\Gamma_2\colon \Gamma_0]\cdot\int_{G_x^+/\Gamma_{1,x}} d\mu_x\cdot\sum_{i=1}^{t}[\Gamma_{1,x_i}\colon \Gamma_{2,x_i}]$$
$$=[\Gamma_1\colon \Gamma_2][\Gamma_2\colon \Gamma_0]\cdot\int_{G_x^+/\Gamma_{1,x}} d\mu_x$$
$$=[\Gamma_1\colon \Gamma_0]\int_{G_x^+/\Gamma_{1,x}} d\mu_x.$$

In view of (4.3) the identity in Lemma 4.1 implies that

$$\mu(ch_{\Gamma_1\cdot x})=\sum_{i=1}^{t}\mu(ch_{\Gamma_2\cdot x_i});$$

this shows that $\mu(f)$ is independent of the decomposition (4.2) to f. Thus μ determines a positive linear functional on $\mathscr{S}(\boldsymbol{V}(\boldsymbol{Q})-\boldsymbol{S}(\boldsymbol{Q}))$.

For a $g\in\boldsymbol{G}(\boldsymbol{Q})$ we put

$$\delta(g)=[g\Gamma g^{-1}\colon \Gamma]$$
$$=[g\Gamma g^{-1}\colon \Gamma\cap g\Gamma g^{-1}]/[\Gamma\colon \Gamma\cap g\Gamma g^{-1}].$$

Then it is easy to check that $\delta(g)$ is independent of the choice of an arithmetic subgronp Γ of $\boldsymbol{G}(\boldsymbol{Q})$ and defines a homomorphism of $\boldsymbol{G}(\boldsymbol{Q})$ into $\boldsymbol{Q}_+^\times$.

For a $g\in\boldsymbol{G}(\boldsymbol{Q})$ and an $f\in\mathscr{S}(\boldsymbol{V}(\boldsymbol{Q})-\boldsymbol{S}(\boldsymbol{Q}))$, put ${}^gf(x)=f(\rho(g^{-1})x)$. Note that ${}^gf$ is also in $\mathscr{S}(\boldsymbol{V}(\boldsymbol{Q})-\boldsymbol{S}(\boldsymbol{Q}))$.

Lemma 4.2. *For any $g\in\boldsymbol{G}(\boldsymbol{Q})$ and any $f\in\mathscr{S}(\boldsymbol{V}(\boldsymbol{Q})-\boldsymbol{S}(\boldsymbol{Q}))$, we have $\mu({}^gf)=\delta(g)\mu(f)$.*

Proof. It is sufficient to prove the lemma for $f=ch_{\Gamma\cdot x}$. Then, by definition, we have

$$\mu({}^{g}f)=\mu(ch_{g\Gamma g^{-1}\cdot gx})=[g\Gamma g^{-1}\colon \Gamma_0]\int_{G_{gx}^{+}/(g\Gamma g^{-1})_{gx}} d\mu_{gx}.$$

Since the pull back of the measure $d\mu_{gx}$ coincides with $d\mu_x$, the right hand side of the identity is equal to

$$[g\Gamma g^{-1}\colon \Gamma][\Gamma\colon \Gamma_0]\int_{G_x^{+}/\Gamma_x} d\mu_x=\delta(g)\mu(ch_{\Gamma\cdot x}).$$

By Lemma 4.2 we may regard μ as a kind of relatively $\boldsymbol{G}(\boldsymbol{Q})$-invariant measure on $\boldsymbol{V}(\boldsymbol{Q})-\boldsymbol{S}(\boldsymbol{Q})$ with multiplier δ. In this sense we often write

$$\mu(f)=\int_{\boldsymbol{V}(\boldsymbol{Q})-\boldsymbol{S}(\boldsymbol{Q})} f(x)d\mu(x) \qquad (f\in\mathcal{S}(\boldsymbol{V}(\boldsymbol{Q})-\boldsymbol{S}(\boldsymbol{Q}))).$$

Moreover for a subset E of $\boldsymbol{V}(\boldsymbol{Q})-\boldsymbol{S}(\boldsymbol{Q})$ such that the characteristic function ch_E of E is in $C^\infty(\boldsymbol{V}(\boldsymbol{Q})-\boldsymbol{S}(\boldsymbol{Q}))$, we write

$$\mu(f\cdot ch_E)=\int_E f(x)d\mu(x) \qquad (f\in\mathcal{S}(\boldsymbol{V}(\boldsymbol{Q})-\boldsymbol{S}(\boldsymbol{Q}))).$$

If ch_E is in $\mathcal{S}(\boldsymbol{V}(\boldsymbol{Q})-\boldsymbol{S}(\boldsymbol{Q}))$, then the value $\mu(ch_E)$ is called the *volume* (or *density*) of E and is simply written as $\mu(E)$.

For an $f\in C^\infty(\boldsymbol{V}(\boldsymbol{Q})-\boldsymbol{S}(\boldsymbol{Q}))$, take a Γ such that f is $\rho(\Gamma)$-invariant and write

$$f=\sum_{x\in\Gamma\backslash\boldsymbol{V}(\boldsymbol{Q})-\boldsymbol{S}(\boldsymbol{Q})} c_x\cdot ch_{\Gamma\cdot x} \qquad (c_x\in\boldsymbol{C}).$$

If the infinite series $\sum_{x\in\Gamma\backslash\boldsymbol{V}(\boldsymbol{Q})-\boldsymbol{S}(\boldsymbol{Q})}|c_x|\mu(\Gamma\cdot x)$ is convergent, then we call the function f an *integrable function* (*with respect to* μ) and put

$$\mu(f)=\int_{\boldsymbol{V}(\boldsymbol{Q})-\boldsymbol{S}(\boldsymbol{Q})} f(x)d\mu(x)=\sum_{x\in\Gamma\backslash\boldsymbol{V}(\boldsymbol{Q})-\boldsymbol{S}(\boldsymbol{Q})} c_x\mu(\Gamma\cdot x).$$

The value $\mu(f)$ is independent of the choice of Γ.

4.3. For $s=(s_1,\cdots,s_n)\in\boldsymbol{C}^n$ and i $(1\leqq i\leqq\nu)$, we define a function $|P|_i^{-s}$ on $\boldsymbol{V}(\boldsymbol{R})-\boldsymbol{S}(\boldsymbol{R})$ by

$$|P(x)|_i^{-s}=\begin{cases}\prod_{j=1}^{n}|P_j(x)|^{-s_j} & (x\in V_i),\\ 0 & (x\notin V_i).\end{cases}$$

Then the restrictions of $|P(x)|_1^{-s},\cdots,|P(x)|_\nu^{-s}$ to $\boldsymbol{V}(\boldsymbol{Q})-\boldsymbol{S}(\boldsymbol{Q})$ define functions in $C^\infty(\boldsymbol{V}(\boldsymbol{Q})-\boldsymbol{S}(\boldsymbol{Q}))$, which we denote by the same symbols.

The restriction $\phi|_{V(Q)-S(Q)}$ of $\phi \in \mathscr{S}(V(Q))$ also defines a function in $C^\infty(V(Q)-S(Q))$.

We define distributions $Z_i(s)$ ($1 \leq i \leq \nu$, $s \in C^n$) on $V(Q)$, which we call the *zeta distributions associated with* (G, ρ, V) *over* Q, by

$$Z_i(s)\colon \mathscr{S}(V(Q)) \ni \phi \longmapsto Z_i(s)(\phi) = \int_{V(Q)-S(Q)} |P(x)|_i^{-s} \phi(x) d\mu(x) \in C.$$

Here we must assume the following condition, which assures the existence of the zeta distributions:

(4.4) *If* $\operatorname{Re} s_1, \cdots, \operatorname{Re} s_n$ *are sufficiently large, then the function* $|P(x)|_i^{-s} \cdot \phi|_{V(Q)-S(Q)}$ *is integrable for all* $i = 1, \cdots, \nu$ *and all* $\phi \in \mathscr{S}(V(Q))$.

It is a conjecture that the assumption (4.1) implies (4.4).

Let Γ be an arithmetic subgroup of $G(Q)$ such that $|P(x)|_i^{-s}$ $(1 \leq i \leq \nu)$ and $\phi(x)$ are Γ-invariant. Then by definition we have

$$Z_i(s)(\phi) = \sum_{x \in \Gamma \backslash V(Q) \cap V_i} \phi(x) \mu(\Gamma \cdot x) \Big/ \prod_{j=1}^{n} |P_j(x)|^{s_j}.$$

In particular, if ϕ is the characteristic function of a Γ-invariant lattice L in $V(Q)$, then

$$Z_i(s)(ch_L) = [\Gamma : \Gamma_0] \sum_{x \in \Gamma \backslash L \cap V_i} \left(\int_{G_x^+/\Gamma_x} d\mu_x \right) \Big/ \prod_{j=1}^{n} |P_j(x)|^{s_j}. \tag{4.5}$$

The right hand side of (4.5) is nothing but the zeta functions associated with (G, ρ, V) introduced in [40] and [35, § 4]. Therefore the condition (4.4) can be rephrased into the usual assumption on convergence of zeta functions (cf. [35, (4.3)]). Notice that, if $Z_i(s)(ch_L)$ is absolutely convergent for an L and sufficiently large $\operatorname{Re} s_1, \cdots, \operatorname{Re} s_n$, then so is $Z_i(s)(\phi)$ for all $\phi \in \mathscr{S}(V(Q))$.

4.4. The above argument clarifies the meaning of zeta functions. Consider the zeta functions (4.5) and assume that $P_1, \cdots, P_n$ take values in Z on L. For an n-tuple $m = (m_1, \cdots, m_n)$ of positive integers, let $A_i(m; L)$ be the set of all solutions in $L \cap V_i$ of the simultaneous Diophantine equation

$$\begin{cases} P_1(x) = \pm m_1, \\ \quad \vdots \\ P_n(x) = \pm m_n. \end{cases} \tag{4.6}$$

In general the set $A_i(m; L)$ of solutions of (4.6) may be an infinite set.

However we can measure the density of $A_i(m; L)$ by using μ and the zeta function (4.5) can be transformed into the following form:

$$Z_i(s)(ch_L) = \sum_{m_1, \cdots, m_n = 1}^{\infty} \mu(A_i(m; L)) m_{m_1}^{-s_1} \cdots m_{m_n}^{-s_n}.$$

From this expression one sees immediately that the zeta function has an intuitive meaning as a generating function of the densities of solutions of the Diophantine equation (4.6).

Remark 1. The idea of measuring the densities of solutions of Diophantine equations by using integrals of the form $\int_{G_x^+/\Gamma_x} d\mu_x$ goes back to Siegel's work on the arithmetic of indefinite quadratic forms ([45], [46]).

Remark 2. Let $\hat{G}(\boldsymbol{Q})$ be the completion of $\boldsymbol{G}(\boldsymbol{Q})$ with respect to the subgroup topology defined by the family of arithmetic subgroups of $\boldsymbol{G}(\boldsymbol{Q})$. Let

$$X = \lim_{\Gamma} \operatorname{proj} \Gamma \backslash (\boldsymbol{V}(\boldsymbol{Q}) - \boldsymbol{S}(\boldsymbol{Q})),$$

where Γ runs through arithmetic subgroups of $\boldsymbol{G}(\boldsymbol{Q})$. Then X is a totally disconnected locally compact topological space and $\hat{G}(\boldsymbol{Q})$ acts topologically on X. Moreover $\mathscr{S}(\boldsymbol{V}(\boldsymbol{Q}) - \boldsymbol{S}(\boldsymbol{Q}))$ can be identified with the space of locally constant functions with compact support on X. The linear functional μ defined above induces a relatively $\hat{G}(\boldsymbol{Q})$-invariant measure on X and δ: $\boldsymbol{G}(\boldsymbol{Q}) \to \boldsymbol{Q}_+^\times$ coincides with the restriction of the module of $\hat{G}(\boldsymbol{Q})$ to $\boldsymbol{G}(\boldsymbol{Q})$. In this sense μ is really a measure. The space X is closely related to the space $\prod'_p (\boldsymbol{V}(\boldsymbol{Q}_p) - \boldsymbol{S}(\boldsymbol{Q}_p))$ of finite adeles; however X does not coincide with it in general, unless $(\boldsymbol{G}, \rho, \boldsymbol{V})$ is universally transitive in the sense of [18].

4.5. In the following we assume that

(4.7) *$(\boldsymbol{G}, \rho, \boldsymbol{V})$ is $\boldsymbol{Q}$-regular*

and

(4.8) *The singular set $\boldsymbol{S}$ is a hypersurface in $\boldsymbol{V}$,*

as well as the conditions (4.1) and (4.4).

We employ the same notation as in Section 1.1 and Section 1.2 for $K = \boldsymbol{Q}$. Then the prehomogeneous vector spaces $(\boldsymbol{G}, \rho^*, \boldsymbol{V}^*)$ dual to $(\boldsymbol{G}, \rho, \boldsymbol{V})$ satisfies automatically the assumptions (4.1), (4.7) and (4.8). Recall that the number of $\boldsymbol{Q}$-irreducible components, (all of which are of codimension 1 by (4.8)), of the singular set $\boldsymbol{S}^*$ of $(\boldsymbol{G}, \rho^*, \boldsymbol{V}^*)$ is equal to

n (= the number of $\boldsymbol{Q}$-irreducible components of $\boldsymbol{S}$). Let $\boldsymbol{S}^* = \boldsymbol{S}_1^* \cup \cdots \cup \boldsymbol{S}_n^*$ be the decomposition of $\boldsymbol{S}^*$ into $\boldsymbol{Q}$-irreducible hypersurfaces and take a $\boldsymbol{Q}$-irreducible relative invariant P_i^* defining $\boldsymbol{S}_i^*$ for each $i=1, \cdots, n$. Let $\boldsymbol{V}^*(\boldsymbol{R}) - \boldsymbol{S}^*(\boldsymbol{R}) = V_1^* \cup \cdots \cup V_\nu^*$ be the G^+-orbit decomposition. Here the number of G^+-orbits in $\boldsymbol{V}^*(\boldsymbol{R}) - \boldsymbol{S}^*(\boldsymbol{R})$ coincides with ν (= the number of G^+-orbits in $\boldsymbol{V}(\boldsymbol{R}) - \boldsymbol{S}(\boldsymbol{R})$).

We are able to define the zeta distributions $Z_i^*(s)$ $(1 \leqq i \leqq \nu)$ on $\boldsymbol{V}^*(\boldsymbol{Q})$ associated with $(\boldsymbol{G}, \rho^*, \boldsymbol{V}^*)$ in the same way as for $(\boldsymbol{G}, \rho, \boldsymbol{V})$:

$$Z_i^*(s)(\phi^*) = \int_{V^*(Q) - S^*(Q)} |P^*(y)|_i^{-s} \phi^*(y) d\mu^*(y) \qquad (\phi^* \in \mathscr{S}(V^*(Q))),$$

where

$$|P^*(y)|_i^{-s} = \begin{cases} \prod_{j=1}^{n} |P_j^*(y)|^{-s_j}, & \text{if } y \in V_i^* \cap V^*(Q), \\ 0, & \text{otherwise.} \end{cases}$$

We assume the condition (4.4) also for $(\boldsymbol{G}, \rho^*, \boldsymbol{V}^*)$.

Let D_0 (resp. D_0^*) be the set of $s \in \boldsymbol{C}^n$, for which $Z_i(s)(\phi)$ (resp. $Z_i^*(s)(\phi^*)$) are absolutely convergent and D (resp. D^*) be the convex hull of $D_0 \cup (D_0^* U^{-1} + \lambda)$ (resp. $D_0^* \cup (D_0 - \lambda) U$).

Now the main result in the theory of zeta functions associated with prehomogeneous vector spaces can be stated in the language of distributions on $\boldsymbol{V}(\boldsymbol{Q})$ and $\boldsymbol{V}^*(\boldsymbol{Q})$ as follows:

Theorem $\boldsymbol{Q}$. (1) *The distributions $Z_i(s)$ (resp. $Z_i^*(s)$) have analytic continuations to distributions depending meromorphically on s in D (resp. D^*).*

(2) *The following functional equations hold for $s \in D$:*

$$Z_i^*((s-\lambda)U)^\wedge = \sum_{j=1}^{\nu} \Gamma_{ji}(s-\delta) Z_j(s),$$

where $(\Gamma_{ij}(s))$ is the Γ-matrix appearing in Theorem $\boldsymbol{R}$ in Section 1.3 and δ is an n-tuple of rational numbers appearing as exponents of $|P_i(x)|$ in the definition of $d_V^\times(x)$ in Section 4.2.

In the present formulation, the analogy of Theorem $\boldsymbol{Q}$ with Theorem $\boldsymbol{R}$ and Theorem $k_\mathfrak{p}$ is quite obvious.

The proof of Theorem $\boldsymbol{Q}$ is the same as in the proof of [35, § 6 Theorem 2]. The only necessary modification is that the Poisson summation formula should be applied in the form given in Section 4.1. It is known that D and D^* coincide with whole of $\boldsymbol{C}^n$, if $\boldsymbol{G}$ is reductive (see [35, § 6 Corollary 1]).

Remark. Hoffman [16] independently introduced the space $\mathscr{S}(V(\boldsymbol{Q}))$ and studied the zeta function $Z_i(s)(\phi)$. His aim is to establish fundamental properties of zeta function of Hurwitz type associated with prehomogeneous vector spaces of parabolic type whose special values contribute trace formulas for Hecke operators (see also § 4.7).

4.6. Our zeta functions $Z_i(s)(\phi)$ (= the value of the zeta distribution) are more general than the zeta functions treated in [40] and [35], because the consideration in these papers was restricted to characteristic functions of lattices in $V(\boldsymbol{Q})$. As is seen in the following, by specializing $\phi \in \mathscr{S}(V(\boldsymbol{Q}))$ suitably, one can obtain several interesting Dirichlet series.

(A) *Zeta functions of Hurwitz type*: Fix an $a \in V(\boldsymbol{Q})$ and a lattice L. Let ϕ be the characteristic function of $a+L$. Then $Z_i(s)(\phi)$ can be regarded as a generalization of the Hurwitz zeta function $\zeta(s, a) = \sum_{n=0}^{\infty}(n+a)^{-s}$ $(0<a\leq 1)$ to prehomogeneous vector spaces. Since the Fourier transform of $\phi = ch_{a+L}$ is given by

$$\hat{\phi}(y) = [L: L_0] e^{2\pi i\langle a, y\rangle} ch_{L^*}(y),$$

the zeta function $Z_j^*(s)(\hat{\phi})$ is a generalization of $\zeta^*(s, a) = \sum_{n=1}^{\infty} e^{2\pi i n a} n^{-s}$.

Aside from the original Hurwitz zeta function, several examples of zeta functions of this type have appeared in literatures. In [3] Bushnell and Reiner proved a functional equation satisfied by "Hurwitz series" associated with a semisimple algebra over $\boldsymbol{Q}$. From our point of view their result can be interpreted as follows.

Let A be a finite dimensional semisimple $\boldsymbol{Q}$-algebra and $A^{\times}$ the group of invertible elements in A. We consider $A^{\times}$ as (a set of $\boldsymbol{Q}$-rational points of) an algebraic group defined over $\boldsymbol{Q}$. Then the action of $A^{\times}$ on A from the left provides us a typical example of prehomogeneous vector spaces defined over $\boldsymbol{Q}$, which we denote by $(A^{\times}, \rho, A)$. It is easy to see that the triple $(A^{\times}, \rho, A)$ satisfies all the assumptions in the present section.

Let $A = A_1 \times \cdots \times A_r$, where A_i is a simple $\boldsymbol{Q}$-algebra. Let

$$P_i(a_1, \cdots, a_r) = \mathrm{Nr}_{A_i}(a_i) \qquad ((a_1, \cdots, a_r) \in A_1 \times \cdots \times A_r),$$

where Nr_{A_i} is the reduced norm of A_i. Then $P_1, \cdots, P_r$ are $\boldsymbol{Q}$-irreducible relative invariants of $(A^{\times}, \rho, A)$ and the singular set $\boldsymbol{S}$ is given by

$$\boldsymbol{S} = \bigcup_{i=1}^{r} \{x \in A;\ P_i(x) = 0\}.$$

Let $Z_1(s_1, \cdots, s_r), \cdots, Z_\nu(s_1, \cdots, s_r)$ (ν = the number of connected component of $A^{\times}(\boldsymbol{R}) = A \otimes \boldsymbol{R} - \boldsymbol{S}(\boldsymbol{R})$) be the zeta distributions associated with

$(A^\times, \rho, A)$. Taking $\phi = ch_{a+L}$ with $a \in A$ and a full $\boldsymbol{Z}$-lattice L in A, we obtain a zeta function $Z(s_1, \cdots, s_r)(ch_{a+L}) = \sum_{i=1}^{\nu} Z_i(s_1, \cdots, s_r)(ch_{a+L})$ of Hurwitz type. Since $A^\times$ is reductive as an algebraic group, the zeta function can be continued meromorphically to whole of $\boldsymbol{C}^r$. The Hurwitz series considered by Bushnell and Reiner is the function $Z(s, \cdots, s)(ch_{a+L})$ obtained by restricting the variable $(s_1, \cdots, s_r)$ to the diagonal $\{(s, \cdots, s); s \in \boldsymbol{C}\}$, In order to make the functional equation explicit, we need the explcit formula for the Γ-matrix $(\Gamma_{ij}(s))$ entering into the functional equation. For the present example the formula for $(\Gamma_{ij}(s))$ is well-known (see e.g. [25, p. 497]).

In their proof of the functional equation in [3], the authors first reduce the things to the case of simple $\boldsymbol{Q}$-algebra, and then two subcases are distinguished according as A satisfies the so-called Eichler condition or not. They posed a problem of finding a method applicable equally to both cases. The theory of prehomogeneous vector spaces supplies the demand.

One can find another discussion on zeta functions of this type in [6].

(B) *L-functions associated with prehomogeneous vector spaces*: Fix a positive integer N and let $\chi_1, \cdots, \chi_n$ be Dirichlet characters defined modulo N. (In the following the characters corresponding to relative invariants do not appear; hence the notation will not cause any confusion.) As usual we set $\chi_i(m) = 0$, if $(m, N) \neq 1$.

Let L be a lattice on which $P_1, \cdots, P_n$ take integral values. Consider the function $\phi_\chi = \phi_{\chi_1,\ldots,\chi_n} \in \mathcal{S}(V(\boldsymbol{Q}))$ defined by

$$\phi_\chi(x) = \sum_{y \in L/NL} \prod_{i=1}^{n} \chi_i(P_i(y)) ch_{y+NL}(x).$$

We define L-functions $L_i(s; \chi)$ $(1 \leqq i \leqq \nu)$ associated with $(\boldsymbol{G}, \rho, \boldsymbol{V})$ by the formula

$$L_i(s; \chi) = Z_i(s)(\phi_\chi) = \sum_{x \in \Gamma \backslash L \cap V_i} \mu(\Gamma \cdot x) \prod_{j=1}^{n} \chi_j(P_j(x)) \Big/ \prod_{j=1}^{n} |P_j(x)|^{s_j},$$

where Γ is an arithmetic subgroup of $\boldsymbol{G}(\boldsymbol{Q})$ satisfying $\rho(\Gamma)(y+NL) = y+NL$ for all $y \in L/NL$.

To write down the functional equation explicitly, we must calculate the Fourier transform $\hat{\phi}_\chi$ of ϕ_χ. It is easy to check the following identity:

$$\hat{\phi}_\chi(x^*) = [NL : L_0] \sum_{y^* \in N^{-1}L^*/L^*} G_{\chi \circ P}(y^*) ch_{y^*+L^*}(x^*), \tag{4.9}$$

where

$$G_{\chi\circ P}(y^*)=\sum_{y\in L/NL}\prod_{i=1}^{n}\chi_i(P_i(y))e^{2\pi i(y,y^*)}.$$

Thus the calculation of $\hat{\phi}_\chi$ is reduced to the calculation of a kind of Gaussian sums $G_{\chi\circ P}$.

This is the problem solved by Gyoja and Kawanaka ([10], [11]) under the conditions

(4.10) (i) $(\boldsymbol{G}, \rho, \boldsymbol{V})$ is a regular irreducible prehomogeneous vector space equipped with a natural $\boldsymbol{Z}$-structure,

(ii) $N=p$ is a sufficiently large prime number.

In the rest of this paragraph we assume these two conditions.

To describe the result of Gyoja and Kawanaka, we need some notational preliminaries. By (4.10) (i) the singular set $\boldsymbol{S}$ is an absolutely irreducible hypersurface defined over $\boldsymbol{Q}$; hence $n=1$. Let $P(x)$ (resp. $P^*(y)$) an irreducible relative invariant of $(\boldsymbol{G}, \rho, \boldsymbol{V})$ (resp. $(\boldsymbol{G}, \rho^*, \boldsymbol{V}^*)$), which is unique up to constant multiple, such that $P(x)$ (resp. $P^*(y)$) takes integral values on the lattice L (resp. L^*) defining the $\boldsymbol{Z}$-structure of $(\boldsymbol{G}, \rho, \boldsymbol{V})$ (resp. $(\boldsymbol{G}, \rho^*, \boldsymbol{V}^*)$).

Let $b(s)$ be the b-function of $(\boldsymbol{G}, \rho, \boldsymbol{V})$, namely the polymomial in s defined by the formula

$$P^*(\mathrm{grad}_x)P(x)^{s+1}=b(s)P(x)^s$$

(cf. § 1.3). It is known that $b(s)$ is a polynomial in s of degree $d=\deg P$ and all the roots of $b(s)=0$ are rational numbers.

Let

$$b^{\exp}(t)=\prod_{j=1}^{d}(t-\exp(-2\pi i\,\alpha_j)),$$

where $\alpha_1, \cdots, \alpha_d$ are the roots of $b(s)=0$. Then $b^{\exp}(t)$ can be expressed as a product of cyclotomic polynomials. We define non-negative integers $m(l)$ $(l\geqq 1)$ by

$$b^{\exp}(t)=\prod_{l=1}^{\infty}\Phi_l(t)^{m(l)},$$

where Φ_l is the l-th cyclotomic polynomial.

Theorem $\boldsymbol{F}_p$ ([11]). *Let χ be a Dirichlet character defined modulo p of order* $\mathrm{ord}\,\chi$. *Then the following identity holds*:

$$G_{\chi\circ P}(p^{-1}y^*)=\varepsilon(\chi)p^{(\dim V-m(\mathrm{ord}\,\chi))/2}(\chi^{-1}\otimes\theta)(P^*(y^*))$$
$$(y^*\in L^*,\ P^*(y^*)\not\equiv 0\ (\mathrm{mod}\,p)),$$

where $\varepsilon(\chi)$ is an algebraic number with absolute value 1 *and*

$$\theta=\begin{cases}\text{trivial character}, & \text{if } \dim V/d \in \boldsymbol{Z},\\ \text{the Legendre symbol } \left(\dfrac{\cdot}{p}\right), & \text{if } \dim V/d \in \dfrac{1}{2}+\boldsymbol{Z}.\end{cases}$$

In order to see the analogy of Theorem $\boldsymbol{F}_p$ to Theorem $\boldsymbol{R}$, k_p and $\boldsymbol{Q}$, it is enough to note that a Dirichlet character defined modulo p can be regarded as a (quasi-)character of the multiplicative group $\boldsymbol{F}_p^\times$ and the Gaussian sum $G_{\chi\circ P}$ is the Fourier transform of $\chi\circ P$ over $V(\boldsymbol{F}_p)$.

Corollary to Theorem $\boldsymbol{F}_p$ ([11]). *If $m(\mathrm{ord}\,\chi)=0$, then $G_{\chi\circ P}(p^{-1}y^*)=0$ $(y^*\in L^*,\ P^*(y^*)\equiv 0 \pmod p)$.*

Now we denote by $L_i^*(s;\chi^{-1}\otimes\theta)$ the L-function associated with $(\boldsymbol{G}, \rho^*, V^*)$.

Theorem L. *Let χ be a Dirichlet character defined modulo p with $m(\mathrm{ord}\,\chi)=0$. Then the L-functions $L_i(s;\chi)$ and $L_j^*(s;\chi^{-1}\otimes\theta)$ are continued analytically to entire functions of s and satisfy the following functional equation*:

$$L_i^*\left(\frac{\dim V}{d}-s,\ \chi^{-1}\otimes\theta\right)=p^{(2ds-\dim V)/2}\varepsilon(\chi)^{-1}[L_0:L]$$
$$\times\sum_{j=1}^{\nu}\Gamma_{ji}\left(s-\frac{\dim V}{d}\right)L_j(s;\chi).$$

Proof. From Theorem $\boldsymbol{F}_p$, Corollary to Theorem $\boldsymbol{F}_p$ and (4.9), it follows that

$$\hat{\phi}_\chi(x^*)=[pL:L_0]\varepsilon(\chi)p^{\dim V/2}\sum_{y^*\in L^*/pL^*}(\chi^{-1}\otimes\theta)(P^*(y^*))ch_{y^*+pL^*}(px^*)$$

for any $x^*\in V^*(\boldsymbol{Q})$. Hence by Theorem $\boldsymbol{Q}$ we get the functional equation of the L-functions. The formula for $\hat{\phi}_\chi$ shows that the points belonging to $\boldsymbol{S}^*$ make no contribution to the Poisson summation formula at the end of Section 4.1. Therefore the L-functions can be continued to entire functions.

Theorem L generalizes Stark's result on L-functions of quadratic forms ([47]). His result corresponds to a $\boldsymbol{Z}$-form of the prehomogeneous vector space $(\boldsymbol{GL}(1)\times\boldsymbol{SO}(n), \Lambda_1, V(n))$ such that the group of real points of $\boldsymbol{SO}(n)$ is compact.

Remark 1. If $m(\mathrm{ord}\,\chi)\neq 0$, $\hat{\phi}_\chi$ does not vanish on $\boldsymbol{S}^*(\boldsymbol{Q})$. Therefore the left hand side of the functional equation in Theorem L does not coincide with $L_i^*\left(\frac{\dim V}{d}-s, \chi^{-1}\otimes\theta\right)$, but involves some mysterious terms in addition. Moreover by the same reason the L-function $L_i(s;\chi)$ may have poles, even if χ is a nontrivial character. This explains the observation of Datskovski and Wright that the L-functions of a nontrivial cubic character associated with the vector space of binary cubic forms have a simple pole ([7, p. 30]). Another example of L-functions with poles was studied by Arakawa [1]. He calculated the residues of L-functions attached to the ternary quadratic form x^2-yz and the character $\left(\frac{*}{p}\right)$ given by the Legendre symbol ([1, Theorem 1.11]).

Remark 2. In many cases the constant $\varepsilon(\chi)$ can be expressed in terms of the classical Gaussian sums. For explicit calculation of $\varepsilon(\chi)$, we refer to [4]. In [5], Christian calculated $\varepsilon(\chi)$ for $(\boldsymbol{GL}(n)\times\boldsymbol{SL}(n), \Lambda_1\otimes\Lambda_1, M(n))$ by using the fact that $\varepsilon(\chi)$ appears as a coefficient of the functional equation satisfied by certain twist of Koecher-Maaß series (= zeta function associated with $(\boldsymbol{SO}(m)\times\boldsymbol{GL}(n), \Lambda_1\otimes\Lambda_1, M(m, n))$ $(m\geqq n)$).

Remark 3. There exists another application of Theorem $\boldsymbol{F}_p$ due to Weissauer ([51]). From Theorem $\boldsymbol{F}_p$ for $\left(\boldsymbol{GL}(n), 2\Lambda_1, V\left(\frac{n(n+1)}{2}\right)\right)$ he derived that certain twist of Siegel modular forms by Dirichlet characters are again Siegel modular forms.

Remark 4. The original motivation of Gyoja and Kawanaka for their study of Gaussian sums of prehomogeneous vector spaces was not L-functions but the fact that this kind of Gaussian sums enters into the character formula for certain irreducible representations of finite reductive groups (see Kawanaka [21]).

4.7. In [41] Selberg pointed out that the cusp contribution to the formula for the dimensions of spaces of automorphic forms (based upon his celebrated trace formula) can be evaluated as special values of certain Dirichlet series at integer argument. The theory of prehomogeneous vector spaces might provide us a convenient tool of formulating Selberg's observation in a more concrete form. Roughly speaking one can expect that the parabolic contributions to the dimension formula for a space of automorphic forms on a real reductive Lie group with respect to a lattice can be expressed in terms of special values of zeta functions of Hurwitz type associated with prehomogeneous vector spaces of parabolic type (in the sense of Rubenthaler [33]). This expectation has been confirmed by several authors mainly in two special cases, namely the case of Siegel cusp forms and the case of lattices of rank 1.

1) *Siegel cusp forms*: In [43] Shintani calculated the the purely unipotent contributions to the dimension of the space of Siegel cusp forms of degree n and of weight k ($k \geqq 2n+3$) and showed that the contributions coincide with certain special values of zeta functions associated with the prehomogeneous vector spaces $\left(GL(m),\ 2\Lambda_1,\ V\left(\frac{m(m+1)}{2}\right)\right)$ $(1 \leqq m \leqq n)$ up to elementary factors. Hashimoto [12] made a detailed investigation in the case of degree 2 and found that the contributions of (not necessarily unipotent) conjugacy classes can also be expressed in terms of special values of certain zeta functions (the Riemann zeta function, the Hurwitz zeta function).

Recently Shintani's calculation was extended in two directions. Arakawa considered in [1] the representation μ_k of the finite symplectic group $Sp(2n, F_p)$ on the space of Siegel cusp forms of degree n and weight k with respect to the principal congruence subgroup $\Gamma_{2n}(p)$, p being a odd prime. For some unipotent elements α in $Sp(2n, F_p)$, he obtained an expression of the purely unipotent contribution to tr $\mu_k(\alpha)$ in terms of special values of L-functions associated with $\left(GL(m),\ 2\Lambda_1,\ V\left(\frac{m(m+1)}{2}\right)\right)$.
In this Arakawa's work we encounter the L-function we have studied in Section 4.6 (B) and some other type of L-functions.

In [27], Murase and Sugano extended Shintani's calculation in another direction, namely to the dimension formula for the space of Jacobi cusp forms. This time in the purely unipotent contribution there appear special values of zeta functions associated with the following prehomogeneous vector space:

$$G=\left\{g(h,x)=\left(\begin{array}{c|c} I_m & x \\ \hline 0 & h \end{array}\right) \in GL(m+n);\ h \in GL(n),\ x \in M(m,n)\right\},$$

$$V=\left\{v=\begin{pmatrix} v_1 \\ v_2 \end{pmatrix};\ v_1 \in M(m,n),\ v_2={}^t v_2 \in M(n)\right\},$$

$$\rho(g(h,x))v=g(h,x)v\,{}^t h.$$

What is remarkable here is that (G, ρ, V) is not a regular prehomogeneous vector space. Murase and Sugano succeeded in proving the existence of analytic continuations and functional equations of the associated zeta functions. Since the existing theory of zeta functions can be applied only to regular prehomogeneous vector spaces, their result goes beyond the general theory and seems to suggest a possibility of further generalization of the theory of prehomogeneous vector spaces.

2) *Lattices of rank* 1: Let G be an admissible reductive Lie group and Γ a lattice of rank 1. In [30] Osborne and Warner established the trace formula for $L^{\mathrm{disc}}_{G/\Gamma}(\alpha)$ $(\alpha \in C_0^\infty(G))$; however in their result it was not clear what kind of zeta functions appear in the parabolic contributions to the trace formula. Previous result of Warner ([49]) showed that the Epstein zeta function is sufficient to describe the parabolic contribution, if G is of $\boldsymbol{R}$-rank 1 (see also Osborne and Warner [29]). Recently Hoffmann [16] extended the work of Osborne and Warner [30] to the trace formula for Hecke operators. Moreover he succeeded in expressing the parabolic contributions, which were still mysterious in [30], in terms of orbital integrals and special values of zeta functions associated with prehomogeneous vector spaces of parabolic type obtained from split parabolic subgroups of centralizers of semisimple elements.

Kato [20] gave an example of explicit determination of the dimensions of automorphic forms on a group of $\boldsymbol{R}$-rank 1, namely on $SU(p, 1)$.

It is quite desirable to develop a method of calculating special values of zeta functions associated with prehomogeneous vector spaces. As for the progress made in this direction, we refer the reader to Satake [34], Kurihara [24] and Arakawa [1]; these works were stimulated by the pathbreaking work of Shintani [44].

Finally we note that the appearence of zeta functions associated with prehomogeneous vector spaces of parabolic type in trace formulas has a strong resemblance to the appearence of Gaussian sums (zeta functions over finite fields) associated with prehomogeneous vector spaces of parabolic type in the formulas for characters (evaluated at unipotent elements) of certain irreducible representations of finite reductive groups (cf. Kawanaka [21] and Remark 4 at the end of Section 4.6).

References

[1] Arakawa, T., Special values of L-functions associated with the space of quadratic forms and the representation of $Sp(2n, \boldsymbol{F}_p)$ in the space of Siegel cusp forms, in this volume.
[2] Böcherer, S. and Sato, F., Rationality of certain formal power series related to local densities, Comment. Math. Univ. St. Pauli, **36** (1987), 53–86.
[3] Bushnell, C. J. and Reiner, I., Functional equations for Hurwitz series and partial zeta functions of orders, J. reine angew. Math., **364** (1986), 130–148.
[4] Chen, Z., Fonction zeta associée à un espace préhomogène et sommes de

Gauss, preprint.

[5] Christian, U., Maaßsche L-Reihen und eine Identität für Gaußsche Summen, Abh. Math. Sem. Univ. Hamburg, **54** (1984), 163–175.

[6] Cogdell, J. W., Congruence zeta functions for $M_2(Q)$ and their associated modular forms, Math. Ann., **266** (1983), 141–198.

[7] Datskovski, B. and Wright, D. J., The adelic zeta function associated to the space of binary cubic forms, Part II: Local theory, J. reine angew. Math., **367** (1986), 27–75.

[8] Denef, J., The rationality of the Poincaré series associated to p-adic points on a variety, Invent. math., **77** (1984), 1–23.

[9] ——, On the evaluation of certain p-adic integrals, Sém. de Théorie des Nombres, Paris 1983/84, Progress in Math. Vol. **59**, 25–47.

[10] Gyoja, A. and Kawanaka, N., Gauss sums of prehomogeneous vector spaces, Proc. Japan Acad., **61** Ser. A (1985), 19–22.

[11] Gyoja, A., Gauss sums of prehomogeneous vector spaces, Preprint.

[12] Hashimoto, K., The dimension of the spaces of cusp forms on Siegel upper half plane of degree two (I), J. Fac. Sci. Univ. Tokyo, **30-2** (1983), 403–488.

[13] ——, Representation of the finite symplectic group $Sp(4, F_p)$ in the space of Siegel modular forms, Contemporary Math., **53** (1986), 253–276.

[14] Hironaka, Y. and Sato, F., Spherical functions and local densities of alternating forms, Amer. J. Math., **110** (1988), 473–512.

[15] ——, Local densities of alternating forms, Preprint.

[16] Hoffmann, W., The non-semi-simple term in the trace formula for rank one lattices, J. reine angew. Math., **379** (1987), 1–21.

[17] Igusa, J., Some results on p-adic complex powers, Amer. J. Math., **106** (1984), 1013–1032.

[18] ——, On functional equations of complex powers, Invent. Math., **85** (1986), 1–29.

[19] Kashiwara, M. and Miwa, T., Micro-local calculus and Fourier transforms of relative invariants of prehomogeneous vector spaces (in Japanese), Surikaiseki Kenkyujo Kokyuroku, **238** (1974), 60–147.

[20] Kato, S., A dimension formula for a certain space of automorphic forms of $SU(p, 1)$, Math. Ann., **266** (1984), 457–477.

[21] Kawanaka, N., Generalized Gelfand-Graev representations of exceptional simple algebraic groups over a finite field I, Invent. Math., **84** (1986), 575–616.

[22] Kimura, T., Theory of prehomogeneous vector spaces, Sugaku, **32** (1980), 97–118 (in Japanese).

[23] Kimura, T., Kasai, S. and Yasukura, O., A classification of the representations of reductive algebraic groups which admit only a finite number of orbits, Amer. J. Math., **108** (1986), 643–692.

[24] Kurihara, A., On the values at non-positive integers of Siegel's zeta functions of Q-anisotropic quadratic forms with signature $(1, n-1)$, J. Fac. Sci. Univ. Tokyo, Sect IA **28** (1982), 567–584.

[25] Muller, I., Décomposition orbital des espaces préhomogènes réguliers de type parabolique commutatif et application, C.R. Acad. Sci. Paris t. **303** (1986), 495–496.

[26] ——, Rubenthaler, H. and Schiffmann, G., Structure des espaces préhomogènes associés à certaines algèbres de Lie graduées, Math. Ann., **274** (1986), 95–123.

[27] Murase, A. and Sugano, T., A note on zeta functions associated with certain prehomogeneous affine spaces, in this volume.

[28] Muro, M., Micro-local analysis and calculations of some relatively invariant hyperfunctions related to zeta functions associated with the vector spaces of quadratic forms, Publ. Res. Inst. Math. Sci., **22** (1986), 395–463.

[29] Osborne, M. S. and Warner, G., Multiplicities of the integrable discrete series: The case of a nonuniform lattices in an **R**-rank one semisimple group, J. Funct. Anal., **30** (1978), 287–310.

[30] ——, The Selberg trace formula I: Γ-rank one lattices, J. reine angew. Math., **324** (1981), 1–113.

[31] Pyasetskii, V. S., Linear Lie groups acting with finitely many orbits, Funct. Anal. Appl., **9** (1975), 351–353.

[32] Reiter, H., Über den Satz von Weil-Cartier, Monatsh. f. Math., **86** (1978), 13–62.

[33] Rubenthaler, H., Espaces préhomogènes de type parabolique, Thèse, Université de Strasbourg, 1982.

[34] Satake, I., Special values of zeta functions associated with self dual cones, Progress in Math. Vol. **14**, Birkhäuser (1981), 359–384.

[35] Sato, F., Zeta functions in several variables associated with prehomogeneous vector spaces I: Functional equations, Tôhoku Math. J., **34** (1982), 437–483.

[36] ——, Zeta functions in several variables associated with prehomogeneous vector spaces III: Eisenstein series for indefinite quadratic forms, Ann. of Math., **116** (1982), 177–212.

[37] Sato, M., Theory of prehomogeneous vector spaces (notes by Shintani in Japanese), Sugaku no Ayumi, **15** (1970), 85–157.

[38] Sato, M., Kashiwara, M., Kimura, T. and Oshima, T., Micro-local analysis of prehomogeneous vector spaces, Invent. Math., **62** (1980), 117–179.

[39] Sato, M. and Kimura, T., A classification of irreducible prehomogeneous vector spaces and their invariants, Nagoya Math. J., **65** (1977), 1–155.

[40] Sato, M. and Shintani, T., On zeta functions associated with prehomogeneous vector spaces, Ann. of Math., **100** (1974), 131–170.

[41] Selberg, A., Discontinuous groups and harmonic analysis, Proc. Intern. Congr. Math., Stockholm, 1962, 177–189.

[42] Serre, J. P., Cohomologie galoisienne, Lect. Notes in Math., **5**, Springer Verlag, 1965.

[43] Shintani, T., On zeta functions associated with the vector space of quadratic forms, J. Fac. Sci. Univ. Tokyo, **22** (1975), 25–65.

[44] ——, On evaluation of zeta functions of totally real algebraic number fields at non-positive integers, J. Fac. Sci. Univ. Tokyo, **23** (1976), 393–417.

[45] Siegel, C. L., Über die analytische Theorie der quadratischen Formen II, Ann. of Math., **37** (1936), 230–263.

[46] ——, Über die Zetafunktionen indefiniter quadratischer Formen, Math. Z., **43** (1938), 682–708.

[47] Stark, H. M., L-functions and character sums for quadratic forms (I), Acta Arith., **16** (1968), 35–50.

[48] Suzuki, T., Fourier transforms of relative invariants of prehomogeneous vector spaces, Master thesis of Nagoya University, 1975 (in Japanese).

[49] Warner, G., Selberg's trace formula for non-uniform lattices: The **R**-rank one case, Advances in Math., Supplementary Studies **6** (1979), 1–142.

[50] Weil, A., Sur certaines groupes d'operateurs unitaires, Acta Math., **111** (1964), 143–211; Collected Papers III, 1–69.

[51] Weissauer, R., Siegel modular forms and Dirichlet series, preprint, 1986.

Department of Mathematics
Rikkyo University
Nishi-Ikebukuro, Tokyo 171
Japan

Advanced Studies in Pure Mathematics 15, 1989
Automorphic Forms and Geometry of Arithmetic Varieties
pp. 509–523

Multi-Tensors of Differential Forms on the Hilbert Modular Variety and on Its Subvarieties, II

Shigeaki Tsuyumine

Dedicated to Prof. Ichiro Satake and Prof. Friedrich Hirzebruch on their sixtieth birthdays

Let Γ_K denote the Hilbert modular group associated with a totally real algebraic number field K of degree $n>1$. Let X_K be the Hilbert modular variety H^n/Γ_K. The present paper is the continuation of a study [8], and our purpose is to extend the known range of K for which an assertion (☆) holds where

(☆) *any subvariety in X_K of codimension one is of general type.*

We show that *if $n\geq 3$, then (☆) holds only with finite exceptions.* It was shown in our previous paper [8] that if the dimension $n\geqq 3$ is fixed, then (☆) holds with finite exceptions. The main theorem of the present paper is as follows:

Theorem. *(☆) holds if $n>26$, or if $n>14$ and the ideal in the maximal order of K generated by 2 is unramified at any prime of degree one.*

As stated in [8], (☆) has the consequent on the property of X_K which we restate here for reader's convenience.

(I) *Let X_K° denote the smooth locus of X_K, and let $\tilde{X}_K$*[1] *be any smooth variety having X_K° as an open subset. Then for any birational morphism φ of $\tilde{X}_K$ to a smooth variety, $\varphi|_{X_K^\circ}$ gives rise to an open embedding.*

(II) *The birational automorphism group of X_K (or equivalently, the automorphism group of the Hilbert modular function field over $\boldsymbol{C}$) is equal to the automorphism group of X_K, which is canonically isomorphic to a semi-direct product $H_K^{(2)}\rtimes \mathrm{Aut}(K/\boldsymbol{Q})$ where $H_K^{(2)}=\{x\in H_K \mid x^2=1\}$, H_K denoting the ideal class group of K in the narrow sense.*

As we see in §2, in order to prove Theorem we need to show

Received February 10, 1987.
1) ~ is missing in [8], Cor. 1, p. 660.

existence of a Hilbert modular form g for every irreducible divisor D of X_K such that (i) $g \not\equiv 0$ on D and that (ii) the quotient by weight (g), of the vanishing order of g at the cusps is at least $n/2(n-1)$. By the dimension formula by Shimizu it follows that there exist modular forms f satisfying the condition (ii), or better one as well, except for a finite number of K. In [8], under a certain condition we have got a modular form g satisfying (i) as well as (ii) by differentiating an "irreducible" factor of f vanishing on D, where non-existence of automorphy factors of very low weight plays an important role, which has been shown by Gundlach [3]. The basic idea of the present paper is to consider a "shifted" modular form $f(\alpha z)$ for α totally positive. Namely, if α will be taken suitably, then $f(\alpha z)$ will still have zeros of large order at cusps and it will not vanish identically on D. $f(\alpha z)$ is generally a modular form for a congruence subgroup, not for Γ_K, and so we must construct desired modular forms from those of congruence subgroups. Combining this method with our previous method, we can find a modular form g satisfying both (i) and (ii) in the case of dim $K \geq 3$ with finite exceptions, where the proof is served by some refinement [9] of [3].

The author completed the present work while staying at Göttingen. He wishes to express his heartfelt gratitude to the Sonderforschungsbereich 170 "Geometrie und Analysis" and especially to Professor U. Christian for hospitality and support during the author's visit.

§ 1. Preliminaries

Let K be a totally real algebraic number field of degree $n>1$. $SL_2(K)$ acts on the product H^n of n copies of the upper half plane $H=\{z_1 \in C \mid \text{Im } z_1>0\}$ by the usual modular substitution

$$z=(z_1, \cdots, z_n) \longrightarrow Mz=\left(\frac{\alpha^{(1)}z_1+\beta^{(1)}}{\gamma^{(1)}z_1+\delta^{(1)}}, \cdots, \frac{\alpha^{(n)}z_n+\beta^{(n)}}{\gamma^{(n)}z_n+\delta^{(n)}}\right)$$

for $M=\begin{pmatrix}\alpha & \beta \\ \gamma & \delta\end{pmatrix} \in SL_2(K)$, where $\alpha^{(1)}, \cdots, \alpha^{(n)}$ denote the conjugates of $\alpha \in K$. Let O_K be the maximal order of K. We put $\Gamma_K=SL_2(O_K)$, which is called the *Hilbert modular group* associated with K and which acts properly discontinuously on H^n. Let $\hat{K}=K \cup \{\infty\}$. We define an equivalence relation in $\hat{K}$ in terms of Γ_K; $\lambda_1, \lambda_2 \in \hat{K}$ are equivalent if $\lambda_1=M\lambda_2=(\alpha\lambda_2+\beta)/(\gamma\lambda_2+\delta)$ for some $M=\begin{pmatrix}\alpha & \beta \\ \gamma & \delta\end{pmatrix} \in \Gamma_K$ where $\zeta/0=\infty$ for $\zeta \in K, \neq 0$, and $\zeta/\infty=0$, $\infty+\zeta=\infty$ for any $\zeta \in K$. Let $\mathfrak{a}=(\rho, \sigma)$, $\rho, \sigma \in K$, be a non-zero fractional ideal. We associate with $\mathfrak{a}$, an element λ of $\hat{K}$ given by $\lambda=\rho/\sigma$. Then if we denote by $C(K)$ the (fractional) ideal class group of

K, then we have a well-defined bijective map of $C(K)$ onto $\hat{K}/\Gamma_K$ by sending $\mathfrak{a}$ to λ (see for instance, Siegel [6, Chap. III, Sect. 2]). $\hat{K}/\Gamma_K$ can be regarded as the set of inequivalent cusps of Γ_K, and so Γ_K has h inequivalent cusps where h denotes the class number of K.

Let $\mathfrak{a}=(\rho, \sigma)$ be a non-zero fractional ideal, and let $\lambda=\rho/\sigma$. We can take the generators ξ, η of an ideal $\mathfrak{a}^{-1}$ for which $\rho\eta-\sigma\xi=1$. Let us put

$$(1) \qquad M_\lambda=\begin{pmatrix}\rho & \xi\\ \sigma & \eta\end{pmatrix} \in SL_2(K).$$

The modular substitution corresponding to M_λ maps a cusp λ to ∞, and $M_\lambda^{-1}\Gamma_K M_\lambda$ equals $\left\{\begin{pmatrix}\alpha & \beta\\ \gamma & \delta\end{pmatrix} \in SL_2(K) \,|\, \alpha, \delta \in O_K, \gamma \in \mathfrak{a}^2, \beta \in \mathfrak{a}^{-2}\right\}$.

Let us fix a subgroup Γ in $SL_2(K)$ commensurable with Γ_K. Let J be the automorphy factor for Γ which is of the form

$$(2) \qquad J(M, z)=v(M)\prod_{i=1}^{n}(\gamma^{(i)}z_i+\delta^{(i)})^{k_i}, \qquad M=\begin{pmatrix}\alpha & \beta\\ \gamma & \delta\end{pmatrix} \in \Gamma,$$

where $k_1, \cdots, k_n \in \boldsymbol{Q}$, and v is the multiplier whose value for $M \in \Gamma$ is a root of unity. v is, of course, depending on the choice of the branches of $(\gamma^{(i)}z_i+\delta^{(i)})^{k_i}$ if $k_i \in \boldsymbol{Q}-\boldsymbol{Z}$ $(i=1, \cdots, n)$. A holomorphic function f on H^n is called a (*Hilbert*) *modular form* for Γ associated with J if it satisfies

$$f(Mz)=J(M, z)f(z) \qquad \text{for any } M \in \Gamma.$$

f is said to be of *vector weight* $(k_1, \cdots, k_n)$, and conversely $(k_1, \cdots, k_n)$ is called the *weight vector* of f. If all the k_i's are equal to k, then f is said to be of *scalar weight* k, and in notation $k=\text{weight}(f)$. A map $\beta \rightarrow v\begin{pmatrix}1 & \beta\\ 0 & 1\end{pmatrix}$ gives a *finite* character of an additive group $\left\{\beta \in K \,|\, \begin{pmatrix}1 & \beta\\ 0 & 1\end{pmatrix} \in \Gamma\right\}$, and moreover it is true also for the unipotent subgroup of the stabilizer subgroup in Γ at every cusp $\lambda \in K$ (for the references, see [8, p. 665]). So a modular form f for Γ has a Fourier expansion centered at λ. Let $\lambda=\rho/\sigma$, $\mathfrak{a}=(\rho, \sigma)$, M_λ be as above, and let

$$(3) \qquad w=M_\lambda^{-1}z, \qquad w_i=(M_\lambda^{-1})^{(i)}z_i \qquad (1\leq i\leq n).$$

Then

$$f_\lambda(w)=\prod_{i=1}^{n}(-\sigma^{(i)}z_i+\rho^{(i)})^{k_i}f(z)$$

is a modular form in w for $M_\lambda^{-1}\Gamma M_\lambda$, and it has a Fourier expansion

$$(4) \qquad f_\lambda(w)=\sum_{\nu} c_\nu \exp\left(2\pi\sqrt{-1}\,\mathrm{tr}(\nu w)\right)$$

where $\mathrm{tr}(\nu w) = \nu^{(1)}w_1 + \cdots + \nu^{(n)}w_n$, and ν runs over the set composed of 0 and totally positive numbers contained in some lattice in K. f_λ is defined independently of $\rho, \sigma \in K$ with $\lambda = \rho/\sigma$, up to a constant multiple. Now we define the vanishing order $\mathrm{ord}_\lambda(f)$ of f at λ to be the minimum of the set of non-negative rational numbers

$$\{\mathrm{tr}(\nu\zeta) \mid \text{totally positive } \zeta \in \mathfrak{a}^{-2},\ \nu \text{ with } c_\nu \neq 0\}.$$

It is easy to see that the above definition is independent of $\rho, \sigma \in K$ for which $\lambda = \rho/\sigma$, and further that it is independent of the choice of λ in one equivalence class of cusps of Γ.

Remark. We give a comment about the vanishing order. For simplicity we suppose that a multiplier v is trivial and $k_1 = \cdots = k_n \in 2\mathbf{Z}$ in (2). Let U be a neighborhood at a cusp λ in the analytic space given by compactifying H^n/Γ. If U is small enough, then $f_\lambda(w)$ can be regarded as a function on U. Let $\varphi: \tilde{U} \to U$ be a desinguralization, and $\{E_i\}$, the exceptional divisors. Then $\mathrm{ord}_\lambda(f)$ has a geometric meaning, namely, it equals the minimum of the vanishing orders of $\varphi^* f$ at the E_i's, provided that $\Gamma = \Gamma_K$. But this is not necessarily true for general Γ. To make general definition, the "width" of cusps of Γ must be taken into account. However as far as we focus only on modular forms for Γ_K, the above definition works well.

Let f, g be modular forms for Γ. The inequalities $\mathrm{ord}_\lambda(fg) \geqq \mathrm{ord}_\lambda(f) + \mathrm{ord}_\lambda(g)$, $\mathrm{ord}_\lambda(f+g) \geqq \min\{\mathrm{ord}_\lambda(f), \mathrm{ord}_\lambda(g)\}$ holds[2)]. In the Fourier expansion (4), we call ν *minimal* if $c_\nu \neq 0$ and if ν cannot be written as $\nu = \nu' + \nu''$ with $c_{\nu'} \neq 0$, $c_{\nu''} \neq 0$. We write

$$g \preccurlyeq f$$

if at every cusp the following holds; for any minimal ν in the Fourier expansion of f, there is a minimal ν' in that of g for which $\nu'^{(i)} \leq \nu^{(i)}$ $(i = 1, \cdots, n)$. In such a case $\mathrm{ord}_\lambda(fh) \geq \mathrm{ord}_\lambda(gh)$ holds for any modular form h[3)]. Finally in this section we define the vanishing order $\mathrm{ord}(f)$ of f at cusps by

$$\mathrm{ord}(f) = \min_{\lambda \in \hat{K}} \{\mathrm{ord}_\lambda(f)\}$$

where λ may actually run over a finite set of representatives of cusps with respect to the equivalence under Γ.

[2)] Although a resulting inequality is true, there is inaccuracy at the last equality in the sequence of inequalities in [8, p668, line 1~2].

[3)] This gives a correct proof of the above.

§ 2. Résumé of [8]

We put

$$\omega_i=(-1)^i dz_1\wedge\cdots\wedge d\check{z}_i\wedge\cdots\wedge dz_n \in \Omega_{H^n}^{n-1} \qquad (1\leq i\leq n),$$

and

$$\omega=\omega_1\otimes\cdots\otimes\omega_n \in (\Omega_{H^n}^{n-1})^{\otimes n}.$$

Then

$$M\cdot\omega=\prod_{i=1}^{n}(\gamma^{(i)}z_i+\delta^{(i)})^{-2(n-1)}\omega \qquad \text{for } M=\begin{pmatrix}\alpha & \beta\\ \gamma & \delta\end{pmatrix} \in SL_2(K).$$

If f is a Hilbert modular form for Γ_K of scalar weight $2r(n-1)$, then $f\omega^{\otimes r}$ is Γ-invariant, and it may be regarded as a multi-tensor of differential forms on the smooth locus X_K° of X_K. It is extendable over all elliptic fixed points except for the cases listed in [8, Lemma 3], and it is always so in particular for $n>6$. It is extendable to a projective non-singular model of X_K, then its restriction to D gives a multi-tensor of canonical differential forms on D. The restriction is not zero unless f vanishes at D.

Proposition 1. *Let $n>6$. Let D be a subvariety in X_K of codimension one. If there is a modular form f for Γ_K of scalar weight satisfying that* (i) $f|_D\not\equiv 0$ *and that* (ii) $\operatorname{ord}(f)/\text{weight}(f)>n/2(n-1)$, *then D is of general type.*

Proof. The proposition was essentially proved in [8, Sect. 6]. Here we give only a sketch. We may assume that weight $(f)=2r(n-1)$, $r \in \boldsymbol{Z}$, replacing f by its power if necessary. Let $g_1, \cdots, g_t$ be modular forms for Γ_K of weight $2r'(n-1)$, $r' \in \boldsymbol{Z}$, by which X_K is embedded into a projective space. If $m \in \boldsymbol{Z}$ is large enough, then $\operatorname{ord}(f^m g_j)/\text{weight}(f^m g_j)$, $1\leq j\leq t$, are at least $n/2(n-1)$, and hence $f^m g_j\omega^{\otimes(mr+r')}$, $1\leq j\leq t$, are extendable to a projective non-singular model of X_K. Since $f|_D\not\equiv 0$, D is of general type. q.e.d.

The proof of Theorem is reduced to find the modular form satisfying the condition in Proposition 1 for any fixed D, which in substance, we carry out in the present paper.

§ 3. Vanishing order

Let μ be a totally positive integer in O_K. Let m be a positive rational integer such that

$$(5) \qquad m\mu^{-1} \in O_K.$$

Let $f(z)$ be a modular form of weight $(k_1, \cdots, k_n)$ for some subgroup in $SL_2(K)$ commensurable with Γ_K. Then such is $f(\mu z)=f(\mu^{(1)}z_1, \cdots, \mu^{(n)}z_n)$. Let $M=\begin{pmatrix}\alpha & \beta \\ \gamma & \delta\end{pmatrix} \in \Gamma_K$. Then

$$g(z) := \prod_{i=1}^{n} (\gamma^{(i)}z_i+\delta^{(i)})^{-k_i} f(\mu Mz)$$

is also such a modular form. We estimate ord (g) in terms of ord (f) and m in (5).

Lemma 1. *Let the notation be as above. Then* ord $(g) \geq m^{-1}$ ord (f).

Proof. Let $\lambda=\rho/\sigma$ be a cusp. Let us put $\lambda'=\mu M(\lambda)$ and $M_{\lambda'}=\begin{pmatrix}\sqrt{\mu} & 0 \\ 0 & \sqrt{\mu}^{-1}\end{pmatrix} M M_\lambda \begin{pmatrix}\sqrt{\mu}^{-1} & 0 \\ 0 & \sqrt{\mu}\end{pmatrix}=\begin{pmatrix}\rho' & * \\ \sigma' & *\end{pmatrix}$, M_λ being as in (1). We have the equalities $\lambda'=M_{\lambda'}(\infty)=\rho'/\sigma'$ and $\rho'=\alpha\rho+\beta\sigma$, $\sigma'=\mu^{-1}(\gamma\rho+\delta\sigma)$. $f(z)$ has a Fourier expansion centered at λ'

$$\prod_{i=1}^{n} (-\sigma'^{(i)}z_i+\rho'^{(i)})^{k_i} f(z)=\sum_{\nu} c_\nu \exp(2\pi\sqrt{-1}\,\mathrm{tr}\,(\nu M_{\lambda'}^{-1}z)).$$

Here $\mathrm{ord}_{\lambda'}(f)=\min\{\mathrm{tr}\,(\nu\zeta) \mid$ totally positive $\zeta \in (\rho', \sigma')^{-2}$, ν with $c_\nu \neq 0\} \geq$ ord (f). We replace z by μMz in the above identity. Then a simple calculation leads to

$$\prod_{i=1}^{n} (-\sigma^{(i)}z_i+\rho^{(i)})^{k_i} g(z)=\sum_{\nu} c_\nu \exp(2\pi\sqrt{-1}\,\mathrm{tr}\,(\nu\mu w)),$$

w being as in (3). Then $\mathrm{ord}_\lambda(g)=\min\{\mathrm{tr}\,(\zeta\mu\nu) \mid$ totally positive $\zeta \in (\rho, \sigma)^{-2}$, ν with $c_\nu \neq 0\} = \min\{\mathrm{tr}\,(\zeta\nu) \mid$ totally positive $\zeta \in \mu(\rho, \sigma)^{-2}$, ν with $c_\nu \neq 0\}$. Since $m^{-1}(\rho', \sigma')=m^{-1}\mu^2(\mu(\alpha\rho+\beta\sigma), \gamma\rho+\delta\sigma)^{-2} \supset m^{-1}\mu^2(\alpha\rho+\beta\sigma, \gamma\rho+\delta\sigma)^{-2} \supset m^{-1}\mu^2(\rho, \sigma)^{-2} \supset \mu(\rho, \sigma)^{-2}$, $\mathrm{ord}_\lambda(g)$ is at least $m^{-1}\,\mathrm{ord}_{\lambda'}(f) \geq m^{-1}$ ord (f).
q.e.d.

Let us construct a modular form for Γ_K from f;

$$(6) \qquad f_{\mu,r}(z) := \sum_{M} \{\prod_{i} (\gamma^{(i)}z_i+\delta^{(i)})^{-k_i} f(\mu Mz)\}^r$$

where r is a positive rational integer, and $M=\begin{pmatrix}\alpha & \beta \\ \gamma & \delta\end{pmatrix} \in \Gamma_K$ runs through a system of right representatives of Γ_K mod $\Gamma \cap \Gamma_K$, Γ denoting the subgroup in $SL_2(K)$ commensurable with Γ_K for which $f(\mu z)$ is modular form.

Corollary. *Let f be a modular form for a subgroup $SL_2(K)$ commensurable with Γ_K, and let μ be a totally positive integer in O_K. Then $f_{\mu,r}$ is* *a modular form for* Γ_K *whose weight vector is an r times the weight vector of f, and* ord $(f_{\mu,r})$ *is at least* $m^{-1}r$ ord (f) *where m is as in* (5).

The common zero of $f_{\mu,r}$, $r=1, 2, \cdots$, is equal to the common zero of $f(\mu Mz)$'s, M being as above. In particular $f_{\mu,r}$ does not vanish identically for infinitely many r unless f vanishes identically.

§ 4. Modular groups

First we introduce several congruence subgroups of Γ_K. Let $\mathfrak{b}$ be a non-zero integral ideal of O_K. We define subgroups of Γ_K associated with $\mathfrak{b}$;

$$\Gamma(\mathfrak{b}) := \left\{M \in \Gamma_K \mid M \equiv \begin{pmatrix} 1 & 0 \\ 0 & 1 \end{pmatrix} \bmod \mathfrak{b}\right\},$$

$$\Gamma_0(\mathfrak{b}) := \left\{\begin{pmatrix} \alpha & \beta \\ \gamma & \delta \end{pmatrix} \in \Gamma_K \mid \gamma \equiv 0 \bmod \mathfrak{b}\right\},$$

$$\Gamma^1(\mathfrak{b}) := \left\{\begin{pmatrix} \alpha & \beta \\ \gamma & \delta \end{pmatrix} \in \Gamma_K \mid \alpha \equiv \delta \equiv 1 \bmod \mathfrak{b},\ \beta \in \mathfrak{b}^2\right\}.$$

Let μ be a totally positive integer in O_K. A simple calculation shows that $M(\mu z) = \mu M_{(\mu)} z$ with

$$M_{(\mu)} = \begin{pmatrix} \sqrt{\mu}^{-1} & 0 \\ 0 & \sqrt{\mu} \end{pmatrix} M \begin{pmatrix} \sqrt{\mu} & 0 \\ 0 & \sqrt{\mu}^{-1} \end{pmatrix} = \begin{pmatrix} \alpha & \mu^{-1}\beta \\ \mu\gamma & \delta \end{pmatrix} \qquad \text{for } M = \begin{pmatrix} \alpha & \beta \\ \gamma & \delta \end{pmatrix}.$$

We note that

$$\Gamma_K \cap \begin{pmatrix} \sqrt{\mu}^{-1} & 0 \\ 0 & \sqrt{\mu} \end{pmatrix} \Gamma_K \begin{pmatrix} \sqrt{\mu} & 0 \\ 0 & \sqrt{\mu}^{-1} \end{pmatrix} = \Gamma_0(\mu),$$

and

$$\begin{pmatrix} \sqrt{\mu}^{-1} & 0 \\ 0 & \sqrt{\mu} \end{pmatrix} \Gamma^1(\mu) \begin{pmatrix} \sqrt{\mu} & 0 \\ 0 & \sqrt{\mu}^{-1} \end{pmatrix} = \Gamma(\mu).$$

Let f be a modular form. We denote by Γ_f, the maximal subgroup of $SL_2(K)$ for which f is a modular form, namely,

$$\Gamma_f = \{M \in SL_2(K) \mid f(Mz)/f(z) \text{ is holomorphic}\}.$$

Then $\Gamma_{f(\mu z)} = \begin{pmatrix} \sqrt{\mu}^{-1} & 0 \\ 0 & \sqrt{\mu} \end{pmatrix} \Gamma_f \begin{pmatrix} \sqrt{\mu} & 0 \\ 0 & \sqrt{\mu}^{-1} \end{pmatrix}$.

Lemma 2. *Let* $f(z)$ *be a non-constant modular form for* Γ_K. *Then* $\Gamma_{f(\mu z)} \cap \Gamma_K = \Gamma_0(\mu)$ *for a totally positive integer* μ *in* O_K.

Proof. By the above fact it is enough to show that $\Gamma_f=\Gamma_K$, where $\Gamma_f\supset\Gamma_K$ is our assumption. By Maass [5] it was shown that Hurwitz's extension $\tilde{\Gamma}_K$ of Γ_K is the unique maximal extension acting properly discontinuously on H^n except for an extension by a group acting trivially on H^n. $\tilde{\Gamma}_K$ is consisting of matrices

$$\begin{bmatrix}\alpha/\sqrt{\omega} & \beta/\sqrt{\omega}\\ \gamma/\sqrt{\omega} & \delta/\sqrt{\omega}\end{bmatrix}$$

where $\alpha, \beta, \gamma, \delta \in K$, and $\omega=\alpha\delta-\beta\gamma$ is totally positive, and $\alpha/\sqrt{\omega}$, $\beta/\sqrt{\omega}$, $\gamma/\sqrt{\omega}$, $\delta/\sqrt{\omega}$ are integral over O_K. $\tilde{\Gamma}_K\cap SL_2(K)$ equals Γ_K, which implies $\Gamma_f=\Gamma_K$. q.e.d.

Let $\mathfrak{p}$ be a prime ideal of O_K. It is easy to check that the set $\Gamma_0(\mathfrak{p}^e)\backslash\Gamma_K$ is naturally isomorphic to

$$\left\{\begin{pmatrix}0 & 1\\ -1 & \delta\end{pmatrix} \,\middle|\, \delta\in\mathfrak{p} \bmod \mathfrak{p}^e\right\}\cup\left\{\begin{pmatrix}1 & 0\\ \gamma & 1\end{pmatrix} \,\middle|\, \gamma\in O_K \bmod \mathfrak{p}^e\right\}.$$

In particular the index $[\Gamma_K:(\mathfrak{p}^e)]$ equals $\mathrm{Nm}(\mathfrak{p}^e)(1+\mathrm{Nm}(\mathfrak{p})^{-1})$ where Nm denotes the norm of K over $\boldsymbol{Q}$. The general case is reduced to the prime power case by the Chinese remainder theorem. We have the following:

Lemma 3. *Let $\mathfrak{b}$ be a non-zero integral ideal of O_K. Then the index $[\Gamma_K:\Gamma_0(\mathfrak{b})]$ equals $\mathrm{Nm}(\mathfrak{b})\prod_{\mathfrak{p}}(1+\mathrm{Nm}(\mathfrak{p})^{-1})$ where $\mathfrak{p}$ runs through prime ideals dividing $\mathfrak{b}$.*

§ 5. Irreducibility

We assume $n\geq 3$ in what follows. Let Γ be a subgroup in $SL_2(K)$ commensurable with Γ_K. Then any holomorphic automorphy factor for Γ is, up to trivial automorphy factors, of the form (2) (Freitag [2]). If there is a non-constant modular form associated with J in (2), then the entries of the weight vector are all positive (Freitag [1, Sect. 2]). A modular form f for Γ is said to be *irreducible in Γ* if its divisor corresponds to an irreducible divisor of the modular variety H^n/Γ. Then any modular form f has a unique irreducible decomposition up to a constant factor;

$$f=f_1^{r_1}\cdots f_t^{r_t}$$

where f_i $(1\leq i\leq t)$ are irreducible in Γ.

Let us suppose that $\Gamma\subset\Gamma_K$. Let g be a modular form for Γ, and let $\mathrm{div}(g)$ be the divisor of g on H^n. Then $\mathrm{div}(g)$ is written as

$$\sum_{j=1}^{t} m_j(\sum_M M\cdot D_j), \quad m_j \in \boldsymbol{Z}, \ >0,$$

where $D_1, \cdots, D_t$ are a finite number of irreducible divisors on H^n inequivalent under Γ and where M runs over a system of left representatives of Γ mod the stabilizer subgroup in Γ at D_j. We note that D_j's may be equivalent under Γ_K. Suppose that in the equivalence relation under Γ_K, D_j's are divided into several classes $\{D_1, \cdots, D_{j_1}\}, \{D_{j_1+1}, \cdots, D_{j_2}\}, \cdots, \{D_{j_s+1}, \cdots, D_j\}$. We define a new divisor invariant under Γ_K by

$$\sum_{k=0}^{s} \max\{m_{j_k+1}, \cdots, m_{j_{k+1}}\} (\sum_k M\cdot D_{j_k+1}) \tag{7}$$

where $j_0=0, j_{s+1}=j$, and where M runs over a system of left representatives of Γ_K mod the stabilizer subgroup in Γ_K at D_{j_k+1}. We denote by $N(g)$ the modular form for Γ_K whose divisor equals (7). $N(g)$ is divisible by g as modular forms for Γ, and moreover it is minimal in the sense that it divides any such modular form for Γ_K. We do not mind the ambiguity in the definition of $N(g)$ up to a constant multiple, which will never cause a trouble in the present paper.

Lemma 4. *Let $n\geq 3$, and let Γ be a normal subgroup of Γ_K. Let g be a modular form for Γ of weight $(k_1, \cdots, k_n)$ which is irreducible in Γ. Then $N(g)$ is given as*

$$\prod_M \prod_i (\gamma^{(i)}z_i+\delta^{(i)})^{-k_i} g(Mz) \tag{8}$$

where $M=\begin{pmatrix}\alpha & \beta\\ \gamma & \delta\end{pmatrix}$ runs through a system of right representatives of Γ_K mod *$\Gamma_K\cap\Gamma_g$. In particular the weight vector of $N(g)$ is a $[\Gamma_K: \Gamma_K\cap\Gamma_g]$ times the weight vector of g.*

Proof. Since (8) is a modular form for Γ_K, it is divisible by $N(g)$. We prove the contrary. Since Γ is normal in Γ_K, for each $M=\begin{pmatrix}\alpha & \beta\\ \gamma & \delta\end{pmatrix} \in \Gamma_K$ $\prod_i (\gamma^{(i)}z_i+\delta^{(i)})^{-k_i}g(Mz)$ is a modular form for Γ which is also irreducible in Γ, and which divides $N(g)$. If M runs through a system of right representatives of Γ_K mod $\Gamma_K\cap\Gamma_g$, then any two of them have no common divisor of zero. This shows our assertion. q.e.d.

Lemma 5. *Let $n\geq 3$, and let Γ be a subgroup of $SL_2(K)$ commensurable with Γ_K. Let g be a modular form for Γ of weight $(k_1, \cdots, k_n)$ irreducible in Γ. Suppose that g is not irreducible in some subgroup in Γ of finite index. Then for any i within $1\leq i\leq n$ and for any positive even r,* *there is a modular form h for* Γ *of weight* $r(k_1, \cdots, k_{i-1}, k_i+2, k_{i+1}, \cdots, k_n)$ *satisfying that* (i) $g^r \preccurlyeq h$ *and that* (ii) g, h *have no common divisors.*

Proof. By assumption, there is a subgroup Γ' of Γ in which g is not irreducible. Replacing Γ' by a smaller one if necessary, we may assume that Γ' is normal in Γ. Let $g = g_1 \cdots g_t$, $t \geq 2$, be an irreducible decomposition in Γ'. Let k, l be distinct integers in $\{1, \cdots, t\}$. Because of the irreducibility of g in Γ, g_k, g_l are transformed into each other for any k, l, more precisely, for some $M = \begin{pmatrix} \alpha & \beta \\ \gamma & \delta \end{pmatrix} \in \Gamma$, $\prod_i (\gamma^{(i)} z_i + \delta^{(i)})^{-e_i} g_k(Mz)$ equals $g_l(z)$ up to a constant factor, $(e_1, \cdots, e_n)$ being the weight vector of g_k. Hence, in particular, g_j's have the same vector weight, namely, $(1/t)(k_1, \cdots, k_n)$. The irreducibility of g in Γ implies also that the multiplicity of each factor is one namely, any two of the g_j's have no common divisors.

$$g_l^2 \frac{\partial}{\partial z_i}(g_k/g_l) = g_l \frac{\partial}{\partial z_i} g_k - g_k \frac{\partial}{\partial z_i} g_l \tag{9}$$

is modular form for Γ' of weight $(2k_1/t, \cdots, 2k_{i-1}/t, 2+2k_i/t, 2k_{i+1}/t, \cdots, 2k_n/t)$. By a matrix M in Γ, (9) is transformed into a similar one. So

$$h := \sum_{k<l} \left(g_l \frac{\partial}{\partial z_i} g_k - g_k \frac{\partial}{\partial z_i} g_l \right)^r \prod_{j \neq k,l} g_j^r \tag{10}$$

is a modular form for Γ of weight $r(k_1, \cdots, k_{i-1}, k_i+2, k_{i+1}, \cdots, k_n)$. A function (9) is $\succcurlyeq g_k g_l$ (see [8, Proof of Proposition 3]), and so each term of (10) is $\succcurlyeq g^r$. This implies that (i) $g^r \preccurlyeq h$. Finally we show that h is not divisible by g, which implies (ii). If otherwise, h is divisible by g_1 in Γ'. $(\partial/\partial z_i)g_1$ does not vanish identically on any component of div(g_1) on H^n by [8, Lemma 5] where the proof is given only for Γ_K, but it is easy to observe validity in general. Expanding (10), we get only one term $((\partial/\partial z_i)g_1)^r \prod_{j \neq 1} g_j^r$ that does not contain g_1. This implies that h is not divisible by g_1, a contradiction. q.e.d.

Corollary. *Let g be as in Lemma* 5. *Then there is a modular form h for* Γ *such that* (i) *the weight vector is* $2r(1, \cdots, 1) + nr(k_1, \cdots, k_n)$, r *being any positive integer, and that* (ii) $g^{nr} \preccurlyeq h$, *and that* (iii) g, h *have no common divisors.*

Proof. We take a product for $j = 1, \cdots, n$ of modular forms given in the lemma. q.e.d.

§ 6. Key proposition

There is a positive rational number k_0 depending only on K for which

$\sum k_i/k_0$ is integral if $(k_1, \cdots, k_n)$ is the weight vector of any automorphy factor for Γ_K. Gundlach [3] (see also [8, p. 666]) shows that $k_0 \geq 1/2$ if n is even. By [9] it was shown that $k_0 \geq 1/2$ for any $n > 1$, and that $k_0 \geq 2$ particularly if the ideal in O_K generated by 2 is unramified at any prime ideal of degree one. We note that if f is a non-constant modular form for Γ_K of scalar weight, then n weight$(f) \geq k_0$.

Now we state the key proposition of the paper, which gives a refinement of [8, Proposition 3].

Proposition 2. *Let $n \geq 3$. Let l be a non-negative real number for which there is a non-constant modular form f for Γ_K of scalar weight such that* ord(f)/weight$(f) > l$. *Let D be any effective divisor on H^n invariant under Γ_K which corresponds to an irreducible divisor of the modular variety H^n/Γ_K. Then there exists a modular form g for Γ_K of scalar weight such that* (i) *$g|_D \not\equiv 0$, and that* (ii) *ord(g)/weight$(g) > \min\{l/(2+1/2^{n-2}k_0)$, $\max\{3l/4(1+1/k_0), l/(1+2/k_0)\}\}$, which is at least $l/4$ and which is at least $l/(2+2^{-n+1})$ particularly if the ideal in O_K generated by 2 is unramified at any prime ideal of degree one.*

Let F denote the modular form for Γ_K defining D, which is obviously irreducible in Γ_K. Let $(k_1, \cdots, k_n)$ be the weight vector of F. If F is not a factor of f, then there is nothing to prove. So we assume that F is a factor of f in what follows.

Suppose that $F(2z)$ does not divide $f(z)$ (in $\Gamma(2)$), or equivalently that $F(z)$ does not divide $f(\frac{1}{2}z)$ (in $\Gamma(2)$). Then $F(z)$ does not divide $f(2z)$ since $f(2z)$ (resp. $F(z)$) is equal to $\Pi(2z_i)^{-\mathrm{weight}(f)} f(\frac{1}{2}(-z^{-1}))$ (resp. $\Pi(z_i)^{-k_i} \times F(-z^{-1})$) up to a constant multiple and since $F(-z^{-1})$ does not divide $f(\frac{1}{2}(-z^{-1}))$. Then by the comment below Corollary to Lemma 1, $F(z)$ is not a factor of $f_{2,r}$ in (6) for some positive integer r. Since

$$\mathrm{ord}(f_{2,r})/\mathrm{weight}(f_{2,r}) \geq \frac{1}{2}(\mathrm{ord}(f)/\mathrm{weight}(f)) > \frac{l}{2}$$

by Corollary to Lemma 1, our assertion follows. So the problem is reduced to the case that

(i) F is a factor of f, and

(ii) $F(2z)$ is a factor of $f(z)$, or equivalently $N(F(2z))$ divides f.

N being as in the preceding section. We consider two cases;

I: F is not irreducible in $\Gamma^1(2)$,

II: F remains irreducible in $\Gamma^1(2)$.

The proof in the case I is given in the next section, and the case II, in the section after next.

§ 7. Proof in case I

Let $f=F^sG$ be the decomposition in Γ_K where G is not divisible by F. At first we assume that $\sum k_i \geq 2k_0$ for the weight vector $(k_1, \cdots, k_n)$ of F. Obviously n weight $(f) \geq s\sum k_i \geq 2sk_0$. By Corollary to Lemma 5 there is a modular form h for Γ_K such that (i) the weight vector is $4(1, \cdots, 1)+2n(k_1, \cdots, k_n)$, and (ii) $F^{2n} \preccurlyeq h$, and that (iii) $h|_D \not\equiv 0$. Then we take as g, h^sG^{2n} which is a modular form for Γ_K of scalar weight $4s+2n$ weight (f), and which does not vanish identically on D. ord (g)/weight $(g) \geq 2n$ ord $(f)/(4s+2n$ weight $(f)) > l/(1+2s/n$ weight $(f)) \geq l(1+1/k_0)$. So in this case our assertion follows, because $l/(1+1/k_0) > \max\{3l/4\,(1+1/k_0), l/(1+2/k_0)\}$.

Lemma 6. *Let F be a non-constant modular form for Γ_K of weight $(k_1, \cdots, k_n)$ irreducible in Γ_K, and let μ be a totally positive integer in O_K other than units. Then $N(F(\mu z))$ has as a factor, at least one modular form for Γ_K different from $F(z)$. If $\sum k_i = k_0$, then it is not irreducible in $\Gamma(\mu)$.*

We note that in the above lemma F is not assumed to be irreducible in a subgroup of Γ_K.

Proof. Let $F(\mu z)=F_1(z)\cdots F_t(z)$ be an irreducible decomposition in $\Gamma(\mu)$. By the irreducibility of $F(z)$ in Γ_K, any two of the $F_j(z)$'s have no common divisor. If $N(F(\mu z))$ has only $F(z)$ as a factor, then $F(z)$ is divisible by each of the $F_j(z)$, and hence by $F(\mu z)$. So $\Gamma_{F(z)}$ is properly larger than Γ_K, which contradicts to Maass [5] (see the proof of Lemma 2). This shows our first assertion. Now suppose that $\sum k_i = k_0$. If $t \geq 2$, then any of $F_1, \cdots, F_t$ is not a modular form for Γ_K by the definition of k_0. The irreducible factor of $N(F(\mu z))$ equals one of $N(F_j)$, up to a constant multiple, which is not irreducible in $\Gamma(\mu)$. If $t=1$, i.e., $F(\mu z)$ is irreducible in $\Gamma(\mu)$, then Lemma 4 together with Lemma 2 shows that $N(F(\mu z))$ is not irreducible in $\Gamma(\mu)$.

We continue to prove Proposition 2. We assume that $k_0=\sum k_i$. Let $f=F^sF'^{s'}G$ be a decomposition in Γ_K, where F' is an irreducible factor of $N(F(2z))$ in Γ_K different from F, and G is divisible by neither F nor F'. Let $(k'_1, \cdots, k'_n)$ be the weight vector of F', and let $k'=\sum k'_i \geq k_0$. Then obviously n weight $(f) \geq sk_0+sk' \geq (s+s')k_0$. Suppose that $s'(k'+4) \geq s(2-k_0)$. Let h be as in the beginning of this section. We take as g, $h^s(F'^{s'}G)^{2n}$ which is a modular form for Γ_K of scalar weight $4s+2n$ weight (f) with order at least $2n$ ord (f) and which does not vanish identically on D. Then ord (g)/weight $(g) \geq 2n$ ord $(f)/(4s+2n$ weight $(f)) > l/(1+2s/n$ weight $(f)) \geq \max\{3l/4(1+1/k_0), l/(1+2/k_0)\}$. Then our assertion follows. Now let us suppose that $s'(k'+4) \leq s(2-k_0)$, which implies

$k_0<2$. By Lemma 6, F' is not irreducible in $\Gamma(2)$. Then by Corollary to Lemma 5, there is a modular form h' for Γ_K such that (i) the weight vector is $4(1, \cdots, 1)+2n(k'_1, \cdots, k'_n)$ and that (ii) $F'^{2n} \preccurlyeq h'$, and that (iii) h', F' have non common divisor. Let $g'=h'^{s'}(F^sG)^{2n}$, which is a modular form for Γ_K of scalar weight $4s'+2n$ weight (f). ord (g') is at least $2n$ ord (f). g' is not divisible by $F(2z)$, because if otherwise, g' is divisible by F' just as f, which is not the case. Then $g'(2z)$ is not divisible by $F(z)$ (see the argument in the preceding section). We take as g, $(g')_{2,r}$ in (6) where r is taken so that $(g')_{2,r}$ is not divisible by $F(z)$. Then by Corollary to Lemma 1, ord (g)/weight $(g) \geqq \frac{1}{2}$(ord (g')/weight $(g')) > l/(2+4s'/n$ weight $(f)) \geq l/(2+(-2k_0+4)/(k'+2k_0)) \geq 3l/4(1+1/k_0) = \max\{3l/4(1+1/k_0), l/(1+2/k_0)\}$, since $k_0<2$. We are done.

§ 8. Proof in case II

By assumption $F(2z)$ is irreducible in

$$\Gamma(2)=\begin{pmatrix}\sqrt{2}^{-1} & 0\\ 0 & \sqrt{2}\end{pmatrix}\Gamma^1(2)\begin{pmatrix}\sqrt{2} & 0\\ 0 & \sqrt{2}^{-1}\end{pmatrix},$$

and by Lemma 2 $\Gamma_{F(2z)} \cap \Gamma_K = \Gamma_0(2)$. By Lemma 4, $F'(z) := N(F(2z))$ is irreducible in Γ_K and its weight vector is $[\Gamma_K : \Gamma_0(2)]\,(k_1, \cdots, k_n)$. Let $f=F^sF'^{s'}G$ be a decomposition in Γ_K, where G is divisible neither F nor F'. We give a similar proof as in the last part of the preceding section. By Corollary to Lemma 5, there is a modular form h' for Γ_K such that (i) the weight vector is $4(1, \cdots, 1)+2n[\Gamma_K : \Gamma_0(2)](k_1, \cdots, k_n)$ and that (ii) $F'^{2n} \preccurlyeq h'$, and that (iii) h' is not divisible by F'. Let $g'=h'^{s'}(F^sG)^{2n}$, which is a modular form for Γ_K of scalar weight $4s'+2n$ weight (f) and with order $\geq 2n$ ord (f). $g'(z)$ is not divisible by $F(2z)$, in other words, $g'(2z)$ is not divisible by $F(z)$. Then we take as g, $(g')_{2,r}$ in (6) which is not divisible by $F(z)$. Then by Corollary to Lemma 1, ord (g)/weight $(g) \geq \frac{1}{2}$(ord (g')/weight $(g')) > l/(2+4s'/(s+s'[\Gamma_k : \Gamma_0(2)]\sum k_i)) > l/(2+4/[\Gamma_K : \Gamma_0(2)]k_0) > l/(2+1/2^{n-2}k_0)$ since $[\Gamma_K : \Gamma_0(2)] > 2^n$ by Lemma 3. We have proved Proposition 2.

§ 9. Proof of Theorem

Let us recall the asymptotic dimension of the space of modular forms for Γ_K of scalar weight $k \in 2Z$ with a trivial multiplier and with ord $(f)/k > l$ ([8], see also [7]). It is

$$\{2^{-2n+1}\pi^{-2n}d_K^{3/2}\zeta_K(2) - 2^{n-1}l^n n^{-n} d_K^{1/2} h R\}k^n + O(k^{n-1})$$

where d_K, h, R denote the discriminant, the class number, the regulator of

K respectively. In particular, if

$$l<2^{-3}\pi^{-2}n\left(\frac{4d_K\zeta_K(2)}{hR}\right)^{1/n},$$

then there are such modular forms for sufficiently large even k.

Theorem 1. *Let $n>6$. If*

$$2^{-4}\pi^{-2}(n-1)\left(\frac{4d_K\zeta_K(2)}{hR}\right)^{1/n}>1, \tag{11}$$

then (☆) holds. If the ideal in O_K generated by 2 is unramified at any prime ideal of degree one and if

$$2^{-3}(1+2^{-n+1})^{-1}\pi^{-2}(n-1)\left(\frac{4d_K\zeta_K(2)}{hR}\right)^{1/n}>1, \tag{12}$$

then (☆) holds.

Proof. By Proposition 2, for any subvariety D in X_K of codimension one, there is a modular form g of scalar weight such that (i) $g|_D\not\equiv 0$, and that (ii) $\operatorname{ord}(g)/\operatorname{weight}(g)>\{2^{-3}\pi^{-2}n(4d_K\zeta_K(2)/hR)^{1/n}\}/4$ which is larger than $n/2(n-1)$ by the condition (11). By Proposition 1, (☆) holds. Also to the second assertion, the similar argument is applicable. q.e.d.

$d_K\zeta_K(2)/hR$ is at least $2^{n-2}\pi^n$ (cf. Lang [4, p. 261]). Hence (11) holds for $n>26$, and (12) holds for $n>14$. This proves our theorem.

References

[1] Freitag, E., Lokale und globale Invarianten der Hilbertschen Modulgruppe, Invent. Math., **17** (1972), 106–134.

[2] ——, Automorphy factors of Hilbert's modular group. In: Discrete subgroups of Lie groups and applications to moduli, pp. 9–19, Tata Institute, 1975.

[3] Gundlach, K.-B., Multiplier systems for Hilbert's and Siegel's modular groups, Glasgow Math. J., **27** (1985), 57–80.

[4] Lang, S., Algebraic number theory. Reading, MA: Addison-Wesley, 1970.

[5] Maass, H., Über die Erweiterungsfähigkeit der Hilbertschen Modulgruppe, Math. Z., **51** (1949), 255–261.

[6] Siegel, C. L., Lectures on advanced analytic number theory, Tata Institute, 1961.

[7] Tsuyumine, S., On the Kodaira dimensions of Hilbert modular varieties, Invent. Math., **80** (1985), 269–281.

[8] ——, Multi-tensors of differential forms on the Hilbert modular variety and on its subvarieties, Math. Ann., **274** (1986), 659–670.

[9] ——, Automorphy factors for a Hilbert modular group, to appear in Glasgow Math. J.

Department of Mathematics
Mie University
Tsu 514
Japan